NAVAIR 01-245FDD-1

NATOPS FLIGHT MANUAL

Navy Model

F-4J

AIRCRAFT

I0809167

Schiffer Military/Aviation History
Atglen, PA

The U.S. Navy F-4J NATOPS flight manual used to print this book is from the collection of Michael L. Roberts. We wish to thank Commander L.N. Carota, USN, Technical Documentation Policy and Programs Office, for his cooperation in ensuring that this manual was cleared for release.

Cover photography from the Doug Siegfried collection.

Library of Congress Catalog Number: 94-62121

Printed in China.
ISBN: 0-88740-781-1

We are interested in hearing from authors with book ideas on related topics.

Published by Schiffer Publishing Ltd.
77 Lower Valley Road
Atglen, PA 19310
Please write for a free catalog.
This book may be purchased from the publisher.
Please include $2.95 postage.
Try your bookstore first.

NATOPS FLIGHT MANUAL

NAVY MODEL

F-4J

AIRCRAFT

McDonnell Aircraft
NOw(A)65–0044–f
N00019–75–A–0403

This manual is incomplete without NAVAIR 01–245FDD–1A and NAVAIR 01–245FDD–1A–1

ISSUED BY AUTHORITY OF THE CHIEF OF NAVAL OPERATIONS AND UNDER THE DIRECTION OF THE COMMANDER, NAVAL AIR SYSTEMS COMMAND

1 MAY 1975
CHANGE 1 - 1 MAY 1976

LIST OF EFFECTIVE PAGES

Insert latest changed pages; dispose of superseded pages in accordance with applicable regulations.

NOTE: On a changed page, the portion of the text affected by the latest change is indicated by a vertical line, or other change symbol, in the outer margin of the page.

Dates of issue for original and changed pages:

Original . . . 0 . . 1 May 75
Change . . . 1 . . 1 May 76

Total number of pages in this manual is 528 consisting of the following:

Page No.	#Change No.	Page No.	#Change No.
Title	1	10-1 — 10-5	0
A	1	10-6 Blank	0
Flyleaf 1	1	10-7 — 10-21	0
Flyleaf 2	0	10-22 Blank	0
i — vii	0	11-1 — 11-76	0
viii	1	1-77 Blank	0
ix	0	11-78	1
x	1	11-79 — 11-138	0
xi — xii	0	A-1	0
1-1 — 1-47	0	A-2 Blank	0
1-48	1	A-3	0
1-49 — 1-128	0	A-4 Blank	0
1-129	1	A-5	0
1-130 — 1-136	0	A-6 Blank	0
2-1 — 2-4	0	A-7	0
3-1 — 3-3	0	A-8 Blank	0
3-4	1	A-9	0
3-5 — 3-9	0	A-10 Blank	0
3-10	1	A-11	1
3-11 — 3-20	0	A-12 Blank	0
3-21	1	A-13	0
3-22 — 3-55	0	A-14 Blank	0
3-56	1	A-15	0
4-1 — 4-29	0	A-16 Blank	0
4-30 Blank	0	A-17	0
5-1 — 5-5	0	A-18 Blank	0
5-6	1	A-19	0
5-7 — 5-19	0	A-20 Blank	0
5-20	1	A-21	0
5-21	0	A-22 Blank	0
5-22	1	A-23	1
5-23 — 5-62	0	A-24 Blank	0
6-1 — 6-6	0	A-25	1
7-1 — 7-8	0	A-26 Blank	0
8-1 — 8-2	0	Index 1 — Index 13	0
9-1 — 9-7	0	Index 14 Blank	0
9-8 Blank	0		

Zero in this column indicates an original page.

INTERIM CHANGE SUMMARY

The following Interim Changes have been canceled or previously incorporated in this manual:

INTERIM CHANGE NUMBER(S)	REMARKS/PURPOSE
1 thru 73	

The following Interim Changes have been incorporated in this Change/Revision:

INTERIM CHANGE NUMBER	REMARKS/PURPOSE

Interim Changes Outstanding - To be maintained by the custodian of this manual:

INTERIM CHANGE NUMBER	ORIGINATOR/DATE (or DATE/TIME GROUP)	PAGES AFFECTED	REMARKS/PURPOSE

INTERIM CHANGE SUMMARY (continued)

Interim Changes Outstanding - To be maintained by the custodian of this manual:

INTERIM CHANGE NUMBER	ORIGINATOR/DATE (or DATE/TIME GROUP)	PAGES AFFECTED	REMARKS/PURPOSE

1 May 75

DEPARTMENT OF THE NAVY
OFFICE OF THE CHIEF OF NAVAL OPERATIONS
WASHINGTON, D.C. 20350

LETTER OF PROMULGATION

1. The Naval Air Training and Operating Procedures Standardization Program (NATOPS) is a positive approach toward improving combat readiness and achieving a substantial reduction in the aircraft accident rate. Standardization, based on professional knowledge and experience, provides the basis for development of an efficient and sound operational procedure. The standardization program is not planned to stifle individual initiative, but rather to aid the Commanding Officer in increasing his unit's combat potential without reducing his command prestige or responsibility.

2. This manual standardizes ground and flight procedures but does not include tactical doctrine. Compliance with the stipulated manual procedure is mandatory except as authorized herein. In order to remain effective, NATOPS must be dynamic and stimulate rather than suppress individual thinking. Since aviation is a continuing, progressive profession, it is both desirable and necessary that new ideas and new techniques be expeditiously evaluated and incorporated if proven to be sound. To this end, Commanding Officers of aviation units are authorized to modify procedures contained herein, in accordance with the waiver provisions established by OPNAVINST 3510.9 series, for the purpose of assessing new ideas prior to initiating recommendations for permanent changes. This manual is prepared and kept current by the users in order to achieve maximum readiness and safety in the most efficient and economical manner. Should conflict exist between the training and operating procedures found in this manual and those found in other publications, this manual will govern.

3. Checklists and other pertinent extracts from this publication necessary to normal operations and training should be made and may be carried in Naval Aircraft for use therein. It is forbidden to make copies of this entire publication or major portions thereof without specific authority of the Chief of Naval Operations.

W.D. Houser

W.D. HOUSER
Vice Admiral, USN
Deputy Chief of Naval Operations
(Air Warfare)

TABLE of CONTENTS

FDD-1-(287)

FOREWORD

SCOPE

The NATOPS Flight Manual is issued by the authority of the Chief of Naval Operations and under the direction of Commander, Naval Air Systems Command in conjunction with the Naval Air Training and Operating Procedures Standardization (NATOPS) Program. This manual contains information on all aircraft systems, performance data, and operating procedures required for safe and effective operations. However, it is not a substitute for sound judgement. Compound emergencies, available facilities, adverse weather, or terrain may require modification of the procedures contained herein. Read this manual from cover to cover. It's your responsibility to have a complete knowledge of its contents.

APPLICABLE PUBLICATIONS

The following applicable publications complement this manual:

NAVAIR 01-245FDD-1A (Supplement)
NAVAIR 01-245FDD-1B (Pocket Checklist)
NAVAIR 01-245FDD-1C (Servicing Checklist)
NAVAIR 01-245FDD-1F (Functional Checkflight Checklist)
NAVAIR 01-245FDB-1T (Tactical Manual)
NAVAIR 01-245FDB-1T(A) (Tactical Manual Supplement)
NAVAIR 01-245FDB-1T(B) (Tactical Manual Pocket Guide)

HOW TO GET COPIES

AUTOMATIC DISTRIBUTION

To receive future changes and revisions to this manual automatically, a unit must be established on the automatic distribution list maintained by the Naval Air Technical Services Facility (NATSF). To become established on the list or to change distribution requirements, a unit must submit NAVAIR Form 5605/3, Part II, to NATSF, 700 Robbins Ave., Philadelphia, Pa. 19111, listing this manual and all other NAVAIR publications required. For additional instructions refer to NAVAIR INST 5605.4 series and NAVSUP Publication 2002, Section VIII, Part C.

ADDITIONAL COPIES

Additional copies of this manual and changes thereto may be procured by submitting Form DD 1348 to NPFC Philadelphia in accordance with NAVSUP Publication 2002, Section VIII, Part C.

UPDATING THE MANUAL

To ensure that the manual contains the latest procedures and information, NATOPS review conferences are held in accordance with OPNAVINST 3510.11 series.

CHANGE RECOMMENDATIONS

Recommended changes to this manual or other NATOPS publications may be submitted by anyone in accordance with OPNAVINST 3510.9 series.

Routine change recommendations are submitted directly to the Model Manager on OPNAV Form 3500-22 shown on the next page. The address of the Model Manager of this aircraft is:

Commanding Officer
Fighter Squadron 121
U.S. Naval Air Station
Miramar, California 92145
Attn: F-4 Model Manager

Change recommendations of an URGENT nature (safety of flight, etc.,) should be submitted directly to the NATOPS Advisory Group Member in the chain of command by priority message.

YOUR RESPONSIBILITY

NATOPS Flight Manuals are kept current through an active manual change program. Any corrections, additions, or constructive suggestions for improvement of its content should be submitted by routine or urgent change recommendation, as appropriate, at once.

NATOPS FLIGHT MANUAL INTERIM CHANGES

Flight Manual Interim Changes are changes or corrections to the NATOPS Flight Manuals promulgated by CNO or NAVAIRSYSCOM. Interim Changes are issued either as printed pages, or as a naval message. The Interim Change Summary page is provided as a record of all interim changes. Upon receipt of a change or revision, the custodian of the manual should check the updated Interim Change Summary to ascertain that all outstanding interim changes have been either incorporated or canceled; those not incorporated shall be recorded as outstanding in the section provided.

NATOPS/TACTICAL CHANGE RECOMMENDATION
OPNAV FORM 3500/22 (5-69) 0107-722-2002

DATE

TO BE FILLED IN BY ORIGINATOR AND FORWARDED TO MODEL MANAGER

FROM (originator)	Unit
TO (Model Manager)	Unit

Complete Name of Manual/Checklist	Revision Date	Change Date	Section/Chapter	Page	Paragraph

Recommendation (be specific)

☐ CHECK IF CONTINUED ON BACK

Justification

Signature	Rank	Title

Address of Unit or Command

TO BE FILLED IN BY MODEL MANAGER *(Return to Originator)*

FROM	DATE
TO	

REFERENCE
(a) Your Change Recommendation Dated ____________

☐ Your change recommendation dated ________ is acknowledged. It will be held for action of the review conference planned for ____________ to be held at ____________

☐ Your change recommendation is reclassified URGENT and forwarded for approval to ____________ ____________ by my DTG ____________

/S/ ____________ MODEL MANAGER | ____________ AIRCRAFT

CHANGE SYMBOLS

Revised text is indicated by a black vertical line in either margin of the page, adjacent to the affected text, like the one printed next to this paragraph. The change symbol identifies the addition of either new information, a changed procedure, the correction of an error, or a rephrasing of the previous material.

WARNINGS, CAUTIONS, AND NOTES

The following definitions apply to "WARNINGS", "CAUTIONS", and "NOTES" found through the manual.

WARNING

An operating procedure, practice, or condition, etc., which may result in injury or death, if not carefully observed or followed.

CAUTION

An operating procedure, practice, or condition, etc., which may result in damage to equipment if not carefully observed or followed.

NOTE

An operating procedure, practice, or condition, etc., which is essential to emphasize.

WORDING

The concept of word usage and intended meaning which has been adhered to in preparing this Manual is as follows:

"Shall" has been used only when application of a procedure is mandatory.

"Should" has been used only when application of a procedure is recommended.

"May" and "need not" have been used only when application of a procedure is optional.

"Will" has been used only to indicate futurity, never to indicate any degree of requirement for application of a procedure.

GLOSSARY

A

AC – Aerodynamic Center

ac – Alternating current

ACK – Acknowledge

ACLS – Automatic Carrier Landing System

ACP – Aircraft Communications Procedures

ADCS – Air Data Computer Set

ADI – Attitude Director Indicator

ADIZ – Air Defense Identification Zone

AFC – Automatic Frequency Control

AFCS – Automatic Flight Control System

agc – Automatic gain control

AI – Airborne Intercept

AJB – Airborne, electro–mechanical, bombing

AMCS – Airborne Missile Control System

AOJ – Acquisition On Jam

APA – Airborne, radar, auxiliary assembly

APCS – Approach Power Compensator System

APN – Airborne, radar, navigational aid

APQ – Airborne, radar, special purpose

AR – Air Refueling

ARC – Airborne, radio, control

ARI – Aileron Rudder Interconnect

ARTC – Air Route Traffic Control

ASA – Airborne, special type, auxiliary assembly

ASE – Allowable Steering Error

ASN – Airborne, special type, navigational aid

ASQ – Airborne, special type, combination of purposes

ATC – Air Traffic Control

AWW – Airborne, armament, control

B

BACSEB – BuWeps Aviation Clothing and Survival Equipment Bulletin

BDHI – Bearing Distance Heading Indicator

BINGO – Return fuel state; Divert

BIT – Built–In–Test

BLC – Boundary Layer Control

Bolter – Hook down, unintentional touch and go (missed wire)

BRC – Base Recovery Course

BST – Boresight

Buster – Full military power

C

CADC – Central Air Data Computer

CAP – Combat Air Patrol

CARQUAL – Carrier Qualifications

CAS – Calibrated Air Speed

CAT – Catapult

CAT – Clear Air Turbulence

CATCC – Carrier Air Traffic Control Center

CCA – Carrier Control Approach

CDI – Command Display Indicator

CG – Center of Gravity

Charlie Time – Expected time over ramp

CIC – Combat Information Center

CIT – Compressor Inlet Temperature

CMD – Command

CNI – Communication Navigation Identification

COT – Cockpit Orientation Trainer

CPLR – Coupler

cps – Cycles per second

CVA – Aircraft Carrier (Attack)

CV – Aircraft Carrier

cw – Continuous Wave

D

dc – Direct current

GLOSSARY (CONT)

DCU – Douglas Control Unit

DDI – Digital Display Indicator

Dead Beat – Causing the object, when disturbed, to return to its original position without oscillation

DL – Data Link

DME – Distance Measuring Equipment

Dog Radial – An assigned radial on which to set up a holding pattern

DR – Dead Reckoning

E

EAC – Estimated Arrival Carrier

EAS – Equivalent Airspeed

EAT – Estimated Approach Time

ECCM – Electronic Counter-Countermeasure(s)

ECM – Electronic Countermeasure(s)

EGT – Exhaust Gas Temperature

F

FAM – Familiarization

FL – Flight Level

FMLP – Field Mirror Landing Practice

FOJ – Fuse on Jam

G

G – Gravity

Gate – Maximum Power

GCA – Ground Control Approach

GCI – Ground Control Intercept

gpm – Gallon per minute

H

Hangfire – A delay or failure of an article of ordnance after being triggered

Hang Start – A start that results in a stagnated rpm and temperature

HOJ – Home on Jam

Hot Start – A start that exceeds normal starting temperatures

HSI – Horizontal Situation Indicator

I

IAS – Indicated Airspeed

IFF – Identification Friend of Foe

IFR – Instrument Flight Rules or In Flight Refueling

ILS – Instrument Landing System

IP – Identification Point

I/P – Identification of Position

IR – Infrared

J

JANAP – Joint Army Navy Air Force Publication

JP – Jet Petroleum

Judy – Radar contact with target taking over intercept

K

KTS – Knots

L

LABS – Low Altitude Bombing System

LBA – Limits of Basic Aircraft

LE – Leading Edge

LID – Limited Instrument Departure

LOX – Liquid Oxygen

lpm – Liters per minute

LSO – Landing Signal Officer (Paddles)

M

MAC – Mean Aerodynamic Chord

Meatball – Glide slope image of mirror landing system

MIL – Military

MIM – Maintenance Instruction Manual

Misfire – A permanent fialure of an article of ordnance being triggered

MLP – Mirror Landing Practice

MSL – Mean Sea Level

GLOSSARY (CONT)

N

NAMT – Naval Air Maintenance Training

NATOPS – Naval Air Training and Operating Procedures Standardization

NMPP – Nautical Miles Per Pound

NOTAMS – Notice to Airmen

NTDS – Naval Tactical Data System

NWIP – Naval Warfare Intercept Procedures

NWP – Naval Warfare Publications

O

OAT – Outside Air Temperature

OMNI – Omni Directional Range

P

Paddles – Landing signal officer

PC – Power Control

Pigeons – Bearing and distance

Platform – 20 miles, 5000 ft. commence descent at 2000 ft. min. level off at 1000 ft.

PMBR – Practice Multiple Bomb Rack

pps – Pulses per seconds

prf – Pulse repetition frequency

psi – Pounds per square inch

Punch – Target detected, aircraft still under ground control

Q

Q – Dynamic Pressure, psf

R

RADAR – Radio Detection and Ranging

RCVG – Replacement Carrier Air Group

RESCAP – Rescue Air Patrol

rf – Radio Frequency

RF – Reconnaissance – Fighter

RIO – Radar Intercept Officer

S

SAR – Search and Rescue

SATS – Short Airfield for Tactical Support

SID – Standard Instrument Departure

SIF – Selective Identification Feature

SPC – Static Pressure Compensator

T

TACAN – Tactical Air Navigation

TAS – True Airspeed

TE – Trailing Edge

TILT – Transmission of Intercept and Landing Information Terminated

TMN – True Mach Number

Trap – Arrested Landing

U

UHF – Ultra High Frequency

UTM – Universal Test Message

V

VFR – Visual Flight Rules

VHF – Very High Frequency

Vn – Velocity Acceleration Relationship

VORTAC – Very high frequency – omni range and tactical air navigation

W

WST – Weapons System Trainer

WSTH – Weapon System Tactical Handbook

10 MILE GATE – 10 mi; transition to landing configuration; maintain 1200 ft.

6 MILE GATE – 6 mi; descend to 600 ft.

TECHNICAL DIRECTIVE SUMMARY

In accordance with BUWEPS Instruction 5215.8, Technical Directive concerning modification, inspection, maintenance or operating procedures and limits of all Naval aircraft and related equipment are titled as follows:

Airframe Change (AFC) or Airframe Bulletin (AFB)
Power Plant Change (PPC) or Bulletin (PPB)
Aviation Armament Change (AAC) or Bulletin (AAB)
Avionics Change (AVC) or Bulletin (AVB)
Accessory Change (AYC) or Bulletin (AYB)
Support Equipment Change (SEC) or Bulletin (SEB)
Photographic Change (PHC) or Bulletin (PHB)
Air Crew System Change (ACC) or Bulletin (ACB)
Air Launched Missile Change (AMC) or Bulletin (AMB)
Target Control System Change (TCC) or Bulletin (TCB)
Clothing and Survival Equipment Change (CSEC) or Bulletin (CSEB)

TECHNICAL DIRECTIVE	ECP	TITLE	PRODUCTION EFFECTIVITY
AAC 483		ADDS ADAPTER ASSEMBLY FOR LAU-7/A	
AAC 509		AUTO HOMING CAPABILITY OF MER/TER - 7 RACKS	
AAC 537		MODIFIES LAU-7/A LAUNCHERS	
AAC 587		MODIFIES LAU-7/A LAUNCHERS	
ACC 74		PROVIDES GRAVITY DROP LIFE RAFT INFLATION	
ACC 157		ADDS THIGH GARTERS	
ACC 169		ADD MK-H7 ROCKET MOTOR SEAR BOOT	
ACC 176		IMPROVED 28 FOOT PERSONNEL PARACHUTE	
ACC 187		ADDS INTERDICTOR PIN AND GUARD	
ACC 217		REPLACES INERTIA REEL	
ACC 224		INSTALLS GAS OPERATED ROCKET MOTOR INITIATOR	
AVC 495		MODIFIES STEERING INFORMATION (RADAR MOD)	
AVC 496		ADDS DATA LINK SYMBOLS (RADAR MOD)	
AVC 498		MODIFIES SCOPE DISPLAY (RADAR MOD)	
AVC 588		ADDS B-SWEEP IN MAP MODE (RADAR MOD)	
AVC 642		MOD OF RADAR TRANSMISSION INTERLOCKS	
AVC 743		ADDS CAPABILITY OF DATA LINK COUPLING IN ACL PHASE	
AVC 750		MOD OF AN/APG-59	
AVC 752		LIMITS USE OF APCS FOR CARRIER LANDINGS	
AVC 801		ADDS GUARD TO ZEROIZE SW ON KY-28 PANEL	
AVC 833		INCREASES ANT SLEW RATE AND MISSILE CONTROL SIGNAL	
AVC 834		MODIFIES FEEDHORN NUTATION	
AVC 838		MOD OF RADAR SCOPE BRT CONTROL IN AFT COCKPIT	
IAVC 852, AVC 1007		MOD OF RADAR BORESIGHT MODE	
IAVC 873		ADDS CAPABILITY OF REMOVING RADAR B PLUS POWER	
AVC 925		CONVERTS MANUAL DATA LINK SYSTEM TO AUTOMATIC SYSTEM	
AVC 926		CONVERTS MANUAL DATA LINK SYSTEM TO AUTOMATIC SYSTEM	
AVC 927		CONVERTS MANUAL DATA LINK SYSTEM TO AUTOMATIC SYSTEM	
AVC 937		ADDS CAPABILITY OF DATA LINK COUPLING IN ACL PHASE	
AVC 940		IMPROVES TUNING AND FIRING CIRCUITS	
AVC 947		ADDS CAPABILITY OF DATA LINK COUPLING IN ACL PHASE	
AVC 1054	911	PROVIDES PILOT LOCKON MODIFICATION.	
IAVC 1198		PROVIDES AUTOMATIC PULSE SWITCHING AT SHORT RANGES.	
AVC 1234		ADDS AN&ALR-45(V) AND AN/ALR-50(V)	
AVC 1590	152	AWG 10A MISSILE CONTROL SYSTEM	
CSEC 4		ADDS HELICOPTER LIFT RING	
CSEC 14		ADDS SUIT MOUNTED PRESSURE SUIT CONTROLLER	
CSEC 15		ADDS OXYGEN MAKE-UP VALVE IN REGULATOR	

TECHNICAL DIRECTIVE SUMMARY

(CONTINUED)

TECHNICAL DIRECTIVE	ECP	TITLE	PRODUCTION EFFECTIVITY
AFC 263	610	INCORP 1/2 FLAP BLC MALFUNCTION LIGHT	153776aa
AFC 312	738	ADDS REMOTE UHF CHANNEL INDICATOR IN AFT COCKPIT	153840ae
AFC 328	792	ADDS THREE POSITION RAIN REMOVAL SWITCH	155867ak
AFC 331	703, S1	INSTALLS KY-28 SPEECH SECURITY UNIT	155529ag
AFC 335	745	REPLACEMENT OF FIRE/OVERHEAT WARNING LIGHT CAPS	153877af
AFC 346		REMOVES ALL POSITION FROM MISSILE JETTISON SELECT SW	153900af
AFC 364	743	INCORP OF STAB INPUT TO APCS	155785ai
AFC 370	739R1	INCORP OF EMERGENCY PNEUMATIC EXTENSION OF AIR REFUEL PROBE	155867ak
AFC 373	770R1	INCORP OF ENGINE SMOKE ABATEMENT	155867ak
AFC 388	840, S1, S2	ADDS B-SWEEP IN MAP MODE (AIRCRAFT MOD)	155529ag
		ADDS DATA LINK SYSTEM	155529ag
		ADDS AN/ALE-29A CHAFF DISPENSER	155529ag
		INCORP RHAW (AN/APR-25) CAPABILITY	155529ag
		ADDS ECM SET (AN/ALQ-51A/100) CAPABILITY	155529ag
		ADDS ECM SET (AN/ALQ-91) CAPABILITY	155529ag
		INSTALLS RADAR BEACON SYSTEM	155529ag
		INCORP INTERROGATOR SET (AN/APX-76) CAPABILITY	155529ag
		JET AIRCRAFT CONVENTIONAL LOW ALTITUDE BOMBING SYSTEM	155529ag
		ISOLATES GENERATOR PMG XFMR-RET OUTPUT	155785ai
		ADDS FAULT LIGHT FOR AN/APX-76	155793ai
		REPLACE ICS CONTROL PANEL	157242an
AFC 392	826	MOD OF APCS (DISENGAGE APCS WITH SPEED BRAKE SWITCH)	155867ak
AFC 399	862	PROVIDES EMER ELEC PWR FOR ENGING IGNITERS	155844aj
AFC 415	687S8	ADDS INTERROGATOR COMPUTER (AN/APX-76)	155844aj
AFC 417	758S5	PROVIDES ADDITIONAL BIT CAPABILITIES	155867ak
AFC 420		ADDS DUAL MODE BELLMOUTH	155785ai
AFC 433	864S7	REPLACE ICS CONTROL PANEL	157242an
AFC 439	754	INSTALLS BLEED AIR LEAKAGE DETECTION SYSTEM	155903ap
AFC 440	1013	ADDS ENGINE BLEED AIR SHUTOFF VALVE AND SWITCH	155903ap
AFC 448	896R1	INCORP FWD COCKPIT CHAFF DISPENSING SWITCH	155890am
AFC 450	912, S3P2	REMOVE PSEUDO AND MODIFIES AIM-7 TUNE-UP SEQUENCE	155844aj
AFC 451	918	RELOCATION OF GUN SWITCH	155890am
AFC 452	906	RELOCATION OF AN/ALQ-91 INDICATOR LIGHTS	155890am
AFC 453	687S9	INCORP FAULT ISOLATION CIRCUIT	155793ai
AFC 463	927	INSTALL RADAR COOLING AIR DIVERTER	155875al
AFC 470	907S1	ADDS ILS AN/ARA-63	157309ar
AFC 472	827S3, 1013	ADDS PROVISIONS FOR ATTACHING ARMOR PLATING	157286aq
AFC 474	928, P1	ADDS SECONDARY EMERGENCY CANOPY JETTISON	157256an
AFC 476	912S2	ADDS TIME DELAY TO AIM-7 SWEEP SELECT CIRCUIT	157261ao
AFC 477	714R1S13	ELIMINATES 3/4 SEC DELAY IN FRONT EJECT SEAT	155903ap
AFC 478	883R1	ADDS GYRO FAST ERRECT CAPABILITY	155903ap
AFC 482	938	MOD OF SEQUENCING SYSTEM FOR AFT SEAT	155903ap
AFC 486	822	ADDS RADAR ANNUNCIATOR LIGHTS DIMMING CONTROL	157242an
AFC 491	949	ADDS REAR COCKPIT CANOPY LEVER GUARD	157286aq

TECHNICAL DIRECTIVE SUMMARY

(CONTINUED)

TECHNICAL DIRECTIVE	ECP	TITLE	PRODUCTION EFFECTIVITY
AFC 496	968P2A2	ADDS COARSE SYNC OPERATION TO ALT ENCODER UNIT	158366au
AFC 497	967, S1	ADDS ELECTRICAL BALLISTIC THRUSTERS (FWD CANOPY)	157258an
AFC506	958, 971	INSTALLS COCKPIT OPT FOR ACM	158355at
		ADDS VISUAL TARGET ACQUISITION SET	158355at
		PROVIDES SIDEWINDER EXPANDED ACQUISITION MODE	158355at
		PROVIDES FIXED MOUNTING FOR REAR COCKPIT RADAR SCOPE	157298ar
		PROVIDES AIM-7E-2 DOGFIGHT MODE SELECTION CAPABILITY	158355at
AFC 508	863R1	ADDS ALL WEATHER CARRIER LANDING CAPABILITY	158355at
AFC 513		PROVIDES MODE SWITCHING LOGIC FOR DUAL MODE APCS	
AFC 514	996	PROVIDES LINE-OF-SIGHT RATE TO AIM-& MISSILES	158355at
AFC 515	993	ADDS RADHAZ WARNING LIGHTS	158355at
AFC 516	850R1S1	ADDS AIM-7 CARRIAGE CAPABILITIES	158355at
AFC 517	970R1	ADDS DOGFIGHT COMPUTER	158355at
AFC 518		ADDS WIRING TO OPERATE LB-30A STRIKE CAMERA POD	
AFC 524	1023A2, S1, S2	ADDS AN/ALR-45(V) AND AN/ALR-50(V)	
AFC 526	1011	REPLACES COMMAND SELECTOR VALVE	
AFC 528		INSTALLS GAS OPERATED ROCKET MOTOR INITIATOR	
AFC 534	1008A1P1, P2	ADDS PILOT OPTION EMERGENCY AILERON DROOP	
AFC 535		30 KVA GENERATOR WIRING ISOLATION	155785ai
AFC 536	1018S1	IMPROVED PRIMARY FLIGHT INSTRUMENT LIGHTING	
AFC 539		INSTALLS AN/APX-76 INTERROGATOR SET	
AFC 541	1035S1	ADDS COUNTERMEASURES SET AN/ALQ-126	
AFC 545		RELOCATION OF REFUEL PROBE CIRCUIT BREAKER	158366au
AFC 550	1033A2	ADDS BLEED AIR OFF LIGHT AND BLEED AIR SWITCH GUARD	
AFC 555	952S3A1	REPLACES AIR COND PACKAGE WITH IMPROVED PACKAGE	158355at
AFC 574	1038	ADDS ALE-39 COUNTERMEASURES DISPENSER SYSTEM AND REMOVES ALE-29 DISPENSER	
AFC 576	325R1	AWG 10A MISSILE CONTROL SYSTEM	

BLOCK NUMBERS

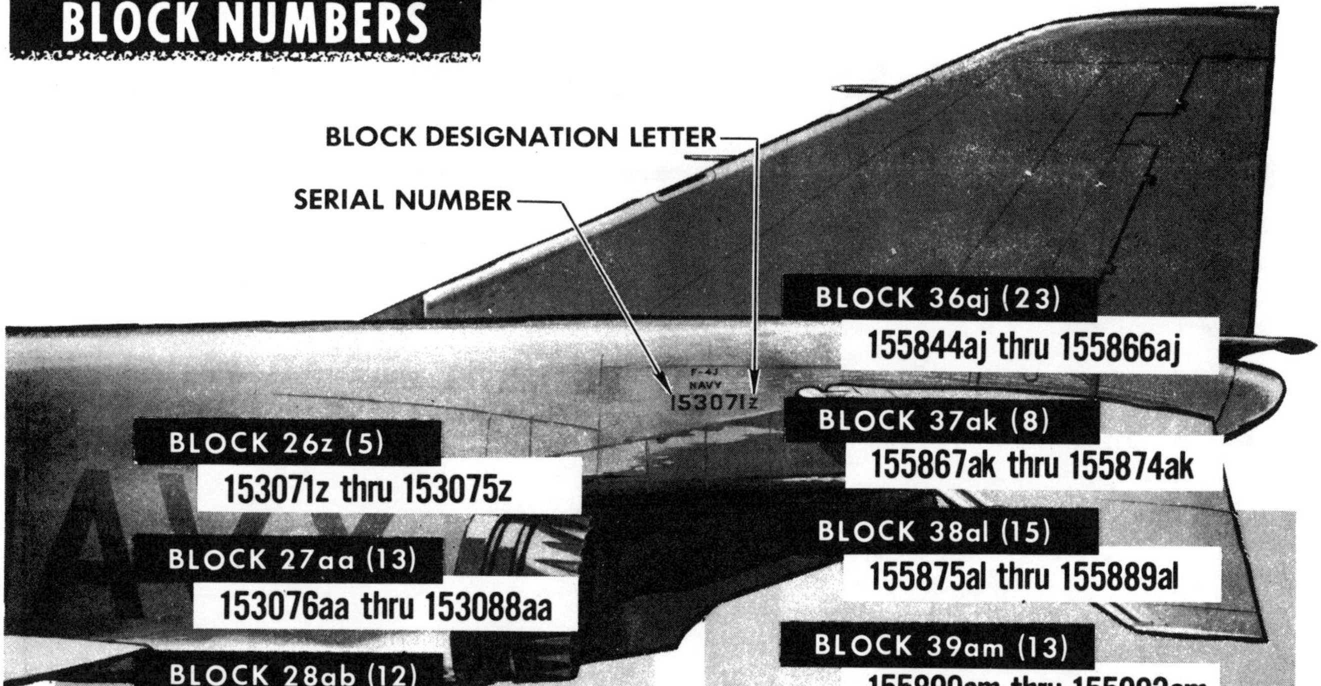

BLOCK 26z (5)
153071z thru 153075z

BLOCK 27aa (13)
153076aa thru 153088aa

BLOCK 28ab (12)
153768ab thru 153779ab

BLOCK 29ac (20)
153780ac thru 153799ac

BLOCK 30ad (40)
153800ad thru 153839ad

BLOCK 31ae (37)
153840ae thru 153876ae

BLOCK 32af (40)
153877af thru 153911af
154781af thru 154785af

BLOCK 33ag (69)
154786ag thru 154788ag
155504ag thru 155569ag

BLOCK 34ah (65)
155570ah thru 155580ah
155731ah thru 155784ah

BLOCK 35ai (59)
155785ai thru 155843ai

BLOCK 36aj (23)
155844aj thru 155866aj

BLOCK 37ak (8)
155867ak thru 155874ak

BLOCK 38al (15)
155875al thru 155889al

BLOCK 39am (13)
155890am thru 155902am

BLOCK 40an (19)
157242an thru 157260an

BLOCK 41ao (13)
157261ao thru 157273ao

BLOCK 42ap (13)
155903ap, 157274ap
thru 157285ap

BLOCK 43aq (12)
157286aq thru 157297aq

BLOCK 44ar (12)
157298ar thru 157309ar

BLOCK 45as (9)
158346as thru 158354as

BLOCK 46at (11)
158355at thru 158365at

BLOCK 47au (14)
158366au thru 158379au

FDD-1-(4)

FDD-1-(288)

Figure 1-0

SECTION I
AIRCRAFT

TABLE OF CONTENTS

PART 1 GENERAL DESCRIPTION

AIRCRAFT

NOTE

Some illustrations referred to within this system writeup may be located in appendix A. These illustrations are referred to, within the the text, as (figure A–, appendix A).

DESCRIPTION

The F-4J aircraft (figure 1-1) is a two-place (tandem) supersonic, long-range, all-weather fighter built by McDonnell Douglas Corporation. The aircraft is designed for intermediate and long range high altitude interceptions using missiles as the principal armament and for intermediate or long range attack missions to deliver airborne weapons/stores. The aircraft is powered by either two single rotor, axial flow, variable stator turbo-jet J79-GE-8 or J79-GE-10 engines with afterburner. The aircraft features a low mounted swept-back wing with anhedral at the wing tips, and a one-piece stabilator with cathedral, mounted low on the aft fuselage. The wings have hydraulically operated leading edge and trailing edge flaps, ailerons, spoilers and speed brakes. All the control surfaces are positioned by irreversible hydraulic power cylinders to provide desired control effectiveness throughout the entire speed range. A self-charging pneumatic system provides normal, emergency, and secondary emergency canopy operation, as well as emergency operation for the landing gear, inflight refueling probe and wing flaps. The pressurized cockpit is enclosed by two clamshell canopies. A drag chute, contained in the aft end of the fuselage, reduces landing roll distances.

AIRCRAFT DIMENSIONS

The approximate dimensions of the aircraft are as follows:

Span (Wings Spread) – 38 feet, 5 inches

Span (Wings Folded) – 27 feet, 7 inches

Length – 58 feet, 3 inches

Height (To Top of Fin) – 16 feet, 6 inches

GENERAL ARRANGEMENT

TYPICAL

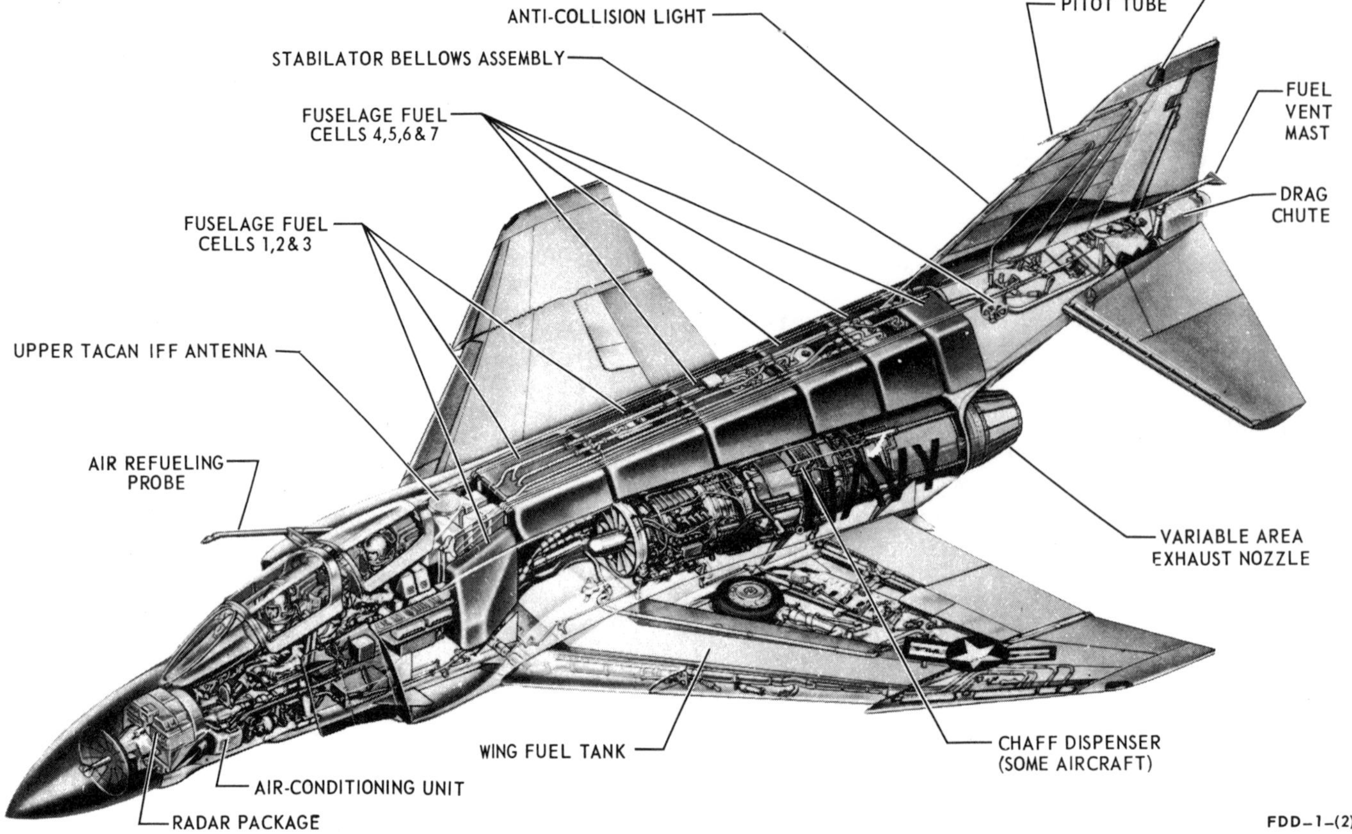

Figure 1-1

ARMAMENT

The aircraft is equipped to carry and deliver an assortment of air-to-air missiles, air-to-ground missiles, rockets, bombs, land mines, leaflet dispensers and airborne weapons/stores. The aircraft is also equipped with gunnery capabilities with the addition of the MK-4 gun pod on the centerline station. Refer to Weapons System, section VIII, for additional information on armament.

TECHNICAL DIRECTIVE SUMMARY

For technical directive incorporation, and main difference between aircraft, refer to Technical Directive Summary in the introduction section of this manual.

BLOCK NUMBERS

Refer to Block Numbers illustration in the introduction section of this manual for block production with corresponding assigned aircraft serial numbers.

ARMOR PLATING

On aircraft 157286aq and up and all others after AFC 472, provisions for attaching armor plating to doors 15, 16, 22, 23, 28 right and 28 left are provided. This armor, when installed, provides protection for the oxygen bay and hydraulic/engine fuel feed compartment. The armor adds approximately 109 pounds to the weight of the aircraft and shifts the CG forward by approximately 0.1% MAC.

AIRPLANE SECURITY REQUIREMENTS

The occasion may arise when it will be necessary to land at a civilian field that does not have a military installation associated with it, or when the airplane is to be presented in a static display. In order to prevent the compromise of classified information, accidental damage to the aircraft, or injury to observers, the following guide lines are provided.

Static Displays

a. Inert airborne weapons/stores may be carried on the aircraft in any combination.
b. External tanks may be carried.

c. The radome must be secured.
d. The front canopy must be secured.
e. The rear canopy, with blackout curtain in place must be secured.
f. The pneumatic system must be bled to prevent the canopies from being opened.
g. No smoking rules must be enforced.

RON at Civilian Fields

a. Make necessary security guard arrangements.
b. Secure radome.
c. Front cockpit is classified confidential and must be secured.
d. Rear cockpit is classified confidential and must be secured.
e. Bleed off pneumatic systems to prevent opening of the canopies.

COCKPITS

NOTE

Some illustrations referred to within this system writeup may be located in appendix A. These illustrations are referred to, within the text, as (figure A–, appendix A).

DESCRIPTION

Although the cockpits are separately enclosed, the cockpit pressure–oxygen environment is the same. Each cockpit incorporates an ejection seat that is adjustable in the vertical plane for comfort. The forward cockpit instrument panel contains the flight and engine instruments. Engine controls, autopilot and fuel management panels are on the left console. Communication, navigation, heating and lighting controls are on the right console. Left and right vertical panels forward of the consoles contain the flight control trim position indicators and the telelight panels. The aft cockpit instrument panel contains the necessary instruments for navigation, plus miscellaneous switches and indicator lights. Radar equipment is below the instrument panel. The right side of the cockpit contains the circuit breaker panels, and the left side contains the communication, and oxygen controls. Refer to figures A–1 and A–2, appendix A, for the instrument panels and consoles.

PART 2 SYSTEMS

AIR CONDITIONING AND PRESSURIZATION SYSTEM

NOTE

Some illustrations referred to within this system writeup may be located in appendix A. These illustrations are referred to, within the text, as (figure A–, appendix A).

DESCRIPTION

Both cockpits are pressurized and supplied with conditioned air from the cockpit air conditioning unit. The same air that pressurizes and heats the cockpits also is used to keep the windshield free of fog, frost and rain. The cockpit air conditioning unit utilizes high temperature, high pressure, engine compressor bleed air from either or both engines. The cockpit air conditioning system (figure 1–2) consists of two air–to–air heat exchangers, an expansion turbine, pressure regulating, mixing and shutoff valves, and temperature controls necessary to select cockpit temperatures, defogging, rain removal and ram air operations. High temperature/high pressure engine compressor bleed air passes through the primary and secondary heat exchanger and is expanded through the cooling turbine. After being mixed with hot compressor bleed air (as required by pilot's temperature selection) it enters the cockpits through several manifolds, one near the RIO's feet, one near the pilot's feet, one along the lower surface of each windshield side panel and one at the base at the flat optical panel of the windshield. Two eyeball type air nozzles are just below the canopy sill on the right and left side of the RIO's cockpit. Air is also routed into the aft cockpit via an open tube duct behind the circuit breaker panels. On aircraft after AFC 328, the inlet bleed air pressure to the air conditioning system can be controlled by selecting the PRESS LOW or PRESS NORMAL position on the rain removal switch. When the PRESS NORMAL position is selected, the bleed air pressure is regulated at 62 $\pm$5 psi. Selection of the PRESS LOW position reduces the bleed air pressure to 40 $\pm$5 psi. The PRESS LOW position reduces the cockpit noise to the same level as produced when the rain removal system is operating.

COCKPIT AIR CONDITIONING

The cockpit air conditioning system operation can best be explained and understood by referring to the cockpit temperature schedule (figure 1-3). The low temperature range, which refers to the curve labeled Foot Heat, produces temperatures from −20°F to 100°F. These temperatures refer to inlet air, and not cockpit temperature - so cockpit temperature will be determined by a combination of inlet air and environmental conditions. The low temperature curve is the governing schedule for all air entering the cockpit while in automatic temperature control with the defog-foot heat lever in the LOW range. A little air is always entering through the defog port and this air increases (while foot heat air decreases) as the defog-foot heat lever is moved forward. But until a switch is made at the HI/LOW position, both defog and foot heat air enter on the low temperature schedule. Thus, full range of the temperature control knob (from 8 o'clock to 4 o'clock positions) will only produce −20°F to 100°F air - unless the defog-foot heat lever is moved into the HI range. When the HI/LOW switch is made, the temperature schedule of all entering air switches to the high temperature curve. Thus, if the temperature control knob was positioned toward HOT (about 3:00 o'clock) 87°F would be the temperature of incoming air in the low temperature range, but when the switch to HI is made, the temperature would change to 137°F. The HI-LOW switch is actuated after approximately 50% of defog-foot heat lever travel to provide more air through the defog nozzles without switching to the high temperature range, thus aiding crew comfort at low altitudes. As the defog-foot heat lever is moved forward through full travel, the foot heat butterfly valves - for both front and rear cockpits - are closing as the defog valve opens. Thus the defog air volume increases on a rather steep slope, and when the lever is closed to full defog position (full forward) the temperature of the air entering the cockpit is quite warm.

NOTE

> The detection of a low pitched howl in many cases indicates icing in the pressure suit heat exchanger. This "organ piping" effect can be eliminated by increasing the cabin air temperature.

Manual Override-Cockpit Temperature Mixing Valve

If the automatic temperature control malfunctions, the manual position of the temp control auto-manual switch can be used to select a full range of temperatures up to 230°F. The HI/LOW switch on the defog-foot heat lever is bypassed. Thus the entire temperature range for both foot heat and defog air is scheduled directly by the mixing valve position, which in turn is moved only when the temperature control switch is held to either HOT or COLD. The switch is spring-loaded to OFF and in the OFF position the mixing valve is held stationary.

CAUTION

> The manual override should be used only if an automatic temperature control system malfunction occurs. To increase the temperature in this mode, the manual control switch should be held toward the HOT position for no more than 1/2 second at a time between pauses of at least 3 seconds until the desired temperature is reached. Actuating the switch for more extended periods does not allow the temperature limitor adequate time to function, and may result in an overheat condition. Detection of smoke in the cockpit after use of manual control is evidence of improper use of the switch and requires the selection of a colder valve position to avoid overheating of the cockpit distribution ducting.

Cockpit Ambient Fog

It is possible, through selection of cold temperature settings, particularly on humid days, for the air conditioning system to deliver air at temperatures well below the dew point, with resultant cockpit fogging. This fog can be rapidly dissipated by selecting a slightly warmer temperature. When operating in high humidity conditions, it is recommended that warmer than normal temperatures be selected prior to starting the takeoff run, to preclude the possibility of cockpit fogging as thrust is increased.

Emergency Vent Knob

The cockpits may be cleared of undesired smoke or fumes and the cabin air conditioning unit may be shut off by pulling up on the emergency ventilating handle (figure A-1, appendix A). Push button on top of knob and then pull up on the knob. The handle may be placed in an intermediate position to obtain desired amount of emergency ventilation. When pulled up, three actions occur simultaneously.

a. All air conditioning and pressurization air from the cabin air conditioning unit to the cockpits, and rain removal system is shut off.

b. The cabin/pressure regulator and safety (dump) valve is opened and the cockpit becomes completely depressurized.

c. A ram air shutoff valve is opened and the atmospheric air is allowed to enter the cockpit through a port just forward of the pilot's feet.

WINDSHIELD DEFOGGING

Fogging of the windshield is prevented by heating the inside surface with incoming cabin air that is diverted into the defogging manifolds along the lower surfaces of the side and center panels. The defog lever (figure A-1, appendix A) on the pilot's right console, outboard of the right utility panel, is provided to select windshield defogging. The lever proportions the cabin airflow between the foot heaters and windshield defogging tubes such that in the full aft (FOOT HEAT) position approximately 90% of the total cockpit airflow is delivered to the pilot's and RIO's air distribution manifolds and 10% through the windshield defog manifold. At the full forward (DEFOG) position approximately 20% of the total airflow is delivered through the foot heat manifold and 80% through the windshield defog manifold. Obtaining adequate defog air is achieved only after the defog-foot heat lever has been moved to the HI range. The pilot should attempt to anticipate fogging conditions and preheat the windshield.

COCKPIT AIR-CONDITIONING and PRESSURIZATION SYSTEM

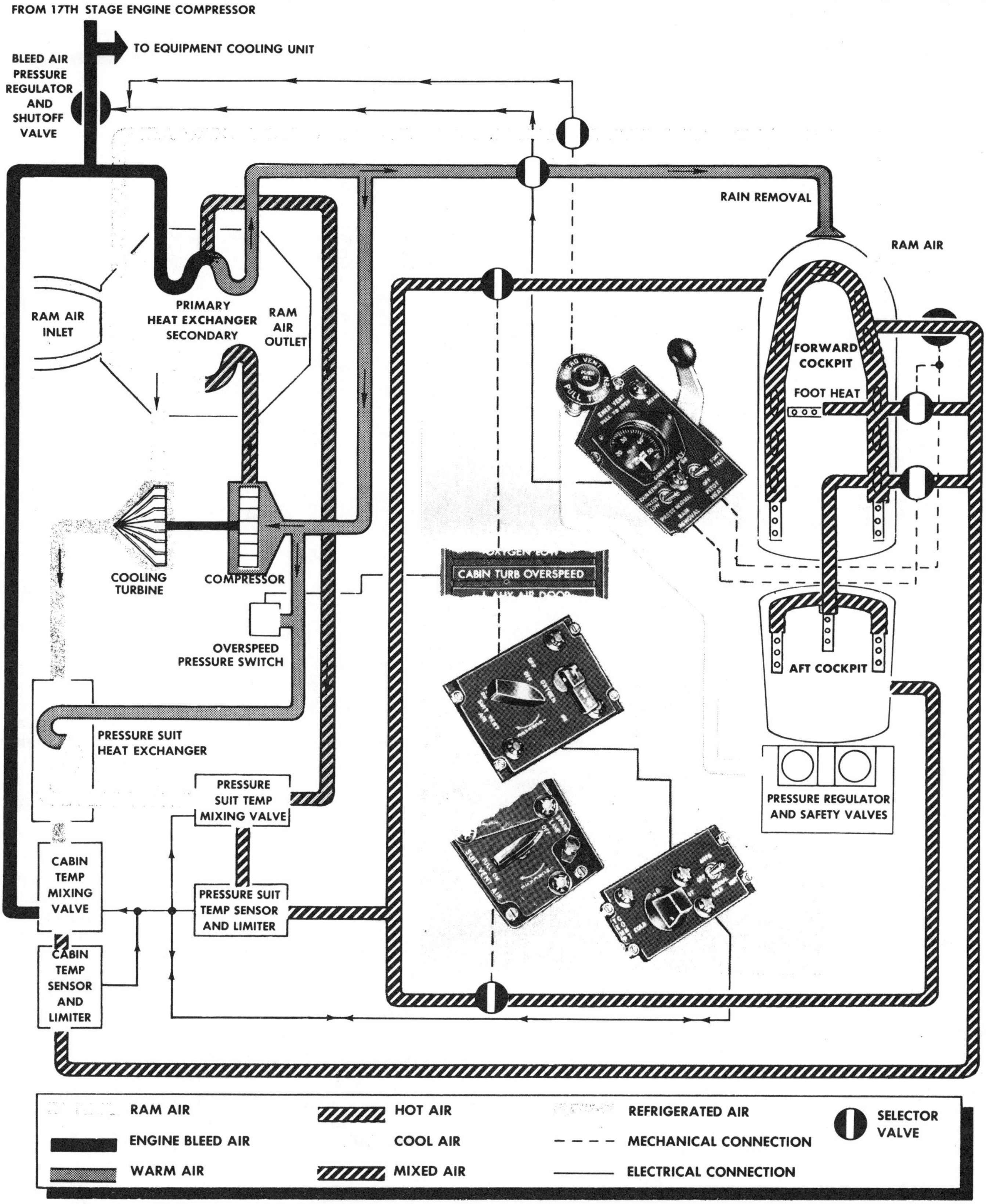

Figure 1-2

COCKPIT TEMPERATURE SCHEDULE

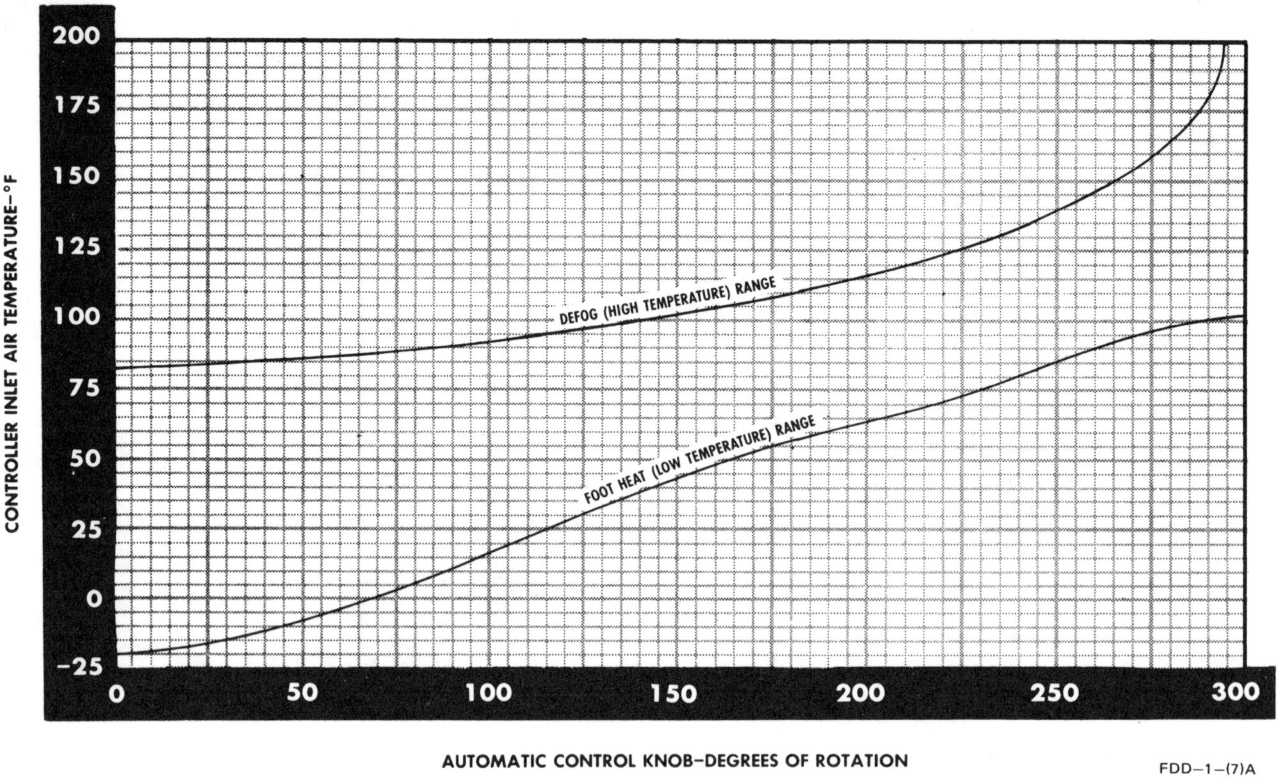

Figure 1-3

NOTE

If the windshield starts to fog over and it is imperative that the pilot maintain visual contact outside the cockpit, the temperature rheostat should be turned full cold and windshield defog air applied. In the HOT position the warm air picks up moisture in the air conditioning system, and when it encounters the relatively cold windshield, condensation (fogging) invariably occurs. Generally, applying hot air to a partially fogged windshield will completely fog it over in a matter of seconds.

WINDSHIELD RAIN REMOVAL

Windshield rain removal is controlled by a rain removal switch (figure A–1, appendix A) on the right utility panel, front cockpit. Placing the switch ON opens a valve causing warm air to flow through nozzles directed up the outer surface of the windshield center panel. This air breaks up the rain drops into small particles and diverts the majority of them over the windshield. A W'SHIELD TEMP HI light on the telelight panel illuminates when the windshield material approaches a temperature which causes optical deterioration. If the light illuminates, the system should be turned off immediately. If the windshield rain removal system cannot be shut down, pull up on the cockpit emergency vent handle. Engine bleed air will be shut off prior to entering the rain removal ducts. If the windshield temperature sensors have not been calibrated properly, the light may illuminate as a result of aerodynamic heating (occurs only near level flight maximum speed with maximum afterburning). In this event, the overheat signal may be disregarded. On aircraft after AFC 328, the rain removal switch is changed to a three position switch with positions of RAIN REM–ON, PRESS LOW, and PRESS NORMAL. This switch provides two methods of reducing cockpit noise. Inlet bleed air pressure is controlled by selecting the PRESS LOW or PRESS NORMAL positions on the rain removal switch. When the PRESS NORMAL position is selected, the bleed air pressure is regulated at 62 $\pm$5 psi. Selection of the RAIN-REM-ON position reduces the bleed-air pressure to 40 $\pm$5 psi, reducing the cockpit noise. This same lower cockpit noise level can be obtained by placing the switch to PRESS LOW position. This position enables the pilot to select a lower pressure setting on the bleed air pressure regulator without actuating the rain removal system.

CAUTION

- For a static ground check, the system must be operated with leading edge flaps in the down position and engines running at or below 88% rpm.
- Do not operate the rain removal system in flight with a dry windshield or at airspeed above Mach 1.0.

COCKPIT PRESSURE SCHEDULE

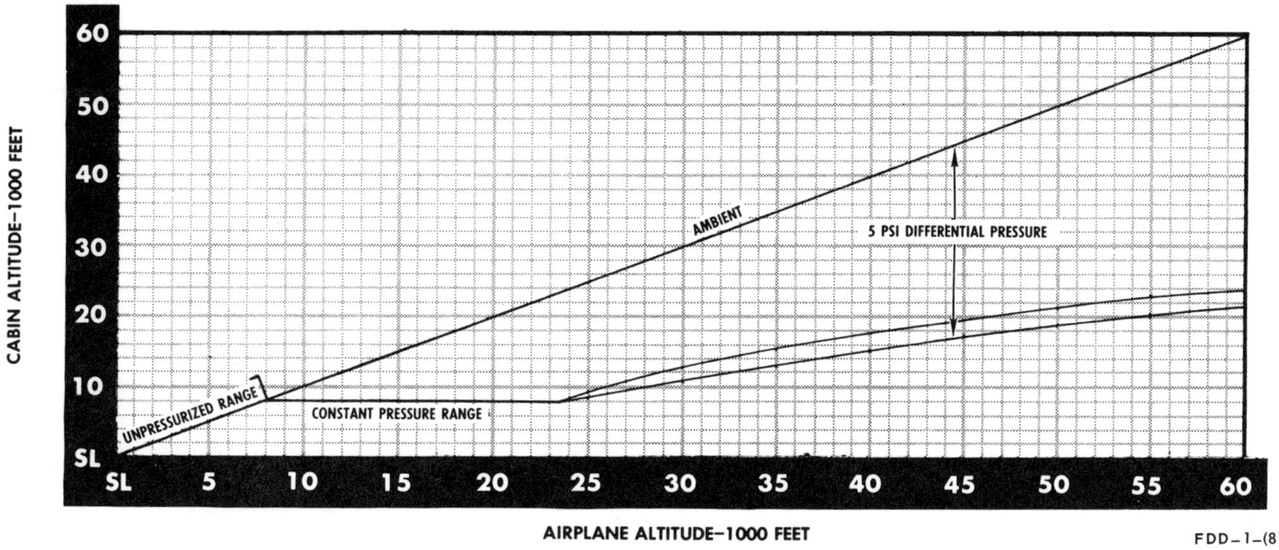

Figure 1-4

- Unless visibility is seriously restricted, the rain removal system should not be used in conjunction with afterburner.

COCKPIT PRESSURIZATION

With the canopy closed, and with the cockpit refrigeration system in operation, the cockpit automatically becomes pressurized at an altitude of 8000 feet and above (figure 1-4). The pressure in the cockpit is maintained by the cockpit pressure regulator (on the cockpit floor aft of the RIO's seat), which controls the outflow of air from the cockpit. Below 8000 feet, the regulator relieves cockpit air at a rate to keep the cockpit unpressurized. Above 8000 feet, the regulator relieves cockpit air as necessary to follow a definite cockpit pressure schedule. Operation of the pressure regulator is completely automatic. The cockpit safety valve prevents the cockpit pressure differential from exceeding positive or negative differential pressure limits in case of a malfunction of the cockpit pressure regulator, and provides an emergency means of dumping the cockpit air. The dump feature of the safety valve is pneumatically connected to a dump feature on the cockpit pressure regulator. Both valves which are operated pneumatically from a single control have sufficient capacity to permit the cockpit differential pressure to be reduced from 5.5 psi to 0.05 psi within 5 seconds or less.

CAUTION

Opening the canopy when the cockpit is overpressurized may result in canopy hinge damage, or a canopy separation. Therefore, any time the cockpit is overpressurized, dump cockpit pressurization prior to operating the canopies.

Cockpit Pressure Altimeters

The pressure altitude of the cockpit is indicated on a pressure altimeter. The pilot's cabin altimeter (figure A-1, appendix A), is on the right console. The RIO cabin altimeter (figure A-2, appendix A) is in a panel on the left side of the aft cockpit. The cabin altimeters are vented directly to cockpit pressure.

Cabin Turbine Overspeed Indicator Light

The cabin turbine overspeed indicator light (figure A-1, appendix A), is on the pilot's telelight panel. The indicator light illuminates when the cooling turbine in the cockpit refrigeration unit is being subjected to pressures and temperatures in excess of normal operation, and is therefore subject to premature failure. If possible, the airplane speed and engine power should be reduced until the light goes out. If the light fails to go out, the pilot should then select ram air by pulling UP on the emergency vent knob, which will divert ram air into the cockpit and at the same time shut off bleed air to the refrigeration unit, thereby stopping the cooling turbine. The cooling turbine may also be shut down by pulling the cockpit heat and vent circuit breakers (L9, M9, before AFC 388; M9, N9 after AFC 388; No. 2 panel).

EQUIPMENT AIR CONDITIONING SYSTEM

The equipment air conditioning system consists of an air-to-air heat exchanger, pressure regulating and shutoff valve, expansion turbine, mixing valve, and temperature controls. It also includes a liquid coolant heater exchanger, coolant pump, coolant reservoir, and pressure and relief valves. The unit provides conditioned air to the nose radar package and radar unregulated power supply, the CNI equipment shelf aft of the nosewheel well, the electronic equipment shelf aft of the rear cockpit, and the liquid coolant heater exchanger. The conditioned air directed to the liquid coolant heat exchanger is used to cool the coolant which in turn cools the radar package in the nose. The equipment air conditioning system also provides high pressure air (Auxiliary air) for fuselage fuel tank pressurization, wing fuel tank pressurization and transfer, anti G suits, electronic equipment pressurization, high pressure pneumatic system air source, ADC, and canopy seal pressure. Operation of the system is entirely automatic with airflow initiated on engine start. Engine bleed air, after flowing through the heat exchanger and pressure regulating valve, is expanded through the cooling turbine and then mixed with warm bleed air as necessary to provide a delivery temperature of 85°F from sea level to 25,000 feet, and 40°F above 25,000 feet. In the event of a system failure such that air temperatures exceed 150 ±10°F, the air conditioning unit is automatically shut off and emergency ram air cooling is provided. A warning light labeled RADAR CNI COOL OFF illuminates on the radar intercept officer's instrument panel and the pilot's right vertical panel whenever ram air is being utilized for cooling. A reset button labeled cooling reset is on the RIO's main instrument panel, and on the console below the pilot's right vertical panel. On aircraft 158355at and up or after AFC 555, the reset button on the RIO's main instrument panel is integrated with the RADAR CNI COOL OFF light If the RADAR CNI COOL OFF light illuminates, attempt to restart the refrigeration unit by reducing speed below that at which the light illuminated, waiting at least 15 seconds and then depressing the cooling reset button. If refrigeration unit fails to restart, no further restart should be made. On aircraft 155875al and up and all others after AFC 463, a diverter valve in the air inlet duct of the liquid coolant heat exchanger diverts air to the nose radar package for pulse radar operation during sea level static conditions. The diverter valve is normally open in flight or during radar ground operations using high power modes. When the aircraft is on the ground and using internal electrical power, the diverter valve is closed (if a radar high power mode is not selected) and the RADAR COOL DIV CLOSED light on the telelight panel illuminates. This light illuminates only as an indicator light and should not illuminate in flight. If the light illuminates during flight, a radar low power mode must be selected. Since the RADAR COOL DIV CLOSED light causes the MASTER CAUTION light to illuminate, these lights normally come on at touchdown if the radar set is in a low power mode. Refer to limitations in NAVAIR 01–245FDD–1A. On aircraft 158355at and up or after AFC 555, the equipment air conditioning system is modified by the addition of three ground cooling fans, a water separator and a RADAR LIQ COOL OVERHT light. The modified system supplies cooling air to the same aircraft units at approximately the same temperatures as the previous system; but ground cooling capabilities and moisture prevention are greatly improved. The ground cooling fans provide a positive air flow to the various systems requiring cooling air and operate automatically whenever the generators are ON and the aircraft weight is on the main gear. If the ambient temperature is below 0°F the fans are not needed for adequate cooling and do not operate. Sufficient cooling is also provided for radar/CNI ground operation with external power applied and the ground cooling shutoff switch, in the nose wheel well, in the ON position. Engine operation or external cooling air is not required with the ground cooling fans operating, provided ambient temperature is below 103°F. If one or more of the fans fail (when operation is required) the RADAR CNI COOL OFF light illuminates and cannot be reset. In this event, a radar low heat mode must be selected, utilization of equipment requiring cooling air must be kept at a minimum, and sustained static engine operation be kept below 80% rpm. After takeoff, the fans are not required and the warning circuit is removed from the RADAR CNI COOL OFF light; at this time the reset button must be pressed to reset the equipment cooling system. The RADAR COOL DIV CLOSED light is removed from the modified system and a RADAR LIQ COOL OVERHT light, on the telelight panel, is added. The RADAR LIQ COOL OVERHT and the MASTER CAUTION lights illuminate if the radar liquid coolant temperature reaches 145° ±5°F. In this event, the radar is not receiving sufficient cooling and a low heat mode must be selected.

CAUTION

- Malfunction of the equipment cooling turbine may be indicated by a high pitch whine and/or vibration in the nose of aircraft. The turbine may be shut off by pulling equipment cooling circuit breakers (L8, N8, No. 2 panel, aircraft 153071z thru 155528ag before AFC 388) (N8, M8, No. 2 panel, aircraft 155529ag and up and all others after AFC 388). This will shut off equipment air conditioning, and turn on emergency ram air cooling. Auxiliary air (fuel tank pressure and transfer, anti G suits, etc.) will not be affected.

- When operating with emergency ram air cooling avoid high speed flight if possible. Maximum allowable cooling temperatures may be exceeded during high speed flight with the result that the following electronic equipment life and/or reliability may be affected: AIM–7 Four Channel Tuning Drive Radar Package, Tacan, IFF, and UHF Comm Radio Receiver–Transmitter.

- If the RADAR CNI COOL OFF light does not extinguish when the reset button is depressed, place the Tacan to OFF, the IFF and radar to STBY, and operate only when necessary.

NOTE

- If the RDAR CNI COOL OFF light illuminates on the ground, it may be due to a failed ground cooling fan. To determine if a fan is at fault pull and immediately reset the EQUIP COOL circuit breaker (N8, No. 2 panel). If the RADAR CNI COOL OFF light illuminates immediately, the equipment cooling system has failed and the mission must be aborted. If the light takes approximately 10 seconds to illuminate, a ground cooling fan has failed and the equipment cooling system can be reset after takeoff.

- Illumination of the RADAR CNI COOL OFF light shall be logged on the yellow sheet (OPNAV Form 3760–2).

EQUIPMENT AUXILIARY AIR SYSTEM

The equipment auxiliary air system utilizes partially cooled 17th stage engine bleed air after it has passed through the equipment air conditioning air-to-air heat exchanger. This partially cooled air is distributed to the anti-G suits, canopy seals, air data computer, fuel system pressurization, pneumatic system air compressor, radio receiver-transmitters (Tacan), radar wave guide, and radar antenna.

CAUTION

On aircraft 158355at and up or after AFC 355, if the RADAR CNI COOL OFF light illuminates on the ground, due to a ground cooling fan failure; the equipment auxiliary system/components may be damaged if sustained static engine runs are made above 80% rpm.

Anti G Suit System

The anti-G system delivers low pressure equipment auxiliary air to the anti-G suits. The air is routed through the anti-G suit control valve and then to the suit. The suit remains deflated up to approximately 1.5 G. As this force is reached or exceeded, air will flow into the suit in proportion to the G forces experienced. When the G force levels off to a constant, the suit remains inflated in proportion to the constant G-force. As the G forces decrease, the suit begins to deflate, again in proportion to the decreasing G forces. A manual inflation button in the anti-G suit control valve allows the crewman to manually inflate his suit for purposes of checking the system or for fatigue relief. A pressure relief valve incorporated within the system is set to relieve at approximately 11 psi and is used as a safety backup in the event of a malfunction. The system is automatic and operates any time an engine is running.

NORMAL OPERATION

Optimum cockpit environment can be achieved by placing the override selector switch on the temperature control panel to AUTO, and adjusting the temperature control knob for the desired cockpit temperature. Adjust the defog control lever on the right utility panel for personal comfort and effective windshield defogging. If the automatic temperature control system fails, a temporary adjustment may be obtained by bumping the override selector switch to the HOT or COLD position. To prevent windshield defogging during letdown into hot humid atmosphere, place the override selector switch to AUTO, and have the defog lever positioned about ¾ of the way forward. Five minutes prior to letdown select the full defog position and adjust the temperature control knob to the 2 o'clock (200° clockwise rotation) and maintain these settings throughout the letdown. If fogging persists, and will not clear up, retract flaps if extended, or increase power (use speed brakes as necessary to maintain airspeed) to provide more engine bleed air to the mixing valves.

EMERGENCY OPERATION

Although there are no provisions made for emergency operation of the cockpit air conditioning system, emergency ventilating air is available. The cockpits may be cleared of undesired smoke or fumes and the cockpit air conditioning unit may be shut off by depressing the button and pulling up on the emergency ventilating knob. The handle may be placed in an intermediate position to obtain the desired amount of emergency ventilation.

LIMITATIONS

There are no specific limitations pertaining to the operation of the Air Conditioning and Pressurization System.

AIR DATA COMPUTER SYSTEM

NOTE

Some illustrations referred to within this system writeup may be located in appendix A. These illustrations are referred to, within the text, as (figure A-, appendix A).

DESCRIPTION

The air data computer system (ADC) receives inputs of static pressure, pitot pressure, total temperature, angle of attack, and 17th stage air. These inputs are supplied by two static ports, one on each side of the aft part of the radome, a total temperature sensor on the left air conditioning inlet duct, a pitot tube on the vertical fin, an angle of attack probe on the left forward side of the fuselage, and bleed air from the equipment air conditioning system. These inputs are corrected in the air data computer to compensate for errors in the sensing equipment installation. The corrected inputs are converted to usable outputs by the ADC and are displayed on the airspeed/mach indicator, the altimeter, and the vertical velocity indicator. Corrected signals are also used by other aircraft systems; see figure 1-5. The instruments and/or systems utilizing the outputs from the ADC are inoperative or in error if a failure or an interruption occurs in the essential ac power supply, essential dc power supply, or engine bleed air system. An interruption or failure in any of the above systems illuminates the STATIC CORR OFF indicator light.

STATIC PRESSURE COMPENSATOR

One of the functions of the ADC is to supply all systems requiring static pressure inputs with a static source position error correction. This correction is accomplished through the static pressure compensator. When operating normally, the compensator utilizes the static air pressure as a balancing force only. The corrected static pressure output is actually auxiliary equipment air, corrected for the static source error as dictated by the instantaneous flight situation. If a malfunction occurs in the compensator, a fail-safe solenoid is deenergized allowing static pressure from the static source to be routed directly to all systems requiring static pressure inputs. With a malfunction, overall accuracy suffers; however, no system dependent on static pressure becomes inoperative.

Static Pressure Compensator Switch

The pilot is alerted to a compensator malfunction by illumination of the STATIC CORR OFF indicator light, on the telelight panel (figure A-1, appendix A). Light illumination may be accompanied by a rapid change in the altimeter reading. A static pressure compensator switch is on the inboard engine control panel (figure A-1, appendix A). This switch is the only control associated with the air data computer system. The switch has positions of RESET CORR, NORM, and CORR OFF, which are used to reset or turn off the compensator. With the STATIC CORR OFF indicator light illuminated, moving the compensator switch to the RESET position returns the compensator to normal operation and extinguishes the indicator light. The switch may then be released and it returns to the NORM position. If the compensator cannot be reset, as evidenced by the indicator light again illuminating when the switch returns to the NORM position, move the switch to the OFF position. The pressure instruments are in error any time the light is illuminated or the switch is in the OFF position. Refer to the airspeed and altimeter position error correction charts in section IV and the NATOPS pocket checklist.

NOTE

The static pressure compensator must be reset after the engines are started. The altimeter reading after reset must not differ from the altimeter reading before reset by more than ±40 feet. This variation is an indication of the accuracy of the compensator and its effect on other instruments. It is possible to experience large errors in both altitude and airspeed if the altimeter jump exceeds 40 feet.

ALTITUDE ENCODER UNIT

Aircraft 153780ac and up, and all others after AFC 388, have an altitude encoder unit installed. The altitude encoder is a dual purpose electronic unit in the rear cockpit that receives static pressure signals from the ADC. The altitude encoder in turn provides a digital output of altitude in 100 foot increments, referenced to a standard day (29.92 inches Hg), to the IFF transponder, and a synchro output to two servoed altimeters. This synchro signal is modified within the altimeters to provide reference to the actual altimeter setting. When mode C is selected on the IFF control panel, automatic altitude reporting, in a coded form, is provided to the air traffic control system, thereby eliminating a need for voice communications. For more information on mode C operation and on the servoed altimeters, refer to Identification System and Instruments, this section.

NORMAL OPERATION

Normal operation of the ADC consists of momentarily placing the static pressure compensator switch in the spring-loaded RESET position, after an engine has been started. This action extinguishes the STATIC CORR OFF light, and the ADC is operating with compensated static pressure. If the STATIC CORR OFF light illuminates in flight and cannot be reset, the AFCS, navigation computer, AMCS, and most flight instruments are in error. The ADC receives power from a three phase, essential 115 volts ac bus, a 28 volts dc essential bus, and a 28 volts ac essential bus. The system is protected by circuit breakers on No. 1 circuit breaker panel in the rear cockpit. The circuit breakers are marked CADC.

NOTE

The variable area inlet ramps may extend while taxiing behind another aircraft's exhaust. Ensure that the variable area inlet ramps are fully retracted prior to takeoff.

EMERGENCY OPERATION

No emergency operations pertain to the ADC. If the STATIC CORR OFF light illuminates in flight and cannot be reset, the AFCS, navigation computer, AMCS, GVR-10, and most of the flight instruments are in error.

LIMITATIONS

After initial altimeter jump, the altimeter variation should not exceed ±40 feet when the SPC is reset. Refer to section XI for SPC OFF airspeed and altimeter position error corrections.

AIR DATA COMPUTER[1]

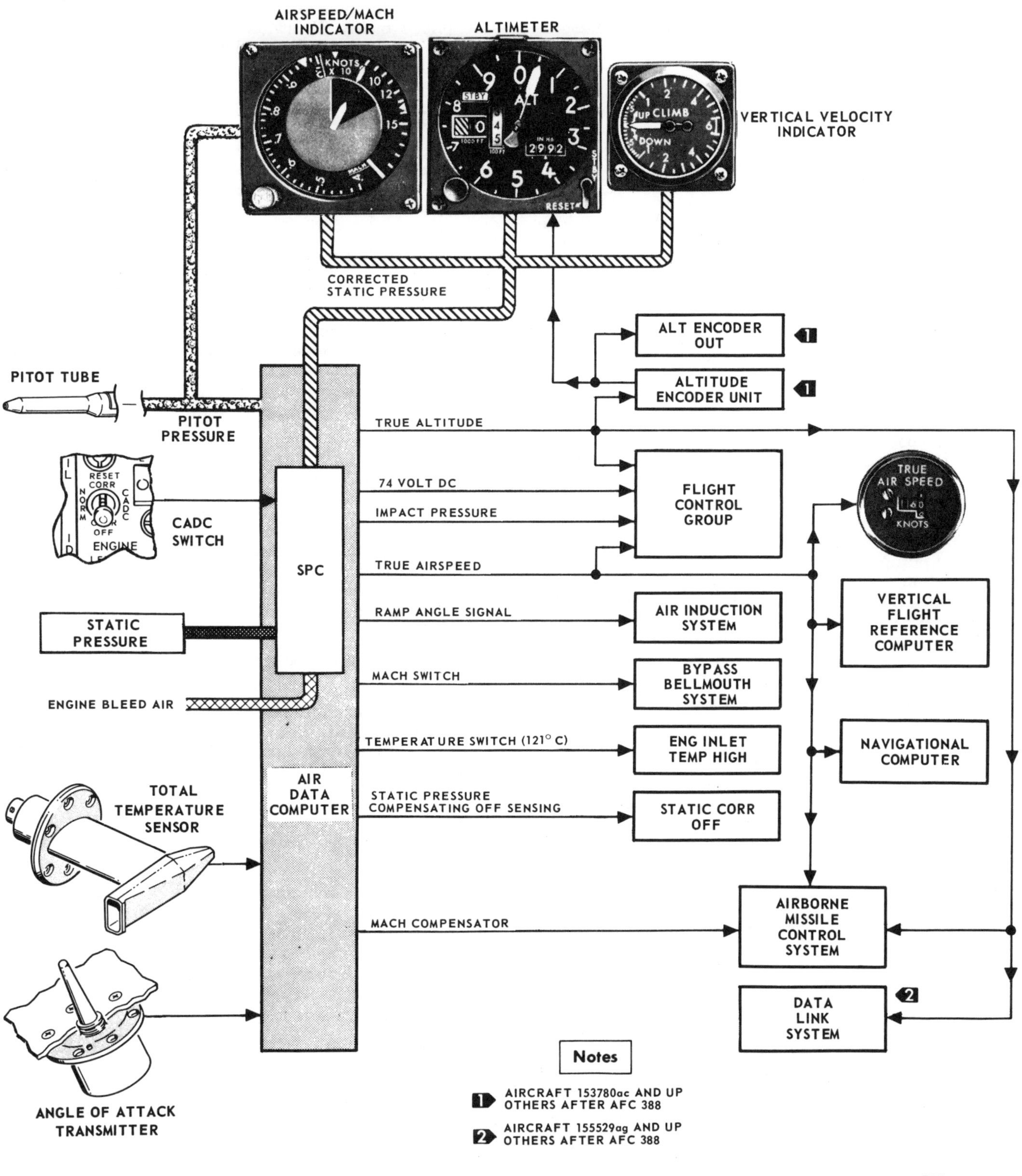

Figure 1-5

AIR REFUELING STORE

NOTE

Some illustrations referred to within this system writeup may be located in appendix A. These illustrations are referred to, within the text, as (figure A-, appendix A).

DESCRIPTION

The airplane can be used as a probe-and-drogue type air refueling tanker upon the installation of a D-704 air refueling store (hereafter referred to as buddy tank). The tank (figure 1-6) is divided into three major sections: power supply, fuel cell, and hose-reel mechanism. The power supply section contains an electrically feathered and governed ram air turbine, a hydraulic pump, a hydraulic fluid radiator and a hydraulic reservoir. The fuel cell contains a hydraulically driven fuel transfer pump, gravity and pressure fueling receptacles, emergency fuel dump and manual drain valves, float actuated fuel shutoff and vent valves, and electrical, fuel and air connections. The hose-reel mechanism section, contains a spool with 50 feet of hose, a drogue refueling receptacle, a level wind mechanism, a hydraulic motor, and a cartridge actuated guillotine for emergency hose and drogue jettisoning. The ram air turbine drives the hydraulic pump which, in turn, drives the fuel transfer pump and hydraulic motor. The fuel transfer pump is capable of pumping 200 gallons per minute at 55 psi. The hydraulic motor is used to snub the hose and drogue during extension, and to rewind the hose and drogue during retraction. The buddy tank contains 300 gallons of fuel, which can be transferred to the tanker airplane for its own use, or to a receiver airplane. The tanker airplane can transfer its internal and external fuel load to the buddy tank, and subsequently to a receiver airplane.

BUDDY TANK CONTROL PANEL

The buddy tank control panel is on the left side of the rear cockpit instrument panel. The panel will normally be installed upon installation of the buddy tank, and removed when the buddy tank is removed. The panel contains six switches and two indicators; a power switch, a hose control switch, a fuel transfer switch, a hose jettison switch, a tank light switch, a ship-tank transfer switch, a drogue position indicator, and a gallons delivered indicator.

Buddy Tank Power Switch

The buddy tank power switch is a guarded toggle lock switch with positions of ON, OFF and DUMP. Placing the switch to ON energizes the buddy tank electrical system which electrically unfeathers the tank's ram air turbine. Placing the switch to OFF feathers the ram air turbine. The DUMP position is guarded. When in DUMP, a solenoid operated dump valve is opened for emergency fuel jettison. If an electrical failure occurs while jettisoning buddy tank fuel, the dump valve automatically closes and any fuel remaining in the tank will be trapped.

NOTE

Before dumping buddy tank fuel, ensure that the pilot's buddy fill switch is in the STOP FILL position to preclude complete depletion of aircraft fuel system.

Hose Control Switch

The buddy tank hose control switch, marked EXT and RET, is used for normal extension and retraction of the hose and drogue.

Transfer Switch

A solenoid operated transfer switch has positions of OFF and TRANS. Selecting TRANS allows automatic fuel transfer to a receiver airplane upon proper drogue engagement. Selecting OFF discontinues fuel transfer. If an electrical failure occurs while transferring fuel to a receiver airplane, the buddy tank transfer switch will revert to OFF.

Hose Jettison Switch

The buddy tank hose jettison switch, marked OFF and CUT, is a guarded toggle lock switch. Placing the switch to CUT electrically fires a cartridge that actuates the hose guillotine mechanism. Also, all electrical power, except power to operate the dump valve, is shut off.

Buddy Tank Light Switch

The buddy tank light switch, marked BRT (bright) and DIM, controls the brilliance of the two indicator lights on the tail cone of the tank. Normally the BRT position would be selected during daylight hours, and the DIM position at night.

Ship-Tank Transfer Switch

This switch is not wired into the airplane electrical system and is inoperative. Fuel is transferred to the buddy tank by placing the buddy fill switch on the fuel control panel (figure A-1, appendix A) to FILL. Buddy tank fuel may be transferred to the airplane fuselage cells by placing the external transfer switch on the fuel control panel to CENTER.

Drogue Position Indicator

The buddy tank drogue position indicator has readouts of RET (retract), EXT (extend), and TRA (transfer). The position of the drogue is indicated by a drum dial viewed through a cutout in the panel. The indicator shows RET when the hose and drogue are completely retracted, EXT when the hose and drogue is extended and ready for engagement, and TRA when the receiver aircraft has

D-704 AIR REFUELING STORE (BUDDY TANK)

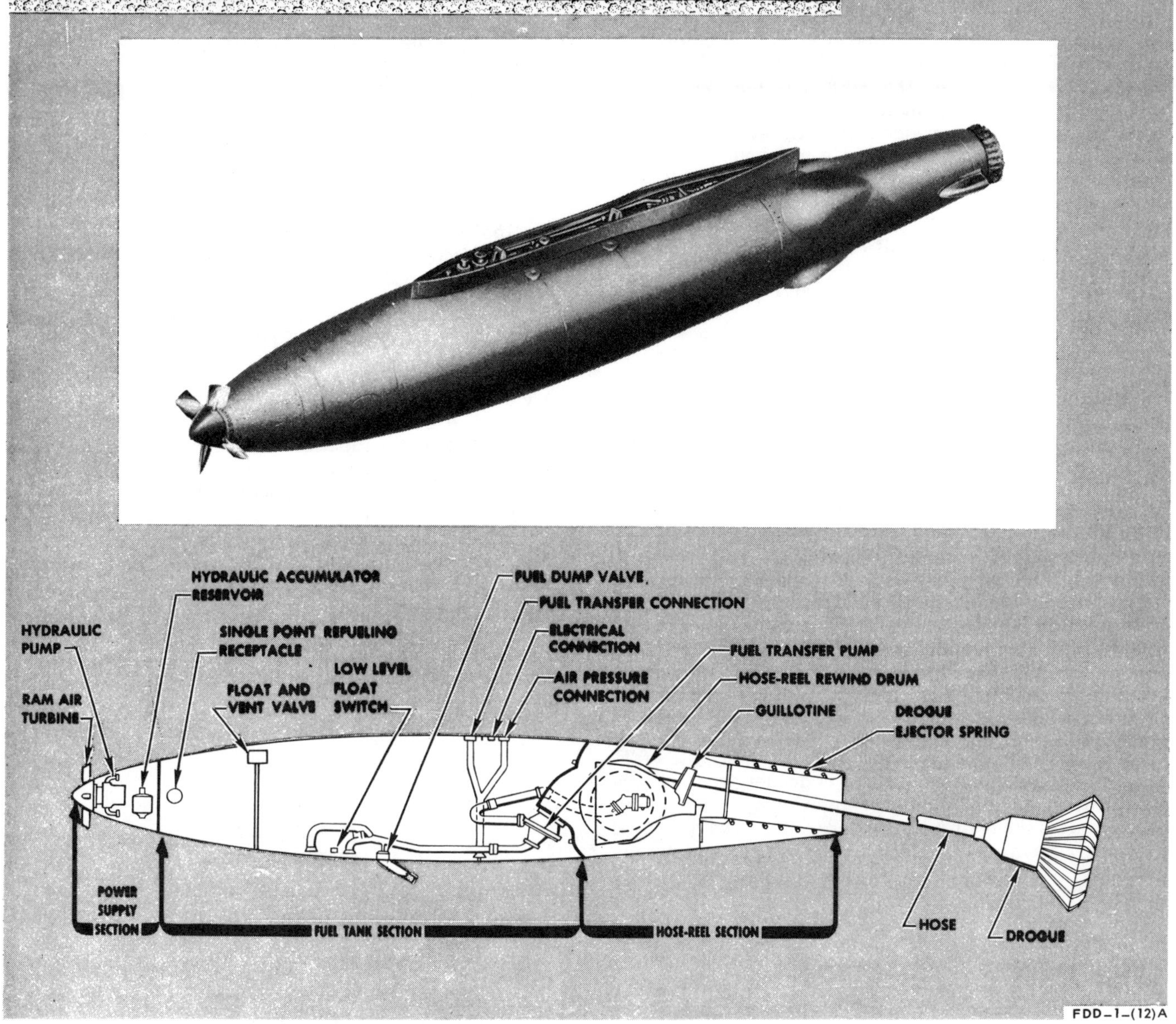

Figure 1-6

completed engagement and retracted hose and drogue a minimum of 2 feet.

NOTE

The drogue position indicator will not show TRA unless the hose and drogue have completely extended prior to its 2-foot rewind.

Gallons Delivered Indicator

A gallons delivered indicator presents a direct reading of total gallons delivered, in 2 gallon increments. A reset knob located immediately adjacent to the indicator windows, permits resetting the gallons delivered indicator to zero.

NORMAL OPERATION

When the buddy tank power switch is placed to ON, 28 volts dc, and 115/200 volts ac power energizes the buddy tank electrical system. Once the electrical system is energized, the ram air turbine unfeathers and drives the hydraulic pump. When the hydraulic pressure builds to approximately 1600 psi, the hose-reel lock mechanism unlocks to allow action of the hose-reel mechanism. When the hose control switch is then placed to EXT, hydraulic pressure to the rewind motor is decreased from 3000 psi to approximately 1000 psi and the drogue ejector spring ejects the drogue out of its receptacle in the tail cone of the tank. Upon ejection, the drogue will blossom to its full target diameter of 24 inches. Air drag on the drogue will then complete the hose extension. The hydraulic rewind motor maintains heavy hose and drogue snubbing during the initial drogue ejection and the last few feet of hose extension. The snubbing reduces shock loading on the hose, drogue, and tank. When the hose and drogue are fully extended the amber READY light on the tail cone will illuminate, and the drogue position indicator will display EXT. For DRY receiver hook-ups, the transfer switch must remain in the OFF position. For WET hook-ups, the transfer switch must be placed in the TRANS position. After engagement of the drogue, the hose must be retracted a minimum of 2 feet before fuel will transfer. When the hose and drogue are retracted 2 feet, the drogue position indicator reads TRA. When fuel is being transferred to the receiver aircraft, the green TRANSFER light on the tail cone illuminates and the amber READY light goes out. If the buddy fill switch is in STOP FILL, only buddy tank fuel (300 gallons) will be transferred.

NOTE

If the buddy fill switch is held in the FILL position, the tanker airplane can transfer its entire fuel supply.

The amount of fuel transferred is indicated on the gallons delivered indicator. Buddy tank fuel transfer may be terminated by three means: by placing the fuel transfer switch to OFF; by emergency disengagement of the receiver; and by low level float switch actuation when the buddy tank empties. A major surge or reduction in fuel pressure will also terminate buddy tank fuel transfer. After probe separation, the hose control switch should be placed in the RET position. The hydraulic rewind motor retracts the hose and drogue. When the drogue position indicator reads RET, the buddy tank power switch may be placed to OFF. When the power switch is placed to OFF, the buddy tank electrical system is deenergized and the ram air turbine will feather, terminating hydraulic pump operation. As the hydraulic pressure decays, the hose-reel lock mechanism mechanically locks the hose and drogue in its retracted position. Also, the buddy tank indicator lights will go out.

EMERGENCY OPERATION

Since an empty buddy tank with its hose and drogue retracted produces the same aerodynamic and/or cushioning effect as an empty external fuel tank, no special emergency procedures are required. Fire, structural damage, uncontrollable fuel loss, etc., may require that the tank be jettisoned. Normally the buddy tank may be safetied by placing the buddy tank power switch to OFF. Placing the power switch to OFF terminates all electrical power to the tank, except for the fuel dump valve, stops hydraulic pump operation, and as a result, mechanically locks the hose and drogue. Buddy tank fuel may be jettisoned at any time by placing the buddy tank power switch to DUMP. Inflight emergencies that do not allow sufficient time for hose and drogue retraction may require hose jettisoning. Placing the hose jettison switch to CUT electrically fires a cartridge which actuates the guillotine cutting blade. The inside segment of the hose is crimped fuel tight by the guillotine blades. Fuel may be dumped after hose and drogue jettison if required.

CAUTION

If the buddy tank hose and drogue is lost unintentionally, the buddy tank electrical system will not be deenergized by placing the power switch to OFF. In this case, the ram air turbine continues to turn and the hydraulic fluid temperature increases causing damage to the O-rings. To deenergize the system and feather the ram air turbine, place the buddy tank power switch to OFF, and pull the buddy tank hyd pump circuit breaker C6, No. 2 panel on aircraft thru 155528ag and C8, aircraft 155529ag and up. Once the buddy tank holding relay is broken, the buddy tank hyd pump circuit breaker may be reset.

HOSE AND DROGUE JETTISONING

A violently whipping hose and drogue, or the inability to retract the hose for any reason, may require hose and drogue jettisoning. Placing the hose jettison switch to CUT deenergizes the buddy tank electrical system, feathers the ram air turbine, mechanically lock the hose reel mechanism, and electrically fires a cartridge which actuates the guillotine cutting blades. A holding relay delays (5 to 20 seconds) the firing of the guillotine hose cutter until the hose-reel mechanism has locked; thus, preventing the hose reel from rotating and the hose from whipping around and spraying fuel inside the tail cone. Fuel may be dumped after hose and drogue jettisoning if required. To jettison the hose and drogue proceed as follows:

1. Hose jettison switch - CUT

CAUTION

Do not change the position of the hose jettison switch after being placed to CUT. If the switch is positioned to NORMAL after jettisoning, the buddy tank electrical system becomes energized.

BUDDY TANK JETTISONING

The buddy tank may be jettisoned individually from the centerline station, or it may be jettisoned along with all external stores. Refer to External Stores Jettison chart, section V.

LIMITATIONS

The following limitations apply for carriage and operation of the buddy tank:

1. With hose retracted and turbine feathered, the maximum airspeed for carriage is 500 knots CAS or 1.1 Mach, whichever is less. The maximum acceleration for carriage is zero G to +4.0 G.
2. With hose extended, the maximum airspeed is 300 knots CAS or 0.8 Mach, whichever is less.

3. The recommended envelope during refueling is 200 knots CAS to 300 knots CAS or 0.8 Mach at any altitude from sea level to 35,000 feet. Maximum acceleration is +2.0 G and no abrupt maneuvers are permitted.
4. The maximum airspeed for hose retraction is 250 knots CAS.

NOTE

If the hose fails to retract fully at 250 knots, a reduction in airspeed to 230 knots CAS should allow complete retraction.

5. The maximum airspeed for dumping fuel from the buddy tank is 250 knots CAS.

CAUTION

Afterburners should not be used while dumping fuel from the buddy tank due to a fire hazard created by the dumped fuel passing through the engine exhaust close to the aft fuselage.

6. The maximum airspeed for jettisoning the buddy tank is 300 knots CAS or 0.9 Mach, whichever is less.

ANGLE OF ATTACK SYSTEM

NOTE

Some illustrations referred to within this system writeup may be located in appendix A. These illustrations are referred to, within the text, as (figure A–, appendix A).

DESCRIPTION

An angle of attack system is incorporated into the airplane to present a visual indication of optimum airplane flight conditions; i.e., stall, landing approach, cruise, etc. Optimum angles of attack are not affected by gross weight, bank angle, load factor, airspeed, density altitude, or airplane configuration. For example, the optimum angle of attack for landing approach is always the same regardless of gross weight. The approach airspeed automatically varies to compensate for the change in weight. The system consists of an angle of attack probe and transmitter, and angle of attack indicator, indexer lights, approach lights, and a stall warning vibrator (rudder pedal shaker). Two electrical heaters, one in the angle of attack probe, and one in the case (adjacent to the fuselage skin), prevent the formation of ice while flying through precipitation. The case heater element is energized when the static compensator switch is placed to RESET CORR, and both case and probe heater elements are automatically energized when weight is off the landing gear. The probe heater and the case heater are both powered by the left main 115 volt ac bus through the angle of attack probe heater circuit breaker (H13 No. 1 panel). Refer to figure 1–7 for angle of attack conversions and displays.

WARNING

With gear up and flaps down (bolter configuration) aircraft angle of attack will be 3 units greater than indicated angle of attack due to a difference in the airflow about the AOA probe with the nose gear door closed vice open. Thus the aircraft will stall at about 21 units indicated angle of attack with gear up and flaps down as compared to 24 units indicated angle of attack with both gear and flaps down.

ANGLE OF ATTACK INDICATOR

This indicator (figure 1–7) displays the angle of attack of the airplane. This is accomplished by means of a probe with parallel slots protruding through the fuselage skin. When the airplane changes its angle of attack, pressure becomes greater in one slot than the other, and the probe rotates to align the probe slots with the airstream. Probe rotation moves potentiometer wiper arms, producing resistance variations which are sent to the angle of attack indicator. The angle of attack indicator is calibrated from 0 to 30 in arbitrary units, equivalent to a range of −10 to +40 angular degrees of rotation of the probe. A reference bug is provided for approach (on speed) angle of attack which is set at 19.0 units (before AFC 388) or 18.3 units (after AFC 388). Approach angle of attack values are only valid with the landing gear down. Gear up angle of attack values are 3 to 4 units lower because of a local flow difference caused by the nose gear door. Additional bugs are provided and can be set at any desired angle of attack. The suggested values for the bug settings are as follows:

Climb (400 KCAS) – 5.5 units

Max Endurance – 8.5 units

Stall Warning (Before AFC 388) – 21.3 units

Stall Warning (After AFC 388) – 20.6 units

The angle of attack indicator also contains a switch that actuates the stall warning vibrator (rudder pedal shaker). On aircraft 155529ag and up, and all others after AFC 388, the approach (on speed) reference bug and the stall warning marker are set at 18.3 and 20.6 units respectively. These two changes on the angle of attack indicator compensate for error induced by extending the corner reflector. If the corner reflector is not extended, the 18.3 on speed indication is actually 0.7 units faster than proper on speed approach. When the indicator is inoperative, the word OFF shows in a small window on the face of the dial. Refer to Airspeed Indicator Failure illustrations, section V

ANGLE OF ATTACK CONVERSION AND DISPLAYS

AIRCRAFT THRU 155528ag BEFORE AFC 388

APPROACH LIGHT	INDICATOR	INDEXER	ANGLE OF ATTACK UNIT	AIRSPEED
			20.1-30	VERY SLOW
			19.5-20.1	SLIGHTLY SLOW
			18.5-19.5	ON SPEED
			17.9-18.5	SLIGHTLY FAST
			0-17.9	VERY FAST

ALL CONFIGURATIONS – FLAPS AS NOTED, GEAR DOWN

1 "g" LEVEL FLIGHT

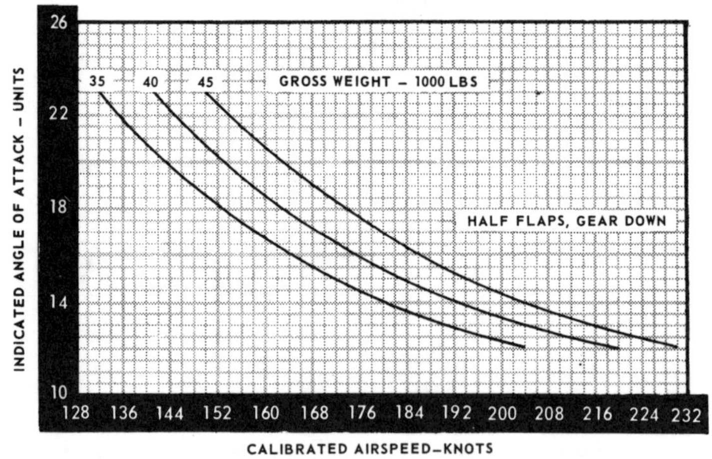

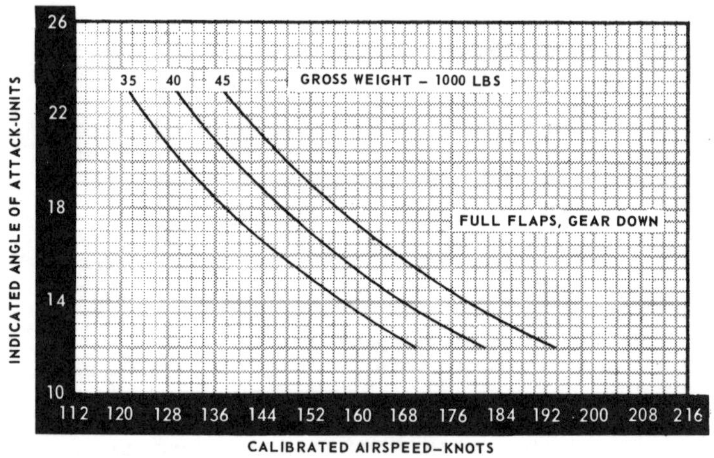

ALL CONFIGURATION-FLAPS RETRACTED, GEAR UP

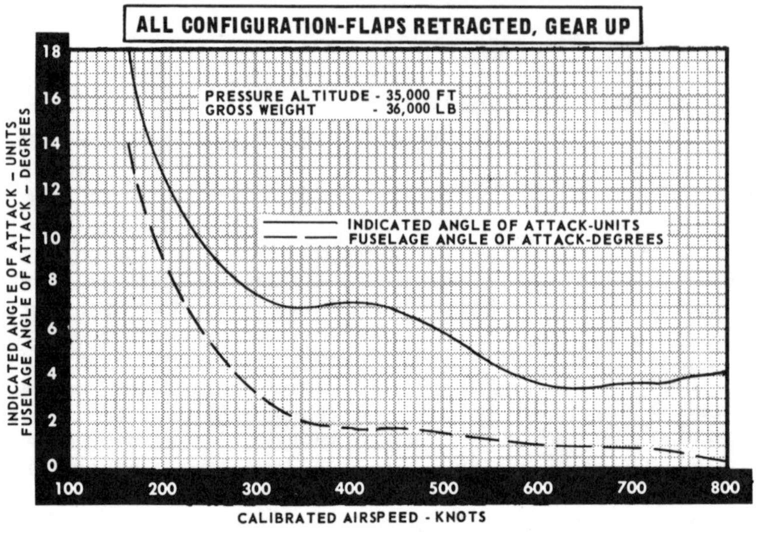

FDD-1-(13-1)

Figure 1-7 (Sheet 1 of 2)

ANGLE OF ATTACK CONVERSION AND DISPLAYS

AIRCRAFT 155529ag AND UP, AND ALL OTHERS AFTER AFC 388

APPROACH LIGHT	INDICATOR	INDEXER	ANGLE OF ATTACK UNIT	AIRSPEED
			19.4-30	VERY SLOW
		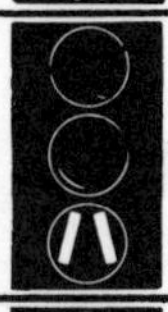	18.8-19.4	SLIGHTLY SLOW
		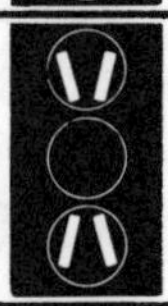	17.8-18.8	ON SPEED
			17.2-17.8	SLIGHTLY FAST
			0-17.2	VERY FAST

ALL CONFIGURATIONS – FLAPS AS NOTED, GEAR DOWN

1 "g" LEVEL FLIGHT

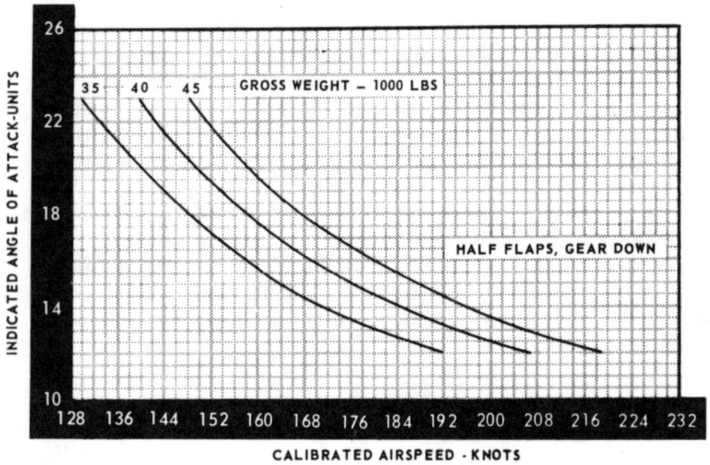

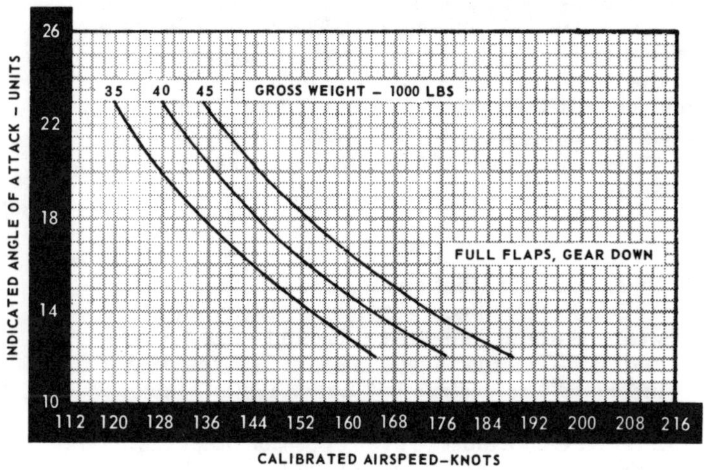

ALL CONFIGURATION-FLAPS RETRACTED, GEAR UP

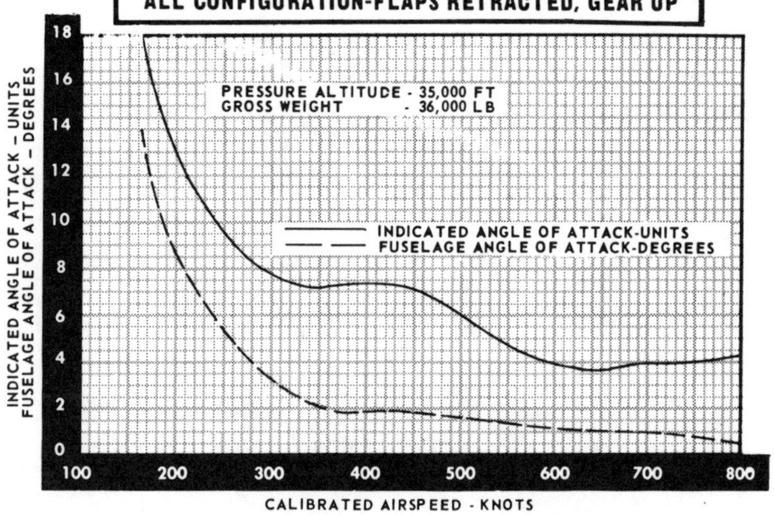

FDD–1–(13–2)

Figure 1-7 (Sheet 2 of 2)

to obtain angle of attack ranges for various drag indexes and flight conditions.

ANGLE OF ATTACK INDEXER

An angle of attack indexer is on the left side of the windshield in the front cockpit. On some airplanes there are two angle of attack indexers, one on each side of the windshield. The indexer presents landing approach angle of attack information by illuminating symbols. The symbols are energized through switches in the angle of attack indicator when the landing gear is down and weight is off the gear. At fast airspeeds (low angle of attack) the lower symbol (an inverted V) is illuminated, at slightly fast airspeed the lower symbol and circular symbol is illuminated. At optimum approach speeds only the center circular symbol illuminates. For slightly slow airspeeds the center symbol and upper symbol illuminate. For slow airspeeds (high angle of attack), only the upper symbol (a V) is illuminated.

Indexer Lights Control Knob

An indexer lights control knob is on the front cockpit right console (figure A-1, appendix A). The indexer lights are automatically illuminated when the landing gear is down and the weight is off the gear. The indexer lights control knob controls the intensity of the indexer lights. When testing the lights on the ground with the warning lights test switch, the knob can be used to adjust the lights intensity.

NOTE

On aircraft 158355at and up, and all others after AFC 506, the knob controls intensity of the weapons status panel (front cockpit), the AIM-9 and AIM-7 status panels, in addition to the AOA indexers. The instrument panel lights control knob (the flight instruments light control knob after AFC 536) must be in any position other than OFF before the ACM/indexer lights control knob (formerly indexer) is operational.

STALL WARNING VIBRATOR

Refer to Rudder Control System, this section.

APPROACH LIGHTS

An approach light system, with red, amber, and green indicator lights, shows symbolic airplane angle of attack during landing approaches. Four switches within the angle of attack indicator determine which of the three lights will be illuminated. The lights are energized through switches in the landing gear, wing flaps and arresting gear systems. A hook bypass switch on the exterior lights control panel, bypasses the wing flap and arresting gear switches. When the hook bypass switch is in NORMAL, the landing gear is down and locked and the arresting hook is down, the applicable approach light glows steady. With the hook by-pass switch in NORMAL, the landing gear down, the wing flaps 1/2 or full down and the arresting gear up, the approach light flashes. When the hook bypass switch is in BYPASS and the landing gear is down, the approach light glows steady regardless of the position of the arresting gear and wing flaps. A green approach light indicates a high angle of attack, low airspeed; an amber light indicates optimum angle of attack; and a red approach light indicates a low angle of attack, high airspeed.

NORMAL OPERATION

No controls pertain to the angle of attack system other than the indexer lights control knob.

EMERGENCY OPERATION

No alternate or emergency provisions pertain to the angle of attack system. However if the indexer lights do not illuminate following a catapult launch, the nose gear catapult extension chamber has failed to deflate. Refer to Nose Gear Strut Extension, this section.

LIMITATIONS

No limitations pertain to the angle of attack system.

APPROACH POWER COMPENSATOR SYSTEM

NOTE

Some illustrations referred to within this system writeup may be located in appendix A. These illustrations are referred to, within the text, as (figure A–, appendix A).

DESCRIPTION

The approach power compensator system (APCS) maintains the aircraft at optimum approach angle of attack by automatically positioning engine throttles. The system consists of the approach power control set, angle of attack transmitter, integrated torque boosters, control switches, and the warning light. The approach power compensator, when engaged, automatically sets the throttles at the calculated thrust, regardless of previous settings, needed to maintain the aircraft at the proper angle of attack. If the pitch attitude is being changed by the pilot to place the aircraft on the desired glide path, the engine speed automatically stabilizes at a thrust level needed to correct for this change in aircraft attitude. However, engine speed does not increase above 99% rpm or decrease below 73%. Manual throttle operation is available with the system turned off, and an emergency override feature allows the pilot to manually position the throttles any time the system is engaged by applying a force of 20 to 25 pounds per throttle. The system may be disengaged by moving the speed brake switch to the IN position if the emergency speedbrake switch is in NORMAL and the speed brake circuit breaker is in, or by moving the APCS system switch to OFF or to STBY. In aircraft 155867ak and up and all others after AFC 392, the system may be disengaged by moving the speed brake switch to IN regardless of the position of the emergency speed brake switch or the speed brake circuit breaker.

APPROACH POWER CONTROL SET AN/ASN-54(V)

The approach power control set AN/ASN-54(V) is connected in parallel with the throttles to provide manual or automatic power control. The set consists of the aircraft accelerometer, throttle control computer, and electronic control amplifier. On aircraft 155785ai and up and all others after AFC 364, the set includes the stabilator position transducer. The APCS compensates for outside air temperature and is capable of controlling one or both engines. The set utilizes normal acceleration and angle of attack error inputs and, in aircraft 155785ai and up and all others after AFC 364, stabilator position change input to position the throttle of the selected engine, thus commanding thrust changes to maintain an optimum angle of attack.

Aircraft Accelerometer

The aircraft accelerometer measures acceleration perpendicular to the glide path and pitch axis. If the aircraft oscillates about the glide path prior to arriving at the proper airspeed, an electrical signal is generated by the accelerometer. The signal is sent to the throttle computer to dampen these excursions off the glide path by positioning the throttles for the proper thrust level.

Stabilator Position Transducer

In aircraft 155785ai and up and all others after AFC 364, a stabilator position transducer is installed to measure the rate of displacement of the stabilator. Stabilator movement from a preset position is sensed by the transducer and sent as an electrical signal to the throttles control computer. The transducer signal allows the computer to anticipate an angle of attack change, thus reducing the time required by the APCS to react to pilot and vertical gust induced attitude changes. This function is operational upon conversion of the APC computer to the CP-974 configuration by incorporation of AVC-743.

Throttle Control Computer

The throttle control computer, computes the throttle position required to maintain an optimum angle of attack. The computer utilizes an error input from the angle of attack transmitter and the accelerometer when modified it will also accept a stabilator position change input from the stabilator position transducer (aircraft 155785ai and up and all others after AFC 364). The computer computes the throttle required for one or both engines as selected by the engine selector switch and compensates for temperature as selected by the temperature switch. The throttle control computer delivers the throttle position signal to the electronic control amplifier.

Electronic Control Amplifier

The electronic control amplifier commands throttle movement necessary to maintain the angle of attack. The amplifier receives a desired throttle position signal from the throttle control computer and the actual throttle position from the torque booster feedback. These two signals are compared by the amplifier and a command signal is applied to the torque boosters.

Integrated Torque Booster

The torque boosters (one per engine) amplify the input signal from the amplifier and position the throttle linkage as required to satisfy the amplifier signal. The torque boosters are hydro-mechanical servo motors which use engine fuel as the controlling medium. Power to the boosters is controlled by the engine selector switch and the APCS power switch. In aircraft 155867ak and up and all others after AFC 392 power is not available to the torque boosters unless both the throttle control computer and the electronic control amplifier are installed in the aircraft, and are modified in accordance with AVC 924.

APCS POWER SWITCH

The APCS power switch is a three-position toggle switch with positions of OFF, STBY, and ENGAGE. In OFF, all power to the system is removed and the throttles must be manually positioned. In STBY, power is supplied to the throttle computer so that the computer may synchronize with the prevailing flight conditions. However, the control amplifier is not active and the throttles must be manually positioned. In the ENGAGE position, the system automatically controls the engine thrust by varying throttle position. The power switch is held in the ENGAGE position by a holding coil. Should the pilot disengage the system or should disengagement occur due to a malfunction, the power switch automatically moves to STBY and the APCS OFF light illuminates. In aircraft 155867ak and up and all others after AFC 392, the APCS power switch does not remain in the ENGAGE position unless both the throttle control computer and the electronic control amplifier are installed in the aircraft and modified in accordance with AVC 924.

ENGINE SELECTOR SWITCH

An engine selector switch, with positions of L (Left), R (Right), and BOTH, allows the pilot to select automatic operation of either or both engines. If a single engine is selected, the manually controlled engine throttle should be positioned so that rpm is between 77 and 82%. The engine selector switch can be positioned at any time; however, it is effective only when the APCS power switch is in the ENGAGE position. This switch will be deactivated upon incorporation of AFC–513. If AVC–743 is incorporated prior to AFC–513, operate engine selector switch in BOTH position only.

AIR TEMPERATURE SWITCH

An air temperature switch, with positions of COLD, NORM and HOT, allows the pilot to select a temperature input which is representative of the ambient conditions. Since thrust developed for any given throttle setting varies with outside air temperature, the pilot should select the correct temperature prior to engaging the APCS. The COLD position is used when temperature is below 40°F and the HOT position when temperature is above 80°F. When the temperature is between 40°F and 80°F the NORM position is utilized. If the incorrect temperature is selected, the APCS throttle command signal is incorrect, but the computer compensates and produces the correct throttle movement. However, less correction time is required if the correct temperature is selected.

APCS OFF LIGHT

The APCS OFF light on the telelight panel and the MASTER CAUTION light illuminates whenever the system becomes disengaged or is in STBY regardless of the cause. On aircraft 158355at and up, and all others after AFC 508, an identical APCS OFF light is added on the left side of the glare shield, near the AOA indicator. The APCS OFF light and the MASTER CAUTION light, when illuminated, can be extinguished by depressing the reset button, by engaging the system, or by turning the system off.

NORMAL OPERATION

It is possible to engage the approach power compensator system anytime the aircraft is in flight. However, it operates properly only in the 73 to 99% rpm throttle range. It is recommended that engagement be made while downwind so that the system operation can be observed prior to commencing the landing approach. Before engagement, ensure that the speed brake switch is in the STOP position and the throttle friction lever is full aft. To engage the system, momentarily place the APCS power switch to STBY and then to ENGAGE. The system then remains engaged until disengagement is accomplished by the pilot, a malfunction, or immediately after touchdown through the landing gear scissors switch. The pilot must assume manual throttle control after disengagement occurs.

WARNING

- Do not engage APCS with gear up and flaps 1/2 or full down. Gear up airspeeds at 19 units angle of attack correspond to gear down airspeeds at 22 to 23 units (10 to 12 knots slow). This is especially critical because landing configuration stalls occur at 24 units angle of attack.
- If AFC–513 is not incorporated concurrent with AVC–743, operate engine selector switch in BOTH position only. If L (left) or R (right) is selected on engine selector switch, the APCS computer will continue to schedule for two–engine operation; consequently, power response will be severely jeopardized.

NOTE

- In aircraft 155897ak and up and all others after AFC 392, the APCS can be disengaged by moving the speed brake switch to IN regardless of emergency speed brake switch or speed brake circuit breaker position.
- Do not engage approach power compensator unless all components are installed in system. With components removed, engagement of system causes throttles to assume any setting aside from 73 to 99%. In aircraft 155867ak and up and all others after AFC 392, the APCS will not engage when the throttle control computer and/or the electronic control amplifier is not installed in the aircraft or the computer and amplifier are not modified in accordance with AVC 924. Absence of a complete system is indicative to the pilot by the APCS power switch failing to remain in the ENGAGE position.

EMERGENCY OPERATION

There are no provisions for emergency operation of the Approach Power Compensator System.

NOTE

When manually retarding the throttles with the approach power compensator engaged, the throttle can be pulled against the APCS minimum-speed stop. The throttles cannot be retarded below this point unless the force against the minimum speed stop is relieved and the APCS is disengaged.

LIMITATIONS

The approach power compensator system is limited to operation within the 73 to 99% throttle range. The APCS should not be utilized for carrier landings until AVC 752 is incorporated in aircraft with J79-GE-8 engines and either AVC-836 or AVC-743 (which also requires incorporation of AFC-364 and AFC-392) is incorporated in aircraft with J79-GE-10 engines.

ARRESTING HOOK SYSTEM

NOTE

Some illustrations referred to within this system writeup may be located in appendix A. These illustrations are referred to, within the text, as (figure A-, appendix A).

DESCRIPTION

The arresting hook system consists of an arresting hook, a combination hydraulic actuator and dashpot, a mechanical uplatch, and a control handle. The forward end of the arresting hook is pivoted in a manner which not only permits up and down movement, but left and right motion as well. Coil springs keep the hook centered for retraction. It is retracted up and aft by a hydraulic actuating cylinder, and is caught and held there by a mechanical uplatch. It is extended by the pneumatic action of the dashpot and its own weight. Hook extension time is approximately 5 seconds, and retraction time is approximately 13 seconds. The arresting hook is controlled by a handle on the right side of the front cockpit instrument panel. The handle and the uplatch mechanism are joined by a control cable and in the event of a cable failure, the arresting hook extends.

ARRESTING HOOK CONTROL HANDLE

The arresting hook is controlled by an arresting hook shape handle (figure A-1, appendix A) on the right side of the front cockpit instrument panel. When the handle is placed in the down position the uplatch is released and the arresting hook extends. Placing the handle in the up position energizes a solenoid valve which directs utility hydraulic pressure to the cylinder. A red warning light inside the control handle illuminates when the control handle is moved to the up or down position and remains illuminated until the arresting hook is fully extended or retracted. On aircraft 155529ag and up, and all others after AFC 388, placing the arresting hook handle down, with the landing gear down, causes the corner reflector to extend. If the arresting hook bypass switch is in the bypass position with the landing gear down the corner reflector will also be extended.

NORMAL OPERATION

Normal operation of the arresting hook system consists of placing the arresting hook control handle down, and observing the warning light to assure the arresting hook is fully extended. The warning light will remain on if the hook is in contact with the ground and prevented from reaching the down limit. To retract the arresting hook, place handle up and observe warning light to assure arresting hook is fully retracted. The hook should extend within 5 seconds and should retract within 13 seconds.

EMERGENCY OPERATION

If the arresting hook fails to extend when the control handle is placed in the down position, the hook-up limit switch has probably failed to deenergize the solenoid operated selector valve. To deenergize the solenoid selector valve, pull the arresting hook control circuit breaker (H7 No. 2 panel on aircraft thru 155528ag and K7 No. 2 panel on aircraft 155529ag and up). Utility hydraulic pressure is then removed from the up side of the arresting hook actuator cylinder and the hook extends. There are no provisions for arresting hook retraction in the event of a utility hydraulic failure or double generator failure.

LIMITATIONS

There are no practical arresting hook limitations for field arrestment; however, there are airspeed gross weight limitations for shipboard recovery. Refer to applicable Recovery Bulletins.

AUTOMATIC FLIGHT CONTROL SYSTEM

NOTE

Some illustrations referred to within this system writeup may be located in appendix A. These illustrations are referred to, within the text, as (figure A-, appendix A).

DESCRIPTION

The automatic flight control system is an electro-hydraulic system designed to provide stable and coordinated flight maneuvers without interferring with manual control. The automatic flight control system is capable of performing two modes of operation: stability augmentation and AFCS. Stability augmentation operation provides airplane stability in pitch, roll, and yaw, in that it opposes any changes of attitude. It does not, however, return the airplane to a given attitude or heading. This mode of operation may be used while the aircraft is under manual control. Stability augmentation can be engaged individually or in any combination for pitch, roll, or yaw axis. However, all three channels must be engaged before complete AFCS operation can be engaged. The AFCS mode of operation maintains any aircraft heading and/or attitude selected within the AFCS limits and corrects for any deviation from the selected heading or attitude of the aircraft within the AFCS limits. The altitude hold mode of operation holds any altitude selected while in the AFCS mode. The AFCS components are the AFCS panel, the control amplifier, force transducer, accelerometers, and rate gyro sensors. Equipment used in conjunction with AFCS operation are the attitude reference bombing and computer set, the air data computer, and the control servos.

STABILITY AUGMENTATION MODE

In the stab aug mode of operation, the system senses motion about the horizontal, vertical, and lateral axis, by means of rate gyro sensors. All attitude, heading, and bank angle changes cause these sensing devices to transmit signals representing the changing motion about their respective axes. These signals are sent to servo valves of the control surface power control cylinders. Therefore, any output signals from the rate gyro sensors, indicating yawing, pitching, or rolling motion, or from the lateral accelerometer indicating side slip, causes the flight controls system to position the appropriate control surfaces to oppose that motion. This action decreases any tendency of the aiplane to oscillate in roll, yaw, or pitch. The rate gyro sensors send signals to the surface controls to oppose any deviations from selected flight attitude but do not return the airplane back to its original heading or attitude. Stability augmentation can be obtained individually or in any combination for pitch, roll, or yaw axis by placing the pitch, roll, and yaw stab aug switches to the ENGAGE position.

AFCS MODE

In the AFCS mode of operation, vertical gyro, directional gyro, and accelerometer signals are used in addition to the rate gyro sensor signals, to maintain the airplane in a desired attitude with pitch, roll, and yaw stability. The autopilot system can be engaged and hold maneuvers and attitudes within a range of ±70° in bank and pitch, and unlimited in azimuth, provided the G limits are not being exceeded. The airplane attitude may be less than 70° roll when the AFCS mode disengages or the roll attitude may be more than 70° when the mode reengages. This difference depends on the rate at which the roll maneuver is performed in the AFCS mode. Autopilot operation is interrupted when ±70° pitch or bank is exceeded. However, the AFCS engage switch remains engaged and the autopilot resumes normal operation when the airplane is returned to within the ±70° limits. If a load factor of plus 4 G or minus 1 G is sensed by the G-limit accelerometer, the autopilot reverts to the stab aug mode and the AFCS engage switch must be reengaged for AFCS operation. On aircraft 158355at and up, and all others after AFC 508, the AFCS engage switch disengages when the right main landing gear scissors switch is actuated upon aircraft touchdown.

CONTROL AMPLIFIER

The control amplifier comprises the control center for the entire automatic flight control system. It receives the signals from the various sensing elements in the system and supplies power to the flight control components.

COUPLER

On aircraft 155529ag and up, and 153768ab thru 155528ag after AFC 388, there is an autopilot coupler installed. With the coupler switch, on the AFCS panel, in the ENGAGE position, the coupler provides additional attitude and heading commands from the data link system to the AFCS control amplifier. Refer to Data Link System, section VIII, classified supplement.

AN/AJB-7

The AN/AJB-7 attitude reference and bombing computer set provides vertical and directional references for the autopilot. The directional reference is controlled by the compass system controller on the pilot's right console. With the compass system controller in the slaved mode, the autopilot receives magnetic heading as a directional reference. The DG mode on the compass system controller provides deviations from a manually set heading as a directional reference to the autopilot. The compass mode is an emergency mode of the AN/AJB-7 and the autopilot cannot be engaged in this mode. The autopilot also disengages when the mode selector is switched from DG to the slaved mode, or when the reference system selector knob is switched between PRIM and STBY while in the slaved position. However, the autopilot can be reengaged after either of the controls has been switched to its new position.

NOTE

The AFCS should not be engaged when the AN/AJB-7 bombing mode is being used. Transients introduced into the attitude input of the AFCS at pull-up can cause pitch and/or roll oscillations.

AIR DATA COMPUTER

The air data computer performs two functions for the autopilot. First, it provides all required gain changes. This is necessary to maintain constant maneuvering rates regardless of changes in airspeed and altitude. Second, the ADC contains a clutched synchro which supplies the autopilot with a signal proportional to the deviation from the barometric altitude which existed when altitude hold was engaged. This signal is used by the autopilot to move the stabilator as necessary to maintain constant barometric altitude.

FORCE TRANSDUCER

The force transducer is a unit which senses the force applied by the pilot to the control stick. This unit actually comprises the visible portion of the control stick with the stick grip mounted on top of it. The force transducer contains pressure sensitivity switches which react to longitudinal and lateral stick forces. A lateral stick force of 2.25 $\pm$0.25 pounds or more closes a force switch. When a roll force switch closes, the roll rate gyro signal in stab aug and the roll rate and attitude gyro signals in AFCS mode are cut out so that pilot initiated maneuvers are not bucked while in the AFCS and stab aug modes. The pilot maneuvers the aircraft by mechanical linkages until the lateral stick force is reduced to less than 2.25 $\pm$0.25 pounds. At this time the roll channel is returned to normal stab aug or AFCS operation. Fore and aft stick forces close a switch to operate certain autopilot components and also cause a force sensing device to send a signal proportional to the applied stick force to the servo amplifier and stabilator actuator, thus causing the desired pitch attitude change. During this operation, the pilot controls stabilator position through the autopilot and the ailerons and spoilers are being manually controlled by the pilot. When the stick forces are released, the autopilot holds the new pitch attitude. This feature is commonly known as control stick steering (CSS). If the pitch or roll limits of the autopilot are exceeded ($\pm$70°), the autopilot is disengaged, although the autopilot switch remains ON. Therefore, when the airplane returns to within the limits of the autopilot, the autopilot once again takes over.

ACCELEROMETERS

During autopilot operation two accelerometers are being utilized to ensure proper functioning of the autopilot system. One of the accelerometers is of the G-limiting type which prevent G loads from occurring as a result of autopilot operation. The other accelerometer is a lateral accelerometer which performs coordinated maneuvers while in autopilot operation.

G-Limit Accelerometer

The normal load factor interlock (G – disengage) feature of the AFCS inhibits the system from commanding excessive load factors on the airplane. The system reverts automatically, from whatever mode is engaged, to the stab aug mode if plus 4 or minus 1 G absolute is sensed by the G – disengage accelerometer switch. This switch is mounted forward on the radar bulkhead so that if the airplane is rotated rapidly into a maneuver, disengagement occurs at lower values of normal load factor because of the anticipation resulting from the forward location sensing a component of pitching acceleration. If, in addition to the G switch being operated, the stab aug servo is hard over in a direction that tends to increase the magnitude of the existing load factor, the stab aug mode also disengages. The G – disengage feature is inoperative outside the $\pm$70° limits of the autopilot.

Lateral Accelerometer

This accelerometer detects airplane skids or slips and produces error signals proportional to the lateral forces developed. These error signals cause the autopilot to take corrective action with the rudder to coordinate the maneuver being performed.

SERVOS

The automatic flight control system contains four control servos which function to operate the aircraft flight controls during stability augmentation and AFCS operation. Two lateral series servos (one in each wing) function to operate the spoilers and ailerons during AFCS and stab aug operation. A directional series servo, on the rudder power control cylinder, functions to operate the rudder during AFCS and stab aug operation. A longitudinal servo, integral with the stabilator power cylinder, functions in a series mode to operate the stabilator during stab aug operation. This same servo functions in a parallel mode to operate the stabilator during AFCS operation.

RATE GYRO SENSORS

Refer to Stability Augmentation Mode, previously discussed in this section.

AFCS CONTROLS

The AFCS panel (figure A-1, appendix A) is on the pilot's left console. This panel contains all the controls for the normal operation of the automatic flight control system.

Stab Aug Switches

The three stab aug switches for pitch, roll, and yaw are two-position toggle switches on the AFCS panel. Placing any one of these switches in the ENGAGE position establishes the stability augmentation mode for the axis selected. These switches can be engaged individually or in any combination for stability augmentation in pitch, roll, or yaw. Yaw stab aug engaged also provides 5° of ARI rudder authority.

AFCS Switch

The AFCS switch is a two–position toggle switch on the AFCS panel. The switch positions are AFCS and ENGAGE. The AFCS switch can be energized with only the pitch stab aug switch engaged. However, for the AFCS mode to be fully selected, all three stab aug switches must be engaged.

NOTE

On aircraft 153075z and up, if the AFCS switch does not readily engage, do not make a further attempt to engage as a pitch sync drive malfunction can exist within the system which could cause violent pitch maneuvers.

Altitude Hold Switch

The altitude hold switch is a two–position toggle switch on the AFCS panel. The switch positions are ALT and ENGAGE. The altitude hold feature functions only if AFCS is engaged. Placing the switch to ENGAGE energizes an altitude sensor in the air data computer which is controlled by barometric altitude. As the altitude varies, an error signal is produced and fed to the pitch servo amplifier. The amplifier then sends a signal to the stabilator actuator which deflects the stabilator as necessary to return the aircraft to its hold altitude. The altitude sensor holds the aircraft within $\pm$50 feet or $\pm$0.3 percent of the reference altitude at speeds up to 0.9 Mach and at speeds greater than 1.0 Mach. When using the AFCS altitude hold mode, the aircraft may experience pitch oscillations while accelerating through the transonic range (0.9 to 1.0 Mach) because of fluctuations in the ADC airspeed system. Engaging the altitude hold mode in climbs or descents greater than 1000 feet per minute may result in a reference altitude other than the engage altitude.

NOTE

When operating in the transonic region in altitude hold mode, the MASTER CAUTION light may flash. However, this is not an indication of failure in the air data computer system.

Coupler Switch

On aircraft 155529ag and up, and 153768ab thru 155528ag after AFC 388, a two–position coupler switch is installed on the AFCS panel. The coupler switch selects data link information as autopilot control signals. The switch positions are COUPLER and ENGAGE. In COUPLER position, data link signals are removed from the autopilot coupler input. The ENGAGE position applies data link signals to the autopilot coupler input. Before engaging the coupler switch, the stab aug pitch, roll, and yaw switches and the AFCS switch must be engaged, and also the CPLR ON (COUPLER ON) light on the command display indicator must be illuminated. The light indicates that the displayed command data is suitable for autopilot control. On aircraft 158355at and up, and all others after AFC 508, the coupler switch disengages when the right main landing gear scissors switch is actuated upon aircraft touchdown.

HEADING HOLD CUTOUT

When operating in the AFCS mode, roll attitudes must be larger than 5° angle of bank in order to disengage the heading hold. Actuating the nose gear steering button provides the means of disengaging the heading hold to allow turns at an angle of bank of 5° or less. Heading hold may be reestablished by again actuating the nose gear steering button.

AFCS/ARI EMERGENCY DISENGAGE SWITCH

A spring–loaded emergency disengage switch (figure A–1, appendix A) is on the control stick. Depressing the lever causes the AFCS and altitude hold switch to return to OFF. The stability augmentation mode and ARI are disengaged as long as the lever is held depressed. When the lever is released, the stability augmentation and ARI are again in operation, but the AFCS is no longer engaged. To permanently disengage the stability augmentation mode, the pitch, roll, and yaw stab aug switches must be placed off. To permanently disengage the ARI, the yaw stab aug switch must be off and the ARI circuit breaker, on the front cockpit left sub panel, must be pulled.

NOTE

In case of suspected flight control malfunction, the pilot should immediately disengage the AFCS by depressing the AFCS/ARI emergency disengage lever.

AUTOPILOT PITCH TRIM

An automatic pitch trim feature is included in the autopilot system which attempts to keep the airplane longitudinally trimmed to the flight conditions experienced while in AFCS mode. Thus, an out of trim condition which would not be sensed while in autopilot mode is prevented, ensuring against an excessive pitch transient when disengaging the autopilot. The auto–pitch trim operates at approximately 40% the speed of the normal trim system resulting in a slight delay, after changing flight conditions, before the basic airplane is properly trimmed. During CSS maneuvering, the auto–pitch trim is inoperative. Auto–pitch trim operation can be observed on the pitch trim indicator after changing flight conditions in the AFCS mode.

NOTE

The pilot shall not use the manual trim button when in AFCS mode except under the conditions specified in paragraph titled Auto–Pitch Trim Light.

AUTO–PITCH TRIM LIGHT

An AUTO PILOT PITCH TRIM light (figure A–1, appendix A) is on the telelight panel. This light illuminates during AFCS operation if the automatic pitch trim is inoperative or lagging sufficiently behind airplane maneuvering to cause an out of trim condition in the basic airplane. Since: (1) auto-pitch trim rate is only 40% of normal trim rate and, (2) auto-pitch trim is inoperative anytime the stick grip transducer switches are made (i.e., during CSS maneuvering), it is possible to develop an out-of-trim condition while maneuvering in the AFCS mode. However, this out-of-trim condition must exist for approximately 10 seconds before the AUTO PILOT PITCH TRIM indicator light illuminates, thus eliminating constant light flickering. Momentary illumination of the light does not necessarily indicate a malfunction, however, if the light remains on and it is apparent from the pitch trim indication that the trim is not working, the pilot should realize that a pitch transient may be experienced when the AFCS mode is disengaged. Airspeed pitch trim indicator relationship should provide an indication of the severity of the transient. If an out-of-trim condition is realized by the steady illumination of the AUTO-PILOT PITCH TRIM light, grasp the stick firmly before disengaging the AFCS mode in anticipation of a pitch bump. However, before disengaging the AFCS following an automatic pitch trim malfunction, the pilot may elect to alleviate the out-of-trim condition by operating the manual trim button and observing the pitch trim indicator. If the out-of-trim condition is thus reduced to within five pounds of trim, the AUTO PILOT PITCH TRIM light is extinguished. Illumination of the AUTO PILOT PITCH TRIM light also illuminates the MASTER CAUTION light. Depressing the master caution reset button only extinguishes the MASTER CAUTION light, leaving the AUTO PILOT PITCH TRIM light illuminated.

AUTOPILOT DISENGAGED INDICATOR LIGHT

The AUTO PILOT DISENGAGE light (figure A-1, appendix A) is on the telelight panel. After initial engagement of the AFCS mode of operation, the AUTO PILOT DISENGAGE light and MASTER CAUTION light illuminate when the AFCS mode is disengaged by any means. Both lights are extinguished by pressing the master caution reset button. The lights remain extinguished until the AFCS is again disengaged.

NOTE

> If PC-1 hydraulic pressure is lost or drops below 500 psi, the pitch axis in stab aug and AFCS is inoperative. If utility hydraulic pressure is lost or drops below 500 psi, the roll axis and yaw axis in stab aug and AFCS are inoperative. In either case, the AUTO PILOT DISENGAGE light and the PITCH AUG OFF light do not illuminate. The CHECK HYD GAGES light and the MASTER CAUTION light illuminate at approximately 1500 psi.

PITCH AUG OFF INDICATOR LIGHT

The PITCH AUG OFF indicator light is on the telelight panel and remains illuminated unless pitch stab aug is engaged. After pitch stab aug engagement, disengagement also illuminates the MASTER CAUTION light.

NORMAL OPERATION

Prior to engaging the AFCS, the following conditions must exist: the stability augmentation mode can be selected individually or in any combination for pitch, roll, or yaw axis; however, all three switches must be engaged for complete AFCS operation; the airplane should be in trim, and an attitude within the AFCS limits must be established. AFCS operation can then be achieved by engaging the AFCS switch. Manual trim during AFCS operation should not be used unless roll reversal is encountered, and then only a small amount of trim should be used to counteract the roll. Autopilot disengagement can be accomplished by placing the AFCS switch to off, the airplane still is in the stab aug mode. To permanently disengage stab aug, the pitch, roll and yaw stab aug switches must be turned off. The autopilot system receives power from the three phase left main 115 volts ac bus. The system also receives power from a left main 28 volt dc bus, a warning lights 28/14 volts ac bus, and a left main 28 volts ac bus. The autopilot system is protected by circuit breakers on No. 1 circuit breaker panel in the rear cockpit. The circuit breakers are marked AUTO PILOT.

OPERATIONAL PRECAUTIONS

Generator Switching

Power to the autopilot, ADC, and AN/AJB-7 may be momentarily interrupted during the starting and stopping of airplane engines or generators. When the left engine or generator is started with the right generator already on the line, the connection between the right and left main busses is momentarily opened to allow the left generator to come on the line. This momentary interruption allows the solenoid held switches to disengage. This necessitates reengaging the autopilot to bring back autopilot operation. The autopilot, ADC, and AN/AJB-7 are not affected by starting or stopping the right engine or generator with the left generator on the line. If failure of the left generator occurs, place the left generator switch to OFF and disengage the stab aug, by depressing the emergency disengage switch, prior to cycling the left generator switch back to ON. This is done to prevent the possible occurrence of control surface transients. Stab aug may be engaged by releasing the emergency disengage switch after the left generator has been cycled to ON, or may be engaged with the left generator switch retained in the OFF position.

Roll Reversal

There is a possibility of a condition called roll reversal occurring when operating the AFCS in the autopilot mode. This condition occurs infrequently and is apparent only when attempting small changes in bank angle. Roll reversal is associated with a small out-of-trim condition in the lateral control system, and is apparent as a slow rolling of the aircraft in the opposite direction of the low lateral force. If, for instance, the airplane is out of trim laterally to the left when the autopilot mode is engaged, roll reversal may occur when low right stick forces are applied. A roll reversal situation may also be caused by operating the lateral trim button while in the autopilot mode, followed by low lateral stick forces being applied opposite to the direction of the trim. There is also a possibility of roll reversal occurring even if the airplane has been trimmed prior to engaging the autopilot mode, and the manual trim button has not been touched. This condition is brought about by changes in aircraft trim accompanying changed flight conditions.

AFCS Operation With Static Correction Off

A malfunction of the static pressure compensator (indicated by the illumination of the STATIC CORR OFF light) has no effect upon autopilot operation. The autopilot operates satisfactorily with the static pressure compensator out; however, the altitude hold mode may be affected. If the altitude hold mode is affected, the reference altitude changes when the static pressure compensator fails.

Pitch Oscillations (Altitude Hold Mode)

When using the AFCS altitude hold mode, the aircraft may experience pitch oscillations in the transonic regions and below due to fluctuations in the ADC airspeed system. The nature of these oscillations vary from stick pumping to divergent pitch oscillations. It is recommended that in the event pitch oscillations occur at subsonic speeds that the following corrective steps be attempted: disengage the AFCS; place the static correction switch OFF; reengage the AFCS; and engage altitude hold. If the oscillations persist after taking corrective action, or if they are encountered at supersonic speeds, disengage the altitude hold mode. In any event, divergent pitch oscillations should not be allowed to develop. If any divergent pitch activity is noted, corrective action should be taken immediately.

NOTE

When using AFCS/CSS or altitude hold, there are no automatic AFCS cut–out features to prevent the aircraft from flying into a stall. Therefore, it is possible for the aircraft to enter uncontrolled flight with AFCS/CSS or altitude hold engaged if airspeed is allowed to dissipate.

EMERGENCY OPERATION

There are no provisions for emergency operation of the Automatic Flight Control System.

LIMITATIONS

Autopilot operation is interrupted when ±70° pitch or bank is exceeded. However, the AFCS engage switch remains engaged and the autopilot resumes normal operation when the airplane is returned to within the ±70° limits. If a load factor of plus 4 or minus 1 G is sensed by the G–limit accelerometer, the auto–pilot reverts to the stab aug mode and the AFCS engage switch must be reengaged for AFCS operation.

BRAKE SYSTEM

NOTE

Some illustrations referred to within this system writeup may be located in appendix A. These illustrations are referred to, within the text, as (figure A–, appendix A).

DESCRIPTION

The main landing gear wheels are equipped with power operated brakes. Two power brake valves are in the nose wheel well and each is operated in a conventional manner from linkage attached to the rudder pedals. The brake control valves are power operated rather than a power boost type. Excessive pedal travel and pumping of the brakes in order to obtain a firm pedal is eliminated since the fluid supply to the wheel cylinders is virtually unlimited. This brake system provides differential wheel brake pressures. With no utility hydraulic pressure available, a 25 cubic inch hydraulic accumulator will provide pressure to actuate the normal brake system. In addition the brake control valves act as master cylinders in a conventional non–power system as long as integrity has been maintained between the control valves and the wheel brakes. Pilot effort in this manual operation is capable of securing the airplane in deck rolls up to 8°. Although it has not been determined by test flight, pilot assumption is that the manual brake system should be capable of stopping the airplane on a typical jet runway provided the drag chute is also used. The manual braking feature is selective and may be used for differential brake steering while the emergency pneumatic brake system is employed in stopping the airplane. Each main landing wheel contains three fuse plugs to protect against tire explosion. If the brakes are used excessively, causing overheating of the wheels and tires, the fuse plugs should melt and let the tire go flat before a tire explosion can occur. On aircraft 157242an and up, an anti–skid system is incorporated into the normal brake system to prevent wheel skid.

NOTE

Only three maximum effort brake applications from the accumulator should be anticipated when utility hydraulic pressure is lost.

WHEEL BRAKE ANTI–SKID SYSTEM

WARNING

The use of anti–skid braking is prohibited during normal field taxi or during shipboard operations.

Aircraft 157242an and up are equipped with an electrically controlled wheel brake anti-skid system which prevents wheel skid. The system detects the start of a skid condition at the wheel(s) and automatically releases the brake pressure from both brakes in proportion to the severity of the skid. Use of the anti-skid system offers protection from skids, and can provide consistently shorter landing rolls on wet, icy, and dry runways. If a wheel is locked up, due to a blown tire, the opposite wheel still has anti-skid protection. The system provides a failure detection circuit which indicates anti-skid failures, but does not deactivate the system. The system is activated by placing the anti-skid control switch ON and lowering the landing gear. It may be disengaged by placing the anti-skid control switch to OFF, or by depressing the emergency quick release lever on the stick force transducer below the stick grip. An ANTI-SKID INOPERATIVE light illuminates when the system is not activated.

NOTE

Whenever electrical power to the anti-skid system is momentarily interrupted, the system fail-detection circuit is automatically rechecked.

Anti-Skid Control Switch

This two-position toggle switch is on the left console, front cockpit, adjacent to the oxygen quantity gage. When the switch is ON and the landing gear handle is down, power is supplied to the system. The anti-skid system may be shut off by placing the anti-skid control switch to OFF.

Anti-Skid Inoperative Light

An ANTI SKID INOPERATIVE light is on the left console, front cockpit. The light illuminates any time the landing gear handle is down and the anti-skid control switch is OFF, the system is inoperative, or when the emergency quick release lever is held depressed. The light flashes momentarily when the landing gear handle is placed in the DOWN position to indicate that the anti-skid circuit has been checked and is operating properly. If the light remains illuminated, the anti-skid system is inoperative and the control switch should be placed to OFF.

Emergency Quick Release Lever (Anti-Skid)

An emergency quick release lever is on the control stick below the stick grip. This lever is provided to disengage the anti-skid system as desired. The lever must be held depressed to disengage the system. Normal wheel braking is immediately available when the lever is depressed and the ANTI SKID INOPERATIVE light illuminates. When the anti-skid emergency quick release lever interrupts electrical power to the system, the circuit to the ANTI-SKID INOPERATIVE light is completed, and the light illuminates. The control stick mounted emergency quick release lever, and the console mounted control switch are connected in series, and actuation of either will deactivate the system. When the landing gear handle is up, all power to the anti-skid system, including the light, is shut off.

EMERGENCY BRAKE SYSTEM

If a hydraulic pressure failure occurs, an emergency air system is provided to accomplish maximum braking. Emergency pressure is provided by a hydraulic accumulator which is pre-charged to 1000 psi and will indicate 3000 psi when normal utility hydraulic pressure is present in the system. Up to ten maximum effort applications may be made by means of the emergency brake handle (figure A-1, appendix A) just inboard of the right console. It is hand operated, and is spring-loaded to the off or brake released position. It is a power brake valve of conventional design and meters air pressure to completely independent pneumatic brake cylinders, in proportion to applied effort. The emergency pneumatic brake system does not provide differential wheel brake pressure; however, by applying the manual portion of the hydraulic brakes in conjunction with the emergency air brakes, differential braking can be accomplished. Hydraulic and pneumatic systems are entirely separate. Actuation of the emergency pneumatic system will not introduce air into the hydraulic system.

ANTI-SPIN SYSTEM

The airplane has an anti-spin system which provides automatic and simultaneous actuation of the left and right wheel brakes during main landing gear retraction. When the gear handle is UP and the weight is off the gear, utility hydraulic pressure from the main landing gear up line actuates the wheel brakes and stops the wheels before stowage. When the gear is up and locked, pressure is automatically released from the anti-spin system.

NORMAL OPERATION

The brakes are conventionally operated by toe action on the rudder pedals. This action meters utility hydraulic pressure to force the brake disks together. Pedal pressure felt by the pilot is proportionate to braking force applied. The pilot is capable of locking the brakes by both normal and emergency braking systems. Caution must be exercised in overbraking, since a fully locked wheel offers less retardation than very light normal braking. If one wheel is locked during application of the brakes, there is a very definite tendency for the airplane to turn away from that wheel and further application of brake pressure offers no corrective action. This produces a rapidly decreasing coefficient of friction between the skidding tire and the runway, while the coefficient of friction between the other tire and the runway remains near optimum for braking effectiveness. It is, therefore, apparent that a wheel once locked never frees itself until brake pressure to that wheel is reduced sufficiently to permit the wheel to rotate. It has been found that optimum braking occurs when the wheel is in a slight skid. The wheel continues to rotate, but at a speed of approximately 80 to 85 percent of its normal free rolling rotational speed. Increasing the rolling skid above approximately 15 to 20 percent only decreases the braking effectiveness. Since no anti-skid system is fitted, recognition of maximum braking force is strictly a matter of pilot sensitivity. As with other examples of operating on the limit, the only sure way of determining maximum braking effort is to exceed it. Since this is seldom a desirable technique, the pilot should attempt to mentally catalogue his body response to normal braking, in order to more readily recognize the maximum, if an emergency should require it. For all conditions, normal and emergency, the most desirable braking technique is a single, smooth application of the brakes with a constantly increasing pedal pressure (to just below the skid point) as the airplane decelerates. In the event of a reduction in retardation being felt while exercising maximum braking, pedal force must be fully released in order to allow the skidding wheels to regain full rolling speed before further application of brakes. Rough runways tend to emphasize the skip or bounce characteristics of the airplane which are caused by relatively stiff struts. To prevent locking a wheel while momentarily off the ground, use light braking until the airplane is solidly on the ground and all skipping has ceased.

On aircraft 157242an and up, the anti-skid system should be utilized at all times to protect against inadvertently locking a wheel, or wheels, during braking. The anti-skid system is completely passive unless the wheel is approaching skid; therefore, under conditions of normal braking, it has no effect on the amount of brake applied. If maximum deceleration is desired, the anti-skid system can be utilized to maintain the wheel at the optimum deceleration point. In this case, apply sufficient brake pressure to skid the wheel and allow the anti-skid system to reduce applied pressure to proper values. Full pedal application, or any amount of pedal which will produce anti-skid action, will provide maximum wheel braking for the existing conditions. A minimum roll landing using the anti-skid system can be accomplished from a normal touchdown and drag chute deployment followed by full brake pedal deflection with the stick full aft and nose gear steering engaged. Less than full pedal can be used, if desired, as long as there is sufficient brake pressure to keep the anti-skid system active. Cycling of the anti-skid system can be detected by a change in longitudinal deceleration; however, cycling of the anti-skid system may not be apparent when braking at high speeds; i.e., immediately after landing, wet runway, etc.

CAUTION

- On aircraft without anti-skid or with anti-skid inoperative, at high speeds, brake pedal deflections as small as 1/16 inch have proven sufficient to blow a tire.
- Anti-skid protection is not available until the wheels have initially come up to speed. Do not land with brake pedals depressed. In addition, anti-skid protection is not available below approximately 10–20 knots.
- If it is suspected that the brakes have been used excessively, and are in a heated condition, the airplane should not be taxied into a crowded parking area. Peak temperatures occur in the wheel brake assembly from 5 to 15 minutes after maximum braking. To prevent brake fire and possible tire explosion, the specified procedures for cooling brakes should be followed. It is recommended that a minimum of 15 minutes elapse between landings where the landing gear remains extended in the airstream, and a minimum of 30 minutes between landings where the landing gear has been retracted to allow sufficient time for cooling. Additional time should be allowed for cooling if brakes are used for steering, crosswind taxiing operation, or a series of landings. To minimize brake fusing when brake-stored energy is known to be high, do not come to a full stop and hold brakes hard on; instead, allow aircraft to roll freely as much as possible.

NOTE

Nose gear steering may be required to maintain directional control when making a minimum roll landing.

EMERGENCY OPERATION

If a utility hydraulic system failure occurs, the airplane can still be stopped by utilizing accumulator brake pressure. This can be accomplished by depressing the brakes and applying a constantly increasing brake pressure. Do not pump the brakes as this rapidly depletes the accumulator pressure. If accumulator brake pressure fails to stop the airplane, utilize the emergency air brake system. The system meters air in proportion to pilot effort and does not provide differential braking. In most cases asymmetrical braking is prevalent when utilizing the air brakes due to runway crown, crosswinds, and unequal brake torque; however, the manual hydraulic brakes (because of fluid trapped between the brake valves and brakes) are still capable of furnishing flow and pressure to accomplish differential braking.

WARNING

Do not use emergency air brakes when a known hot brake condition exists. Some combustible substance such as oil, grease, hydraulic fluid, etc, may be present in the wheel assembly. The combination of heat and the introduction of 3000 psi compressed air sets up a highly explosive or combustible situation hazardous to personnel.

CAUTION

There is a noticeable time lag between pulling the emergency brake handle and braking action. Failure to take the system delay into consideration often results in blown tires.

LIMITATIONS

Maximum effort anti-skid braking (pedals depressed to the limits) may be employed as follows: (assuming a cool brake 75°C at brake application velocity).

If brake/tire service life is the prime consideration:

a. 130 knots maximum with drag chute deployed at a

BRAKE ENERGY LIMITATIONS

NOTE

BASED UPON 10 MILLION FT-LBS. OF ENERGY PER BRAKE FOR THE STOP WITH A 75°C MAXIMUM BRAKE STACK TEMPERATURE AT START OF FLIGHT.

CAUTION

BRAKE LININGS WILL FUSE IF THESE LIMITS ARE REACHED AND THE AIRCRAFT IS BROUGHT TO A COMPLETE STOP ON THE RUNWAY.

EXAMPLE: A 38,000 LB AIRPLANE WITH DRAG CHUTE:
MAXIMUM EFFORT BRAKING MAY BE APPLIED FROM 144 KTS. MAXIMUM.

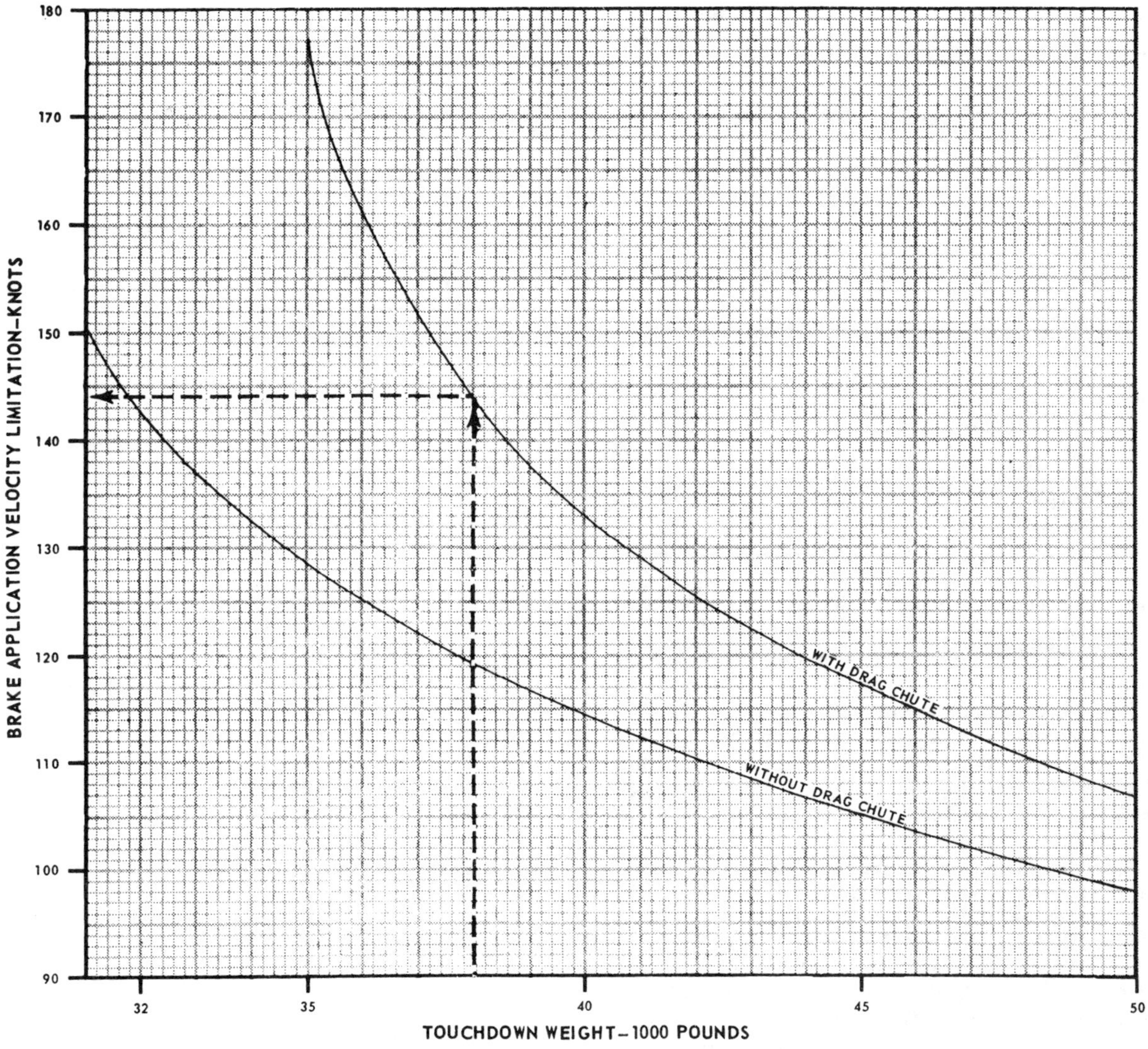

FDD-1-(14)

Figure 1-8

maximum landing weight of 38,000 pounds.

b. 110 knots maximum without drag chute deployed at a maximum landing weight of 38,000 pounds.

If short field performance is the prime consideration:

a. Braking to the limits shown on figure 1-8 may be employed.

CAUTION

When brake energy reaches the defined limits of figure 1-8, brake fusing may result if heavy pedal effort is applied with the wheels stopped.

CANOPY SYSTEM

NOTE

Some illustrations referred to within this system writeup may be located in appendix A. These illustrations are referred to, within the text, as (figure A–, appendix A).

DESCRIPTION

Each cockpit area is enclosed by a separate transparent, acrylic plastic, clam shell type canopy, hinged aft of each cockpit enclosure. Canopy operation, both normal and emergency, is accomplished pneumatically. Clean dry air is supplied by the basic pneumatic system for use in the normal canopy system. Individual manual controls are provided for each cockpit. In addition, external operation is provided for each cockpit by individual push buttons on the exterior of the fuselage. In normal operation, the canopy operating time is set at 5 seconds. Each cockpit employs an inflatable canopy seal for cockpit pressurization. The canopy seals are automatically inflated and deflated upon opening and closing of the canopies. Canopy emergency operation is also accomplished pneumatically. The emergency canopy system utilizes clean dry air supplied from the basic pneumatic system to two 15 cu. in. (forward canopy and aft canopy) air bottles. On aircraft 157256an and up, and all others after AFC 474, a secondary emergency system also provides jettisoning of the forward and aft canopies. This secondary source of emergency air, the emergency brake air bottle, is available, should the normal and emergency pneumatic systems experience battle damage. Each canopy has individual emergency manual control valves that operate independently of each other. These valves can be operated by pulling on the face curtain handles, lower ejection handles, or by actuating the emergency canopy jettison lever. In addition, one emergency control is provided on the fuselage to open both canopies simultaneously for ground crew rescue purposes. The canopy emergency system also actuates the forward and aft cockpit flooding doors in order to reduce the time required to equalize the cockpit pressure under water. Design time for deck operation or testing of canopy emergency system is 2 seconds. Inflight operation of the canopy emergency system takes 1 second. Design time for underwater pressure equalization of the cockpit is 7 seconds. On aircraft 157258an and up, and all others after AFC 497, electrically initiated cartridge thrusters under the forward canopy sills assist the canopy actuating cylinder in jettisoning of the forward canopy. The thrusters accomplish this operation by producing a thrust on the forward part of the canopy to lift it quickly into the airstream as soon as the canopy is unlocked. The thrusters are fired electrically by a thermal battery which is actuated by gases from either the seat mounted or cockpit mounted initiators.

EXTERNAL CANOPY CONTROLS

External Canopy Buttons

The forward cockpit external canopy control buttons are on the left side of the airplane just forward of the engine air intake duct. The aft cockpit external canopy control buttons are on the left side of the airplane just above the engine air intake ducts between the forward and aft canopies. The forward and aft canopy push buttons operate independently of each other but their functions are the same. Push OPEN button to open the canopy; push CLOSE button to close the canopy. The push buttons operate the same valves as the internal canopy controls.

External Manual Unlock Handles

The forward canopy external manual unlock handle is on the left side of the fuselage below the aft end of the canopy. The aft canopy external manual unlock handle is in the same position below the aft canopy. Each handle operates independently, but their functions are the same. Operating a push-type latch causes the handle to pop out about 1¾ inches. A forward rotation of the handle unlocks the canopy down-lock mechanism and permits the canopy to be lifted open manually.

INTERNAL CANOPY CONTROLS

Canopy Control Handle

The forward cockpit canopy control handle is on the left side of the cockpit above the flap control panel and just below the canopy sill. The aft cockpit canopy control handle is in the same position in the aft cockpit. Each canopy control operates independently, but their functions are the same. Pull control handle aft to OPEN the canopy; push forward to CLOSE the canopy. When closing the canopy, to determine that the canopy is locked by the canopy hooks engaging the rollers, check that the alignment tape on the canopy lock push rod aligns with the alignment mark on the bracket hanging from the left canopy sill. If canopy alignment tape is not installed, assure rollers have engaged canopy hooks by observing approximately 1-inch aft travel of canopy push rod mechanism. On some aircraft a guard is incorporated on the rear cockpit normal canopy lever to prevent inadvertent actuation.

Manual Canopy Unlock Handles

The forward cockpit manual canopy unlock handle (figure A-1, appendix A) is on the right side of the cockpit above the arresting hook control handle. The aft cockpit manual canopy unlock handle is in the same position in the aft cockpit. Each cockpit manual canopy unlock handle operates independently, but their functions are the same. The handle when pulled aft, releases the canopy downlock linkage, so the canopy may be pushed open to permit exit from the cockpit. Before manual unlocking of the canopy, the (normal) canopy control handle must be placed in the open position. When the canopy has been opened manually it should be held open while entering or leaving the cockpit, since without power the actuator may not hold the canopy in the open position. The manual unlock handle is used normally when the airplane pneumatic systems are depleted.

Canopy Unlocked Indicator Light

An amber CANOPY UNLOCKED light (figure A-1, appendix A) in the forward and aft cockpits is used to notify the crew that a canopy is unlocked. The pilot's CANOPY UNLOCKED light illuminates when either canopy is unlocked. The RIO's light illuminates only when his canopy is unlocked.

EXTERNAL EMERGENCY CANOPY CONTROLS

Emergency Jettison Lanyard

The external emergency canopy jettison lanyard is on the left lower side of the fuselage, just forward of the nose wheel door. Pulling the handle jettisons both canopies.

INTERNAL EMERGENCY CANOPY CONTROLS

Emergency Canopy Release Handle

The forward cockpit emergency canopy jettison handle (figure A-1, appendix A) is on the left side of the cockpit above the landing gear control. The aft cockpit emergency canopy jettison handle is in the same relative position. Each handle is painted with black and yellow stripes for ease of identification. The canopy emergency jettison handles operate independently of each other but function the same. Pulling the handle aft jettisons the canopy and opens the cockpit flooding doors. In addition, operation of the aft handle stows the rear cockpit radar scope, the radar set controls and the hand control to provide more room for egress from the cockpit. On aircraft 157298ar and up, and all others after AFC 506 the radar scope is moved forward and no longer stows when the canopy jettisons. On aircraft 158355at and up, and all others after AFC 506 the radar set controls and radar antenna hand control will no longer stow when the canopy is jettisoned. A downward pull on the face curtain handle or an upward pull on the lower ejection handle: jettisons the canopy, opens the flood doors, and when an aft seat ejection handle is pulled, stows the applicable radar equipment. The applicable aft cockpit radar equipment will stow when a dual ejection is initiated from the forward cockpit.

NORMAL OPERATION

Normal operation of the canopies is accomplished through the use of the push buttons on the exterior of the aircraft and a lever in the cockpit. Refer to applicable canopy control, Canopy Systems.

NOTE

> If canopy closure is attempted with engines running, the engines should not be running above a stabilized idle rpm. Attempted canopy closure with engine rpm above idle may result in canopy not fully locking due to back pressure caused by the aircraft pressurization system.

EMERGENCY OPERATION

Emergency operation of the canopies is accomplished through the use of an emergency jettison lanyard on the exterior of the airplane and an emergency canopy jettison handle in each cockpit. Refer to applicable emergency canopy control, Canopy Systems.

LIMITATIONS

The canopy is designed to remain in the full open position up to 60 knots and to separate from the airplane at approximately 100 knots.

COMMUNICATION-NAVIGATION-IDENTIFICATION (CNI) EQUIPMENT

NOTE

Some illustrations referred to within this system writeup may be located in appendix A. These illustrations are referred to, within the text, as (figure A–, appendix A).

DESCRIPTION

The communication–navigation–identification (CNI) equipment consists of an intercom, a UHF comm receiver–transmitter and auxiliary receiver, a tacan set, an automatic direction finder system, and an identification system. The functions and control locations for each of the above equipment are listed in figure 1-9. CNI equipment operating on external power without cooling air, is limited to 10 minutes of accumulated operation in a 1 hour span. This limitation applies to all CNI equipment except the intercom and UHF comm receiver–transmitter. Cooling air is provided to the receiver–transmitter unit by two internal blowers. The maximum permissible altitude with the CNI equipment on is 70,000 feet. Flight above 70,000 feet with CNI equipment on may result in damage to the equipment because of arcing. Operation of the UHF comm receiver–transmitter is limited to a transmit–receive ratio of one third or less transmitting time to two thirds receiving time.

EXTERNAL GROUND POWER OPERATION

With external ground power connected to the airplane and neither of the two generators connected to the line, the CNI equipment does not operate unless the CNI ground power switch is placed to the ON position. The CNI ground power switch, in the left wheel well, applies external power to the CNI equipment for ground operation. The switch is manually operated and electrically held and must be reset after each interruption of external power.

INTERCOM SYSTEM

An intercom panel (figures A–1, appendix A and A–2, appendix A) in both cockpits provides inter–cockpit communication in the airplane. An external station, connected in parallel with the aft cockpit microphone and headset, is used by ground personnel. An additional function of the intercom system is to amplify audio frequency signals received from various sources. Transmission over the UHF receiver–transmitter may be accomplished in either cockpit in conjunction with the intercom system.

INTERCOM CONTROLS

The intercom controls consist of volume controls, function selector switches, emergency amplifier selector switches, intercom switches, and microphone buttons.

Volume Control

The intercom volume control knobs are on the left side of the pilot's intercom panel, and at the right side of the RIO's intercom panel. The input level of the intercom signals to the headsets is increased by rotating the respective volume control knobs in a clockwise direction. The signals received from the radio receivers are not affected by operation of these intercom volume controls.

Function Selector Switch (Pilot's)

A three–position toggle switch, with positions of RADIO OVERRIDE, HOT MIC and COLD MIC, is on the right side of the pilot's intercom control panel. The RADIO OVERRIDE position of the switch is momentary, the HOT MIC and COLD MIC positions are fixed. The HOT MIC position is used when duplex operation of the intercom system is desired. The RADIO OVERRIDE position is identical with HOT MIC except that all radio gain is reduced, communication between cockpits then override radio reception. When the switch is set to COLD MIC, normal radio reception and transmission is still available, but the pilot can no longer communicate with the RIO over the intercom system without using the ICS switch on the throttle. However, the RIO can still transmit to the pilot by either placing the radio override switch in the RADIO OVERRIDE position, or by pressing the foot operated intercom switch.

Function Selector Switch (RIO)

A two–position toggle switch, with positions of RADIO OVERRIDE and NORMAL, is on the left side of the RIO's intercom panel. The NORMAL position is used for normal duplex operation at the same time the pilot's switch is set at HOT MIC. Radio signals are received at normal volume when the RIO's switch is set at NORMAL. The RADIO OVERRIDE position is used to reduce the reception of radio signals in both cockpits. This switch position may also be employed to accomplish intercockpit communication if the pilot's switch is set at COLD MIC. The RADIO OVERRIDE position is a momentary switch position. If the RIO has selected either EMER ICS or EMER RAD and then selects RADIO OVERRIDE, a one–way conversation results. The RIO can talk to the pilot, but the pilot cannot talk to the RIO until the RIO releases his function selector switch and allows it to return to NORMAL position.

Intercom Foot Switch (RIO)

A foot operated switch is on the left foot ramp in the RIO's cockpit. This switch is wired in parallel with the pilot's intercom switch. By depressing his foot switch, the RIO may override any of the positions selected on the pilots function selector switch, allowing intercom transmission without the necessity of releasing other manual controls.

COMMUNICATION NAVIGATION IDENTIFICATION EQUIPMENT

TYPE DESIGNATION	FUNCTION	RANGE	CONTROL LOCATION
INTER COM SYSTEM	Intercockpit and cockpit-to-ground communications.	N/A	Front cockpit left console. Rear cockpit left console.
UHF RADIO COMMUNICATION SYSTEM	UHF radio communication between airplane and ground, airplane and ship, or between airplanes.	Up to line of sight, depending upon frequency and antenna coverage.	Front cockpit right console. Rear cockpit left console.
AUTOMATIC DIRECTION FINDER SYSTEM	Indicates relative bearing of and homes on Radio signal sources.	Up to line of sight depending upon frequency and antenna coverage.	Front cockpit, right console. Rear cockpit, left console.
TACAN NAVIGATIONAL SET AN/ARN-86	Indicates bearing and distance to ground station. Determines identity and dependability of beacon. Also determines distance to other aircraft.	Line of sight distance up to 300 miles depending on attitude and altitude.	Front cockpit, right console. Rear cockpit, left console.
IDENTIFICATION SYSTEM	Identifies aircraft in which equipment is installed as friend or foe. Provides selective identification of a single airplane within a group.	0-200 miles or line of sight.	Front cockpit, right console.
ATTITUDE REFERENCE AND BOMBING COMPUTER SET AN/AJB-7	Provides continuous information of the aircrafts attitude and azimuth condition.	N/A	Front cockpit, right console.
VERTICAL FLIGHT REFERENCE SET GVR-10	Provides continuous information of the aircrafts attitude (roll and pitch only).	N/A	N/A, system operative with power on aircraft buses.
NAVIGATION COMPUTER AN/ASN-39	Provides continuous information of the aircrafts navigation situation.	N/A	Rear cockpit, instrument panel.
FLIGHT DIRECTOR GROUP	Provides an integrated display of the navigation situation of the airplane.	N/A	Front cockpit, instrument panel.

FDD-1-(15)

Figure 1-9

Emergency Amplifier Selector Knobs

The emergency amplifier selector knobs are three–position rotary type switches and are in the center of both intercom control panels. The operator uses these controls to bypass a defective amplifier. Both operators may have occasion to switch to one of the emergency settings at the same time. In certain instances of amplifier failure, this arrangement is necessary in order to maintain intercockpit communication. There are three possible settings for each control. The NOR position is used when both amplifying stages in the respective control boxes are functioning properly. The other two positions for each control are EMER RAD and EMER ICS, which are used when it is desired to bypass a defective amplifier. On aircraft 157242an and up, and 153768ab thru 155528ag after AFC 388, and 155529ag thru 155902am after AFC 433 (Part 1), the emergency amplifier selector knobs have positions of B/U, NORM, and EMER. If the headset amplifier in either ICS station fails, place the switch to the B/U (back–up) position in the cockpit with the defective station. This switches from the normal headset amplifier to the back–up amplifier and restores normal operation. If selecting B/U does not restore ICS operation, select EMER (emergency). Audio from the operative station is then connected directly to the back–up headset amplifier in the defective station. The volume control on the station with EMER selected has no effect on the audio level. The switch is left in NORM (normal) if the amplifiers in both ICS stations are operating normally.

NOTE

> If both amplifier selector knobs are in an emergency position (EMER RAD and/or EMER ICS), and both intercom volume control knobs are above 75% of their volume range, a loud squeal is heard in both headsets. To eliminate this squeal, turn either volume control knob to a position below 75% of its volume range.

Microphone Buttons

The microphone buttons are used to connect microphone outputs to the UHF transmitter. The pilot's microphone button is on the inboard throttle grip, the RIO's microphone button is a foot–operated switch on the right foot ramp. When either crewmember wishes to transmit, he depresses the microphone button, and the output from the microphone is fed into the transmitter. The positioning of other controls in no way affects the transmitting operation from either cockpit.

COMM RECEIVER–TRANSMITTER (UHF)

The main receiver–transmitter transmits and receives UHF frequencies in a range of 225.0 to 399.0 mc for air–to–air or air–to–ground communications. Complete control over the operation of the main radio receiver–transmitter can be maintained by either the pilot or the RIO through the comm–nav group control panels (figures A–1, appendix A and A–2, appendix A), located in each cockpit. The comm–nav group control panel provides controls for the operation of the main receiver–transmitter on any of 3500 manually selected frequencies, or any of 18 preset channels plus a guard channel, and for ADF operation with associated direction finder equipment.

NOTE

> On aircraft after AFC 331, provisions for the installation of the speech security system (KY–28) are provided. This system can have a direct effect on UHF transmission and reception. Refer to NAVAIR 01–245FDB–1T(A) for detailed description of the system and its operational application.

COMM RECEIVER–TRANSMITTER CONTROLS

The controls for operation of the comm receiver–transmitter are on each comm–nav group control panel (figures A–1 and A–2, appendix A). The controls consist of the UHF volume control, the comm–aux pushbuttons, the mode selector switch, the comm channel knob, the set channel pushbutton, and the comm frequency thumbwheels.

UHF Volume Control

The UHF volume control is a thumbwheel type control which turns on the UHF communications and ADF systems, and controls the volume of the comm receiver–transmitter. The thumbwheel also has an on–off switch, with the 0 position being off. When adjusting the volume, the higher the number the control is set to, the stronger is the incoming audio signal.

COMM–AUX Pushbuttons

The comm–aux pushbuttons control the mode of operation of the UHF comm receiver–transmitter, aux receiver, and ADF systems. The comm receiver, comm transmitter, comm guard receiver, aux receiver and aux guard receiver are controlled by the pushbuttons as indicated below (only the comm receiver–transmitter shall be discussed at this time):

T/R–ADF	Comm receiver –communication reception. Comm transmitter –communication transmission. Comm guard receiver –not used.
T/R+G–ADF	Comm receiver –communication reception. Comm transmitter –communication transmission. Comm guard receiver –guard reception.
ADF+G–CMD	Comm receiver –ADF reception. Comm transmitter –communication transmission with interuption of ADF during transmission. Comm guard receiver –guard reception.

ADF-G	Comm receiver -ADF reception. Comm transmitter -communication transmission with interruption of ADF during transmission. Comm guard receiver -not used.

The pushbuttons are only effective from the cockpit with comm cmd.

Mode Selector Switch

The mode selector switch is a three position toggle switch which controls the mode of channel selection as indicated below.

CHAN	When used in conjunction with the comm chan knob, provides selection of preset channels 1 thru 18.
GUARD	Channels the comm receiver and transmitter to the guard frequency with the T/R-ADF pushbutton depressed. With the T/R+G-ADF pushbutton depressed the comm receiver and transmitter are channelled to the guard frequency and the guard receiver is on. This control is effective only from the cockpit with comm cmd.
MANUAL	Permits manual selection of the comm receiver and transmitter frequency as indicated by the manual frequency dials. This control is effective only from the cockpit with comm cmd.

COMM Channel Knob

The comm channel knob is a rotary switch used to select preset comm channels 1 through 18. The mode selector switch must be in the CHAN position for the comm channel knob to be operative. The selected channel is indicated in the channel indicator adjacent to the comm channel knob.

Set Channel Pushbutton

The set channel pushbutton is used to preset comm channels 1 through 18. To set in a frequency into a channel, the mode selector switch must be placed to the CHAN position, the comm channel knob must be rotated to the desired channel and the comm frequency thumbwheels must be positioned to the desired frequency. When the set channel pushbutton is then depressed, the frequency is set into the desired channel.

COMM Frequency Thumbwheels

The comm frequency thumbwheels are four rotary controls used to set a frequency into the comm receiver-transmitter or to preset frequencies into the comm receiver-transmitter channels. Frequencies can be set into the comm receiver-transmitter with the mode selector switch in the MANUAL position, and frequencies can be set into the comm receiver-transmitter channels with the mode selector switch in the CHAN position while depressing the set channel pushbutton. Frequencies from 225.0 to 399.0 mc, in increments of 0.05 mc, can be set using the comm frequency thumbwheels. The frequency selected is displayed on the manual frequency dials located above the thumbwheels. The thumbwheels are effective only from the cockpit with comm cmd, except when used for presetting comm channels 1 through 18.

UHF Remote Channel Indicators

A remote channel indicator is on the forward cockpit instrument panel. Aircraft 153840ae and up after AFC 312 have a remote channel indicator in the aft cockpit on the instrument panel. The indicators provide a secondary means of indicating the preset channel selected with the communication channel knob.

AUXILIARY RECEIVER (UHF)

An auxiliary receiver operates in conjunction with the comm receiver-transmitter under normal conditions, and operates as an emergency receiver in the event of failure of the main radio receiver-transmitter. The auxiliary receiver can be used as a conventional radio receiver for reception of AM radio signals in the frequency range of 265.0 to 284.9 mc or as an ADF receiver, for reception of radio signals in the same frequency range. A guard receiver, preset to 243.0 mc is contained within the auxiliary receiver. Guard channel can be monitored when the ADF-G pushbutton is depressed. The auxiliary receiver can be placed in either function by operation of the controls on either comm-nav group control panel. The auxiliary receiver contains 20 channels preset to frequencies within the receiver's range. Channel selection is also accomplished by operation of the aux chan knob on one of the comm-nav group control panels. The comm-aux pushbuttons control the functions of the auxiliary receiver equipment and provides for either the auxiliary receiver or the main radio receiver-transmitter to be operating as an ADF receiver while the other equipment is operating as a voice receiver. The direction finder group of the auxiliary receiver provides the pilot with continuous indication of the direction of radio frequency signals intercepted by either the radio receiver-transmitter or the auxiliary receiver which is used in conjunction with the ADF system. These receivers function to intercept amplitude modulated and unmodulated signals in the frequency range of 225 through 400 mc and 265 through 284.9 mc depending on which receiver is used. Continuous indication of the bearing of the intercepted signals relative to the airplane heading is presented on the horizontal situation indicator. Necessary primary power, 115 volt ac, is applied to the auxiliary receiver when the electrical system is energized.

AUXILIARY RECEIVER CONTROLS

The controls for operation of the auxiliary receiver are on each comm-nav group control panel (figures A-1 and A-2, appendix A). The controls consist of the aux volume control, the comm-aux pushbuttons, and the aux channel knob.

AUX Volume Control

The aux volume control is a thumbwheel type control which is used to control the volume of the aux receiver.

COMM-AUX Pushbuttons

The comm-aux pushbuttons control the mode of operation of the UHF comm receiver-transmitter, aux receiver and ADF systems. The comm receiver, comm transmitter, comm guard receiver, aux receiver and aux guard receiver are controlled by the pushbuttons as indicated below (only the aux receiver shall be discussed at this time):

T/R-ADF	Aux receiver -ADF reception Aux guard -not used
T/R+G-ADF	Aux receiver -ADF reception Aux guard receiver -not used
ADF+G-CMD	Aux receiver -communication reception within aux receiver frequency range Aux guard receiver -not used
ADF-G	Aux receiver -not used Aux guard receiver -guard reception

The pushbuttons are effective only from the cockpit with comm cmd.

AUX Channel Knob

The aux channel knob is a rotary switch used to select preset aux channels 1 through 20. The channel frequencies cannot be reset from the cockpit. This control is effective only from the cockpit with comm cmd.

ANTENNA SELECTOR SWITCH

A two-position antenna selection switch is in each cockpit. The pilot's antenna switch is on the left console outboard of the throttles; the antenna switch for the RIO is on the RIO's instrument panel. The switch positions are UPR and LWR and are used to select one of two communication antennas to be used with the command communications set. The circuitry of the antenna selection switch is wired through the take command relay. Therefore, whoever has command of the comm-nav group control panel has command of the antenna selection switch. Aircraft 155529ag and up, and all others after AFC 388 have the data link system installed. In these aircraft, the antenna selection switch is on the cockpit lights/data link control panel in the rear cockpit on the left side. When the antenna selection switch is placed to the UPR position, the upper communications antenna is selected for UHF communications, and the lower antenna is selected for the data link system. Placing the antenna switch to the LWR position selects the lower communications antenna for UHF communications, and the upper antenna for the data link system.

TACAN (TACTICAL AIR NAVIGATION) SYSTEM

The tacan system functions to give precise geographical bearing and distance information at ranges up to approximately 300 miles (depending on aircraft altitude) from an associated ground or shipboard radio beacon. It also determines the identity of the beacon and indicates the dependability of the beacon signal. It also provides deviation from a selected course. In the air-to-air mode, line-of-sight distance can be obtained between two aircraft operating their tacan sets 63 mc apart. Up to five aircraft can determine line-of-sight distance from a sixth lead aircraft in the A/A mode, providing their tacan sets are 63 mc apart from the lead aircraft. The lead aircraft indicates distance from one of the other five, but it cannot readily determine which one. Before operating in the A/A mode, the frequencies used by each aircraft must be coordinated. The tacan system employs UHF radio frequencies, the propagation of which is virtually limited to line of sight distances. The maximum distances from the beacon at which reliable tacan signals can be obtained depends on the altitude of the aircraft and the height of the beacon antenna. Tacan information is presented on the HSI and ADI in the pilot's cockpit (refer to Flight Director Group, this section), and on the No. 2 pointer of the BDHI in the RIO's cockpit. The BDHI is a conventional RMI display with the additional feature of displaying distance to a tacan station. The No. 2 bearing pointer of the BDHI is the tacan needle and indicates magnetic bearing to the tacan station. The No. 1 bearing pointer is the ADF needle and provides relative bearing to the selected UHF station. The BDHI indicator is also capable of displaying distance information to a tacan station. When a usable signal is not being received the red warning flag partly obscures the distance indicators from view and the word OFF in black letters appears in the window. The units digit indicator dial is divided into ½ mile increments. The BDHI is controlled strictly from the aft cockpit by the CNI-NAV COMP switch on the aft instrument panel. Switch positions selected on the mode-bearing/distance selector panel in the pilot's cockpit have no effect on BDHI operation.

TACAN CONTROLS

The controls for tacan operation are on the tacan panel on the pilot's right console (figure A-1, appendix A) and the RIO's left console (figure A-2, appendix A), the comm-nav group control panel on the RIO's left console (figure A-2, appendix A), and the comm-nav group panels on the pilot's right console. The controls on the tacan control panel are the function selector knob, the channel selector control and the volume control. The controls on the HSI are the course set knob and the heading set knob. The controls on the navigation function selector panel are the mode selector knob and the brg/dist switch. The control on the cockpit lights/data link panel is the CNI-NAV COMP switch. The control on the comm-nav group control panels is the nav cmd pushbutton.

Function Selector Knob

The function selector knob is a four position rotary knob used for selecting tacan modes of operation as follows:

OFF	Deenergizes tacan system
REC	Tacan receives bearing signals from the tacan ground station for display on the HSI and BDHI.
T/R	Tacan receives bearing signals from the tacan ground station, and in addition, the tacan interrogates the tacan ground station to establish distance from the aircraft to the ground station. The bearing and distance information is displayed on the HSI and BDHI.
A/A	Tacan interrogates other aircraft which contain a tacan in the A/A mode and tuned 63 channels apart from the channel setting of the interrogating aircraft. The interrogation provides line–of–sight distance information for display on the HSI and BDHI of both (or up to six) aircraft.

The function selector knob is only operative in the cockpit having nav com.

Channel Selector Knob

The channel selector knob provides for selection of 126 tacan channels. The controls consist of an outer knob used to select the hundred and tens digit of the desired channel number, and an inner knob used to select the units digit of the channel number. Positions 0, 00, 127, 128 and 129 are not used.

Volume Control Knob

The volume control knob is used for volume adjustment of the station identity tone signal.

Course Set Knob

The course set knob is used for manually selecting desired tacan magnetic course.

Heading Set Knob

The heading set knob is used for manually selecting desired magnetic heading. Its setting is initially the same as that set in by the course set knob, but then is normally reset to compensate for wind.

Mode Selector Knob

With the mode selector knob in the tacan position, the tacan supplies the HSI information used to determine the course of the selected tacan radial, and whether the course is to or from the tacan station. A steering signal which aids the pilot in making an asymptotic approach to the selected tacan radial is displayed on the vertical director pointer of the attitude director indicator.

Brg/Dist Switch

With the brg/dist switch in the tacan position, bearing and distance information from the tacan set is supplied to the HSI for display.

CNI–Nav Comp Switch

With the CNI–nav comp switch in the CNI position, bearing information from the ADF system, and bearing distance information from the tacan system is displayed on the BDHI.

Nav Cmd Button

The navigation command push button transfers control of the tacan function from one cockpit to the other. When the green light illuminates, command of the tacan has been obtained in that cockpit. Once having obtained control, it can be relinquished to the other cockpit by pressing the nav cmd button. The navigation volume control is effective in both cockpits regardless of the take command situation.

ANTENNA SELECTOR SWITCH

The antenna selector switch is in the forward cockpit on the emergency floodlights control panel (figure A–1, appendix A). The switch is used to give the pilot a choice of either manual or automatic selection of the upper or lower tacan antenna. The switch has positions of UPPER, AUTO and LOWER. When the AUTO position is selected the tacan coax switch is used for automatic antenna selection. Operation of the AN/APR–27 or after AFC 524, the AN/ALR–50, causes the tacan to utilize the upper antenna regardless of antenna selector switch position.

IDENTIFICATION SYSTEM (IFF)

NOTE

The following discussion of the identification system includes incorporation of AFC 388 (AIMS). Aircraft without AIMS will not have mode C (altitude reporting) or mode 4 (secure IFF) capability. However, the other functions are the same.

The identification system is capable of automatically reporting coded identification and altitude signals in response to interrogations from surface or airborne stations. These coded replies allow the stations to establish aircraft identification, control air traffic, and maintain vertical separation. The system has five operating modes (1, 2, 3/A, C and 4). Modes 1 and 2 are IFF modes, mode 3 (civil mode A) and mode C (automatic altitude reporting) are primarily air traffic control modes, and mode 4 is secure (encrypted) IFF mode. (Mode 4 is not operational unless the system includes a KIT–1A/TSEC transponder computer.) The basic identification system consists of a transponder and IFF control panel. The altitude reporting function requires an altitude encoder unit and the air data computer (ADC). The altitude encoder unit receives log pressure synchro signals from the ADC. These signals correspond to aircraft altitude corrected for static position error. The altitude encoder provides an output of altitude information in both synchro and digital encoded form. The synchro output is supplied to the AAU-19/A servoed altimeters and the digital output is supplied to the transponder for transmission in mode C. The servoed altimeters are discussed under Instruments, later in this section. An ALT ENCODER OUT light on the telelight panel monitors operation of the altitude encoder unit. When this light is illuminated the transponder is not receiving digital altitude information for mode C operation. Refer to figure 1-24 for additional information on effect of ALT ENCODER OUT light.

IFF CONTROL PANEL

The controls for operation of the identification transponder are on the IFF control panel (figure A-1, appendix A). The REPLY light and controls on the left side of the panel are concerned with mode 4. The TEST light and the remaining controls are associated with modes 1, 2, 3/A, and C; except that the master control knob controls all modes of operation.

Master Control Knob

The master control knob applies power to the transponder but does not control power to the components necessary for altitude reporting (altitude encoder unit and ADC). These units are supplied power when aircraft power is available. The master control knob has positions of OFF, STBY, LOW, NORM, and EMER. The knob must be lifted over a detent to select EMER or OFF. STBY should be selected for two minutes prior to selecting LOW or NORM to allow the transponder to warm up. NORM initiates transponder operation at normal receiver sensitivity. LOW initiates transponder operation at a reduced sensitivity. In EMER the transponder transmits emergency replies to mode 1, 2, or 3/A interrogations. The mode 3/A emergency reply includes code 7700. Modes C and 4 are automatically enabled when EMER is selected, regardless of the position of the selector switches.

NOTE

Emergency operation of the transponder is automatically activated whenever either crewmember ejects.

Mode Selector Switches

The four mode selector switches (M-1, M-2, M-3/A, and M-C each have OUT, ON, and spring loaded TEST positions. The center ON position of each switch enables that mode. To test the transponder, press the mode selector switch of each mode to the TEST position. Illumination of the TEST light indicates proper operation of that mode. The master control knob must be in NORM for the test function to operate. The mode selector switches of the modes not being tested should be OUT when testing on the ground to prevent unnecessary interference with nearby ground stations.

Mode 1 and Mode 3/A Code Selectors

The two mode 1 code selectors allow selection of 32 mode 1 codes. The four mode 3/A code selectors allow selection of 4096 mode 3/A codes. The mode 2 code selector is on the transponder and cannot be changed in flight.

Identification of Position Switch

The identification of position switch has positions of IDENT, OUT, and MIC. The spring loaded IDENT position adds an identification of position pulse to mode 1, 2, and 3/A replies for a period of 15 to 30 seconds. In MIC the identification of position function is activated for 15 to 30 seconds each time the UHF microphone button is pressed. OUT disables the identification of position function.

Monitor - Radiation Test Switch

The switch has positions of RAD TEST, MON, and OUT. RAD TEST (radiation test) is used in conjunction with ground test equipment for testing modes 3/A and 4. It has no function during flight. MON (monitor) is used to monitor operation of modes 1, 2, 3/A, and C. With MON selected, the TEST light illuminates for 3 seconds each time an acceptable response is made to an interrogation on a selected mode. OUT disables the monitor function.

Mode 4 Selector Switch

Mode 4 is selected by placing the mode 4 selector switch to ON, provided that the master control knob is in NORM or LOW. OUT disables mode 4.

Mode 4 Code Selector

The mode 4 code selector has positions of ZERO, B, A, and HOLD. The knob must be lifted over a detent to select ZERO. It is spring-loaded to return from HOLD to the A position. Position A selects the mode 4 code for the present code period and position B selects the mode 4 code for the succeeding code period. Both codes are mechanically inserted into the KIT-1A/TSEC transponder computer by a single insertion of the KIK-18/TSEC code changing key. The codes are mechanically held in the KIT-1A/TSEC computer, regardless of the position of the master control knob or status of aircraft power, until after weight is off the landing gear. Thereafter, mode 4 codes will automatically zeroize anytime the master control knob or aircraft power is turned off. The code settings can be mechanically retained after the aircraft has landed by turning the code selector to HOLD and releasing it at least 15 seconds before the master control knob or aircraft power is turned off. The codes again will be held, regardless of master control knob position or aircraft power status, until the next time weight is off the landing gear. Mode 4 codes can be zeroized, when power is on the aircraft and the master control knob is in any position but OFF, by placing the code selector knob to ZERO.

Mode 4 Indication Switch

The switch has positions of AUDIO, OUT, and LIGHT. In AUDIO, an audio tone indicates that valid mode 4 interrogations are being received and the REPLY light illuminates if mode 4 replies are transmitted. In LIGHT, the REPLY light illuminates as mode 4 replies are transmitted, but no audio tone is present. In OUT, the audio and light indications are inoperative, and the REPLY light will not press-to-test.

IFF Light

The IFF light, on the telelight panel, illuminates to indicate that mode 4 is inoperative. The light is operative with power on the aircraft and the master control knob out of OFF. However, the light will not operate unless the KIT-1A/TSEC transponder computer is installed in the aircraft. Illumination of the IFF light indicates that (1) the mode 4 codes have zeroized, (2) the self test function of the KIT-1A/TSEC computer has detected a fault, or (3) the transponder is not replying to proper mode 4 interrogations. If the IFF light illuminates, place the master control knob to NORM (if in STBY) and ensure that mode 4 selector switch is ON. If illumination continues, employ operationally directed flight procedures for an inoperative mode 4 condition.

NORMAL OPERATION

NORMAL OPERATION OF INTERCOM SYSTEM

The intercom system is placed in operation without additional switching as soon as the aircraft receives electrical power. The controls should be set in the following manner in order to check the equipment before takeoff:

Pilot's Controls

1. Function selector switch - HOT MIC
2. Emergency amplifier selector switch - NOR
3. Volume - Rotate clockwise

Radar Intercept Officer's Controls

1. Function selector switch - NORMAL
2. Emergency amplifier selector switch - NOR
3. Volume - Rotate clockwise

With controls positioned as stated, check the duplex operation of the equipment by talking into the microphones. Rotate the volume controls to ensure that they are operating properly. Switch to the EMER ICS - NOR - EMER RAD positions to ensure that they are mechanically sound. The radio override functions of the interphone should be checked by each operator. To check the equipment properly, the operators should not switch to RADIO OVERRIDE at the same time, since reduction in the volume for the radio receivers is accomplished in both headsets when only one of the operators selects RADIO OVERRIDE. Therefore, each control must be positioned at different times to check the radio override circuitry in each unit. With all four of the amplifying stages working and the intercom system functioning normally, the pilot and RIO should place their function selector switches in the HOT MIC and NORMAL positions respectively. Both the pilot's and RIO's emergency amplifier switches should be placed in the NOR positions and the volume controls on each panel should be set as desired. No further switching is necessary to operate in duplex. The system is turned off when aircraft electrical power is removed.

NORMAL OPERATION OF COMMUNICATIONS TRANSMITTER AND RECEIVERS

With aircraft power activated, the UHF communications and ADF systems are placed into operation by rotating the UHF volume control thumbwheel to the ON position. Either comm-nav group control panel may be used in tuning the radio equipment to the comm and ADF channels required for operation; however, comm cmd must be established on whichever control panel is used for tuning. To enable the T/R-ADF and T/R+G-ADF modes, depress the T/R-ADF or the T/R+G-ADF push button and set the UHF volume control thumbwheel to its mid-position. The comm receiver-transmitter is set by use of the comm frequency thumbwheels with the mode selector switch in the MANUAL position; or, if a preset channel is available it can be set into the comm receiver-transmitter channels with the mode selector switch in the CHAN position and depressing the set channel push button. Voice communication is now possible, and ADF bearing information can be displayed on the BDHI and HSI. To enable the ADF+G-CMD mode, depress the ADF+G-CMD pushbutton and set the UHF volume control and the aux volume control to the approximate mid-position. Set the ADF channel to be used by rotating the comm frequency thumbwheels with the mode selector switch in the MANUAL position. Or, if a preset channel is available which is the same frequency as the ADF channel, it can be selected by rotating the comm channel knob with the mode selector switch in the CHAN position. Select the channel to be used for communication reception by rotating the aux channel knob. Voice reception is now possible and ADF information can be displayed on the BDHI and HSI. To enable the ADF-G mode, depress the ADF-G pushbutton and set the UHF volume control to approximately its mid-position. To set ADF channel to be used, rotate the comm frequency thumbwheels with the mode selector switch in the MANUAL position; or, if a preset channel is available which is the same frequency as the ADF channel, it can be selected by rotating the comm channel knob with the mode selector switch in the CHAN position. ADF bearing information can now be displayed on the BDHI and HSI.

NOTE

When in the gear down configuration, the ADF antenna pattern is distorted due to the close proximity of the nose landing gear door to the antenna. Therefore, the ADF system should not be relied upon as a primary navigational aid in the gear down configuration.

NORMAL OPERATION OF TACAN

The tacan system is made operational by placing the function selector switch to any position other than OFF. To operate the tacan receiver and transmitter, set the function selector knob to REC to allow the system to warmup, then select T/R if both bearing and distance information is desired. Allow a warmup period of approximately 2 minutes. Set the channel selector control to the channel of a tacan station within operating range. Place the brg/dist switch on the pilot's instrument panel to TACAN, and the cni-nav comp switch in the RIO's cockpit to the CNI position. Bearing and distance to the tacan station is displayed on the HSI and BDHI. The identification signal tone for the selected tacan station should be heard in the headsets. For air-to-air ranging, set the function selector knob to the A/A position. The tacan interrogates the other aircraft which contain a tacan in the A/A mode and are tuned 63 channels apart from the channel setting of the interrogating aircraft. Place the brg/dist switch to TACAN and cni-nav comp switch to CNI and the interrogation provides line-of-sight distance information for display on the HSI and BDHI of both (or up to six) aircraft. To utilize the tacan set for steering purposes place the brg/dist switch to the TACAN position and set the function selector knob to either REC or T/R mode. To fly directly to or from a tacan station, position the HSI course pointer and command heading marker to the desired tacan station radial (referenced to magnetic north). The HSI course deviation bar indicates the deviation from the desired course. The HSI to-from pointer indicates whether the selected radial will lead the aircraft to or away from the tacan station. The ADI displays a steering signal, which aids in making an asymptotic approach on the desired radial. Should the tacan information be unreliable, the vertical director warning flag on the ADI comes into view.

CAUTION

To prevent blowing fuses in tacan R/T, set tacan function selector knob to REC for 2 minutes warmup prior to selecting T/R or A/A.

NOTE

Operation of the AN/APR-27 or after AFC 524, the AN/ALR-50 causes the tacan system to utilize the upper antenna only. Tactical Manual Supplement, NAVAIR 01-245FDB-1T(A) provides additional details concerning ECM/tacan system integration.

NORMAL OPERATION OF IDENTIFICATION SYSTEM-IFF

To operate the system rotate the master control knob to the NORM position and set the mode 1, mode 2 and mode 3/A switches ON unless otherwise directed. On aircraft 153071z thru 153779ab, mode 1 (security identification feature) is automatically made operative when the master control knob is in an operation mode. On aircraft 153780ac and up, and all others after AFC 388, the ON and OUT positions are operative. Set the mode 1 and mode 3/A code selector dials as directed. The system is now ready for interrogation and response signals. In the event of an emergency, rotate the master control knob to the EMER position. The reply for modes 1 and 2 are special emergency signals of the codes selected on the applicable dials while mode 3/A reply is special emergency signals of code 7700. The same special emergency signals are replied when either or both crewmembers eject. For I/P switch operation, place the I/P switch in the IDENT position, or place it in the MIC position and key the UHF microphone. The IFF system responds with special I/P signals.

EMERGENCY OPERATION

CAUTION

Operation of the CNI while utilizing ram air with the RADAR CNI COOL OFF light illuminated, could affect equipment life and/or reliability. If the RADAR CNI COOL OFF light does not go out when the reset button is depressed, place the TACAN to OFF, and the IFF/SIF to STBY and operate only when necessary.

EMERGENCY OPERATION OF INTERCOM SYSTEM

Each cockpit ICS unit is equipped with two amplifiers, both of which are used during normal duplex (hot mic) operation. An emergency selector knob with EMER ICS, EMER RAD, and NOR positions is provided on the ICS panels. These selections enable an operator to bypass a faulty or dead amplifier in his unit. Assuming the pilot has selected HOT MIC, the operation during the various emergency selections is as follows:

Pilot's Switch Position	RIO's Switch Position	Resulting ICS Operation
Nor	Nor	Normal hot microphone
Emer ICS	Nor	Normal hot microphone
Emer RAD	Nor	Normal hot microphone
Nor	Emer ICS	Pilot's microphone is hot. RIO must actuate radio override switch to transmit on ICS. Foot button inoperative.
Nor	Emer RAD	Pilot's microphone is hot. RIO must actuate radio override switch to transmit on ICS. Foot button inoperative.

If both pilot's and RIO's intercom systems are in NOR operation, and the pilot then selects the COLD MIC position of the function switch, actuation of either front or rear seat radio override switches or front or rear seat ICS mic switches opens the system to HOT MIC operation from both cockpits. If it is necessary for both front and rear seat operators to select an emergency position, rear seat emergency conditions prevail. In addition, under NOR operation, both front and rear cockpit ICS microphone switches perform the same function as the radio override switches, reducing UHF volume.

NOTE

Even though the pilot's function selector switch is set at COLD MIC, the RIO may talk and listen to the pilot if he switches to RADIO OVERRIDE, or depresses the foot operated ICS switch. This is the only instance where duplex operation may be maintained when the pilot is not at HOT MIC. This switching arrangement is not normally used for intercockpit communication, since the pilot would usually be at HOT MIC position regardless of the settings of the other controls and the RADIO OVERRIDE settings are momentary switch positions.

On aircraft 157242an and up, and 153768ab thru 155528ag after AFC 388, and 155529ag thru 155902am after AFC 433 (Part 1), the emergency amplifier selector knob has positions of B/U, NORM, and EMER. If the headset amplifier in either ICS station fails, place the switch to the B/U (back-up) position in the cockpit with the defective station. This switches from the normal headset amplifier to the back-up amplifier and restores normal operation. If selecting B/U does not restore ICS operation, select EMER (emergency). Audio from the operative station is then connected directly to the back-up headset amplifier in the defective station. The volume control on the station with EMER selected has no effect on the audio level. The switch is left in NORM (normal) if the amplifiers in both ICS stations are operating normally.

EMERGENCY OPERATION OF AUX RECEIVER

The auxiliary receiver works in conjunction with the comm receiver-transmitter under normal conditions and operates as an emergency receiver in the event of power failure to the main radio receiver-transmitter. A guard receiver, contained within the auxiliary receiver, can also be monitored.

EMERGENCY OPERATION OF IFF SYSTEM

In an emergency replies may be transmitted for modes 1, 2, 3/A and 4, but the modes must be operating. The system begins transmitting an emergency reply when the master control knob is placed to the EMER position. In the emergency mode the system generates additional signals which readily identify the aircraft as being in distress. The reply is discontinued when the knob is removed from the emergency position.

LIMITATIONS

CNI equipment operating on external power without cooling air is limited to 10 minutes of accumulated operation in one hour span. This limitation applies to all CNI equipment except the intercom and UHF com receiver-transmitter. The maximum permissible altitude with CNI equipment ON is 70,000 feet. Flight above 70,000 feet with CNI equipment ON may result in damage to the equipment due to arcing.

DRAG CHUTE SYSTEM

NOTE

Some illustrations referred to within this system writeup may be located in appendix A. These illustrations are referred to, within the text, as (figure A–, appendix A).

DESCRIPTION

The airplane is equipped with a 16 foot, ring slot type parachute which is deployed after touchdown to aid in reducing landing roll distances. The drag chute may also be utilized for out of control/spin recovery. The chute is carried in a compartment within the empennage at the base of the vertical stabilizer, and is pulled into the airstream by a pilot chute when the spring-loaded compartment door is opened. The design of the attaching mechanism is such that should the compartment door open, without operating the cockpit control handle, the chute is released and falls free of the airplane. The drag chute is retained to the airplane structure upon normal deployment. There is no breakaway fitting within the attaching mechanism.

DRAG CHUTE HANDLE

The drag chute is deployed by means of a control handle (figure A-1, appendix A) beside the left console. A cable joins the handle, the release and jettison mechanism, and the door latch mechanism. Rotating the handle aft without depressing the button on the handle releases the door latch mechanism. The spring-loaded actuator then opens the drag chute door, and at the same time the hook lock is positioned over the drag chute attach ring. The spring-loaded pilot chute pops out, opens, and pulls out the drag chute. The drag chute is jettisoned by pulling aft on the handle to clear the detent, depressing the thumb button and lowering the handle. The release and jettison mechanism then returns to the normal position, permitting the drag chute to pull free.

NORMAL OPERATION

Normal operation of the drag chute system consists of deploying and jettisoning the drag chute. The drag chute is deployed by grasping the drag chute handle and rotating the handle aft. When landing with a known crosswind component equal to or greater than 20 knots, the drag chute should not be used because of excessive weather-cocking. To jettison the drag chute, rotate handle further aft to clear detent, depress the thumb button and then rotate handle full forward.

EMERGENCY OPERATION

There are no specific emergency operations pertaining to the drag chute system.

LIMITATIONS

Maximum airspeed for drag chute deployment is 200 knots CAS.

EJECTION SEATS

NOTE

Some illustrations referred to within this system writeup may be located in appendix A. These illustrations are referred to, within the text, as (figure A-, appendix A).

DESCRIPTION

The MK-H7 ejection seat system (figure 1-10) provides the occupant with a means of safe escape from the aircraft at practically all altitudes. The ejection seat is an automatic device that primarily regulates the opening of the personnel parachute at a predetermined altitude or after the seat has decelerated to within a safe limit for the occupant. The seat is equipped with two separate firing controls which are used to fire the seat during the ejection sequence. These two controls are the face curtain and lower ejection handle. The face curtain, located above the head, provides head protection and promotes proper ejection posture during the ejection sequence. The lower ejection handle, located between the occupant's knees on the seat bucket, provides an alternate method of firing the seat. Either control will jettison the canopy and fire the catapult gun. A rocket pack on the underside of the seat fires as the seat leaves the cockpit to provide a higher seat trajectory. A sequencing system (figure 1-11) provides an automatic ejection sequence if either of the crewmembers actuates his ejection seat. This allows both crewmembers to eject in a shorter period of time, and precludes collisions between seats or a seat with a canopy. A command selector valve, installed in the aft cockpit, provides the capability of selecting single ejection or dual automatic ejection. Should the automatic sequencing system malfunction, single ejection from either cockpit can be accomplished by pulling an ejection handle after canopy removal. A shoulder harness snubber incorporates an inertial reel lock with a velocity retention system, and a powered retraction device to retract the crewmember to the back of his seat during ejection. The velocity retention system locks the inertial reel whenever the crewmember pitches forward faster than a predetermined rate. The seat is linked to the canopy by the interlock mechanism to prevent accidental ejection through the canopy plexiglas. Ejection through the plexiglas is not possible due to the high strength qualities of the plexiglas used on the canopy.

The drogue chutes on the top of the seat stabilize the seat after ejection and deploy the personnel parachute. A hard-shell container is provided for the 29.7 foot skysail personnel parachute. On aircraft 157242an and up, and all others after ACC 176, the Skysail chute is replaced by a 28 foot parachute which gives better performance at low airspeeds. The container separates from the crewmember after chute deployment. Three handles are incorporated on the seat bucket, they are the emergency restraint release handle, the shoulder harness lock lever, and the leg restraint manual release handle. Two finger rings, located adjacent to the leg restraint snubber units, provide a means of releasing the snubbers on the leg restraint cords. The emergency restraint release handle permits manual release of the harness during manual separation procedures. The shoulder harness lock lever controls the upper harness movement. The leg restraint manual release handle releases the leg restraint cords. A seat position switch on the forward right side of the seat bucket, controls the seat adjusting motor.

CANOPY INTERLOCK BLOCK

NOTE

Although an ejection handle is pulled, ejection cannot occur if the canopy is fully opened. Since the canopy actuator is at the top limit of its travel, the canopy will not jettison to remove the interlock block and thus will not allow the ejection gun to fire.

The canopy interlock block and various lever arrangements within the interlock mechanism prevent firing of the ejection seat before the canopy has been jettisoned from the aircraft. The interlock block is connected to the canopy by a cable and is pulled from the interlock mechanism by the canopy during the ejection sequence. On aircraft 158346as and up, and all others after ACC 187, link connects the canopy interlock block to the ejection gun firing mechanism (interdictor) safety pin. The interdictor safety pin remains inserted in the ejection gun firing mechanism until it is pulled by the interlock block as the canopy is jettisoned. This gives added protection against inadvertent ejection initiation.

EJECTION SEAT

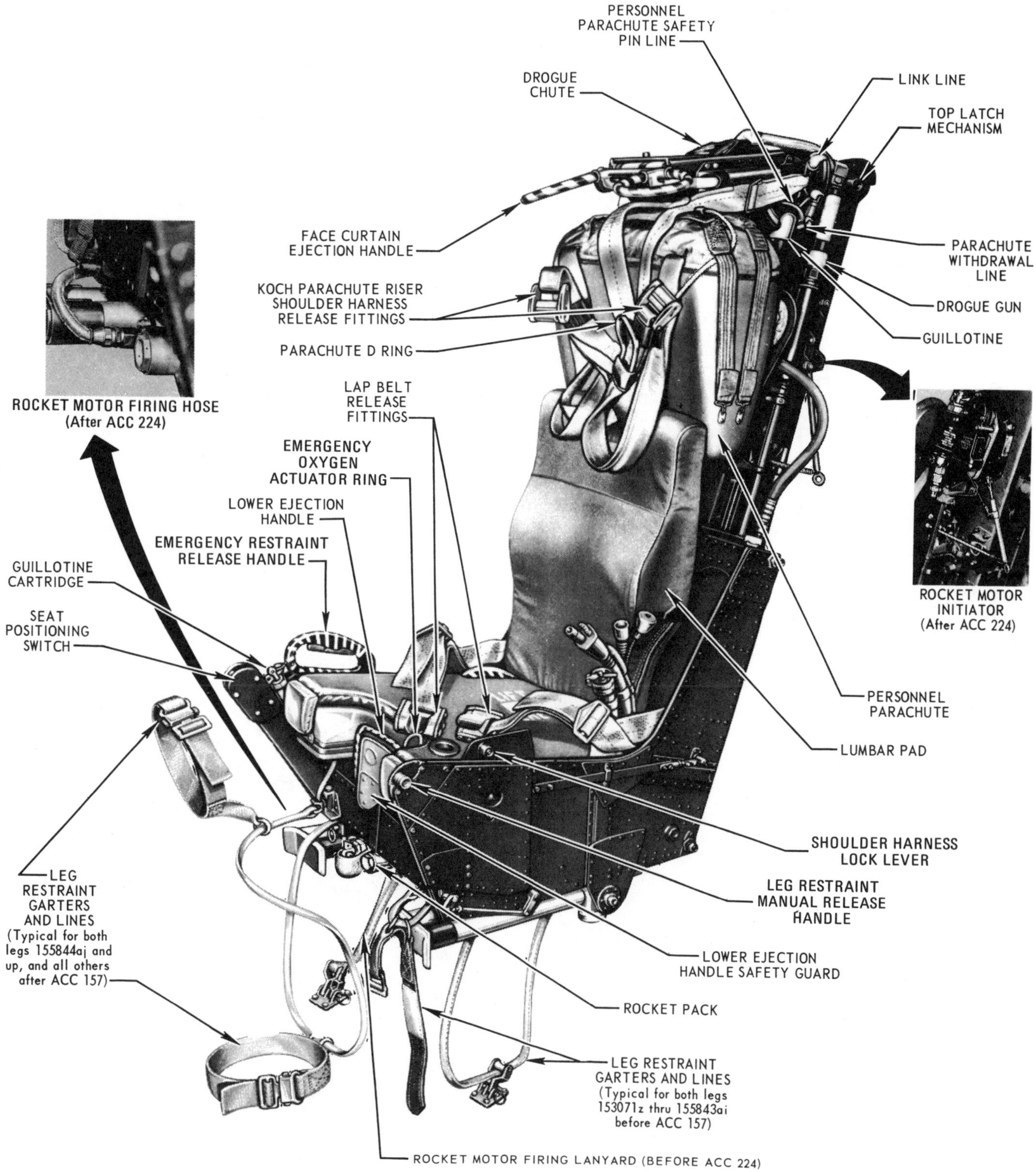

Figure 1-10

AUTOMATIC SEQUENCING SYSTEM

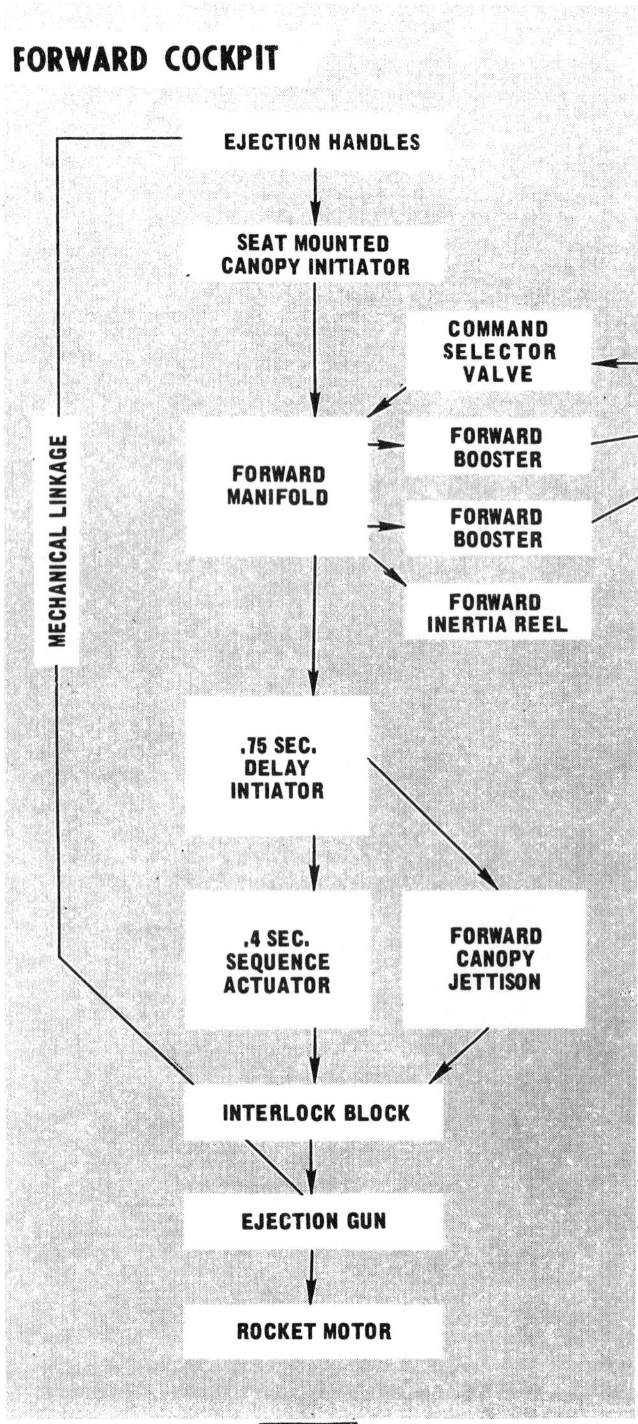

Note

ELAPSED TIME TO EJECTION GUN FIRING – 1.392 SECONDS.

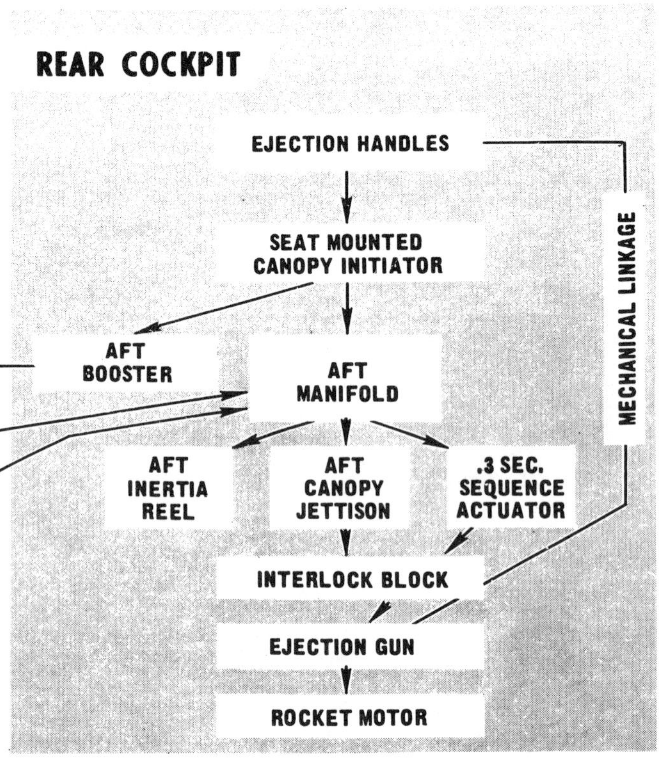

Note

- ELAPSED TIME TO EJECTION GUN FIRING – 0.54 SECONDS.
- PRIOR TO 155903ap BEFORE AFC 482, .3 SEC. SEQUENCE ACTUATOR IS INOPERATIVE WITH SINGLE EJECTION SELECTED BY COMMAND SELECTOR VALVE, TO FIRE SEAT, EJECTION HANDLE MUST BE PULLED AFTER CANOPY REMOVAL.

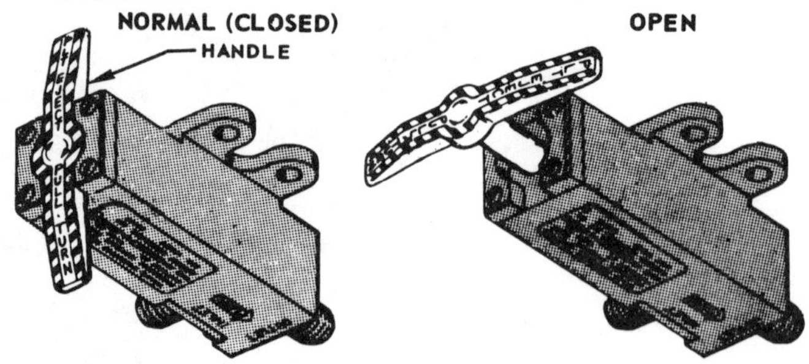

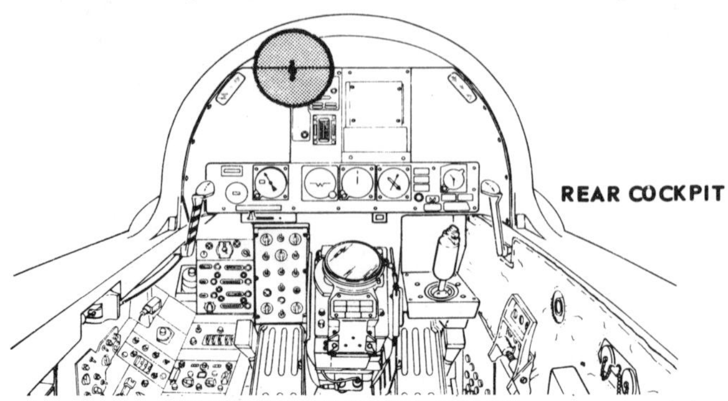

COMMAND SELECTOR VALVE

FDD-1-(17)A

Figure 1-11

CANOPY ACTUATING CYLINDER GUARD

The canopy actuating cylinder guard protects the banana links if the canopy shear pin fails and the actuating cylinder falls against the seat. The guard is installed on the seat mounted canopy initiator in the front cockpit only. On aircraft 158346as and up, and all others after ACC 187, a guard is installed over the seat mounted initiator linkages in the aft cockpit to give added protection against inadvertent initiation due to foreign objects.

CATAPULT GUN

The catapult gun mounted on the back of the ejection seat is used to propel the seat from the cockpit during the ejection sequence.

ROCKET MOTOR

A rocket motor is incorporated to provide a propulsion system for the rocket thrust phase of the ejction sequence. The rocket motor is on the bottom of the seat bucket and consists of a number of small diameter combustion tubes containing solid propellant, screwed into a manifold containing nozzles. As the ejection seat nears the end of the ejection gun stroke a static line attached to the cockpit floor withdraws the sear from the rocket motor firing mechanism causing ignition of the propellant. Since seat CG location is a function of seat bucket position, the rocket motor contains a thrust angle adjustment mechanism that changes the rocket motor thrust angle to compensate for the new CG location whenever the seat is raised or lowered. The ejection seat contains leg guards to prevent inadvertent contact between the crewmember's legs and the hot rocket tubes. On aircraft 157242an and up, and all others after ACC 169, a protector is installed around the rocket sear and sear cable to prevent accidental pulling of the rocket sear.

Rocket Motor Initiator - After ACC 224

After ACC 224, the rocket motor sear and lanyard are removed from beneath the seats. Rocket motor ignition is then provided by a gas rocket motor initiator mounted on the left side of the seat (figure 1-10). The initiator contains a cable lanyard which fires the initiator when the seat reaches a predetermined height. The initiator when fired produces a ballistic gas which activates the firing pin in the firing body to fire the rocket motor. Removing the rocket motor sear and lanyard from the bottom of the seat makes the motor less susceptible to inadvertent firing.

DROGUE GUN

The drogue gun is on the left side of the ejection seat headrest. Upon ejection, a trip rod fixed to the airplane structure pulls a sear from the drogue gun to initiate a 0.75 seconds time delay, at which time a cartridge is fired and the resultant gas pressures propel a piston out of the drogue gun barrel. Attached to this piston is a lanyard which pulls the controller drogue chute from its container. When deployed, the controller drogue chute pulls the stabilizer drogue chute from its containers. A cocking indicator is installed on the bottom of the drogue gun. When the gun is cocked, the indicator extends approximately ½ inch below the gun housing with the indicator shaft showing. If the indicator is flush with the bottom of the gun housing without the shaft showing, the drogue gun is not cocked and will not fire during ejection.

DROGUE CHUTE RESTRAINING SCISSORS

The drogue chute restraining scissors are on the top of the seat, and are used to connect the drogue chutes to the top of the seat when they are deployed during ejection. A movable jaw of the scissors releases the drogue chutes from the seat when the time release mechanism actuates.

TIME RELEASE MECHANISM

The time release mechanism is on the right side of the ejection seat headrest. Its function is to delay deployment of the personnel parachute and seat separation until the occupant has descended from the upper atmosphere, and/or has slowed enough to prevent excessive opening shock of the personnel parachute. The time release mechanism is armed upon ejection by a trip rod secured to the airplane. Initiation of the timing sequence follows immediately, providing the altitude is below 13,000 $\pm$1500 feet. If above 13,000 $\pm$1500 feet, initiation is delayed until the altitude condition is met. Two and one quarter (2.25) seconds after initiation of the timing sequence, the time release mechanism releases the drogue chutes from the restraining scissors allowing the personnel parachute to be pulled from its container. At the same time, it unlocks the harness and leg restraint lines to allow the occupant to be pulled from the seat when the personnel parachute deploys.

STICKER CLIPS

The sticker clips are on each side of the inner seat bucket. Each clip is made of spring steel with a detent point to hold the harness sticker strap lugs. The sticker clips retain the occupant in the seat until the personnel parachute blossoms and pulls the occupant clear of the seat.

FACE CURTAIN EJECTION HANDLE

A face curtain ejection handle (figure 1-10) is provided for seat ejection. The ejection handle, at the top of the seat, projects forward and provides a gripping surface for the crewmember. When ejection is desired, a forward and downward pull on the handle initiates operation of the sequencing system. During dual ejection from either cockpit, when the face curtain ejection handle is pulled to the first position, the automatic ejection sequence is initiated. After canopy jettison the catapult gun can be fired by either the automatic sequencing system or the ejection handle. By continuing the pull on the handle it is possible to beat the sequencing system in firing the catapult gun. This pulls the face curtain over the face and removes the wedge-shaped sear from the ejection gun, firing the main charge and ejecting the seat. For single ejection from the rear cockpit, an additional pull on the handle is required, since the catapult gun will not fire automatically. On aircraft 155903ap and up, and all others after AFC 482, the sequencing system is modified so that the rear seat will fire automatically during single ejection, and that an additional pull on the ejection handle is no longer required.

LOWER EJECTION HANDLE

The lower ejection handle (figure 1-10) is on the seat bucket between the occupant's legs. This handle is connected to a cable assembly, routed under and behind the seat, that initiates the same sequences and firing functions as the face curtain. The lower ejection handle is guarded (by a plate) to prevent inadvertent actuation. The guard provides a mechanical stop to prevent the handle from moving, and must be rotated to the down position before ejection.

COMMAND SELECTOR VALVE HANDLE

The command selector valve handle (figure A-2 appendix A) above the instrument panel on the left side of the aft cockpit is used to select single or dual ejection from the aft cockpit. The vertical position of the handle is the single ejection position and the handle is normally kept in this position during flight. To select dual ejection, the handle is pulled directly out (without applying torque to handle) from the valve. During the pull, the handle rotates 90° clockwise to the locked open position through cam action. On airplanes 158355at and up, and all others after AFC 506, the command selector valve is moved to the left console in the aft cockpit.

NOTE

A cycle of the command selector valve handle is defined as movement from the vertical to horizontal to vertical position. Inspection of valve will be in accordance with applicable maintenance directives.

Command Selector Valve Handle – After AFC 526

After AFC 526, the command selector valve is replaced by a more durable valve capable of 20,000 cycles. The new valve operates essentially like the old valve except for the following: the new valve is opened by applying torsion to the handle instead of a pulling action; the valve handle does not move away from the valve body when opening or closing; and the handle, if released in an intermediate position, will not always return to the vertical position, but to the vertical or 90° position depending on which side of the center of travel it is released.

SEAT POSITIONING SWITCH

The ejection seats may be adjusted vertically only. Fore and aft seat positioning is compensated for by adjusting the rudder pedals (front cockpit only). Vertical seat positioning is accomplished by actuating a momentary contact switch (figure 1-10) on the right forward side of the seat bucket. Each seat can be adjusted up or down through a total distance of six inches. It is not necessary to adjust the seat height before ejection; however, the seat should be low enough to afford adequate clearance between the helmet and face curtain ejection handle.

SHOULDER HARNESS SNUBBER AND POWER RETRACTION UNIT

The seat incorporates a shoulder harness snubber utilizing a velocity retention system, and a power retraction device. The snubbing unit prevents any forward shoulder movement whenever it is locked. When unlocked, the snubber automatically prevents rapid forward shoulder movement when the aircraft is subject to forces which tend to pitch the seat occupant forward, such as longitundinal crash forces or violent maneuvers. When forward motion of the seat occupant causes the harness straps to unwind faster than a safe predetermined rate of speed, the shoulder harness automatically locks. This feature is called the velocity retention system. The powered retraction device retracts the shoulder harness loop straps to position and locks the crewmember's shoulder harness for ejection. The device is gas powered and can only be initiated by pulling the face curtain or lower ejection handles.

Shoulder Harness Lock Lever

The shoulder harness lock lever (figure 1-10) has two positions, a forward or locked position, and an aft or unlocked position. Selecting the locked position locks the shoulder harness. Selecting the unlocked position will not prevent the snubber from locking when the velocity retention system detects a high rate of velocity change of the crewmember in a forward direction. Once the shoulder harness is automatically locked, it must be manually unlocked by cycling the release handle full forward then full aft. It is noteworthy that G forces on the aircraft by itself will not lock the snubber. On airplanes 155844aj and up and all others after ACC 217, the shoulder harness, after being locked automatically, can be unlocked by the crewmember relaxing tension on the harness, provided that the shoulder harness release handle is in the unlocked position.

LEG RESTRAINERS

A leg restraint assembly on the seat holds the occupant's legs in place and prevents them from flailing during ejection. The leg restraint assembly consists of garters worn by the crewmembers, leg restraint lines with lock pins, snubber units, and shear fittings secured to the floor. The garters are strapped on to the legs level with the leg restraint lock-in hole, but not below the midcalf position. Each leg line runs from the shear fitting beneath the seat, passes through the snubber unit, through the garter, and then the lock pin on the leg lines plugs into the leg restraint lock-in mechanisms on the front of the seat pan. When the seat is ejected, the slack in the leg restraint line is taken up by the upward travel of the seat, pulling the occupant's legs to the front face of the seat pan. When all the slack has been removed in the leg restraint lines, the tension of the line will cause the shear fitting to fail. The occupant's legs are firmly held against the seat pan by the snubbing unit until the harness is released and the occupant is separated from the seat. Leg restraint disengage rings (figure 1-10) on the face of the seat pan are used to adjust the amount of slack in the leg restraint lines. This slack may be adjusted by the occupant by pulling out on the appropriate finger ring. This allows more restraint line to be pulled out to provide sufficient slack. To take in excess slack, the occupant need only reach under the seat bucket and pull in the excess restraint line through the snubber unit. A spring is contained in each of the lock-in mechanisms on the front of the seat pan. The spring ejects the lock pins whenever any of the following actions occur: the time release mechanism actuates, the emergency harness release handle is pulled, or the leg restraint manual release handle is moved to the unlocked position. On airplanes 155844aj and up, and all others after ACC 157, the leg restrainers (figure 1-12) utilize two garters on each restraint line, a calf garter worn above the flight boot and a thigh garter worn on the thigh just above the knee. Each garter contains a quick release device which allows the garter to be released and left in the aircraft without disturbing garter adjustment. The restraint line routing under the seat is changed to provide a slower leg withdrawal during ejection. When routing the restraint lines through the garters be certain the lines are not twisted and route first through the calf garter (route first through outboard ring of calf garter then inboard ring) then through the thigh garter before inserting the lock pins in the lock—in mechanisms (figure 1—12).

Leg Restraint Manual Release Handle

The leg restraint manual release handle (figures 1-10) is on the left forward side of the seat bucket. When the handle is moved to the aft (unlocked) position, the lock pins on the leg lines are released from the leg lock mechanism. This allows the occupant to thread the leg lines back through the garter, enabling him to leave the seat without removing the garters.

GUILLOTINE ASSEMBLY

The guillotine (figures 1-10) is a pyrotechnically operated device used to sever the personnel parachute withdrawal line during manual separation from the seat. When the emergency restraint release handle is pulled, the guillotine cartridge fires, and gas pressure forces the guillotine blade upward, the withdrawal line is severed and the personnel parachute is disconnected from the drogue chutes.

INTEGRATED HARNESS

The integrated harness (figure 1-13) is a vest-like garment or a series of web straps worn by the crewmember. The harness, when used with the C/MBEU/615/PA integrated type parachute, takes the place of a lap belt and shoulder harness. Both of the harness configurations have four buckles for attaching the parachute of the crewmember. The lower two buckles, when connected to the lap belt release fittings, which in turn is fastened to the seat, serve as the lap belt. The upper two buckles, when connected to the parachute riser-shoulder harness release fittings, which in turn is fastened to the locking reel assembly, serve as the shoulder harness. The integrated harness eliminates the need for the crewmember to wear his parachute to and from the airplane, and it eliminates a separate lap belt and shoulder harness with its inherent limited restraint capabilities.

EMERGENCY RESTRAINT RELEASE HANDLE

The emergency restraint release handle (figure 1-10) is located on the right side of the seat bucket. In landing emergencies, in ditching, and in the event of the drogue gun or automatic time release failing to function during an ejection, the occupant can release himself, his parachute, and his survival equipment kit from the seat by squeezing the trigger of the restraint release handle and pulling the handle up and aft. The initial 1 inch of travel of the handle fires the guillotine and severs the connecting line between the stabilizer drogue parachute and the personnel parachute. Once the connecting line is severed the automatic parachute must be manually deployed. Continued travel of the handle up and aft causes the handle to lock in the open position and release the leg restraint lines, shoulder harness attach point and lap belt attach points. The occupant is then held in the seat only by the sticker clips; it is necessary to push free of the sticker clips to release from the seat. When the emergency restraint release handle is actuated it should be in a continuous motion to the full up and locked position. It is possible to partially rotate the handle and fire the guillotine and not release the leg lines, shoulder harness, or lap belt. The handle is spring loaded and returns to the down position, when released, unless locked in the fully up and aft position.

SURVIVAL KIT

A modified PK-2 survival kit is packed within a two piece fiberglas container (figure 1-14) which, in turn, is attached to the occupant by strap-harnessing. The content of the survival kit is the same as for a normal issue PK-2, but the packing arrangement has been changed to suit the requirements of the container. The following is a list of contents of this kit:

Pararaft with inflation bottle, sleeve type sea anchor and lanyard
Shark repellant
Survival radio
De-salter kit (tablets)
Water storage bag
Signal mirror
Bailing sponge
50 feet of nylon line
2 packs dye marker
Mark 73 Mod 0 smoke and flare signal
Canned rations
Can of sunburn ointment
Emergency code instruction sheets

The emergency provisions included in the PK-2 survival kit are subjected to local option and may be altered at the discretion of the area commander. All items except the pararaft and its associated gear are packed in a zipper enclosed bag which is attached to the pararaft by a lanyard. Both the pararaft and bag are packed into the survival kit container and the pararaft is attached by a lanyard to the upper half of the container. The upper half

LEG RESTRAINERS

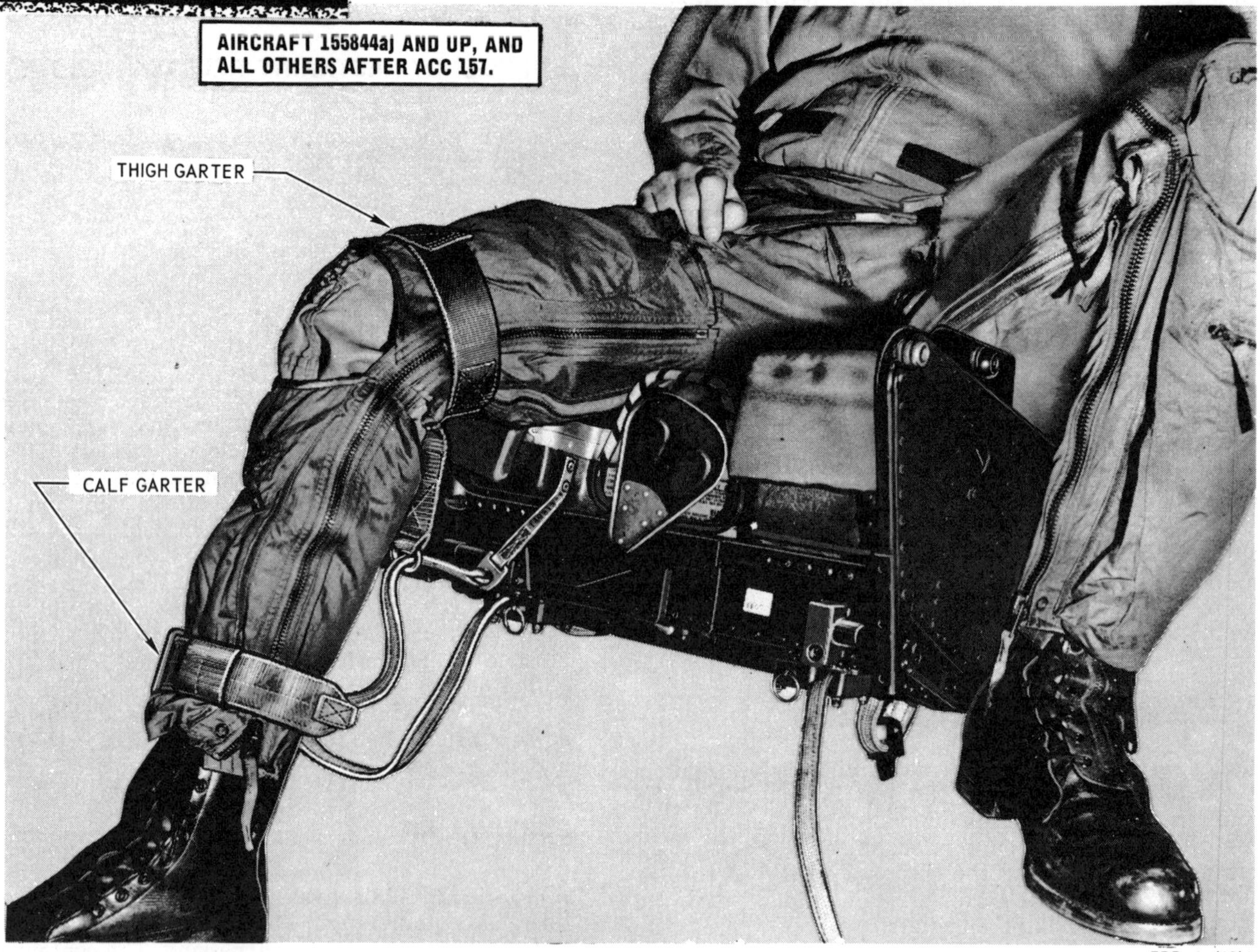

Figure 1-12

FDD-1-(18)A

of the container contains the emergency oxygen. A receptacle for the composite disconnect is located in the left rear corner of the survival kit container, and a kit release handle is on the right rear side of the survival kit container. Pulling up on the kit release handle unlocks the container which causes the lower half of the container and the life raft to drop below the crewmember on a drop line. The dropping action initiates inflation of the life raft. In the event of an ejection over water, the life raft should be inflated prior to entering the water since the kit release handle is accessible while still in the parachute harnessing and all survival equipment is secured to the crewmember. Should the crewmember enter the water before the survival kit release handle is pulled, the life raft can only be inflated by pulling the release handle and then reaching into the opened kit and pulling the life raft inflation bottle cable. The survival kit contains four negative G straps. These straps, located on each corner of the kit, prevent the kit from raising above the seat pan during negative G flight. The front straps are held in place by the leg restraint line lock pins. Care should be taken whenever the lock pins are removed from the leg restraint line leg lock-in mechanism to resecure the front negative G straps by re-threading the lock pin through the strap lugs.

COMPOSITE DISCONNECT

The composite disconnect is designed to connect aircraft oxygen, ventilating air, anti-G air and communications lines to the crewmember. The composite disconnect assembly consists of a lower block, an intermediate block, and an upper block. The intermediate block is fastened to the upper part of the survival kit and contains the tie-in between the crewmember and the emergency oxygen supply. The disconnect is so designed that the crewmember, during normal aircraft entrance or departure, is capable of quickly attaching or detaching all hoses and electrical lines leading from the aircraft to the survival kit and man. The lower block contains check valves in the ventilating air, anti-G and oxygen ports, that are open when the three selections of the disconnect are plugged in, and closed when either the upper or lower blocks are disconnected from the intermediate block. The check valves prevent gas leakage in the normal direction of flow when the valves are closed. The lower block is provided with a lanyard operated unlocking device, the free ends of the lanyard being attached to the airplane structure. As the seat is ejected, tension in the lanyard unlocks the device and separates the lower block from the

INTEGRATED HARNESS

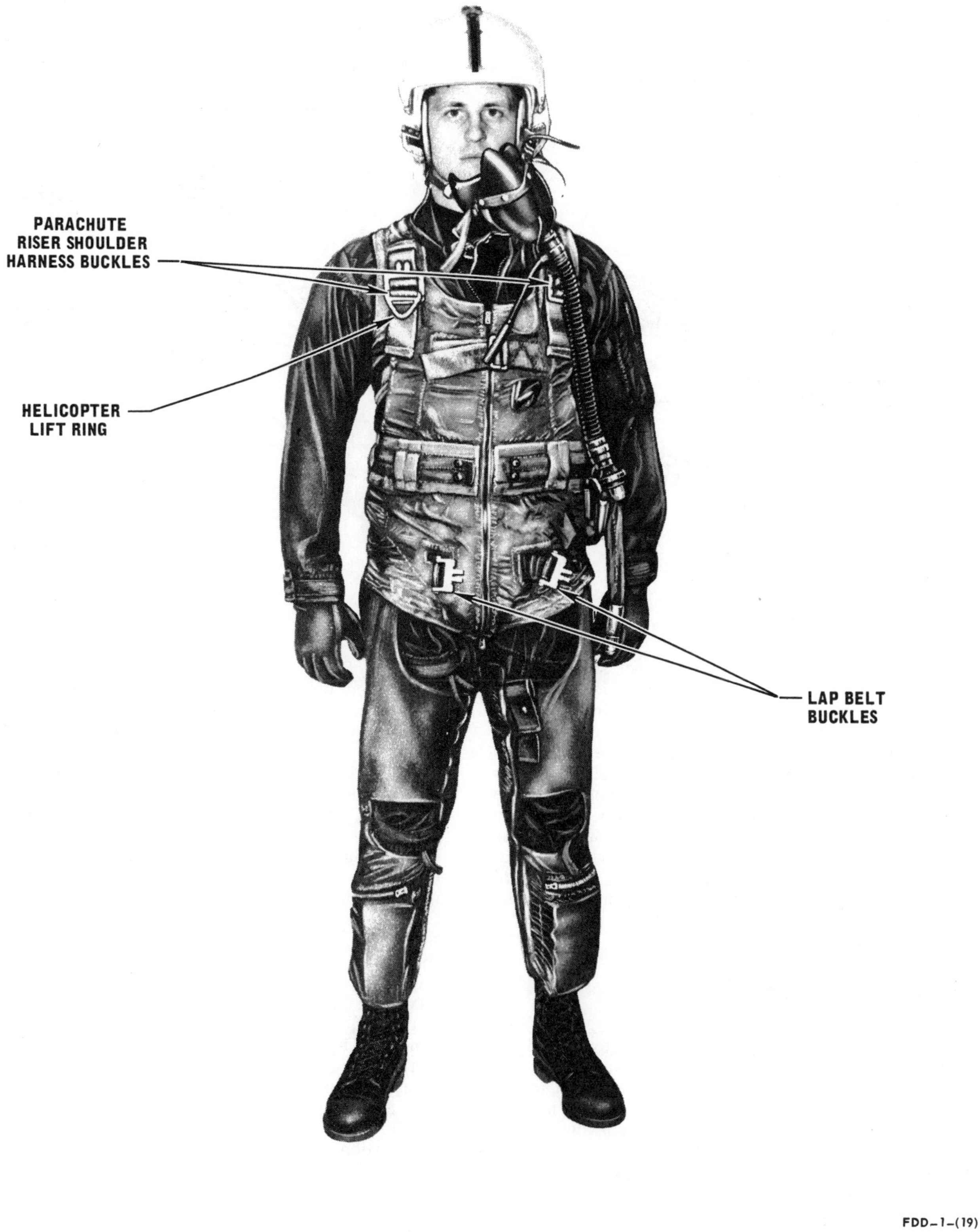

Figure 1-13

SURVIVAL KIT

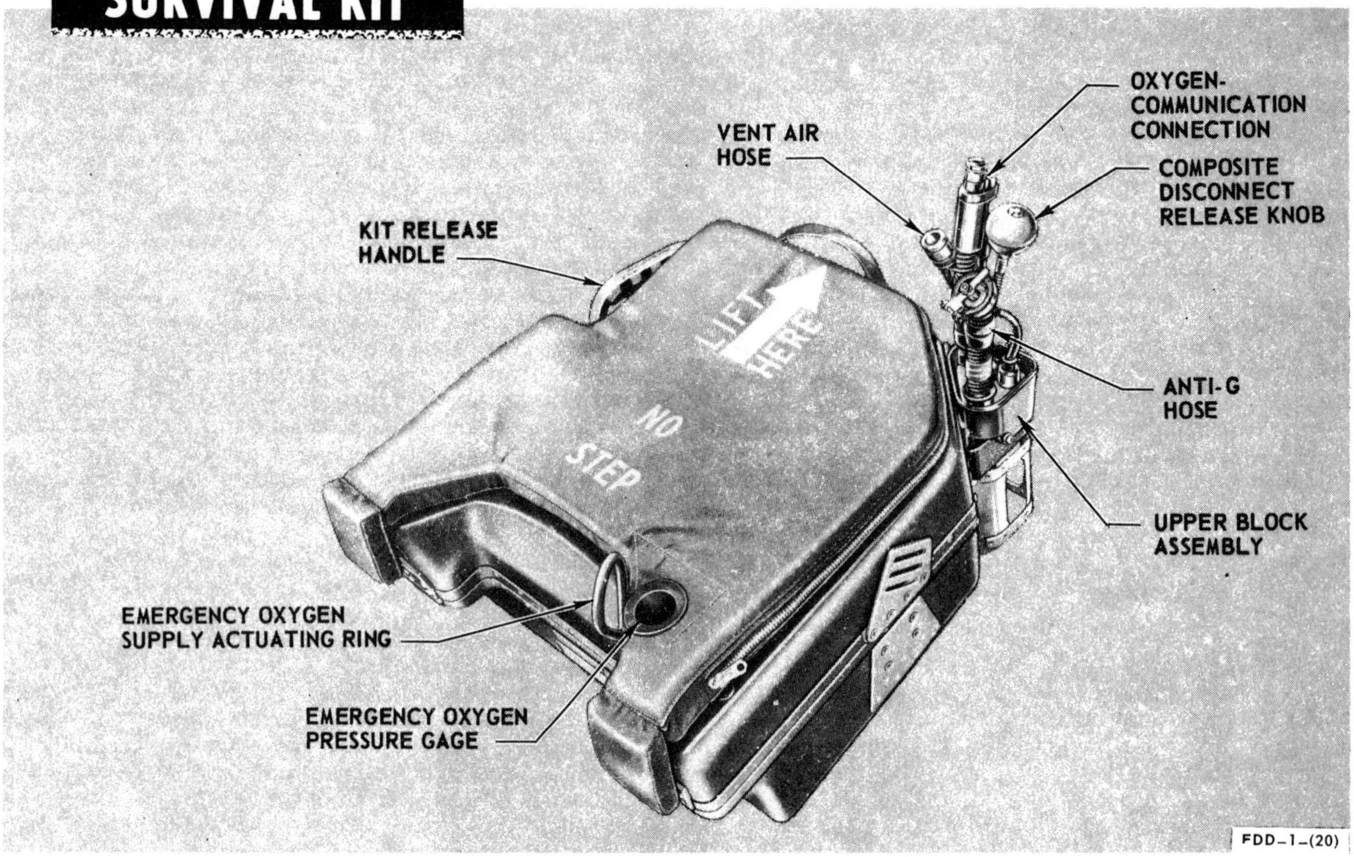

Figure 1-14

intermediate block. The intermediate block serves as the connecting link between upper and lower blocks and, in addition, by means of a Tee in the oxygen line, connects the emergency oxygen to the system. The upper block provides the means for attaching all service lines corresponding to those leading to the lower block. This is accomplished by in-line connections between the crewmember and the upper block. It also contains a manual disconnect device that permits the crewmember to free himself from all kit connections during normal aircraft departure by a single pull on the normal composite disconnect release knob. This action unlocks the upper block from the intermediate block.

NORMAL OPERATION

Ejection from the aircraft is accomplished by propelling the seat from the aircraft with a pyrotechnically energized catapult gun, followed by the firing of the rocket pack on the bottom of the seat. Three ejection sequences may be selected. Dual ejection may be initiated from the forward cockpit and dual or single ejection may be initiated from the aft cockpit (figure 1-11) A command selector valve is provided in the aft cockpit to select single or dual ejection. Ejection is initiated by pulling the face curtain or the lower ejection handle. Once an ejection handle is pulled, the sequencing system automatically fires the seat(s). Should the sequencing system malfunction, the automatic features of the system can be overidden, once the canopy is removed, by a continued pull on the ejection handle(s). Actuation of either handle fires an initiator which subsequently jettisons the canopy from the aircraft. When the canopy separates from the aircraft, the canopy interlock block which is attached to the canopy is pulled from the seat. Once the canopy is removed, the primary cartridge of the catapult gun is fired. Gas pressure generated by the cartridge causes the inner and intermediate tubes of the gun to extend upward. The upward travel of the inner tube actuates the top latch mechanism, which releases the seat from the aircraft. Continued movement of the inner tube propels the seat up the tracks. During upward travel of the seat, the sears of the drogue gun and time release mechanism are pulled by trip rods. Also during upward seat movement, the auxiliary cartridges are fired when they become exposed to the hot propellant gasses within the gun. Gas pressure generated by the auxiliary cartridges adds additional force to the gun during upward travel. Staggered firing of the catapult gun cartridges furnishes even pressure within the gun during the power stroke eliminating high acceleration forces during ejection. Separation of the inner tube from the gun occurs when the inner and intermediate tubes are fully extended in the outer tube. Upward seat travel after separation from the catapult gun continues by the momentum of the seat mass, and the rocket pack fires to propel the seat to a still greater height. Approximately 0.75 second after ejection, the drogue gun fires, deploying the controller drogue, which subsequently deploys the stabilizer drogue. The seat is stabilized and decelerated by the drogue chutes and the seat and occupant descend rapidly through the upper atmosphere. When an altitude of approximately 13,000 ±1500 feet is reached, a barostat initiates the timing sequence of the time release mechanism which, after 2.25 seconds, actuates to release the occupant's harnessing, leg restraint lines and chute restraint straps. The drogue chute pulls the parachute link line and the parachute safety pin line to deploy the personnel parachute. The purpose of the safety pin is to secure the parachute from premature opening due to windblast during descent prior to time release mechanism actuation. The occupant is held to the seat by sticker clips until the opening shock of the parachute snaps him out of the seat. If an ejection is made below 13,000 ±1500 feet the preceding sequence of events will occur approximately 2.25 seconds after the time release mechanism is tripped. The dual ejection sequence is initiated whenever the crewmember pulls either the face curtain handle or the lower ejection handle to fire the forward seat mounted initiator. Gas pressure from the seat mounted initiator is routed to the sequencing system stowing the aft cockpit equipment and operating the forward seat inertial reel, the aft seat inertial reel, aft canopy pressure operated valve (jettisoning aft canopy and opening flooding doors), aft pressure operated sequence actuator (ejecting aft seat), forward canopy pressure operated valve (jettisoning forward canopy), and forward pressure operated sequence actuator (ejecting forward seat), in that order. It is possible to save part of the 0.4 second sequence actuator delay time by maintaining a continuous pull on the ejection handle so that the catapult gun sear is pulled as soon as the canopy interlock block is removed.

Single Ejection From The Front Cockpit After Rear Cockpit Has Ejected Independently

Even though the rear cockpit has ejected independently, when the front seat occupant initiates the ejection sequence, the 0.75 second interlock delay is still present. This delay prevents collision between the front canopy and the rear seat in a sequenced ejection. When the ejection sequence is initiated by the crewmember, after a 0.75 second delay, gas pressure from the seat mounted initiator is routed to the forward canopy pressure operated valve (jettisoning the forward canopy), and then 0.4 seconds after canopy jettison the forward pressure operated sequence actuator operates to eject the forward seat. It is possible to save part of the 0.4 second sequence actuator delay time by maintaining a continuous pull on the ejection handle so that the catapult gun sear is pulled as soon as the canopy interlock block is removed. On aircraft 155903ap and up, and all others after AFC 477, the system is modified such that the 0.75 second delay will be bypassed after the rear cockpit has ejected independently. That is, if the rear cockpit has ejected, the front seat will eject approximately 0.4 seconds after a front ejection handle is pulled. However, in order for the 0.75 second delay to be completely bypassed, the rear ejection handle must be pulled at least 0.75 of a second before ejection is initiated from the front seat. If the front ejection handle is pulled sometime within 0.75 seconds of the rear handle, the delay will be reduced proportionally. That is, if the front seat initiates ejection 0.5 second after the rear seat, the 0.75 second delay is reduced to 0.25 second and the front seat then ejects approximately 0.65 second after the front ejection handle is pulled.

Dual Ejection Initiated From Rear Cockpit

The aft crewmember initiates a dual ejection by opening the command selector valve and pulling either the face curtain or lower handle to fire the seat mounted initiator. Gas pressure generated by the initiator is routed to the sequencing system which operates the aft cockpit equipment stowage (before AFC 506), the aft seat inertial reel, aft canopy pressure operated valve (jettisoning aft canopy and opening the flooding doors), aft pressure operated sequence actuator (ejecting aft seat), forward canopy pressure operated valve (jettisoning forward canopy), and forward pressure operated sequence actuator (ejecting forward seat), in that order. It is possible to save part of the 0.3 second sequence actuator delay time by maintaining a continuous pull on the ejection handle so that the catapult gun sear is pulled as soon as the canopy interlock block is removed.

NOTE

> When opening the command selector valve, pull directly out on the valve handle without applying torque. The handle will rotate 90° clockwise through cam action. After AFC 526, the handle is turned by the application of torque only, and there is no requirement to pull the handle.

Single Ejection Initiated From Rear Cockpit

Single ejection occurs when the aft cockpit crewmember pulls either the face curtain handle or lower ejection handle with the command selector valve in the normal (closed) position. Gas pressure generated by the aft seat-mounted initiator is routed to the sequencing system which operates the aft seat inertia reel, the aft cockpit equipment stowage (after AFC 506), and the aft canopy pressure operated valve (jettisoning the aft canopy and opening the flooding doors). An additional pull on the ejection handle then fires the seat catapult. On aircraft 155903ap and up, and all others after AFC 482, the sequencing system will fire the seat and there is no need for the additional pull on the ejection handle. However the ejection handle may be still pulled to save part of the 0.3 second sequence actuator delay time so that the catapult gun sear is pulled as soon as the canopy interlock block is removed.

Ejection With Front/Both Canopies Missing

Should the front canopy or both canopies be lost, the front canopy interlock block will also be lost. If ejection is then initiated from the front seat, this could expose the rear crewman to the front seat's rocket blast, and if conditions are right, a collision between seats could result. Should loss of the front canopy or both canopies occur, the rear crewman should be ordered to eject individually, allowing the front crewman to eject once the rear seat leaves, or have the rear crewman select command ejection by actuation of the command selector valve, and eject both seats by the automatic sequencing. The automatic sequence when initiated from the aft seat with the front canopy missing is not affected by the fact that the front interlock block is gone. With the loss of the rear canopy only, normal sequenced ejection can be initiated from either cockpit.

EJECTION HANDLE SELECTION

Due to its greater accessibility and shorter travel when compared to the face curtain, the lower ejection handle should be used during situations requiring an expeditious ejection. Some of these situations are insufficient flying speed from catapult, ramp strike, parting of cross deck pendants during carrier arrestment, low altitude, uncontrolled flight, and under high G during spin or air combat maneuvers.

EMERGENCY OPERATION

There are no provisions for emergency operation of the ejection seats; however, if the ejection seats fail to eject, the crewmember can abandon the airplane by following the procedures outlined in Emergency Procedures.

LIMITATIONS

Assuming wings level and no aircraft sink rate, the ejection seats provide safe escape within the following parameters:

a. Ground level (zero altitude) –zero airspeed (canopy must be closed)
b. Ground level and up –400 KCAS maximum (based on human factors) – 500 KCAS or Mach 0.92 maximum, whichever is less (based on seat limitations)

At airspeeds greater than 400 KCAS, appreciable forces are exerted on the body which makes escape more hazardous.

ELECTRICAL POWER SUPPLY SYSTEM

NOTE

Some illustrations referred to within this system writeup may be located in appendix A. These illustrations are referred to, within the text, as (figure A–, appendix A).

DESCRIPTION

The airplane electrical power supply system consists of a primary ac electrical system powered by two ac generators, a secondary dc electrical power system composed of two dc transformer–rectifiers for conversion of ac power to dc, a power distribution (bus) system, an emergency power supply system, and a receptacle for plugging in external power. The generators supply ac power to the left main 115/200 volt ac bus, the left main 28 volt ac bus, the essential 115/200 volt ac bus, the essential 28 volt ac bus, the right main 115/200 volt ac bus, the right main 14 volt ac bus, and the right main 28 volt ac bus. The transformer–rectifiers convert 115/200 volt ac power to 28 volt dc power, which is supplied to the essential 28 volt dc bus, the right main 28 volt dc bus, the left main 28 volt dc bus, the armament 28 volt dc bus, and the radar 28 volt dc bus. Refer to Electrical System (figure A–3, appendix A) for the individual bus loading.

AC ELECTRICAL POWER

The primary source of electrical power is derived from two 400 Hz, three–phase, 115/200 volt ac generators. The generators are engine driven and are rated at 30,000 volt–amperes. The generators are driven by constant speed drive (CSD) unit, which utilizes engine oil as a coolant and as the hydraulic media to regulate the generators at constant speed of 8000 rpm. The CSD incorporates a mechanical shaft disconnect feature which disconnects the CSD unit from the engine when a malfunction occurs which causes an excessive temperature within the CSD unit housing. The right engine generator supplies power to the right main 115/200 volt ac bus, and the right transformer–rectifier. The left engine generator supplies power to the left main 115/200 volt ac bus, the essential 115/200 volt ac bus, and the left essential transformer–rectifier, and on aircraft 155844aj and up and all others after AFC 399, the 115 volt ac ignition bus. Either generator is capable of supplying electrical power to the entire bus system through a bus tie relay contactor. The bus tie relay contactor is opened when both generators are operating and connected to their buses, resulting in split bus operation. Over/under excitation and voltage protection is provided. Should any of these conditions occur, the system automatically detects it and removes the affected generator from its buses. The bus tie relay contactor then closes and the full bus load is assumed by the remaining generator. Each generator may be manually disconnected from the bus system by placing its generator control switch to OFF. Under maximum generator loads, a generator drops off the line at an engine rpm of approximately 53%. At less than maximum generator loads, a generator drops off the line at a lower engine rpm than for maximum generator loads. When a generator is disconnected from its buses by the generator control switch or the automatic fault protection circuits, the appropriate generator warning light, either L.H. GEN OUT or R.H. GEN OUT illuminates. Operating checks consist of monitoring the above warning lights to determine if the generators are connected to the line. With both primary generators off the line, or the left generator off the line with the bus tie contactor open, the emergency ac generator can be operated to supply power to the essential 115 volt ac bus, the ignition 115 volt ac bus (after

AFC 399), the essential 28 volt ac bus, and the left transformer-rectifier. External power can be used to energize the ac buses in place of the engine driven generator. The required power is three phase, 400 cycle, 115/200 volt ac and is distributed to the bus system in the same manner as the generator output.

AC Power Distribution

The ac power distribution system consists of the following ac buses: right main 115/200 volt ac bus, essential 115/200 volt ac bus, left main 115/200 volt ac bus, and on aircraft 155844aj and up and all others after AFC 399, the 115 volt ac ignition bus. Auto transformers, powered by the right main 115/200 volt ac bus, the essential 115/200 volt ac bus, and the left 115/200 volt ac bus, power the right main 28 volt ac bus, the essential 28 volt ac bus and the left main 28 volt ac bus, respectively. The right main 14 volt ac bus receives its power from the same auto-transformer as the right main 28 volt ac bus. See figure A-3, appendix A. The circuit breakers protecting ac powered equipment are located on the essential circuit breaker panel in the front cockpit and the No. 1 and No. 2 circuit breaker panels in the rear cockpit. See figure A-7, appendix A for circuit breaker location and figure A-3, appendix A for circuit breakers associated with each ac bus.

Emergency Generator

An emergency three phase, 115/200 volt ac generator is provided as a source of electrical power in the event both generators go off the line or the left generator is off the line with the bus tie contactor open. The emergency generator is capable of supplying the essential ac and dc buses with 3000 volt-amperes of electrical power. The generator is powered by a ram air turbine which is extended into the airstream pneumatically. When operating, the emergency generator energizes an ac relay which connects generator output to the essential 115/200 volt ac bus, the essential 28 volt ac bus, and the left transformer-rectifier which supplies power to the essential 28 volt dc bus. The emergency generator ac relay will not become energized unless both engine driven generators are off the line, or the left generator is off the line with the bus tie contactor open. When operating with emergency power, the emergency generator does not deenergize until airspeed is reduced to approximately 90 KCAS. On aircraft 155844aj and up and all others after AFC 399, operation of the emergency generator energizes the ignition bus relay to connect the 115 volt ac ignition bus directly to an output phase of the emergency generator. The ignition bus can be energized by the RAT without energizing the essential buses, should the essential buses be overloaded, etc. To extend the ram air turbine, place the ram air turbine control handle to RAT OUT (push-down). If the engine driven generators are restored, the emergency generator is automatically disconnected from the essential buses. The ram air turbine may then be retracted by placing the ram air turbine control handle to RAT IN (pull-up). Based on NATC tests, the emergency generator delivers fully rated power for a minimum continuous period of three hours.

DC ELECTRICAL POWER

Two 100 ampere transformer-rectifiers receive 400 cycle, three phase, 115/200 volt ac power, and supply 28 volt dc power. The right transformer-rectifier supplies power directly to the right main 28 volt dc bus. The left transformer-rectifier supplies power directly to the essential 28 volt dc bus, and through an essential dc line relay to the left main 28 volt dc bus. The output of both transformer-rectifiers is connected in parallel through a 60 ampere bus tie current limiter. If one of the transformer-rectifiers fails, the remaining transformer-rectifier supplies power to the entire dc bus system. The emergency ac generator will supply power through the left transformer-rectifier to the essential 28 volt dc bus.

DC Power Distribution

The dc power distribution system consists of the following dc buses: left main 28 volt dc bus, right main 28 volt dc bus, essential 28 volt bus, armament 28 volt dc bus, and radar 28 volt dc bus. The left main 28 volt dc bus is connected to the right main dc bus through a 60 ampere bus tie current limiter. The essential 28 volt dc bus is connected to the left main 28 volt dc bus through contacts of the essential dc line relay. The essential 28 volt dc bus is connected to the left 28 volt dc bus whenever the left generator is connected to the line. The essential 28 volt dc bus becomes disconnected from the left main 28 volt dc bus if the left generator fails and the bus tie relay remains open. The armament 28 volt dc bus and the radar 28 volt dc bus are connected to the tie between the right and left main 28 volt dc buses. The armament 28 volt dc bus is energized by closure of the armament safety switch in the landing gear control handle when the handle is moved to the UP position. The radar 28 volt dc bus is energized by the missile power switch on the missile control panel being placed to the RADAR STBY or PWR ON position. The circuit breakers protecting dc powered equipment in the front cockpit are on the left utility panel, and the essential circuit breaker panel. Most of the aft cockpit circuit breakers are on circuit breaker panels No. 1 and No. 2. See figure A-7, appendix A.

EXTERNAL ELECTRICAL POWER RECEPTACLE

To provide adequate power for ground operation of electrical equipment, an external power receptacle (figure 1-1) is on the bottom of the left air duct. The external power required is three-phase, 400 cycle, 115/200 volt ac and it is distributed through the entire electrical system in the same manner as generator output (figure A-3, appendix A).

GENERATOR CONTROL SWITCHES

Two generator control switches, one for each generator are on the generator control panel (figure A-1, appendix A). The switches, labeled R GEN - ON, OFF, ON - EXT and L GEN - ON, OFF, ON - EXT, are utilized to select the source of electrical power for the airplane bus system. With external electrical power applied, and with both generator control switches in EXT ON, electrical power is supplied to the entire bus system. When either engine driven generator is operating, its output may be connected to the entire bus system by placing its control switch to GEN ON.

NOTE

If either generator indicator light illuminates as a result of a temporary generator malfunction, the generator may be reset. To reset the generator, place the generator control switch to OFF, and reposition it to GEN ON.

CIRCUIT BREAKERS

Most of the circuits are protected by circuit breakers in the aft cockpit. The circuit breakers in essential circuits are on the left utility panel and a panel on the right console in the forward cockpit (figure A-1, appendix A). The majority of the remaining circuit breakers are on panels in the rear cockpit (figure A-7, appendix A).

GENERATOR INDICATOR LIGHTS

The LH GEN OUT and RH GEN OUT indicator lights, and the BUS TIE OPEN light are on the generator control panel (figure A-1, appendix A) in the front cockpit. One of the generator lights will illuminate any time its generator is not on the line. The pilot may attempt to reset the generator by placing the generator switch to OFF and then placing it back to ON. If a double generator failure occurs in airplanes 153071z thru 155784ah prior to AFC 388 or AFC 535, both generator lights should be illuminated by dc control voltage generated from the permanent magnet generators contained within the ac generators. However, typical double generator failure in airplanes 153071z thru 155784ah prior to AFC 388 or AFC 535 is caused by loss of control voltage from the permanent magnet generators. The generator warning lights will not illuminate since the control voltage is utilized to illuminate the lights. Therefore, detection of double generator failure without illumination of the generator warning lights must be determined by other means such as the appearance of OFF flags on the instrument panels and the sudden quiet caused by loss of electrical control power to the air conditioning system. The generator warning lights cannot be restored by RAT power. On airplanes 155785ai and up, and all others after AFC 388 or AFC 535, double generator failure due to permanent magnet generator control voltage failure is minimized, but the warning lights never illuminate, unless the RAT is operating. As above, the double generator failure is determined by the appearance of OFF flags and the quiet caused by shutdown of the air conditioning system. The warning lights will be restored on RAT power. Also, on aircraft 155785ai and up, and all others after AFC 388 or AFC 535, a left generator failure accompanied by a failure of the bus tie will not be noted by the illumination of the LH GEN OUT and BUS TIE OPEN lights if the pilots instrument lights knob (flight instrument lights knob after AFC 536) is OFF. Failure can be detected initially by noting the OFF flag appearing on the ADI, at which time the instrument lights should be turned on and the generator lights monitored for verification of the failure. If the pilots instrument lights are not in the OFF position when a right generator failure accompanied by a bus tie failure occurs, the RH GEN OUT and BUS TIE OPEN lights will not illuminate. The failure may be detected by loss of the front cockpit console floodlights if in the MED position, loss of the rear cockpit floodlights if in the DIM position, or loss of the rear cockpit utility light. In this case the failure is not obvious and may not be detected until attempting to land, at which time the APCS and indexer lights will be inoperative. Since a right generator failure accompanied by a bus tie failure is difficult to detect when using instrument lights, it is recommended that anytime instrument lighting is being used, the console floods should be in the MED position. The MED position of the console floods is powered by the right generator, therefore, the console floods can be used as a failure indicator. Warning lights power will be available at all times on the RAT, regardless of the position of the pilot's instrument lights knob (flight instrument lights knob after AFC 536).

ESSENTIAL DC TEST BUTTON

An essential dc test button and indicator light are on the right side of the rear cockpit (figure A-2, appendix A). When the dc test button is depressed, the essential dc line relay is deenergized. Power will then be supplied to the essential dc test light by the left (essential) transformer-rectifier. If the indicator light illuminates, the left transformer-rectifier is receiving ac power and supplying dc power. If the indicator light does not illuminate, the left transformer-rectifier is either malfunctioning or not receiving ac power. With the dc test button depressed, the right transformer-rectifier can be checked by actuating the warning lights test switch. If the warning lights illuminate, the right transformer-rectifier is operating properly.

AUTOPILOT GROUND TEST SWITCH

With external power applied to the aircraft buses and the autopilot ground test switch in the NORM position, no electrical power is applied to the AFCS circuits. To apply power to the AFCS, the autopilot ground test switch must be placed to the TEST, solenoid held, position. The autopilot ground test switch, just aft of No. 2 circuit breaker panel (figure A-2, appendix A), will remain in the TEST position with power applied to the autopilot until either external power is removed, or the switch is manually placed to the NORM position. When either generator switch is placed to a position other than EXT, the autopilot ground test switch can no longer be used to remove power to the AFCS. The purpose of the switch is to prolong AFCS component life by removing power to the system while external power is applied to the aircraft for maintenance of other systems.

A-A/IFF GROUND TEST SWITCH

On aircraft 153071z, 153851ae, and 155529ag and up, and all others after AFC 388, an A-A/IFF ground test switch is installed just below the autopilot ground test switch in the rear cockpit. The manually actuated, electrically held switch permits ground operation of the AN/APX-76 interrogator set by by-passing circuitry provided to inhibit inadvertent ground operation. With external power on the aircraft and the switch in NORM, no power is applied to the AN/APX-76 circuits. To apply power, the switch must be placed to TEST, solenoid held, position. The switch will remain in TEST with power applied to the AN/APX-76 until either external power is removed or the switch is placed to NORM. When either generator switch is placed to a position other than EXT, the A-A/IFF ground test switch can no longer be used to remove power to the AN/APX-76.

NORMAL OPERATION

Normal operation of the electrical system commences when external power is applied to the aircraft and the generator switches are in EXT ON, or when the engines are running and the generator control switches are in GEN ON.

EMERGENCY OPERATION

SINGLE GENERATOR FAILURE

Failure of one generator is indicated by illumination of either the LH or RH GEN OUT light. The light indicates which generator has failed. One generator in normal operation is sufficient to support the entire electrical demand or load. If generator failure occurs, cycle the affected generator control switch from ON to OFF, and back to ON. If the generator fault has been corrected, the generator is brought back on the line and the light goes out. If the light remains illuminated, monitor engine oil pressure and variable nozzle operation. If oil starvation is indicated, secure the affected engine if practicable. On airplanes 153088aa and up, and all others after PPC 62, illumination of the applicable engine oil low warning light can be used as an indication of oil starvation.

CAUTION

- Upon illumination of a generator indicator light, immediately check the corresponding oil pressure gage. The generator failure could have been caused by oil starvation which will also affect the engine oil system.
- On airplanes 153088aa and up, and all others after PPC 62, illumination of the applicable engine oil low warning light shall be used as an indication of oil starvation.

DOUBLE GENERATOR FAILURE

Although a double generator failure is highly remote, the possibility of a double failure is still present. As previously stated, one generator out light will be illuminated when one of the generators fail. On airplanes 153071z thru 155784ah prior to AFC 388 or AFC 535, when the other generator fails, both generator out lights should be illuminated by dc control voltage generated from the permanent magnet generators contained within the ac generators. However, typical double generator failure in airplanes 153071z thru 155784ah prior to AFC 388 or AFC 535 is caused by loss of control voltage from the permanent magnet generators. The generator warning lights will not illuminate since the control voltage is utilized to illuminate the lights. Therefore, detection of double generator failure without illumination of the generator warning lights must be determined by other means such as the appearance of OFF flags on the instrument panels and the sudden quiet caused by loss of electrical control power to the air conditioning system. The generator warning lights cannot be restored by RAT power. On airplanes 155785ai and up, and all others after AFC 388 or AFC 535, double generator failure due to permanent magnet generator control voltage failure is minimized but the warning lights never illuminate, unless the RAT is operating. As above, the double generator failure is determined by the appearance of OFF flags and the quiet caused by shutdown of the air conditioning system. The warning lights will be restored on RAT power. Upon loss of both generators, extend the ram air turbine and turn off all electrical equipment not necessary to maintain flight. Attempt to return the generators to the line by cycling the generator controls switches to OFF, then back to ON. If the fault has been corrected, the generator lights will be extinguished. With the loss of all electrical power in flight, the emergency pneumatic system should be utilized to extend the landing gear. To preclude the loss of the utility hydraulic system, do not blow down the flaps unless a utility hydraulic system failure is indicated, or a carrier landing is being anticipated. The utility hydraulic system gage does not operate under RAT power.

NOTE

If the flaps had been lowered prior to electrical failure, air loads will return the flaps to a low drag position.

BUS TIE OPEN

Under normal conditions, both generators operate independently of each other, with the bus tie relay open and without the BUS TIE OPEN light illuminated. But with certain electrical faults present in the system, it is possible to lose either of the generators while retaining the bus tie open. The result is a single generator operating to provide power to only part of the electrical buses with illumination of the BUS TIE OPEN light and either the LH GEN OUT light or the RH GEN OUT light. An important example of this kind of fault is the loss of the essential buses due to a short in the generator system. A short on one of the left generator buses will be noted by the illumination of the LH GEN OUT light followed in 2 seconds by the illumination of the BUS TIE OPEN light. All the buses powered by the left generator, including the essential buses, will be lost. An attempt shall be made to regain the left generator by placing the left generator control switch OFF, and placing the switch back ON. If the left generator comes on the line, the short is no longer present and the generators will resume normal operation. Should the lost generator fail to be restored, cycle the right generator control switch in an attempt to close the bus tie. If none of these procedures succeed the short is probably still present, and the emergency generator must be used to restore the essential buses. Care, however, must be exercised after extending the RAT as the short might be on one of the essential buses. A short on one of the buses associated with the right generator is not of such a serious nature because the left main and essential buses will still be in operation. The same procedure shall be followed, however, to attempt to restore the lost generator and, failing that, to close the bus tie. On aircraft 155785ai and up, and all others after AFC 388 or AFC 535, a left generator failure accompanied by a failure of the bus tie will not be noted by the illumination of the LH GEN OUT and BUS TIE OPEN lights if the pilots instrument lights knob (flight instrument lights knob after AFC 536) is in the OFF position. Failure can be detected initially by noting the OFF flag appearing on the ADI, at which time the instrument lights should be turned on and the generator lights monitored for verification of the failure. If the pilots instrument lights knob is not in the OFF position when a right generator failure accompanied by a bus tie failure occurs, the RH GEN OUT and BUS TIE OPEN lights will not illuminate. The failure may be detected by loss of the front cockpit console floodlights if in MED position, loss of the rear cockpit floodlights if in DIM position, or loss of the rear cockpit utility light. If the failure is not detected before attempting to land, the APCS and indexer lights will be inoperative. Since a right generator failure accompanied by a bus tie failure is difficult to detect when using instrument lights, it is recommended that anytime instrument lighting is being used, the console floods should be in the MED position. The MED position of the console floods is powered by the right generator, therefore, the console floods can be used as a failure indicator. Warning lights power will be available at all times on the RAT, regardless of the position of the pilots instrument lights knob (flight instrument lights knob after AFC 536). Refer to Emergency Procedures for action to be taken upon illumination of the BUS TIE OPEN light.

ELECTRICAL FIRE

If an electrical fire occurs, extend the ram air turbine and turn the generator control switches and all electrical switches OFF. When the fire subsides place the generator control switches ON. If the fire still persists after turning the generator control switches ON, turn the generator control switches OFF. Operate essential equipment only and land as soon as practicable. If no fire is apparent when the generator control switches are placed ON, individually reposition the electrical equipment switches to ON beginning with the most essential equipment first. If the malfunctioning piece of equipment is found, turn the equipment OFF, and pull the applicable circuit breaker.

LIMITATIONS

Based on NATC tests, the emergency generator delivers fully rated power for a minimum continuous period of 3 hours.

EMERGENCY EQUIPMENT

NOTE

Some illustrations referred to within this system writeup may be located in appendix A. These illustrations are referred to, within the text, as (figure A–, appendix A).

DESCRIPTION

The airplane emergency equipment consists of a pneumatically extended and retracted ram air turbine, various jettison switches, and a comprehensive set of warning and indicator lights.

RAM AIR TURBINE

A ram air turbine, in the upper left side of the center fuselage, is provided as a power source for an emergency ac generator. The turbine assembly consists of a housing that contains two variable pitch turbine blades, a governing unit that controls the pitch of the blades, and the gearing to transfer blade rotation to a vertical drive shaft. The gear box then drives the generator. When the ram air turbine is extended into the airstream, the turbine blades are at a maximum angle of attack. This results in a rapid acceleration of the turbine blades and governing unit. As the regulating speed of the governing unit is approached, the turbine blades decrease their angle of attack. Turbine blade angle of attack (as directed by the governing mechanism) then varies with respect to the velocity of the airstream in order to maintain a constant 12,000 rpm. In effect, the ram air turbine functions as a constant speed drive unit for the emergency generator, providing the airspeed is above approximately 180 knots. As the airspeed drops below approximately 180 knots, the velocity of the airstream across the turbine blades is not sufficient to maintain 12,000 rpm, and as a result, electrical power output will decrease in proportion to the decrease in rpm of the turbine. Electrical power output ceases at approximately 90 KCAS. Maximum airspeed for operation of the ram air turbine is 515 KCAS or Mach 1.1, whichever is less. Acceleration limits with RAT extended are −1.0 to +5.2G. Ram air turbine operating time shall be logged on the yellow sheet (OPNAV Form 3760–2).

Ram Air Turbine Control Handle

The ram air turbine is extended and retracted pneumatically by a ram air turbine handle (figure A-1, appendix A) in the forward cockpit. Air pressure for extension and retraction is taken from a 4.2 cubic inch air bottle (in reality, not a bottle but an enlarged air line) which is charged by the pneumatic system. If the pneumatic system loses pressure, the air line retains its pressure through the action of a check valve. However, after loss of pneumatic system pressure there is only enough charge left in the air line for a single actuation (extension or retraction). Pushing down on handle extends the turbine, pulling up on the handle retracts the turbine.

WARNING

Minimum RAT extension pressure is 1000 psi. Therefore to ensure RAT extension in the event of an air bottle check valve failure, the RAT should be extended whenever pneumatic pressure begins dropping and RAT utilization is anticipated.

NOTE

- During ram air turbine extension, it is normal for the FIRE and OVERHT lights to flicker until the RAT comes up to full speed. Disregard these lights unless they remain on after RAT has obtained full speed.
- Except for illumination resulting from RAT extension, illumination of the FIRE or OVERHT lights shall be logged on the yellow sheet (OPNAV FORM 3760-2).

WARNING AND INDICATOR LIGHTS

Warning and indicator lights have been incorporated throughout the cockpits to reduce instrument surveillance to a minimum. The majority of the lights are in the front cockpit, with most of them being grouped on the telelight panel.

Telelight Panels

Telelight panels, on the front cockpit right vertical panel and the generator control panel (figure A-1, appendix A), contain the telelight bars. When a condition exists that requires corrective action, or is worthy of note, a telelight bar corresponding to the condition illuminates. Most of the lights on the telelight panels illuminate in conjunction with the MASTER CAUTION light. Indicator lights that do not illuminate in conjunction with the MASTER CAUTION light are: SPEED BRAKE OUT, L EXT FUEL, CTR EXT FUEL, R EXT FUEL and REFUEL READY. Electrical power to the telelights is supplied by the 28/14 volt ac warning lights bus. The light(s) will be extinguished when the condition that cause the light(s) to illuminate is corrected. Amplification of conditions that exist upon illumination of a telelight, and its corrective action can be found in Warning/Indicator Light Analysis, section V.

Master Caution Light

A MASTER CAUTION light operates in conjunction with lights on the telelight panel. It is only necessary to monitor the MASTER CAUTION light for an indication of a condition requiring attention, and then referring to the telelight panel for the specific condition. The MASTER CAUTION light may be extinguished by depressing the master caution reset button, located on the generator control panel. The illuminated lights on the telelight panel will not be extinguished by the master caution reset button, with the exception of the AUTO PILOT DISENGAGE light and the APCS OFF light, until their respective faults have been cleared. After the MASTER CAUTION light is cleared, and an additional condition exists that requires attention, the MASTER CAUTION light will again illuminate, except when initial illumination is caused by a hydraulic system failure and the subsequent failure is also a hydraulic system failure.

CAUTION

When the MASTER CAUTION light illuminates with no other indications, activate the warning lights test switch to check for a burned out bulb in the telelight panel.

Warning Light Test and Dimmer Circuit

The warning light test and dimmer circuit provides a means of testing the operation of the bulbs in the warning and indicator lights. All warning and indicator lights are included in the test and dimmer circuit which is powered by the warning lights 28/14 volt ac bus. The circuit does not provide an operational check of any warning or indicator system, it checks only the light bulbs. The warning and indicator lights may be illuminated by actuating the warning lights test switches on the interior lights control panels.

Eject Light

An EJECT LIGHT system (removed from airplanes 158355at and up, and all others after AFC 506) provides for a positive visual ejection command from the pilot to the RIO. The lights can be actuated only from the front cockpit. The pilot's switch and monitor light are incorporated into a single unit, mounted under the left canopy sill just forward of the flap switch (figure A-1, appendix A). The switch is a push ON, push OFF type, with the push button being the lens of the light. The lens is recessed sufficiently to preclude an accidental actuation. The light in the rear cockpit is a rectangular press to test unit mounted at the bottom right of the instrument panel (figure A-2, appendix A). Pressing the lens of the rear light will test the rear light bulb and circuitry only. When the switch in the front cockpit is depressed, both EJECT lights illuminate. Depressing the switch again will extinguish both lights. Both lights incorporate red lenses, the rear light with black lettering. Power for the system is supplied by a separate dry cell battery mounted in the front cockpit.

External Stores Emergency Release Button

The external stores emergency release button (figure A-1, appendix A) is on the left vertical panel. This button when depressed, will jettison all external stores (including missiles and pylons) provided the gear handle is up or, the main gear struts are extended. On airplanes 158355at and up and all others after AFC 506, the jettison select switch limits the external stores jettisoned by the external store emergency release button.

Jettison Select Switch

On airplanes 158355at and up and all others after AFC 506, the jettison select switch (figure A-1, appendix A) is on the air-to-air panel in the front cockpit. The switch is used to limit the stores jettisoned by the external stores emergency release button. Generally, the switch can be used to retain all air-to-air missiles (Sidewinders and Sparrows), and if required, all air-to-air missiles and whatever store is on the centerline station (fuel tank, etc.), while jettisoning the remaining external stores by depressing the external stores emergency release button (Refer to figure 5-17). The jettison select switch has three positions: ALL, AIR-GRD, and AIR-GRD + CL. In ALL, the external stores emergency release button jettisons all external stores. In AIR-GRD + CL, all external stores are jettisoned except the air-to-air missiles, and also various armament attachments when carried in combination with air-to-air missiles on the inboard wing stations as listed below. In AIR-GRD , all external stores are jettisoned except air-to-air missiles and whichever store is carried on the centerline station, and also various armament attachments when carried in combination with air-to-air missiles on the inboard wing stations as listed below. With any combination of air-to-air missile(s) and other stores on the inboard wing stations (Sparrow on station 2 and TER on 8, Sidewinders and TER carried simultaneously on stations 2 and 8, etc.), and either AIR-GRD or AIR-GRD + CL selected on the jettison select switch, the inboard pylons, regardless of whether they carry an air-to-air missile, will not be jettisoned when the external stores emergency release button is depressed. However, if TER are carried on either or both inboard wing stations, the stores carried on the TER stations will automatically be bombed off unarmed, although the TER is retained on the pylon.

ENGINES

NOTE

Some illustrations referred to within this system writeup may be located in appendix A. These illustrations are referred to, within the text, as (figure A-, appendix A).

DESCRIPTION

The airplane is powered by two General Electric J79 engines. Aircraft 153071z thru 153087aa have J79-GE-8 engines installed, with a thrust rating of 10,900 pounds each. Afterburner operation increases the maximum thrust to 17,000 pounds. Aircraft 153088aa and up have J79-GE-10 engines installed, with a thrust rating 11,870 pounds each. Afterburner operation increases the maximum thrust to 17,900 pounds each. The engine uses a variable stator (first six stages), a 17-stage compressor, 10 annular combustion chambers, a three-stage turbine, a variable area exhaust nozzle and a variable thrust afterburner. An impingement type starter, supplied with air from an external auxiliary power unit, is used to crank the engine during starting. During operation, air enters the inlet of the engine and is directed into the compressor rotor by the variable inlet guide vanes. As it is compressed, the air is forced through the compressor rear frame into the combustion chambers. Fuel nozzles, projecting into the combustion chambers, eject a fuel spray which mixes with the compressed air. Ignition is provided by a spark plug in the number 4 combustion chamber; the remaining nine combustion chambers are ignited through cross fire tubes. The gases resulting from combustion are directed onto the turbine. The three turbine wheels move as a unit on a common shaft which is directly splined to the compressor rotor. After passing through the turbine section, the exhaust gases flow into the afterburner where their flow is stabilized and then ejected through the variable exhaust nozzles. Additional fuel may be injected into hot exhaust gages for afterburner combustion, producing considerable thrust augmentation. The engine oil system is a dry sump type completely contained on the engine. The compressor inlet guide vanes and the stators in the first six compressor stages are variable, and are controlled by the variable stator system. The variable stators and inlet guide vanes are interconnected externally and are positioned by two actuators which utilize high pressure engine fuel as the hydraulic medium. The variable nozzle system is hydromechanically controlled and schedules nozzle area by positioning the nozzle opening to obtain optimum thrust with respect to altitude and airspeed conditions. The purpose of the variable exhaust nozzles is to control the operating temperature of the engine as governed by the engine amplifier during Military and Maximum engine operation. Air bled from the 17th stage of each engine compressor is used by the auxiliary equipment cooling system, the boundary layer control system, the central air data computer system, the cockpit air conditioning and pressurization system, the engine anti-icing system, the fuel tank pressurization system, the pneumatic system (air source), and the windshield rain removal system.

ENGINE FUEL CONTROL SYSTEM

The fuel control system (figure A-4, appendix A) for each engine is complete in itself and the two systems are identical. For simplicity of discussion, only one system or engine shall be discussed. The engine fuel control system transports fuel from the engine fuel inlet to the combustion chambers. This fuel is discharged in the proper state of atomization for complete burning. Varying engine power settings and conditions demand changes in fuel flow; therefore the engine fuel control system must also control fuel flow to obtain maximum engine efficiency within the design limits of the engine. Only the engine fuel system is discussed in the following paragraphs. The afterburner fuel system is discussed separately in this section.

Fuel Pump Unit

The engine fuel pump unit consists of a low pressure impeller-type pumping element, a high pressure gear-type main pumping element, a low pressure fuel filter, a fuel filter bypass, and an output pressure relief valve. The centrifugal impeller performs a pressure boost function which assures pump operation at low inlet fuel pressures. Airplane boosted fuel from the main fuel manifold passes through the impeller-pump. The impeller boosted fuel then passes through the fuel filter to the main gear-pump, which delivers it to the main engine fuel control at approximately 1000 psi. If the fuel pressure differential across the fuel filter exceeds approximately 33 psi, the fuel filter bypass opens, and a CHECK FUEL FILTERS light, on the pilot's right vertical panel, illuminates. If the discharge pressure of the gearpump exceeds approximately 1125 psi, the output pressure relief valve will open to maintain safe fuel pressures. The output pressure relief valve will reseat when discharge pressures reduce to approximately 1025 psi.

Engine Fuel Control

The main engine fuel control governs engine speed by controlling fuel flow. Fuel is delivered to the fuel control under pressure by the main fuel pump. The fuel enters the control at the main inlet or the servo inlet. Fuel entering the servo inlet will be used for operation of the variable vanes and power lever control. Fuel entering at the main inlet will be divided into excess or metered fuel. Excess fuel is returned to the main fuel pump, metered fuel is directed to the fuel nozzles. The main fuel control uses 5 signal inputs for operation - throttle positions, engine speed, compressor inlet air temperature, compressor discharge air pressure, and position of the variable vanes.

Fuel Oil Heat Exchanger

Metered fuel from the main fuel control, passes through the coolant tubes of the fuel-oil heat exchanger and then to the fuel nozzles. The fuel serves as the coolant for the scavenge oil which flows around the heat exchanger tubes. The heat exchanger incorporates a bypass valve to regulate the flow of oil which controls the temperature of the oil and fuel. There are two fuel-oil heat exchangers on the engine; one utilizes normal engine fuel as a coolant while the other uses core afterburner fuel. Both fuel-oil heat exchangers serve the same purpose and their operation is the same.

Fuel Pressurization and Drain Valve

The fuel pressurization and drain valve prevents fuel flow to the engine until sufficient fuel pressure is attained in the main fuel control to operate the servo assemblies, which are used to compute the fuel flow schedules. It also drains the fuel mainfold at engine shutdown to prevent post shutdown fires, but keeps the upstream portion of the system primed to permit faster starts. On the J79-GE-8 engines when the fuel pressure differential across the pressurization valves drops below 80 psi, the pressurization valve closes, cutting off fuel flow to the engine and the drain valve opens to drain the fuel manifold. On the J79-GE-10 engines, when the fuel pressure differential across the pressurization valve drops below 90 psi, the pressurization valve closes, cutting off fuel flow to the engine and the drain valves open to drain the fuel manifold.

Fuel Nozzles

A flow-divider type fuel nozzle in each inner combustion chamber liner delivers metered fuel, in the proper state of atomization for maximum burning, into the compressor discharge air entering the combustion chamber. The nozzles produce a uniformly distributed, cone shaped, hollow fuel spray. High velocity compressor air is directed around the nozzle by an air shroud to provide a cooling action around the nozzle orifice and to reduce carbon deposits.

Check Fuel Filter Indicator Light

The CHECK FUEL FILTERS indicator light is on the telelight panel. The CHECK FUEL FILTERS indicator light, and MASTER CAUTION light illuminate when the fuel filter on either engine is being bypassed. The light circuit is completed through a pressure differential switch which senses filter inlet and filter outlet pressure. In the case of a partially clogged fuel filter, the CHECK FUEL FILTERS indicator light may be extinguished by reducing power on both engines. Check each engine individually by adding power to see if the light illuminates. In this manner, it can be determined if one or both engines are affected.

OIL SYSTEM

Each engine is equipped with a completely self contained, dry sump, full pressure oil system. Oil is stored in a 5.3 gallon pressurized reservoir, located at the 1:00 o'clock position on the engine compressor front casing. The oil pump is a positive displacement, dual element, rotary vane type unit. The lubrication element is capable of delivering 11.8 gpm at 60 psi. Each element contains a filter through which the oil is pumped before distribution. Engine oil is used for engine lubrication, variable exhaust nozzle operation, generator lubrication and cooling, and constant speed drive unit control. After distribution to various points throughout the engine the oil is picked up by three scavenge pumps, routed through a scavenge filter, through an air/oil cooler and two fuel/oil heat exchangers and then back to the tank. The pressurizing system maintains the proper relationship between ambient air pressure and air pressure in the bearing sumps, gear boxes, damper bearing and reservoir to ensure effective oil seal operation, and to prevent damage to the reservoir and sumps during high speed climbs and descents. Oil is also supplied directly from the reservoir to the constant speed drive unit, where it is used as both the control and final drive medium for controlling generator speed.

Oil Quantity Indicator Lights

A two point oil level indicating system is provided to indicate servicing and operation level. A L ENG OIL LOW and R ENG OIL LOW light is on the telelight panel and illuminates when a low level condition exists in either or both engine oil tanks. A detector, in the right missile well, detects a low level condition and provides a signal to illuminate the appropriate indicator light. There is an oil quantity sensor switch on the utility panel on the left console. During flight, with the switch in the NORM position, the indicator light will illuminate when the oil quantity drops to 3.1 ±0.1 gallons. During ground checking operations, with the switch in the SERV CHK position, the indicator will illuminate if the oil level in either or both tanks is 4.6 ±0.1 gallons or less (provided the aircraft is in a level ground attitude, external electrical power is applied, both engines are off and the check is performed within 20 minutes of engine shutdown). If more than 20 minutes has elapsed since engine shutdown, the engines may be windmilled to scavenge the oil system prior to making the ground check.

NOTE

On aircraft with J79-GE-8 engines installed, and before PPC 62, the indicating lights are installed; however, the oil level sensors are not installed in the engine oil tank. On these aircraft, the system will be inoperative; however, the indicating lights may flicker when the warning lights test switch is actuated.

Lubrication

The lubrication element of the oil pump supplies oil to cool and lubricate bearings, gears and other rubbing or moving parts in the engine. Lubricating oil is also circulated through the engine driven generator for cooling purposes. Oil is drawn from the lowest standpipe in the reservoir by the constant displacement, rotary vane type lubrication element of the oil pump, which is capable of delivering 11.8 gpm at 60 psi. By supplying the lubrication element from the lowest standpipe, oil will be available for lubrication should leakage occur in the nozzle control or constant speed drive unit. Lubricating oil is routed from the lubricating element, through the filter to three branch lines. The first branch distributes oil forward to the transfer gear box, intermediate damper bearing, number 1 bearing, front gear box and the afterburner fuel pump. The second branch distributes to the number 2 and 3 bearings and the rear gear box. The third branch distributes to the pressure transmitter and the pressure relief valve. The pressure relief valve protects the system and is set to relieve at approximately 95 psi. The oil is picked up from the bearing sumps by the scavenge pumps, which pumps the oil through the scavenge filter, the air/oil cooler, the two fuel/oil heat exchangers and returns it to the tank. Oil temperature is maintained by temperature regulators on the two fuel/oil heat exchangers. A check valve, located in the lubrication element outlet, prevents gravity flow of oil from the oil tank when the engine is not running.

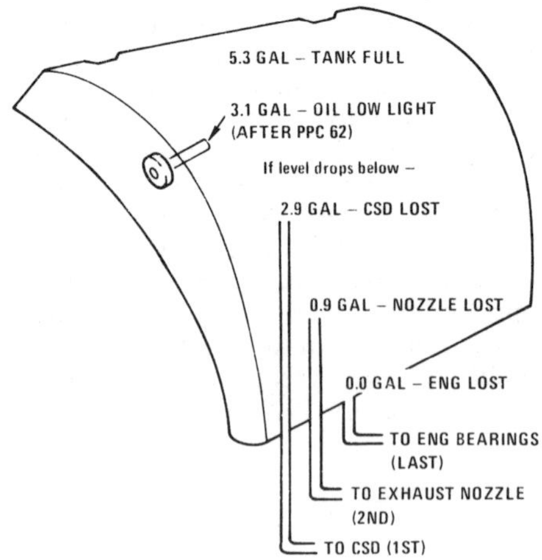

FDD 1 (317)

Figure 1-15

Variable Nozzle Control

Engine oil is used as the hydraulic medium for positioning the variable nozzle flaps. During normal flight attitudes, oil flows through a gravity valve and into an accumulation compartment in the reservoir. During inverted flight, the gravity valve closes and oil for nozzle positioning is available for approximately 30 seconds. From this compartment, oil is drawn through a weighted, flexible standpipe, which stays submerged regardless of flight attitude, to the hydraulic element of the oil pump. This element is of the rotary vane type, and is capable of delivering 4.1 gpm at 70 to 110 psi. A relief valve in the pump protects the system by opening at approximately 70 psi. Oil is routed from the pump, through the hydraulic element filter to the nozzle pump. From here it may return directly to the scavenge system, or be boosted and directed to the nozzle actuators on command from the nozzle area control. From the actuators, oil is routed back to the nozzle pump and into the scavenge system. The scavenge system is the same as described previously.

ENGINE AIR INDUCTION SYSTEM

There are two independent but identical air induction systems, one for each engine. The component units are fixed and variable ramps, which make up the primary air system, a variable bypass bellmouth and an auxiliary air door which make up the secondary air system.

Variable Duct Ramp

The variable duct ramp system provides primary air at optimum subsonic airflow to the compressor face throughout the range of aircraft speeds. The ramp assembly consists of a fixed forward ramp and two variable ramps. The forward variable ramp is perforated to allow boundary layer air to be bled off and exhausted overboard. The aft variable ramp is solid. Movement of the aft ramp positions the forward variable ramp through mechanical linkage. The central air data computer supplies a total temperature input to the ramp control amplifier. The amplifier sends a signal to a utility hydraulic system servo unit which positions the ramps for optimum airflow at Mach numbers of 1.5 and above. The system has a fixed forward ramp angle of 10° and a variable ramp angle of 0° to 14° relative to the fixed forward ramp.

Engine Inlet Temperature High Indicator Light

The engine inlet temperature high indicator light, marked ENG INLET TEMP HIGH, is on the telelight panel (figure A–1, appendix A). The light illuminates when the temperature within the engine intake duct (compressor inlet) is beyond allowable limits for steady state engine operation. The maximum steady state operation of the engine is limited to 121° Centigrade (250°F) compressor inlet temperature. Operating the engine at high altitudes, with the compressor inlet temperature above the prescribed limit, causes the life of the gears, bearings, and carbon seals to be reduced because the lubricating oil will exceed its design temperature. Exceeding the temperature will also cause structural components of the engine (compressor rear frame and combustion casings) to exceed their design limit because of high temperatures and pressures.

Variable Bypass Bellmouth

The variable bypass bellmouth is an automatic unit which allows excess induction air from the compressor face to flow into the engine compartment. Air diverted in this fashion is referred to as the Secondary Air System. The variable bellmouth is a perforated ring located between the airplane duct structure and the engine compressor inlet. Between 0.4 to 0.98 Mach the bellmouth is closed, however, a limited amount of bypass air flows into the engine compartment through the perforations in the bypass bellmouth and the engine air–oil cooler bleed. On aircraft 155785al and up and all others after AFC 420, the bellmouth is also closed below 225 KCAS. Above 0.98 Mach the bypass bellmouth controller senses the optimum airflow (based on duct air velocity) for induction into the engine. When this airflow is exceeded, (rapid throttle retardation) the controller signals a utility system hydraulic actuator which opens the bypass until the optimum airflow to the engine is established.

Auxiliary Air Doors

Two auxiliary air doors, one for each engine compartment, are on the center underside of the fuselage. They are normally controlled by the landing gear handle and actuated open or closed by utility hydraulic pressure. When the landing gear handle is placed in the down position, the doors open, making additional air available to the engine compartments for cooling purposes. When the landing gear handle is placed in the up position, the doors close. If the engine compartment pressures exceed the designed limits, the door will be forced open by an amount proportional to the overpressure. As soon as the overpressure is relieved, the actuator pulls the door closed.

CAUTION

> In the event either auxiliary air door indicator light illuminates (other than momentary), corrective action should be taken as soon as possible. Refer to Auxiliary Air Door Malfunction, section V.

Auxiliary Air Door Indicator Lights

The auxiliary air door indicator lights, on the telelight panel (figure A–1, appendix A) are marked L AUX AIR DOOR and R AUX AIR DOOR respectively. The lights illuminate if the auxiliary air doors operate out of phase with the landing gear handle or momentarily when engine compartment overpressures are relieved. Illumination of the auxiliary air door indicator lights causes the MASTER CAUTION light to illuminate. Refer to part 2, section III, for operating instructions during auxiliary air door malfunction.

VARIABLE AREA EXHAUST NOZZLE SYSTEM

Two sets of cylindrical nozzles, operating together, make up the variable area exhaust nozzle system. The primary nozzle, hinged to the aft end of the tail pipe, controls the convergent portion of the nozzle, while the secondary nozzle, hinged to a support ring, controls the divergent portion of the nozzle. The two sets of nozzles are linked together and maintain a scheduled area and spacing ratio which is infinitely variable between full open and full closed. The nozzles are regulated by the nozzle area control. Movement of the nozzles is accomplished automatically by four synchronized hydraulic actuators. The exhaust gases leave the primary flaps at sonic velocity and are accelerated to supersonic velocity by the controlled expansion of the gases. Control of this expansion is provided by the cushioning effect of the secondary airflow through the annular passage between the two sets of nozzles.

Exhaust Nozzle Control Unit

Throttle position, nozzle position feedback, and exhaust gas temperature are the paramters utilized to schedule the correct nozzle area. During engine operation in the sub–military region, nozzle area is primarily a function of throttle angle and nozzle position feedback. The nozzle is scheduled full open at idle and the area is decreased as the throttle is advanced toward the military position. However, during a rapid throttle burst from below 79% rpm to 98% rpm, a control alternator supplies engine speed information to the temperature amplifier which in turn schedules engine off speed inputs as a function of temperature limiting. This signal prevents the primary nozzle from closing down past a preset position, permitting a rapid increase in engine rpm. During engine operation in the Military and afterburner region, it becomes necessary to limit the nozzle schedule as established by throttle angle and nozzle feedback to prohibit exhaust gas temperature from exceeding the design limits. Exhaust gas temperature is sensed by 12 thermocouples loops and the resulting millivoltage is transmitted to the magnetic temperature amplifier. The amplifier which receives its power supply from the control alternator, compares the thermocouple signal to a preset reference voltage, representing desired engine temperature. The difference is amplified and transmitted to the nozzle area control. Nozzle area control output signal directs the operation of the variable pressure, variable displacement nozzle pump.

NOTE

Spasmodic exhaust nozzle operation shall be logged on the yellow sheet (OPNAV Form 3760-2).

AFTERBURNER SYSTEM

The engine is equipped with an afterburner, where additional fuel may be injected into the hot exhaust gases for afterburner combustion, producing considerable thrust augmentation. The main components of the afterburner system are the afterburner fuel pump, afterburner fuel control, afterburner fuel manifold and spray bars, and the torch igniter.

Afterburner Fuel System

The afterburner fuel system (figure A-4, appendix A) provides the fuel for augmentation of the thrust produced by the engine. A separate, constant pressure drop, variable fuel control meters afterburner fuel. Ignition is by a separate AB ignition system. In operation, the airplane boost pumps supply fuel to the inlet of the afterburner fuel pump. The inlet valve is opened by a fuel pressure signal from the main fuel control when engine speed is above approximately 91 percent rpm. On the J79-GE-8 engines, the throttle angle is 76.5°. On the J79-GE-10 engines the throttle angle is 78°, (in minimum afterburner range). Fuel passing through the check valve continues to the afterburner fuel control, is metered, separated into core and annulus flows, and directed to the afterburner pressurizing valve. The core fuel passes through the afterburner fuel oil heat exchanger on the way to the control and pressurizing valve. The core and annulus flows are further subdivided into primary and secondary flows by the pressurizing valve. The flow sequence, as the throttle is advanced in the afterburner range, is to the primary core, secondary core, primary annulus and secondary annulus.

Afterburner Fuel Pump

The afterburner fuel pump is an engine-driven centrifugal pump. It operates continuously, but discharges fuel to the afterburner fuel system only when the inlet valve on the pump is open. To open the inlet valve to the afterburner fuel pump, the pilot must move the throttle into the afterburner modulation range and engine speed must be sufficiently high (above 91% rpm) to support combustion.

Afterburner Fuel Control

The afterburner fuel control is linked mechanically to the main fuel control through the use of teleflex cabling. Any movement of the throttle moves the main fuel control teleflex and also moves the teleflex to the afterburner fuel control. Fuel entering the afterburner fuel control is metered and separated into core and annulus flows in response to throttle movement and changes in compressor discharge pressure. The control varies fuel flow between the minimum necessary for afterburner combustion for any flight condition and the maximum fuel flow allowable at the flight condition. The afterburner fuel control is designed to hold a constant pressure drop across an orifice while the area of that orifice is varied in accordance with throttle position and compressor discharge pressure.

Afterburner Fuel Distribution

The afterburner fuel pressuring valve delivers fuel to four separate fuel manifolds: primary annulus, primary core, secondary annulus, and secondary core. The fuel is distributed by these manifolds to 21 multi-jet afterburner fuel nozzles which are equally spaced around the perimeter of the afterburner section. Each multi-jet nozzle contains four tubes, one for each manifold, and holes in the sides of the tubes spray the fuel into the exhaust gases. When the throttle is first placed into the afterburner position, the pressurizing valve directs fuel to the primary core manifold. Further advancement directs fuel to the secondary core manifold which joins the primary core manifold in delivering fuel for afterburner operation. When the throttle is advanced still further, the pressurizing valve directs fuel to the primary annulus manifold. As the throttles are advanced to the maximum afterburner position the fuel is directed to the secondary annulus thus joining the other three manifolds in delivering fuel to the nozzles; this is full afterburner operation. The afterburner fuel manifolds and multi-jet nozzle system gives a smooth afterburner operation, with no appreciable acceleration surge between full military and minimum afterburner, or between minimum afterburner and maximum afterburner.

IGNITION SYSTEM

The ignition system consists of an ignition button on each throttle, a low voltage high energy ignition unit on the engine, a spark plug in No. 4 combustion chamber and the necessary wiring. The main ignition system ignites the atomized fuel-air mixture in the No. 4 combustion can. The remaining nine combustion cans are ignited through the cross fire tubes. The afterburner ignition system includes the torch igniter, an afterburner ignition switch, and a torch igniter fuel metering valve. The metering valve supplies fuel to the torch igniter only during the time that afterburner operation is selected with the throttle. When the throttle is moved into the afterburner detent, fuel from the main engine fuel control is directed to the pressure actuated afterburner ignition switch. Electrical power from the right main 115/200 volt ac bus is then supplied to the afterburner sparkplug which emits a continuous arc to ignite the fuel–air mixture in the torch igniter. Continuous ignition is provided as long as the throttle is in the afterburner detent. The torch igniter produces an intense flame which ignites the afterburner fuel.

Ignition Buttons

The ignition buttons are spring–loaded push button type switches on each throttle directly below the throttle grips. Depressing the ignition button causes the spark plug to discharge, igniting the fuel–air mixture as the throttle is moved from OFF to IDLE during engine start. The spark plugs fire only when the ignition button is depressed. The ignition duty cycle is 2 minutes on, 3 off, 2 on, and 23 off. The ignition circuits are completed anytime aircraft power is on, and ignition button is depressed.

STARTING SYSTEM

The impingement starting system consists of an assembly of ducting and valves which are airframe mounted and a manifold assembly which is mounted on the turbine frame of the engine. The single receptacle for connecting the air supply line is on the bottom left side of the fuselage aft of the main gear wheel well. Air from the external source is directed to the left or right selector valve which distributes the air to either the left or right engine, depending on cockpit selection. The engine manifold assembly distributes the starting air to seven impingement nozzles, which direct the air against the second stage turbine blades of the turbine wheel.

Engine Start Switch

The engine starter switch (figure A–1, appendix A) is on the left console in the pilot's cockpit just inboard of the throttles. The starter switch is a three–position switch and is marked L, OFF, and R. The switch receives power from the essential 28 volt dc bus. With APU air connected, actuating the starter switch to L energizes the left engine selector valve and permits air to flow to the left engine impingement nozzle. Selecting R energizes the right engine selector valve and permits air to flow to the right engine impingement nozzles. The OFF position closes both selector valves and stops airflow to the engines. For impingement starting a 5.5:1 pressure ratio gas turbine starting unit is desired.

ENGINE CONTROLS

Engine Master Switches

The toggle lock type two–position engine master switches (figure A–1, appendix A) are on the left console in the pilot's cockpit on the inboard engine control panel. Placing the switch in the ON position directs power to the fuel boost pumps and fuel transfer pumps. The circuits for the fuel shutoff valves, which are normally operated by the throttles, are such that either valve is closed when its respective engine master switch is placed OFF regardless of the throttle position.

Throttles

A throttle (figure A–1, appendix A) for each engine is on the left console of the forward cockpit. Mechanical linkage and teleflex cables transmit mechanical motion from the throttle to those accessories requiring coordination to obtain the degree of thrust desired. Movement of the throttle is transmitted by mechanical linkage to the main fuel control. The main fuel control unit incorporates a throttle booster which reduces the amount of effort needed to move the throttles. The boost power is supplied by fuel from the engine driven fuel pump. Teleflex cables from the main fuel control link the nozzle area control and afterburner fuel control to throttle movement so that fuel flow and nozzle area are compatible throughout the full range of engine operation. A friction adjusting lever is mounted between the throttles which permits adjustment of throttle friction to suit individual requirements. The throttle mechanism is a gear shift type. Included on the throttles are the ignition buttons (one for each engine on the applicable throttle), speed brake switch and microphone button on the right throttle, and master lights control switch on the outboard side of the left throttle. On aircraft 155890am thru 158354as and all others after AFC 448, a singles chaff switch is added to the right throttle below the microphone button. On aircraft 158355at and up and all others after AFC 506, the singles chaff switch is replaced by a flare/chaff switch. Limit switches which control the main fuel shutoff valves are built into the throttle quadrant. Advancing the throttle from OFF to IDLE (with the engine master switch ON) actuates electrical switches which open the main fuel shutoff valve corresponding with the throttle moved. With further advancement of the throttle from IDLE to MIL, engine thrust increases proportionally. At the MIL position of the throttles, the engine should be delivering its rated Military power. Afterburner light–off can be initiated anywhere within the afterburner modulation range by shifting the throttles outboard and moving forward toward MAX position. Movement of the throttles from IDLE to OFF actuates a switch which closes the main fuel shutoff stopping fuel flow to the engine. Throttle movement through the cutouts is as follows: To move throttles from OFF to IDLE, push forward and then shift throttles inboard. To move from MIL to MAX shift throttles outboard, throttles can now be moved forward in the afterburner range.

Catapult Throttle Grips

Catapult throttle grips secured to the pilot's cockpit structure above the MIL throttle detent and MAX throttle detent may be hinged upward to line up with the throttle grips at the MIL and MAX throttle positions. The grips and throttles may then be held together during catapulting to prevent inadvertently throttling back. The grips are automatically stowed when released.

ENGINE INSTRUMENTS

Engine Fuel Flow Indicators

The engine fuel flow indicators (figure A–1, appendix A) are on the right side of the pilot's instrument panel. The fuel flow indicating system indicates the amount of main fuel system flow, in pounds per hour, of fuel the engines are using at a particular power setting. The rate of fuel flow is shown i 1000 pounds per hour by a pointer moving over a scale calibrated from 0 to 12. Maximum fuel flow fluctuation is: 100 pph for indicator readings of 0–3000 pph; 750 pph for readings of 3001–12,000 pph. The flow is measured by transmitters mounted on the engines. Afterburner fuel flow bypasses the fuel flow transmitters and therefore is not shown on the indicators.

Tachometers

The electric tachometer system is composed of two tachometer indicators (figure A–1, appendix A) on the pilot's instrument panel and one engine–driven tachometer generator on each engine. The system is completely self–contained in that it requires no external source of power. The tachometer generator develops a poly–phase alternating current which is used to indicate percentage of maximum engine rpm. The indicator dials are calibrated from 0 to 110. Each indicator includes two pointers, a large one operating on the 0 to 100 scale and a small one operating on a separate scale calibrated from 0 to 10.

Exhaust Gas Temperature Indicators

The exhaust gas temperature indicators (figure A–1, appendix A) are on the pilot's instrument panel. The scale range on the indicators is 0 to 11 with the reading multiplied by 100° centigrade. The system indicates the temperature of the exhaust gas as it leaves the turbine unit during engine operation. Twelve dual loop thermocouples are installed on each engine and are connected in parallel. The millivoltages produced by one of the sets of dual loop thermocouples is directed to an amplifier for temperature limiting. The millivoltages produced by the other set of 12 thermocouples is directed to the cockpit indicator. The indicator is a null–seeking potentiometer type. It balances a thermocouple voltage against a constant voltage source with a small servo simultaneously balancing a bridge circuit and operating the indicator pointers.

Exhaust Nozzle Position Indicators

Exhaust nozzle position indicators (figure A–1, appendix A) which show the exit area of the exhaust nozzle, are on the pilot's instrument panel. The instruments are placarded Jet Nozzle Position and are calibrated from CLOSE to OPEN in ¼ increments. The nozzle position indicators enable the pilot to make a comparison of nozzle position between engines, and is also used to establish a relationship between nozzle position and exhaust gas temperature and nozzle position and throttle settings.

Oil Pressure Indicators

The oil pressure indicators (figure A–1, appendix A) are on the pedestal panel. The scale range on the indicators is 0 to 10 with readings multiplied by 10. The oil pressure indication system senses oil pressure downstream of the main lube pump in the main lube discharge line.

ENGINE ANTI–ICING SYSTEM

The engine anti–icing system is a compressor discharge air bleed type system, controlled by an on–off pressure regulating valve. Air for anti–icing purposes is supplied from the 17th stage of the engine compressor at pressures up to 275 psi, and temperatures up to 593°C. A regulator in the anti–icing valve reduces the incoming air to a pressure of approximately 14–20 psi. Air from the anti–icing valve is distributed through the first stage stator vanes, the compressor front frame struts, the inlet guide vanes, and to a port in the front gear box for engine nose dome anti–icing.

CAUTION

During supersonic flight, the anti–icing system should only be used when actual icing is noted.

NOTE

During subsonic flight, the anti–icing system should be used when icing conditions are anticipated.

Engine Anti–Icing Switch

A two–position engine anti–icing switch (figure A–1, appendix A) is on the outboard engine control panel. The switch is marked engine anti–icing and the switch positions are DE–ICE and NORMAL. Placing the switch to DE–ICE opens the regulator valve which starts anti–icing air flow. With the switch in NORMAL, no anti–icing operation is being performed.

Engine Anti–Ice Lights

Engine anti–ice lights are on the telelight panel (figure A–1, appendix A). The lights marked L ANTI–ICE ON and R ANTI–ICE ON, operate from a pressure sensitive switch which is actuated by the pressure of engine bleed air when the system is turned on. There are two situations forseen in which light illumination may be observed with the switch in the NORMAL position. One situation occurs in high Mach flight where impact pressure will trip the pressure switch and illuminate the light. The other situation occurs at any speed when the regulator valve leaks and allows sufficient anti–icing airflow to trip the pressure switch and illuminate the light. During high Mach flight, if the light illuminates while the anti–icing switch is in NORMAL, reduce speed to subsonic. If the light goes out, the indication is caused by impact pressure and high Mach flight may be continued, regardless of the light indication. If a speed reduction will not extinguish the light, then the cause is probably a leaking regulator valve and high Mach flight may not be continued unless actual icing is present on the airplane. Operation of the system is not limited in subsonic flight, and therefore, flight may be conducted in this region regardless of light indication. System output may be checked by placing the anti-icing switch to the DE-ICE position and observing a 10° rise in EGT. This check will not constitute clearance to operate at high Mach with the light illuminated if a speed reduction failed to extinguish it.

CAUTION

If the L ANTI-ICE ON and/or R ANTI-ICE ON lights illuminate during high Mach flight, reduce speed. If a speed reduction will not extinguish them operate at a subsonic speed. Continued operation in the high Mach range may cause engine damage.

FIRE AND OVERHEAT DETECTOR SYSTEMS

Engine Fire/Overheat/Bleed Air Leakage Warning Lights

The fire and overheat detection system consists of three separate and independent systems: Engine Fire Detection System, Aft Fuselage Overheat system, and on aircraft 155903ap and 157274ap and up, and all others after AFC 439, a Bleed Air Leakage Detection System. The FIRE and OVERHT warning lights, one for each engine, are on the upper right portion of the main instrument panel. The BLEED AIR OVERHT lights (3 each) are on the telelight panel. In addition to the lights, each system consists of a control unit, and series of continuous sensing elements. The fire warning sensing elements are routed throughout the engine compartments. The right or left FIRE warning light illuminates if a temperature of approximately 765°F occurs in the corresponding engine compartment. The aft fuselage overheat sensing elements are routed in vertical recesses of the skin fairing on each side of the keel. These recesses are opposite the aft end of the secondary engine nozzle fingers. The left or right OVERHT warning light illuminates if a temperature of approximately 1050°F occurs at the corresponding aft fuselage skin. Do not use afterburner if an aft fuselage OVERHT warning light illuminates. This indicates a safety of flight condition such as an open engine compartment door or a damaged engine nozzle. Either of these conditions can lead to the loss of flight control if afterburner is used. The bleed air leakage sensing elements are routed along the bleed air ducts, and illuminate three different lights, depending on where the leak occurs. The elements routed under fuel cell 2 through the fuel/hydraulic bay (door 22) and then through the left and right refrigeration packages illuminates the FUS BLEED AIR OVERHT warning light if a temperature of approximately 410°F occurs in this area. The elements routed along each wing leading edge outboard to the BLC shutoff valve illuminates the WING BLEED AIR OVERHT light if a temperature of approximately 410°F occurs in this area. The element routed along the engine bleed air duct, in the keel web, illuminates the ENG BLEED AIR OVERHT light if a temperature of approximately 575° occurs in this area. Illumination of any of the fire or overheat lights is a warning to initiate the appropriate emergency procedure(s). The MASTER CAUTION light does not illuminate in conjunction with the FIRE or OVERHT warning lights; however, it will illuminate in conjunction with any of the bleed air leakage warning lights. The engine fire and aft fuselage overheat lights operate on RAT power; however, the bleed air overheat lights do not.

NOTE

- During ram air turbine extension , it is normal for the FIRE and OVERHT lights to flicker until the RAT comes up to speed. Disregard these lights unless they remain on after RAT has obtained full speed.
- Except for illumination resulting from RAT extension, illumination of the FIRE or OVERHT lights shall be logged on the yellow sheet (OPNAV Form 3760-2).

Fire Detector Check Switch

The engine fire, aft fuselage overheat, and on aircraft 155903ap and 157274ap and up, and all others after AFC 439, the bleed air overheat warning light system may be checked by depressing the fire detector check switch. The lights should first be checked by placing the warning light test switch, on the right console, to test. This action only checks the light bulbs. Depressing the fire detector check switch illuminates the FIRE, OVERHT, FUS BLEED AIR OVERHT, WING BLEED AIR OVERHT, and ENG BLEED OVERHT lights and also checks the continuity of their sensors and the operation of each system's control panel. The engine fire and aft fuselage overheat warning test circuit receives power from the right main 28 volt dc bus. Therefore, when operating on RAT, these lights cannot be tested. The bleed air overheat warning lights do not operate on the RAT.

ENGINE BLEED AIR SYSTEM

The bleed air system supplies high temperature, high pressure air from the engines to the boundary layer control system, the cabin air conditioning system, the equipment air conditioning system, and the fuel pressurization system. The functional control of the bleed air is initiated by the requirements of each individual system and the flow, temperature, and pressure is regulated by the system. The system utilizes engine compressor bleed air tapped off the 17th stage compressor. Normally, both engines supply the air for the operation of these systems, but when necessary, single engine operation supplied sufficient air for their operation. On aircraft 155903ap and 157274ap and up, and all others after AFC 440, during single engine operation, trailing edge BLC will be lost on the wing adjacent to the inoperative engine. The system ducting routes the flow of bleed air from the engines to the systems and is insulated to protect the airframe structure from heat radiation. Check valves are installed in the ducting to prevent back flow into the nonoperating engine during starting and single engine operation.

Engine Bleed Air Switch

On aircraft 155903ap, and 157274ap and up, and all others after AFC 440, the lever-locked engine bleed air switch is installed on front cockpit right utility panel. The switch has two positions, NORM and OFF, and controls a bleed air shutoff valve installed in the keel Y duct between the engines. After AFC 550, the lever-locked engine bleed air switch is replaced by a simple toggle switch and a red guard. Actuation of the switch to OFF shuts off engine bleed air to all systems except trailing edge BLC. CNI equipment (Tacan, ADF, UHF, IFF, and SIF), and radar will lose cooling air from the equipment refrigeration package, but will still have ram air cooling available. Ram air is also available for cabin temperature control. Under RAT power, bleed air will be delivered to all systems regardless of switch position.

With the engine bleed air switch off, operation of the following equipment will be lost.

1. External fuel and internal wing fuel transfer
2. Leading edge BLC
3. Cockpit air conditioning and pressurization
4. Rain removal air
5. Defog/Foot heat
6. Equipment cooling air
7. Normal pneumatic compressor charging
8. SPC inputs to all systems
9. Automatic altitude reporting
10. Anti G suit

NOTE

With the engine bleed air switch in OFF, 17th stage air is not available for pneumatic compressor charging; however, the pneumatic compressor will compress ram air (the rate of charging is dependent on altitude and airspeed).

CAUTION

When the engine bleed air switch is positioned to OFF, radar and CNI equipment should be placed in the OFF position unless safety of flight/operational necessity requires their use.

Bleed Air Off Light

After AFC 550, a BLEED AIR OFF light is added to the caution light panel. The light illuminates any time the bleed air shutoff valve is in the off position. The MASTER CAUTION light illuminates in conjunction with the BLEED AIR OFF light. The BLEED AIR OFF light is powered by the warning lights 28/14 volt ac bus through the BLC warning light circuit breaker.

NORMAL OPERATION

STARTING ENGINES

Refer to Starting Procedures, section III, part 3.

ENGINE OPERATING CHARACTERISTICS

T2 Reset

During high compressor inlet temperatrue operation (high speed flight) engine idle speed is rescheduled upward to maintain sufficient airflow to prevent compressor stall. When the inlet temperature increases from +57°C to +108°C, engine idle speed is raised from normal idle (65 percent) to 100 percent regardless of the throttle setting. To reduce engine idle speed, once it has been reset, compressor inlet temperature must be reduced. This is effected by retarding the throttles out of afterburner to reduce thrust. Thrust can be further reduced by retarding the throttles below the military position so that the exhaust nozzles open, lowering exhaust gas velocity and temperature. As thrust decreases, compressor inlet temperature decreases as a result of lower airspeed, and engine speed control is returned to thethrottle.

T2 Cutback

When the compressor inlet temperature (T2) falls below a predetermined level, the maximum engine rpm is limited to prevent excessive mass air flow through the engine. On the J79-GE-8 engines, T2 cutback starts to occur at +4°C and is reduced until at -54°C the maximum rpm is approximately 91.5 percent. On the J79-GE-10 engines, T2 cutback starts to occur at +45°C and reduced until at -54°C the maximum rpm is approximately 90 percent.

T5 Reset (J79-8 Engines)

The engines incorporate an exhaust gas temperature (T5) reset during military and full AB operation. This T5 reset occurs at the same point as T2 cutback, and reduces EGT at the same time that T2 cutback is reducing rpm. As a result of T5 reset, the engines run at lower EGT's, operate with larger nozzle areas, provide less net thrust and consume less fuel while operating in the speed cutback region at low compressor inlet temperature conditions.

T5 Reset (J79-10 Engines)

The engines incorporate an exhaust gas temperature reset during afterburner operation. When the compressor discharge pressure exceeds 290 ± 5 psia while in afterburner operation, the exhaust gas temperature limit is lowered to 571°C ± 9°C. As a result of T5 reset, the engines run at a lower exhaust gas temperature, operate with larger nozzle areas, and consume less fuel. The purpose of T5 reset is to prevent certain components from exceeding their operating temperatures.

RAMP SCHEDULING

Ramps begin scheduling at +52°C total temperature and stop at +146°C total temperature. The following schedule is representative of ramp opening at 40,000 feet:

OAT°C	Ramps Begin To Open At Approx. Mach
-40	1.40
-45	1.45
-50	1.50
-55	1.55
-60	1.60
-65	1.65
-70	1.70

Engine Operating Envelope

The engine operating envelopes (figure 1-16) show pertinent engine operating data for an ICAO Standard Day. The various envelopes are plotted to show an approximate area of operation; therefore, airstarts, afterburner light-offs, minimum airspeed operation, etc. may occur, depending on prevailing flight conditions, on either side of the plotted operational area. However, under 1 G level flight conditions, satisfactory engine operations can be expected within the plotted envelopes. The transient operation zone (Mach 2.0 to 2.4) and the maximum engine operation curve are standard day airspeed restrictions and are shown for reference only. In all cases the Engine Airspeed Limit Chart, section I, part 2 and the Airplane Speed Restriction Chart, section I, part 4 of this manual shall take precedence over any or all operations shown herein.

EMERGENCY OPERATION

ENGINE FAILURES

Jet engine failures in most cases are caused by improper fuel scheduling due to malfunction of the fuel control system or incorrect techniques used during certain critical flight conditions. Engine instruments often provide indications of fuel control system failures before the engine actually stops. If engine failure is due to a malfunction of the fuel control system or improper operating technique, an airstart can usually be accomplished, providing time and altitude permit. If engine failure can be attributed to some obvious mechanical failure within the engine proper, do not attempt to restart the engine.

RUNAWAY ENGINE

There is no provision made on the main fuel control for stabilized engine rpm in the event the throttle linkage becomes disconnected from the control. If a disconnect occurs, vibration may cause the fuel control to hunt or assume any setting from idle to maximum power. Therefore, at the first indication of a runaway engine while on the ground, secure the engine with the engine master switch. If a runaway engine occurs in flight, use of the engine may be regained in the landing configuration by engaging the auto throttles, provided the disconnected throttle linkage is not binding the engine fuel control and the RPM has stabilized between 73 and 99%. If the auto throttle will not function properly, shutdown the affected engine before commencing the approach, or sooner if necessary.

AIRSTARTS

In general, airstart capability is increased by higher airspeeds and lower altitudes; however, airstarts can be made over a wide range of airspeeds and altitudes. Depending on airspeed and altitude, the engine starts at various low rpm indications. Above 12% rpm, however, is considered optimum. An airstart is accomplished by depressing the ignition button with the throttle at IDLE. A start is indicated by a rapid increase in EGT followed by an increase in rpm. If light-off does not occur within 30 seconds after ignition; the engine does not continue to accelerate after light-off; the EGT exceeds maximum limitations; or the oil pressure does not attain 12 psi minimum at idle, retard the throttle to OFF. Wait 30 seconds before initiating a restart.

NOTE

- If airstarts are attempted outside the airstart envelope (figure 1-16) a hung acceleration condition may result. RPM will hang between 65% and 71% while EGT will continue to rise and the engine will not respond to throttle movement. To terminate this condition the throttle may have to be retarded to OFF. Restart attempt may be initiated as soon as RPM drops below 60%.

- If one or both engines flame out, do not delay the airstart. If a mechanical failure is not immediately evident, depress and hold the ignition button(s) to restart the engine(s) before excessive rpm is lost.

OIL SYSTEM FAILURE

The standpipes which supply the three systems utilizing engine oil are in the reservoir such that the pipe for the constant speed drive unit is the highest, the one for the nozzle control is the next highest, and the lubricating system pipe is the lowest. Therefore, a leak in the constant speed drive unit would probably cause a failure of that system only, while a leak in the nozzle control system may cause failure of that system and the constant speed drive unit. A leak in the lubricating or the scavenging system causes failure of the constant speed drive unit and the nozzle control system and ultimately, engine bearing failure results. A GEN OUT light illumination, followed

ENGINE OPERATING ENVELOPES

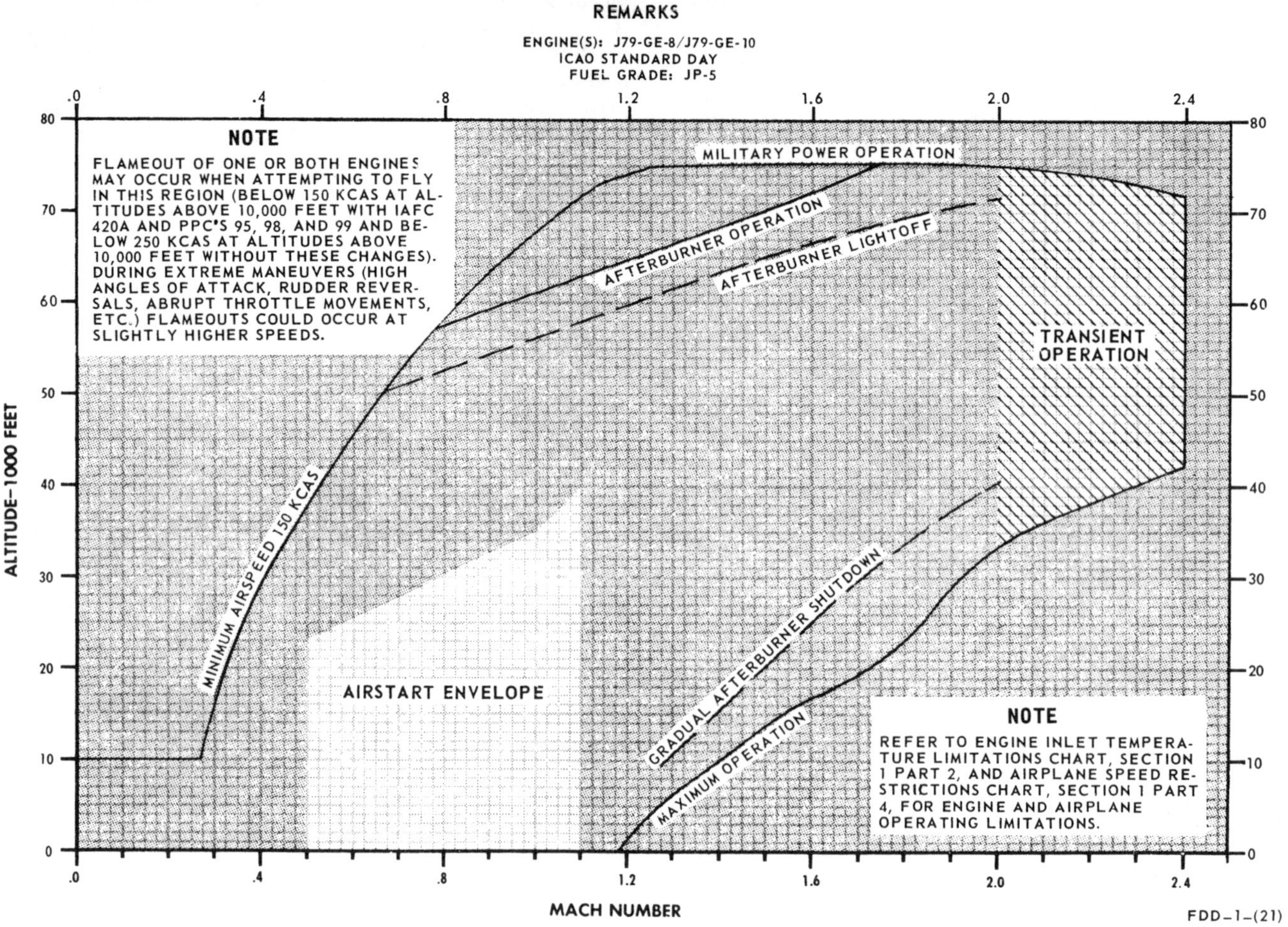

Figure 1-16

by sluggish exhaust nozzle action, are early indications of impending engine oil starvation. The engine oil pressure gage should be mon'tored closely subsequent to a generator failure. In general, it is advisable to shut down the engine as early as possible after a loss of oil supply is indicated, to minimuze the possibility of damage to the engine and the constant speed drive unit. The engine operates satisfactorily at military power for a period of one minute, with an interrupted oil supply, However, continuous operation, at any engine speed, with the oil supply interrupted, results in bearing failure and eventual engine seizure. The rate at which a bearing fails, measured from the moment the oil supply is interrupted, cannot be accurately predicted. Such rate depends upon the condition of the bearing before oil starvation, temperature of the bearing and loads on the bearing. Malfunctions of the oil system are indicated by a shift (high or low) from normal operating pressure, sometimes followed by a rapid increase in vibration. A slow pressure increase may be caused by partial clogging of one or more oil jets; while a rapid increase may be caused by complete blockage of an oil line. Conversely a slow pressure decrease may be caused by an oil leak; while a sudden decrease is probably caused by a ruptured oil line, or a sheared oil or scavenge pump shaft. Vibration may increase progressively until it is moderate to severe before the pilot notices it. At this time complete bearing failure and engine seizure is imminent. Limited experience has shown that the engine may operate for 4–5 minutes at 80–90 percent speed before a complete failure occurs. In the event of a drop in oil pressure or a complete loss of pressure, shut down the affected engine if power is not required or, set the engine speed at 86–89% if partial power is required. If partial power is required on the affected engine, avoid abrupt maneuvers causing high G forces and avoid unnecessary or large throttle bursts.

AUTO-ACCELERATION

If the auxiliary air doors fail to open when the landing gear is lowered, there is a possibility that the engines may automatically accelerate up to 100% rpm. A utility hydraulic system failure renders the variable bypass bellmouth and auxiliary air doors inoperative. Operation of an engine with an open variable bypass bellmouth and closed auxiliary air doors allows engine compartment secondary air to recirculate to the engine inlet. During low altitude or ground operation, the temperature of the recirculating air may be high enough to initiate T2 reset.

When T2 reset is initiated the engine(s) auto-accelerates. The auto-accelerated engine(s) can be shutdown if on the ground, by placing the throttle to OFF. If engine operation is required the thrust output can be regulated by modulation of the engine throttle. Modulation of the engine throttles repositions the exhaust nozzles. However, the engine rpm is not affected.

VARIABLE AREA INLET RAMP FAILURE

There are no provisions made for emergency operation of the inlet ramps. Malfunctions of the inlet ramp control or actuating system may cause the ramp to assume the fully retracted (maximum duct area) position or the fully extended (minimum duct area) position. A failure of the ramp to the retracted position has no effect on engine operating characteristics or performance below approximately 1.5 Mach. Engine compressor stalls may occur above 1.7 Mach. A failure of the inlet ramps to the extended position below approximately 1.5 Mach will cause a substantial loss of thrust, and engine compressor stall and flameout may occur above approximately 18,000 feet altitude. Above 1.5 Mach, maximum airspeed attainable is reduced, and engine compressor stalls may occur. A failure of the inlet ramps to the extended position at any power setting from idle to max AB at airspeeds from 400 knots to landing approach speed does not cause unstable engine operation. Extended inlet ramps may be detected by; observing the ramp position in the rear view mirror; significantly reduced fuel flow at power settings above 85% rpm; high pitched howl at airspeeds above 300 knots; and significantly reduced thrust (approximately 35%) at power settings above 90% rpm. Engine acceleration time and response to throttle movement are not affected by the extended ramp. No special procedures are required for throttle manipulation under these conditions below 18,000 feet altitude. Slam accelerations to military and max AB power, stabilized high power operation, sideslips and airstarts may be performed without overtemperature or compressor stalls. Power settings above 94% rpm with the ramps extended produce increased fuel consumption without increasing engine thrust output. For this reason, cruising altitudes should be selected at which the recommended maximum range Mach number for existing configuration and gross weight can be maintained with 94% rpm or less. Refer to section V for flight procedures with ramps extended.

AFTERBURNER IGNITION FAILURE

If for any reason afterburner ignition is not available, afterburner light-offs can generally be obtained through turbine torching. If afterburner thrust is required and/or desired, slam accelerate the engine into afterburner range from 90% rpm or higher. If the first selection is not successful, an immediate slam reselection should achieve successful results.

LIMITATIONS

RPM DROP

When entering afterburner from throttle settings less than military, the allowable rpm drop is 14%. When entering afterburner from stabilized military power, allowable rpm drop is 10%. All exceeded engine speed limitations must be recorded on the flight forms (yellow sheets).

ENGINE SPEED

Engine speed limitations are listed in figures 1-17 and 1-19.

ENGINE TEMPERATURE

Engine temperatures are limited by degree and time as shown in figures 1-18 and 1-20.

THROTTLE BURST

When operating with maximum engine compressor bleed air (flaps down and cockpit pressurized) in outside air temperatures of -37°C and below, rapid throttle bursts may result in an rpm hang-up. If a throttle burst into maximum afterburner is made, cyclic engine operation may result. When rapid throttle bursting is necessary under these conditions, it is recommended that the throttle be advanced to minimum afterburner first and the engine rpm be allowed to stabilize before advancing further into the afterburner range.

WINDMILLING

Except for emergency shutdown, do not allow the engine to windmill below 7% rpm below 40,000 feet for periods greater than 10 minutes. Extended windmilling may result in engine damage from inadequate lubrication or oil

ENGINE SPEED LIMITATIONS –8 and –10 ENGINE

CONDITION	STATIC	INFLIGHT
* NORMAL OPERATION	100 ± 0.5% RPM	100 ± 0.5% RPM
*ALLOWABLE OVERSPEED NON TIME LIMITED	103% RPM	102% RPM
ALLOWABLE OVERSPEED TIME LIMITED	103-105% RPM FOR 3 MINUTES	102-103.6% RPM FOR 1 MINUTE
IDLE	65 ± 1% RPM	————

* Refer to power limitations, this section.

Note

ANY RPM IN EXCESS OF THE ABOVE LIMITATIONS SHALL BE LOGGED ON THE YELLOW SHEET (OPNAV FORM 3760-2)

FDD-1-(22)

Figure 1-17

ENGINE EXHAUST TEMPERATURE LIMITATIONS

CONDITION	TEMP	TIME
Starting from light-off	1000°C	3 sec
	980°C	10 sec
	930°C	1 min
	900°C Down to 733°C	1 min 30 sec
During all engine operation other than starting	Above 750°C	3 sec
	(-8 Eng) 635°C (-10 Eng) 668°C	NO TIME LIMIT

Note

If any of the above limits are exceeded the aircraft will be aborted and written up. In addition, if 705°C is exceeded during start for any period of time, the engine will require corrective action to prevent recurrence. This is not an abort item.

FDD-1-(23)A

Figure 1-18

depletion, and may cause internal engine conditions that are conducive to sump fires when relighting. Prior to non-emergency conditions, the engine should be decelerated to the coolest operating point (lowest EGT), and this speed maintained long enough to stabilize EGT. The number of 10 minute intervals below 7% rpm is not limited, providing the engine is operated above 7% rpm for a minimum of 10 minutes between intervals.

POWER SETTING

Military Power

Military power is obtained with full non-afterburning thrust and is limited to 30 minutes below 35,000 feet and 2 hours above 35,000 feet.

Maximum Power

Maximum power is obtained with full afterburning thrust and is time limited to 30 minutes below 35,000 feet, and 2 hours above 35,000 feet.

ENGINE OIL PRESSURE

The oil pressure limitations for the primary engine lubricating oil, MIL-L-23699, are as follows: 12 psi at idle, 45 to 70 psi during steady state ground operations at military power, 35 to 70 psi during steady state flight operation at 100% rpm. Oil pressure fluctuations of ±2.5 psi are allowed around a known steady state pressure. Below 20,000 feet during steady state operation, any erratic pressure change which exceeds 5 psi for more than 1 second must be investigated. At 20,000 feet altitude and above, pressure fluctuations are limited to the following: maximum of 20 psig below normal pressure for a duration

MILITARY POWER OPERATING LIMITS

REMARKS
ENGINE(S): J79-GE-10

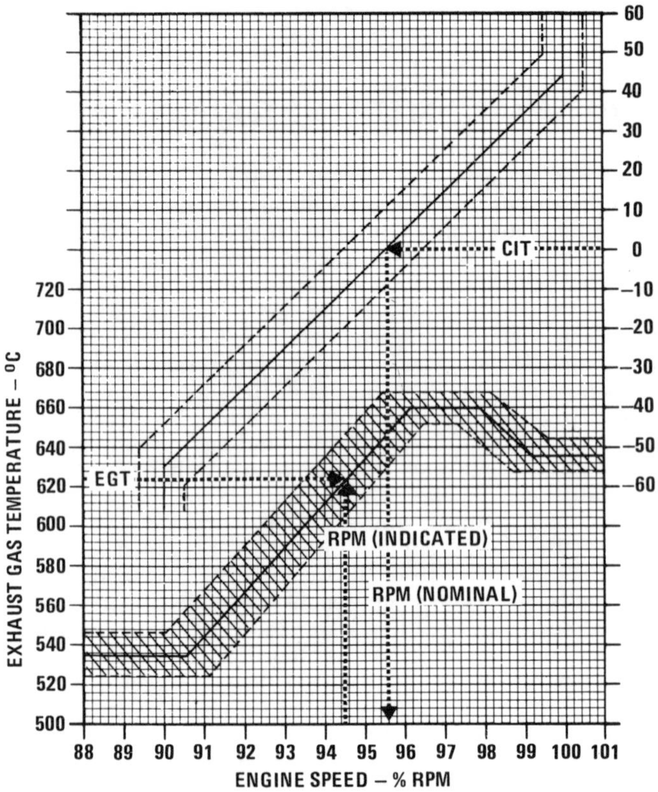

Note

- This graph depicts OAT vs RPM and RPM vs EGT when aircraft is in static condition.
- Disregard OAT for inflight reference between RPM and EGT.
- Enter chart with CIT (OAT) to find military RPM.
- Use actual engine RPM to find EGT.

FDD-1-(24)A

Figure 1-19

of 3 seconds, occurring no more than 4 times per minute, or a maximum of 10 psig below normal pressure for a duration of 1 second, occurring no more than 15 times per minute. During any engine speed reduction, indicated oil pressure will decrease approximately 1 psi per 1 percent reduction in rpm from 100%. Pressure changes resulting from airspeed increases or going ON/OFF afterburner are acceptable down to 35 psi minimum. From flight to flight, indicated pressure must repeat within 5 psi of the known normal pressure for a particular aircraft/engine combination. When an alternate lubricating oil is utilized, refer to section I, part 4, of this publication for oil pressure limitations.

ENGINE IGNITION

The engine ignition duty cycle is as follows:

2 minutes ON – 3 minutes OFF
2 minutes ON – 23 minutes OFF

NOTE

In an emergency use the ignition system as required. Exceeded limits must be entered in the flight forms (yellow sheets).

EMERGENCY FUEL

The engines may be operated on MIL–G–5572B 115/145 AVGAS if JP–4 or JP–5 is not available. When AVGAS is used the aircraft is restricted to one flight of no more than 5 hours duration at subsonic speeds. AVGAS has a specific gravity range between 0.730–0.685. The fuel control should be set to correspond to these values. The engine top speed should be adjusted as necessary. If the fuel control adjustments cannot be made, the aircraft may be flown; however, the pilot should be aware that the following degradation in engine performance will occur:

a. Longer time to start and accelerate, with possible missed–starts or start–stalls.
b. Maximum engine RPM and EGT may not be attained.
c. Slow acceleration throughout the operating range.
d. Lower than normal afterburner thrust.
e. Reduced aircraft range.

GRADUAL AFTERBURNER SHUTDOWN

Afterburner shutdown must be gradual in certain areas of the airplane flight envelope, and is intended to allow the airplane to decelerate to a lower Mach number before the engine exhaust nozzles close. This prevents the nozzle from becoming overpressurized due to peak transient pressures between Max and Mil power.

CAUTION

If a compressor stall occurs above approximately 630 knots, retard the throttle to idle immediately to prevent engine over temp.

ENGINE G

Due to limited oil distribution to all systems utilizing engine oil for lubrication or operation during negative G or zero G flight, the airplane is limited to the following:

a. 30 seconds of negative G flight.
b. 10 seconds of zero G flight.

FLAPS

NOTE

Some illustrations referred to within this system writeup may be located in appendix A. These illustrations are referred to, within the text, as (figure A–, appendix A).

DESCRIPTION

The wing flap system comprises two–position leading edge flaps, three position trailing edge flaps and drooped ailerons. The system is actuated by the utility hydraulic system. The leading edge flaps are mounted on the center and outer wing panels. Trailing edge flaps are mounted on the inboard portion of the wing adjacent to the fuselage. Each flap and aileron has its own hydraulic actuator. The leading edge flaps are locked in the retracted position by overcenter linkages. Trailing edge flaps and ailerons are locked in the retracted position by internal locks in the cylinders. A check valve is provided as in integral part of the selector valve to prevent unlocking of overcenter mechanisms and internal locks by back pressure in the return lines. A flow divider is provided to synchronize the trailing edge flaps. There is no synchronization between leading edge flaps or between leading and trailing edge flaps and ailerons. On aircraft 154786ag and up, the aileron droop is actuated by electrical droop actuators powered by the right 115 volt ac bus. The actuator is locked in the extended or retracted position by an integral brake.

WING FLAP SWITCH

The leading and trailing edge flap switch (A–1, appendix A) on the wing flap control panel, is mounted above the left console outboard of the throttles. The three–position toggle switch is marked UP, 1/2, and DN and is shaped like an airfoil for ease of identification. Selecting the 1/2 position moves the center and outboard leading edge flaps to the full down position (60°, 55° respectively), droops the ailerons (16–1/2°) and moves the trailing edge flaps 1/2 (30°) down. Selection of the DN position moves the trailing edge flaps to the fully extended position (60°). Selecting the 1/2 position after the flaps have been fully extended raises the trailing edge flaps to the 1/2 (30°) position. Placing the flap switch in the UP position simultaneously returns all the flaps and ailerons to the full retracted position. There is no individual selecting of flaps.

EMERGENCY AILERON DROOP SWITCH

On aircraft after AFC 534, an emergency aileron droop switch is provided. This lever–locked switch replaces the emergency speed brake switch located on the left utility panel and allows selection of aileron droop during emergency flap operation. The switch has positions of NORMAL and DISABLE; in NORMAL, the ailerons will droop when emergency flap is selected; in DISABLE, the ailerons will remain in the non–drooped position when emergency flap is selected or return the ailerons to a non–drooped position if already drooped during emergency flap operation. If emergency flap is selected during a utility hydraulic system failure with both PC systems operating normally, and the switch is in NORMAL, the ailerons will droop normally; however, if the utility and a single PC system fail simultaneously, an asymmetric aileron droop condition will result. In this condition, or anytime a non-drooped condition is preferred during emergency flap operation, place the switch to DISABLE and the ailerons will return to a non-drooped condition. Power for the emergency aileron droop control is supplied by the right 115 volt ac bus (on aircraft with electro-mechanical actuators) and by the essential 28 volt dc bus (on aircraft with hydraulic/pneumatic actuators).

EMERGENCY FLAP EXTENSION

Emergency extension of the wing flaps is accomplished pneumatically. High pressure air (approximately 3000 psi), stored in a 300 cubic inch air bottle, is released to extend the flaps by pulling the flap circuit breaker and pulling full aft on the emergency flap extension handle (figure A1, appendix A). The handle is airfoil shaped and is painted in alternating black and yellow stripes for ease of identification. Actuation of the emergency flap extension handle extends the leading edge flaps to the full down position, the trailing edge flaps to the one-half down position, and the aileron droop as follows: On aircraft before AFC 534, the ailerons remain in the non-drooped position (if normal flap extension preceded emergency extension, the ailerons would return to non-drooped condition). On aircraft after AFC 534, with the emergency aileron droop switch in NORMAL, the ailerons move to the drooped position. With the switch in DISABLE before emergency flap operation, the ailerons remain in the non-drooped position; however, if during emergency flap operation aileron droop is not desired, placing the switch to DISABLE will return the ailerons to a non-drooped condition.

NOTE

The flaps, once extended pneumatically, will not be retracted through action of the flap airspeed blowup switch.

ENGINE INLET TEMPERATURE LIMITATIONS

CONDITION	TEMP	TIME
ENG INLET TEMP HIGH Warning light Illumination	121°C	Prohibited operation with light illuminated below 30,000 FT.
Transient Temperature Operation	121°C–193°C (max. temperature occurs approximately .4 Mach above illumination of ENG INLET TEMP HIGH Warning light)	5 min. per hour (noncumulative) above 30,000 FT.

FDD-1-(25)

Figure 1-20

FLAP POSITION INDICATOR

The leading edge and trailing edge flap indicators (figure A-1, appendix A) are on the left vertical panel in the pilot's cockpit. The indicators work in conjunction with position switches on the leading and trailing edge flaps. The position of the flaps is indicated by drum dials viewed through cutouts in the instrument panel. With flaps up, the word UP appears on the indicators; flaps in transit is indicated by a barber pole; half flaps is indicated by the fraction 1/2 appearing on the drum dial for the trailing edge flaps only; with flaps down, the letters DN appear on the indicators. On aircraft before AFC 534, the trailing edge flap indicator will show barber pole following pneumatic flap extension. On aircraft after AFC 534,if the emergency aileron droop switch is in DISABLE, the trailing edge flap indicator will show barber pole following pneumatic extension but if the switch is in NORMAL, the trailing edge flap indicator will show 1/2.

CAUTION

Should the leading edge flaps indicate barber pole throughout all flap positions, this should be treated as a bleed air duct failure.

WHEELS WARNING LIGHT

The WHEELS warning light is on the upper left corner of the main instrument panel. The light flashes any time the flaps are down and the landing gear handle is UP. Failure of the flasher may cause the light to illuminate steady.

BOUNDARY LAYER CONTROL SYSTEM

The boundary layer control system utilizes air bled from the 17th stage of the engine compressor. This air passes through ducts attached to the rigid part of the wing between leading edge flaps and the spar and between the trailing edge flap and the flap closure beam. Slots along the ducts behind the outboard and center panel leading edge flaps and in front of the trailing edge flaps direct laminar air over the wing and flaps when the flaps have deflected sufficiently to expose the slots. The high temperature and high velocity laminar air directed over the wings and flaps delay flow separation over the airfoil, hence reducing turbulence and drag. This results in a lower stall speed and therefore a reduction of landing speed. Leading edge BLC is operative in the 1/2 or full flap position. Trailing edge BLC is operative only when the flaps are in the full down position. BLC air is controlled by four valves, one in each leading edge duct and one in each trailing edge duct. The BLC valves are actuated by mechanical linkages connecting the valves to the flaps. The leading edge flap BLC valves open with the flap switch in the 1/2 or full flap position, and the trailing edge flap BLC valves open with the flap switch in the full flap position. On aircraft 155903ap, and 157274ap and up, and all others after AFC 440, leading edge BLC can be shut off by placing the engine bleed air switch OFF.

CAUTION

- Operation of the engine(s) with the flaps down and wings folded should be avoided, to protect the leading edge BLC bellows from unrestrained heat expansion. If engine operation must be accomplished, power settings should be kept at a minimum.
- On aircraft 155903ap and 157274ap and up, and all others after AFC 440, if an engine is shut down trailing edge BLC will be lost on the wing adjacent to the shut down engine.

BLC Malfunction Indicator Light

A BLC MALFUNCTION light (figure A-1, appendix A) is on the telelight panel. The purpose of the light is to indicate a BLC valve malfunction in the 1/2 flaps or flaps up condition. The light illuminates when any one of the four BLC valves is not fully closed in the flap up condition, or when one of the trailing edge BLC valves open in the 1/2 flaps condition. It must be remembered that the illumination of the BLC MALFUNCTION light only indicates that a BLC valve has failed to close when the flaps are up, or that a trailing edge BLC valve has opened with the flaps 1/2 down. No indication is provided for a completely inoperative system, nor is there an indication provided for a BLC valve failing to open when the flaps are down. On aircraft 153071z thru 153075z before AFC 263, the BLC MALFUNCTION light does not indicate a condition where the trailing edge BLC valves are open at half flaps.

NORMAL OPERATION

The leading edge flaps and aileron droop actuators are operated by the use of a manifold-mounted selector valve and single-acting actuators, while the trailing edge flaps employ the same manifold-mounted selector valve, a wing-mounted selector valve, and dual-acting actuators. Placing the flap switch in the 1/2 position energizes the manifold-mounted selector valve allowing utility hydraulic pressure to lower the leading edge flaps full travel, the ailerons drooped 16½° down and the trailing edge flaps to 1/2 down. Further movement of the switch to the DN position energizes the wing-mounted selector valve resulting in complete extension of the trailing edge flaps. Immediate movement of the switch from the UP position to the DN position causes both selector valves to become energized simultaneously, thereby completely extending the leading and trailing edge flaps and ailerons. The limit switches, provided on each flap, are all connected in parallel to deenergize the electrical circuits to the selector valves after all flaps are retracted. The electrical circuits are continuously energized to maintain hydraulic pressure on flaps down. Should the cockpit switch inadvertently be left in the down position, the leading and trailing edge flaps and ailerons are protected from structural damage by an airspeed pressure switch which operates the common solenoid selector valve. This switch is set so as to limit the maximum speed before automatic retraction between 230 and 244 knots. During deceleration, the flaps automatically extend (providing the flap switch is down) between 234 and 210 knots. Normal flap and aileron extension is accomplished in approximately 8 seconds and retraction in approximately 6 seconds. On aircraft 154786ag and up, the aileron droop is actuated electrically; however, the normal operation is the same.

CAUTION

The airspeed switch receives its sensing pressure through the pitot system. If the pitot tube becomes clogged, erroneous indications will be sensed by the flap pressure switch as well as the ADC. It is therefore possible to lower the flaps by the normal means at excessive airspeeds.

EMERGENCY OPERATION

If normal wing flap operation fails, the flaps can be lowered by pulling the flap circuit breaker and pulling full aft on the emergency extension handle. The flap circuit breaker must be pulled prior to lowering the flaps by the emergency system. This causes the flap hydraulic selector valve to return to its full trail position, blocking hydraulic pressure to the flap actuators and insuring that hydraulic fluid will not be forced into the actuators on top of pneumatic pressure. Once the emergency wing flap extension handle has been pulled, it should be left in the full aft position. Returning the handle to its normal position allows the compressed air from the flap down side of the actuating cylinder to be vented overboard, and the flaps are blown up by the airstream. If the flaps are inadvertently extended in flight by emergency pneumatic pressure, they must be left in the extended position until post flight servicing. If retraction in flight is attempted, rupture of the utility reservoir could occur with subsequent loss of the utility hydraulic system. When the flaps are lowered by the emergency system the aileron droop action is as follows:

On aircraft before AFC 534, the ailerons remain in the non-drooped position.

On aircraft after AFC 534, the ailerons will droop if the emergency aileron droop switch is in NORMAL. If the emergency aileron droop switch is in DISABLE, the ailerons remain in the non-drooped position.

CAUTION

Pull the flap circuit breaker prior to extending the flaps by the emergency system.

NOTE

- Any pneumatic extension of the wing flaps shall be logged on the yellow sheet (OPNAV FORM 3760-2).
- On aircraft before AFC 534, the trailing edge flap indicator will show barber pole following pneumatic flap extension. On aircraft after AFC 534, the trailing edge flaps will show barber pole following pneumatic flap extension if the

emergency aileron droop switch is in DISABLE but will show 1/2 if the switch is in normal.

LIMITATIONS

Do not attempt to lower flaps above 250 knots CAS.

FLIGHT CONTROLS

NOTE

Some illustrations referred to within this system writeup may be located in appendix A. These illustrations are referred to, within the text, as (figure A–, appendix A).

DESCRIPTION

The airplane primary flight controls consist of the stabilator, rudder, ailerons, and spoilers, The stabilator, ailerons and spoilers are actuated by irreversible, dual power cylinders. The rudder is actuated by a conventional, irreversible power cylinder. Artificial feel systems provide simulated aerodynamic control stick and rudder pedal forces due to the lack of aerodynamic feedback forces from the power control cylinders. The feel systems have trim actuators which, through the power cylinders, move the entire control surface. Secondary controls are leading edge flaps, trailing edge flaps, and wing mounted speed brakes. See figure 1–22.

LATERAL CONTROL SYSTEM

The lateral control system (figure 1–22), an aileron–spoiler combination, basically consists of the control stick; left and right push–pull rod systems; left and right walking beam bellcranks; aileron dual power cylinders with integrated control valves; spoiler dual control valves; left and right autopilot series servos; left and right lateral feel trim actuators; and a left and right aileron droop actuating cylinder. The ailerons travel downward 30 degrees from a full trail position. Upward travel is limited to 1 degree. The spoilers travel upward, 45 degrees from a flush contour position in the upper wing surface. Lateral movement of the control stick is transmitted mechanically by the push–pull rods through the walking beam bellcranks to the spoiler and aileron control valves. The control valves meter hydraulic fluid to their respective dual power cylinders in proportion to the mechanical displacement. An override spring cartridge is incorporated into the left and right push–pull rod systems. In the event one side becomes jammed, the override spring will deflect under force, allowing operation of the other lateral control surfaces. The walking beam bellcranks receive control surface movement inputs from three sources: the control stick, the lateral trim system, and the autopilot series servos. A self–serviced hydraulic damper, attached to the aileron backup structure, is utilized as an up–stop for the aileron as well as a flutter damper. The control system uses dual power cylinders to provide the ailerons and spoilers with three independent sources of hydraulic pressure. PC–1 and utility systems provide hydraulic pressure to the left aileron and spoiler, and PC–2 and utility systems provide hydraulic pressure to the right aileron and spoiler. If a single hydraulic system fails, the remaining system(s) supply adequate power for control.

Aileron Control

The ailerons are controlled by dual, irreversible, power cylinders that receive metered hydraulic fluid from dual integrated control valves. The control valves, in turn, are controlled by the push–pull rods, through the walking beam bellcranks, and control stick. Each power cylinder contains four parallel inner cylinders with rods and pistons. The piston rods are joined at one end by a yoke that is attached to the airplane structure. The cylinder portion of the power cylinder is attached to the aileron. The two outer cylinders of the left aileron and spoiler receive hydraulic fluid from the utility system, and the two inner cylinders receive hydraulic fluid from PC–1. The two outer cylinders of the right aileron and spoiler receive hydraulic fluid from PC–2, and the two inner cylinders receive hydraulic fluid from the utility system. This arrangement provides symmetrical loading of the yoke should one of the systems fail. The ailerons deflect 16½° down when 1/2 or full flaps are selected. This is accomplished by utilizing an aileron droop actuating cylinder which repositions a bellcrank pivot point when flaps are selected. As the bellcrank pivot point is repositioned, linkage to the aileron control cylinder is deflected which, in effect, tells both ailerons that 16½° of travel is required. Even though the ailerons are drooped, they will continue to function as originally designed, except that the ailerons neutral point will be 16½° down, and any aileron movement will take place around this new neutral point. Aileron deflection up for a particular maneuver will be as much as 16½° back to the streamlined position, instead of 1° up as is the case without the ailerons drooped. For instance, if the control stick is moved 5° to the right with the flaps half or full down, requiring 6⅞° of aileron, the right aileron will raise 8¼° from the 16½° position, while the left aileron will deflect an additional 6⅞° and assume a 23⅜° down position. Therefore, the ailerons will move essentially in the same manner as the non–drooped ailerons; however, the aileron neutral point will be 16½° lower. The aileron droop cylinder is positioned by utility hydraulic pressure. Aircraft 154786ag and up, the aileron droop cylinder is positioned by electromechanical droop aileron actuator.

Spoiler Control

Each wing contains two spoiler surfaces, two spoiler power cylinders, and a dual spoiler control valve. Each surface has a dual, irreversible power cylinder, with a feedback linkage to a dual spoiler control valve. The spoiler control valve divides each power control system input into equal parts which is then distributed to each spoiler dual power cylinder. One portion of each power cylinder of the right wing receives hydraulic pressure from PC–2, and the other portion receives hydraulic pressure from the utility system. One portion of each power cylinder of the left wing receives hydraulic pressure from PC–1, and the other portion receives hydraulic pressure from the utility system. If one of the systems fails, the other(s) supply adequate pressure for spoiler control.

Lateral Control Feel and Trim System

The lateral trim system consists of the trim switch (figure 1–27), a rotary power unit, two flexible drive shafts, and two screwjack actuators. When the trim switch is energized, the rotary power unit and flexible drive shafts position the screwjack actuators. The screwjack actuators are connected to the airplane structure on one end, and the walking beam bellcranks on the other end. As the screwjack actuators extend and retract, the lateral controls are repositioned and the control stick follows the trim movements. Lateral control artificial feel is provided by double–action spring cartridges connected in tandem with the screwjack actuators. When the control stick is moved from neutral, the springs are compressed. The farther the control stick is moved from neutral, the greater the force required to compress the springs. The spring cartridges return the control stick to neutral when the force on the control stick is removed.

Aileron Position Indicator

An aileron position indicator (figure A–1, appendix A) is on the left vertical panel in the front cockpit. A transmitter is mechanically connected to the lateral control linkage in the left wing. As the control linkage moves, the mechanical input is converted into electrical impulses which are sent to the position indicator. The indicator, marked in units of percent of system travel, represents actual control surface position. A wings level indication is zero trim, and a full down left or right indication is maximum trim travel. The maximum lateral trim available is 33%.

STABILATOR CONTROL SYSTEM

Longitudinal control is provided by a single unit horizontal tail surface (stabilator), that is actuated by an irreversible dual power cylinder. A slotted stabilator is provided to increase stabilator effectiveness, and thereby counter the nose down pitching moment caused by the drooped ailerons. The stabilator control system components include the control stick; push–pull rods; cables; bellcranks, integrated control valve; and an irreversible dual power cylinder. Additional components include a ram air bellows and 5 pound per G bob weight for system artificial feel, a trim actuator, and an AFCS servo that is integral with the control valve. When the control stick is moved longitudinally, the motion is transmitted by push–pull rods to a bellcrank. It is then transmitted by a cable assembly to another push–pull rod set. The second push–pull rod set actuates the control valve which meters hydraulic fluid to the dual power cylinder. Hydraulic pressure to the stabilator power cylinder is supplied by both power control hydraulic systems. If one of the power control hydraulic systems should fail, the remaining system will provide adequate control response. A hydraulic AFCS servo is integrated into the stabilator dual servo valve. It positions the dual servo valve in the same manner as control stick inputs. As a result, when the autopilot signals for a pitch attitude change, the control stick will follow the movement. The bob weight in the control linkage also increases stick forces proportionately to increases in G forces.

Stabilator Control Feel and Trim System

Artificial feel is provided by a dynamic (ram air) pressure bellows acting through a variable bellcrank on the stabilator trim actuator and a 5 pound per G bob weight. When the airplane is in trim, the ram air force on the bellows is balanced by the bob weight. As the aircraft increases or decreases in airspeed, the pressure on the bellows changes causing the bellows bob weight assembly to become off balance. The off balance condition is then transmitted through the trim actuator, control cables, and push–pull rods back to the control stick. Actuating the trim switch causes the stabilator trim actuator to move, balancing the forces between the bellows and the bob weight, thereby eliminating force on the control stick. A viscous damper, attached to the trim actuator, prevents abrupt control surface movements by increasing control stick forces with rapid stick movements. An override spring cartridge allows the feel and trim portion of the stabilator control system to be bypassed in the event of a nose up trim malfunction (runaway trim and/or bellows diaphragm failure). A heater is installed in the bellows ram air inlet probe and venturi to prevent freezing of moisture which causes restriction of airflow in these units. The heaters are controlled by the pitot heat switch on the right console (figure A–1, appendix A).

Stabilator Trim Position Indicator

The stabilator trim indicator (figure A–1, appendix A) is on the left vertical panel in the front cockpit. It is directly controlled by a transmitter which is integral with the stabilator feel trim actuator. The indicator, marked in units of percent of trim, represents trim actuator position

CONTROL STICK

The control stick, consisting of a stick grip and motional pickup transducer, is mounted in a yoke to permit left, right, fore and aft movement. The control stick grip contains five controls: a four–way trim switch, a bomb and centerline stores release button, a nose gear steering/heading hold cutout button, a missile trigger switch, and an emergency disengage switch. On aircraft 158355at and up and all others after AFC 506, a weapon select switch and a target slave and acquisition switch is added; the nose gear steering/heading hold cutout button is relocated to the lower part of the stick grip. The motional pickup transducer is utilized in conjunction with the automatic flight control system to provide control stick steering. The nose gear steering button also functions as a heading hold cutout button for the automatic flight control system. Refer to figure 1-21 for the location of the control stick grip controls.

RUDDER CONTROL SYSTEM

The rudder control system consists of the rudder pedals, push-pull rods, cable assemblies, bellcranks, a rudder feel trim system, an aileron-rudder interconnect actuator, a rudder damper, and an irreversible power cylinder with integral control valve. When the rudder pedals are moved, the motion is transmitted by the push-pull rods, bellcranks and cable assemblies to the control valve of the power cylinder. The control valve meters utility system hydraulic fluid to the power cylinder which positions the rudder. It is possible to have limited mechanical authority over the rudder if a utility hydraulic system failure occurs. A bypass valve in the power cylinder opens when system pressure is lost, allowing fluid to pass from one side of the cylinder to the other. Total amount of rudder deflection available is then a function of air loads on the rudder; however, under all speed conditions it requires a considerable amount of pilot effort to manually deflect the rudder. A hydraulic servo for yaw damping and AFCS operation is incorporated into the control valve of the power cylinder. Operation of the AFCS, however, does not move the rudder pedals.

Rudder Feel Trim System

Artificial feel is supplied to the rudder pedals by an artificial feel trim system. A hydraulic cylinder with utility system hydraulic pressure on both sides of a differential area piston, provides a pedal force of approximately 2.6 pounds per degree of rudder deflection below an airspeed between 228 and 252 knots during acceleration and below an airspeed between 232 and 218 knots during deceleration. At the accelerating airspeed, a pressure switch in the pitot static system cuts off hydraulic pressure to the low area side of the piston, and the pedal force becomes approximately 11.5 pounds per degree of rudder deflection. Use of trim switch on the console, in conjunction with an electric trim actuator, removes loads from the pedals after the rudder has been positioned to the proper flight attitude. Normal trim range is 7.5 $\pm$1 degrees of rudder deflection left and right.

CAUTION

In the event of a loss of the essential main 28 volt dc bus, while above approximately 235 knots CAS, the rudder feel force of approximately 11.5 pounds per degree of rudder deflection automatically reverts to approximately 2.6 pounds per degree of rudder deflection. As a result, rudder pedal forces become more sensitive, and excessive structural loads can be imposed on the airplane if full rudder deflection is commanded.

Rudder Trim Switch

The rudder trim switch (figure A-1, appendix A) is in the front cockpit on the inboard engine control panel. This switch controls the trim actuator in the rudder feel and trim system to trim the airplane directionally.

Rudder Position Indicator

A rudder position indicator (figure A-1, appendix A), is on the left vertical panel in the front cockpit. A transmitter is mechanically connected to the rudder control linkage. As the control linkage moves, the mechanical input is converted into electrical impulses which are sent to the indicator. The indicator is only marked for takeoff trim, which is zero degrees of rudder deflection.

Rudder Pedals

The rudder pedals are conventional type suspended units which are coupled to the rudder push-pull rod system by individual screwjacks. The screwjacks provide adjustment of the rudder pedals for comfort and are adjusted simultaneously by turning a crank on the pedestal panel. The pedals are also coupled to the power brake valves so that toe pressure on the pedal will apply the brakes. The rudder pedals are also used to control the nose gear steering unit when the nose gear steering button on the control stick grip is depressed.

Stall Warning Vibrator

A stall warning vibrator is on the front cockpit left rudder pedal to warn of approaching stall conditions. The vibrator is electrically connected to a switch in the angle of attack indicator, which is set at 21.3 units. On aircraft 155529ag and up and all others after AFC 388, the switch is set at 20.6 units. The stall warning vibrator motor is powered from the right main 28 volt dc bus through the angle of attack probe heater circuit breaker in the rear cockpit G8, No. 1 panel (on aircraft 155529ag and up K14, No. 1 panel). If the vibrator runs continuously, it may be rendered inoperative by pulling this circuit breaker.

AILERON RUDDER INTERCONNECT (ARI)

The aileron-rudder interconnect system causes rudder displacement proportional to aileron displacement which provides coordinated turns at low airspeeds. The limits of the system are 15° of rudder displacement when the automatic flight control system is in the yaw stab aug mode, and 10° rudder displacement when the yaw stab aug switch is disengaged. Components of the system include the control amplifier, the 10° servo actuator, acting through a walking beam, an airspeed pressure switch and two aileron transducers. The ARI circuit is completed through the flap blowup airspeed pressure switch. When the flap switch is in either the 1/2 or DN position with the

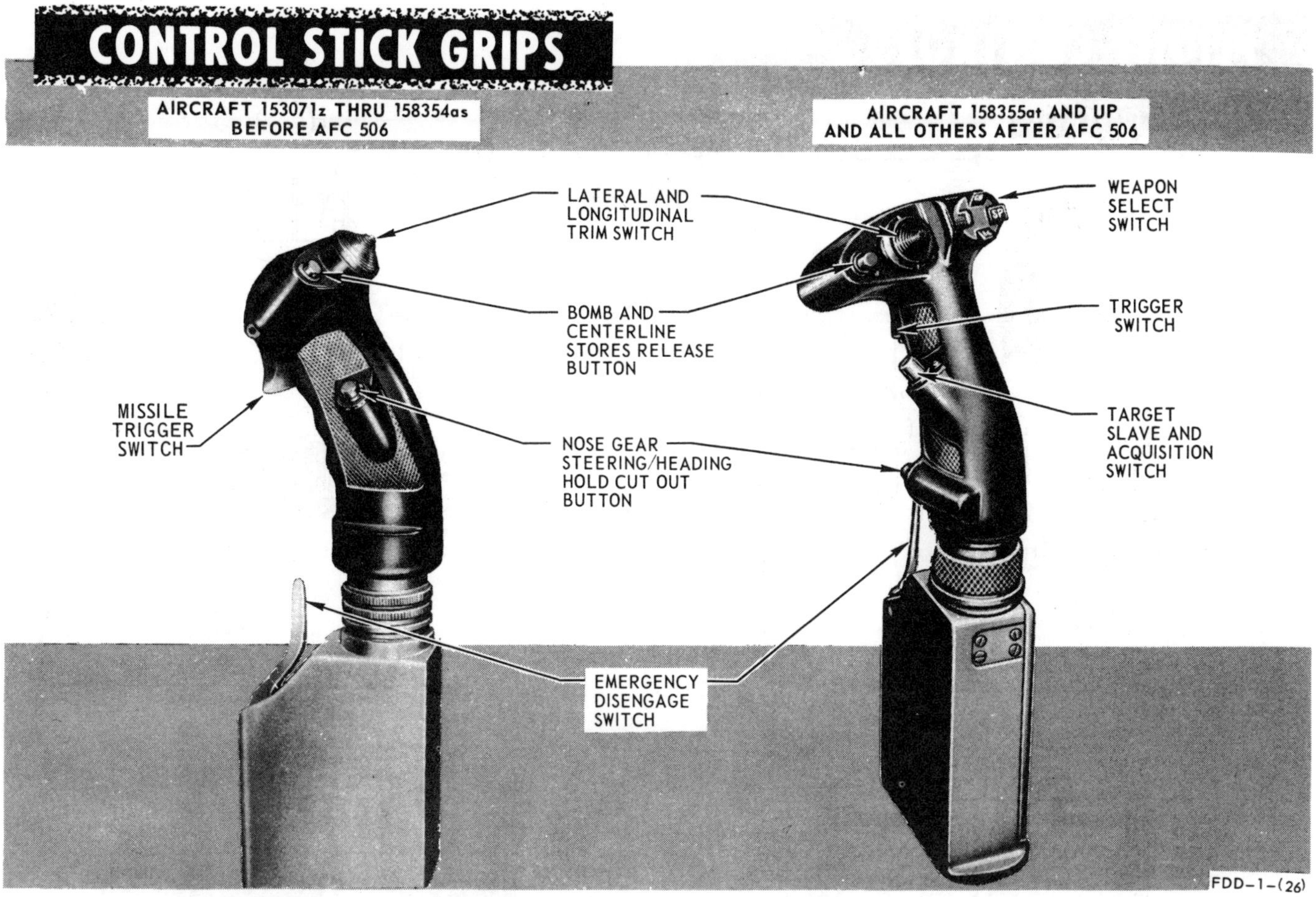

Figure 1-21

airspeed pressure switch in the low airspeed position, 28 volts dc is applied to the engage relay solenoids of the ARI system. This allows the hydraulic 10° servo actuator to move the control linkage (if aileron displacement is present) and cause rudder displacement. The system can be disengaged by depressing the emergency disengage switch on the control stick; this will disengage yaw stab aug and the ARI only as long as it is held depressed. When the switch is released the ARI (10°) and the yaw stab aug (5°) rudder authority will be regained. Regardless of the amount of ARI rudder authority engaged, the pilot can easily override the ARI system by pushing on the rudder pedals.

NOTE

There are various inflight situations where rudder jump will be experienced when the ARI system cuts in or out with a lateral input to the control stick. Rudder jumps are most apt to occur in situations where the flaps are put up or down during a turn, such as retracting the flaps during a climb out after takeoff or during a go-around. Assuming no manual rudder inputs, it is possible that after the flap switch is placed to the UP position during a go-around, for example, the rudder can jump from a deflected position to neutral after the flap switch is actuated. Another jump displacing the rudder back from neutral will then occur when the right hand trailing edge one half down limit switch closes. When the flaps go above the limits of the one half down limit switch the rudder will again deflect to neutral. Sometimes the first jump just described will not occur because the one half down limit switch will not be open when flaps up is selected. Rudder jumps will also occur whenever the flap airspeed switch is actuated when the flaps limit speed is exceeded, or by placing the flap switch to the DN or 1/2 position.

NORMAL OPERATION.

Normal operation of the flight controls is accomplished through the use of the control stick for longitudinal axis (ailerons) and lateral axis (stabilator) control, and the rudder pedals for vertical axis control.

EMERGENCY OPERATION

STABILATOR FEEL TRIM FAILURE

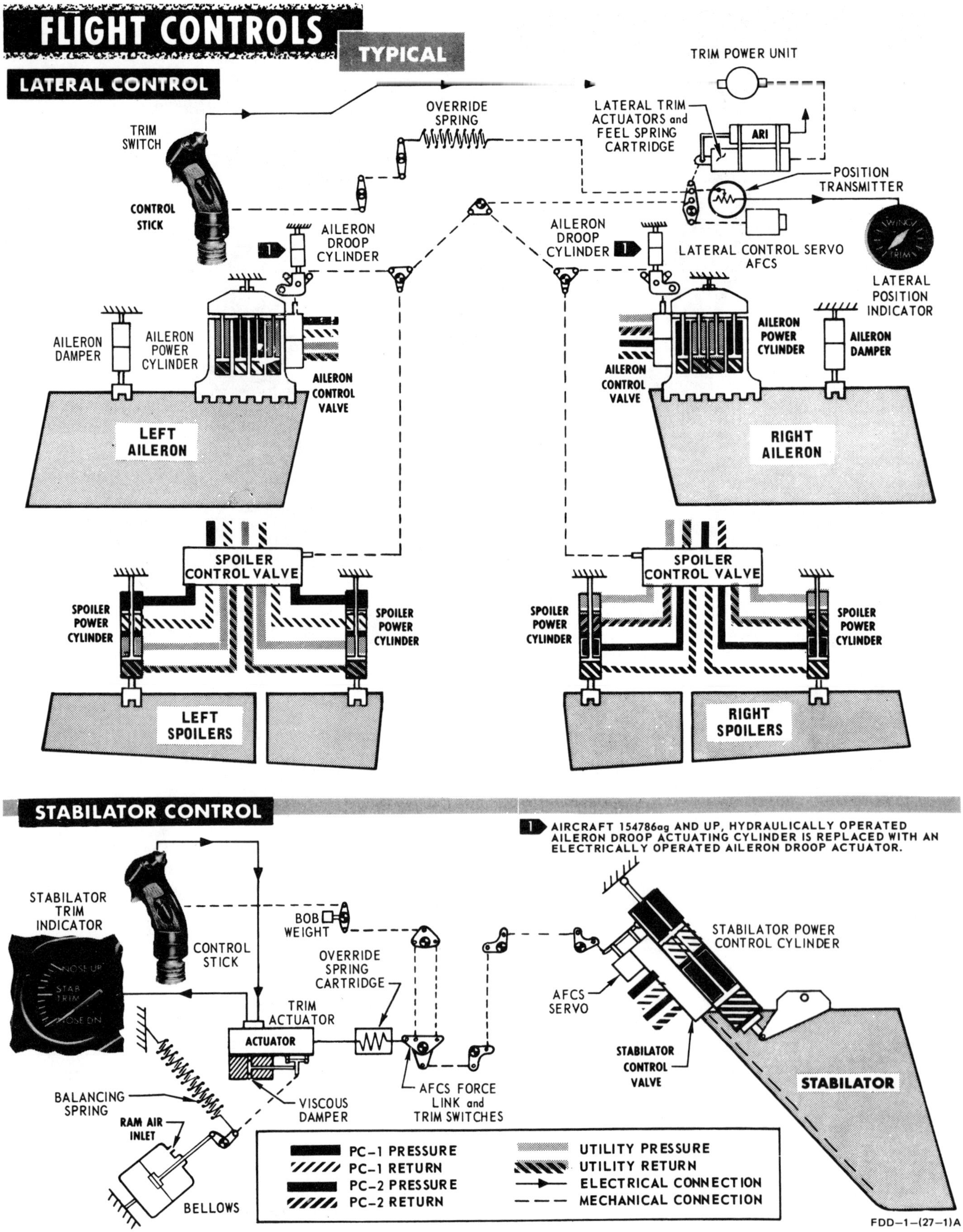

Figure 1-22 (Sheet 1 of 2)

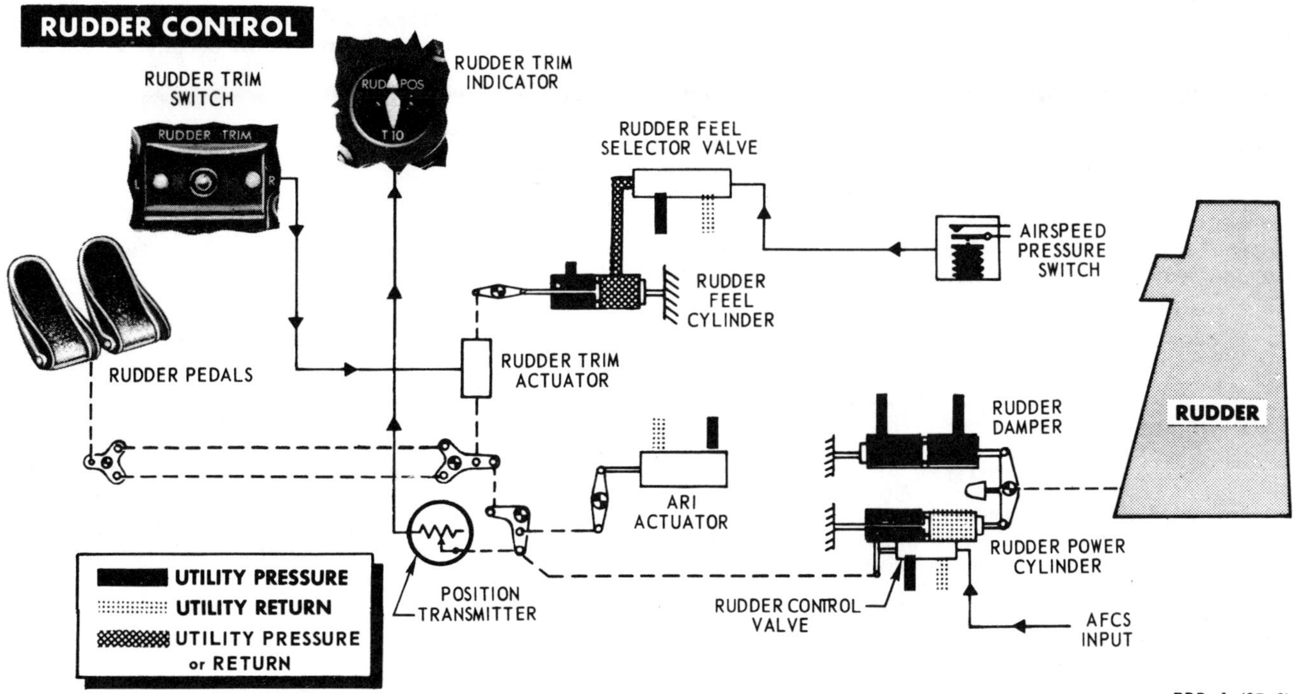

FDD-1-(27-2)

Figure 1-22 (Sheet 2 of 2)

Partial Bellows Failure

Partial bellows failure is recognized by a mild nose down stick force proportional to the airspeed, unless the failure occurs during maneuvering flight at which time it may not be noticeable. Reduction of stick centering and pitch stability will result. Should this failure occur, reduce airspeed to 250–300 knots CAS; retrim the airplane; avoid abrupt fore and aft stick movements; and land as soon as practicable.

Complete Bellows Failure

A complete bellows failure is recognized by a somewhat heavier nose down feel force at the control stick. Stick force will never exceed approximately 5 pounds/G and this force cannot be trimmed out. Should a complete bellows failure occur, reduce airspeed to 250–300 knots CAS, avoid abrupt fore and aft stick movements; and land as soon as practicable.

Ice/Water Blockage of Ram Air Line

The ram air bellows line is equipped with a heater which operates in conjunction with the pitot heat switch. With this arrangement, bellows line icing should not be encountered. If, however, the bellows line is allowed to ice up, the pilot will experience a situation similar to a complete bellows failure. If ice or water blockage is suspected, insure that the pitot heat switch is in the ON position, and do not apply longitudinal trim to relieve control stick forces. The intermittent nature of this condition and the suddenness of return to normal can cause violent pitch transients. When the ram air line is blocked, no stick force gradient will be felt by the pilot should a change in stick position be required. If ice or water blockage of the ram air line occurs, reduce airspeed to 250–300 knots CAS, maintain attitude by pilot effort; and if practicable descent to air that is above freezing. If the condition persists, land as soon as practicable.

Runaway Stabilator Trim

If stabilator trim appears to be running away, it is possible under certain conditions to lessen the situation. Runaway stabilator trim can be alleviated by engaging the autopilot, providing, the stab trim circuit breaker has been pulled immediately upon detection of runaway trim; runaway trim is in the nose up direction; nose down runaway trim has not exceeded 2–1/2 units; and airspeed is reduced to 300 knots CAS or less. If the above conditions are met, engage the autopilot. When the autopilot is used to alleviate a runaway trim condition, and excessive out of trim forces are present (full nose down runaway trim), the autopilot will alternately disengage and re–engage. If this occurs, discontinue use of the autopilot and plan to land as soon as practicable. If the autopilot is still engaged when in the landing configuration (gear and flaps extended), grasp the control stick firmly and disengage the autopilot at 180 to 190 knots CAS. Depending upon the severity of the malfunction, the airplane may or may not be in trim; if out of trim the forces should not be too high and the airplane can be landed with the out of trim condition, or the autopilot can be reengaged, and the landing made with control stick steering. If the landing is made with autopilot engaged, disengage the autopilot immediately after touchdown to prevent damage to autopilot components.

ARI System Disengagement

The ARI system can be temporarily disengaged by depressing the emergency disengage switch; this will disengage the ARI and stab aug only as long as it is held depressed. To permanently disengage the ARI system, the circuit breaker on the left utility panel must be pulled and the yaw stab aug switch must be disengaged. Pulling the circuit breaker only, and keeping the yaw stab aug engaged still provides 5° of ARI rudder authority. To permanently disengage the ARI while retaining complete stab aug, pull the rudder feel trim circuit breaker G15, No. 1 panel (on aircraft 155529ag and up B13, No. 1 panel). Pulling the rudder trim circuit breaker is the preferred method.

LIMITATIONS

There are no specific limitations pertaining to the flight controls.

FUEL SYSTEM

NOTE

Some illustrations referred to within this system writeup may be located in appendix A. These illustrations are referred to, within the text, as (figure A–, appendix A).

DESCRIPTION

NOTE

All fuel weights in this manual are based on JP–5 at 6.8 pounds per gallon.

The fuel system (figure A–4, appendix A) consists of seven interconnected fuel cells in the fuselage, and two integral wet wing cells located in the wing torque boxes. Provisions are made for two externally mounted droppable wing tanks and a droppable fuselage centerline external tank, which is interchangeable with a refueling tanker external store (hereafter referred to as Buddy Tank). Provisions are also made for an air refueling system. The function of fuselage cells 2, 3, 4, 5, 6, and 7 is to keep cell No. 1 supplied with fuel. See figure 1–22, fuel Quantity Data Table, for fuel quantities. An air pressure fuel transfer system is provided to transfer wing and external tank fuel to the fuselage cells. Hydraulic and electric transfer pumps plus gravity feed are utilized to transfer fuel from the fuselage cells to No. 1 tank which is the engine feed tank. Single point ground pressure fueling at the rate of approximately 1700 pounds per minute may be accomplished. Two point ground pressure fueling is available by using the air refueling probe. There are no gravity fueling or defueling provisions made for the internal or external fuel systems. Single point defueling is accomplished by using the single point fueling receptacle. All internal fuel cells incorporate capacitance type fuel gaging units which continuously indicate the total fuel quantity in pounds in all internal cells. The fuel system is equipped with refueling level control valves which are float type valves that shut off the pressure fueling when predetermined fuel levels are reached. All internal and external fuel tanks are pressurized in flight by regulated engine bleed air which is also utilized to transfer wing or external fuel to the fuselage cells or to dump wing fuel. The internal cells and external centerline tank or Buddy Tank are all vented to a common manifold which dumps overboard from the fuel vent mast located immediately below the rudder. The external wing tanks are vented to the wing cell dump lines. With the Buddy Tank installed, the airplane becomes a tanker with the capabilities of transferring; in flight, a predetermined amount of its internal fuel supply (plus the Buddy Tank fuel supply) to a receiver airplane, or return transfer from the Buddy Tank to its own internal fuel supply.

FUEL BOOST SYSTEM

Fuel is supplied to the engine during all flight attitudes by two submerged electric motor–driven centrifugal type boost pumps. The left pump is a two–speed unit. During normal operation, both pumps operate at high speed. If a complete electrical failure or double engine failure occurs, extending the ram air turbine automatically switches the left pump from high to low speed and shuts off the right pump, thereby reducing a high amperage load and conserving electrical power and at the same time maintaining positive fuel pressure at the engine inlet. The boost pumps are in the engine feed (No. 1) tank. Both pumps are mounted on the bottom of the tank and provide for negative G requirements. Due to internal tank baffling and check valves, which trap approximately 905 pounds of fuel in the lower third of the tank during negative G flight, the boost pumps will always remain submerged and provide a continuous fuel flow to the engines. The two boost pumps operate when either engine master switch is ON, provided ac power is supplied to the system.

NOTE

- When the electrical fuel boost pumps are inoperative, gravity fuel is sufficient to maintain full military power at altitudes below 20,000 feet, provided no unusual attitudes and/or negative G conditions are present.
- In the event of a double engine failure and loss of electrical power, extending the ram air turbine automatically switches the left boost pump to low speed and shuts off the right pump. The low speed boost pump plus gravity feed will supply enough fuel pressure to the engine driven fuel pumps to enable the engines to be started.

FUEL QUANTITY DATA TABLE

	FULLY SERVICED		USABLE FUEL	
	(GALLONS)	(POUNDS)	(GALLONS)	(POUNDS)
FUSELAGE CELL NO. 1			231	1571
CELL NO. 2			221	1503
CELL NO. 3			164	1115
CELL NO. 4			221	1503
CELL NO. 5			201	1367
CELL NO. 6			235	1598
CELL NO. 7			95	646
TOTAL FUSELAGE FUEL	1393	9473	1368	9303
INTERNAL WING TANKS	638	4338	630	4284
TOTAL INTERNAL	2031	13811	1998	13587
EXTERNAL WING TANKS	744	5059	740	5032
INTERNAL FUEL PLUS EXT. WING TANKS	2775	18870	2738	18619
EXTERNAL CENTER TANK	602	4094	600	4080
INTERNAL FUEL PLUS EXT. CENTER TANK	2633	17905	2598	17667
MAX. FUEL LOAD TOTAL INTERNAL AND ALL EXTERNAL FUEL	3377	22964	3338	22699
TOTAL UNUSABLE	39	265	–	–

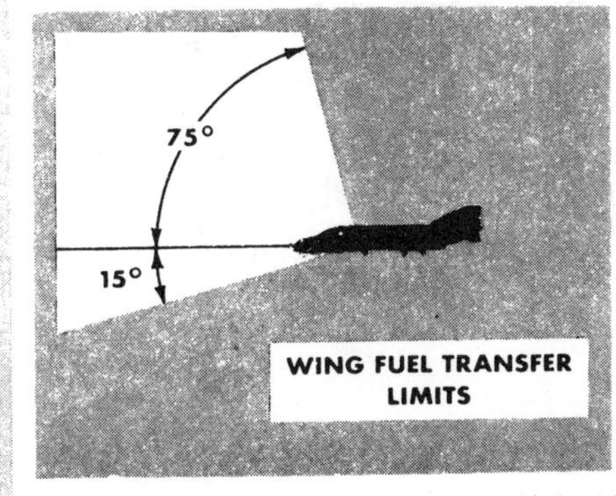

Note

FUEL WEIGHTS ARE BASED ON THE JP-5 AVERAGE WEIGHTS OF 6.8 POUNDS PER GALLON AT 60 DEGREES FAHRENHEIT.

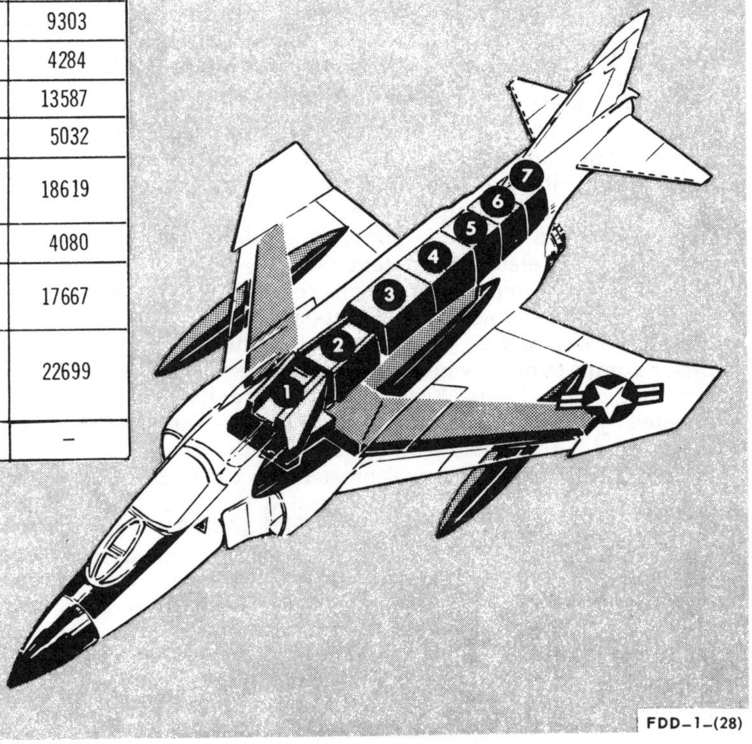

FDD-1-(28)

Figure 1-23

Boost Pump Pressure Indicators

The boost pump pressure indicator (figure A-1, appendix A) are on the left console in the front cockpit. The gage dials are calibrated from 1 to 5 with readings multiplied by 10. Pressure transmitters on the airplane keel in the engine compartment measure pressure in the aircraft fuel system as it enters the engine fuel pump. This signal is transmitted to the indicators in the cockpit.

Fuel Boost Pump Check

It is possible for the pilot to check the operation of the fuel boost pumps through use of the fuel pump check switches. The left and right boost pump check switches, with a CHECK position and a spring-loaded NORMAL position, are on the fuel control panel. On aircraft 155903ap and aircraft 157274ap and up, the left and right boost pump check switches are red press-to-check switches, on the left utility panel (figure A-1, appendix A). A boost pump check may be made only with external power applied to the aircraft, both engine master switches OFF and the refuel/defuel switch (in the right wheel well) OFF. Holding either check switch in the CHECK position operates the corresponding left or right engine shut-off valve, allowing a pressure transmitter to pick up boost pump pressure. Fuel boost pump pressure transmitters transmit an electrical signal to the applicable pressure indicator on the left vertical panel. To perform a boost pump pressure check, operate each boost pump check switch individually and check for a reading of 30 $\pm$5 psi on the applicable pressure indicator. Should fuel in cell 1 be less than approximately 1000 pounds when the boost pump check is being performed, a low boost pump pressure reading may be experienced. In flight with the fuel tanks pressurized, if a reading of more than 4 psig above the reading is noted, the fuel cells are overpressurized and a malfunctioning pressure regulator and/or fuel vent valve should be suspected. The system should be vented by slowing down and extending the probe. If it is not practicable to slow down, pull the refuel probe circuit breaker (G5, No. 1 panel on aircraft before AFC 388 and AFC 545; J14, No. 1 panel on aircraft after AFC 388 before AFC 545; J5, No. 1 panel after AFC 545) and place the refuel probe switch to REFUEL.

FUSELAGE FUEL TRANSFER SYSTEM

Fuel from the fuselage cells (3 thru 7) will transfer into cells 1 and 2 by gravity feed and transfer pumps. Cell 3 will gravity feed cell 4; cells 5 and 7 will gravity feed cell 6. The fuel is then transferred from cells 6 and 4 to cells 1 and 2 by electrical and/or hydraulic transfer pumps located within the cells. The electrical transfer pumps (2 each) will commence transferring fuel when ac electrical power is available and either engine master switch is ON. The hydraulic transfer pumps (2 each) will commence transferring when hydraulic power is available with no electrical power (electrical system failure), when either engine is in afterburner or when the fuel low level warning circuit is energized. Float-type level control valves, in cells 1 and 2, will open to allow fuel from the transfer pumps to enter when the level in these cells drops below that of the floats. Fuselage cell 7 gravity feed is controlled by the pilot valve in cell 2. When the fuel level in cells 1 and 2 drops below 2050 $\pm$200 lbs, the valve will open and direct air and/or fuel pressure to the dual (air and fuel) actuator transfer valve in cell 7, thus allowing cell 7 fuel to gravity feed cell 6. On aircraft 155903ap and aircraft 157274ap and up, cell 7 has an electric transfer pump which transfers fuel into cell 6. The pump is controlled by two thermistors; one in the upper portion of cell 5, and one on the bottom of cell 7. When the fuel level drops below the thermistor in cell 5 the pump is energized and fuel transfer is initiated. After all the fuel in cell 7 is transferred the exposed thermistor in cell 7 interrupts the pump circuit and energizes the TANK 7 EMPTY light. If the transfer pump fails the fuel will gravity feed into cell 6.

INTERNAL WING TRANSFER SYSTEM

Wing fuel will transfer (if selected) to cells 1 and 3 only as soon as the internal wing tanks are pressurized. The internal wing tanks are pressurized when the gear handle is UP and an engine is running. The internal wing tanks incorporate an automatic transfer feature that will transfer wing fuel into cells 1 and 3 when energized (regardless of switch position). Wing fuel normally enters cells 3 and will enter cell 1 when the fuel level in the cell drops low enough to permit the refuel level control valve to open. Internal wing fuel can be transferred when operating on the emergency generator.

Internal Wing Transfer Switch

The internal wing transfer switch is a two-position toggle switch on the fuel control panel. The switch positions are marked NORMAL and STOP TRANSFER. In NORMAL, internal wing fuel is transferred to fuselage cell 3 as soon as the internal wing tanks are pressurized, and the refueling level control valve is open. Internal wing fuel will also transfer to cell 1 if space is available in the cell. Selecting STOP TRAN closes the internal wing valves, thus preventing internal wing fuel transfer. If the automatic fuel transfer circuit is energized, the internal wing fuel transfer valves will open, regardless of the internal transfer switch position, and all internal wing fuel will transfer.

Wing Transfer Pressure Switch

The wing transfer pressure switch (figure A-1, appendix A) is a two position switch on the fuel control panel. The switch positions are marked NORMAL and EMERG. When the landing gear handle is UP and the wing transfer pressure switch is in NORMAL, all internal and external tanks become pressurized by the pressure regulator valves being deenergized open and the pressure relief valves energized closed. If the landing gear is down, internal wing or external fuel will not transfer unless the wing transfer pressure switch is in the emergency position. Placing the switch in EMERG performs the same function as the landing gear handle switch; all pressure regulators open and all pressure relief valves close; the tanks are thereby pressurized and ready to transfer. To prevent the external tanks from collapsing during a high altitude descent with wheels down, place the wing transfer pressure switch to EMERG before lowering the landing gear. If the tanks have been depressurized in level flight, place the wing transfer pressure switch to EMERG and continue in level flight for approximately 30 seconds to ensure adequate repressurization before commencing descent. Place wing transfer pressure switch to NORMAL prior to landing.

EXTERNAL FUEL TRANSFER SYSTEM

External fuel will transfer (if selected) to cells 1, 3 and 5 as soon as the external tanks are pressurized. The external tanks become pressurized when the gear handle is UP and an engine is running. The external tanks will automatically transfer fuel to cells 1 and 3 if the automatic fuel transfer circuit is energized, regardless of external tank switch position. External fuel will normally enter cells 3 and 5 and will enter cell 1 when the fuel level in the cells drop low enough to permit the refuel level control valve to open. External fuel can be transferred when operating on the emergency generator.

External Tank Transfer Switch

The external transfer switch (figure A-1, appendix A) is a three-position toggle switch on the fuel control panel. The switch positions are marked CENTER, OFF, and OUTBD. Upon selection of the CENTER position, the centerline tank fuel transfers to cells 3 and 5, providing the external tanks are pressurized. Placing the switch to OUTBD, transfers outboard wing tank fuel to cells 3 and 5, providing the external tanks are pressurized. When the switch is OFF, no external fuel can be transferred. When selecting fuel transfer with the transfer switch, only the tanks selected will transfer. However, if the automatic fuel transfer circuit is energized, all external tanks will transfer simultaneously.

CAUTION

Catapult launching acceleration can force fuel out of the external tanks through the transfer lines to the fuselage cells at a rate beyond tank venting capability, thus creating a partial vacuum in the external tanks. Therefore, to prevent external tank collapse during a catapult launch, ensure that the external transfer switch is OFF prior to launch.

NOTE

If external tanks are being carried, internal wing fuel will not transfer (prior to the automatic level) if the external transfer switch is in any position other than OFF.

FUEL TRANSFER SELECTOR KNOB

On aircraft 155903ap and aircraft 157274ap and up, a four position fuel transfer selector knob replaces the individual internal wing transfer switch and the external tank transfer switch. This makes it impossible to select external fuel transfer and internal wing fuel transfer simultaneously. The knob positions are marked STOP, OUTBD, CTR, and INT WING. Selecting INT WING, transfers internal wing fuel to fuselage cell 3. Selecting OUTBD or CTR, transfers fuel from the external wing tanks or the centerline tank, respectively, to cells 3 and 5. Selecting STOP, provides only fuselage fuel transfer. Selecting STOP until TANK 7 EMPTY light illuminates is the fastest way to transfer cell 7 fuel. If the automatic fuel transfer circuit is energized, all external tanks and the external wing tanks will transfer regardless of the position of the fuel transfer selector knob.

AUTOMATIC FUEL TRANSFER

Inflight, when the fuel level in cells 1 and 2 drops below 2615 $\pm$200 pounds, all internal wing fuel, external wing fuel and centerline fuel will simultaneously transfer to cells 1 and 3 regardless of fuel transfer switch or landing gear positions. However, the buddy tank (if installed) will not automatically transfer if FILL is selected. This fuel will not enter cells 5, 6 or 7. On aircraft 155903ap and aircraft 157274ap and up, the automatic fuel transfer sensor is in cell 1 and the circuit is energized when the fuel level drops below 1465 $\pm$200 pounds. On these aircraft, during landing approach or loiter, the automatic transfer circuit may activate early due to high AOA and a partially full cell 1. The automatic fuel transfer system is completely independent of the fuel quantity indicating system. The automatic fuel transfer system is inoperative when the air refuel is in the refuel position.

EMERGENCY FUEL TRANSFER

Transfer system failure can usually be attributed to failure of the fuel system to become pressurized. If the fuel system fails to become pressurized, place the wing transfer switch in the EMERG position. This performs the same functions as the landing gear handle switch, all pressure regulators open and all pressure relief valves close.

PRESSURIZATION AND VENT SYSTEM

The pressurization and vent system provides regulated engine bleed air pressure to all internal and external tanks for pressurization, fuel transfer, and wing fuel dump. The system also provides venting of external tanks to prevent collapse during rapid descents.

Wing Tank Pressurization and Vent

The wing cells and external tanks pressurization system utilizes pressure regulators and pressure relief valves which are set respectively at 15 $\pm$0.5 psi and 17.5 $\pm$0.5 psi. The wing cell pressure relief valves, which provide fuel tank pressure and vacuum relief, dump into a common manifold which is vented overboard under the rudder. The external wing tanks are vented through their pressure relief valves to the wing cell dump lines. The wing cells and external wing tanks are vented to the atmosphere when the landing gear is extended.

Fuselage Tank Pressurization and Vent

The fuselage tank pressure regulator, in conjunction with the flow limiter and pressure relief vent valve, will maintain regulated air pressure at 2 $\pm$0.5 psi and pressure relief at 3.5 $\pm$0.5 psi. The fuselage cells and the Buddy Tank or centerline external tank are vented to the common fuel vent manifold and then vented overboard through the fuselage pressure relief valve. When the aircraft is on the ground all pressure relief valves are open, venting all tanks to the atmosphere.

WING FUEL DUMP

Wing fuel may be dumped in flight at any time regardless of any other transfer position by selecting the DUMP position on the internal wing dump switch (figure A-1, appendix A). The two-position toggle switch marked NORMAL and DUMP is on the fuel control panel on the left console of the pilot's cockpit. A hex-head is installed on the internal wing dump switch to make it more easily recognized. Selecting DUMP, opens the left and right wing dump shutoff valves and closes the wing transfer and vent valves (if not previously closed). The wing air pressure regulators open (if not previously open) allowing the wing tanks to pressurize and force fuel out the dump lines at the wing fold trailing edge. At 85% rpm in level flight, the fuel dumping capability is approximately 680 pounds per minute. The dumping rate varies directly with rpm and pitch attitude; i.e., lower rpm and/or nose pitch down will decrease the dumping rate. Air pressure will continue to bleed out the dump line until the internal wing dump switch is placed in the NORM position to close the dump valve.

NOTE

- Wing fuel dump cannot be initiated on RAT power.

- Since the internal wing dump switch will function with the engine master switch ON or OFF, and the landing gear UP or DOWN, wing fuel will be dumped ON THE DECK when internal wing dump switch is placed in the DUMP position and external power is applied to the airplane.

SMOKE ABATEMENT SYSTEM (FUEL ADDITIVE)

On aircraft 155867ak and up, and all others after AFC 373, an engine exhaust smoke abatement system is incorporated to improve fuel combustion. The improved fuel combustion reduces the black smokey exhaust emitting from the engine exhaust while operating at military power settings at low altitudes. The system, when energized, injects a combustion improver CI-2 into the engine fuel manifolds. The system consists of two interconnected reservoirs, a pressure regulator and vent valve, a shutoff valve, flow restrictors, and a control switch. The reservoirs have a total capacity of approximately 2 gallons (approximately 30 minute duration) and are pressurized by 17th stage engine bleed air. Pressure within the reservoir is maintained between 46 and 55 psi by the pressure regulator and vent valve. The valve receives pressure from the fuel tank pressurization line and vents through the centerline tank vent line. The shutoff valve and flow restrictors, control and regulate flow from the reservoirs to the fuel manifolds. During afterburner operation the system is deactivated by the afterburner sensing relays regardless of the control switch position.

Smoke Abatement Control Switch

A two position smoke abatement control switch (figure A-1, appendix A) with positions of OFF and ON is on the engine control panel. When the switch is in the ON position, the shutoff valve is energized open and additive is injected into the fuel manifolds. Additive will continue to be injected into the fuel manifolds until the switch is placed to the OFF position, the reservoirs are depleted, or until afterburner is selected.

FUEL QUANTITY INDICATING SYSTEM

The fuel quantity indicating system is of the capacitance type and provides a reading in pounds of total internal fuel. The system components include the fuel quantity indicators, feed tank check switch and a fuel level low indicator light. There are 14 fuel gaging units located throughout the internal tanks which register at the one cockpit fuel quantity gage.

Fuel Quantity Indicator

A combination (counter-sector) fuel quantity indicator (figure A-1, appendix A) is in the upper right corner of the pilot's instrument panel. On aircraft 158355at and up and all others after AFC 506, a fuel quantity indicator is installed on the left side of the RIO's main instrument panel. The forward cockpit indication is more dependable than the aft cockpit indication. Therefore, if a discrepancy occurs between the two indications, the forward indicator is more likely to give the best reading. The counter unit of the gage(s) continuously indicates the total usable fuel quantity (with readings multiplied by 10) in all internal tanks. The sector portion of the indicator(s) simultaneously indicates the total usable fuel quantity in fuselage cells 1 through 6 only, with readings multiplied by 1000. Fuselage cell 7 and internal wing fuel quantities are not indicated on the sector portion of the indicator(s). After all wing fuel has transferred, the counter should read approximately 640 pounds higher than the sector when fuselage cell 7 is full. On aircraft 155903ap and aircraft 157274ap and up, cell 7 feeds before the internal wings transfer. With both internal wing tanks and fuselage cell 7 empty, the counter and sector should read within 350 pounds of each other. There is a possibility that fuel quantity variations will be noted on the fuel quantity indicator during aircraft accelerations and decelerations. These variations are due to the high acceleration and deceleration rates which can be achieved with the airplane. Transient increases in fuel quantity readings may be noted during deceleration, and transient decreases in fuel quantity readings may be noted during acceleration. Therefore, an optimum fuel quantity gage indication is achieved with the aircraft in a straight and level attitude in conjunction with moderate stabilized power settings.

WARNING

At the low end of the fuel scale, the counter portion of the fuel quantity gage has a tolerance of ± 200 pounds, and the sector portion has a tolerance of ± 150 pounds. Therefore, if the FUEL LEVEL LOW light illuminates above an indicated 2080 pounds, the warning light should be used as the primary indication of a low fuel state, and continued aircraft operation should be judiciously considered. The feed tank check switch should be actuated in an effort to gain additional information as to the location of the fuel and time of flight remaining.

Feed Tank Check Switch

The two-position feed tank check (figure A-1, appendix A), with switch positions of CHECK and NORMAL permits the pilot to check the fuel quantity in the engine feed tank. When the switch is placed in the spring-loaded CHECK position, the sector portion and the counter portion of the fuel quantity gage(s) indicates engine feed tank fuel quantity. In addition to checking feed tank fuel quantity, it is also an indication that there is power to the fuel quantity circuits and that the gage is functioning properly.

Fuel Level Low Indicator Light

The FUEL LEVEL LOW indicator lights, on the telelight panel in the forward cockpit and on the instrument panel in the aft cockpit, illuminate when the combined usable fuel in the engine feed tank and cell 2 is reduced to approximately 1880 ±200 pounds. The FUEL LEVEL LOW lights illuminate at the above fuel quantities only if the airplane is in a perfectly level attitude and moderate stabilized power settings are being used. However, due to the various attitudes and power settings required during a normal flight, the illumination of the FUEL LEVEL LOW indicator lights is not an accurate indication of the amount of fuel remainining in cells 1 and 2. The illuminated lights only indicate that the fuel is low. In this system, the unit which operates the low level indicator lights is a thermistor sensing switch which is located in cell 2. When the fuel level in cells 1 and 2 is above the thermistor sensing switch, its resistance is increased and no current flows to the fuel level low light relay. At this time the FUEL LEVEL LOW light is OFF. When the fuel level in cells 1 and 2 is below the thermistor sensing switch, its resistance is decreased and current flows to the fuel level low light relay. At this time the FUEL LEVEL LOW light illuminates. On aircraft 155903ap and aircraft 157274ap and up, the low level sensor is in cell 1. On these aircraft, during landing approach or loiter, the FUEL LEVEL LOW light may illuminate early due to high AOA and a partially full cell 1. The fuel level low light is completely independent of the fuel quantity indicating system.

Tank 7 Fuel Light

The TANK 7 FUEL light, on the telelight panel, illuminates in conjunction with the FUEL LEVEL LOW indicator light if cell 7 fuel is not being transferred. The light indicates that the dual actuator transfer valve in cell 7 did not open when the fuel level in the engine feed tank and cell 2 reached 2050 pounds.

NOTE

With the TANK 7 FUEL light illuminated, the counter portion of the fuel quantity indicator indicates approximately 640 pounds more fuel remaining than is actually available.

Tank 7 Empty Light

On aircraft 155903ap and aircraft 157274ap and up, the TANK 7 EMPTY light replaces the TANK 7 FUEL light and illuminates when cell 7 is empty. After cell 7 has transferred the light may be extinguished by depressing the master caution reset button.

Left and Right External Fuel Lights

The L EXT FUEL or R EXT FUEL indicator lights on the telelight panel are provided to indicate an empty left or right outboard external tank with OUTBD position selected on the external transfer switch. The corresponding external fuel indicator light will illuminate when fuel flow from its external wing tank ceases. This indicates to the pilot that the tank indicated is empty or flow is interrupted. Since external fuel transfer is intermittent rather than continuous, the L EXT FUEL, and the R EXT FUEL indicator lights will come on during a temporary halt of fuel flow. Although the external tanks fuel indicator lights have gone out, the lights will again illuminate (approximately 10 to 30 seconds later) during the next interruption of fuel flow. Intermittent external fuel transfer is desired since this means the transfer rate is greater than engine consumption and fuselage fuel is being maintained at its highest possible volume. The L EXT FUEL or R EXT FUEL indicator lights will illuminate when selecting the OUTBD or REFUEL position on the fuel control panel and during automatic transfer when the fuel flow is less than 5 gpm. Each external fuel indicator light will also illuminate when the tanks are full during refueling operations. When the fuel level in cells 1 and 2 drops below automatic fuel transfer, and the outboard tanks are empty or MERS are installed, the L EXT FUEL and R EXT FUEL warning lights illuminate.

Centerline External Tank Fuel Light

The CTR EXT FUEL indicator light is provided to indicate an empty centerline tank with CENTER position selected on the external transfer switch. The CTR EXT FUEL indicator light illuminates when fuel flow ceases. When selecting the CENTER, BUDDY, FILL, or REFUEL positions on the fuel control panel or during automatic fuel transfer, the CTR EXT FUEL light will illuminate any time fuel flow is less than 5 gpm. The CTR EXT FUEL indicator light will also illuminate when the tank flow stops during fuel transfer to the Buddy Tank. When the fuel level in cells 1 and 2 drops below automatic fuel transfer, and the centerline tank is empty, the CTR EXT FUEL warning light illuminates.

TK Light

The TK light, on the missile status panel, is located on the lower left side of the main instrument panel. The light illuminates when the centerline tank or any other store is installed on the centerline station. When the TK light is illuminated, the forward fuselage missiles cannot be fired. On aircraft 158355at and up and all others after AFC 506, the TK light is located on the AIM 7 status panel which is relocated to the upper right side of the main instrument panel. The light illuminates only when a weapon is selected with the weapon select switch (on the stick grip) and the centerline tank or other store is installed on the centerline station. If the weapon lights reset switch (on the right console) is actuated, the TK light will go out.

EXTERNAL TANK JETTISON SYSTEM

External Tank Jettison Switch

The external wing tanks can be jettisoned by selecting the JETT position on the fuel control panel (figure A-1, appendix A) on the left console in the pilot's cockpit. On aircraft 158355at and up and all others after AFC 506, the external tank jettison switch is on bomb control panel of the pilot's main instrument panel. The tanks can be jettisoned before or after the flow indicating light illuminates. Illumination of the flow light indicates flow has ceased and external tanks are empty. If the external transfer switch has been inadvertently left in either the OUTBD or CENTER position and external tanks are not installed on the airplane, or the tanks have been jettisoned, the external wing tanks fuel shutoff valve will close and the switch will be ineffective, allowing wing fuel to transfer in its normal manner.

CAUTION

The external wing tanks can be jettisoned by the external wing tank jettison switch any time power is on the airplane and the external tanks safety pins are removed. This circuit is not wired through the landing gear handle.

NOTE

Refer to External Stores Limitations chart (part 4 of this section) for external tanks jettison restrictions.

External Centerline Tank Jettison

Centerline external stores may be made ready for jettison by accomplishing the following: landing gear handle, UP; centerline station switch, READY; bomb control switch, DIRECT; weapons switch CONV-OFF NUCL-ON. With these switches set, the centerline store may be jettisoned by depressing the bomb release button on the control stick grip.

External Stores Emergency Release Button

The external stores emergency release button (figure A-1, appendix A) is on the left vertical panel. This button may be used to jettison the external tanks. The external tanks can be jettisoned before or after the flow indicating light illuminates. Illumination of the flow light indicates flow has ceased and external tanks are empty. If the external transfer switch is in the CENTER position at time of jettison, the centerline tanks fuel shutoff valve will close and the switch will be ineffective, allowing wing fuel to transfer in its normal manner. For a complete description of the external stores emergency release button refer to Emergency Equipment this section.

AIR REFUELING (RECEIVER) SYSTEM

The air refueling probe is on the right side of the fuselage above the engine air inlet duct. The probe is equipped with an MA-2 refueling nozzle which is capable of receiving fuel from any drogue type refueling system. The refueling operation is actuated by the refuel probe switch on the fuel control panel. The refuel probe switch has three positions: REFUEL, EXTEND and RETRACT. The REFUEL position conditions the airplane fuel system for inflight refueling of all tanks and extends the inflight refueling probe. The EXTEND position retains the probe in the extended position, but returns the airplane fuel system to normal operation with the exception that fuselage cells 1 and 3 will accept fuel from the tanker at full flow until full, then at a rate equal to fuel consumption. This position is used when it is necessary to replenish the fuel in the engine feed tank, either by normal fuel transfer or from the tanker. It is also used if the probe is damaged and cannot be retracted. The RETRACT position returns the fuel system to normal transfer operation and retracts the probe. When the refuel probe switch is placed in the REFUEL position, a REFUEL READY light, on the telelight panel, illuminates if the fuselage pressurization and vacuum relief valve opened properly. This assures that the fuselage tanks are properly vented for refueling. On aircraft 155867ak and up, and all others after AFC 370, the air refueling probe can be extended by pneumatic pressure from the canopy air bottle when the normal extension system fails. Pneumatic extension is initiated by the emergency refuel probe control switch on the outboard engine control panel. The switch, marked NORM and EMER EXT, extends the probe when placed in the EMER EXT position. When extension is accomplished by pneumatic pressure, the probe remains extended for the remainder of the flight. The refuel selection switch is on the fuel control panel. This is a two-position guarded switch with ALL TANKS and INT ONLY position. The ALL TANKS position opens the external tank fuel shutoff valves when refueling. The INT ONLY position closes the external tank fuel shutoff valves and allows only the internal tanks to be refueled during air refueling. On aircraft 155903ap and aircraft 157274ap and up, the emergency refuel probe switch is removed from the engine control panel and installed (with a red guard) on the fuel control panel. The refuel selection switch is also replaced with a four position rotary switch that provides options of refueling internal only and all tanks, with or without refueling cell 7.

IFR Probe Unlock Indicator Light

The IFR PROBE UNLOCKED indicator light on the telelight panel will illuminate when the air refueling probe is not fully retracted. The illumination of the IFR PROBE UNLOCKED light also energizes the MASTER CAUTION light. The indicator light circuit is completed through a limit switch located within the air refueling probe latching actuator.

Air Refueling Probe Light

An air refueling probe light is on the right side of the fuselage forward of the air refueling probe. The light is used during night air refueling operations to illuminate the refueling probe and the drogue from the refueling airplane. The light is controlled by the IFR switch and variable intensity control knob, both are on the exterior lights control panel.

GROUND REFUELING

The airplane is capable of either single point or two point pressure refueling. The single point refueling receptacle is on the right underside of the fuselage in the area below the aft cockpit. Single point pressure fueling at the rate of approximately 250 gallons per minute may be accomplished. Two point pressure fueling at the rate of approximately 480 gallons per minute may be accomplished by utilizing the inflight refueling probe with a special fitting attached. The system allows a controlled partial refueling capability. If desired, fuel is locked out of the left and right wing tanks, and cells 5, 6 and 7. This allows the airplane to be partially refueled to an amount of 837 gallons (approximately 5,692 pounds) without creating an undesirable CG condition.

Cockpit Switch Positions

The switches on the fuel control panel located on the left console in the pilot's cockpit should be in the following position before single point pressure fueling; external transfer switch OFF, wing transfer pressure switch NORMAL, refuel selection switch ALL TANKS, buddy fill switch STOP FILL, refuel probe switch RETRACT. Refueling of the internal tanks only, with any or all external tanks installed may be accomplished by selecting INT ONLY on the refuel selection switch. The Buddy Tank is interchangeable with the centerline external tank and is refueled in the same manner as the centerline tank. The landing gear control handle must be in the gear down position and master switches and throttles should be in the OFF position. The generator control switches should be in the EXT ON position. If two-point pressure refueling is desired, the refueling probe switch should be placed in the REFUEL position. On aircraft 155903ap and aircraft 157274ap and up, the fuel transfer selector knob replaces the external transfer switch and should be placed to STOP. Also, the refuel selection switch is replaced by a refuel selection knob and should be placed to ALL WITH 7 or ALL W/O 7 as required.

Refueling Operation

Apply external ac power to the airplane and place the generator control switches to EXT ON. Open filler door and attach fueling nozzle to service inlet valve. Set ground fueling switch in the right wheel well to REFUEL position. (The ground fueling switch is only effective with the engine master switches OFF and ground electrical power applied.) With the REFUEL position selected on the ground fueling switch, all valves in the fuel system are closed with the following exception. The fuselage air pressure regulators will be open, all internal tank vent valves will be open, all external tank vent valves will be open if their respective tanks are installed and the refuel selection switch is on ALL TANKS. All fuel level control valves are open to receive fuel until their respective tanks are filled at which time floats rise in the valves to shut off fuel. Outboard and centerline external tanks motor-operated shutoff valves are open allowing fuel to fill the external tanks installed. A fuel flow transmitter in each refueling line to any external tank energizes a caution indicator light in the cockpit corresponding to the tank not accepting fuel. Partial refueling is accomplished by actuating and holding the left and right wing tank and No. 5 fuselage tank fuel level control valve switches. This prevents fuel from entering the left and right wing tanks and fuselage cells 5, 6 and 7. Fuel can be prevented from entering fuselage cell 7 only by depressing and holding the air and fuel actuator checkout plungers which are located in the right wheel well. On aircraft 155903ap and aircraft 157274ap and up, the air and fuel actuator checkout plungers are removed and refueling of cell 7 is controlled by a switch on the refuel-defuel control panel in the right wheel well.

NOTE

No. 7 fuel tank will not fill until the interconnect valve opens. The valve will not open until fuselage tanks 1 thru 6 are full and fuel pressure builds up to approximately 12 psi in the interconnect valve actuator line. No. 7 fuel tank will take approximately 5 minutes to fill after the interconnect valve opens.

Functional Precheck of Electric Transfer Pumps

Individual momentary type check switches for each electrically operated transfer pump and a pressure indicator light are on a panel in the left wheel well to provide a functional check for each electric transfer pump. The check can be performed only while external power is connected to the aircraft. When either switch is placed in CHECK, the primary circuit shuts off both fuselage fuel level control valves in fuselage cells 1 and 2, energizes the pressure transmitter switch and operates the transfer pumps in cells 4 and 6. The pressure transmitter switch energizes the green indicator light if the discharge pressure of the selected pump is normal. No light, while the pump is being checked, indicates a malfunction in the pump.

Functional Precheck of Hydraulic Transfer Pumps

A momentary type check switch is provided on a panel in the left wheel well to check the operation of the hydraulic transfer pumps. The switch operates in conjunction with two indicator lights. The check switch, when placed in CHECK, energizes closed the transfer pump level control valves in fuselage cells 1 and 2, opens the hydraulic shutoff valve to allow both pumps to operate and energize each pressure transmitter switch. Each pressure transmitter switch illuminates the green indicator light for each pump if their discharge pressure is normal. No light, with the switch in the CHECK position, indicates a malfunction of that pump. Hydraulic and electrical ground power must be connected to the airplane to conduct the above check.

Functional Precheck of Refueling Level Control Valves

A double throw momentary type master check switch and seven individual momentary type check switches are located in a panel in the right wheel well. The master switch has positions of CHECK NO. 1 and CHECK NO. 2. With fuel flow started from the fueling source, hold the master check switch to the CHECK NO. 1 position. This position closes the motor operated shutoff valves of any external tank installed and energizes a solenoid in the primary float unit of the refueling level control valves, causing the primary floats to rise and shut off fuel flow to all internal tanks. Placing the master check switch in CHECK NO. 2, closes the motor operated shutoff valves of any external tank installed and energizes a solenoid in the secondary float unit of the refueling level control valves which causes the secondary floats to rise and shut off fuel flow to all internal tanks. Continuation of fuel flow with the master switch in CHECK NO. 1 or CHECK NO. 2 indicates a malfunction of one or more of the refueling level control valves and/or motor operated shutoff valves.

When a malfunction occurs in the primary or secondary system, that respective position on the master switch shall be held. Malfunction of any refueling level control valve can then be isolated by operating the individual momentary type check switches one at a time to their respective position until fuel flow is stopped. The respective position of the individual switches energizes the solenoid in the circuit of each valve opposite to the circuit checked on the master switch. The individual switch that stops fuel flow indicates a malfunction of that valve in the primary or secondary unit respective to the circuit checked. On aircraft 155903ap and aircraft 157274ap and up, cell 7 refuel switch (on refuel–defuel panel in the right wheelwell) must be in the STOP position when making CHECK NO. 1 and CHECK NO. 2.

NOTE

If the master check switch in the CHECK NO. 1 or CHECK NO. 2 position does not stop fuel flow and operational necessity dictates, refuel normally except for the following: Pull the fuel transfer pump circuit breakers on circuit breaker panel No. 1, at approximately 6000 pounds total reading on the counter. Cut down the source pressure to 30 psig maximum until a counter reading of 9000 pounds is reached. Then cut the source pressure down to 4–5 psig until the airplane is fully fueled.

NORMAL OPERATION

Operation of the fuel system is normally controlled through the fuel control panel. However, if the proper switch settings for fuel transfer have not been selected, the unselected fuel will automatically be transferred when the automatic fuel transfer circuit is energized. With no external tanks aboard, all switches on the fuel control panel should be in the inboard position, with the exception of the external transfer switch which should be OFF. With this switch arrangement, the fuel system is set up for proper fuel transfer and no further switching is required. If external tanks are carried, switch positions are the same as with no external tanks except that the external transfer switch is placed in an appropriate external tank position. In this case it will be necessary to switch to another external tank position, or place the external transfer switch off when the fuel in the selected tank(s) is depleted. After all external fuel is expended and the external transfer switch is OFF, internal wing fuel will transfer automatically and no further switching will be required. The L EXT FUEL, R EXT FUEL and CTR FUEL warning lights will illuminate when flow from the selected tank is interrupted; therefore, the only indication of completed external fuel transfer is the illumination of the external fuel warning lights accompanied by a decrease in internal fuel. Upon depletion of external tank fuel, the fuselage cells will continue to supply fuel to the engine feed tank; however, internal wing fuel will not commence transferring until the external transfer switch is turned OFF. On aircraft 155903ap and aircraft 157274ap and up, the internal wing fuel, centerline fuel and external wing fuel is controlled by a common four position fuel transfer selector knob. Therefore, normal operation is as follows: STOP, until TANK 7 EMPTY light illuminates; OUTBD or CTR, until their corresponding L EXT FUEL, R EXT FUEL or CTR FUEL light(s) illuminate (with a decrease in internal fuel quantity); INT WING, when internal wing fuel is required. During carrier operation, manage internal wing fuel so as to arrive at the carrier with the maximum trap weight.

EMERGENCY OPERATION

FUEL BOOST PUMP

If fuel boost pumps fail, fuel is still supplied to the engines by gravity feed. If both boost pumps fail above 20,000 feet and/or at a high power setting, flameout or an unstable rpm indication on one or both engines may occur. During gravity feed, high fuel flow rates required by afterburner operation cannot be met. A boost pump pressure indication of 0 psi indicates that both boost pumps are inoperative. If both engines have flamed out, reduce airspeed to 515 knots CAS or Mach 1.1 whichever is less and extend the ram air turbine. Extending the ram air turbine operates the left fuel boost pump at low speed. This will supply enough fuel to either engine to accomplish an airstart. If an airstart has been accomplished or the engines have not flamed out, reduce power and/or descend until stable engine operation can be maintained. Since the boost pumps feed into a common manifold before branching off to the engines and boost pump pressure transmitters, an operative pump will be noted on both boost pump indicators. Therefore, a boost pump pressure reading below normal (30 $\pm$5 psi) is a good indication that one of the boost pumps is inoperative. The power settings on each engine should be reduced as necessary until boost pump pressure reading of 5 psi or greater is obtained.

INTERNAL TANKS TRANSFER SYSTEM

Transfer system failure can usually be attributed to failure of the fuel system to become pressurized. This will only affect external and internal wing fuel transfer. Fuselage fuel is transferred by two electrically driven transfer pumps that run continuously whenever electrical power is applied. In case of complete electrical failure, two hydraulically driven pumps take over. The hydraulically driven pumps will commence transferring fuel when hydraulic power is available with no electrical power (electrical system failure), when either engine is in afterburner or when the fuel low level warning circuit is energized. If the fuel system fails to become pressurized, place the wing transfer pressure switch in the EMER position. This performs the same functions as the landing gear handle switch, all pressure regulators open and all pressure relief valves close.

AIR REFUELING PROBE

On aircraft 155867ak and up, and all others after AFC 370, the air refueling probe can be extended with pneumatic pressure by pulling the refuel probe circuit breaker (G5 before AFC 388, J14 after AFC 388 and before AFC 545, J5 after AFC 388 and AFC 545, No. 1 panel), placing the probe switch to REFUEL, and placing the emergency refuel probe switch to EMER EXT. This extends the refueling probe with canopy air, and cycles the fuel system for refueling. After air refueling is completed, the fuel system is pressurized by placing the refuel probe switch (on the fuel control panel) to EXTEND. The emergency refuel probe switch must stay in EMERG EXT. On aircraft 155903ap and aircraft 157274ap and up, the emergency refuel probe switch is removed from the engine panel and installed on the fuel control panel with a red guard.

CAUTION

To prevent possible damage to the utility reservoir and/or loss of utility pressure, the air refueling probe must be left in the extended position, once it is extended with pneumatic pressure.

CENTERLINE TANK JETTISONING

Jettisoning the centerline tank when it is partially full may result in aircraft damage and severe control problems. If release of the centerline tank is necessary proceed as follows:

1. Maintain flight integrity or attempt to rendezvous with an available aircraft if a single plane flight and conditions permit.
2. Maintain an altitude at or above 5,000 feet if possible.
3. Maintain a wings level attitude and 1 G flight.
4. Establish appropriate release airspeed determined from NATOPS publications.
5. Jettison tank according to the type of situation as defined below.

a. If the tank is ruptured:
   (1) Wingman advise the extent of streaming fuel.
   (2) If fire hazard from streaming fuel is imminent, jettison the tank and be prepared for control difficulties.
   (3) If wingman indicates fire danger is not imminent, allow the tank to drain and land ashore with it empty. If landing ashore is not practicable, jettison the tank after it is empty.
   (4) Do not select afterburner power with fuel streaming from a ruptured tank.

b. If the tank is not ruptured:
   (1) If the tank is partially full, attempt to fill it with fuselage fuel if total fuel available permits. If time or total fuel available preclude such procedures be prepared for aircraft damage and control difficulties after jettisoning the tank.
   (2) Completely empty or completely full tank – establish proper jettison airspeed and jettison the tank.

LIMITATIONS

WING FUEL TRANSFER

Internal wing fuel will not transfer above 75° nose up attitude, or below 15° nose down attitude.

HYDRAULIC POWER SUPPLY SYSTEM

NOTE

Some illustrations referred to within this system writeup may be located in appendix A. These illustrations are referred to, within the text, as (figure A–, appendix A).

DESCRIPTION

Hydraulic power is supplied by three, completely independent, closed center hydraulic systems. They are Power Control System One (PC–1), Power Control System Two (PC–2), and Utility System. The systems have operating pressure of approximately 3000 psi, and are pressurized any time the engines are running. The power control systems supply hydraulic pressure to the dual power control cylinders of the ailerons, spoilers, and stabilator. The utility system supplies hydraulic pressure to the power control cylinder of the rudder, ailerons, spoilers, and to all other hydraulically operated systems. Each system can be pressurized by an external hydraulic power source.

POWER CONTROL SYSTEM ONE (PC–1)

PC–1 (figure A–5, appendix A) is pressurized to 3000 $\pm$250 psi, by a variable volume (18 to 26 gpm), constant pressure hydraulic pump mounted on the left engine. This system supplies hydraulic pressure to one side of the dual power control cylinders of the stabilator and one side of the left aileron and left spoiler. Fluid is supplied to the pump by an airless, pressure loaded, piston-type hydraulic reservoir that has a usable capacity of 0.83 gallons. The reservoir ensures positive hydraulic pressure and fluid supply to the pump suction port, regardless of airplane altitude or flight attitude. A 50 inch accumulator, precharged to 1000 psi, is utilized as a pump surge suppressor, and as a limited source of hydraulic fluid and pressure when system demands exceed pump output. A pressure relief valve protects the system from pressure surges, and limits pressure build-up, by dumping excessive pressures to return. The relief valve begins to relieve at 3250 psi and fully opens at 3850 psi. A pressure transmitter for the PC-1 hydraulic pressure indicator (figure A–1, appendix A), is in a main pressure line. In the event of a loss of system pressure, a CHECK HYD GAGES indicator light and MASTER CAUTION light illuminate. The hydraulic fluid is maintained at a usable temperature by a fuel–hydraulic fluid heat exchanger.

POWER CONTROL SYSTEM TWO (PC-2)

PC-2 (figure A-5, appendix A), is pressurized to 3000 ±250 psi, by a variable volume (18 to 26 gpm), constant pressure hydraulic pump mounted on the right engine. This system supplies hydraulic pressure to one side of the dual power control cylinders of the stabilator and one side of the right aileron and right spoiler. Fluid is supplied to the pump by an airless, pressure loaded, piston-type hydraulic reservoir that has a usable capacity of 0.83 gallons. The reservoir ensures positive hydraulic pressure and fluid supply at the pump suction port regardless of airplane altitude or flight attitude. A 50 cubic inch accumulator precharged to 1000 psi, is utilized as a pump surge suppressor, and as a limited source of hydraulic fluid and pressure when system demands exceed pump output. A pressure relief valve protects the system from pump surges, and limits pressure build-up, by dumping excessive pressures to return. The relief valve begins to relieve at 3250 psi and fully opens at 3850 psi. A pressure transmitter, for the PC-2 hydraulic pressure indicator (figure A-1, appendix A), is in a main pressure line. In the event of a loss of system pressure, a CHECK HYD GAGES indicator light and MASTER CAUTION light illuminate. The hydraulic fluid is maintained at a usable temperature by a fuel-hydraulic fluid heat exchanger.

UTILITY SYSTEM

The utility hydraulic system (figure A-5, appendix A), is pressurized to 3000 ±250 psi by two variable volume (22 to 30 gpm combined), constant pressure hydraulic pumps, one on each engine. To prevent the utility hydraulic pumps from resonating, check valves with different cracking pressures are installed on the pump output lines. As a result, the right engine utility hydraulic pump delivers 2775 ±225 psi at idle rpm, while the left engine utility hydraulic pump delivers 3000 ±250 psi at idle rpm. Fluid is supplied to the pumps by an airless, pressure loaded, piston-type hydraulic reservoir that has a usable capacity of 1.84 gallons. The reservoir insures positive hydraulic pressures and fluid supply at the suction ports of the pumps regardless of airplane altitude or flight attitude. A 50 cubic inch accumulator, precharged to 1000 psi, is utilized as a pump surge suppressor, and as a limited source of hydraulic fluid and pressure when system demands exceed the output of the pumps. A pressure relief valve protects the system from pump surges, and limits pressure build-up by dumping excessive pressures to return. The relief valve begins to relieve at 3250 psi and is fully open at 3850 psi. A pressure transmitter, for the utility hydraulic pressure indicator, is in a main pressure line. If either pump fails, a CHECK HYD GAGES indicator light and MASTER CAUTION light illuminate. The hydraulic fluid is maintained at a usable temperature by two fuel-hydraulic fluid heat exchangers. The utility hydraulic system supplies hydraulic pressure to the:

Aileron Power Control Cylinders
Aileron Dampers
Aileron Droop Cylinders
Aileron-Rudder Interconnect
Air Refueling Probe
Anti-Skid - Aircraft 157242an and up
Arresting Hook (retraction)
Auxiliary Air Doors
Chaff Dispenser Doors
Corner Reflector
Flaps (leading and trailing edge)
Forward Missile Cavity Doors
Fuel Transfer Pumps (hydraulic)
Landing Gear
Lateral Control Servo (autopilot)
Nose Gear Steering
Pneumatic System Air Compressor
Radar Antenna Drive
Rudder Feel/Trim
Rudder Power Control Cylinder
Speed Brakes
Spoiler Power Control Cylinders
Variable Engine Bellmouth
Variable Engine Intake Duct Ramps
Wheel Brakes
Wing Fold
Yaw And Roll Augmentation

HYDRAULIC PRESSURE INDICATORS

Three hydraulic pressure indicators (figure A-1, appendix A), are on the pedestal panel in the front cockpit. Pressure transmitters, one for each system, convert pressure impulses into electrical impulses which, in turn, are supplied to the indicators. The indicators are identical, and cover a pressure range of 0 to 5000 psi and are marked from 0 to 5 with readings multiplied by 1000. Electrical power for PC-1 and PC-2 indicators is supplied by the essential 28 volt ac bus. Electric power for the utility hydraulic pressure indicator is supplied by the right main 28 volt ac bus.

NOTE

Electrical power for the utility pressure indicator is supplied by the right main 28 volt ac bus; therefore, the utility hydraulic pressure gage will be inoperative on RAT power or when the right generator is inoperative and the bus tie is open.

HYDRAULIC SYSTEM INDICATOR LIGHT

An amber CHECK HYD GAGES indicator light is on the telelight panel (figure A-1, appendix A). This single light is utilized by both the power control system and the utility system to indicate loss of hydraulic system pressure and direct the pilot's attention to the hydraulic pressure indicators. Illumination of the CHECK HYD GAGES indicator light is controlled by the hydraulic system pressure switches. The CHECK HYD GAGES light illuminates when the pressure in any one system drops below 1500 ±100 psi and/or when one of the utility hydraulic pumps fail. In all cases a loss of system pressure is noted on the applicable hydraulic pressure indicator, but a failed utility hydraulic pump may not register a significant pressure drop on the utility pressure indicator. However, it can be generally concluded that an illuminated CHECK HYD GAGES light with no noted pressure drop on any of the hydraulic pressure indicators signifies that the right utility hydraulic pump has failed. An illuminated CHECK HYD GAGES light in conjunction with a utility hydraulic pressure drop of 200 psi signifies that the left utility pump has failed. The MASTER CAUTION light will illuminate in conjunction with the CHECK HYD GAGES indicator light. The MASTER

CAUTION light may be extinguished by depressing the reset button. The CHECK HYD GAGES light remains illuminated until the pressure in the faulty system increases beyond 1750 psi. If a failure occurs in one of the remaining hydraulic systems while the CHECK HYD GAGES light is already illuminated, the MASTER CAUTION light does not illuminate again and the pilot is not alerted to the second failure.

NOTE

The MASTER CAUTION light, AUX AIR DOOR light, and CHECK HYD GAGES light may illuminate momentarily when the landing gear is being lowered because of high system demands.

NORMAL OPERATION

Normal operation of the hydraulic system commences with engine operation. Hydraulic pressure indicators are on the pilot's pedestal panel.

EMERGENCY OPERATION

The loss of a hydraulic pressure in either power control system, or in the utility hydraulic system, is indicated by the illumination of the CHECK HYD GAGES light. This single light serves all three systems, and the pilot should check the hydraulic gages to determine which system has malfunctioned. A hydraulic failure in either PC–1 or PC–2 presents no problem in that both systems are independent of each other and the normal operating hydraulic supply system assumes the full demand of the stabilator control cylinder. A hydraulic failure in either PC–1, PC–2, or utility presents no problem in that the systems are independent and the two remaining systems will assume the full demand of the power control cylinders of the ailerons and spoilers. Thus, each flight control hydraulic supply system serves as an emergency system for the other(s). If simultaneous loss of the utility system and one of the power control systems, the operable aileron and spoiler provides adequate lateral control for an emergency landing.

COMPLETE POWER CONTROL SYSTEM FAILURE

The pilot should, upon initial detection of hydraulic power loss, note the trend of failure as to whether the gages show a definite steady drop, or gage fluctuations. With a steady drop indication, hydraulic power will probably not recover. In the event of complete power control hydraulic failure, the aircraft becomes uncontrollable.

UTILITY HYDRAULIC SYSTEM FAILURE

Failure of the utility system will prevent/degrade the hydraulic operation of the following essential items:

Anti–Skid (some aircraft)
Auxiliary air doors
Drooped ailerons (some aircraft)
Flaps
Fuel transfer pumps
Landing gear
Nose gear steering
Rudder
Speed brakes
Spoilers
Variable by–pass bellmouth
Variable engine intake ramps
Wheel brakes

Of the above items, emergency pneumatic operation is provided for the following:

Landing gear
Wheel brakes
Wing flaps
Drooped ailerons (some aircraft)

In addition to emergency (pneumatic) operation of the landing gear, wheel brakes, wing flaps and air refueling probe, back–up or alternate operation is provided for the rudder and fuel transfer pumps, the ailerons and the spoilers. The rudder can be manually operated; however, deflection is entirely dependent upon air loads on the rudder surface. The electrical fuel transfer pumps, the primary means of fuel transfer, continue to operate even though the hydraulic transfer pumps may have failed. The speed brakes can be retracted to a low drag trail position by placing the emergency speed brake switch to RETRACT. The air refueling probe can be extended pneumatically by selecting EMER EXT on the air refuel switch. The power control hydraulic systems act as a back–up for the ailerons and spoilers.

LIMITATIONS

No specific limitations pertain to the hydraulic power supply system.

INSTRUMENTS

NOTE

Some illustrations referred to within this system writeup may be located in appendix A. These illustrations are referred to, within the text, as (figure A–, appendix A).

DESCRIPTION

Most of the instruments are electrically operated by power from the electrical system, see figure A–3, appendix A. Some instruments, such as the accelerometer, are self–contained and do not require external power. Only the instruments which are not covered under another system are discussed herein.

TRUE AIRSPEED INDICATOR

A true airspeed indicator (figure A–1, appendix A), is on the pilot's and RIO's instrument panels. The airspeed is indicated by a small counter which rotates to show a row of numbers through a window on the indicator face. The airspeed indicator is read directly in knots TAS; the range of the instrument is from 0 to 1500 knots to the nearest knot. The true airspeed system calibrated range is 150 to 1500 knots. Therefore, true airspeed readings below 150 knots are not reliable. The true airspeed indicator may indicate between 60 and 180 knots while the airplane is motionless on the ground. The true airspeed outputs are produced from the signal from the total temperature sensor of the ADCS by routing this signal through a potentiometer driven by one of the Mach number function cams. Thus, Mach number is translated into true airspeed.

ALTIMETER

An altimeter (figure A–1, A–2, appendix A), on the pilot's and RIO's instrument panel indicates the altitude of the airplane above sea level. This unit is of the counter pointer type which displays the whole thousands numbers in a counter window and indicates the increments of the whole number with a pointer which rotates on the face of the instrument. The pointer scale is graduated in 50 foot units with major 100 foot scale divisions from 1 to 10. The range of the altimeter is 0 to 80,000 feet. An adjustable barometric scale is provided so that the altimeter may be set to a sea level pressure. This scale range is from 28.50 to 30.90 inches mercury. The altimeter is positioned by a corrected static pressure supplied by the ADC, and no position error correction need be made to the indicated altitude as long as the static correction source is operative. In the event static correction monitors off, (illumination of STATIC CORR OFF light) the altimeter indications will be in error.

NOTE

Even though the ADCS supplies corrected static pressure to the altimeters, it is normal for the altimeter to fluctuate when passing through the transonic speed range. Refer to NAVAIR 01–245FDB–6 for allowable tolerances.

SERVOED ALTIMETERS (AN/AAU–19/A)

NOTE

On aircraft 158366au and up, and all others after AFC 496, when operating aircraft with the servoed altimeter installed, the RESET position can be used. On aircraft before AFC 496 use the STBY mode only.

Aircraft 153780ac and up, and all others after AFC 388 have a servoed altimeter in each cockpit (figure A–1, A–2, appendix A), instead of the conventional pressure altimeter. The altimeter consists of a precision pressure altimeter combined with a servomechanism. Altitude is displayed in direct digital form by a 10,000–foot counter, a 1000–foot counter, and a 100–foot drum. A single pointer also indicates hundreds of feet on a circular scale, with center graduations of 50 feet. Below 10,000 feet a diagonal warning symbol appears on the 10,000–foot counter. A barometric pressure setting (baroset) knob is provided to insert barometric pressure in inches of Hg. The altimeter has a primary (servoed) mode and a standby (pneumatic) mode, controlled by a mode switch on the altimeter. The mode switch is spring–loaded to the center position and has momentary positions of RESET and STBY. The primary mode is selected by positioning the mode switch to RESET for 1 to 3 seconds, provided aircraft power is available. Standby operation can be selected by placing the mode switch to STBY. The altimeter will automatically switch to standby mode if electrical power is interrupted longer than 3 seconds, or a failure occurs in the altimeter, altitude encoder unit, or air data computer. During standby operation a red STBY flag appears on the dial face. In primary (servoed) mode the altimeter is corrected automatically by signals from the air data computer. Synchro signals representing pressure altitude corrected for position error are supplied from the air data computer to the altimeter servomechanism via the altitude encoder unit. The servomechanism drives the counter, drum, and pointer to provide an accurate altitude readout. In standby (pneumatic) mode the altimeter functions the same as a standard pressure altimeter. Whether the altitude readout is corrected for position error depends on the status of the static pressure compensator in the air data computer. Refer to figure 1–24 for status of altimeter and altitude reporting function of the transponder (mode C) with various failures. The altimeter is set and checked in the same manner as a conventional pressure altimeter. Refer to section III for altimeter check.

CAUTION

To prevent stripping of barometric setting gear, do not force settings beyond 28.50 inches minimum and 30.90 inches maximum.

NOTE

During normal use of the baroset knob, if momentary locking of the counters is experienced do not force the setting. Application of force may cause internal gear disengagement and result in

ALTITUDE REPORTING FAILURE INDICATIONS

AIRCRAFT 153780ac AND UP; OTHERS AFTER AFC 388

TYPE OF FAILURE	ALTIMETER (AAU-19/A)	ALT ENCODER OUT LIGHT	STATIC CORR OFF LIGHT	MASTER CAUTION LIGHT	RESULTING SYSTEM OPERATION
ALTIMETER SERVO FAILURE OR MANUAL STBY SELECTION	STBY	OFF	OFF	OFF	ALTIMETER REVERTS TO PNEUMATIC OPERATION ON CORRECTED STATIC PRESSURE.
ALTITUDE ENCODER UNIT FAILURE OR AIR DATA COMPUTER LPC MODULE FAILURE	STBY	ON	OFF	ON	ALTIMETER REVERTS TO PNEUMATIC OPERATION ON CORRECTED STATIC PRESSURE, AND NO ALTITUDE INFORMATION SUPPLIED TO ALTITUDE REPORTING TRANSPONDER.
AIR DATA COMPUTER SPC FAILURE OR MANUAL OFF SELECTION	STBY	ON	ON	ON	ALTIMETER REVERTS TO PNEUMATIC OPERATION ON UNCORRECTED STATIC PRESSURE, AND NO ALTITUDE INFO SUPPLIED TO ALTITUDE REPORTING TRANSPONDER.

FDD-1-(11)

Figure 1-24

excessive altitude error in both primary and standby modes. If locking occurs, rotate the knob a full turn in the opposite direction and approach the setting again with caution.

Altimeter Operating Characteristics

The counter-drum-pointer display is designed so that the 100-foot drum and the pointer rotate continuously during altitude changes, while the 10,000-foot and 1000-foot counters remain in a fixed position. When each 1000-foot increment is completed the counter(s) abruptly index to the next digit. However, when operating in the standby mode, there may be a noticeable pause or hesitation of the pointer caused by the additional friction and inertia loads involved in indexing the counter(s). The momentary pause is followed by a noticeable acceleration as the counters change. This occurs in the "9" to "1" sector of the scale and is more pronounced when the 10,000-foot counter changes. It is also more pronounced at high altitudes and high rates of climb and descent. During normal rates of descent, at low altitudes, the effect is minimal.

NOTE

If the altimeter internal vibrator is inoperative, the pause and accelerate effect may be exaggerated in the standby mode. If operating in standby mode, be alert for this behavior when minimum approach altitude is within the "8" to "2" sector of the scale; i.e., 800-1200 feet, 1800-2200 feet, etc.

AIRSPEED AND MACH NUMBER INDICATOR

The combination airspeed and Mach number indicator (figure A-1, appendix A) shows airspeed reading at low speeds and includes Mach number readings at high speeds. Both readings are provided by a single pointer moving over a fixed airspeed scale, graduated from 80 to 850 knots, and a rotatable Mach number scale graduated from Mach 0.4 to Mach 2.5. A movable bug is included as a landing speed reference and can be positioned by the knob on the face of the instrument. The same knob can position another bug on the Mach number scale for maximum indicated airspeed reference. The airspeed indicator pointer and the Mach number scale are synchronized so that a proper relationship between the two is assured throughout all altitude changes. Thus, at sea level and under standard conditions, the pointer indicates Mach 1 approximately 660 knots. Under the same conditions, but at 50,000 feet, if the same true airspeed is maintained the pointer indicates approximately 292 knots and a Mach number of 1.15. The airspeed and Mach number indicator utilizes a corrected static pressure source from the ADC which eliminates the need for instrument position error correction. Therefore, the airspeed and Mach number indicator displays calibrated airspeed (CAS) and true Mach number (TMN) when the static pressure correction portion of the ADC is operative. In the event the static correction source is lost (illumination of STATIC CORR OFF light) the airspeed and Mach number indicator reverts to displaying indicated airspeed (IAS) and indicated Mach number (IMN).

VERTICAL VELOCITY INDICATOR

A vertical velocity indicator (figure A–1, appendix A), is on the pilot's instrument panel. The indicator shows the rate of ascent or descent of the airplane, and is so sensitive that it can register a rate of gain or loss of altitude which would be too small to cause a noticeable change in the altimeter reading. The upper half of the indicator face is graduated in 500 foot units from 0 to 6000 feet with 100 foot scale divisions from 0 to 1000 feet. The upper half of the instrument indicates rate of climb in thousands of feet per minute. The lower half of the indicator face is identical to the upper half except that it indicates rate of descent in thousands of feet per minute. The vertical velocity indicator is connected to the corrected static pressure system of the airplane and measures the change in atmospheric pressure as the airplane climbs or descends. Since the vertical velocity indicator utilizes corrected static pressure, a failure of the corrected static pressure source (illumination of STATIC CORR OFF light) may result in a slightly erroneous vertical velocity indication.

TURN–AND–SLIP INDICATOR

A slip inclinometer and rate of turn needle is incorporated into the attitude director indicator (figure A–1, appendix A), on the front cockpit instrument panel of all airplanes. A conventional turn and slip indicator (figure A–2, appendix A), is on the rear cockpit instrument panel of dual control airplanes. The rear cockpit indicator displays a 90° per minute turn rate for a single needle width deflection (4 minute turn). The turn indicator consists of an electrically driven gyro, linked to a pointer needle. When the needle is off center, it indicates that the airplane is turning in the direction shown by the needle. The amount the needle is off center is proportional to the rate of turn. The slip indicator is a ball type inclinometer. The ball is in a curved, fluid filled tube. When the airplane is flying straight and level, the ball is centered in the tube by its own weight. When the airplane is making a turn, the ball is acted upon by centrifugal and gravity forces. During a coordinated turn, both forces balance out to hold the ball centered in the tube. The ball, therefore, indicates proper lateral attitude for any rate of turn as well as for straight and level flight. The front cockpit turn and slip indicator, however, displays turn rate about the vertical axis of the airplane, and as a result, does not provide an accurate indication of actual airplane rate of turn. When the airplane bank angle is increased, the rate of turn needle shows increased turn rate up to a point, beyond which the rate of turn around the vertical axis of the airplane decreases. This characteristic will normally preclude obtaining a full needle width deflection at high altitudes and/or high airspeeds. Therefore, the front cockpit rate of turn needle should not be used as a primary turn rate instrument. At low airspeeds, however, a single needle width deflection is approximately equivalent to a two minute turn. The turn needle is deflected electronically, and therefore an electrical failure renders the needle immediately inoperative, despite the fact that the gyro is still spinning.

ACCELEROMETER

An accelerometer, to measure and record positive and negative acceleration G loads, is on the front cockpit instrument panel (figure A–1, appendix A). The indicator has three movable pointers. One pointer moves in the direction of the G load being applied, while the other two (one for positive G, and one for negative G) follow the indicator pointer to its maximum travel. These recording points remain at their respective maximum travel position of the G load being applied. Depressing a PUSH TO SET button, in the lower left corner of the instrument allows the recording pointers to return to the one G position.

NOTE

Accelerometers may read as much as 0.5 G low; possibly lower if the pull–in rate is high.

STANDBY ATTITUDE INDICATOR

A standby attitude indicator, on the pilot's main instrument panel (figure A–1, appendix A), and a remote attitude indicator, on the RIO's instrument panel (figure A–2, appendix A), are identical except for size. Attitude information is supplied by the AN/AJB–7, and is limited to pitch and roll. After AFC 478, a gyro fast erect switch on the main instrument panel may be used to increase the erection rate of both attitude indicators. The indicators display an OFF flag when power is interrupted.

STANDBY COMPASS

A conventional magnetic compass on the cockpit windshield frame is provided for navigation in event of instrument or electrical malfunction. Compass deviation cards are above the canopy sill on the right side of both cockpits.

RADAR ALTIMETER SET

The radar altimeter set is a pulsed range–tracking radar, that provides the pilot with accurate terrain clearance information from 0 to 5000 feet, within ±5 feet or ±5 percent of the indicated altitude, whichever is greater. The set functions normally up to 30° bank angles and 50° pitch angles. The set consists of two identical antennas, a receiver–transmitter unit, a r–f switching unit, and a height indicator. The receiving antenna is on the lower left front fuselage near the left inboard leading edge flap, and the transmitting antenna is on the corresponding spot on the starboard side of the fuselage.

Height Indicator

The height indicator, on the left side of the pilot's main instrument panel (figure A–1, appendix I) provides the read–out for the set. The face of the indicator contains a dial scale, an altitude pointer, a movable low altitude index pointer, and an OFF flag. The dial scale is logarithmic throughout its range. The OFF flag indicates that power is not supplied to the set, that the 5000–foot altitude range has been exceeded, or that the altitude indication is unreliable.

Function Control Knob

The functional control knob, on the lower left side of the height indicator, provides complete control of the set. By rotating the knob clockwise past the off detent, power is supplied to the set. Rotating the knob further clockwise positions the low altitude pointer. Pushing in on the knob activates a self test function, when airborne, provides the set with an artificial return signal. The altitude pointer moves to 5 ±5 feet. The test function is activated continuously on the ground by a scissors switch on the right main landing gear.

Low Altitude Warning Light

A red, low altitude warning light is directly below the front cockpit true airspeed indicator. The light illuminates any time the aircraft descends below the altitude set on the low altitude index pointer.

NORMAL OPERATION

RADAR ALTIMETER SET

With electrical power supplied to the airplane, rotate the function control knob clockwise to turn the set on. Move the knob further clockwise to set the low altitude index pointer as desired. After a 3 minute warm-up period, the set is ready for operation. As the airplane ascends through 5000 feet (approximately), the OFF flag becomes visible and above 14,000 feet the transmitter is disabled by a barometric pressure switch. The set test function may be activated any time the airplane is airborne below 14,000 feet.

EMERGENCY OPERATION

There are no specific emergency operations pertaining to the instruments.

LIMITATIONS

RADAR ALTIMETER SET

The set is limited to 5000 feet altitude range and 30° bank angle/50° pitch angle.

WARNING

High frequency radar waves can penetrate snow and ice fields. When operating in areas covered with snow and ice, the radar altimeter may indicate a greater terrain clearance than actually exists.

INSTRUMENT LANDING SYSTEM (AN/ARA-63)

DESCRIPTION

Aircraft 157309ar and up and all others after AFC 470, Pt I, contain an AN/ARA-63 instrument landing system (ILS). This system can be used for primary manual instrument landing approach, or it can be used to monitor the automatic landing performance of an automatic landing approach system. The ILS is used with carrier based azimuth and elevation transmitters. The major components of the AN/ARA-63 system are a receiver, decoder, and a control panel.

SURFACE TRANSMITTERS

The AN/ARA-63 ILS receives continuous coded angular guidance azimuth and elevation signals from the carrier based AN/SPN-41 transmitters. These signals maintain the aircraft on the approach path toward the touchdown point. The transmitted elevation scan pattern is from ground 0 to 10° looking up. Proportional angle steering in elevation is displayed ±1.4° from the 3° glide slope. The azimuth channel sweeps ±20° from an established reference. The reference used is the carrier deck centerline. For azimuth, proportional angle steering is displayed between ±6°. At a range of 20 miles, the azimuth scanning beam sweeps an area approximately 8 miles wide, while the elevation or glide path is about 4 miles high.

ILS RECEIVER

The ILS receiver receives coded transmissions of azimuth and elevation guidance data from the surface transmitters. The receiver transforms these coded signals to video pulses suitable for processing in the decoder.

ILS DECODER

The ILS decoder receives the azimuth and elevation video pulses from the receiver, and converts them to azimuth and elevation command signals which drives the pitch and bank steering bars and their associated warning flags on the ADI.

Azimuth Command Signals

If the decoder receives azimuth signals of sufficient strength for tracking, the signals control the movement of the bank steering bar, and the vertical director warning flag (at the 12 o'clock position) deflects out of view. If the decoder receives weak azimuth signals or no signals, the bank steering bar deflects to the right side of the ADI, and the warning flag comes into view.

Elevation Command Signals

If the decoder receives elevation signals of sufficient strength for tracking, the signals control the movement of the pitch steering bar, and the vertical displacement warning flag (at the 9 o'clock position) deflects out of view. If the decoder receives weak elevation signals or no signals, the pitch steering bar deflects to the upper portion of the ADI, and the warning flag comes into view.

ILS CONTROLS AND INDICATORS

The controls and indicators for the ILS are on the AN/ARA-63 control panel. The ADI and the navigation function selector panel on the front cockpit main instrument panel are used but are not part of the AN/ARA-63 system.

ILS CONTROL PANEL

The ILS control panel is on the right console in the front cockpit (figure A-1, appendix A). The panel contains an on-off power switch, a power on indicator light, a built-in-test pushbutton, and a channel selector knob.

Power Switch

The power switch is a lock-type toggle switch with positions of ON and OFF. Placing the switch to ON applies power to the ILS.

Power On Indicator Light

The power on indicator light illuminates green when the power switch is placed to ON.

Built-In-Test

The BIT pushbutton is depressed to check for correct system operation. If the system is operating properly the bank steering bar on the ADI slowly oscillates one half scale to the left and then to the right. The pitch steering bar indicates the glide slope (3°).

Channel Selector Knob

The channel selector knob is a rotary type knob that can select any one of 20 channels.

NAVIGATION FUNCTION SELECTOR PANEL (ILS)

On aircraft 157309ar and up and all others after AFC 470 Pt I, the navigation function selector panel has a DL/ILS position added. Placing the mode selector knob to the DL/ILS position enables the ILS system to operate in a manual instrument landing approach, or to monitor an automatic landing approach during which the aircraft is receiving signals from the data link and AN/SPN-42 systems. At the time the automatic approach is being monitored the pitch and bank steering bars on the ADI are responding to signals from the AN/ARA-63 system.

NORMAL OPERATION

The AN/ARA-63 ILS is operational when the ILS power switch is placed to ON. To receive the transmitted azimuth and elevation signals from the surface transmitters, set the mode selector knob to DL/ILS and set the channel selector knob to the correct channel for receiving the incoming signals. When the aircraft is within the range of the transmitters, the bank steering bar and the pitch steering bar on the ADI indicate in which direction the aircraft must be flown to line up with the fixed approach path to the carrier deck. If the pitch steering bar is below the miniature wings on the ADI and the bank steering bar is to the right of center, the aircraft must be flown down and to the right to attain optimum glide path. When the aircraft is lined up in azimuth and elevation, both steering bars on the ADI are centered. When the aircraft reaches a point within ½ mile of the carrier deck at a 200-foot altitude, use the Fresnel Lens Optical System, or go visual with the deck as a reference.

LANDING GEAR SYSTEM

NOTE

Some illustrations referred to within this system writeup may be located in appendix A. These illustrations are referred to, within the text, as (figure A-, appendix A).

DESCRIPTION

The airplane is equipped with fully retractable tricycle landing gear which are completely covered by flush doors when retracted. The gear is electrically controlled by the right main 28 volt dc bus and hydraulically actuated by the utility hydraulic system. The airplane is not equipped with a tail skid. Accidental retraction of the landing gear when the airplane is on the ground is prevented by safety switches on the main gear torque scissors, and ground safety locks.

MAIN GEAR

Each main gear is hydraulically retracted and extended. When the gear handle is in the UP position and the weight is off the gear, the gear will retract. As the main gear retracts, the wheels are automatically braked to a stop by the antispin system and the struts are mechanically compressed. When the gear is up and locked, pressure is automatically released from the antispin system. The struts automatically return to the normally extended position when the gear is extended. The gear is locked down by an internal finger-type latch in the side brace actuator. The main gear retracts inboard and is enclosed by fairing doors that are flush with the underside of the wing. The gear is locked up by a hydraulically actuated mechanism. All main gear doors remain open when the gear is extended.

NOSE GEAR

The nose gear is hydraulically retracted and extended. The gear is locked in the down position by an integral downlock mechanism within the gear actuating cylinder. A hydraulically operated nose gear uplock cylinder is located in the nose gear wheel well, and is employed in the system as part of the nose gear up latch mechanism. The nose gear retracts aft into the fuselage and is covered by mechanically operated doors that close flush with the underside of the fuselage. The forward door is attached to the nose gear strut, and closes with retraction; the aft door is operated and latched closed by the gear uplatch mechanism. The nose gear is equipped with dual nose wheels, a combination shimmy damper steering actuator and a self-centering mechanism. The nose gear can be steered by differential braking of the main gear wheels in the event nose wheel steering is not utilized.

LANDING GEAR CONTROL HANDLE

Operation of the landing gear is controlled by a handle (figure A-1, appendix A) at the left side of the main instrument panel. The handle has a wheel shaped knob for ease of identification. Placing the handle in the UP or DOWN position energizes a solenoid valve to connect system pressure to the landing gear. Placing the handle in the gear UP position energizes switches in the fuel tank vent and pressurization, jettison and armament circuit. A red warning light is located in the landing gear control handle knob. This light comes on whenever the control handle is moved to retract or extend the gear and it remains on until the gear completes its cycle and locks.

Emergency Landing Gear Control

Two 100 cubic inch air bottles provide sufficient compressed air to extend the gear pneumatically in the event of a hydraulic system failure. Pulling the landing gear control handle full AFT, when it is in any position, operates an air valve which directs 3000 psi compressed air to open all gear doors, release the uplocks, and extend all gear.

Landing Gear Position Indicators

The landing gear position indicators (figure A-1, appendix A) are on the left vertical panel in the front cockpit. The indicators operate in conjunction with position switches on the landing gear. The position of the landing gear wheels is indicated by drum dials viewed through cutouts in the instrument panel. With gear up, the word UP appears on three indicators; gear in transient is indicated by a barber pole; and with gear down, a picture of a wheel is seen through the cutouts.

Landing Gear Warning Lights

The WHEELS warning light flashes any time the flaps are down and the landing gear handle is UP. The landing gear handle light illuminates any time any gear is unlocked.

NOSE GEAR STRUT EXTENSION

The nose gear strut extension system is utilized to increase the airplane angle of attack for catapult launches. A nose gear extension switch in the left main gear wheel well operates a solenoid valve that ports high pressure air into a chamber above the nose gear shock strut piston. The shock strut piston is then forced down to extend the nose gear. The high pressure air is dumped when the left main gear strut extends after launching, or when the landing gear handle is moved to the UP position. The nose gear strut extension chamber may also be deflated by momentarily placing both generator control switches to OFF.

CAUTION

- If the angle of attack indexer lights and approach lights are not illuminated during a landing approach, it may mean that the nose gear catapult extension is still pressurized. To preclude the possibility of exploding the nose gear strut upon touchdown, relieve the catapult extension pressure by cycling the landing gear.
- During normal operations, the airplane's pneumatic system must be fully charged (2750 psi minimum) before extending the nose gear strut and must indicate a minimum of 1350 psi after extension. Pneumatic system pressures below these values may not be capable of extending the strut enough to give the aircraft the proper angle of attack for a catapult takeoff. Insufficient pneumatic pressure may allow the strut to bottom out causing damage to the strut or fuselage structure.
- Do not allow the pneumatic system pressure to exceed 2300 psi with the nose gear strut extended. If the pneumatic pressure approaches this value, actuate the emergency air brakes as necessary to maintain the pressure below 2300 psi. If the pneumatic pressure exceeds 2300 psi the emergency brakes will not release the pressure in the nose strut. To release this excess pressure the nose strut will have to be deflated and then

re-inflated. Allowing the pneumatic system pressure to exceed 2300 psi subjects the nose strut to excessive loads during catapulting.

NOTE

See Nose Strut Extension Pressure Minimums Chart, section III, for lower than normal launch weights.

NOSE WHEEL STEERING

An electrically controlled, hydraulically operated nose gear steering system is installed in the aircraft. It provides directional control of the aircraft during ground operation in two modes; nose gear steering and shimmy damping. For nose gear steering, the rudder pedals, through mechanical linkage, control a variable gain electrical output from the command potentiometer. Low gain (for fine, more precise steering) occurs near rudder pedal neutral and increases non-linearly in the command potentiometer to high gain (for coarse, quick steering) near full rudder pedal deflection. The first 5° of rudder pedal deflection from neutral deflects the nose wheel approximately 3°, but the last 5° of rudder pedal deflection deflects the nose wheel approximately 40°. Steering is limited to approximately 70° either side of center. The control unit receives the signal from the command potentiometer and electrically selects a servo valve setting in the utility hydraulic system. This electrical sub-system is energized and the servo setting is continually following the rudder pedals when there is electrical power in the aircraft, the nose gear is not locked up, and there is weight on the right main landing gear. Hydraulic pressure to turn the nose wheel is provided when the nose gear steering button on the control stick (figure 1-21) is depressed. This button energizes a relay which opens the selector valve in the hydraulic system. Hydraulic pressure is supplied to the servo valve which moves to the selected setting and ports hydraulic pressure to the power unit, a rotary vane hydraulic motor, on the nose gear strut. The power unit turns the nose wheel through a geared strut torque collar. As the nose wheel turns, the follow-up potentiometer on the power unit balances the electrical circuit in the control unit so that the servo valve closes as the nose wheel reaches the position commanded by the rudder pedals. Releasing the nose gear steering button also closes the servo valve and removes hydraulic pressure from the power unit. A failure detection circuit will shut off hydraulic power to the system upon detection of an electrical short or open. However, the failure detection circuit requires a definite time to operate and may not preclude some degree of nose deflection. Should the electrical fault be removed, nose gear steering will again be operative by depressing the nose gear steering button. Shimmy damping is automatically activated whenever the nose gear steering button is released. A restrictor, bypassing the selector valve, allows a balanced pressure of 275 psi (regulated by the compensator) to remain in the power unit. This pressure prevents cavitation of the rotary vane hydraulic motor as the nose wheel swivels. Fluid flow is metered through one-way restrictors in the power unit to damp wheel shimmy. The nose wheel can swivel to any direction in this mode.

CAUTION

To prevent the landing gear struts from being subjected to abnormal side loads, do not use nose wheel steering and brakes simultaneously while in a turn.

NORMAL OPERATION

Operation of the landing gear is controlled by the wheel shaped landing gear control handle. To lower the landing gear, push the handle down. A red warning light in the control handle knob illuminates and stays illuminated until the gear is fully extended and locked. To raise the gear, move the landing gear handle up; the warning light again illuminates until the landing gear is up and locked.

FLIGHT WITHOUT MAIN LANDING GEAR DOORS

If maintenance or operational consideration require flight without main landing gear doors, the airplane is limited as follows:

a. Below 20,000 feet – 250 knots CAS.
b. Between 20,000 and 35,000 feet – Mach 0.85.
c. Above 35,000 feet – 250 knots CAS or Mach 0.85, whichever is greater.
d. Descent – 250 knots CAS or onset of any buffet.
e. After each flight, wheel wells should be inspected for evidence of cracks or malformed lines and fittings.

EMERGENCY OPERATION

If normal gear operation fails, the gear can be lowered by pushing the landing gear handle down pulling the landing gear circuit breaker and then pulling aft on the landing gear handle. The landing gear circuit breaker must be pulled prior to lowering the gear by the emergency system. This causes the landing gear hydraulic selector valve to return to its full trail position, blocking hydraulic pressure to the landing gear and insuring that hydraulic fluid will not be forced into the actuators on top of the pneumatic pressure. Should this occur, system hammering may result, with possible eventual rupture and loss of system integrity. Hold the handle aft until the gear indicates down and locked. Do not retract the landing gear following an emergency extension. If the landing gear is inadvertently extended in flight by emergency pneumatic pressure, it must be left in the extended position until post-flight servicing.

CAUTION

Hold handle in full aft position until gear indicates down and locked, and then leave the landing gear handle in the full aft position. Returning the handle to its normal position allows the compressed air from the gear down side of the actuating cylinder to be vented overboard. In this condition the main landing gear side brace integral mechanical latch will be the only device preventing the landing gear from collapsing upon landing. Pull the landing gear circuit breaker prior to extending the landing gear by the emergency system.

NOTE

Any pneumatic extension of the landing gear shall be logged on the Yellow sheet (OPNAV Form 3760–2).

LIMITATIONS

Maximum permissible airspeed for lowering of the landing gear is 250 knots CAS.

LIGHTING EQUIPMENT

NOTE

Some illustrations referred to within this system writeup may be located in appendix A. These illustrations are referred to, within the text, as (figure A–, appendix A).

DESCRIPTION

EXTERIOR LIGHTING

The exterior lights consists of the position lights (wing and tail), join–up lights (wings only), fuselage lights, anti–collision light, angle–of–roll light, approach lights, air refueling probe light and taxi light. An exterior lights master switch on the outboard left throttle grip which controls some of these lights is also utilized in the operation of night catapult launches. The exterior lights control panel on the right console in the front cockpit contains all of the manual controls, with the exception of the master switch, for the exterior lights.

Exterior Lights Master Switch

The exterior lights master switch provides a master control for the following exterior lights: position lights, join up lights, fuselage lights and the anti–collision light. Before any of these lights can be operated, the master switch must be in either the ON or the SIGNAL position. Further control of these lights is accomplished from the exterior lights control panel. The switch, on the outboard throttle grip, has three positions: ON, OFF and SIGNAL. Placing the switch to ON energizes the switches on the exterior lights control panel for the above lights. The SIGNAL position functions identically to the ON position, except that it is spring–loaded to OFF. In addition to controlling the above exterior lights, the ON and SIGNAL positions dim the approach lights should they be illuminated. The master switch and exterior lights panel receive power from the right main 28 volt dc bus.

Position and Join Up Lights

The position lights include the wing tip position lights and the tail lights. The join–up light consists of a red or green light on the trailing edge of the applicable wing tip. The wing tip position lights and join–up lights are both operated by the exterior lights master switch on the outboard throttle and the WING switch on the exterior lights control panel. With the exterior lights master switch to ON or SIGNAL, the lights are controlled by BRT, DIM and OFF positions of the WING switch. The wing and join–up lights do not have flash capabilities. The tail light is controlled by the exterior lights master switch, TAIL switch, and FLASHER switch on the exterior lights control panel. With the exterior lights master switch to ON or held in SIGNAL, the tail light can be controlled by the BRT, DIM and OFF position of the TAIL switch. The tail light will flash or illuminate steady depending on whether the flasher switch is in the STEADY or FLASH position. The dim circuit of the position and join–up lights is powered by the right main 14 volt ac bus. The bright lights circuit is powered by the right main 28 volt ac bus.

Fuselage and Anti–Collision Lights

Three semi–flush white lights are installed on the fuselage, one above the number two fuel cell and one below each of the engine air inlet ducts. In addition to the fuselage lights, one red anti–collision light is installed in the leading edge of the vertical stabilizer. The anti–collision and fuselage lights are controlled by the fuselage lights switch labeled FUS, with positions marked DIM, MAN and BRT. The three fuselage lights illuminate in the DIM, and BRT positions in conjunction with the exterior lights master switch. They illuminate steady or flashing depending on the position of the flasher switch. The anti–collision light illuminates only when the fuselage switch is in BRT position and the flasher switch is in the FLASH position. The light does not function in the steady condition or with the FUS switch in MAN or DIM positions. The MAN position of the switch allows the fuselage lights to be energized by the manual key button. The fuselage lights are powered by the right main 28 volt ac bus. Both of the lamps in the anti–collision light are powered by the right main 28 volt ac bus.

Exterior Lights Flasher Switch

The exterior lights flasher switch on the exterior lights control panel has two positions, FLASH and STEADY. If STEADY is selected the tail light and fuselage lights produce a steady illumination, provided the FUS light switch and TAIL light switch are in DIM or BRT positions. Placing the flasher switch to FLASH causes the fuselage lights and tail lights to flash, and anti–collision lights to flash if the FUS switch is in BRT. The flasher unit for this switch receives power from the right main 28 volt dc bus.

Manual Key Button

The manual key button, on the exterior lights control panel, energizes the fuselage lights when the fuselage lights switch is in the MAN position. With the key button depressed and the exterior lights flasher switch in STEADY, the fuselage lights illuminate steady; with the FLASHER switch in FLASH, the fuselage lights flash. An indicator light on the exterior lights control panel glows when the manual key button is depressed. The manual key button receives its power from the right main 28 volt ac bus.

Angle of Roll Light

The low, swept wing design of the airplane prevents the landing signal officer from observing the right wing tip light during a normal carrier approach until the airplane is almost on final; therefore, a green angle of roll light is installed aft of the left engine air intake duct just above the trailing edge of the wing. This light illuminates with a steady glow and serves as a roll reference for the landing signal officer until such time that the right wing tip is visible. The angle of roll light illuminates steady, in flight, with the landing gear down, the flaps 1/2 or full down, and the arresting hook down. With the landing gear down and locked, the flaps 1/2 or full down, and the arresting hook up, the angle of roll light will flash (unless the hook bypass switch is in the BYPASS position). The angle of roll light receives power from the right main 28 volt ac bus.

Hook Bypass Switch

The hook bypass switch on the exterior lights control panel has two positions: NORMAL and BYPASS. This is the only operating control in the cockpit for the approach lights and the angle of roll light systems. The switch, when placed in the BYPASS position, completes a circuit which causes the approach lights to illuminate steady without having the arresting hook extended. With the switch in the NORMAL position, and with the gear down and the flaps in full or half down, the approach lights flash unless the arresting hook is down. With the arresting hook down, the approach lights illuminate steady. Also, with utility hydraulic and electrical power applied, if the landing gear control handle is in the DOWN position and the hook bypass switch is in the BYPASS position, the corner reflector extends.

Approach Lights

Refer to Angle–of–Attack System, this section.

Air Refueling Probe Light

The inflight refueling probe light is on the right side of the fuselage forward of the air refueling probe. The light is used during night air refueling operations to illuminate the refueling probe, and the drogue from the refueling airplane. The light is controlled by the IFR switch and variable intensity control knob, both of which are located on the exterior lights control panel. The inflight refueling probe light receives power from the right main 115 volt ac bus.

Taxi Light

The taxi light is adjacent to the approach light assembly on the nose gear door. The light is controlled by the taxi light switch on the exterior light control panel. Power to the light is supplied by the right 28 volt ac bus.

PILOT'S INTERIOR LIGHTING

Interior lighting in the airplane is powered by the ac electrical system, either from the engine driven generators or by the emergency generator. Most of the pilot's interior lighting controls are on the cockpit lights control panel (figure A-1, appendix A).

Instrument Lights

The instruments are illuminated by integral instrument lights. Variations in instrument lighting intensity on all airplanes is controlled by the instrument panel lights control knob on the forward inboard corner of the cockpit lights control panel. The control knob varies the brilliance of the instrument lights from OFF to BRT. Also, as the control is rotated from OFF to BRT, a switch within the the control knob energizes the warning lights dimming relay, reducing the brilliance of the warning lights in both cockpits. After AFC 536 integral lighting for the following flight instruments is no longer controlled by the instrument panel lights control knob: altimeter, airspeed/mach indicator, vertical velocity indicator, angle of attack indicator, horizontal situation indicator, attitude director indicator and standby attitude indicator. Individual intensity control of the above lights is provided by seven controls on the flight instrument lights balance control panel added above the right console under the canopy sill. Prior to this change the intensity of the standby attitude indicator lighting is less than the other flight instrument lights. With the flight instrument lights balance controls installed, the standby attitude indicator lighting can be balanced with the other flight instrument lights. AFC 536 also installs the flight instrument lights knob on the engine control panel on the left console. The flight instrument lights knob simultaneously varies the intensity of the above seven lights so that they can be balanced with the remaining instrument lighting still controlled by the instrument lights knob on the right console. The flight instrument lights knob varies the flight instrument lights from OFF to BRT. In addition, a switch within the knob dims the HSI mode lights and also takes over the function of dimming the warning lights from the instrument lights knob. In the event of normal instrument lighting system failure, secondary instrument lighting is provided by red floodlights on the glare shield. These floodlights are controlled by the INSTR PANEL EMER FLOOD switch mounted above the cockpit lights control panel. The three position switch labeled OFF, DIM and BRT provides only bright or dim positions.

Console Lights

Console lighting is combination edge and floodlighting. Variation in edge lighting intensity is controlled by the console lights control knob on the cockpit lights control panel. This knob controls all edge lighting on the left and right console, the pedestal panel and the armament control panel. The console control knob varies the brilliance of the console edge lights from OFF to BRT. Also, as the control is rotated from OFF to BRT, a switch within the control knob energizes the DIM position of the console floods switch thus providing console floodlight illumination and edge–lighting. The console floods switch above the console control knob selects BRT, DIM or MED brilliance for the red console floodlights. The console floodlights are off only when the console floods switch is in the DIM position and the console control knob is rotated to the OFF position.

White Floods Switch

One white floodlight is provided above each console under the canopy sill. Control is by the white flood switch in the forward outboard corner of the cockpit lights control panel. This switch is of the lever–lock type to prevent inadvertent operation. No intensity variation is provided on these lights.

Standby Compass Switch

The standby compass switch on the cockpit lights control panel turns the standby compass light ON and OFF. The console lights control knob must be turned ON before the standby compass switch is energized.

Warning Lights Test Switch

The warning lights test switch is a two position, spring–loaded toggle switch on the cockpit lights control panel. The switch is spring–loaded to NORMAL and when placed to TEST illuminates all front cockpit warning lights simultaneously, and the FUEL LEVEL LOW light in the rear cockpit. In aircraft 158355at and up and all others after AFC 506, the warning lights test switch is a three position, spring–loaded toggle switch. The switch is spring–loaded to the mid or off position, the other two positions are WARN LT TEST and WPN LTS RESET. The new WPN LTS RESET position is used to cancel the air–to–air weapon selections and to deenergize the lights on the weapon status panel, and AIM–9 and AIM–7 status panels (including the TK light). This is the only way to reset once the lights are selected. The test circuit for these warning lights receives power from the right main 28 volt dc bus and the warning lights 14/28 volt ac bus.

Utility Light

A utility spot and floodlight on the right side of cockpit, above right console includes an integral ON–OFF intensity control. Its color may be changed from red to white by depressing the latch button and rotating the lens housing. An additional plug–in socket for the light is provided on the right windshield sill aft of the instrument panel.

Spare Edge Lamps

Spare edge lamps are in a spring–loaded cylindrical container on the wing fold control panel on the right console.

Indexer Lights Control Knob

Refer to Angle–of–Attack System, this section.

Radar Annunciator Lights

On aircraft 157242an and up and all others after AFC 486, the interior lighting system includes the radar annunciator lights control knob on the indicator lights control panel. With the instrument lights control knob (flight instrument lights knob after AFC 536) out of the OFF position, the intensity of the radar annunciator lights is independently controlled by the radar annunciator lights control knob in both cockpits.

RIO'S INTERIOR LIGHTING

The RIO's interior lighting is controlled from the RIO's main instrument panel (figure A–2, appendix A) and consists of an instrument lights control knob, an equipment lights control knob, a cockpit floods switch, and a warning light test switch. In addition to these controls, a utility light is provided for auxiliary lighting. On aircraft 155529ag and up and all others after AFC 388, the RIO's interior lighting controls are on the cockpit lights/data link control panel under the RIO's left canopy sill.

Instrument Lights

The instrument lights are controlled by a variable intensity control knob. As the control knob is rotated from OFF towards BRT, instrument light intensity increases.

Equipment Lights

The equipment lights utilize a variable intensity type control knob to control the intensity of the equipment panel edge lights. As the knob is rotated from OFF towards BRT, equipment light intensity increases.

Cockpit Floods Switch

The cockpit floods switch is a three position switch, with switch positions OFF, DIM and BRT. The switch is used to select operation and intensity of the four red floodlights in the RIO's cockpit. On aircraft 155529ag and up and all others after AFC 388, three floodlights have been added to the RIO's cockpit.

Warning Lights Test Switch

The warning lights test switch is a two-position switch with positions OFF and TEST. The switch is spring-loaded to the OFF position. When placed in the TEST position, the warning lights illuminate, except the FUEL LEVEL LOW light. The FUEL LEVEL LOW light will test only when the test switch in the front cockpit is placed to the TEST position. On aircraft 155529ag and up, the warning lights test switch is on the cockpit lights/data link control panel. The warning lights test circuit receives power from the right main 28 volt dc bus and the warning lights 14/28 volt ac bus.

Utility Light

The utility light is above and to the left of the RIO's instrument panel. An additional plug-in socket for the light on the upper structure aft of the instrument panel provides an alternate location from which to illuminate the chartboard when it is being used. This light may be changed from red to white by rotating the lens housing. The light has an integral ON-OFF and intensity control.

Radar Annunciator Lights

On aircraft 157242an and up and all others after AFC 486, with the instrument lights knob (flight instrument lights knob after AFC 536) in the front cockpit out of the OFF position, the intensity of the radar annunciator lights is independently controlled by the radar annunciator lights control knob. This knob is on the radar annunciator lights control panel, outboard and adjacent to the nav computer control panel.

Spare Edge Lamps

Spare edge lamps are in a spring-loaded cylindrical container on the utility panel in the left forward section of the cockpit.

NORMAL OPERATION

Normal operation of the exterior lights and pilot's and RIO's interior lights are controlled by the various switches on their control panels.

EMERGENCY OPERATION

There are no provisions for emergency operation of the exterior or interior lighting. However, the pilot's instrument panel and consoles, and the RIO's cockpit floodlights illuminate when the ram air turbine is extended, by placing the respective floodlight switches in the BRT position. The taxi lights, pilot's and RIO's instrument panel lights, and the warning lights also operate on RAT power. However, the warning lights cannot be dimmed.

LIMITATIONS

No limitations pertain to the lighting equipment.

NAVIGATION EQUIPMENT

NOTE

Some illustrations referred to within this system writeup may be located in appendix A. These illustrations are referred to, within the text, as (figure A-, appendix A).

DESCRIPTION

The navigational systems on the airplane consist of the attitude reference and bombing computer set AN/AJB-7, the vertical flight reference set GVR-10, the flight director group, and the navigation computer AN/ASN-39A. Each of these systems provides its share of information to supply the crewmembers adequate navigational assistance.

ATTITUDE REFERENCE AND BOMBING COMPUTER SET AN/AJB-7

The AN/AJB-7 is an all attitude reference platform which supplies standby attitude and azimuth information to the attitude director indicator, the horizontal situation indicator, and the bearing distance heading indicator in the rear cockpit. The rear cockpit remote attitude indicator, the forward cockpit standby attitude indicator, the autopilot system, and the bombing computer receive attitude information from the AN/AJB-7 at all times. The bombing computer functions of the AN/AJB-7 are described in the special purpose bombing system, this section. The major components of the AN/AJB-7 system are the compass adapter-compensator, displacement gyroscope, and the compass transmitter. The compass adapter-compensator obtains heading information from the compass transmitter and the displacement gyroscope. It processes this information and provides heading outputs to the attitude director indicator, horizontal situation indicator, and to other aircraft systems. The displacement gyro also sends pitch and roll signals to the attitude director indicator and the AN/APG-59 radar set. The compass transmitter is a direction sensing device. It accurately detects its alignment relative to the lines of force of the earth's magnetic field and transmits this information electrically to the compass adapter-compensator.

COMPASS SYSTEM CONTROLLER

The compass system controller (figure A-1, appendix A) provides the controls and indicator necessary for proper operation of the azimuth system and the selection between the primary and standby attitude sources. The mode selector knob initiates the proper relay switching in the compass adapter compensator to select the operating modes (compass, DG, and slaved). The sync position of the mode selector knob, spring-loaded to return to the SLAVED position, is used for fast synchronization of the compass transmitter and the azimuth reference system. The degree of synchronization is indicated by the sync indicator meter. The set heading control knob, spring-loaded to return to the center or zero position, provides a means to manually adjust the azimuth setting of the ADI, HSI, and BDHI. Compensation is provided by the hemisphere switch (N-S) and the latitude control, when they are set to the local hemisphere and latitude respectively. The attitude reference selector knob, with positions of PRIM and STBY, provides selection between the vertical flight reference set GVR-10, and the displacement gyroscope of the AN/AJB-7 system as the source of attitude information. When switching from STBY to PRIM or vice versa, attitude information appears almost immediately but may be accompanied by some unusual but normal gyrations of the attitude director indicator. This phenomenon is a simultaneous 180° flop about all three axes after which normal attitude reference is displayed. Whether or not this occurs is determined by which side of the AJB-7 gyro is uppermost during the initial erection. Accurate heading information may not be immediately available after completing a turn with a rate of more than 15° per minute. To correct this situation, the aircraft must be flown straight and level (rate of turn less than 15° per minute) for approximately 20 seconds and then manually synchronized by placing the mode selector knob to the SYNC position.

Compass Mode

The compass mode is considered an emergency mode. If the displacement gyroscope fails, the attitude reference selector knob may be placed on PRIM and the mode selector knob on COMP, resulting in display of compass transmitter magnetic heading on the attitude director indicator. However, the interlock with the AFCS mode of operation of the automatic flight control system is automatically opened in the compass mode to prevent erratic magnetic heading signals from being applied to the autopilot. Also, if the attitude reference selector knob is on STBY and the mode selector knob is on COMP the attitude director indicator azimuth indications should not be used since it is still connected to the malfunctioning gyroscope. To place the attitude reference system in the compass mode place the bomb control switch on the bomb control panel to OFF, and the mode selector knob on the compass controller to COMP.

DG Mode

The DG mode is used in north and south latitudes greater than 70° and in areas where the earth's magnetic field is appreciably distorted. When the DG mode is initially selected, the magnetic heading of the aircraft must be set into the system with the set heading control on the compass controller. The system then uses this reference for subsequent heading indication. Apparent and real drift compensating voltages are inserted by use of the hemisphere switch (N-S) and latitude control on the compass system controller. To place the reference system in the DG mode place the bomb control switch on the bomb control panel to OFF and the mode selector knob on the compass controller to DG. Then set the hemisphere switch to the local hemisphere and the latitude control knob to the local latitude. Set the aircraft magnetic heading on the ADI with the set heading knob, and readjust the latitude control knob for each 5° change in latitude.

Slaved Mode

The slaved mode is the mode used under normal conditions. In the slaved mode the azimuth system is primarily controlled by signals from the compass transmitter (flux valve). Since system accuracy is now dependent upon the earth's magnetic field, the slaved mode should not be used in latitudes greater than 70° and in areas where the earth's magnetic field is distorted. To place the reference system in the slaved mode place the bomb control switch on the bomb control panel to OFF and then place the mode selector knob on the compass controller to the SLAVED position. Allow 10 seconds for automatic fast synchronization and check the sync indicator meter for a center-scale indication. Slight deviation of the needle from the center position is corrected by normal sync. Adjust the pitch trim control knob on the ADI for zero pitch attitude.

GYRO ERECTION SWITCH

After AFC 478, there is a gyro fast erect switch on the front cockpit main instrument panel with positions of FAST and NORMAL. When placed to the momentary FAST position a high erection voltage is applied to the AN/AJB-7 displacement gyro roll and pitch torques. This provides fast erection after aircraft maneuvering. However, the aircraft must be in straight and level flight approximately 20 seconds before going to fast erect. In NORMAL, low erection voltage is utilized and the gyro erects normally. During the time that the erection switch is held in the FAST position, the power fail flag on the attitude director indicator, standby attitude indicator, and remote attitude indicator is in view, and the auto-pilot is disengaged.

CAUTION

Do not hold the gyro erection switch in the FAST position longer than 1 minute as damage to the gyro could result.

VERTICAL FLIGHT REFERENCE SET GVR-10

The function of the vertical flight reference set is to furnish pitch and roll attitude to the velocity computer of radar set AN/APG-59 for stabilization functions, and pitch and roll to the attitude director indicator for display on the sphere. It also provides flight path angle and vertical acceleration information to the terrain clearance tracker of radar set AN/APG-59 for use in computations of climb/dive signals to be displayed by the pitch steering bar in the attitude director indicator. The vertical flight reference set is comprised of the attitude sensing unit and the electronic computer unit. These components are on the left side of the aft cockpit on the equipment shelf assembly. The attitude sensing unit is basically a vertical gyro which generates pitch and roll signals that are sent to the attitude director indicator and the radar set. An accelerometer is incorporated within the unit to measure vertical acceleration. The electronic computer unit contains operational circuits which are necessary for proper functioning of the attitude sensing unit. The electronic computer unit furnishes flight path angle information to the terrain clearance tracker of radar set AN/APG-59. Vertical acceleration, barometric altitude, and true air speed information are routed to the flight angle circuits for computing climb/dive signals.

Attitude Reference Selector Knob

The attitude reference selector knob on compass system controller (figure A-1, Appendix A) is a two position knob which selects the source of attitude reference information for the attitude director indicator and the AN/APG-59 radar set. When the knob is placed in the PRIM position, pitch and roll from the GVR-10 system, and heading from the AN/AJB-7 is displayed on the attitude director indicator. Pitch and roll information is also furnished by the GVR-10 to the AN/APG-59 radar set. When the knob is in the STBY position, the AN/AJB-7 transmits heading, pitch, and roll signals to the attitude director indicator, and pitch and roll signals to the AN/APG-59 radar set. However, the remote attitude indicator in the rear cockpit and the standby attitude indicator in the front cockpit always display the pitch and roll information from the AN/AJB-7, regardless of the position of the attitude reference selector knob. Prior to taxiing ensure the attitude reference selector knob is in the primary position. If the standby mode is selected and a failure occurs in the AN/AJB-7, there is no accurate attitude information displayed on any gyro including the AN/APG-59 radar set.

NOTE

On aircraft 153071z thru 154785af, before AFC 388, with the left generator failed, bus tie open, and the right generator and ram air turbine operating, pull the GVR-10 circuit breaker on the no. 1 panel (zone A16). Switch the attitude reference selector knob from the PRIM to the STBY position. Pitch and roll information is then obtained from the AJB-7 system. On aircraft 154786ag and up, or after AFC 388, with the left generator failed, bus tie open, and the right generator and the ram air turbine operating, the pitch and roll information is automatically obtained from the STBY mode instead of the PRIM mode. However, when the ram air turbine is extended, switch the attitude reference selector knob to the STBY position, so that if power from the left generator is restored, the attitude information to the ADI is not interrupted.

BUILT-IN-TEST (BIT) CHECKS

A BIT switch is on the AN/AJB-7 test system panel in the aft cockpit. In aircraft 155529ag and up this switch is in the nose wheel well. The switch has positions of INT TEST, OPEN, and EXT TEST. Place the switch to EXT TEST, and if there is a malfunction in the GVR-10 system or in the air data computer, the vertical displacement warning flag on the ADI comes into view (with radar operating or on STBY), the OFF flag on the ADI appears, and the PRIMARY GYRO OFF warning light illuminates. These are all steady indications. If the GVR-10 is functioning properly, the flags cycle in and out of view and the warning light flashes on and off at a steady rate. If no malfunction is indicated, it is not necessary to proceed further with the BIT checks. If a malfunction is indicated in EXT TEST, place the switch to the INT TEST. If there is a malfunction indication, the vertical flight reference system (GVR-10) is unreliable. If there is no malfunction indication, the airspeed inputs from the air data computer to the GVR-10 system are unreliable. When the BIT switch is in OPEN, the BIT circuit is inoperative.

NAVIGATION COMPUTER AN/ASN-39A

The navigation computer is a great circle computer and consists of a control panel and an amplifier-computer. The system furnishes the following information during flight:

a. The aircraft's present position latitude and longitude based on dead reckoning computations from an initial fix.

b. The continuous great circle magnetic bearing and distance to either of two preset targets. This is an instantaneous spherical trigonometric solution based on true north.

The dead reckoning computations are only as accurate as the wind vectors and magnetic variation references furnished the system by the RIO, or the wind vectors furnished by the radar set. Inputs of magnetic heading and true airspeed are automatically supplied by the AN/AJB-7 and ADCS. For purposes of discussion, the navigation computer can be divided into two functional sections. The first is the present position computer section which performs the basic dead reckoning computation. The second is a course and distance computer section which provides great circle course and distance to a target (or base).

PRESENT POSITION COMPUTER

The present position computer is a group of servo-mechanisms contained in the computer control panel with their amplifiers in the amplifier-computer. The present position computer receives magnetic heading from the AN/AJB-7 and true airspeed from the ADCS. Magnetic variation and the latitude and longitude of the starting position, the base, and destination are set manually by means of control knobs on the computer control panel. Wind speed and wind direction can be set manually on the computer control panel, or they can be obtained automatically from the AN/APG-59 radar set. The present position computer resolves true airspeed and wind velocity to their north-south and east-west components and adds them algebraically to derive the components of aircraft ground speed. Ground speed is integrated with respect to time to attain distance which is then converted to a change of latitude and longitude. The north–south mileage covered is converted directly into degrees and minutes of latitude by the position latitude counter on the computer control panel, since one nautical mile is always equal to one minute of latitude. In the longitude channel, an additional step is necessary since a direct conversion can only be made at the equator. At other latitudes, the east–west mileage covered is multiplied by the secant of the latitude at which the aircraft is traveling. The modified east–west mileage is then converted directly to degrees and minutes longitude by the position longitude counter on the computer control panel. The present position computer continuously computes the change in latitude and longitude from the aircraft's starting position (initial fix). These coordinate changes are applied to the corresponding position counters, both of which have been manually set to the coordinates of the starting position. Since the position counters add the change in coordinates to the starting coordinates, they provide continuous indication of the aircraft's present position.

COURSE AND DISTANCE COMPUTER

The basis of course and distance computation is the solution of the spherical triangle (each side of which is a segment of a great circle) formed on the earth's surface by the geographic north pole (true north), the present position, and the preselected target or base. The latitude and longitude of the base and the target are manually inserted into the system by means of the position and target counters on the computer control panel. The base coordinates are retained by memory circuits of the position counters. Since this information is known within the system, as is present position of the aircraft which is available continuously from the present position computer, two sides of the spherical triangle and the angle between them are known. This makes it possible to solve for the third side and angle using the information available. The third side represents the great circle distance and the angle represents the great circle bearing or course angle.

COMPUTER CONTROL PANEL

The computer control panel (figure A–2, appendix A) has been modified to accept radar derived wind inputs from the radar set. The panel contains the operating controls and counters which provide a readout of inserted information. The position counters also provide a continuous readout of present position during flight.

Function Switch

The function switch is a five–position rotary switch with positions of OFF, TARGET, BASE, STBY, and RESET. The OFF position removes all power from the navigation computer. The TARGET position selects output displays of target range, ground track relative to magnetic heading, and bearing relative to magnetic heading to the preselected target coordinates set on the target counters. The BASE position selects the same output displays as the TARGET position but is referenced to the preselected base or alternate target coordinates which are retained in the memory circuits of the position counters. The STBY position supplies filament power to the amplifier–computer and the latitude and longitude integrator channels of the system are inoperative. The RESET position is used to set base alternate target, or return point coordinates in the memory circuits of the position counters. Placing the switch to the RESET position causes the original memorized coordinates to be lost. A restriction on the switch prevents accidental switching to the RESET position. The switch must be pulled outward slightly to override the restriction when switching to the OFF or RESET positions.

Wind Control Knobs

The wind control knobs consist of the wind velocity control knob and the wind from control knob. The wind velocity control knob is used to manually insert the wind velocity affecting flight into the system and is displayed on the wind velocity counter. The wind from control knob is used to manually insert the true wind direction. The true wind direction is expressed as an angle measured clockwise from true north and is presented in degrees on the wind from counter.

NOTE

> The wind control knobs are inoperative once the radar set has supplied wind signals to the navigation computer. To remove radar winds and regain manual winds, see section VIII classified supplement.

Magnetic Variation Control Knobs

The magnetic variation control knob is used to manually insert the magnetic variation angle into the system. The magnetic variation is the angular difference between true north and magnetic north.

Position Control Knobs

The position control knobs are initially used to insert the base, alternate target, or return point latitude and longitude into the system. To insert these coordinates the function switch must be in the RESET position. With the function switch in any position but RESET, the position control knobs are used to manually insert present position latitude and longitude to establish an initial fix for the dead reckoning function of the navigation computer. The base, alternate target, or return point latitude and longitude are not displayed anywhere during flight; a memory of these coordinates is retained by the system so long as the function switch is not placed to the RESET position. The position latitude and longitude counters continuously indicate the aircraft present position in degrees and minutes during flight.

Target Control Knobs

The target control knobs are used to manually insert the target latitude and longitude into the system. These coordinates are displayed on the target latitude and

ATTITUDE DIRECTOR INDICATOR AND HORIZONTAL SITUATION INDICATOR

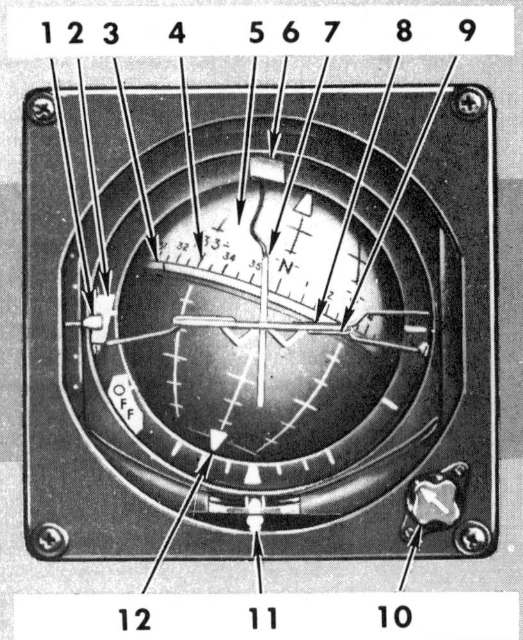

1. VERTICAL DISPLACEMENT POINTER
2. VERTICAL DISPLACEMENT WARNING FLAG (TACAN, ILS, GVR-10, RADAR)
3. HORIZON BAR
4. HEADING REFERENCE SCALE
5. 3-AXIS SPHERE
6. VERTICAL DIRECTOR WARNING FLAG (TACAN, DATA LINK, INS)
7. BANK STEERING BAR
8. PITCH STEERING BAR
9. MINIATURE AIRCRAFT
10. PITCH TRIM KNOB
11. TURN AND SLIP INDICATOR
12. BANK POINTER

1. COMPASS CARD
2. BEARING POINTER
3. LUBBER LINE
4. COURSE DEVIATION INDICATOR
5. COURSE ARROW
6. HEADING MARKER
7. COURSE SET KNOB
8. COURSE SELECTOR WINDOW
9. TO-FROM INDICATOR
10. AIRCRAFT SYMBOL
11. RANGE INDICATOR
12. HEADING SET KNOB

FDD-1-(29)A

Figure 1-25

longitude counters. The system provides output displays to fly to these coordinates when the function switch is placed to TARGET.

FLIGHT DIRECTOR GROUP

The purpose of the flight director group is to provide an integrated display of the navigation situation of the airplane. The flight director group consists of a flight director computer, the horizontal situation indicator (HSI), and a selector panel. Although the attitude director indicator (ADI) is not a component of the flight director group, it does receive some signals from the flight director computer and shall be discussed along with the flight director group.

Flight Director Computer

The flight director computer provides navigation information to the HSI, and steering information to the ADI. Except for the bearing and distance display on the HSI, all signals for the HSI, and signals for portions of the ADI pass through or originate in the computer. The flight director computer has no control over the 3 axis sphere portion of the ADI. Steering signals are computed to provide the pilot with flight direction information when flying either manually or remotely set headings and manually selected tacan radials. These computed signals, together with the required flag signals and off scale signals, are supplied by the computer to the ADI.

Horizontal Situation Indicator

The HSI (figure 1-25) provides the horizontal or plan view of the aircraft with respect to the navigation situation. That is, the HSI is a platform picture of the aircraft's present situation, as seen from above the airplane. The aircraft symbol, in the center of the HSI is the airplane superimposed on a compass rose. The compass card rotates so the aircraft's magnetic heading is always under the lubber line. Index marks are provided every 45 degrees around the perimeter of the compass card, which can be used in holding patterns, procedure turns, etc. The bearing pointer provides magnetic bearing to a selected tacan or navigation computer station, depending on which is selected on the bearing distance selector switch. When ADF is selected on the bearing distance selector switch, the bearing pointer indicates relative bearing to the UHF station. The heading marker is controlled by the heading set knob, except when navigation computer steering is desired at which time it is automatically set. The needle which reflects the steering asked for by the heading marker is the bank steering bar on the ADI. The heading marker is centered under the lubber line when ATT is selected on the mode selector knob, but indicates a command heading for the bank steering bar to steer to in any other mode. The course arrow has two functions. For tacan steering, it must be set (using the course set knob) to coincide with the desired tacan track. The course selector window always agrees with the course arrow when used for tacan. Just as the heading marker provides steering information to the bank steering bar on the ADI, the course arrow provides the course deviation indicator on the HSI with displacement information. When the navigation computer is being used for steering information, the course arrow automatically indicates the magnetic ground track presently being flown. The to-from indicator indicates whether the aircraft is approaching or going away from the tacan station on the course selected, providing tacan azimuth is locked-on. It doesn't indicate whether the aircraft is actually heading toward the selected station. If the to-from indicator points toward the head of the course arrow (equivalent to a TO indication), it indicates the tacan course selected will steer the aircraft toward the station and not away from it. The to-from indicator only operates when tacan mode is selected. The course deviation indicator operates in conjunction with the course arrow and is only operative when in the tacan mode. The course deviation indicator represents the selected tacan radial. The relationship between the aircraft symbol and the course deviation indicator is the same as an actual plan view of the selected tacan radial and the aircraft. The course deviation indicator indicates direction and angular relationship to the tacan radial. In addition, angular error from the tacan radial can be read up to 5°. The two dots, on each side of center, each indicate 2.5° of angular error from the selected tacan radial. Once the course deviation indicator is fully deflected, angular error cannot be read directly, but up to 5° on either side of the selected tacan radial can be read. Four mode-of-operation word messages are shown around the HSI display. These lights are illuminated internally to indicate the selected operating mode, provided the instrument panel lights control knob is in the ON position. The intensity of the mode lights is also controlled by this knob.

Attitude Director Indicator

The primary function of the ADI (figure 1-25) is to provide aircraft attitude reference. The black and gray sphere is movable and stabilized through all attitudes so that the miniature aircraft wings, against the moving horizon line on the sphere, give the pilot attitude reference. Pitch angle increments of 10° are marked on the sphere and can be set using the pitch trim knob. Heading indications are also provided around the sphere's horizon line. Ten degree bank increments, up to 30° are marked on the bottom of the instrument. The bank steering bar is the only bar used in conjunction with the flight director group. The bank steering bar indicates corrective action necessary to intercept the selected heading, tacan radial or navigation computer destination. The bank steering bar receives its bank information from the heading marker on the HSI, through the flight director computer. When the heading marker is positioned either manually or automatically, the bank steering bar deflects right or left to direct the pilot to the new heading. The bank steering bar does not indicate direction or displacement from the desired heading, but rather the corrective action required. The maximum bank angle that is commanded by the bank steering bar is 30°. This is because the maximum error the computer commands is 90°. If the heading marker on the HSI is set at 90° or more from the present position, the bank steering bar indicates a maximum bank of 30°. Any heading error of less than 90° produces a bank angle indication of something less than 30°. Obviously, there are times when more than 30° of bank are desired to intercept the new heading or tacan radial and in these cases, the bank steering bar must be disregarded. The bank steering bar information is only reliable for tacan steering if the selected tacan track is within ±60° of the present inbound track. The course deviation indicator on the HSI is reliable regardless of the tacan track selected. The ±60° limitation applies only to the bank steering bar. If when on the selected tacan radial, it becomes necessary to establish a crab angle due to wind draft (aircraft heading different from selected radial), the bank steering bar indicates a heading error. To eliminate this apparent heading error, the heading marker on the HSI should be manually set to correspond to the new heading. Do not expect the bank steering bar to automatically correct for wind drift. The vertical director warning flag on the ADI (12 o'clock position), normally out of view, appears if an unreliable signal is received from the tacan, data link (some aircraft), and ILS systems (some aircraft). The vertical displacement warning flag (9 o'clock position), appears if an unreliable signal is received from the tacan, ILS, GVR–10, or the range terrain clearance tracker (if the radar is turned to ON or STBY). The OFF power fail flag on the ADI appears if there is a failure of the GVR–10 or AJB–7 systems, if the gyro is in the start cycle, if the gyro fast erect switch is held on, or if there is a power failure to the indicator.

NAVIGATION FUNCTION SELECTOR PANEL

The navigation function selector panel (figure A–1, appendix A) is on the pilot's main instrument panel. The panel contains a mode selector knob and a bearing/distance selector switch. The mode selector knob and bearing/distance selector switch control separate functions of the ADI and HSI, and do not necessarily have to be set up as a pair.

Mode Selector Knob

The mode selector knob is a rotary–type switch with positions of ATT, HDG, TACAN, NAV COMP, and DL. The knob is used to select the source of information to be displayed on the HSI and ADI as shown in figures 1–26, 1–27, and 1–28.

Bearing/Distance Selector Switch

The bearing/distance selector switch is a three–position toggle switch with switch positions of NAV COMP, ADF, and TACAN. The bearing/distance switch activates only the bearing pointer and range indicator (and the mode word which reflects what is selected) on the HSI. The information displayed for the three positions of the bearing/distance switch are shown in figure 1–29.

NORMAL OPERATION

AN/AJB–7 OPERATION

There is no on–off switch for the AN/AJB–7 system. Power is applied when the circuit breakers are engaged and the aircraft bus system is energized. The AN/AJB–7 system receives three phase, 115 volts ac, essential bus power, and 28 volts dc essential bus power. The system also receives power from the single phase, 115 volts ac essential bus, and 28 volts ac essential bus. The system is protected by circuit breakers on the number one circuit breaker panel in the rear cockpit. The circuit breakers are marked AJB–7. After the system is energized, the azimuth modes in which the compass adapter compensator may operate are the compass, directional gyro, and the slaved modes.

GVR–10 OPERATION

The GVR–10 system becomes operational when electrical power is applied to the aircraft. The system receives three phase power for the left main 115 volts ac bus, and the left main 28 volt dc bus. The GVR–10 system is protected by circuit breakers on No. 1 circuit breaker panel in the rear cockpit. The circuit breakers are marked GVR–10.

FLIGHT DIRECTOR GROUP OPERATION

The flight director group receives power when the circuit breakers for the AN/AJB–7, CNI, and navigation computer systems are engaged. To operate the flight director group, the mode selector knob and bearing/distance selector switch should be positioned to the mode desired for proper display on the ADI and HSI.

NAVIGATION COMPUTER OPERATION

All controls required to operate the navigation computer are on the computer control panel. To simplify the procedure, it will be understood that where a counter setting is specified, the control knob associated with the counter will be rotated to set the counter. The position latitude and longitude control knobs must be pressed in to engage them with their associated counter. The initial flight plan may be set into the navigation computer at any time. Electrical power is not required to set in the flight plan. However, for the navigation computer system to be operative the appropriate protective circuit breakers must be engaged. The system receives power from the left main single phase 115 volt ac bus, and the left main 28 volts dc bus. The AN/ASN–39A system is protected by circuit breakers on No. 1 circuit breaker panel in the rear cockpit. The circuit breakers are marked NAV CMPTR.

Initial Flight Plan (Electrical Power Not Required)

a. Set magnetic variation counter to local magnetic variation.

b. Set wind from counter and wind velocity counter to wind direction and wind speed that affects the flight.

NOTE

Weather stations may provide the magnetic wind direction which must be corrected for variation to obtain the true wind direction.

c. If memory base information is desired, place function switch to RESET and set latitude and longitude of alternate target, base, or return point on position counters.

d. Place function switch to any position but RESET and set local latitude and longitude in position counters.

e. Set latitude and longitude of the target or destination on the target counters.

f. Place function switch to OFF until after starting engines.

ATTITUDE AND HEADING DISPLAYS

ATT (Attitude)

THE COURSE ARROW, COURSE DEVIATION INDICATOR AND HEADING MARKER ARE SLAVED TO THE MAGNETIC HEADING OF THE AIRPLANE (I. E. VERTICAL ON THE FACE OF THE HSI). NO MODE LIGHT IS ILLUMINATED.

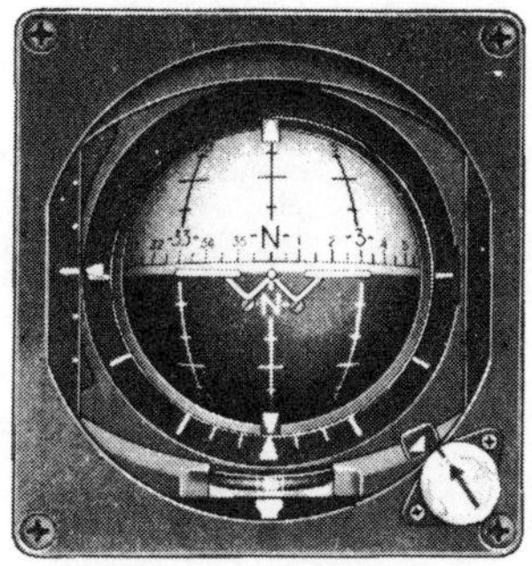

ALL POINTERS ARE DEFLECTED OUT OF VIEW. ONLY ATTITUDE AND AZIMUTH INFORMATION IS DISPLAYED.

HDG (Heading)

THE HEADING MARKER IS POSITIONED BY THE HEADING SET KNOB TO PROVIDE THE ADI BANK STEERING BAR WITH BANK AND AZIMUTH INFORMATION IN ORDER TO TURN TO THE SELECTED HEADING. THE MAN MODE LIGHT IS ILLUMINATED.

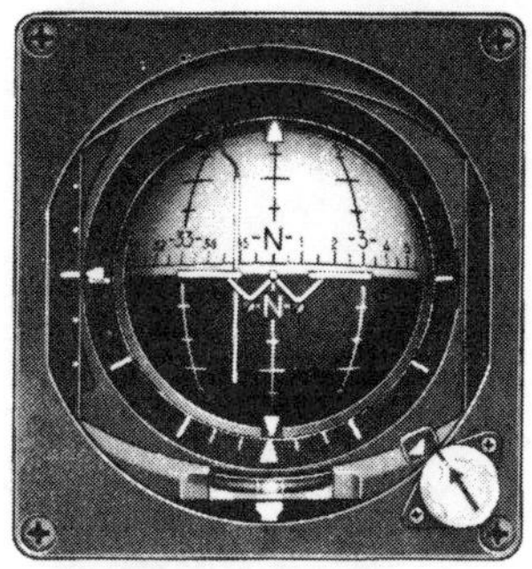

THE BANK STEERING BAR INDICATES BANK ANGLE STEERING UP TO 30° OF BANK TO APPROACH THE HEADING SELECTED BY THE HEADING SET KNOB ON THE HSI.

FDD-1-(30)

Figure 1-26

NOTE

Select an operating mode (TARGET or BASE) prior to takeoff roll.

Output Displays

The navigation computer display information in the front cockpit is shown in figure 1-27. To display navigation computer information on the BDHI in the rear cockpit, select the NAV COMP position on the navigation selector switch. The displayed information of the BDHI is as follows:

a. Magnetic bearing to target or base displayed on No. 1 needle when read against the compass card. A relative bearing to the target or base can also be read by noting the number of degrees from the index clockwise to No. 1 needle.

b. Magnetic ground track displayed on No. 2 needle when read against the compass card. Left or right drift angle can also be read by noting the number of degrees left or right from the index.

c. Distance to the target or base displayed on the range counter.

To travel the great circle route to the target or base, the aircraft should be flown on a course that causes the two needles to be coincident. However, it is not necessary to fly the course shown by the coincidental needles. Departure from the route may be made, as a part of evasive maneuvers or to fly a search pattern, without affecting the operation of the system. Since computations are being made continuously, the current position of the airplane is shown at all times on the position counters regardless of

NAVIGATION COMPUTER DISPLAYS

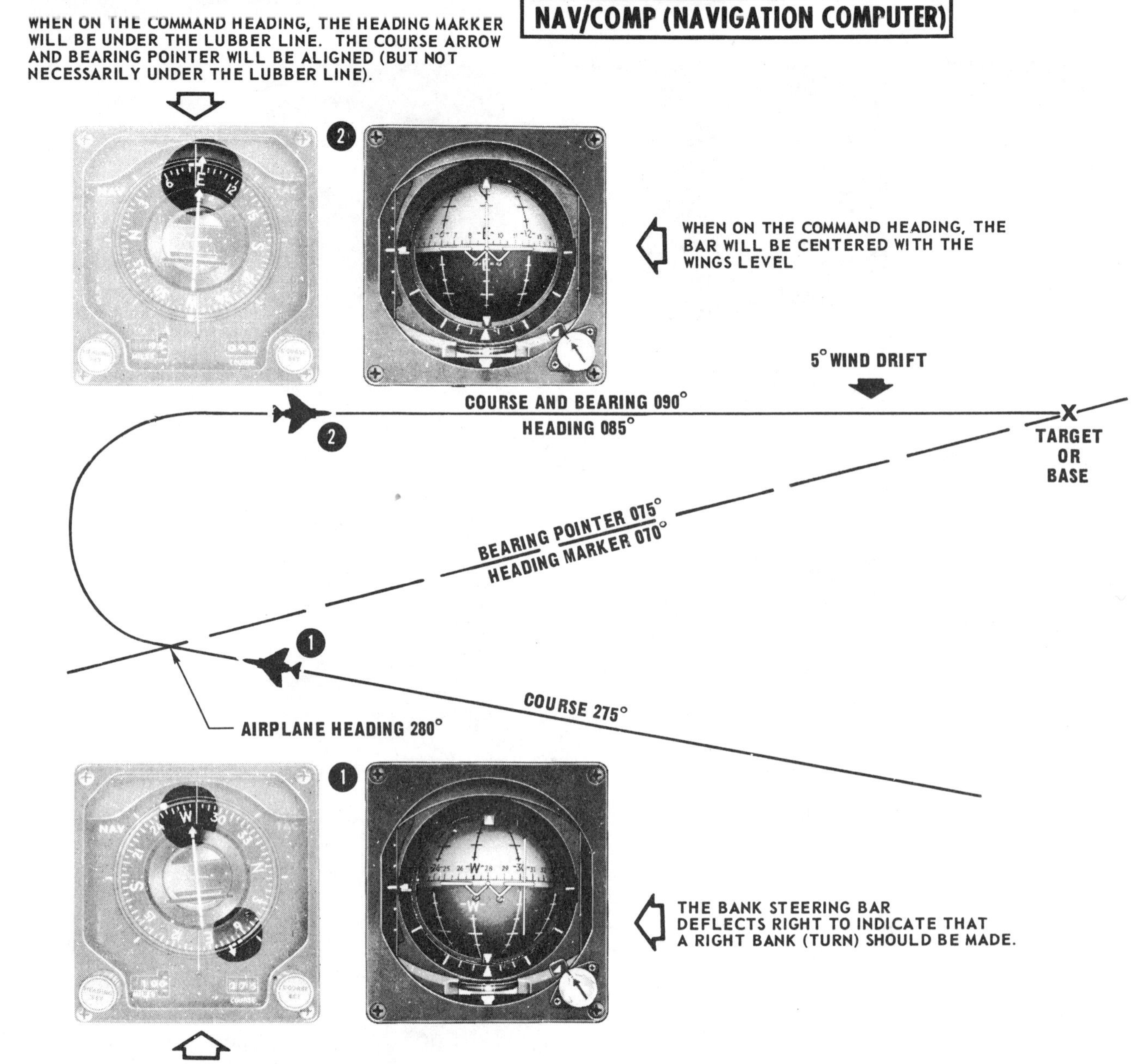

THE HEADING MARKER INDICATES THE MAGNETIC HEADING THAT MUST BE FLOWN TO MAKE GOOD A COURSE DIRECT FROM THE PRESENT POSITION OF THE AIRCRAFT TO THE DESTINATION (TARGET OR BASE) SELECTED ON THE NAV COMPUTER. WHETHER THE HEADING IS CORRECT OR NOT IS DEPENDENT UPON THE ACCURACY OF THE WIND DIRECTION AND VELOCITY, THE VARIATION, AND THE ACCURACY OF THE PRESENT POSITION. THE COURSE ARROW INDICATES THE TRACK THAT IS CURRENTLY BEING MADE GOOD, ALSO DEPENDENT UPON THE ACCURACY OF THE NAV COMPUTER SETTINGS. THE COURSE DEVIATION INDICATOR IS SLAVED TO THE COURSE ARROW. THE COURSE SELECTOR WINDOW WILL INDICATE THE SAME TRACK AS THE COURSE ARROW. THE BEARING POINTER INDICATES MAGNETIC BEARING TO DESTINATION. IN ORDER TO OBTAIN NAV COMPUTER INFORMATION FROM THE BEARING POINTER AND THE RANGE INDICATOR, THE BRG/DIST SWITCH MUST BE IN THE NAV COMP POSITION.

FDD-1-(31)

Figure 1-27

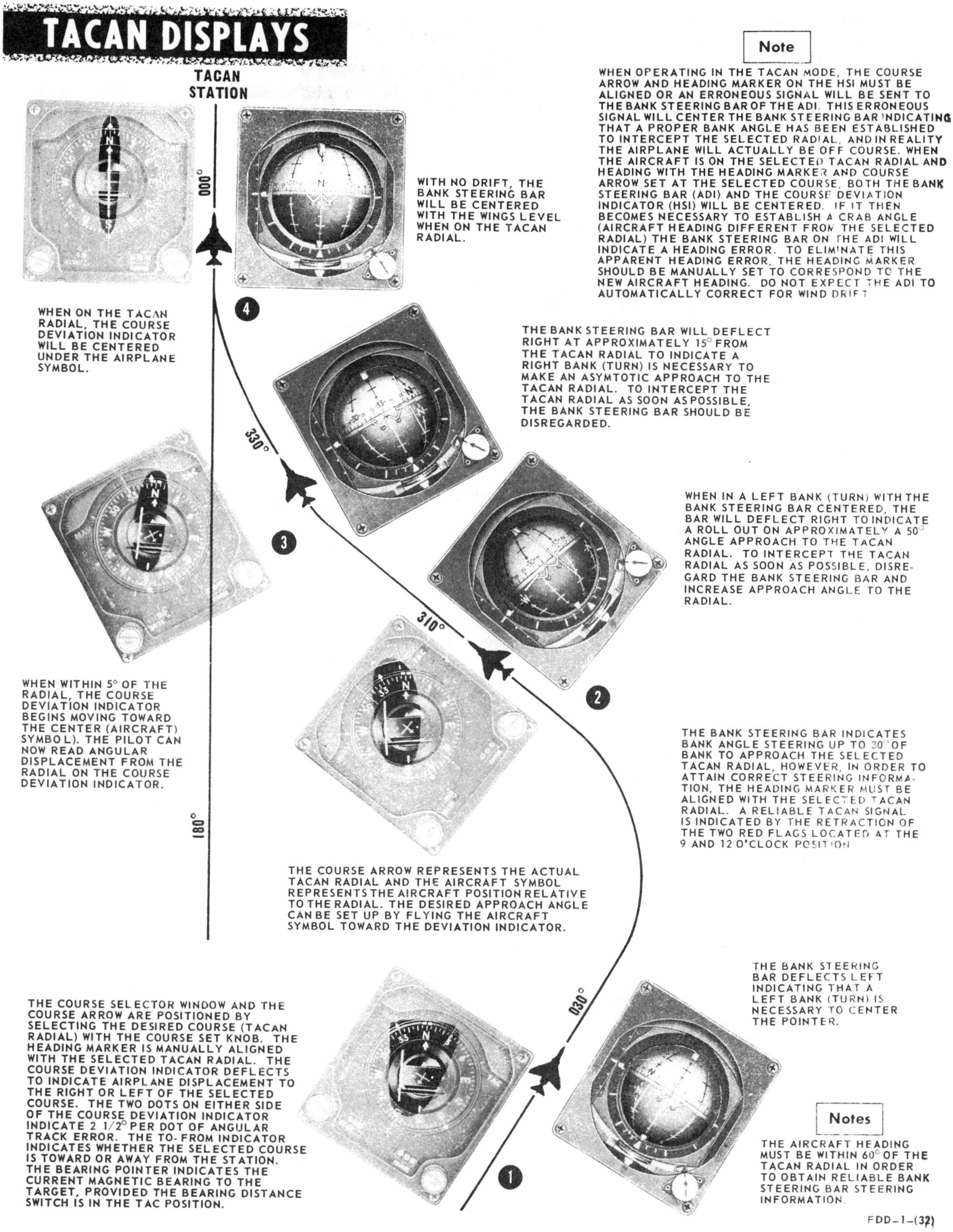
TACAN DISPLAYS
TACAN STATION
Note
WHEN OPERATING IN THE TACAN MODE, THE COURSE ARROW AND HEADING MARKER ON THE HSI MUST BE ALIGNED OR AN ERRONEOUS SIGNAL WILL BE SENT TO THE BANK STEERING BAR OF THE ADI. THIS ERRONEOUS SIGNAL WILL CENTER THE BANK STEERING BAR INDICATING THAT A PROPER BANK ANGLE HAS BEEN ESTABLISHED TO INTERCEPT THE SELECTED RADIAL, AND IN REALITY THE AIRPLANE WILL ACTUALLY BE OFF COURSE. WHEN THE AIRCRAFT IS ON THE SELECTED TACAN RADIAL AND HEADING WITH THE HEADING MARKER AND COURSE ARROW SET AT THE SELECTED COURSE, BOTH THE BANK STEERING BAR (ADI) AND THE COURSE DEVIATION INDICATOR (HSI) WILL BE CENTERED. IF IT THEN BECOMES NECESSARY TO ESTABLISH A CRAB ANGLE (AIRCRAFT HEADING DIFFERENT FROM THE SELECTED RADIAL) THE BANK STEERING BAR ON THE ADI WILL INDICATE A HEADING ERROR. TO ELIMINATE THIS APPARENT HEADING ERROR, THE HEADING MARKER SHOULD BE MANUALLY SET TO CORRESPOND TO THE NEW AIRCRAFT HEADING. DO NOT EXPECT THE ADI TO AUTOMATICALLY CORRECT FOR WIND DRIFT.
000°
WITH NO DRIFT, THE BANK STEERING BAR WILL BE CENTERED WITH THE WINGS LEVEL WHEN ON THE TACAN RADIAL.
4
WHEN ON THE TACAN RADIAL, THE COURSE DEVIATION INDICATOR WILL BE CENTERED UNDER THE AIRPLANE SYMBOL.
THE BANK STEERING BAR WILL DEFLECT RIGHT AT APPROXIMATELY 15° FROM THE TACAN RADIAL TO INDICATE A RIGHT BANK (TURN) IS NECESSARY TO MAKE AN ASYMTOTIC APPROACH TO THE TACAN RADIAL. TO INTERCEPT THE TACAN RADIAL AS SOON AS POSSIBLE, THE BANK STEERING BAR SHOULD BE DISREGARDED.
330°
3
WHEN IN A LEFT BANK (TURN) WITH THE BANK STEERING BAR CENTERED, THE BAR WILL DEFLECT RIGHT TO INDICATE A ROLL OUT ON APPROXIMATELY A 50° ANGLE APPROACH TO THE TACAN RADIAL. TO INTERCEPT THE TACAN RADIAL AS SOON AS POSSIBLE, DISREGARD THE BANK STEERING BAR AND INCREASE APPROACH ANGLE TO THE RADIAL.
310°
2
WHEN WITHIN 5° OF THE RADIAL, THE COURSE DEVIATION INDICATOR BEGINS MOVING TOWARD THE CENTER (AIRCRAFT) SYMBOL). THE PILOT CAN NOW READ ANGULAR DISPLACEMENT FROM THE RADIAL ON THE COURSE DEVIATION INDICATOR.
THE BANK STEERING BAR INDICATES BANK ANGLE STEERING UP TO 30° OF BANK TO APPROACH THE SELECTED TACAN RADIAL, HOWEVER, IN ORDER TO ATTAIN CORRECT STEERING INFORMATION, THE HEADING MARKER MUST BE ALIGNED WITH THE SELECTED TACAN RADIAL. A RELIABLE TACAN SIGNAL IS INDICATED BY THE RETRACTION OF THE TWO RED FLAGS LOCATED AT THE 9 AND 12 O'CLOCK POSITION.
180°
THE COURSE ARROW REPRESENTS THE ACTUAL TACAN RADIAL AND THE AIRCRAFT SYMBOL REPRESENTS THE AIRCRAFT POSITION RELATIVE TO THE RADIAL. THE DESIRED APPROACH ANGLE CAN BE SET UP BY FLYING THE AIRCRAFT SYMBOL TOWARD THE DEVIATION INDICATOR.
THE BANK STEERING BAR DEFLECTS LEFT INDICATING THAT A LEFT BANK (TURN) IS NECESSARY TO CENTER THE POINTER.
030°
THE COURSE SELECTOR WINDOW AND THE COURSE ARROW ARE POSITIONED BY SELECTING THE DESIRED COURSE (TACAN RADIAL) WITH THE COURSE SET KNOB. THE HEADING MARKER IS MANUALLY ALIGNED WITH THE SELECTED TACAN RADIAL. THE COURSE DEVIATION INDICATOR DEFLECTS TO INDICATE AIRPLANE DISPLACEMENT TO THE RIGHT OR LEFT OF THE SELECTED COURSE. THE TWO DOTS ON EITHER SIDE OF THE COURSE DEVIATION INDICATOR INDICATE 2 1/2° PER DOT OF ANGULAR TRACK ERROR. THE TO-FROM INDICATOR INDICATES WHETHER THE SELECTED COURSE IS TOWARD OR AWAY FROM THE STATION. THE BEARING POINTER INDICATES THE CURRENT MAGNETIC BEARING TO THE TARGET, PROVIDED THE BEARING DISTANCE SWITCH IS IN THE TAC POSITION.
Notes
THE AIRCRAFT HEADING MUST BE WITHIN 60° OF THE TACAN RADIAL IN ORDER TO OBTAIN RELIABLE BANK STEERING BAR STEERING INFORMATION.
1
FDD-1-(32)

Figure 1–28

HORIZONTAL SITUATION INDICATOR BEARING AND DISTANCE DISPLAYS

THE BEARING DISTANCE SWITCH CONTROLS ONLY THE INDICATIONS DISPLAYED BY THE BEARING POINTER AND RANGE INDICATOR OF THE HSI.

BEARING DISTANCE SWITCH

NAV COMP (Navigation Computer)

THE BEARING POINTER INDICATES MAGNETIC BEARING TO THE DESTINATION SELECTED ON THE NAV COMPUTER. THE RANGE INDICATOR INDICATES THE NAUTICAL MILES TO THE DESTINATION SELECTED ON THE NAV COMPUTER (TARGET OR BASE). THE NAV LIGHT WILL BE ILLUMINATED.

BEARING DISTANCE SWITCH

ADF (Automatic Direction Finder)

THE BEARING POINTER INDICATES RELATIVE BEARING TO THE UHF STATION SELECTED ON THE COMMUNICATION CONTROL PANEL (EITHER COMM OR AUX ADF POSITION MUST BE SELECTED). THE RANGE INDICATOR WINDOW WILL BE BLANK. THE UHF LIGHT WILL BE ILLUMINATED.

BEARING DISTANCE SWITCH

TACAN

THE BEARING POINTER INDICATES MAGNETIC BEARING TO THE SELECTED TACAN STATION. THE RANGE INDICATOR WILL INDICATE TACAN RANGE WHEN TACAN IS IN THE A/A OR T/R MODE. THE TAC LIGHT WILL BE ILLUMINATED.

FDD-1-(33)

Figure 1-29

the path flown. As the target (or base) is approached, the distance on the range counter decreases. When the target is reached, uncertainty is exhibited by No. 1 needle which turns 180° as the target is passed in order to indicate bearing to target. At any time during flight, the present position may be compared with a known check point or a fix obtained from radar, tacan, or GCI and the position counters changed accordingly. This corrects errors in computation that have previously occured but does not disturb the memory of the base or target position.

Updating Methods

The position counters should be checked and updated at each opportunity by one of the following methods:

a. Visual reference to a geographical position. While in one of the operating modes or temporarily in STBY, adjust the position latitude and longitude to agree with the latitude and longitude of the visual fix. This latitude and longitude may be obtained from maps, charts, or publications.

b. Radar reference to a geographical position. Use radar mapping to obtain a bearing and distance to a known geographical position. Set the latitude and longitude of this position in the target counters or in base memory. Adjust the position counters so that No. 1 needle and the range counter on the BDHI agree with the radar bearing and distance.

c. Tacan fix. Set the latitude and longitude of the acquired tacan station in the target counters or in base memory. Adjust the present position counters so that the No. 1 needle and the range counter on the BDHI agree with the tacan readout.

d. GCI or radar monitored fix. Set the latitude and longitude of the controlling agency in the target counters or in base memory. Adjust the position counters so that No. 1 needle and range counter on the BDHI agree with the bearing and distance provided by the controller.

Radar Wind Updating

For information on updating the radar derived east–west, and north–south wind velocity, refer to NAVAIR 01–245FDD–1A.

Wind Finding Technique

To find true wind utilizing a tacan fix, it is assumed that a constant altitude is maintained while the wind is being found. The wind velocity counter is set to zero; the latitude and longitude of the reference tacan station are set on the target counters; and the position latitude and longitude counters are set so that the tacan range and bearing and the computer range and bearing on the BDHI are identical. After a convenient time interval has elapsed, the tacan range and bearing and the computer range and bearing should be plotted, either on a chart, a grid such as that on an MB–4 computer, or pictured on the face of the BDHI itself. The difference between these two plotted positions gives the wind vector for the elapsed time interval. Applying variation, the direction of the wind vector is from the computer or fix to the tacan fix. Multiply the distance by the factor necessary for an hourly rate to produce an accurate wind speed setting. For best results repeat this process as necessary and add the new vector to the original vector.

EMERGENCY OPERATION

There is no emergency operation for the GVR–10, the flight director group, or the navigation computer system.

AN/AJB–7

Compass Mode

The compass mode is considered an emergency mode in the AN/AJB–7 system and should be used only if the displacement gyro fails. To place the system in the compass mode, place the attitude reference selector knob to the PRIM position, and the mode selector knob to the COMP position. Magnetic heading is displayed on the ADI.

LIMITATIONS

There are no limitations for the AN/AJB–7, the GVR–10, or the flight director group.

NAVIGATION COMPUTER

Beyond 70°N or 72°S latitude, the navigation computer continues to operate but with a progressive loss of accuracy as either geographic pole is approached.

OXYGEN SYSTEM

NOTE

Some illustrations referred to within this system writeup may be located in appendix A. These illustrations are referred to, within the text, as (figure A–, appendix A).

DESCRIPTION

The liquid oxygen (LOX) system consists of a ten liter capacity vacuum insulated container, build–up coils, check valves, vent valves and quantity gages. The system is designed to deliver gaseous oxygen to the crew at a continuous rate of up to 120 liters per minute at 60 to 90 psi. From the container, the liquid oxygen flows to the build–up coils which are predetermined lengths of tubing wrapping around the outside bottom of the container. When the liquid flows through the build–up coil, it absorbs heat from the surrounding area and becomes gaseous. The gaseous oxygen is then increased in pressure to assure 40 to 90 psi to the regulators. From the build–up coils, the oxygen flows into the warm–up plate in the rear cockpit aft of the ejection seat. The warm–up coils warm it to a temperature no colder than 20°F under the cockpit temperature. From the warm–up plate the oxygen is now ready for crew consumption. A system relief valve set at 110 psi vents excessive pressures that may occur due to the boil–off of the LOX when the system is not being used. A blow out patch in the oxygen container provides added safety if a relief valve should fail. An electrical capacitance type indicator provides the pilot and RIO with an accurate means of determining the amount of LOX remaining at any time. An OXYGEN LOW indicator light operated by the indicator circuit and on the telelight panel alerts the pilot when the liquid oxygen gage indicates one liter of oxygen remaining in the system

OXYGEN SUPPLY LEVERS

A two–position ON–OFF oxygen supply lever is on a panel in each cockpit. The pilot's supply lever (figure A–1, appendix A) is on the forward end of the left console. The RIO's supply lever (figure A–2, appendix A) is on the utility panel which is in the left forward section of the cockpit.

OXYGEN QUANTITY GAGE

An oxygen quantity gage is in each cockpit which reads 10 liters maximum. The pilot's oxygen gage (figure A-1, appendix A) is on the forward end of the left console. The RIO's oxygen gage (figure A-2, appendix A) is on the utility panel which is in the left forward section of the cockpit.

OXYGEN REGULATOR

The oxygen breathing regulator is personnel mounted and is used both in the inflight and bailout or emergency conditions. The regulator is so designed that with an inlet pressure of 40 to 90 psi, it will deliver 100% oxygen automatically to the user, between the altitudes of 0 to 50,000 feet. In addition, the regulator incorporates automatic safety pressure build-up to a maximum of 2 inches of water below 35,000 feet and automatic pressure breathing for altitudes above 35,000 feet and is designed to integrate with the A13/A oxygen breathing mask.

OXYGEN CONNECTION

The oxygen connection utilizes a universal upper block which remains in the airplane. The oxygen-communication lead is an integrated single line. It is fastened to the back of the torso harness at the left waist and center shoulder areas with velcro fasteners. The line is then routed forward over the left shoulder and fastened permanently to the oxygen regulator. The line is clipped to the left side of the helmet. A communications lead plug is attached to the line where it is clipped to the helmet. The universal upper block features a locking indicator in the release handle. When securely locked, the indicator is flush with the handle.

WARNING

The over the shoulder oxygen-communication integrated line must be routed under the shoulder harness to preclude the possibility of the crewmember's helmet being jerked from his head after ejection. This action is caused by the left parachute riser (shoulder harness strap) snapping up against the integrated line with the opening of the personnel parachute.

EMERGENCY OXYGEN

Emergency oxygen is stored in a cylinder in the upper half of the survival kit container. The emergency oxygen cylinder is a coil assembly constructed of steel tubing closed at both ends with a volume of 100 cubic inches. The cylinder is normally charged to 1800 psi and supplies gaseous oxygen in emergencies for breathing. The flow of oxygen from this coil is controlled and regulated by the pressure reducer manifold which is actuated either manually or automatically. The pressure reducer manifold is within the survival kit and is attached to the forward left corner of the upper half of the container. It is used to reduce the oxygen pressure within the emergency oxygen cylinder to 65 $\pm$15 psi with a flow up to 140 lpm. Components of the manifold include a toggle arm, pressure gauge, relief valve, filler valve, and safety plug. When the toggle arm of the manifold is in the cocked position, flow of oxygen from the emergency oxygen cylinder is prevented by action of the pressure reducer valve within the manifold. When the toggle is tripped, emergency oxygen flows through the manifold at a reduced pressure to the intermediate block for breathing. The emergency oxygen is delivered through the miniature demand regulator and duration of the supply will depend upon altitude and demand. Under a normal workload, there is sufficient oxygen for 10 minutes at sea level. At 40,000 feet this supply would last approximately 50 minutes. The relief valve, attached to the manifold, prevents excessive pressure buildup in the system when manifold pressure regulation fails. The emergency oxygen filler valve is accessible through a hole in the upper half of the container which permits ease of servicing. The safety plug of the manifold prevents excessive pressure within the emergency oxygen cylinder due to over-servicing or thermal expansion. A pressure gauge attached to the pressure reducer manifold provides pressure indication for the cylinder and is visible through a hole in the kit cushion.

Emergency Oxygen Supply Actuating Ring

The emergency oxygen supply actuating ring is on the left forward end of the survival kit container. The ring is colored green and is in the ready position during all normal flight conditions. When emergency oxygen is needed for breathing, the actuating ring is pulled from the kit to actuate the emergency oxygen. The ring, when pulled, separates from the kit after the emergency oxygen has actuated.

WARNING

When pulling the emergency oxygen supply actuating ring, apply the force of the pull at an angle perpendicular to the horizontal plane of the survival kit. A pull at any other angle may cause the ring to malfunction, resulting in no emergency oxygen available to the crewmember.

NORMAL OPERATION

Operation of the oxygen system consists of turning the oxygen supply lever from OFF to ON.

EMERGENCY OPERATION

Emergency oxygen is obtained by pulling up on the emergency oxygen manual release ring until the ring separates from the seat.

OXYGEN DURATION CHART

● DURATION TIME IS HALVED WHEN TWO CREW MEMBERS ARE USING OXYGEN

OXYGEN DURATION-HOURS

Cabin Pressure Altitude Feet	Gage Quantity-Liters						
	10	8	6	4	2	1	Below 1
40,000 and Above	60.6	48.5	36.4	24.2	12.0	6.0	-EMERGENCY- DESCEND TO ALTITUDE NOT REQUIRING OXYGEN
35,000	37.0	29.6	22.2	14.8	7.4	3.6	
30,000	27.2	21.8	16.4	10.8	5.4	2.8	
25,000	20.4	16.4	12.4	8.2	4.0	2.0	
20,000	16.0	12.8	9.6	6.4	3.2	1.6	
15,000	12.8	10.2	7.6	5.2	2.6	1.2	
10,000	10.0	8.0	6.0	4.0	2.0	1.0	
5,000	8.4	6.6	5.0	3.2	1.6	0.8	
SEA LEVEL	7.0	5.6	4.2	2.8	1.4	0.6	

FDD-1-(34)

Figure 1-30

LIMITATIONS

There are no limitations pertaining to the operation of the oxygen system.

PITOT STATIC SYSTEM

NOTE

Some illustrations referred to within this system writeup may be located in appendix A. These illustrations are referred to, within the text, as (figure A-, appendix A).

DESCRIPTION

A conventional pitot-static system is used in the airplane, with a single pitot tube near the top of the vertical stabilizer and two static ports, one on each side of the radome. The purpose of the pitot-static system is to supply both impact and atmospheric pressure to various instruments and system components. The pitot-static system is composed of two separate systems. Both pressures may be utilized by the same instruments but at no time do the pressures intermingle. Both pitot and static pressures are supplied to airspeed pressure switches that retract the flaps and actuate the rudder feel system. The pitot and static pressures are also directed to the air data computer where they are calibrated and corrected (static pressure only) and then sent to the various instruments and systems requiring pitot-static pressures.

Pitot Heat Switch

A pitot heat switch, on the right utility panel, in the front cockpit right console (figure A-1, appendix A), has positions of OFF and ON. When placed to the ON position, the heating element in the pitot tube is energized, thereby preventing formation of ice during flight.

EMERGENCY OPERATION

No emergency operations pertain to the pitot-static system.

LIMITATIONS

Extended ground operation with the pitot heat switch ON should be avoided.

PNEUMATIC SYSTEM

NOTE

Some illustrations referred to within this system writeup may be located in appendix A. These illustrations are referred to, within the text, as (figure A–, appendix A).

DESCRIPTION

The pneumatic system (figure A-6, appendix A), provides high pressure air for the normal and emergency operation of the canopies, normal operation of the ram air turbine (extension and retraction), the nose gear strut extension, emergency operation of the landing gear, wing flaps, cockpit flooding doors, wheel brakes, and after AFC 370, emergency extension of the inflight refuel probe. On aircraft 155890am and up and all others after AFC 474, a secondary emergency canopy pneumatic system provides canopy jettison capability. Air for the pneumatic system is drawn from the engine bleed air supply, via the electronic equipment air conditioning system, and is compressed to approximately 3100 psi by a hydraulic motor–driven air compressor. A pneumatic pressure sensor in the system moisture separator opens a hydraulic shutoff valve, to activate the air compressor; when the system pressure falls below 2750 +50 −0 psi minimum. When the pneumatic system pressure builds to 3100 +100 −50 psi, the pneumatic pressure sensor closes the hydraulic shutoff valve which deactivates the air compressor. The air compressor discharges through a moisture separator and chemical air dryer to pneumatic system air bottles. Check valves prevent the air bottles from discharging back to the air compressor. Shutoff valves isolate the air bottles from their component systems until they are manually discharged. A pressure transmitter, for the pneumatic pressure indicator, is installed in a main pressure line.

PNEUMATIC PRESSURE INDICATOR

The pneumatic pressure indicator (figure A-1, appendix A) is on the pedestal panel and operates in conjunction with the pneumatic pressure transmitter. The indicator has a range of 0 to 5000 pounds with calibrations of 0 to 50 and readings multiplied by 100. Normal system pressures range from 2680 – 3270 psi due to pressure transmitter and pressure gage tolerances.

NORMAL OPERATION

Normal operation of the pneumatic system is accomplished automatically when the engines are running or by the application of external pneumatic power. A check of the pneumatic system cockpit pressure indicator or the basic system pressure gage denotes only the pressure in the supply line. Operating pressures for the emergency subsystems are indicated by their individual pressure gages. To deactivate the air compressor, pull out the pneumatic system control circuit breaker on no. 2 panel.

EMERGENCY OPERATION

There is no emergency operation of the pneumatic system air compressor. However, all the normal and emergency systems have air storage bottles that assure adequate air pressure to the individual pneumatic subsystems. Operation of the normal and emergency subsystems is discussed under the applicable individual systems.

LIMITATIONS

The normal pneumatic system pressure as read on the cockpit indicator is 2680 to 3270 psi.

RADAR BEACON SYSTEM (AN/APN-154)

DESCRIPTION

The radar beacon system AN/APN-154 may be installed on aircraft 155529ag and up or on aircraft after AFC 388. The system enhances range tracking capabilities of certain ground-based radars. These ground-based radars are transported to forward areas for vectoring the aircraft to a target area or a specific target. The radar transponder equipment emits X-band signals which interrogate the radar beacon equipment in the aircraft. The radar beacon system's transmitter, in turn, transmits a pulse reply to the interrogating radar. Hence, the radar site receives a transmitted signal which is considerably stronger than a radar echo. This improves the radar site's acquisition capabilities at the maximum radar ranges, especially in adverse environmental conditions. With the radar site's accurate knowledge of aircraft and target positions, it can precisely direct the aircraft to the target. The radar beacon system is comprised of the following components: the receiver-transmitter, which includes receiving, transmitting, decoding, and power supply assemblies; the duplexer, which permits signal reception and transmission with a single antenna; the antenna (X-band), which is installed on the surface of the aircraft; and the control panel, which contains the operating controls.

RADAR BEACON CONTROLS

The radar beacon control panel (figure A-2, appendix A) is in the rear cockpit on the left console. The panel contains a power switch and a mode selector knob. The OFF position of the power switch removes radar beacon system power. The STDBY position places the radar beacon system in a standby (warm-up) condition. The POWER position places the radar beacon system in full operation. For optimum performance, leave the switch in the STDBY (warm-up) position for at least 5 minutes before selecting the POWER position. The mode selector knob allows selection of single or double pulse radar interrogation reception. There are five DOUBLE positions (labeled 1, 2, 3, 4, and 5). With the knob in the SINGLE position, the system responds only to single pulse interrogations. The DOUBLE positions (modes 1 thru 5) enable the set to respond to double pulse interrogations. The mode the RIO selects should be determined either during mission briefing or by direct voice communication with the radar site. When operating the radar beacon system in conjunction with the data link system, the RIO receives a signal on the digital display indicator as to whether SINGLE or DOUBLE should be selected, but the mode of the DOUBLE position is still established in the mission briefing or by voice communications.

NORMAL OPERATION

X-band pulse signals from an interrogating radar are received by the antenna and directed by the duplexer to the receiver portion of the receiver-transmitter. These signals may be single or double pulse trains, depending on the selected operating mode at the radar site. The signals are amplified and decoded, and when the mode of incoming signals matches the mode selected by the RIO, the decoded signals trigger the transmitter. The transmitter output is directed by the duplexer to the antenna for transmission to the radar site. There are no indicators that indicate system operation. The RIO simply energizes the system and selects the operating mode. Vectoring information is obtained by normal voice communications between the aircraft and the interrogating radar site. When operating the radar beacon system in conjunction with the data link system, vectoring information is automatically displayed on various cockpit indicators.

EMERGENCY OPERATION

There is no emergency operation pertaining to the radar beacon system.

LIMITATIONS

There are no limitations pertaining to the radar beacon system.

SPEED BRAKES

NOTE

Some illustrations referred to within this system writeup may be located in appendix A. These illustrations are referred to, within the text, as (figure A-, appendix A).

DESCRIPTION

The hydraulically operated speed brakes are on the underside of the inboard wing panels and are hinged on the forward side permitting the brakes to open downward. The speed brakes may be positioned at any point in their travel and are controlled by a switch on the inboard throttle grip. If the throttle mounted switch fails the speed brakes can be retracted with the emergency speed brake retract switch on the oxygen panel. On aircraft after AFC 534, the emergency speed brake switch is replaced by the emergency aileron droop switch. The speed brakes are controlled by the right main 28 vdc bus and powered by the utility hydraulic system.

SPEED BRAKE SWITCH

The speed brake switch (figure A-1, appendix A) on the throttle grip has three positions: IN, STOP and OUT. The STOP position is the normal position of the switch. Only the OUT position of the switch is momentary. Placing the switch in the OUT position operates the speed brakes toward the extend position. When the switch is released, it returns to the STOP position. The STOP position deenergizes the selector valve and blocks all ports giving a hydraulic lock for holding the speed brakes to any desired position. Selecting the IN position of the switch closes the speed brakes flush with the wing. The speed brakes take 2–3 seconds to fully open and 2–3 seconds to fully close. The speed brake switch IN position also serves as an emergency disengage switch for the approach power compensator system.

NOTE

The STOP position of the speed brake switch may not hold the speed brakes completely closed. This is indicated by the illumination of the SPEED BRAKE OUT light. If this occurs, position the speed brake switch to IN and leave in that position. However, with the speed brake switch IN, the APCS system is inoperative.

SPEED BRAKE OUT LIGHT

An amber SPEED BRAKE OUT light (figure A-1, appendix A) on the telelight panel illuminates when either or both of the speed brakes are not fully closed.

NOTE

SPEED BRAKE OUT light does not light the MASTER CAUTION light.

NORMAL OPERATION

Normal operation of the speed brakes is accomplished through the three position throttle mounted speed brake switch. The IN position retracts the speed brakes, the OUT position extends the speed brakes, and the STOP position holds the speed brakes in any intermediate position. A SPEED BRAKE OUT light on the telelight panel illuminates any time the speed brakes are not closed.

EMERGENCY OPERATION

The speed brakes automatically close if an electrical failure occurs. If a hydraulic failure occurs, air loads close the speed brakes to a low drag trail position. If a failure occurs in the throttle mounted speed brake switch, the speed brakes can be closed by placing the emergency speed brake switch in the RETRACT position. On aircraft after AFC 534, the emergency speed brake retract switch is removed and emergency retraction is accomplished by pulling the SP BK (speed brake) circuit breaker, on the essential circuit breaker panel.

NOTE

In aircraft 155897ak and up and all others after AFC 392, the APCS can be disengaged by moving the speed brake switch to IN regardless of emergency speed brake switch or speed brake circuit breaker position.

LIMITATIONS

There are no specific limitations pertaining to the operation of the speed brakes.

TOW TARGET SYSTEMS

NOTE

Some illustrations referred to within this system writeup may be located in appendix A. These illustrations are referred to, within the text, as (figure A–, appendix A).

DESCRIPTION

The airplane can be equipped to tow the TDU-22/B or TDU-22A/B Tow Target with the RMU-8/A Reel Launcher, or to launch the AQM-37A Missile Target with LAU-24/A Launcher.

RMU-8/A REEL LAUNCHER

This jettisonable reel launcher, mounted on the centerline station, is used to tow the TDU-22/B or TDU-22A/B tow targets. The reel launcher (figure 1-32) is a ram air turbine driven power unit, with reversible pitch blades; capstans; a towline storage spool; a level wind mechanism; a target launcher boom and saddle; an emergency cable cutter, and an automatic control system. The automatic control system provides for automatic target launching and reel-out and reel-in to a preset towline length. The tow reel is capable of reel-out and reel-in speeds of 5200 feet per minute. The towline length normally carried on the tow reel is 40,000 feet, but it is capable of carrying up to 100,000 feet. Carriage of the RMU-8/A does not cause any adverse flight characteristics; however, the pilot should expect to encounter the following phenomena:

a. Minor vibration during reeling operation, particularly during reel acceleration and deceleration.

b High pitched whine (like a siren) during high speed reeling.

c Torque effect during high speed reeling.

The launcher is down for launch and recovery only, and is up during target tow. The reel contains a nitrogen bottle to actuate the launcher and the pneumatic brakes. It uses the aircraft emergency bus to provide power for stopping the reel and cutting the cable if aircraft ac power is lost.

Towline Length Sensing Unit

The towline length sensing unit (figure 1–31) contains pulsing switches used in sensing the amount of towline reeled out, three counter operated limit switches used in automatic sequencing during the launch operations, and a reset mechanism. The switches and switch functions are listed below:

OUT STOP	The out stop switch is used to set the maximum towline length required (minus stopping distance).
LAUNCH STOP	The launch stop switch determines the point beyond which high speed reel–out action is permitted.
IN STOP	The in stop switch determines the point at which deceleration from high speed reel–in occurs, within which high speed reel–in may not be started, and within which manually controlled target recovery may be started.
CABLE IN OUT/OFF	The cable switch is a three position switch with positions of IN, OUT, and OFF. This switch allows the ground crew to simulate a complete operating cycle. For inflight tow target operation, the switch is positioned to OFF.

RMU-8/A CONTROL PANEL

FDD-1-(35)

Figure 1–31

Blade Pitch Change Actuator

The blade pitch change actuator consists of a reversible dc motor and a switch package containing six limit switches. The dc motor, through a gear train, rotates the power unit blades clockwise during recovery and counterclockwise during launching. The six limit switches are cam operated and determine the number of degrees rotation the blades turn and the direction of the turn.

Speed Monitor Switch Assembly

The speed monitor switch assembly consists of four switches. The operation of each switch is governed by the power unit speed. The switches are set to operate at 8600 to 9400 rpm (normal governing band), at 9900 rpm (ten percent overspeed), and at 10,800 rpm (twenty percent overspeed). The 8600 and 9400 rpm switches perform target reel–out/reel–in speed monitoring functions. The 9900 and 10,800 rpm switches are for emergency control.

Central Control Box

The central control box contains function control relays and an acceleration monitor. The relays are used to control the overall operation of the RMU–8/A. The acceleration monitor is used to automatically control change–rate of the towline speed during the operating cycle.

Power Unit

The power unit consists of eight reversible–pitch blades, each of which attaches to a central hub. The reversible–pitch blades are arranged in two rows of four blades each. The blade angle is controlled through the blade pitch change actuator which is mechanically coupled to the reversible–pitch blades and drives the blades on their individual axis. The power unit is driven by the wind force created by the forward speed of the aircraft. The unit rotates counterclockwise (as viewed from the rear) during reel–out and functions as a brake against the trailing towline and target. The power unit rotates clockwise (as viewed from the rear) during reel–in and functions as a turbine to furnish drive power to the transmission and spool during the recovery operation.

Transmission Assembly

The transmission assembly couples the driving power developed by the power unit through a gear train assembly to the capstan and through a slip–clutch mechanism to the towline spool. The transmission assembly houses the gear train, clutches, oil pump, scavenger oil–pump, and the oil dip stick.

Clutch Mechanism

The clutch mechanism consists of two, over-running sprag-type clutches and a constant torque slip-clutch driven through a gear train. The arrangement of the clutch mechanism maintains towline tension between the capstan and the towline spool to prevent towline slippage on the capstans.

Spool and Levelwind

The spool and levelwind provides storage of the towline which is reeled-out or reeled-in. For convenience in rewinding, the spool is removable. The levelwind is driven by the spool and may be readily disengaged for purposes of synchronization and threading.

Lubrication

The lubrication system is a wet-sump type with automatic reversing, rotary, internal gear pumps located in the transmission assembly. Positive lubrication is furnished to the main drive gear mesh, capstan bearing, clutch assembly, and power unit. A separate pump scavenges the power unit and discharges the oil into the transmission assembly. The transmission assembly and clutch mechanism are capable of operating approximately 10 seconds with an interrupted oil supply. During the target launch and recovery operations, all oil is routed to the clutch assembly for cooling. There are also two oil pressure switches, an oil pressure indicator, and a temperature indicator in the system. The oil pressure switches and the oil temperature indicator are safety devices which provide a visual indication of low oil pressure and excessive oil temperature to the operator.

Pneumatic System

The pneumatic system provides the necessary power to extend and retract the launcher and to apply the brakes to the reel-launcher. A storage bottle contains compressed nitrogen at 3000 psi. The pressure is reduced to 300 psi before it is delivered to any of the operating mechanisms of the reel launcher. This supply of nitrogen permits approximately four complete operating cycles. When the storage bottle pressure drops below 700 psi, a light in the control panel gives a low air indication. The latch that holds the launcher in the retracted position is also operated by nitrogen pressure. This latch is in series with the actuating cylinder that forces the launcher down. This is to assure that the latch is opened before the launcher boom is extended. The pneumatic system is designed so that towline tension is maintained at all times during launcher extension and retraction and so that the brakes are applied gradually to prevent snapping of the towline. During an emergency the brakes are applied instantly.

Launcher

A pneumatically operated target launcher is provided to launch and retrieve the target through the region of disturbed air flow around the RMU-8/A and the airplane. The launcher also acts as a shock absorber upon initial target contact during recovery. The launcher contains two mechanical clamps which hold the target when the launcher is fully retracted. These automatically open as the launcher is lowered.

Towline Cutter

An explosive cartridge operated towline cutter is mounted on the launcher support structure and provides for cutting the towline under any possible flight operating condition. The cutter is fired manually by actuation of the emergency stop and towline cut switch and is automatically fired under certain emergency conditions. The conditions under which the towline is cut automatically are: a 20% overspeed (towline reel rate of 6000 ft/min.) during any of the four operating cycles, or target approaching within 200 feet of the aircraft after failure to slow down and stop after passing through the preset in stop distance during the reel-in cycle (towline cutter is actuated by an anti-collision device).

Tension System

A strain gage bridge in the high tension sheave works in conjunction with a regulated power supply to present an indication of towline tension on the towline tension indicator.

Towline Speed Indicating System

A tachometer generator in the reel launcher provides an indication of towline speed to the operator and a signal indicating the rate of change of speed to the acceleration monitor.

Control Circuit Sequencing

The normal stop and the emergency stop and towline cut functions can be initiated at all times. The normal stop function must be initiated to interrupt any cycle before initiating a new cycle. The following table illustrates cycles that can be initiated after the completion of a manual or automatic normal stop at various towline lengths relative to the towline length limit switch.

TARGET POSITION	FUNCTIONS POSSIBLE
Target Stowed	Launch, Reel-Out
Target between stowed position and in-stop (2500 ft)	Reel-Out, Recover
Target between in-stop (2500 ft) and out-stop	Reel-Out, Reel-In
Past out-stop	Reel-In

RMU-8/A REEL LAUNCHER

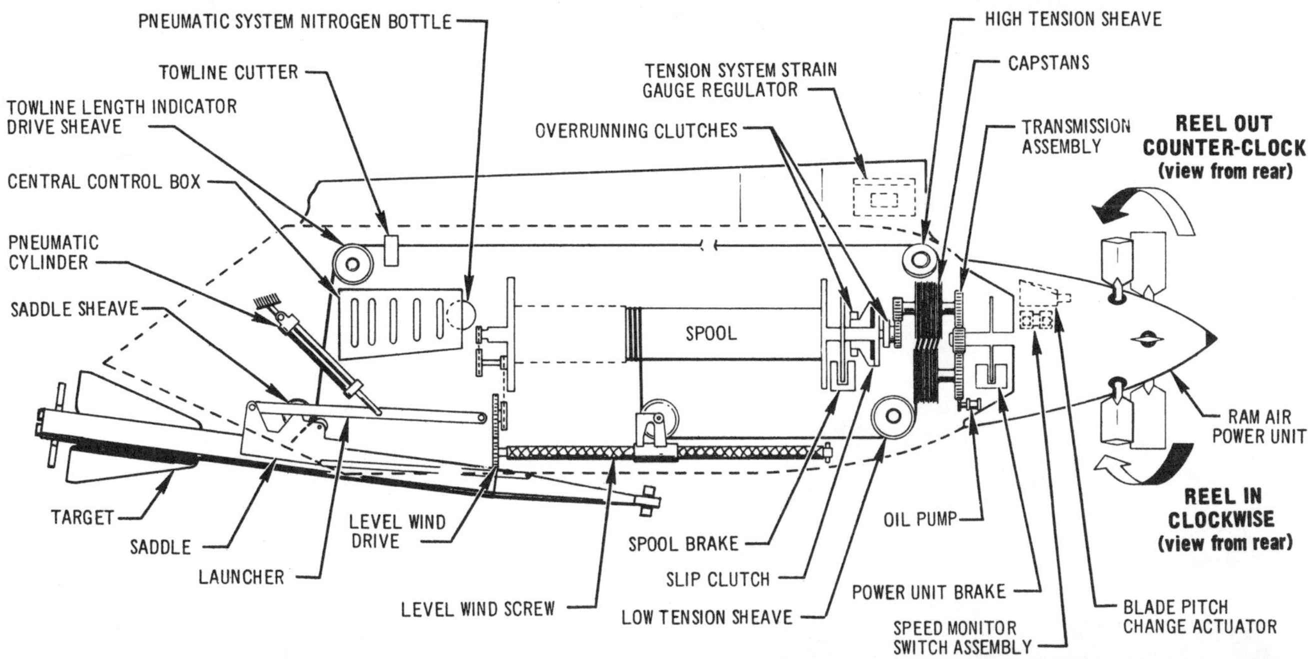

DRAG INDEX	**6.4**
LENGTH	**160 inches**
DIAMETER	**20 inches**
WEIGHT with tow cable and target	**approx. 1800 pounds**
(Add 51 pounds for the Aero 27A Ejector Bomb Rack.)	

FDD-1-(36-1)

Figure 1-32 (Sheet 1 of 2)

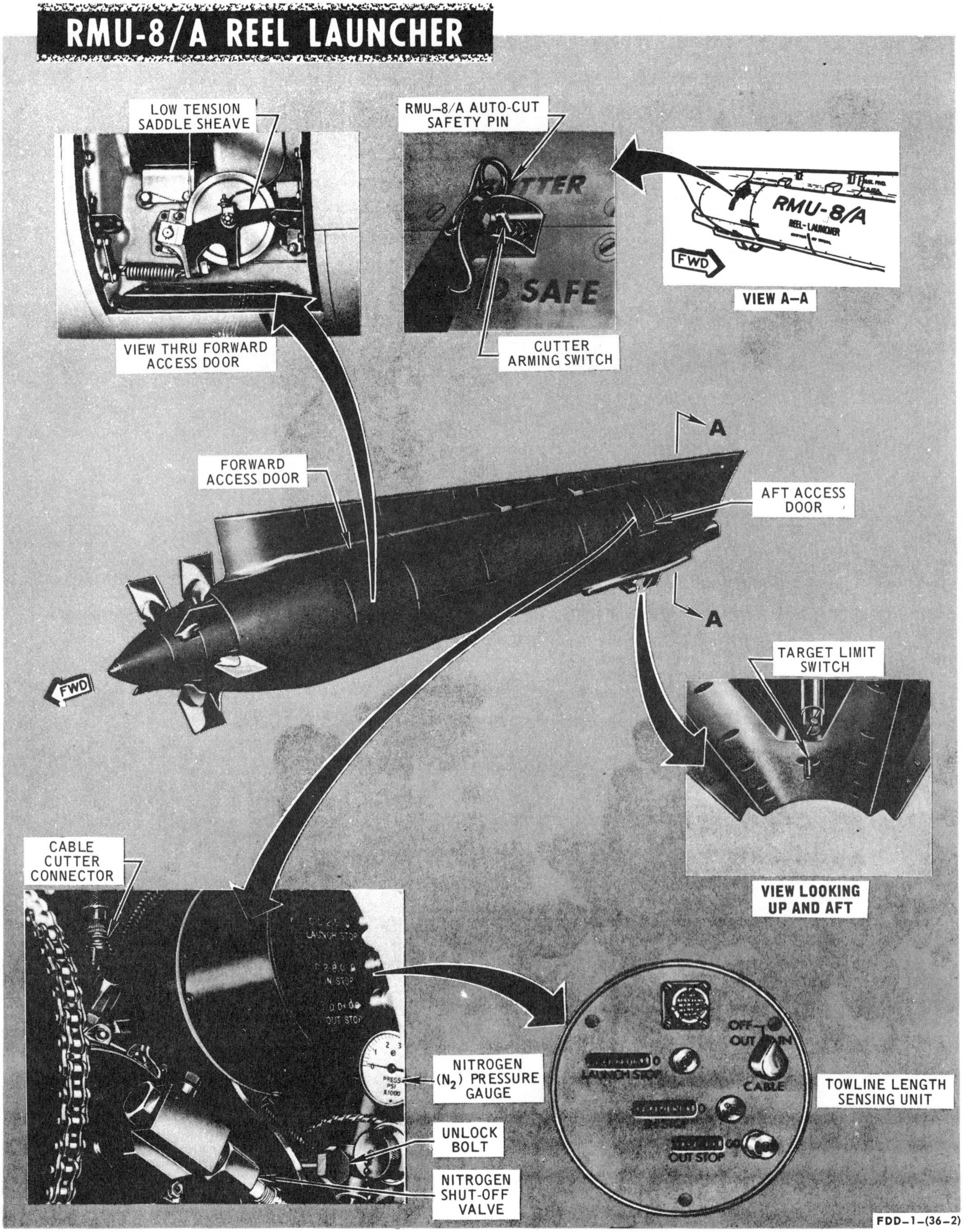

Figure 1-32 (Sheet 2 of 2)

TDU-22/B and TDU-22A/B Tow Targets

Both targets are center-of-gravity towed targets which measure 78 inches in length, 7 inches in diameter, and have a fin span of 22 inches. The TDU-22B weighs approximately 32 pounds and is radar augmented. The TDU-22A/B weighs approximately 40 pounds and is radar and infrared augmented. Radar augmentation consists of Luneberg lenses, and infrared augmentation consists of four flares with command radio receiver and battery pack. Flares are fired by a small tone generator mounted in the RIO's radio mike circuit.

RMU-8/A Control Panel

The reel launcher control panel (figure 1-31) may be mounted adjacent to the RIO's seat, or other accessible location. It has all of the instruments and switches required to control and monitor tow reel functions. Three instruments indicate reel-in or reel-out speed, tow cable tension, and tow cable length. The switches include: a master power switch, a target launch switch, a tow reel control switch, a turbine pitch control switch, a tow reel stop switch and a stop and cut switch. Indicator lights include an EMERG PWR light, a LOW AIR light, a TOW CUT light, a LAUNCHER DOWN light, a TARGET OUT light and a SAFETY ARM light.

Master Power Switch

The master power switch, marked ON and OFF, is utilized to energize the tow target system.

Target Launcher Switch

The target launcher switch, with positions of LAUNCH and RECOVERY, is used during the first stage of target launch, and the last stage of target recovery. Placing the switch to LAUNCH extends the launching boom and allows the target to withdraw from the saddle. The LAUNCHER DOWN and TARGET OUT lights illuminate and the target begins to reel-out to approximately 100 feet. At 100 feet, the launching boom retracts, the LAUNCHER DOWN light goes out, and the tow reel is braked to a stop. During recovery, the reverse process will take place.

Tow Reel Control Switch

The tow reel control switch has positions marked REEL-IN and REEL-OUT, and is used after the target has been launched. Placing the switch to REEL-OUT releases the brake on the tow reel. The target then begins to reel-out at an increasing speed to a maximum of 5000 feet per minute. At a preselected cable length, the tow reel brake is applied to stop the target. Placing the switch in REEL-IN returns the target to a recovery position at reel-in speeds up to 5000 feet per minute.

Turbine Pitch Control Switch

Turbine pitch control switch, marked INCREASE SPEED and FEATHER, is used to manually control target recovery speeds. After the target has completed reel-in, actuating the turbine pitch control switch to INCREASE increases the target recovery speed approximately 150 feet per minute. Actuating the switch to FEATHER stops recovery procedures.

Tow Reel Stop Switch

The stop switch has two positions: STOP, and a spring-loaded OFF position. The switch allows any operation in progress to be halted.

Stop & Cut Switch

The stop and cut switch has two positions: STOP & CUT and a guarded OFF position. The switch provides an emergency means of stopping and cutting the towline at any time.

Towline Length Indicator

The towline length indicator indicates the distance (in ten foot increments) the target is from the reel launcher assembly.

Towline Speed Indicator

The towline speed indicator provides an indication of the speed the towline is entering or leaving the RMU-8/A. The indicator is calibrated in feet-per-minute and consists of two scales, a lower scale with a range of 0 to 700 and an upper scale with a range of 700 to 6500. The upper scale has a green arc to indicate the normal operating range and a red line at 6000 feet-per-minute to indicate the maximum turbine speed. A neon light on the dial indicates which speed range is operative. The lower scale is operative with the light off and the upper scale is operative with the light on.

Towline Tension Indicator

The towline tension indicator provides an indication of the tension of the towline. The indication presented is calibrated in hundreds of pounds.

Target Out Light

The TARGET OUT light provides an indication that the target has been launched or is loose in the launcher because of lost towline tension.

Launcher Down Light

The LAUNCHER DOWN light provides an indication that the target launcher assembly is in the down position or i not locked.

Low Air Pressure Light

The LOW AIR PRESS light illuminates when the air storage bottle is depleted to an air pressure of 700 psi or lower.

Towline Cut Light

The TOWLINE CUT light functions in conjunction with the stop and cut sequence. When the stop and cut sequence is activated, the TOWLINE CUT light shows that the action of cutting the towline has been accomplished.

Emergency Power Light

The EMERG PWR light provides an indication of failure of the primary 28 volt dc supply.

Safety Arm Light

The SAFETY ARM light provides the operator with an indication that the anti-collision safety device is armed. The SAFETY ARM light illuminates during reel-in procedures and goes out during recovery.

AQM-37A MISSILE TARGET SYSTEM

The AQM-37A missile target system consists of a missile target, an LAU-24/A launcher, and a PEU-56/A missile firing control panel. The target and launcher are mounted only on the aircraft fuselage centerline station. Controls and indicator lights necessary for selecting, launching, and/or jettisoning the target, are in the forward cockpit. The target system installation utilizes components from existing aircraft systems as well as those specifically installed for the target.

AQM-37A Missile Target

The target is cylindrically constructed with a streamlined nose section and clipped delta wings which have a vertical fin mounted on each tip. Control of the target is maintained through full span ailerons on the trailing edge of the wings, and by moveable canard surfaces on the forward nose section of the target. The target is propelled by a rocket engine that utilizes dual thrust chambers. The larger chamber is used for initial boost and acceleration, while the smaller chamber is used for sustained power during target cruise operations. Stabilization and control of the target is maintained through a three axis gyro reference to pitch, roll, and yaw, regulated by a self-contained preset flight programmer. Mission capabilities of the target include supersonic flight operations in cruise altitude range up to 70,000 feet; however, operations are normally accomplished at altitudes between 30,000 and 40,000 feet.

PEU-56/A Missile Firing Panel

This panel is in the forward cockpit on the right console. The panel contains a PWR ON light, a READY light, and a target function selector switch with positions marked SELECT, OFF, and JETTISON. When the function selector switch is placed to SELECT, power is supplied to the target and the PWR ON light illuminates. THe READY light illuminates when power is applied to the target for five minutes. Placing the function selector switch to the JETTISON position immediately jettisons the target. Separation of the jettisoned target from the launcher provides a destruct signal to the target's destruct system. The target may also be jettisoned by actuating the external stores emergency release button.

NOTE

The READY light indicates only that 5 minutes have elapsed since power was applied to the target and does not indicate target operational readiness. The target may be launched without the READY light illuminated if the PWR ON light has been on for five minutes.

NORMAL OPERATION

Refer to section IV, Tow Target Procedures.

EMERGENCY OPERATION

Refer to section IV, Tow Target Procedures.

LIMITATIONS

Refer to section IV, Tow Target Procedures.

WING FOLD SYSTEM

NOTE

Some illustrations referred to within this system writeup may be located in appendix A. These illustrations are referred to, within the text, as (figure A–, appendix A).

DESCRIPTION

Each outer wing panel is folded upward to a vertical position by a conventional hydraulic actuator that receives hydraulic pressure from the utility hydraulic system. A mechanical locking system is used to lock wing pins in hinge fittings when wings are spread. A flush mounted control lever (figure A–1, appendix A) on the right console in the pilot's cockpit, is connected by push rods and push–pull cables to a pin locking device in the wing fold area. Pulling UP on the lever unlocks wing pins, extends warning flags on the upper wing surfaces, illuminates amber L WING PIN UNLOCK, and R WING PIN UNLOCK lights in both cockpits and energizes the wing fold. Wing fold is actually accomplished by a two–position toggle switch underneath the wing pin release lever and is exposed when the lever is raised. The switch is marked FOLD and SPREAD. As an added safety precaution, the wing fold hydraulic system receives its hydraulic pressure from the landing gear down pressure line; this prevents pressurizing the wing fold system when the landing gear is UP. When folding or spreading the outer panels, observe the following precautions:

a. Jury struts removed.
b. Do not fold or spread wings broadside of the blast of an aircraft's engines.
c. Do not fold or spread wings in winds over 60 knots.
d. Ensure wings are spread and locked prior to taxiing, or prior to being spotted directly aft of the jet blast deflector during launch operations.

NORMAL OPERATION

Normal operation consists of folding and spreading the wings, and is accomplished through the wing fold panel on the right console. To fold the wings, pull UP on the wing lock lever, and place the wing fold switch in the FOLD position. To spread the wings, remove the jury struts, and place the wing fold switch to SPREAD, visually determine that both outer wing panels are in the full spread position prior to placement of the manual wing pin lock lever in the down (locked) position. Red warning flags which are attached to the wing pin locks are flush with the wing skin if the wing pin locks are fully inserted. The warning flags extend above the wing surface, inboard of the wing fold line, when the wing pin locks are not inserted. When the wing pin locks are fully inserted the L. WING PIN UNLOCK and R. WING PIN UNLOCK warning lights are extinguished.

EMERGENCY OPERATION

There is no emergency operation pertaining to the wing fold system.

LIMITATIONS

Whenever the aircraft is parked or towed with wings folded, jury struts are to be installed. Taxiing with wings folded and jury struts not installed should be held to a minimum. Aboard ship, jury struts must be inserted any time wings are folded.

PART 3 SERVICING

For information pertaining to servicing (i.e., authorized AGE, consumable materials, capacities, pressures, and cockpit procedures), refer to NAVAIR 01–245FDD–1C, NATOPS Servicing Checklist.

PART 4 OPERATING LIMITATIONS

AIRCRAFT

GENERAL

All airplane/systems limitations that must be observed during normal operation are covered or referenced herein. Some limitations that are characteristic only of a specialized phase of operation (emergency procedures, flight through turbulent air, starting procedures, etc.) are not covered here; however, they are contained along with the discussion of the operation in question.

INSTRUMENT MARKINGS

The limitation markings appearing on the instrument faces are shown in figure 1–33 and noted in the applicable text.

INSTRUMENT FLUCTUATION

FUEL FLOW

100 PPH maximum for indicator readings of 0 – 3000 PPH. 750 PPH maximum for indicator readings of 3001 – 12,000 PPH.

RPM

±0.2% from steady–state condition.

EGT

±5°C maximum for steady–state operation from IDLE through MIL power settings. ±8°C maximum for steady–state afterburning.

EXHAUST NOZZLE

Limited by EGT fluctuation.

OIL PRESSURE

±2.5 PSI from steady–state pressure.

ENGINE LIMITATIONS

Refer to Engines, Part 2 of this section.

AIRSPEED LIMITATIONS

The maximum permissible airspeeds in smooth or moderately turbulent air with arresting hook, landing gear, and wing flap retracted, and with speedbrakes retracted or extended are as shown in figure 1–35. Airspeed limitations for operation of various systems are presented in Operations/Limitations, figure 1–33.

CAUTION

When flying at airspeeds in excess of 350 knots CAS below 10,000 feet with CG location aft of 32.0% MAC, avoid abrupt control motions. The CG location aft of 32.0% MAC will normally occur whenever full internal fuel is maintained in conjunction with external stores. Prior to takeoff, refer to the Handbook of Weight and Balance Data (AN 01–1B–40) to determine CG location for the specific configuration.

PROHIBITED MANEUVERS

a. Full–deflection aileron rolls in excess of 360°.

b. Intentional spins.

c. Abrupt control movements when carrying the 600–gallon external fuel tank or the D–704 air refueling store.

d. Lateral control deflections in excess of ½ of the total stick travel when carrying the MK 104, the 600–gallon external fuel tank, the 370–gallon external wing tanks or the D–704 air refueling store. This restriction does not apply when in the takeoff or landing configuration.

e. With the AFCS engaged, intentional maneuvers that exceed the automatic disengagement limits of the system.

f. Lateral control deflections in excess of 1/3 of the total stick travel when carrying the RCPP–105 starter pod. This restriction does not apply when in the takeoff or landing configuration.

g. Negative G in excess of 30 seconds.

h. Zero G in excess of 10 seconds.

i. Airborne deployment of the drag chute except for emergency out of control/spin recovery.

CG LIMITATIONS

The center of gravity for all currently permissible gross weights and configurations must be kept between 27.0% and 36.0% of the mean aerodynamic chord (MAC). However, the maximum allowable aft CG will be forward of 36.0% MAC with certain external loading configurations. To maintain minimum acceptable longitudinal stability, the allowable aft CG must be moved forward as wing–mounted stores are added to the airplane.

INSTRUMENT MARKINGS

BASED ON JP-4 or JP-5 FUEL

AIRPLANES 153071z thru 153087aa

J79-GE-8

EXHAUST GAS TEMPERATURE	
635°C	▬ MAXIMUM STEADY STATE TEMP.

AIRPLANES 153088aa and Up

J79-GE-10

EXHAUST GAS TEMPERATURE	
668°C	▬ MAXIMUM STEADY STATE TEMP.

MIL-L-7808 OIL		MIL-L-23699 OIL
60 PSI	MAXIMUM	70 PSI
35 PSI	MINIMUM AT MILITARY RPM STEADY-STATE GROUND OPERATION	45 PSI
12 PSI	MINIMUM AT IDLE POWER	12 PSI
30–60 PSI	INFLIGHT NORMAL OPERATION AT MIL POWER AND ABOVE	35–70 PSI

TACHOMETER	
102%	▬ MAXIMUM

Notes

- PLACARD PRESSURE ON THE OIL PRESSURE GAGES IS CORRECTED TO 100 PERCENT RPM. AT MILITARY THROTTLE SETTING DURING T_2 CUTBACK, OR ANY OTHER SPEED REDUCTION, INDICATED PRESSURE WILL DECREASE BELOW PLACARD PRESSURE APPROXIMATELY 1 PSI PER 1 PERCENT REDUCTION IN RPM BELOW 100 PERCENT RPM.
- INDICATED OIL PRESSURE CORRECTED TO 100 PERCENT RPM MUST REPEAT WITHIN 5 PSI OF THE PLACARDED OIL PRESSURE.
- BELOW 20,000 FEET DURING STEADY STATE OPERATION, ANY ERRATIC PRESSURE CHANGE WHICH EXCEEDS 5 PSI FOR MORE THAN 1 SECOND MUST BE INVESTIGATED. AT AND ABOVE 20,000 FEET, PRESSURE FLUCTUATIONS ARE LIMITED TO THE FOLLOWING: MAXIMUM OF 20 PSI BELOW NORMAL PRESSURE FOR A DURATION OF 3 SECONDS, OCCURRING NO MORE THAN 4 TIMES PER MINUTE; OR A MAXIMUM OF 10 PSI BELOW NORMAL PRESSURE FOR A DURATION OF 1 SECOND, OCCURING NO MORE THAN 15 TIMES PER MINUTE.
- WHEN OPERATING WITH MIL-L-23699 OIL AND THE OUTSIDE AIR TEMPERATURE IS 100°F OR ABOVE, A MINIMUM OF 40 PSI AT MILITARY RPM STEADY STATE GROUND OPERATION IS PERMISSIBLE.

FDD-1-(38)B

Figure 1-33

OPERATIONS/LIMITATIONS

- SYSTEMS -

REFUELING PROBE	EXTENDING–RETRACTING	300 KT/.9 IMN
	EXTENDED	400 KT/.9 IMN
RAM AIR TURBINE	MAX SPEED	515 KT/1.1 IMN
	G LIMIT EXTENDING	0 – +3G
	G LIMIT OUT	–1.0 – +5.2G
	MINIMUM AIR PRESSURE EXTENSION	1000 PSI
	MINIMUM AIR PRESSURE RETRACTION	2000 PSI
	WORKS TO	90 KT
LANDING GEAR	EXTENSION–RETRACTION	250 KT
FLAPS	LOWERING–RAISING	250 KT
	AUTO BLOW UP	230–240 KT
	AUTO BLOW DOWN	234–210 KT
CORNER REFLECTOR	EXTENSION–RETRACTION	250 KT
RUDDER CHANGEOVER	ACCELERATING	228–252 KT
	DECELERATING	232–218 KT
CANOPY	MAX NORMAL OPENING	60 KT
	DEPARTS AIRCRAFT	100 KT
RAMPS	PROGRAM	1.4–1.7 IMN
DRAG CHUTE	MAX DEPLOYMENT	200 KT
	CROSSWIND LANDING	20 KT @ 90°
EJECTION LIMITATIONS	HUMAN	400 KT
	SEAT	500 KT/.92 IMN
BAROSTAT	INITIATION	13,000 ± 1500 FT
AIRCRAFT SPEED RESTRICTIONS (CLEAN)	0–30,000	750 KT
	ABOVE 30,000	710 KT/2.1 IMN
AFCS LIMITS	±70° PITCH & ROLL	–1 +4G
RADAR ALTIMETER	±30° BANK ±50° PITCH	5,000 ALTITUDE
CG LIMITATIONS	27.0% THRU 36.0% MAC	

- ACCELERATION LIMITS -

CLEAN (37,500)	–3.0 +8.5 (LESS THAN .72 IMN)	+6.5 (MORE THAN 1.05 IMN)

- GROSS WEIGHT/AOA -

MAXIMUM ALLOWABLE GROSS WEIGHT	FIELD TAKEOFF	56,000 LB	ANGLE OF ATTACK AIRSPEED (DRAG COUNT 0–60)	CLIMB (400–.88)	5.5 UNITS
	FIELD LANDING	46,000		MAX ENDURANCE	8.5–9.0
	CATAPULTING	56,000		MAX RANGE	7.5–8.0
	ARRESTED LDG	40,000		GEAR EXTENSION	9.0
	TOUCH & GO	40,000		FLAP EXTENSION	13.0
	FMLP	40,000		APPROACH	18.3
	BARRICADE	34,000		NO FLAP	17.0
	SINGLE ENGINE	36,000			

- FUEL -

FUEL QUANTITIES (USEABLE)	MAX FUSELAGE		9,303 LB
	MAX INTERNAL		13,587 LB
	MAX INTERNAL & ℄		17,667 LB
LOW FUEL/TRANSFER	LOW FUEL	AUTO XFR	TK 7 XFR
THRU BLK 41	1880 ± 200 LB	2615 ± 200 LB	2050 ± 200 LB
BLK 42 & UP	1880 ± 200 LB	1465 ± 200 LB	WHEN ROOM AVAIL.
WING FUEL TRANSFER LIMITS	+75° TO –15° PITCH		
DUMP CAPACITY	680 POUNDS/MINUTE		

FDD-1-(37-1)A

Figure 1-34 (Sheet 1 of 2)

OPERATIONS/LIMITATIONS

- PRESSURES -

		PC 1 PSI	PC 2 PSI	UTILITY PSI
HYDRAULIC PRESSURE	R ENGINE ONLY	0	2750–3250	2550–3000
	L ENGINE ONLY	2750–3250	0	2750–3250
	BOTH ENGINES	2750–3250	2750–3250	2750–3250
	STAB AUGS	PITCH AUG	———	YAW & ROLL AUG
	CHK HYD LIGHT	ON BELOW 1500 ±100	OUT ABOVE 1750	
	ACCUM PREFLIGHT	1000 ±50	1000 ±50	1000 ±50
		WHEEL BRAKE ACCUMULATOR		1000 ±50

PNEUMATIC SYSTEM GAGE PRESSURE	NORMAL COCKPIT INDICATION		2680–3270 PSI
	PRESSURES FOR NOSE GEAR EXTENSION	BEFORE	2750 PSI MIN
		AFTER	1350 PSI MIN
			2300 PSI MAX
	ALL GAGES ON PREFLIGHT		2750–3100 PSI

- ENGINE -

	CONDITION	TEMP (°C)	TIME (SEC)
EXHAUST TEMPERATURE LIMITATIONS	STEADY STATE FOR CONTINUOUS OPS	660 ±8	NO LIMIT
	TEMPERATURE DURING STARTING	1000	3
		980	10
		930	60
		900 DOWN TO 733	90
	ABOVE 705° DURING START REQUIRES DISCREPANCY CARD		
	DURING ALL ENGINE OPERATIONS OTHER THAN STARTING	750°	3
		668	NO LIMIT
INLET TEMPERATURE LIMITATIONS	ENGINE INLET TEMP HIGH WARNING LIGHT	121°	OPS PROHIBITED BELOW 30,000 FT
	TRANSIENT TEMP OPERATION	121°–193°	5 MINUTES PER HR ABOVE 30,000 FT

		STATIC (RPM)	INFLIGHT (RPM)
SPEED LIMITATIONS	NORMAL OPERATION	100 ±0.5%	100 ±0.5%
	ALLOWABLE OVERSPEED NON TIME LIMITED	103%	102%
	ALLOWABLE OVERSPEED TIME LIMITED	103–105% 3 MIN	102–103.6% 1 MIN
	IDLE	65 ±1%	

		BELOW 35,000 FT	ABOVE 35,000 FT
POWER TIME LIMITS	MILITARY	30 MIN	2 HRS
	MAXIMUM	30 MIN	2 HRS

G (LUBRICATION)	NEGATIVE G		30 SEC
	ZERO G		10 SEC
BLEED AIR CHECK VALVE FAILURE (RUN-UP)	RPM ABOVE 67.5%	EGT DOWN 25°C	FUEL FLOW DOWN 100 pph
IGNITION	2 MIN ON 3 OFF 2 ON 23 OFF		

OIL PRESSURE (MIL-L-23699)	MAXIMUM	70 psi
	MIN. AT MILITARY STEADY STATE GROUND OPS	45 psi
	MIN. AT IDLE	12 psi
	NORMAL INFLIGHT MILITARY AND ABOVE	35–70 psi
	FLUCTUATION ALLOWABLE	±2.5 psi
	OIL PRESSURE BELOW 50 PSI PRIOR TO ENGAGING GEN.	

NORMAL FUEL FLOW (PPH)	LIGHT OFF/15 SEC 225–800	IDLE 800–1400	MILITARY (APPROX) 7000
	THROTTLE CHOP 330		
BOOST PUMP PRESSURE	30 ±5 psi		

FDD-1-(37-2)B

Figure 1-34 (Sheet 2 of 2)

AIRPLANE SPEED RESTRICTIONS

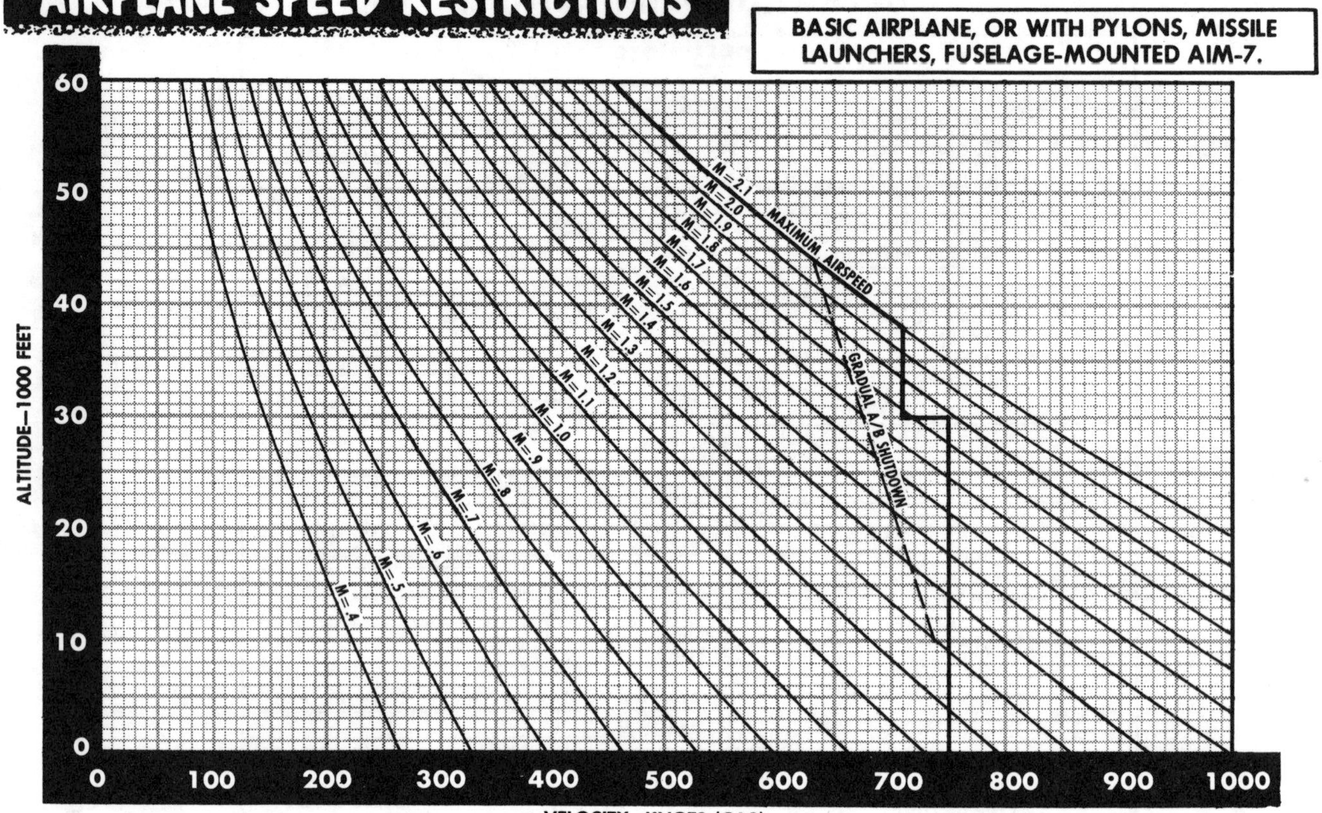

NOTE: ● UNDER SOME CONDITIONS, MAXIMUM AIRSPEEDS ARE DETERMINED BY INLET TEMPERATURE LIMITATIONS REFER TO ENGINE INLET TEMPERATURE LIMITATIONS CHART, PART 2 OF SECTION I.

● FOR AIRSPEED RESTRICTIONS CONCERNING OTHER CONFIGURATIONS, REFER TO EXTERNAL STORE LIMITATIONS CHART.

FDD-1-(39)

Figure 1-35

The minimum acceptable level of stability is based on airplane controllability. Refer to Longitudinal Stability, section IV, for a discussion of flight characteristics near the aft CG limit. Stability numbers for individual stores are contained in part 1 of section XI. After compiling the airplane stability index (sum of stability numbers), refer to the Aft CG Limits chart (figure 1-36) to determine maximum allowable aft CG. For precise loading and CG data, refer to the Handbook of Weight and Balance Data (AN 01-1B-40) for the specific airplane.

WEIGHT LIMITATIONS

The maximum allowable gross weights are as follows:

Field takeoff	56,000 pounds
Field landing (flared)	46,000 pounds
Catapulting	56,000 pounds
Arrested landing, touch-and-go, and FMLP	40,000 pounds
Barricade engagement	34,000 pounds

ACCELERATION LIMITATIONS

The maximum permissible accelerations shown in figures 1-37 and 1-38 are for flight in smooth air. Moderate and heavy buffet should be avoided whenever possible. In conditions of moderate turbulence, it is essential that accelerations resulting from deliberate maneuvers be reduced 2.0 G below that shown in figure 1-37. This is to minimize the possibility of overstressing the airplane as a result of the combined effects of gust and maneuvering loads. Acceleration limitations for ram air turbine operation is 0.0 to +3.0 G for extension/retraction and -1.0 to +5.2 G with the RAT fully extended. Carriage and release acceleration limitations for the various external stores are shown in the External Stores Limitations charts, this section and appendix A of NAVAIR 01-245FDB-1T.

CAUTION

Normal accelerations of 8.5 G are permissible only when gross weight and airspeed are equal to or less than 37,500 pounds and 0.72 Mach respectively. Acceleration to 8.5 G at conditions outside these limitations will impose excessive stresses on aircraft structure. Since aircraft fatigue life depends largely on the number and magnitude of G application, accelerations above 6.5 G should be used only as necessary in mission performance. Care should be taken to avoid G overshoot beyond authorized limits under all conditions.

FLIGHT STRENGTH DIAGRAM

The flight strength diagram (figure 1–38) is a composite presentation of the airplane's operating envelope at three different gross weights. Parameters of each envelope include maximum allowable Mach number, wings level stall speed at sea level, and the positive and negative load factor limits. This diagram further restricts allowable negative load factors at speeds above 1.5 Mach, and allowable positive load factors at speeds above 1.8 Mach.

CARRIER OPERATIONS

For carrier arrested landings, total fuel in fuselage cells 1 through 6 shall not exceed 5100 pounds. This restriction applies to cells 1 through 6 (sector reading on fuel quantity gage) only. Fuselage cell 7 may be either full or empty; however, internal wing tanks should be empty. This restriction does not apply to touch–and–go landings. For carrier approach and arrestment limitations, refer to applicable recovery bulletins.

CAUTION

If it is necessary to make an arrested landing with fuel in the internal wing tanks, a notation to this effect must be made on the yellow sheet.

EXTERNAL STORES

GENERAL

NOTE

Most of the operating limitations data for air–to–ground weapons have been moved to appendix A of the F–4 Tactical Manual, NAVAIR 01–245FDB–1T.

Only the external stores listed in this section and in appendix A of NAVAIR 01–245FDB–1T may be carried and released, singly or in combination, by the aircraft. The External Stores Limitations charts (figure 1–42 and appendix A of NAVAIR 01–245FDB–1T) illustrate all authorized loading configurations; maximum permissible airspeeds and accelerations for carriage, launch, release, and jettison; maximum dive angles for delivery; and pertinent remarks/notes for each authorized store. Catapult launches and arrested landing with the external stores depicted in figure 1–39 are authorized unless otherwise specified.

WARNING

Ordnance shall be jettisoned above the minimum fragmentation clearance altitude when possible, even though jettison safe is selected.

CARRIER OPERATIONS

Normal carrier operations are not permitted with external store loadings in excess of 60,000 inch–pounds of static moment. Under emergency landings only, twin–engine arrestments with asymmetric loading up to 212,000 inch–pounds are permitted, and asymmetric loadings up to 60,000 inch–pounds are permitted for single–engine arrestments.

NOTE

For carrier landing capability with hung ordnance refer to appendix A of NAVAIR 01–245FDB–1T.

AFT CG LIMITS

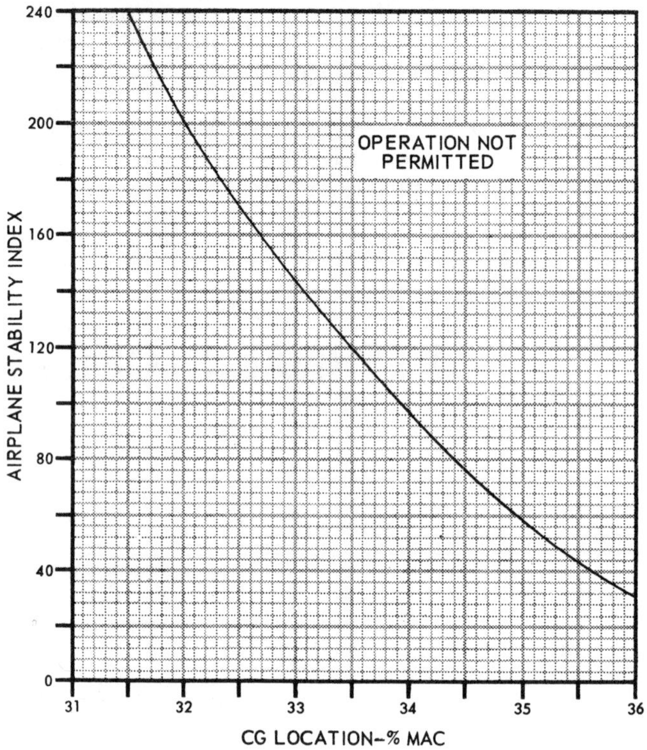

Note

The aft CG limit curve is based on inflight conditions. Before starting engines, initial CG positions up to 0.4% MAC aft of the limit curve are permissible, but shall not exceed 36% MAC. This assumes a corresponding forward shift in CG during ground operation.

FDD-1-(40)

Figure 1-36

ACCELERATION LIMITATIONS

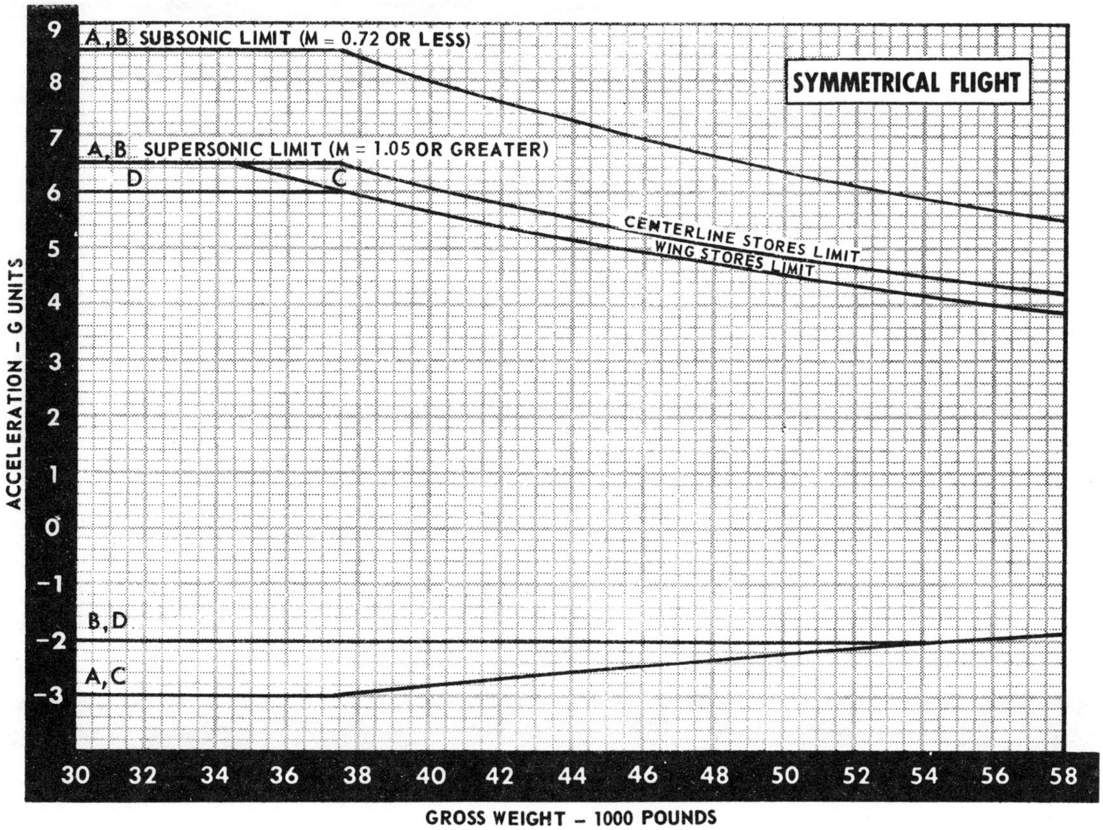

CONFIGURATIONS:

A. BASIC AIRPLANE, OR WITH PYLONS, MISSILE LAUNCHERS, FUSELAGE-MOUNTED AIM-7.
B. EMPTY 600-GALLON TANK (MCDONNELL WELDED TANK ONLY).
C. AIM-7 OR AIM-9 MISSILES AT STATIONS 2 AND 8.
D. EMPTY 370-GALLON TANKS (MCDONNELL OR SARGENT FLETCHER).

Notes

- FOR STORES NOT SPECIFICALLY LISTED HEREON, REFER TO EXTERNAL STORES LIMITATIONS CHART. DO NOT EXCEED APPLICABLE CENTERLINE OR WING STORES LIMIT CURVE.
- REFER TO FLIGHT STRENGTH DIAGRAM FOR ADDITIONAL LOAD FACTOR LIMITS AT SPEEDS ABOVE 1.5 MACH.

CAUTION

THE ENVELOPE FOR A SYMMETRICAL FLIGHT (FLIGHT IN WHICH ROLL OR YAW ACCELERATION IS APPLIED) IS FROM 0.0 G to 80% of SYMMETRICAL FLIGHT LIMIT.

FDD-1-(41)A

Figure 1-37

FLIGHT STRENGTH DIAGRAM

BASIC AIRPLANE, OR WITH PYLONS, MISSILE LAUNCHERS, FUSELAGE-MOUNTED AIM-7.

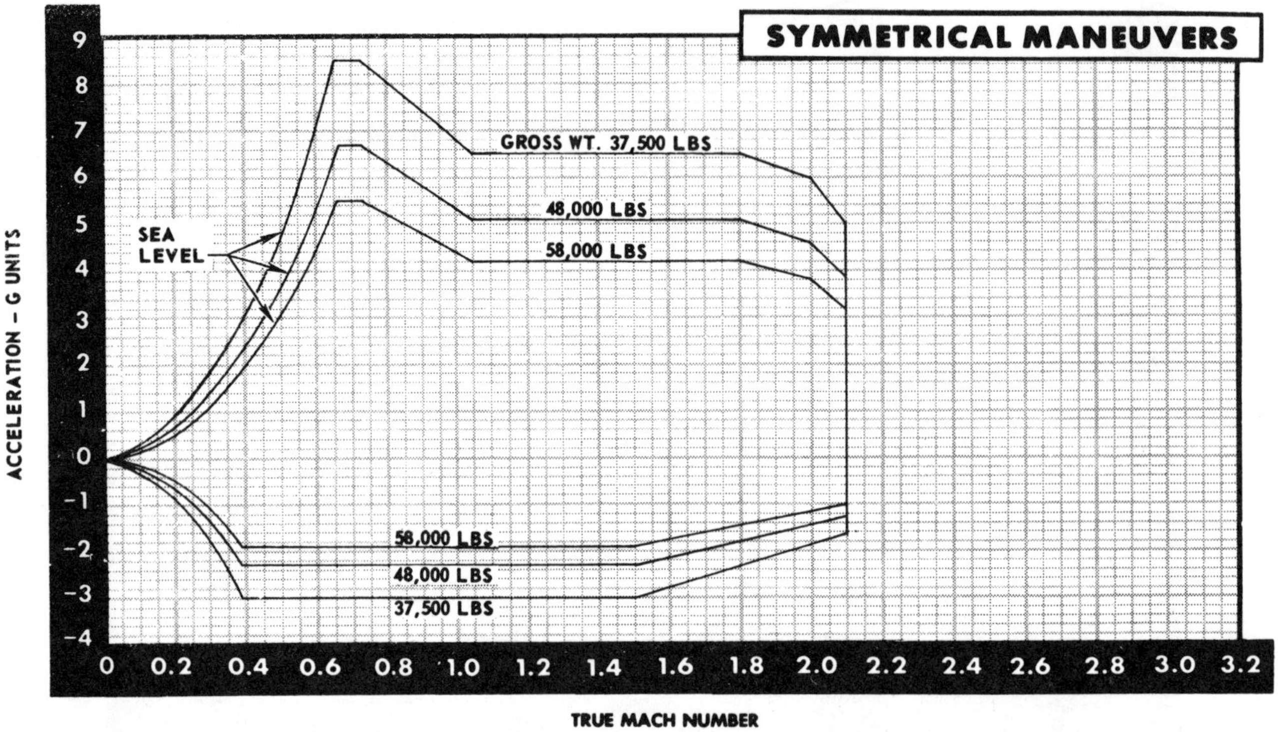

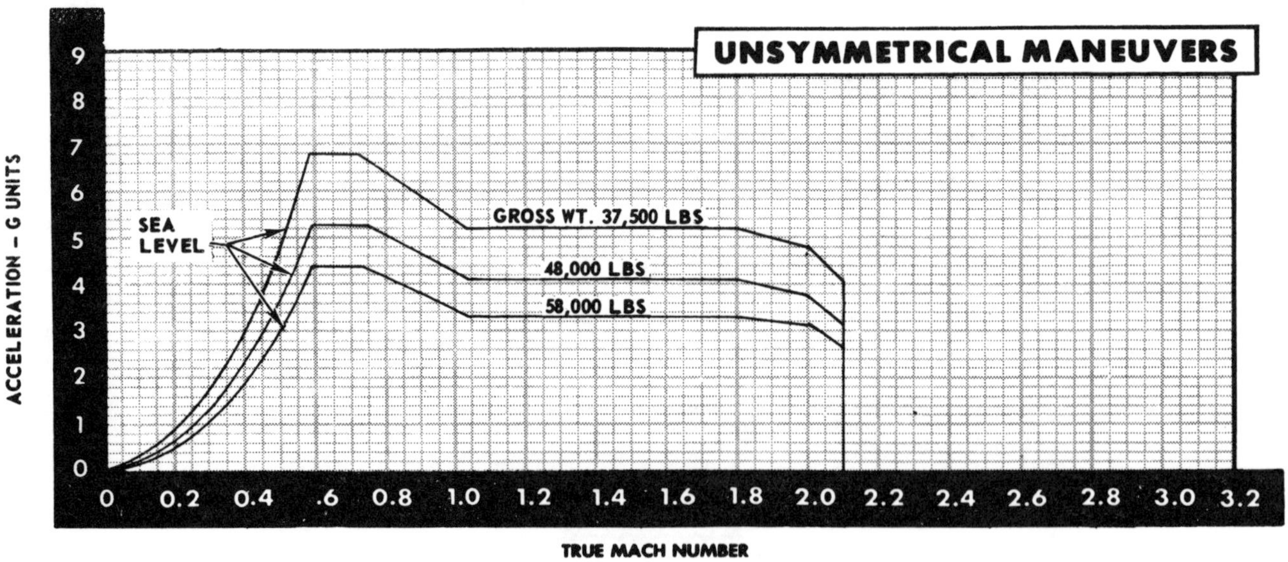

FDD-1-(42)

Figure 1-38

EXTERNAL STORES LIMITATIONS

STORES			STATIONS 1	2	3	4	5	6	7	8	9	CARRIAGE MAX KCAS/MACH	ACCELERATION SYM	ACCELERATION UNSYM	JETTISON MAX KCAS/MACH	REMARKS
370-Gal. Wing Tanks	McDonnell or Sargent Fletcher	EMPTY	X								X	750/1.6	-2.0 to +6.0	0.0 to +4.8	375 Below 30,000 Ft.	See Notes 1, 2, and 3. Arrested landings, both shipboard and shore-based, are permitted only with empty external tanks. Catapulting with partially full external tanks is not permitted
		EMPTY TO 3/4 FULL	X								X	550/1.6	-2.0 to +5.0	0.0 to +4.0		
		FULL	X								X	550/1.6	-1.0 to +4.0	0.0 to +2.0		Full or empty only 370-gallon external wing tanks can be jettisoned at 450 KCAS in symmetrical maneuvers between 2G and 3G below 5000 feet; however, this may result in tank/airplane contact and minor aircraft damage.
	Royal Jet	EMPTY TO 3/4 FULL	X								X	550/1.6	-2.0 to +5.0	0.0 to +4.0	410 Above 30,000 Ft.	
		FULL	X								X	550/1.6	-1.0 to +4.0	0.0 to +2.0		*Jettisoning the 600-gallon tank between 375 KCAS and 425 KCAS below 15,000 feet may result in airplane contact and minor damage.
600-Gal. Centerline Tanks	McDonnell Short Fairing	EMPTY TO 3/4 FULL					X					600/1.8 Above 30,000 Ft. 0.8 Below 30,000 Ft.	0.0 to +5.0	+1.0 to +3.0	300/0.9	**The symmetrical acceleration limitations for the empty Royal Jet centerline tank are 0.0 to +6.5G.
		FULL					X						+0.5 to +3.0	+1.0 Only		
	McDonnell Nestable (Accessory Bulletin 18-62 Incorporated)	EMPTY TO 3/4 FULL					X					600/1.8	0.0 to +5.0	+1.0 to +3.0		
		FULL					X						+0.5 to +3.0	+1.0 Only		
	McDonnell Welded	EMPTY					X					LBA	-2.0 to LBA	LBA	425/1.8 *	
		EMPTY TO 3/4 FULL					X						-2.0 to +6.5	0.0 to +5.2		
		FULL					X						-2.0 to +5.0	0.0 to +4.0		
	Royal Jet	EMPTY TO 3/4 FULL					X					600/1.8	** 0.0 to +5.0	+1.0 to +3.0		
		FULL					X						+0.5 to +3.0	+1.0 Only		
RCPP-105 Starter Pod							X					550/0.9	-1.0 to +5.0	0.0 to +3.0	300/0.9	See Notes 1 and 2. Jettison in level flight.
CNU-169/A FERRY MISSION EQUIPMENT STORE (MODIFIED AERO 1A, 150 GALLON EXTERNAL FUEL TANK, P/N 12E1034-1 OR -2)				X						X		550/1.0	-2.0 to +6.5	0.0 to +5.0	400 Max. 200 Min.	See Notes 1, 2 and 3. **WARNING** The CNU-169/A ferry mission equipment store must be loaded so that the CG is maintained between the attachment lugs. Maximum equipment loading is 250 pounds. The CNU-169/A is carried captive to the LAU-17/A pylon.
ASDC (Air-Ship Delivery Container System)	Delivery Container on TER STA 1; Wire Container on TER STA 2.		X	X								400/0.9	+1.0 to +3.0	+1.0	Release 250/0.40	Release altitude - 400 ft. Release mode - Single only; Bomb sight setting - 290 mils. Carrier operations with ASDC permitted. Release in 1G level flight.
	Delivery Container on TER STA 1; Wire Container on TER STA 3.									X	X					

FDD-1-(43-1)C

Figure 1-39 (Sheet 1 of 2)

EXTERNAL STORES LIMITATIONS

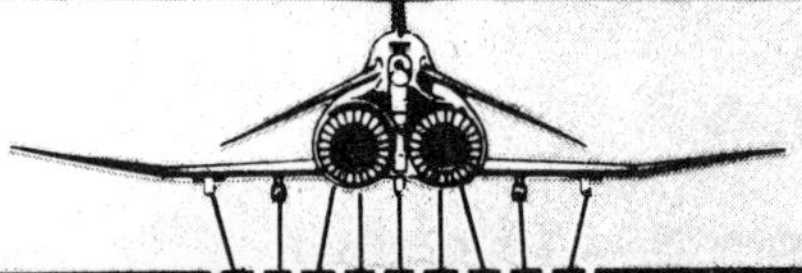

LBA – Limits of Basic Airplane
NE – Not Established

STORES		STATIONS 1	2	3	4	5	6	7	8	9	CARRIAGE MAX KCAS/MACH	CARRIAGE ACCELERATION SYM	CARRIAGE ACCELERATION UNSYM	JETTISON MAX KCAS/MACH	REMARKS
D-704 Air Refueling Store						X					500/1.1	0.0 to +4.0	0.0 to +3.2	300/0.9	See Notes 1 and 2. Buddy tank adapter pylon kit is required. Arrested landing, both shipboard and shore-based, are permitted only with empty tank. Catapulting with partially full tank is not permitted. AB operation is not permitted while dumping fuel. Maximum acceleration for operation is +2.0G, and no abrupt maneuvers are permitted. Maximum airspeeds for operation are 300/0.8 with the hose extended and 250 KCAS for fuel dumping and hose retraction. Jettison in level flight.
AQM-37A Missile Target (LAU-24A Launcher)						X					550/0.95	0.0 to +4.0	+1.0 to +3.0	550/0.92	See Notes 1 and 2. Maximum bank angle not to exceed 60 degrees. Abrupt maneuvers not permitted. Avoid slipping and skidding. Buffet boundary operation not permitted except in landing configuration. Release in level flight.
RMU-8A Reel/ Launcher with TDU-22 Target	Target reel-in, reel-out, or being launched					X					500/0.9	NE	NE	0.95	See Notes 1 and 2. Refer to Tow Target procedures in Section IV for additional limitations.
	Target reel-in, reel-out, or fully streamed					X					500/1.6				
	Target being recovered					X					350/0.85				
RMU-8A Reel Launcher Without Target						X					500/1.6	NE	NE	0.95	

Notes

1. Below 10,000 feet, the following airspeed limitations shall be observed when carrying external stores other than Sparrow missiles:
   - Airplanes with C G location aft of 34% MAC – 0.70 Mach or external store limit, whichever is less.
   - Airplanes with C G location forward of 34% MAC – Basic aircraft limit as shown in Airplane Speed Restrictions Chart or the external store limit, whichever is less.
2. Refer to Acceleration Limitations Chart for additional acceleration limitations while carrying external stores.
3. Jettison, empty or full, in 1.0 G level flight with speed brakes, landing gear, and flaps retracted.

FDD-1-(43-2)B

Figure 1-39 (Sheet 2 of 2)

SECTION II
INDOCTRINATION

TABLE OF CONTENTS

GROUND TRAINING

MINIMUM GROUND TRAINING SYLLABUS

The overall ground training syllabus for each activity will vary according to local conditions, field facilities, requirements from higher authority, and the immediate Unit Commander's estimate of squadron readiness. The minimum ground training syllabus (pilot/RIO) for each phase is set forth below:

FAMILIARIZATION

Flight physiological training as appropriate

NAMT pilot's course

COT/WST (within 10 days)

FLIGHT SUPPORT LECTURES

J79 engine

Air induction system

Flight controls, flaps, BLC, and AFCS

Aircraft systems and emergency procedures

Aircraft operating limitations

Flight Characteristics

No less than two preflight inspections utilizing the preflight and daily maintenance requirements cards

Cockpit/Pressure suit air conditioning and MK IV pressure suit

Ejection seat and survival kit

Cockpit procedures/checklists

BIT checks (RIO only)

Climb, loiter, and cruise performance

Fuel management/mission planning

Single-engine performance

CNI equipment

NATOPS Flight Manual (open and closed book) stressing normal and emergency procedures and aircraft/engine limitations.

INTERCEPT FLIGHT SUPPORT LECTURES

NAMT AMCS course

Naval Air Maintenance Training group instructions, AIM-7 missile

Naval Air Maintenance Training group instructions, AIM-9 missile

AN/APG-59 missile control system functions and computations, AIM-7 missile

AN/APG-59 missile control system computations, AIM-9 missile

NTDS/ATDS operating procedures

Tactical employment of the weapons system

Basic intercept procedures

Voice procedures

High altitude-high Mach intercepts

Low altitude intercepts

AIM-7 beam intercepts with AIM-9 reattack

Forward hemisphere intercepts

Electronic Counter-countermeasures

Air intercept control techniques and procedures - broadcast control

Multiple intercept procedure

Missile firing procedures

Fighter vs fighter combat maneuvering

WEAPONS FIRING FLIGHT SUPPORT LECTURES

Arming/dearming procedures

Firing procedures

Safety procedures

Jettison/dump areas

FMLP/CARQUAL FLIGHT SUPPORT LECTURES

Mirror and Fresnel Lens optical landing system

Day landing pattern and procedures

Night landing pattern and procedures

Shipboard procedures and landing patterns

CCA procedures

Air refueling (Day/Night)

WAIVING OF MINIMUM GROUND TRAINING REQUIREMENTS

All F/RF-4 qualified crewmembers shall be instructed on the differences from model in which qualified and comply with those items listed below, as directed by the unit commanding officer.

Where recent crewmember experience in similar aircraft models warrant, Unit Commanding Officers may waive the minimum ground training requirements provided the crewmember meets the following mandatory qualifications:

Has obtained a current medical clearance

He is currently qualified in flight physiology

Has satisfactorily completed the NATOPS Flight Manual open and closed book examinations

Has completed at least one emergency procedure period in the COT/WST (if available)

Has received adequate briefing on normal and emergency operating procedures

Has received adequate instructions on the use/operation of the ejection seat and survival kit.

FLIGHT TRAINING SYLLABUS

AIRCREW FLIGHT TRAINING SYLLABUS

Prior to flight, all crewmembers will have completed the Familiarization and Flight Support Lectures previously prescribed. A qualified instructor pilot will be assigned for the first familiarization flight. The instructor pilot will occupy the rear seat. The geographic location, local command requirements, squadron mission, and other factors will influence the actual flight training syllabus and the sequence in which it is completed. The specific phases of training are:

FAMILIARIZATION

Military and afterburner power takeoffs

Buffet boundary investigation

Approach to stalls

Slow flight

Acceleration run to Mach 2.0

Subsonic and supersonic maneuvering

Investigate all features of the AFCS/Stab Aug

Formation flight

Aerobatics

Single engine flight at altitude and airstarts

Landings with and without drag chute

Use of nose gear steering

Simulated single-engine

Landing with full, 1/2 and no flaps

Acceleration runs at various altitudes

INSTRUMENTS

Basic instrument work

Penetrations and approaches

Local area round robin (day and night) flights

WEAPON SYSTEM EMPLOYMENT

In accordance with existing training and readiness directives

FIELD MIRROR LANDING PRACTICE AND CARRIER QUALIFICATION

Slow flight

Field mirror landing practice

Carrier qualifications flights

OPERATING CRITERIA

CEILING/VISIBILITY REQUIREMENTS

Prior to the pilot becoming instrument qualified in the airplane, field ceiling/visibility and operating area weather must be adequate for the entire flight to be conducted in a clear air mass according to Visual Flight Rules. After the pilot becomes instrument qualified, the following weather criteria apply:

TIME IN MODEL (Hr)	CEILING (Ft)/VISIBILITY (Mi)
10–20	800/2; 900/1–1/2; 1,000/1
20–45	700/1; 600/2; 500/3
Over 45	Field minimums or 200/1/2 whichever is higher

Where adherence to these minimums unduly hampers pilot training, Commanding Officers may waive time–in–model requirements for actual instrument flight, provided pilots meet the following criteria:

Have a minimum of 10 hours in model

Completed 2 simulated instrument sorties

Completed 2 satisfactory Tacan penetrations.

MINIMUM FLIGHT QUALIFICATIONS

Where recent crewmember experience in similar aircraft models warrant, Unit Commanding Officers may waive the minimum flight training requirements for basic qualifications. Minimum flight hour requirements to maintain pilot and RIO qualifications after initial qualification in each specific phase will be established by the Unit Commanding Officer. Crewmembers who have more than 45 hours in model are considered current subject to the following criteria:

Must have a NATOPS evaluation check with the grade of Conditionally Qualified, or better, within the past 12 months and must have flown 5 hours in model and made two takeoffs and landings within the last 90 days.

Must have satisfactorily completed the ground phase of the NATOPS evaluation check, including COT/WST emergency procedures check (if available), and be considered qualified by the Commanding Officer of the unit having custody of the aircraft.

REQUIREMENTS FOR VARIOUS FLIGHT PHASES

NIGHT

Pilot

Not less than 10 hours in model as first pilot

RIO

Not less than 3 hours in model as crewmember

CROSS COUNTRY

Pilot

Have a minimum of 15 hours in model as first pilot

Have a valid instrument card

Have completed at least one night familiarization flight

RIO

Have completed at least one night familiarization flight

AIR–TO–AIR MISSILE FIRING

Pilot

Have a minimum of 15 hours in model

Be considered qualified by C.O.

RIO

Have a minimum of 15 hours in model as crewmember

Have satisfactorily completed a minimum of two intercept flights on which simulated firing runs were conducted utilizing the voice procedures and clear to fire criteria to be utilized in live firings

Be considered qualified by the Commanding Officer

CARRIER QUALIFICATION

Each crewmember will have a minimum of 50 hours in model, and meet the requirements set forth in the LSO NATOPS manual.

MINIMUM CREW REQUIREMENTS

The pilot and RIO (or two pilots) constitute the normal crew for performing the assigned mission for all flights. Unit commanders may authorize rear seat flights for personnel other than qualified pilots and RIOs provided such personnel have received thorough indoctrination in the use of the ejection seat and oxygen equipment and in the execution of rear seat checklist and emergency procedures. Where operational necessity dictates, unit commanders may authorize flights with the rear seat unoccupied provided the requirements for such flights clearly overrides the risk involved and justifies the additional burden placed on the pilot. Although a rear seat occupant is required in order to properly comply with the procedures in section V of the NATOPS manual, the infrequency of occurrence of these situations tends to justify juduciously selected instances of solo flight. Ferry squadron commanders may authorize solo ferry flights of F–4 aircraft. In no case is solo flight authorized for shipboard operations, combat or combat training missions.

FERRY SQUADRONS

Training requirements, check–out procedures, evaluation procedures, and weather minimal for ferry squadrons are governed by the provisions contained in OPNAVINST 3710.6.

CREWMEMBER FLIGHT EQUIPMENT

MINIMUM REQUIREMENTS

In accordance with OPNAVINST 3710.7, the flying equipment listed below will be worn by crewmembers on every flight. All survival equipment will be secured in such a manner that it will be easily accessible and will not be lost during ejection or landing. This equipment shall be the latest available as authorized by Aircrew Personal Protective Equipment Manual (NAVAIR 13–1–6).

Anti–buffet helmet modified in accordance with current aviation clothing and survival equipment bulletins

Oxygen mask

Anti–G suit (required on all flights where high G forces may be encountered)

Fire retardant flight suit

Steel–toed flight safety boots

Life Preserver

Harness assembly

Shroud cutter

Sheath knife

Flashlight (for all night flights)

Strobe light

Pistol with tracer ammunition, or approved flare gun

Fire retardant flight gloves

Identification tags

Anti–exposure suit in accordance with OPNAVINST 3710.7

Personal survival kit

Other survival equipment appropriate to the climate of the area

Full pressure suit and MK 4 life preserver on all flights above 50,000 feet MSL

SECTION III
NORMAL PROCEDURES

TABLE OF CONTENTS

PART 1 BRIEFING/DEBRIEFING

BRIEFING

GENERAL

The mission commander/flight leader is responsible for briefing all crewmembers on all aspects of the mission to be flown. The pilot/RIO will assist the mission commander/flight leader in preparing required flight or briefing forms and may, if applicable, brief that portion of the mission pertaining to the other crewmember. A briefing guide or syllabus card, as appropriate, will be used in conducting the briefing. Each crewmember will maintain a kneepad and will record all flight numbers, call signs, and all other data necessary to assume the lead and complete the assignment. However, this does not relieve the flight leader of the responsibility for briefing all crews in the operation and conduct of the flight. The briefing guide will include the following:

ASSIGNMENTS

Aircraft assigned, call sign, and deck spot when appropriate.

Engine start, taxi, and takeoff times

Visual signals and rendezvous instructions

MISSION

Primary

Secondary

Operating area

Control agency

Time on station or over target

WEAPONS

Loading

Safety

Arming, dearming

Duds

Special routes with ordnance aboard

Minimum pull-out altitude

Jettison area

COMMUNICATIONS

Frequencies

Radio procedure and discipline

Navigational aids

Identification and ADIZ procedures

WEATHER

Local area

Local area and destination forecast

Weather at alternate

High altitude weather for the jet stream, temperature, and contrail band width

NAVIGATION AND FLIGHT PLANNING

Takeoff speed

Takeoff distance

Abort distance

Crosswind effects

Climb out

Mission route, including ground controlling agencies

Fuel/oxygen management

Marshal

Penetration

GCA or CCA

Recovery

EMERGENCIES

Aborts

Divert fields

Bingo and low state fuel

Waveoff pattern

Ready deck

Radio failure

Loss of visual contact with flight

Ejection

SAR procedures

System failures

AIR INTELLIGENCE AND SPECIAL INSTRUCTIONS

Friendly and enemy force disposition

Current situation

Targets

Safety precautions

ECM and ECCM

OPERATING AREA BRIEFINGS

Prior to air operations in and around a new area, it is mandatory that a comprehensive area briefing be given including, but not limited to, the following:

Bingo Fields

Instrument approach facilities

Runway length and arresting gear

Terrain and obstructions

Emergency Fields

Fields suitable for landing but without required support equipment

Include information under Bingo fields

SAR Facilities

Type

Frequencies

Locations

DEBRIEFING

GENERAL

Postflight debriefing is an integral part of every flight. The mission commander/flight leader should review the entire flight from takeoff to landing, including not only errors and poor techniques, but also the methods of correcting them. Also, the mission commander/flight leader shall cover completely any deviations from standard operating procedures. All intercepts should be reviewed using scope camera and controller information when available.

PART 2 MISSION PLANNING

GENERAL

The pilot and RIO shall be responsible for the preparation of required charts, flight logs, navigation computations including fuel planning, checking weather and NOTAMS, and for filing required flight plans. Refer to section XI, Performance Data, to determine fuel consumption, correct airspeed, power settings, and optimum altitude for the intended flight mission. Planning data for specialized missions will be contained in the F-4 Tactical Manual.

FLIGHT CODES

The proper Kind of Flight classification and codes to be assigned individual flights are established by OPNAVINST 3710.7.

PART 3 SHORE-BASED PROCEDURES

PREFLIGHT

GENERAL

The yellow sheet must be checked for flight status, configuration, armament loading, and servicing prior to manning the aircraft. At least the ten previous B sections should be reviewed for discrepancies noted and the corrective action taken. Weight and Balance clearance is the responsibility of the maintenance department.

EXTERIOR INSPECTION

1. Approaching aircraft - CHOCKS, TIEDOWNS, DOWNLOCKS, LEAKS, DOORS AND PANELS SECURE, STRUTS, TIRES, HOOK UPLOCK REMOVED, COVERS REMOVED
2. Left intake - DUCT CLEAR OF FOD, INLET GUIDE VANES SECURE, BELLMOUTH BYPASS CLEAR, PITOT AND STATIC PORTS CLEAR, RAMP SECURE, RAMP BLEED AIR HOLES CLEAR
3. Nosewheel well - CNI SHELF DOOR SECURE, WHEEL BRAKE ACCUMULATOR PRESSURE GAGE 1000 $\pm$50 PSI, VERTICAL REFERENCE BIT SWITCH OFF AND COVER CLOSED; RAT, LANDING GEAR, AND FLAP GAGES 2750–3100 PSI; DOOR ROLLER MECHANISM FREE
4. Nose gear - GENERAL CONDITION, NGS UNIT ELECTRICAL AND HYDRAULIC CONNECTIONS, TIRES INFLATION AND CONDITION, WHEEL NUTS SECURE, JACKING PAD SAFETY WIRED, APPROACH AND TAXI LIGHTS SECURE, ANTENNAS SECURE
5. Left forward fuselage - ACCESS PANELS SECURE, AOA PROBE COVER REMOVED, AOA PROBE SECURE AND FREE TO ROTATE AND ADJUST TO CENTER RANGE, TOTAL TEMPERATURE SENSOR SECURE, EQUIPMENT COOLING INTAKE CLEAR, RADOME LOCKING LUGS SECURE, STATIC PORTS CLEAR, STATIC SYSTEM DRAIN DRY
6. Right forward fuselage - RADOME LOCKING LUGS SECURE, STATIC PORTS CLEAR, CABIN COOLING INLET CLEAR, ACCESS PANELS SECURE
7. Right intake - DUCT CLEAR OF FOD, INLET GUIDE VANES SECURE, BELLMOUTH BYPASS CLEAR, PITOT AND STATIC PORTS CLEAR, RAMP SECURE, RAMP BLEED AIR HOLES CLEAR
8. Bottom fuselage - UTILITY RESERVOIR FULL AND O PRESSURE, LOX CONVERTER CONNECTIONS SECURE, CANOPY JETTISON GAGES 2750–3100 PSI, FUEL CELL CAVITY DRAINS DRY, MAIN SYSTEM PNEUMATIC GAGE 2200 PSI MINIMUM, KEEL WEB INSPECTION PLATES SECURE
9. Centerline tank - TANK SECURE, SIGHT GAGE CHECK, FILL CAP SECURE
10. Right aux air door - DOOR ACTUATING CYLINDER SECURE, DOOR LIMIT SWITCH CHECK, THROTTLE LINKAGE SECURE, OIL CAPS SECURE, CSD AND FUEL CONTROL CONDITION
11. Lower right wing - PC–2 RESERVOIR FULL AND O PRESSURE; PYLONS, SUSPENSION EQUIPMENT, AND ORDNANCE SECURE
12. Right wheel well - INNER DOOR SECURE AND DOWN, PC–2 PRESSURE GAGE 1000 $\pm$50 PSI, GROUND REFUEL SWITCH OFF, NUMBER 7 CELL REFUEL A & F BUTTONS OUT TO REFUEL, HYDRAULIC SERVICING CAPS SECURE
13. Right main gear - DOOR SECURE, UPPER STRUT GAGE IN GREEN, SHRINK LINK STRAIGHT AND SECURE, BRAKE LINES CONDITION
14. Right main wheel - TIRE INFLATION AND CONDITION, JACK PAD AND DOOR SECURE
15. Right wing tank - SECURE AND QUANTITY CHECKED
16. Right wing fold area - WING LOCKING LUGS FREE OF CRACKS, MANUAL LOCK PIN SECURE, WING LIGHTS SECURE
17. Right aft wing area - DUMP MAST SECURE, SPEED BRAKE SECURE AND FREE OF FLUID
18. Bottom aft fuselage - FUEL CELL CAVITY DRAINS DRY, CATAPULT HOLDBACK SECURE, ENGINE ACCESS DOORS AND PANELS SECURE
19. Left and right engine exhaust area - TURBINE BLADES, AB SPRAY BARS, FLAME HOLDER, TORCH IGNITER, SECURITY AND CONDITION; NOZZLE CONDITION; OVERHEAT CIRCUIT SECURE; AB AND NOZZLE AREA FREE OF FLUIDS AND FOD.
20. Aft tail area - DASH POT PRESSURE GAGE, ANTI–COLLISION LIGHT SECURE, PITOT TUBE AND BELLOWS RAM AIR INLET SECURE AND COVERS REMOVED, HOOK POINT CONDITION AND WEAR, STABILATOR SECURE AND FREE OF CRACKS, DRAG CHUTE DOOR CLOSED AND SECURE, VENT MAST SECURE, RUDDER SECURE, TAIL LIGHT SECURE
21. Left aft wing area - DUMP MAST SECURE, SPEED BRAKE SECURE AND FREE OF FLUID
22. Left wing fold area - WING LOCKING LUGS FREE OF CRACKS, MANUAL LOCK PIN SECURE, WING LIGHTS SECURE
23. Left wing tank - SECURE AND QUANTITY CHECKED
24. Left main wheel - TIRE INFLATION AND CONDITION, JACK PAD AND DOOR SECURE
25. Left main gear - DOOR SECURE, UPPER STRUT GAGE IN GREEN, SHRINK LINK STRAIGHT AND SECURE, BRAKE LINES CONDITION
26. Left wheel well - INNER DOOR SECURE AND DOWN, PC–1 PRESSURE GAGE 1000 $\pm$50 PSI,

EJECTION SEAT AND CANOPY CHECK POINTS

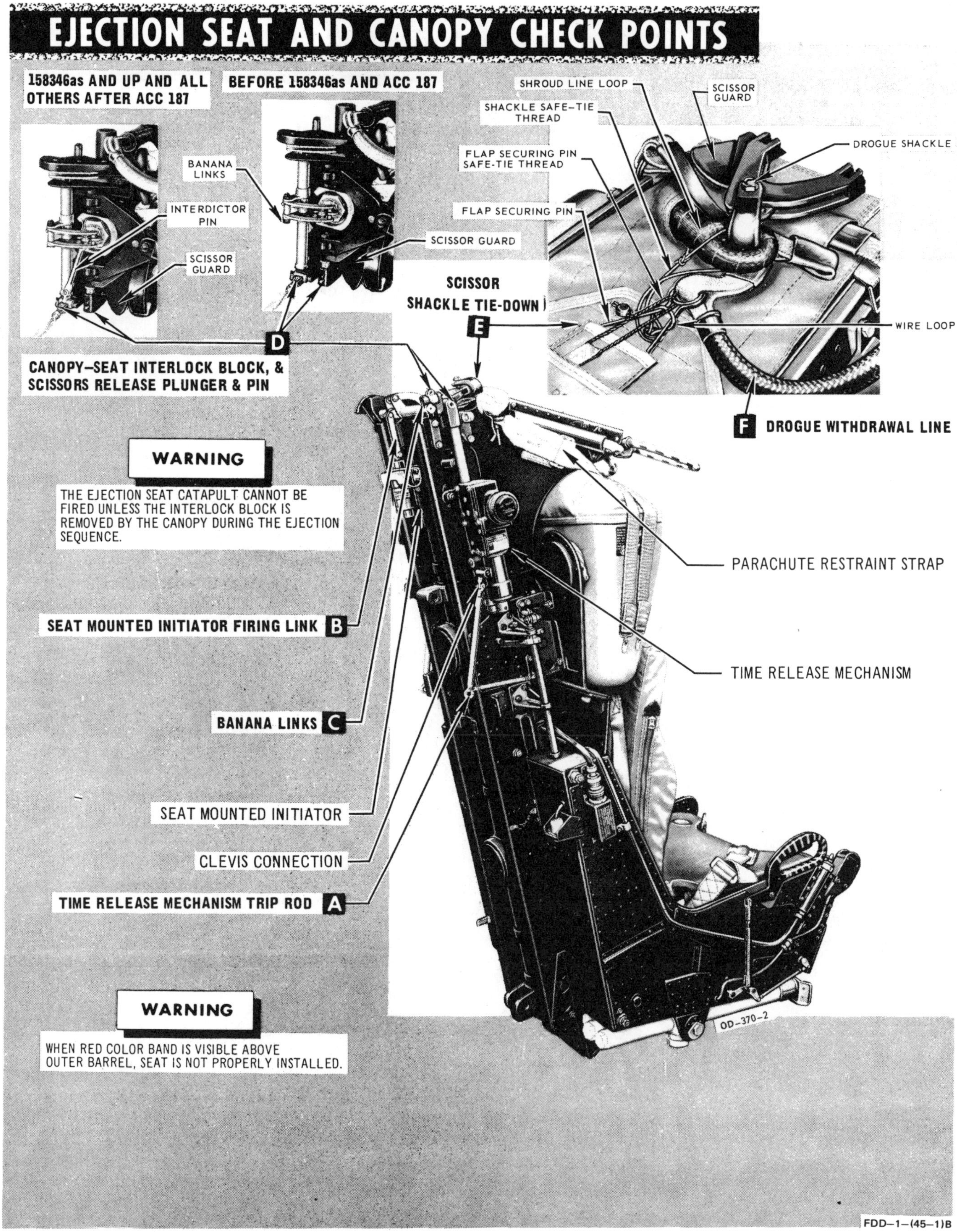

Figure 3-1 (Sheet 1 of 2)

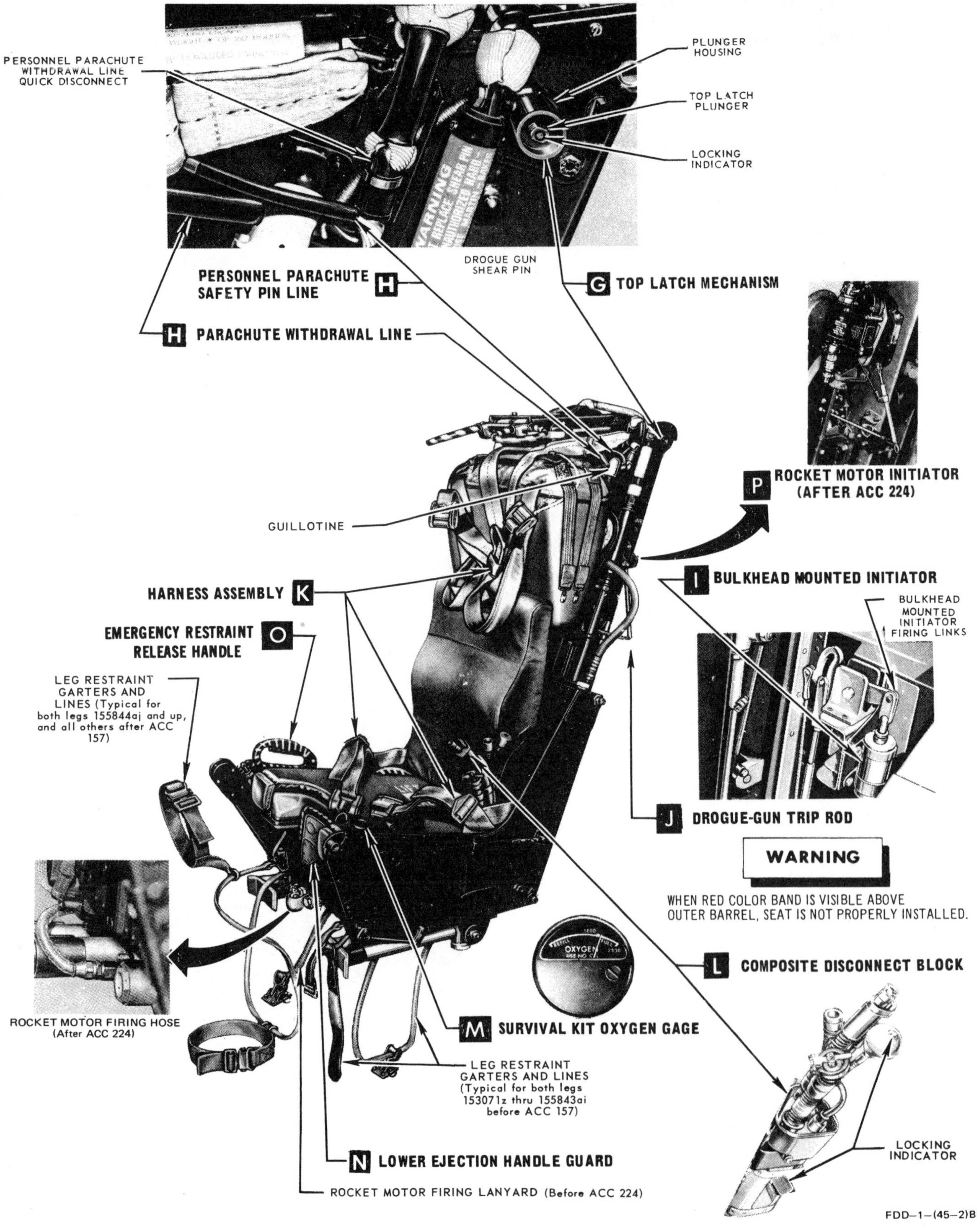

Figure 3-1 (Sheet 2 of 2)

HYDRAULIC SERVICING CAPS SECURE

27. Lower left wing - PC-1 RESERVOIR FULL AND O PRESSURE; PYLONS, SUSPENSION EQUIPMENT, AND ORDNANCE SECURE
28. Left aux air door - DOOR ACTUATING CYLINDER SECURE, DOOR LIMIT SWITCH CHECK, THROTTLE LINKAGE SECURE, OIL CAPS SECURE, CSD AND FUEL CONTROL CONDITION, CENTERLINE JETTISON CARTRIDGES CAPS SECURE
29. Upper fuselage - RAT DOOR SECURE, ACCESS DOORS AND PANELS SECURE

AFT COCKPIT INTERIOR CHECK FOR SOLO FLIGHT

1. Canopy initiator safety pin (bulkhead mounted) REMOVED

   Check pin removed to insure operation of forward initiator.

CAUTION

Exercise caution regarding hand movements in the vicinity of the airplane mounted canopy initiator linkages. Also, do not stow flight equipment or personal items in this area. Failure to comply could result in inadvertent jettisoning of the canopy.

2. Seat safety pins - INSTALLED
3. Circuit breakers - IN
4. Navigation computer function switch - OFF
5. Radar function switch - STBY
6. Pressure suit vent air valve - OFF
7. Oxygen supply lever - OFF
8. Cockpit light switches - OFF
9. Seat harness - STOWED
10. All loose gear - STOWED
11. Aft cockpit electrical test receptacle - INSURE CAP TIGHTENED

NOTE

It is possible to trip both generators off the line if the electrical test receptacle 3P325 under the right canopy sill is loose. The generators can not be restored until the cap is tightened.

12. All armament switches - OFF OR SAFE

After electrical power -

13. Essential dc test button - DEPRESS

    Depress button and check that essential dc bus indicator light illuminates. If light illuminates, the left transformer-rectifier is delivering dc power. If the light does not illuminate, the left transformer-rectifier is inoperative or not receiving power. With the dc test button depressed, the right transformer-rectifier can be checked by actuating the warning lights test switch. If the warning lights illuminate, the right transformer-rectifier is operating properly.
14. Canopy - LOCKED

BEFORE ENTERING COCKPIT

1. Normal opening canopy handle - AFT/OPEN
2. Gear handle - DOWN
3. Emergency canopy release handle - FORWARD and SHEAR WIRED
4. Radar scope - SECURE
5. Center rear view mirror - CHECK

   Ensure the mirror will clear the windshield bow when the canopy is closed.
6. Ejection seat and canopy rigging - CHECK

WARNING

On aircraft before ACC 224, the rocket motor sear is under the seat. Do not use the area for stowage, and exercise caution when performing any function in the vicinity of the rocket pack; e.g., pulling rocket motor safety pin, adjusting leg restraint lines, etc.

a. Time release mechanism trip rod - Check time release mechanism trip rod secured to anchor beam and engaged in time release mechanism. Check for correct distance between the outer sleeve and the point where it bottoms against the clevis connection (approximately ½ inch). If the clevis connection (point where trip rod connects to the time release mechanism sear) on the inner shaft of the trip rod is bottomed down on the outer sleeve of the trip rod, the time release mechanism sear is probably not seated properly. In aircraft with locking indicator type top latch plunger, check that the red color band on the trip rods is not visible.

b. Canopy initiator firing link - Check canopy initiator firing link installed.

c. Banana links - Check banana links pin engaged in firing mechanism sear.

d. Canopy-seat interlock block - Check canopy-seat interlock block in place, and interlock block cable secure to the canopy. On aircraft 158346as and up, and all others after ACC 187, check interdictor pin inserted in firing mechanism sear. The ejection seat will not fire if the interlock block is not removed by the canopy during the ejection sequence.

e. Scissor shackle tie-down and scissor guard - Check that the scissor shackle tie-down passes under the flap securing pin, around the wire loop, then aft under the shroud line loop, through the drogue shackle, over the top of the shroud line, and then forward and tied to the other end of the cord. Check drogue shackle engaged in scissors, and scissors plunger extended against moveable scissors arm with plunger pin visible on top of scissors plunger housing. Check the scissors guard on the lower right side of the scissor shackle assembly to ensure that the guard is not bent.

WARNING

A bent scissor guard could prevent the scissors jaw from opening during ejection and thereby preclude man/seat separation.

f. Drogue withdrawal line – Check drogue withdrawal line (in wire braid sleeve) passes over the lays on top of all other lines.
g. Top latch mechanism – Check that top latch plunger is flush with the end of the top latch mechanism housing. The locking indicator must be flush with the end of the top latch plunger.

WARNING

If the top latch mechanism check does not meet the outlined requirements, an inadvertent ejection could result.

h. Parachute withdrawal line – Check that the parachute withdrawal line passes through the guillotine and is routed underneath the parachute restraint strap. In addition, check the withdrawal line quick disconnect for proper connection and swivel action. Check personnel parachute safety pin line not routed through guillotine and both the parachute safety pin line and parachute withdrawal line loop routed through the alignment ring on the parachute pack.
i. Canopy initiator (bulkhead mounted) firing line – Check that bulkhead mounted canopy initiator firing link is installed.

CAUTION

Exercise caution regarding hand movements in the vicinity of the airplane mounted canopy initiator linkages. Also, do not stow flight equipment or personal items in this area. Failure to comply could result in inadvertent jettisoning of the canopy.

j. Drogue gun trip rod – Check drogue gun trip rod secured to anchor beam and engaged in drogue gun. Check that red color band on the trip rod is not visible. Check drogue gun cocking indicator protruding approximately ½ inch from the bottom of the drogue gun.
k. Harness assembly – Check that the shoulder harness retaining pin is installed at the retraction reel. Check that the pins securing the lap belt and survival kit to the seat are in place.
l. Composite disconnect block – Check that lower block locking indicator (yellow metal flag) is tight. Check that the upper block is positively locked to the intermediate block by depressing the yellow knob on the composite disconnect release cable and noting that the indicator button is flush with the top of the knob. If indicator button is protruding from the top of the knob, the upper block assembly is not positively locked to the intermediate block. Check upper block properly inserted and locked into intermediate block by exerting an upward pull on block assembly after composite disconnect release knob is locked to cable housing.

NOTE

If unlocked, the composite disconnect should be carefully inserted with a downward force parallel to the seat ejection plane. After the composite disconnect is fully inserted, push down on composite disconnect release knob to lock knob to cable housing and prevent the release knob from laying over and dangling.

m. Survival kit oxygen gage – Check survival kit oxygen gage for at least 1800 psi pressure.

WARNING

If an RSSK-1A survival kit is installed in the ejection seat, the seat must also have a lumbar pad installed. If the RSSK-1A survival kit and lumbar pad compatibility is not present and an ejection becomes necessary, the seat occupant will most likely suffer a fractured vertebrae.

n. Lower ejection handle guard – Check lower ejection handle guard is in the up (vertical) position.
o. Emergency restraint release handle – Check emergency restraint release handle down, and firing sear installed in guillotine initiator.
p. Rocket motor initiator and rocket motor firing body (if applicable) – Check initiator cable lanyard connected to drogue gun trip rod without excessive cable hanging from initiator housing. Initiator sear installed with cable lever assembly link inserted. Initiator hose connected with hose pin in hose connection (not in initiator sear). After ACC 224 amendment 1, hose pin is replaced with a nut and bolt combination. On bottom of seat, rocket motor firing body installed and hose connected.

7. (Before ACC 224) Check under seat for foreign objects, for rocket motor firing lanyard connected to floor fitting and ensure rocket motor sear is clear. Ensure area beneath seat is clear of foreign objects before pulling rocket motor sear safety pin.
8. Seat safety pins, except face curtain pin and after ACC 187, interdictor pin – REMOVE

   Check seat safety pins – ejection gun (before ACC 187), guillotine, canopy initiator (seat mounted), drogue gun, rocket motor (before ACC 224), rocket motor initiator (after ACC 224), canopy initiator (bulkhead mounted) removed.
9. Aft cockpit electrical test receptacle – INSURE CAP TIGHTENED

NOTE

It is possible to trip both generators off the line if the electrical test receptacle 3P325 under the right canopy sill is loose. The generators cannot be restored until the cap is tightened.

10. Command selector valve – CLOSED (for cat shot, as pilot directs)

AFTER ENTERING COCKPIT

PILOT

Before electrical power –

1. Generator switches – OFF
2. Oxygen, communications, anti–G lines and helmet mounted sight (if applicable) – CONNECT
   Connect oxygen, communications and anti–G line disconnects, as applicable, and check for position lock. When the helmet mounted sight is utilized, verify that the VTAS egress disconnect will operate normally. After connection, an upward motion from the snap location (seated) should effect release. Reconnect and verify that the movement flange is seated to preclude inadvertent disconnection during maneuvers. During disconnection, a taut lanyard moves the flange to release the lock.

WARNING

An improperly operating VTAS egress disconnect could cause serious injury during ejection.

3. Oxygen – CHECK
   Turn oxygen lever on, check normal flow with mask held away from face. Put mask on, check normal breathing. Turn oxygen off, check no breathing.

WARNING

Do not pull emergency oxygen actuator ring prior to actual use. If the emergency oxygen actuator ring is actuated prior to intended use, the pressure reducer manifold may not prevent emergency oxygen from flowing to the oxygen regulator. If this happens, the crewmember has no way of knowing how much, or if any, emergency oxygen remains and has no way of replenishing the depleted supply.

4. Leg restraint lines – CONNECTED
   For single garter restrainers, pass leg lines through garters so that the D rings are on the aft inboard side of the calf and plugged into the seat pan. Assure that the leg lines are not twisted. For dual garter restrainers, route untwisted lines first through lower garter (route first through outboard ring of calf garter then through inboard ring) and then through thigh garter before inserting lock pins in snubber boxes.

WARNING

It is imperative that leg restraint system be hooked up at all times during flight to ensure legs are pulled aft upon ejection. This will prevent leg injury and enhance seat stability by preventing legs from flailing following ejection. An unhooked leg restraint system necessitates pulling legs aft against seat to preclude hitting canopy bow. This action will cause spine to flex and will increase the possibility of spinal injury during ejection. For dual garter systems, be certain to route lines through lower garters before thigh garters since incorrect routing could result in serious leg injury during ejection.

5. Harnessing – FASTENED

WARNING

- Ensure harness assembly is securely fastened to the seat. The pins must be in their proper receptacles, one pin on each side of the bucket seat and one pin on the harness locking reel assembly. The emergency restraint release handle must be down.
- The over the shoulder oxygen–communication integrated line must be routed under the shoulder harness to preclude the possibility of the crewmember's helmet being jerked from his head after ejection. This action is caused by the parachute riser (shoulder harness strap) snapping up against the integrated line with the opening of the personnel parachute.

6. Shoulder harness lock lever – CHECK OPERATION
7. Face curtain pin – REMOVE

WARNING

- To prevent inadvertent firing of seat or canopy, all ejection seat safety pins must be either installed or removed and properly stowed prior to operating the canopy.
- When removing or checking for the removal of the face curtain safety pin, make sure that the safety pin shank has been removed from the hole. The safety pin collar has been known to separate from the pin shank upon attempted safety pin removal leaving the pin shank in the hole and the face curtain safetied.

- Do not pull down on the face curtain ejection handle. Seat and canopy ejection systems are fully armed when safety pins are removed.

8. Lower ejection handle safety guard - AS DESIRED

WARNING

The lower ejection handle safety guard, when lowered, can rebound to the safe position if it is lowered too rapidly. This could consume critical time when lowering guard for a low altitude ejection.

9. Rudder pedals - ADJUST
10. Stick grip - CHECK SECURITY
11. Gun switch - OFF (some aircraft)
12. Aural tone control - AS DESIRED
13. Dogfight computer training/test switch - NORM (some aircraft)
14. Strike camera mode switch - OFF (some aircraft)
15. Armament safety override switch - OUT

WARNING

If armament safety override switch is IN with landing gear handle down, and electrical power is applied to the aircraft, landing gear armament safety override feature is bypassed and power is supplied to armament bus relay. This could result in inadvertent dropping or firing of ordnance while aircraft is on the ground.

16. Intercom control panel - CHECK
    a. Volume selector knob - AS DESIRED
    b. Function selector switch - HOT MIC
17. Fuel control panel - SET
    a. External tank jettison switch - NORM (GUARD DOWN) (some aircraft)
    b. Buddy fill switch - STOP FILL
    c. Refuel selection switch - AS DESIRED
    d. Refuel probe switch - RETRACT
    e. Internal wing dump switch - NORMAL
    f. Wing transfer pressure switch - NORMAL
    g. Internal wing transfer switch - NORMAL (some aircraft)
    h. External transfer switch - OFF (some aircraft)
    i. Emergency Probe extend switch - GUARD DOWN (some aircraft)
    j. Fuel transfer selector knob - STOP (some aircraft)
18. RAT handle - RAT IN
19. Flap switch - UP
20. Flap emergency handle - UP
21. Stab aug switches - OFF
22. Smoke abate switch - OFF
23. UHF antenna selector switch - UPR
    Electromagnetic interference (EMI) radiating from lower UHF antenna may interfere with nose wheel steering system. Therefore use of lower UHF antenna is restricted to inflight operation.
24. Engine anti-icing switch - NORMAL
25. Throttles - OFF
26. Master lights switch - OFF
27. Speed brake switch - STOP (NEUTRAL)
28. Throttle friction lever - SET AS DESIRED
29. Engine master switches - OFF
30. Engine start switch - NEUTRAL
31. Flight instrument light control knob - AS REQUIRED (some aircraft)
32. Drag chute handle - DOWN & SECURE
33. ARI circuit breaker - IN
34. Emergency speed brake switch - GUARD DOWN (some aircraft)
35. Emergency aileron droop switch - DISABLE (some aircraft)
36. Approach power compensator switch - OFF
37. Gear handle - DOWN
38. True airspeed indicator - 60-180 KNOTS
39. Radar altimeter - OFF
40. Accelerometer — SET

40a. Servoed optical sight mode knob — OFF (some aircraft)

41. Missile jettison selector switch — OFF
42. Missile power switch - OFF
43. Missile (gun/missile) arm switch - SAFE
44. Missile select switch - RADAR (some aircraft)
45. Missile interlock switch - IN
46. AIM-9 coolant switch - OFF (some aircraft)
47. Aural tone control - AS DESIRED (some aircraft)
48. Jettison select switch - ALL (some aircraft)
49. VTAS power switch - OFF (some aircraft)
50. External wing tank jettison switch - GUARD DOWN (some aircraft)
51. Centerline station safe switch - SAFE
52. Bomb control switch - OFF
53. Weapons switch - AS REQUIRED
54. Data link bomb switch - MANUAL (some aircraft)
55. Bomb arming switch - SAFE
56. Multiple weapons station selector switch - OFF
57. Vertical velocity indicator - CHECK
58. Clock - WIND & SET
59. Navigation function selector panel - SET
60. Arresting hook control handle - UP
61. Manual canopy handle - FWD
62. Telelight panel - CHECK SECURE
63. Emergency vent knob - IN
64. Defog footheat handle - AS REQUIRED
65. Pitot heat switch - OFF
66. Rain removal switch - OFF/LOW
67. Bleed air switch - NORM (some aircraft)
68. UHF volume control - OFF
69. Comm-aux pushbutton - TR + G - ADF
70. TACAN function selector switch - OFF
71. Circuit breakers - IN
72. IFF master switch - OFF/MODE SET
73. Compass system controller - SET
    a. Latitude compensator - SET
    b. Mode switch - SLAVE
74. Cockpit temperature control panel - CHECK
    a. Heat knob - SET AS DESIRED
    b. Temperature control switch - AUTO

WARNING

The MAN position of the temperature control switch should not be used except as a back-up in the event of a failure in the automatic system. The full hot manual position can produce temperatures in excess of 300°C at Mil power settings.

75. Instrument panel emergency floodlights switch - OFF
76. TACAN antenna selector switch - AUTO
77. ECM warning lights switch - OFF (some aircraft)
78. Cockpit lights - OFF
79. Exterior lights - AS DESIRED
80. Wing fold switch - COINCIDES WITH WING POSITION
81. Spare lamps - CHECK
82. KY-28 power switch - OFF (some aircraft)
83. KY-28 cipher switch - P (some aircraft)
84. ARA-63 (ILS) power switch - OFF (some aircraft)
85. Flight instrument balance control panel - AS REQUIRED (some aircraft)
86. Flashlight, charts and reference material - CHECK

After electrical power -

CAUTION

Do not place generator control switch to the EXT position until external power has been connected and has had time to reach rated voltage and frequency.

1. Generator control switches - EXT ON
2. KY-28 (if installed) - AS DESIRED
3. Intercom system - CHECK
4. Seat - ADJUST
5. Data link light control - AS DESIRED (some aircraft)
6. Radar annunciator lights - AS DESIRED (some aircraft)
7. Cockpit lights - AS DESIRED
8. ECM warning lights - AS DESIRED (some aircraft)
9. Warning lights - CHECK
   Depress warning light test switch and note master caution light, warning lights panel, command and digital display indicator lights (some aircraft), ECM warning lights (some aircraft), radar scope warning lights, missile status panel lights, arresting hook warning light and landing gear warning lights are illuminated. Check warning lights dimming circuit by holding warning lights test button depressed and rotating instrument panel lights control knob from OFF to BRIGHT. Warning lights should dim and revert to bright when knob is returned to OFF.
10. Fire warning lights - CHECK
    Depress the fire check button and note engine FIRE/OVERHT and on some aircraft the three BLEED AIR OVERHT warning lights illuminated.
11. Fuel quantity gage - CHECK
    actuate feed tank check switch and fuel gage will stabilize at1571 pounds. Check aft gage sector and counter within 100 and 200 pounds, respectively, of the fwd gage.
12. Fuel quantity - CHECK
    Check fuel quantity indicators against known fuel quantity. Check aft gage sector and counter within 100 and 200 pounds, respectively, of the fwd gage.
13. Flap position indicator - UP
    Check flap position indicator corresponds with flap position.
14. Landing gear indicator - DOWN
    Check landing gear position indicators indicate gear down.
15. Liquid oxygen gage - CHECK
16. Anti-skid switch - OFF; LIGHT ON (some aircraft)
17. Eject light system - CHECK (some aircraft)
    Depress the light switch and check that both front rear cockpit EJECT lights illuminate. Depress the switch again and check that lights extinguish.
18. Boost pumps and engine fuel shutoff valve - CHECK
    Observe boost pump pressure indicators while actuating boost pump check switches one at a time. Normal pressure (30 $\pm$5 psi) on side being checked indicated engine fuel shutoff valve open and boost pump running. Concurrent pressure on other indicator indicates other valve faulty (not properly closed). Lack of pressure on side being checked indicates faulty valve (not properly open) or pump inoperative. Also, note that zero fuel flow is registered on the fuel flow indicator.

CAUTION

Allow minimum of 3 seconds between release of one fuel boost pump test switch and actuation of other. Failure to do so could result in burning of switch contacts and subsequent engine flameout.

19. Master lights switch - AS REQUIRED

RIO

Before electrical power -

1. Oxygen, communications, and anti-G lines - CONNECT
   Connect oxygen, communications, and anti-G lines, as applicable, and check for positive lock.
2. Oxygen - CHECK
   Turn oxygen selector on, check normal flow with mask held away from face. Put mask on check normal breathing. Turn oxygen off, check no breathing.

WARNING

Do not pull emergency oxygen actuator ring before actual use. If emergency oxygen actuator ring is actuated prior to intended use, pressure reducer manifold may not prevent emergency oxygen from flowing to suit controller and/or oxygen regulator. If this happens, crewmember has no way of knowing how much, or if any, emergency oxygen remains and has no way of replenishing depleted supply.

3. Leg restraint lines - CONNECTED

For single garter restrainers, pass leg lines through garters so that the D rings are on the aft inboard side of the calf and plugged into the seat pan. Assure that the leg lines are not twisted. For dual garter restrainers, route untwisted lines first through lower garter (route first through outboard ring of calf garter then through inboard ring) and then through thigh garter before inserting lock pins in snubber boxes.

WARNING

It is imperative that leg restraint system be hooked up at all times during flight to ensure legs are pulled aft upon ejection. This will prevent leg injury and enhance seat stability by preventing legs from flailing following ejection. An unhooked leg restraint system necessitates pulling legs aft against seat to preclude hitting canopy bow. This action will cause spine to flex and will increase the possibility of spinal injury during ejection. For dual garter systems, be certain to route lines through lower garters before thigh garters since incorrect routing could result in serious leg injury during ejection.

4. Harnessing – FASTENED

WARNING

- Ensure harness assembly is securely fastened to the seat. The pins must be in their proper receptacles, one pin on each side of the bucket seat and one pin on the harness locking reel assembly. The emergency restraint release handle must be down.
- The over the shoulder oxygen–communication integrated line must be routed under the shoulder harness to preclude the possibility of the crewmember's helmet being jerked from his head after ejection. This action is caused by the parachute riser (shoulder harness strap) snapping up against the integrated line with the opening of the personnel parachute.

5. Shoulder harness lock lever – CHECK OPERATION
6. Face curtain pin – REMOVE

WARNING

- To prevent inadvertent firing of seat or canopy, all ejection seat safety pins must be either installed or removed and properly stowed prior to operating the canopy.
- When removing or checking for the removal of the face curtain safety pin, make sure that the safety pin shank has been removed from the hole. The safety pin collar has been known to separate from the pin shank upon attempted safety pin removal leaving the pin shank in the hole and the face curtain safetied.
- Do not pull down on the face curtain ejection handle. Seat and canopy ejection systems are fully armed when safety pins are removed.

7. Lower ejection handle safety guard – AS DESIRED

WARNING

The lower ejection handle safety guard, when lowered, can rebound to the safe position if it is lowered too rapidly. This could consume critical time when lowering guard for a low altitude ejection.

8. Tacan – OFF
9. UHF – OFF
10. Integrated control panel – SET (some aircraft)
    a. Dispenser selector knob – OFF
    b. ECM mode selector knob – OFF
    c. Receiver power switch – OFF
    d. Payload selector knob – OFF (some aircraft)
    e. Warning receiver interface switch – OFF (some aircraft)
    f. Dispenser mode select switch – STBY (some aircraft)
11. Data link power switch – OFF (some aircraft)
12. Data link BIT switch – NORM (some aircraft)
13. Radar beacon power switch – OFF
14. ALQ-91 function selector switch – OFF (some aircraft)

CAUTION

Failure to ensure that AN/ALQ-91 function selector switch is turned OFF and remains in the OFF position until aircraft is switched to internal power may result in damage to the equipment or a blown fuse.

15. Destruct circuit arm switch – SAFETY PIN INSTALLED (some aircraft)
16. Command selector valve handle – POSITION IN ACCORDANCE WITH TYPE COMMANDER/SQUADRON POLICY

CAUTION

When actuating the command selector valve from the closed (vertical) to open (horizontal) position, pull the handle straight out allowing the cam action of the valve to rotate the handle to the open position. This procedure will prolong the service life of the selector valve. After AFC 526, the handle is turned by the application of torque only, and there is no requirement to pull the handle.

17. Communication antenna selector switch - UPPER
18. Cockpit lights - AS DESIRED
19. Altimeter - SET
20. Clock - CHECK
21. Band disable panel - SET (some aircraft)
    a. Band switches - DISABLE
22. Antenna correlation disable panel - SET (some aircraft)
    a. Correlation switch - DISABLE
    b. Antenna switch - DISABLE
23. Manual canopy unlock handle - FORWARD
24. Circuit breakers - CHECK
25. DRSC power switch - OFF (some aircraft)
26. Eject light - CHECK (some aircraft)
27. Antenna handle control panel - EXTEND (some aircraft)
28. Radar scope - AS DESIRED (some aircraft)
29. Radar set control panel - EXTENDED (some aircraft)
30. Radar function switch - OFF
31. Navigational computer - SET/STBY

After electrical power -

32. Warning lights - TEST
33. Perform data link BIT 3 and 4 checks (some aircraft).
    Data link BIT 3 and 4 checks must be performed with external power connected and the autopilot ground test switch in TEST.
34. Nav selector switch - NAV COMP
    Check DME runoff to zero, note error and return switch to CNI
35. Oxygen quantity gage - CHECK
36. Nav command - AS DESIRED
37. Comm command - As BRIEFED
38. Nav channel - AS BRIEFED
39. Aux channel - AS BRIEFED
40. Comm channel - AS BRIEFED
41. Radar set control switches - AS REQUIRED
42. Radar annunciator lights - AS DESIRED (some aircraft)
43. Seat - ADJUST
44. Essential dc bus test button - DEPRESS
    Depress button and check that essential dc bus indicator light illuminates. If light illuminates, the left transformer-rectifier is delivering dc power. If the light does not illuminate, the left transformer-rectifier is inoperative or not receiving power. With the dc test button depressed, the right transformer-rectifier can be checked by actuating the warning lights test switch. If the warning lights illuminate, the right transformer-rectifier is operating properly.
45. Circuit breakers - CHECK IN
46. Notify pilot - PRE-START CHECKS COMPLETE - CIRCUIT BREAKERS IN

BEFORE STARTING ENGINES

PILOT

1. Wheels - CHOCKED
2. Fire bottle - MANNED
3. Intake and exhaust areas - CLEAR
4. Boarding steps - UP
5. External air supply - CONNECTED & PRESSURE UP
6. Rudder - CHECK MOVEMENT

CAUTION

The center mirror on the forward canopy can be tilted sufficiently to prevent canopy closing; therefore, ensure the mirror will clear the windshield bow before closing the canopy.

7. Inform RIO - READY TO START.

STARTING ENGINES

The pilot starts engines and keeps the RIO informed of any unusual occurrences. The RIO remains alert for any emergency signals from the ground crew and informs the pilot if such signals are observed. When practicable, start and run up engines on paved surface to minimize the possibility of foreign objects being drawn into the compressor with resultant engine damage. Start the engines with the nose into or at right angles to the wind as exhaust temperatures may be aggravated by tail wind.

WARNING

- Suction at the intake is sufficient to kill or severly injure personnel drawn into or pulled suddenly against the duct.
- Danger areas aft of the airplane are created by high exhaust temperature and velocities. The danger increases with afterburner operation.

PILOT

CAUTION

With flaps extended, the BLC ducts are open and the loss of engine bleed air while attempting to start the engines may result in a hot or false (no-ignition) start. If it is imperative that a start be made with flaps down, the starting procedure recommended for starting with low air pressure should be utilized with the exception that the start should be initiated at normal starting rpm.

NOTE

The following engine start procedure establishes the right engine as being started first and was adopted in order to ascertain that both utility hydraulic system pumps are operating. The right engine pump delivers 2775 $\pm$225 psi at idle rpm, and the left engine pump delivers approximately 3000 $\pm$250 psi at idle rpm. Therefore, the single-needle utility hydraulic pressure indicator cannot be used to determine pump operation unless the right engine is started first.

1. Throttles – OFF
2. External compressed air source – CONNECTED
3. Engine master switches – ON

NOTE

When attempting an engine start, there is a possibility that the starter air valve will not open when the start switch is actuated. This does not mean that the solenoid operated valve is completely inoperative, it may be that the valve is only sticking in the closed position. If the valve will not open, have the air shut off at the starter cart and then actuate the start switch. The stuck valve, unopposed by air pressure, may open.

4. Engine start switch – RIGHT

CAUTION

If there is no indication of engine rpm within 15 seconds, or no indication of oil pressure within 30 seconds after start cycle begins, shut down immediately and investigate.

5. At 10% rpm, right engine ignition button – DEPRESS
   At approximately 10% rpm, depress right engine ignition button and simultaneously advance the throttle half–way up the quadrant and then snap it back to the idle stop position while monitoring fuel flow. If the throttle is properly rigged, the snap–back will not affect initial starting fuel flow. However, if a momentary drop of more than 75 pph below minimum starting fuel flow is indicated, the throttle is out of rig. This can be confirmed by snapping the throttle forward and back several times while keeping ignition button depressed.

CAUTION

- Do not attempt to start the engine before reaching 10% rpm. If the starting procedure is initiated at a lower rpm, additional heat distress of the engine hot suction is anticipated. Over–temperature of the turbine will generally occur during a low rpm start if starter air is inadvertently interrupted during the start cycle. However, starting below 10% rpm may be helpful with a hard to start engine in an emergency situation.
- If the engine does not light off by the time fuel flow reaches 800 pph or within 15 seconds after fuel flow or pressure is indicated chop throttle to full OFF position, release ignition button.

NOTE

The engine usually fires at approximately 10 to 16% rpm.

6. Release ignition button when light–off is indicated by a sudden increase in EGT.

CAUTION

If engine does not continue to accelerate after light–off, discontinue start.

7. Start switch – NEUTRAL
   When the engine is operating at a self–sustaining rpm (usually about 45%), move the starter switch to the neutral position.
8. Exhaust temperature gage – CHECK WITHIN LIMITS (980°C max for 10 sec).

CAUTION

- At no time should EGT exceed maximum starting limits.
- With only one engine in operation, do not move control stick (surface controls) excessively. If the stick is moved rapidly with hydraulic pressure on only one side of the tandem power cylinders, the fluid that is the other side of the cylinder is forced back through the return line to the reservoir, filling the reservoir, and either rupturing the reservoir or forcing the excess fluid overboard. The seals within the tandem power cylinders may also be damaged due to the ingesting and expelling of air and lack of lubrication. The power control hydraulic systems must be reserviced and checked.

NOTE

After the engine reaches idle rpm the EGT should recede and stabilize.

9. Fuel flow indicator – CHECK
   Fuel flow should not exceed 800 pph at light–off, up to 2400 pph during the transition to idle, and 800 to 1400 pph at IDLE.

CAUTION

If fuel flow is in excess of 800 pph, a hot start will likely result.

NOTE

Fuel consumed while starting engines is approximately 65 pounds.

10. Oil pressure gage – CHECK
    Check oil pressure 12 psi minimum at idle rpm.

CAUTION

After any wet start or false (no–ignition) start, allow 1 minute or longer for the combustion system to drain before starting the engine.

NOTE

- With the right engine started, the PC–2 and utility hydraulic pressure indicators should read within normal. The CHECK HYD GAGES warning light will remain illuminated until the other engine is started and all four hydraulic pumps (PC-1, PC-2 and utility) are operating properly.

- If the throttles cannot be returned to OFF, the engine may be shut down from any throttle setting by placing the respective engine master switch to the OFF position. This will close the corresponding fuel shutoff valve, thus depriving the engine of fuel. The engine(s) will flame-out in approximately 15 seconds from MIL power.

11. Right generator switch - GEN ON

NOTE

Oil pressure should be below 50 psi before placing the generator control switches to GEN ON.

Check LH GEN OUT warning light illuminated and BUS TIE OPEN warning light out.

12. Left spoiler - CHECK
    With right engine operating, PC—1 pressure zero, slowly deflect control stick approximately one inch to the left. Have ground crew/RIO verify that the left spoiler does not fully deflect and if returns to a flush condition when the stick is returned to neutral.
13. Check fuel boost pump gages for normal indications.
    If right generator phase reversal is present, the boost pumps will run in reverse and the gages will indicate substantially lower than normal.
14. Start the left engine as per items 4 thru 10.
15. Left generator switch - GEN ON
16. External power and air - DISCONNECT
17. Right generator switch - OFF
    Check RH GEN OUT warning light illuminated and BUS TIE OPEN warning light out.

WARNING

If both generator switches are placed to the OFF position with engines running, auxiliary air doors and speed brakes will close violently.

18. Check both boost pump pressure gages for normal indications.
    If left generator phase reversal is present, the boost pumps will run in reverse and the gages will indicate substantially lower than normal.
19. Right generator switch - GEN ON
    Check RH GEN OUT and BUS TIE OPEN warning lights out.
20. Notify RIO BUS TIE OPEN light out.

NOTE

- Non-start or abnormal starts shall be logged on the yellow sheet (OPNAV FORM 3760-2).
- Fuel consumption at idle rpm is approximately 42 ppm.
- After satisfactory starts are accomplished, the engines do not require any warm-up time prior to placing throttles in any position.

RIO

1. Notify pilot of any emergency signals noted from ground crew.

STARTING WITH LOW AIR PRESSURE

If low air pressure units are employed for operational necessity and starting rpm seems to be hanging up, proceed as follows:

At any rpm over 5% -

1. Ignition button - DEPRESS, THROTTLE IDLE
2. Exhaust gas temperature - MONITOR
3. If EGT starts to move up rapidly as it passes 650°C - Discontinue start, THROTTLE OFF
4. Let engine coast until EGT drops to 250°C - DEPRESS IGNITION BUTTON, THROTTLE IDLE

At 250°C the engine rpm should be approximately 12% rpm, so the second relight should be successful. If it is not successful, repeat the procedure, cutting the engine when it starts to overtemp and relighting when it cools to 250°C. A little rpm will be gained each time.

CAUTION

Do not attempt to manually meter fuel flow with the throttles at a position between OFF and IDLE. This results in a premature hot section deterioration without any abnormal EGT indication.

BEFORE TAXIING

PILOT

After switching to internal power, the pilot shall inform the RIO ON INTERNAL POWER: BUS TIE CLOSED/OPEN. The pilot will then complete all Before Taxiing checks. When checks are completed, he will inform the RIO READY TO TAXI. The pilot will turn the missile power (CW) switch ON when requested by the RIO and will acknowledge by POWER SWITCH ON.

1. IFF - STBY/ON
2. Compass system controller - SET
   a. Sync button - PUSH
   b. Sync indicator - CHECK

NOTE

The wings must be spread and locked prior to compass sync to prevent false bearing information.

3. Attitude reference selector knob – PRIM–STBY–PRIM
   Switch to STBY and note shift of ADI to AJB–7 system. Switch back to PRIM.
4. Tacan – REC
5. Radios – ON
6. Altimeter and SPC – SET & CHECK
   a. Place the SPC switch in the RESET CORR position; the STATIC CORR OFF light must go out and remain extinguished. After reset, the altimeter should indicate within ±40 feet of the before reset indication. Altimeter oscillations of any magnitude are unacceptable.

NOTE

On some aircraft, the altimeter may momentarily jump to 90 feet before settling to the actual engagement error. Disregard this initial momentary jump.

   b. Obtain field barometric pressure from tower and set altimeter barometric pressure correction dial. Altimeter pointer should indicate the field elevation within ±75 feet.
   c. With the static correction on (SPC engaged), set the altimeter pointer to field elevation. Check the difference between front and rear cockpit altimeters. This difference should not exceed 0.06 in Hg.
   d. Reset altimeter to field barometric reading.
7. ADI and standby attitude indicator – SET
8. Radar altimeter – ON & SET
   Turn the altimeter on and set the low altitude limit. Ensure that the altitude pointer moves to 5 ±5 feet after warm–up.
9. AFCS – CHECK

NOTE

Before engaging stab aug, neutralize the controls by using the force transducer to place the control stick in the vertical position. Use of the stick grip to neutralize the controls will introduce erroneous signals into the system. If stick is not off fwd or aft stops, AFCS may not engage.

   a. Stab aug switches – ENGAGE
   b. AFCS switch – ENGAGE
   c. AFCS/ARI disengage switch – DEPRESS
      Check that AFCS switch returns to OFF.

NOTE

Only the AFCS switch will move to OFF when the AFCS/ARI emergency disengage switch is depressed. The stab aug switches must be individually disengaged.

   d. Stab aug switches – DISENGAGE
10. Trim switches – CHECK & SET FOR TAKEOFF.
    Check operation of trim switches and set rudder and aileron trim to neutral. Set stabilator trim to 2 units nose down.

On signal from Plane Captain –

11. Wings – SPREAD & LOCKED
12. Speed brakes – CYCLE
    Ascertain from ground crew that speed brakes are fully closed and warning light is out.
13. Flap switch – CYCLE & CHECK
    Actuate the flaps to the 1/2 and DN position. While cycling flaps, have ground crew confirm that both trailing edge flaps extend together at the same rate and that both ailerons droop together at the same rate. Check that the BLC system is operating. Return flaps to 1/2.
14. Flight control surfaces and hydraulic systems – CHECK TRAVEL. NOTE PRESSURE DROP
    a. Cycle the flight controls and check corresponding movement of the control surface. While cycling the ailerons, check ARI by noting corresponding rudder movement.
    b. Set takeoff pitch trim (2 units nose down) and release the control stick from the full aft stop. The control stick should move forward to at least the ½ travel position, and further movement toward the forward stop should require no more than approximately 1 pound push force. Moderate stick movement should be smooth and free of any restrictions.
15. Flap switch – UP
    Actuate the flaps to up and monitor the BLC MALFUNCTION light for a valve malfunction.
16. Arresting hook/corner reflector – CYCLE
17. Canopy rigging – CHECK & PREFLIGHT
    a. Engines – IDLE
    b. Cabin pressure vent knob – IN
    c. Defog footheat handle – AFT
    d. Cockpit temp control knob – SET AT 2 O'CLOCK
    e. Aft canopy – CLOSE, LIGHT OUT
    f. Front canopy – CLOSE (closing and locking not to exceed 9 seconds)
       The canopy is closed when warning light is out, alignment marks aligned, and overcenter links moved over center.
    g. Open aft canopy, keep forward canopy closed, and repeat timing check on aft canopy.
    h. If either canopy fails timing check, inflight loss of canopy may result.
18. Probe and RAT – CYCLE (if practicable) & CHECK
19. Perform VTAS BIT (some aircraft)
    Refer to section VIII, Classified Supplement.
20. Perform Dogfight Computer BIT (some aircraft)
    Refer to section VIII, Classified Supplement
21. Tune AIM–7 missiles (if aboard)
22. Perform digital and command display indicator BIT checks (some aircraft)
    (Refer to Data Link System, section VIII, Classified Supplement.)
23. Perform ARA–63 (ILS) BIT (some aircraft)
    Refer to section I for BIT procedures.
24. Anti–collision light – ON (day or night) and all external lights ON (dusk to dawn) during shore based operations.
25. Report – READY TO TAXI

RIO

1. Ascertain from the pilot that the generator control switches are ON and the bus tie and generator warning lights are out.
2. Remote attitude indicator (VGI) - SET
3. Radios - TR +G
4. Tacan - REC
5. Vent air - AS DESIRED
6. Altimeter - SET & CHECK (compare with pilot)
7. Radar function switch - STBY (safety/operational necessity only, otherwise OFF)
   The radar function switch should be in the OFF position, but may be placed on STBY or ON if safety of flight or operational necessity requires use of the radar.
8. Perform BIT as required
   Refer to section VIII, Classified Supplement for BIT procedures.
9. Data link system operation - CHECK (some aircraft)
   Refer to Data Link System, section VIII, Classified Supplement for universal test message checks, data link BIT checks, and digital display indicator BIT check.
10. Data link BIT switch - NORM (some aircraft)

WARNING

Do not select BIT checks with the data link BIT switch, on the cockpit lights/data link control panel, while in flight. This could result in application of deflection signals to the aircraft control surfaces.

11. ECM equipment operation - CHECK (some aircraft)
    Refer to section V, supplement to Tactical Manual for BIT procedures.

CAUTION

Assure that the lower UHF antenna is off prior to taxi, takeoff, and landing. Upper UHF antenna should be used during these phases because lower UHF antenna radiation may cause the nose wheel to go left or right.

NOTE

If, after engagement of nose gear steering with upper UHF antenna selected, no response is noted or unscheduled steering commands are detected, disengage nose gear steering, and do not reengage.

12. Clock - SET
13. Report - READY FOR TAXI

TAXIING

PILOT

High takeoff gross weight combined with the small wheels and tires dictate that a positive technique be used while taxiing this aircraft. After the chocks have been pulled, add power as required on both engines and engage nose gear steering. After the aircraft has started rolling, check the brakes and reduce power. Taxi at the lowest practicable rpm and use nose gear steering for directional control, where possible, to minimize brake heating. Do not ride or pump the brakes; use a steady pressure when needed. Keep the taxi speed slow and make as few stops as possible. Slow the aircraft before entering a turn in order to reduce side loads while in the turn. Make turns as wide as practicable, 75-foot radius if possible, at 12 to 13 knots. See figure 3-2 for minimum turning radius and ground clearance.

RIO

Complete BIT checks as required if not previously completed. Return the radar function switch to STBY/OFF (as required) after completion of BIT checks.

TAKEOFF

BEFORE TAKEOFF

When in the run-up area, allow the aircraft to roll straight ahead to align the nose wheel. Apply the brakes with a firm, steady pressure and assure the flaps are up. Note the idle RPM, EGT and fuel flow of both engines. Check the engines individually at MIL power and observe that the RPM, EGT, exhaust nozzle, fuel flow, oil pressure and hydraulic pressure are within their normal operating ranges on the engine being checked; also check that the RPM, EGT and fuel flow on the idling engine remains stable and that the ramps are fully retracted.

CAUTION

During engine run-up with flaps full up, a rise in rpm above 67.5%, a drop in EGT of more than 25°C, or a drop of more than 100 pph in fuel flow on the idling engine indicates a defective bleed air check valve on that engine. In cases where this check cannot be made at full military power, a valid check for an inoperative valve may be made at 80% rpm. Failure indications with such a check would be proportionally lower and should be verified at full military power if possible. This

MINIMUM TURNING RADIUS and GROUND CLEARANCE

AIRCRAFT BEING TAXIED

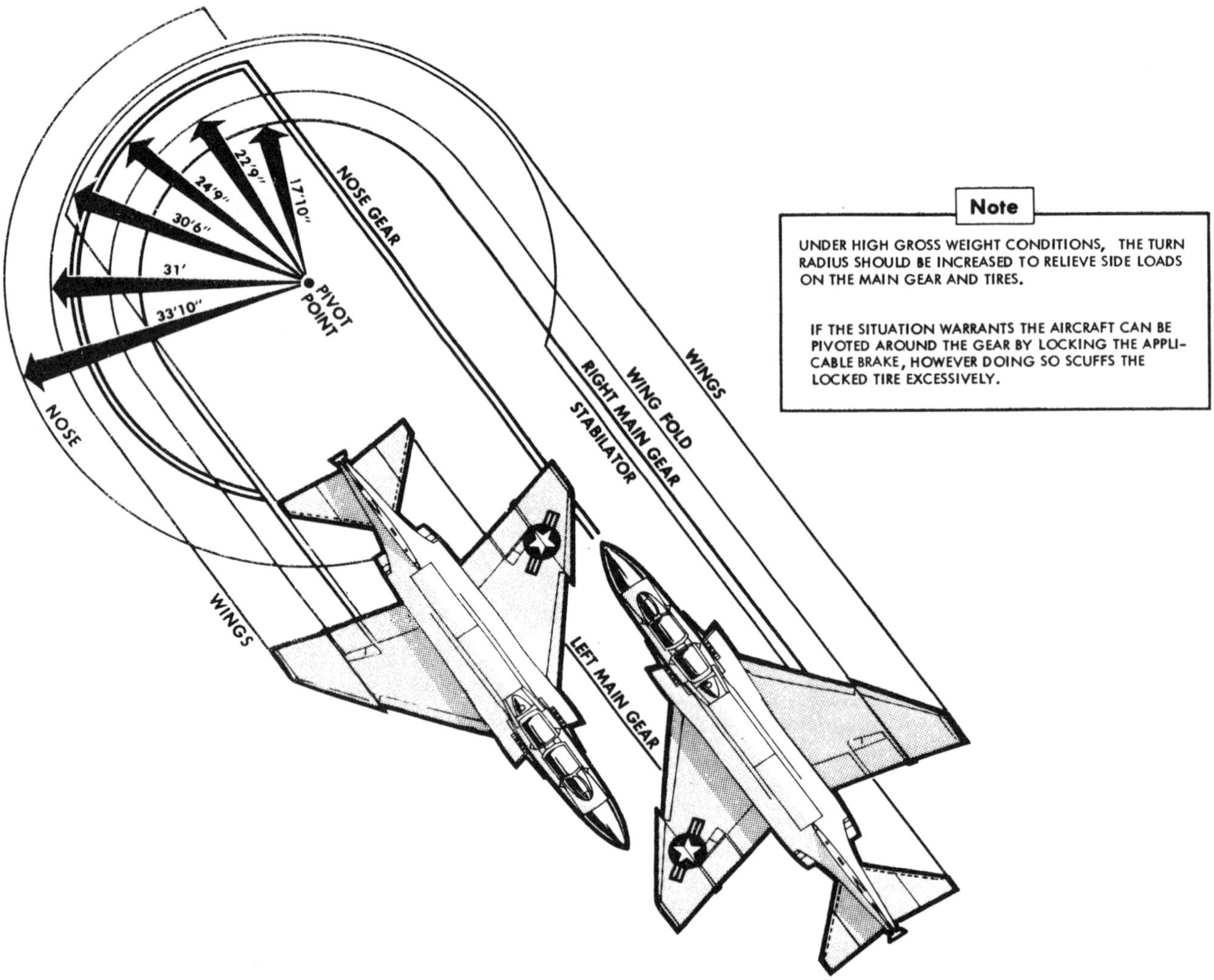

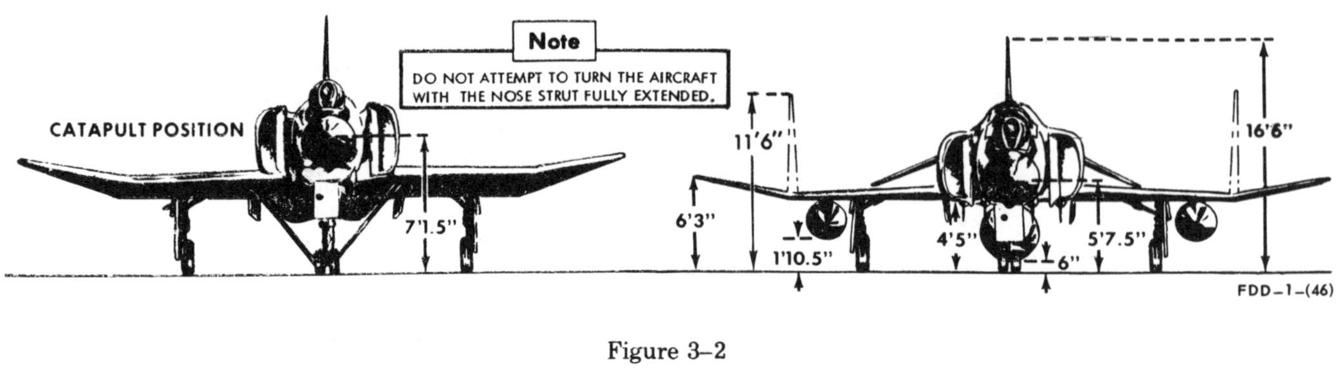

Figure 3-2

check performed with the flaps in any position other than full up is invalid.

To guard against possible engine flame-out during throttle chops at low altitude, check the throttle rigging and fuel control behavior by abruptly retarding each throttle from MIL power to IDLE. Monitor the fuel flow indicator. The momentary minimum acceptable fuel flow during this check is 330 pph. Observe that engine rpm returns to its originally noted value. If fuel flow drops below 330 pph, but the engine recovers to original rpm, proceed with flight; however, do not snap decelerate these engines below 10,000 feet. If engine rpm fails to recover to the original idle rpm value, regardless of fuel flow reading, the flight should be aborted. It is mandatory that an entry be made in OPNAV Form 3760-2 (Yellow Sheet) on all engines which drop below the minimum fuel flow during snap deceleration and/or fail to recover to the original idle rpm. Do not attempt to check the engine in the MAX power range and do not operate the engine at MIL power with the flaps down for a period longer than 1 minute. When the engine checks are completed, complete the remainder of the takeoff checklist. If canopy closure is attempted with engines running, the engines should not be operating above a stabilized idle rpm. Attempted canopy closure with engine rpm above idle may result in canopy not fully locking due to back pressure caused by the aircraft pressurization system.

FLAP POSITIONS

Three flap positions are available for takeoff; half-flaps, full-flaps, and no-flaps. However, the half-flap configuration is recommended for all takeoffs. Full-flaps is not an acceptable takeoff configuration for field operations since it affords no advantages, and several disadvantages (increased drag, reduced thrust, reduced stabilator effectiveness and large trim change during transition to climb) over any other flap configuration. No flaps is not a recommended takeoff configuration. If a no-flap takeoff is attempted in a heavy or draggy aircraft in the same distance as a half-flap takeoff, the attitude of the aircraft at lift-off will result in the airplane flying closer to the stall margin and aircraft control will be e more critical. In order to achieve the same takeoff attitude as that obtained with half-flaps, the airplane takeoff speed must be further increased. By increasing the takeoff speed, the takeoff distance and the airplane's kinetic energy are also increased proportionally, thereby making an abort more difficult. In addition, the increased takeoff speed begins to approach the rotational speed limitations of the tires, making the possibility of tire failure more probable. Stabilator effectiveness during a no-flaps takeoff is considerably greater than during a half-flaps takeoff, therefore, the stick must be programmed forward more rapidly to prevent overrotation beyond the desired 10° to 12° takeoff attitude. Since ARI is available only when flaps are used, increased adverse yaw may be expected in the no-flaps configuration.

PILOT

1. Engines - RUN UP (one at a time)
   a. Variable area inlet ramps - CHECK FULLY RETRACTED
   b. Engine instrument check - NORMAL (relay reading to RIO)
2. Engine anti-ice - AS DESIRED
3. Anti-skid switch - ON, LIGHT-OUT (some aircraft)
4. Radar horizon - SET
5. External transfer switch - CHECK
6. Stab aug/pitch aug - ENGAGE
7. Defog footheat handle - AS DESIRED
8. Pitot heat - ON
9. Tacan - T/R

   Allow a minimum of 2 minutes receive time before selecting T/R.
10. Compass - CHECK SYNC
11. IFF - AS DESIRED
12. Takeoff checklist - COMPLETE WITH RIO
    a. Controls - CHECKED

       Check controls for freedom of movement, normal pressure drop, and direction of movement.

    b. Wings - LOCKED

       Check wing pin unlock handle down, and WING PIN UNLOCK lights out.

    c. Trim - SET
    d. Flaps - 1/2
    e. Hook - UP
    f. Harness - LOCKED & LAP BELT SECURE
    g. Warning lights - OUT
13. Seat pins - REMOVE
14. Lower ejection handle guard - AS DESIRED
15. Command selector valve - AS DESIRED

RIO

1. Tacan - T/R

   Allow a minimum of 2 minutes receive time before selecting T/R.
2. Compass heading sync - CHALLENGE PILOT
3. Circuit breakers - CHECK IN
4. Radar antenna elevation strobe - 17° DOWN
5. Equipment - STOWED
6. Lower ejection handle guard - AS DESIRED
7. Command selector valve - INFORM PILOT OF POSITION
8. Takeoff checklist - COMPLETE
   a. Controls - CHECK
   b. Wings - LOCKED

      Visually check each wing lock pin down, and both WING PIN UNLOCK lights out.

   c. Trim - CHECK SETTING
   d. Flaps - VISUALLY CHECK
   e. Hook - CHECK UP
   f. Harness - LOCKED & LAP BELT SECURE
   g. Warning Lights - OUT
   h. Seat pins - PULLED
9. Report flaps, ramps, ready for takeoff
10. Nav computer - TGT or BASE when airspeed reaches 160 knots.

NORMAL TAKEOFF TECHNIQUE

For individual takeoff, the centerline of the runway should be used as a directional guide. When in position, roll forward slightly to align the nose wheel. If nose gear steering is desired, engagement must be made prior to commencing takeoff roll. Do not engage after the takeoff roll has started. The takeoff roll may be started with the engines in IDLE or the brakes can be applied until 80% rpm and 450–470°C EGT is reached on each engine. After the takeoff roll has begun, the throttles are advanced to MIL power and EGT and RPM are checked. If an afterburner takeoff is desired, afterburner is selected by moving both throttles into the afterburner detent and advancing smoothly to MAX power. If one afterburner fails to light, sufficient directional control is available with the rudder to continue the takeoff with asymmetric power. Very light braking or nose gear steering can be used to maintain directional control until the rudder becomes effective at approximately 70 knots. Nose gear steering should be disengaged when the rudder becomes effective. In any case, nose gear steering must be disengaged prior to lift–off to ensure nose wheel centering and nose gear retraction. Optimum lift–off speeds are contained in the NATOPS Pocket Checklist and in section XI of this publication. Location of the main landing gear a good bit aft of the normal CG prevents this aircraft from being rotated early in the takeoff roll. In the normal rotation technique, position the stick slightly aft of neutral until reaching 120 knots. While the exact stabilator position required to achieve 10° - 12° rotation at lift–off will vary with gross weight and CG, stick position slightly aft of neutral will insure initial rotation for field take–off. As flying speed and stabilator effectiveness are gained, stick position may be smoothly adjusted to achieve the desired 10° - 12° attitude for lift–off. The takeoff attitude of 10° - 12° is identical in the half–flap and no–flap configurations. No–flap takeoff speed will be 10–20 knots faster than half–flap speed. The AUX AIR DOOR, and MASTER CAUTION lights may illuminate momentarily as the landing gear and flaps controls are actuated. This is normal and should be no cause for alarm.

MINIMUM RUN TAKEOFF

For a minimum run takeoff full aft stick is applied at brake release. As the aircraft starts to rotate the stick should be adjusted forward to maintain 10° to 12° of pitch. Concentrate on a smooth rotation and do not exceed 22 units angle of attack. This will allow the airplane to fly off at optimum lift–off speed. Takeoff data is based on minimum run takeoff.

WARNING

From 30 knots below takeoff speed until aircraft is normally airborne, rapid full aft stick movement may cause aircraft overrotation with resultant stalled flight condition, lift–off prior to reaching safe flying speed, or the stabilator striking the runway. With gear down and flaps down, do not exceed 22 units angle of attack. After gear retraction, do not exceed 18 units angle of attack, since the angle of attack system indicates 3 to 4 units low with gear retracted.

CAUTION

Because of the effectiveness of the stabilator, rapid nose–up pitching moment will occur during flaps–up takeoff if the stick is held full aft at rotation.

CROSSWIND TAKEOFF

If nose gear steering is to be used, it must be engaged before commencing takeoff roll. Release brakes evenly, do not ride or keep pressure on the brakes during the initial part of the roll. The brakes should be used sparingly to prevent overheating. Excessive braking will increase the takeoff roll. The rudder will become effective at approximately 70 knots. Hold the nose wheel down until flying speed is reached. Fly the airplane off the runway at optimum lift–off speed. Do not assume an immediate wing low attitude in order to counteract for wind drift; the pilot cannot properly judge the wing tip ground clearance on a swept wing airplane.

FORMATION TAKEOFF

For formation takeoff, all aspects of the takeoff must be prebriefed by the flight leader. This should include flap settings; use of nose gear steering; power changes; power settings; and signals for actuation landing gear, flaps, and afterburner. The leader will take position on the downwind side of the runway with other aircraft in tactical order maintaining normal parade bearing (normal parade is minimum safe aircraft separation). After pretakeoff checks are completed and the flight is in position, engines are run up to approximately 85%, instruments checked and nose gear steering engaged (procedures for nose gear steering technique are the same as for single aircraft takeoff technique). On signal from leader, brakes are released, throttles are advanced to military power minus 2% rpm. (If afterburner is desired, the lead pilot may go into midburner immediately without stopping at military power, or he may select afterburner during the takeoff roll at a later time.) During the takeoff roll, the leader should maintain stick between center and three quarters aft position until reaching 120 knots, then smoothly rotate the aircraft to a 10°–12° nose high attitude. The lead should maintain this attitude until the flight is airborne. The wingman should strive to match the lead aircraft's attitude as well as maintain a position in parade bearing with wingtip separation. The gear and flaps are raised on signal. Turns into the wingman will not be made at altitudes less than 500 feet above ground level. The first section must be airborne before the second section commences its takeoff roll. Visual communication procedures are contained in section VII

CAUTION

- In the event of an aborted takeoff, the aircraft aborting must immediately notify the other aircraft. The aircraft not aborting should add max power and accelerate ahead and out of the way of the aborting aircraft. This will allow the aborting aircraft to steer to the center of the runway and engage the arresting gear if required.
- It is imperative that the wingman always be alert for an over–running situation and take timely steps to preclude such an occurrence. Should an

over-running/over-shooting situation develop after becoming airborne, the wingman should immediately move laterally away from the lead and, if feasible, reduce power in order to maintain wing position, safe flight of both aircraft must not be jeopardized in an attempt to maintain formation. The leader should detach the wingman if he is experiencing loss of thrust and flying speed. The wingman should detach and add power if unable to maintain a safe wing position on the lead.

AFTER TAKEOFF

When the aircraft is definitely airborne, perform the following:

PILOT

1. Ensure that aircraft is definitely airborne before retracting the landing gear.
2. Raise landing gear.
3. Raise flaps at 300 feet or 200 knots while maintaining a 10° to 12° nose-up attitude.
4. Deselect AB at 250 knots minimum
5. External transfer switch - AS DESIRED
6. IFF - AS DESIRED
7. Compass - SLAVED/SYNC

RIO

The RIO will challenge the pilot on the following:

1. Landing gear - UP
2. Flaps - UP
3. External transfer switch - AS DESIRED
4. IFF - AS DESIRED
5. Compass - SLAVED/SYNC
6. Lower ejection handle - AS DESIRED
7. Command selector valve - INFORM PILOT OF POSITION

TRANSITION TO CLIMB

When the airplane is definitely airborne, raise the landing gear. Raise the flaps at 300 feet or 200 knots while maintaining a 10° to 12° nose-up attitude.

CLIMB

A simplified MIL power climb at normal gross weights can be made by maintaining a 10° to 12° nose—up attitude until reaching 350 knots. Vary the pitch attitude as necessary to maintain 350 knots until reaching final cruise Mach. Then Vary the pitch attitude as necessary to maintain cruise Mach until reaching cruise altitude. A simplified MAX power climb at normal gross weights can be made by maintaining a 10° to 12° nose—up attitude until reaching 250 knots. At 250 knots smoothly rotate to a 20° to 25° nose—up attitude and hold until reaching Mach 0.9. Vary the pitch attitude as necessary to maintain Mach 0.9 until reaching cruise altitude. For optimum climb performance, refer to section XI, part 3.

NOTE

The possibility exists that engine flame-outs may occur while flying at altitudes above 35,000 feet in cirrus clouds. Such incidents have occurred and are generally believed to have been caused by excessive ingestion of ice crystals. Under such conditions, ice buildup on the duct lips or other parts of the aircraft are not likely to occur and flame-outs can, therefore, occur without warning. However, in all known incidents of this type, relights have been accomplished and maintained at lower altitudes. Therefore, if flame-out occurs at high altitudes in clouds, it is recommended that relight attempts be deferred until descent to a lower altitude and, if possible, to a less dense part of the cloud.

INFLIGHT

Refer to Inflight Procedures, section IV, and Performance Data, section XI.

LANDING

DESCENT/INSTRUMENT PENETRATION

PILOT

In all descents, care will be taken not to exceed any airframe limitations. See section I, part 4. In any descents from altitude, 5 minutes prior to letdown, select the desired DEFOG position on the defog lever and place the temperature control at the 2 o'clock (200° of clockwise rotation) position. Since rapid descents cannot always be anticipated, the maximum comfortable interior temperature should be maintained. This will aid in defrosting the windshield. Refer to section XI, part 7, for recommended descent. Before starting descent, perform the following:

1. Engine anti-ice – AS DESIRED
2. Radar gyro horizon – CHECK
3. SPC – ENGAGED
4. Radar altimeter – SET & CHECK
   When below 5000 feet, ensure unit is ON, and that the OFF flag is masked. Depress and hold function control knob and check that altitude pointer goes to 5 $\pm$5 feet. Release knob and ensure desired low altitude warning is set.
5. Tacan and UHF homer – CROSS CHECK
6. ARA-63 (ILS) power switch – ON and CHANNEL SET (some aircraft)
7. Defog footheat handle – DEFOG
8. Pitot heat – ON
9. Rain removal – AS DESIRED
10. Cabin heat – SET
11. Compass – SYNC (check against standby).
12. Missile power switch – OFF (STBY with AIM-7 missiles aboard)
13. IFF – AS DIRECTED

NOTE

If it becomes necessary to dump fuel during a descent, thrust settings in excess of 85% rpm may be required to ensure rapid inflight dumping.

RIO

1. Altimeter – SET
2. Flight instruments – CHECK
3. Radar function switch – STBY (safety/operational necessity only, otherwise OFF).
   The radar function switch should be in the OFF position, but may be placed in STBY or ON if safety of flight or operational necessity requires use of the radar.
4. Radar gyro horizon – CHECK (Inform pilot if horizon does not match aircraft attitude and place gyro switch to OUT if so directed by the pilot)
5. Challenge pilot as required for cabin heat, pitot heat, engine de-icing, compass sync, and radar altimeter – ON.
6. Challenge pilot for all armament switches – OFF or SAFE.
7. Determine fuel on board (visual check)

PATTERN ENTRY

Enter the traffic pattern at the altitude and airspeed prescribed by the local course rules. Whenever possible, pattern entry will be made in accordance with figure 3-3.

LANDING

PILOT

1. Landing checklist – COMPLETE
   a. Wheels
   b. Flaps
   c. Hook
   d. Armament
   e. Harness
2. APCS – AS DESIRED
3. UHF antenna – UPR
4. Lower ejection handle guard – AS DESIRED
5. Command selector valve – AS DESIRED
6. Bleed air switch – NORM (some aircraft)
7. Anti-skid – ON (some aircraft)
8. Wheel brakes – CHECK
9. Upon touchdown, throttles – IDLE
10. Drag chute – DEPLOY (as required)
11. Brakes – APPLY

RIO

1. Pilot's checklist – MONITOR
2. Communication antenna – UPR
3. Equipment – STOWED
4. Harness – LOCKED
5. Lower ejection handle guard – AS DESIRED
6. Command selector valve – INFORM PILOT OF POSITION
7. Landing checklist – COMPLETE
8. Report – READY FOR LANDING

FIELD LANDING PATTERN TYPICAL

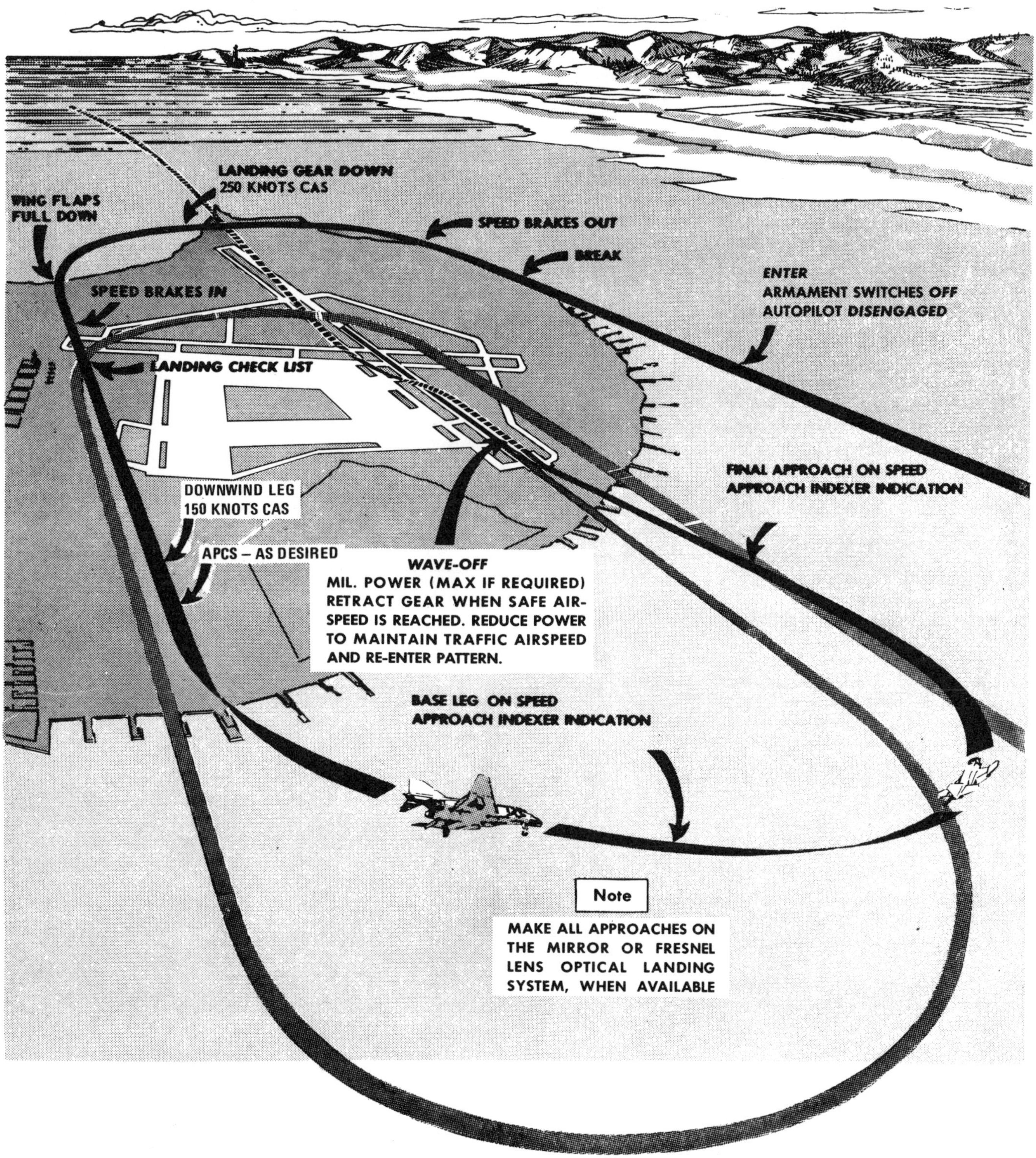

Figure 3-3

LANDING TECHNIQUE

APPROACH

Refer to figure 3-3. Enter the pattern as local course rules dictate. At the break, reduce thrust and extend the speed brakes (if required). As the airspeed decreases through 250 knots, lower the landing gear and extend the wing flaps on the downwind leg. Continue to decelerate to, and maintain, 150 knots, crosschecking angle of attack and airspeed. After completing the landing checklist, roll into the base leg and establish a rate of descent, maintaining an ON SPEED angle of attack. On final approach, maintain an ON SPEED angle of attack and a rate of descent of approximately 700 fpm. This will result in a 2½ – 3° glide slope. Avoid overcontrolling the throttles as power response is immediate. If the APCS is utilized, engage the system after completing the landing checklist with the throttles above 75% and the aircraft at approximately approach airspeed.

Compensate for crosswind in the traffic pattern to guard against undershooting or overshooting the final turn. Fly the final approach course with the aircraft ground track properly aligned with the runway. The crosswind may be compensated for either by using the wing–low method, the crab method, or a combination of the two. When using the wing–low method, the ARI can be overpowered by use of the rudder pedals. If the crab method is employed, the aircraft heading should be aligned with the runway just prior to touchdown. On a wet runway crosswind landing, the aircraft shall be flown to touchdown in a crab.

TOUCHDOWN

Maintain approach attitude and power setting, and touchdown utilizing the mirror or 500 to 700 feet past the runway threshold. On touchdown, place the speedbrake switch IN, retard the throttles to idle, and deploy the drag chute. A firm touchdown will absorb energy, decreasing the landing roll. A minimum sink rate landing should not be made unless required. After touchdown, the nose drops almost immediately due to the airplane center–of–gravity and stabilator location. When the nose gear is on the runway, hold full back stick to increase drag.

CAUTION

- Do not chop power prior to touchdown as the sudden loss of boundary layer control air causes the airplane to settle immediately.
- Do not deploy the drag chute before touchdown. Sink rate and AOA increase rapidly resulting in large deviations from optimum airspeed and glide path.

LANDING ROLLOUT

DIRECTIONAL CONTROL

After touchdown maintain runway track with aerodynamic controls. Lateral stick as well as rudder should be used. In light crosswinds, aerodynamic controls are effective in maintaining track down to very low speeds.

CAUTION

Directional control is a primary consideration during the landing rollout. The most important aspect of directional control is keeping the aircraft precisely tracking down the runway rather than trying to correct back to the runway centerline after it has deviated.

DECELERATION

The airplane is aerodynamically clean, and even with fairly low residual thrust it will tend to roll down the runway with little deceleration. Leave the flaps down to increase aerodynamic drag and to decrease residual thrust by utilizing BLC air. As the drag chute is the most effective means of deceleration early in the landing roll, it should normally be deployed on all landings except for specified no–drag chute landings during the familiarization phase, or landings made with a known crosswind component greater than 20 knots. Use of the drag chute intensifies the weather vane effect for any given condition. The weather vane effect increases as the forward velocity of the aircraft decreases. If the drag chute is to be used, it should be deployed immediately after touchdown.

All landings should be planned and flown as no–drag chute landings. In case of drag chute non–deployment, a waveoff

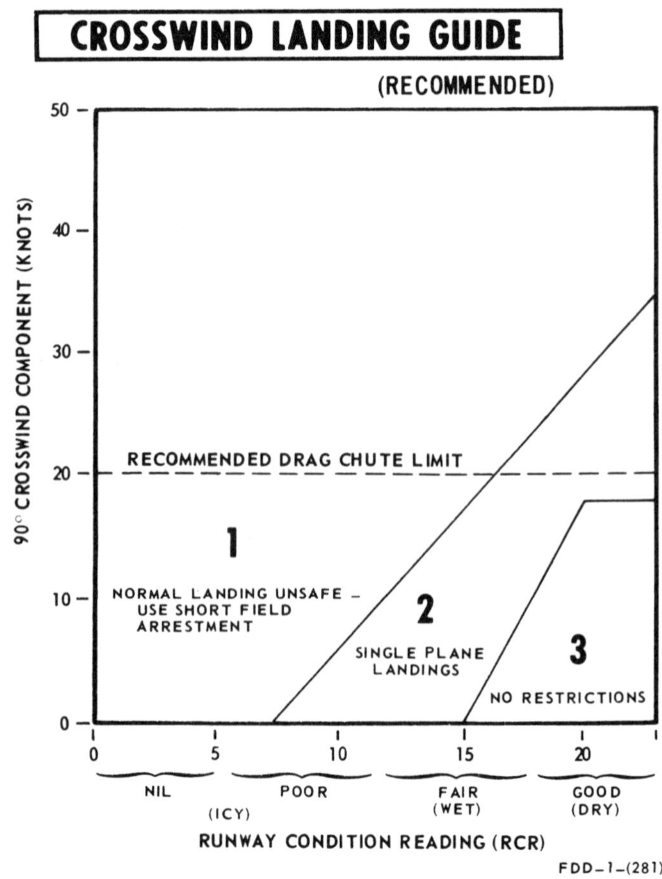

Figure 3–4

shall be initiated if conditions are not ideal to stop the aircraft. If a waveoff is initiated, the drag chute handle should be stowed immediately to preclude inadvertent chute deployment/jettison in the landing pattern. If committed for a no-drag chute landing and there is any possibility that speed or runway conditions will preclude stopping the aircraft on the runway, the pilot must be prepared to drop the hook and engage available arresting gear.

NOTE

- The drag chute should not be used under normal circumstances with a known crosswind component greater than 20 knots. It should not be used at low RCR with a known crosswind component greater than 10 knots. See figure 3-4.
- If the drag chute is used and excessive weathervaning is encountered, jettison the drag chute.

BRAKING TECHNIQUE

With Anti-Skid

The anti-skid system should be utilized at all times to protect against inadvertently locking a wheel or wheels, during braking. The anti-skid system is completely passive unless the wheel is approaching skid; therefore, under conditions of normal braking, it has no effect on the amount of brake the pilot applies. If maximum deceleration is desired, the anti-skid system can be utilized to maintain the wheel at the optimum deceleration point. In this case, the pilot must apply sufficient brake pressure to ensure anti-skid cycling and allow the anti-skid system to reduce applied pressure to proper values. Full pedal application, or any amount of pedal which will produce anti-skid action, will provide maximum wheel braking for the existing conditions. A minimum roll landing using the anti-skid system can be accomplished from a normal touchdown and drag chute deployment followed by full brake pedal deflection with the stick full aft. Less than full pedal can be used, if desired, as long as there is sufficient brake pressure to keep the anti-skid system active. Cycling of the anti-skid system can be detected by a change in longitudinal deceleration; however, cycling of the anti-skid system may not be apparent when braking at high speeds; i.e., immediately after landing, wet runway, etc.

CAUTION

Anti-skid protection is not available until the wheels have initially come up to speed. Do not land with brake pedals depressed. In addition, anti-skid protection is not available below approximately 20 knots.

Without Anti-Skid

Wheel brakes are the primary means of deceleration and directional control when the drag chute and flight control surfaces effectiveness is reduced. The brakes are fully powered rather than boosted, and there is very little feel at the pedals. The tire pressures are high and the tires tend to break loose and skid even with light applications. Normally, wheel brakes should be used only below 100 knots since the probability of blowing a tire decreases significantly with a reduction in ground speed. The most desirable braking technique is a single, smooth application of the brakes with a constantly increasing pedal pressure as the aircraft decelerates. Maintain directional control by easing pressure on the brake opposite the desired direction of turn.

CAUTION

- At high speeds, brake pedal deflections as small as 1/16 inch have proved sufficient to blow a tire.
- Release brake pressure just prior to crossing arresting gear cables. Maintaining brake pressure across arresting gear cables will cause damage which may be sufficient to blow the tire.

CROSSWIND/WET RUNWAY/RUNWAY SURFACE CONSIDERATIONS

The problem of maintaining directional control on a wet runway is greatly intensified with an increase in crosswind component. Characteristics of the runway surface also have a great bearing on directional control capability. A grooved concrete runway provides a good braking surface in most conditions, while a smooth asphalt runway becomes slippery when wet. Heavy rubber marks on the runway surface are very slippery, especially when wet. Standing water greatly decreases braking effectiveness, and may cause total hydroplaning under some conditions. Intermittent puddles of water may cause wheels to lock. Without anti-skid as the locked wheel leaves the puddle, and encounters a good braking surface, it will remain locked, skid, and blow unless brake pressure is released. Brakes should be reapplied only after wheel spin-up has occurred. The following procedures are recommended when landing on a wet runway.

a. Determine field condition prior to approach.
   (1) Braking action.
   (2) Crosswind component
   (3) Type, status, and location of arresting gear.

b. Reduce landing weight as much as possible/practicable.

c. Land on runway centerline, using normal FMLP landing with no flare.

d. When a crosswind or adverse RCR exists, refer to the Crosswind Landing Guide (figure 3-4) to determine the recommended drag chute deployment and arrested landing parameters.

e. If adverse wind and runway conditions exist, plan to make a no drag chute arrested landing being sure to keep ground speed high enough to maintain effective directional control with aerodynamic control surfaces (ailerons, spoilers, and rudders). Braking and nose gear steering may not be effective due to hydroplaning. In the event of a hook skip/bolter, execute a waveoff.

f. If a rollout landing is desired or short field arresting

gear is not available, a wet runway landing should be made at the lowest practicable gross weight. Plan the pattern to be well established on final in a wings-level crab and with an ON SPEED indication. Plan to touchdown on centerline within the first 500 feet. Make a firm touchdown while maintaining the wings-level crab. Touching down in the crab results in a continuation of a straight track down the runway. Immediately after touchdown retard throttles to IDLE and deploy the drag chute. As wheel cornering capability overcomes aerodynamic effects, the nose of the aircraft will gradually assume the track down the runway. Be ready to jettison the drag chute if the weathervaning effect begins to interfere with maintaining desired track. When directional control is firmly established, utilize normal braking. Be prepared to engage the arresting gear if the aircraft is not slowing down properly. During high speed portion of the landing roll, particularly under wet or icy conditions little deceleration will be felt because the braking potential is very low. Unless the pilot is familiar with the variables in braking potential of the aircraft, this low deceleration might be mistakenly interpreted as brake failure.

CAUTION

Do not allow the aircraft to deviate from a straight track down the runway. Jettison the drag chute if necessary. If aerodynamic controls lose effectiveness before effective differential braking can be initiated, engage nose gear steering, as required, to maintain directional control.

Loss of directional control results from poor pilot technique, blown tires, wet runway, crosswinds, nose gear steering malfunctions or a combination thereof. In all but the most extreme crosswind/low RCR conditions, directional control can be maintained through application of proper technique.

A blown main tire on landing rollout may result in a swerve which can be more severe at higher speeds. If a tire should blow above 100 knots, a waveoff is optional. Below this speed a waveoff is not recommended. See section V for emergency procedures for a blown tire on landing rollout.

Hydroplaning may cause severe directional control problems on a wet or flooded runway, especially if a significant crosswind component exists. The best way to avoid this situation is to make an arrested landing, keeping speed high enough to maintain effective aerodynamic directional control until engagement or making a waveoff in the event of a hook skip/bolter.

MAINTAINING/REGAINING DIRECTIONAL CONTROL

Directional control can be maintained/regained by using any or all of the following techniques. Corrective action must be immediate.

a. Necessary aerodynamic controls (stick and rudder) should be applied immediately.

b. Differential wheel braking should be used as necessary. Of course, it is desirable not to blow a tire; however, the primary consideration is to avoid departing the runway. In an extreme situation, if hard differential braking has caused a tire to blow and directional control problems persist, braking may be continued on the wheel with the blown tire. The metal-to-runway contact of a wheel with a blown tire can be effective for deceleration and directional control, especially on a flooded runway. This procedure should be used only to prevent departing the runway. Do not brake on a blown tire under normal circumstances.

c. Nose gear steering should be used if required. Rudder pedals should be neutralized prior to engaging nose gear steering. Lateral stick deflection will provide up to 15° of rudder deflection through the ARI while rudder pedals are neutralized to engage nose gear steering.

d. Jettison the drag chute if adverse weathervaning is encountered.

The nose gear steering system includes a failure detection system/network which detects hardover commands and reverts the nose gear steering system to the shimmy damper mode. The timing device in the failure detection network resets when the nose gear steering button is released. It requires a finite time to operate and may not preclude some degree of nose wheel displacement. Do not use nose gear steering unless required to maintain directional control. If the nose gear steering system fails to respond, or responds unfavorably, release the nose gear steering button and do not reengage.

CAUTION

Do not use differential thrust. The possible benefits are outweighed by the increase in energy that will be undesirable should the aircraft depart the paved surface.

If runway departure cannot be prevented, secure both engines prior to departing the runway.

SECTION LANDING

The leader should transition to optimum approach speed when the runway is sighted, touching down 500–1000 feet down the runway on his side. The wing-man should avoid getting "sucked" and maintain a normal wing position except that as he approaches the runway, he moves out to give additional wingtip clearances at touchdown. The wingman will call "Good Chute" or "No Chute" as the case may be.

MOREST LANDING

The techniques for engaging MOREST are essentially the same as for other types of arresting equipment and are as follows:

1. Notify control tower as soon as possible of intention of engaging MOREST, and transmit estimated gross weight for touchdown.
2. At the 180 position, receive clearance for a MOREST landing.

3. Approach on mirror.
4. Touchdown on centerline or runway and deploy drag chute as required.
5. Lower arresting hook 1000 feet in front of MOREST gear.
6. Longitudinal controls neutral prior to engagement.
7. Engage wire with feet off the brakes.

WAVEOFF

The decision to take a waveoff should be made as early as possible. Advance the throttles to MIL or MAX as required to stop the sink rate. The landing gear should be raised only after the sink rate has been stopped and there is no possibility of the airplane contacting the ground. At a safe airspeed and altitude, raise the flaps.

POST FLIGHT

POSTFLIGHT PROCEDURES

Care must be exercised while taxiing with the drag chute deployed to ensure that the drag chute does not become entangled in the taxi lights, other aircraft, or obstructions. The drag chute will be released on signal from the taxi-signalman in an area where the possibility of interference with other aircraft turning up or taxiing is least. The pilot must advise tower personnel if the drag chute is released elsewhere on the field.

Before engine shutdown, it is recommended (but not required) that the engines be operated at IDLE power for 3 to 5 minutes in order to allow engine temperatures to stabilize. Landing roll and taxi time may be included. Carrier landings may require that the engines be shut down almost immediately after touchdown from high power settings. If the engines are shut down before the recommended idle time, a notation should be made on the yellow sheet. To shut down an engine, move the throttle to OFF, the engine master switch to OFF and the generator control switch to OFF. With only one engine operating, do not move the control stick excessively. Excessive stick movement with hydraulic pressure on only one side of the tandem power control cylinders will cause the hydraulic fluid that is in the unpressurized side of the cylinder to be forced back through the pressure lines to the reservoir, filling the reservoir, and causing the excess fluid to be dumped overboard. The seals within the power cylinders may also be damaged by air ingestion and lack of lubrication. If the above situation occurs, the power control hydraulic systems must be reserviced and checked. Perform the postflight checks as listed in the NATOPS Pocket Checklist, with the exception that during operations where the temperature is below freezing or expected to drop below freezing the aircraft may be parked with wings spread and flaps in the full down position.

POST LANDING

WARNING

- After flight, especially if negative G has been encountered, account for all loose items before opening canopy. Inadvertent seat ejection may occur if any foreign object in the cockpit becomes jammed between the canopy actuator and primary seat-mounted initiator or the ejection gun firing mechanism sear. If all known loose objects cannot be accounted for, leave the canopy closed until the inspection of the banana link area is made by a knowledgeable person.
- After flight, except in those instances where emergency ground egress is the primary consideration, remain completely strapped in until the canopy is fully raised.

PILOT

1. Flaps - UP (when clear of active runway)
2. Anti-skid switch - OFF (some aircraft)
3. Speed brakes - IN
4. Lower ejection handle guard - UP
5. Stab aug - OFF
6. Radar altimeter - OFF
7. Missile power switch - OFF/STBY
8. VTAS power switch - OFF (some aircraft)
9. Pitot heat - OFF
10. Temperature control knob - FULL HOT
    Place the temperature control knob to HOT to evaporate any water that may have collected in the air conditioning system.
11. Defog footheat handle - DEFOG
12. IFF - OFF
13. ARA-63 (ILS) power switch - OFF (some aircraft)
14. Notify RIO - READY FOR SHUTDOWN
15. Ejection seat - RAISE (Before ACC 224)
    Elevate seat to gain clearance for insertion of rocket motor safety pin.
16. Right throttle - OFF
17. Right engine master switch - OFF
18. Right generator switch - OFF
19. Right spoiler - CHECK
    With left engine operating, PC—2 pressure zero, slowly deflect control stick approximately one inch to the right. Have ground crew/RIO verify that the right spoiler does not fully deflect and it returns to a flush condition when the stick is returned to neutral.

With only the left engine operating, PC-2 pressure zero, and stick neutral, verify right spoiler remains in flush position. Any deflection indicates a right spoiler actuator failure.

20. Left throttle – OFF
21. Left engine master switch – OFF
22. Left generator switch – OFF
23. Face curtain pin – INSTALLED
24. All switches, levers and personal equipment – OFF or DISCONNECTED

RIO

1. Lower ejection handle guard – UP
2. Ejection seat – RAISE (Before ACC 224)
   Elevate seat to gain clearance for insertion of rocket motor safety pin.
3. Radar – PERFORM BITS/OFF (if safety/operational necessity requires, otherwise OFF)
4. Nav computer – OFF
5. Data link system – OFF (some aircraft)
6. TACAN – OFF
7. Oxygen – OFF
8. UHF/COMM – OFF
9. AN/ALQ-91A function selector switch – OFF (some aircraft)

CAUTION

Failure to ensure AN/ALQ-91A function selector switch is turned OFF may result in damage to equipment or a blown fuse when external power is reconnected to aircraft.

10. Destruct circuit arm switch – SAFE (safety pin inserted)
11. Face curtain pin – INSTALLED (after engine shutdown)
12. All switches, levers and personal equipment – OFF or DISCONNECTED

HOT REFUELING

Prior to entering refueling pit –

1. Post Landing checklist steps 1 thru 14 – COMPLETED.
2. Air Refueling checklist – COMPLETED.

WARNING

Stop short of refueling pit for tire inspection. If notified of hot brakes, taxi clear of refueling area.

3. Monitor ground control frequency during refueling operation.

WARNING

- Do not operate any transmitter during refueling operations, except in an emergency.
- If fuel starts running from the wing dump masts or the fuselage vent mast, place the refuel selection switch to INT ONLY. If fuel continues to vent on INT ONLY, discontinue fueling. Log either occurance on the yellow sheet.

When refueling is complete –

4. Air refuel switch – RETRACT

REFUELING

ENGINES OFF, WITHOUT ELECTRICAL POWER

If operational expediencies dictate, the aircraft fuel system may be set up for refueling without electrical power. However, the transfer pumps and fuel level shutoff valves cannot be checked using this procedure.

Prior to engine shutdown –

1. Refuel probe circuit breaker – PULL (G5 before AFC 388, J14 after AFC 388, No. 1 panel).
2. Refuel probe switch – REFUEL.
3. Refuel selector switch – AS REQUIRED.
4. Throttles – OFF.

After engine shutdown –

5. After generators drop off the line, engine master switches – OFF.
6. Refuel probe switch – RETRACT.
7. Refuel probe circuit breaker – RESET.

SCRAMBLE OPERATION

SCRAMBLE INTERIOR CHECK

PILOT

1. AIM-9 tone control – 1/4 TURN
2. ICS – HOT MIC
3. Fuel switches – SET FOR NORMAL OPERATION
4. Smoke abate switch – OFF
5. Flap switch – UP
6. Anti-skid switch – ON (some aircraft)
7. Speed brake switch – STOP (NEUTRAL)
8. Engine master switches – ON
9. Engine start switch – RIGHT
10. Anti-ice switch – AS REQUIRED
11. Missile power switch – RADAR STBY
12. Radar altimeter – ON & SET
13. Altimeter – SET
14. Generator control switches – OFF
15. Radio – ON (TR+G)
16. Tacan – ON (TR)
17. IFF – NORM
18. Pitot heat switch – ON
19. Bleed air switch – NORM (some aircraft)
20. Light switches – AS REQUIRED

RIO

1. Radar function switch – OFF
2. All other radar switches for immediate use after normal warm up.
3. Radio – ON (TR+G)
4. Tacan – ON (TR)
5. ICS – NORMAL
6. Light switches – AS REQUIRED
7. Black out curtain – AS DESIRED

SCRAMBLE ENGINE START

PILOT

1. Starting unit up to power – EXTERNAL POWER CONNECTED
2. Generator switches – EXT ON
3. Signal plane captain to turn CNI ground power switch – ON
4. At 10% rpm, right ignition button – DEPRESS WHILE ADVANCING THROTTLE TO IDLE
5. At 35% on right engine, engine start switch – LEFT
6. At 53% on right engine, generator switches – GEN ON
7. Signal to disconnect external ac power
8. Stab aug – ENGAGE
9. At 10% on left engine, left ignition button – DEPRESS WHILE ADVANCING THROTTLE TO IDLE
10. At 35% on left engine, engine start switch – NEUTRAL
11. Static pressure compensator – RESET
12. Complete takeoff checklist

RIO

1. Notify pilot of any emergency signals noted from ground crew.

SCRAMBLE TAKEOFF

Aircraft scrambles invariably take place under various conditions of radio silence (refer to NWP–41A, Chapter 2). The following procedures will be followed for an alert which will probably result in the actual launching of the airplane. Normal preflight, start, and poststart checks will be conducted in accordance with the NATOPS Flight Manual and the NATOPS Pocket Checklist. Shut down the engines but leave the airplane as prepared as possible for takeoff. Remove all seat pins except the face curtain pin. If awaiting the scramble order requires the use of the aircraft radio, observe ground operating limitations. The ground equipment will be positioned to provide rapid removal after starting. When the scramble order is received, start the engines, establish radio communications, determine that all ground locks and safety pins are removed, and that the ground crew and equipment are clear before taxiing. Taxi safely but expeditiously and energize all necessary electrical–electronic equipment. Complete the scramble checklist prior to scramble.

NIGHT FLYING

EXTERNAL LIGHT MANAGEMENT

During night operations, the external lights should be set as follows:

1. On the line – BRIGHT and STEADY
2. When ready for taxiing – BRIGHT and FLASH
3. In flight
   a. Single aircraft – BRIGHT and FLASH (or as weather conditions dictate)
   b. Formations – AS REQUESTED BY WINGMAN
      The last aircraft in formation flight should have his external lights on BRIGHT and FLASH unless tactical situation demands otherwise (actual penetrations etc.)

TAXIING

Night operation demands extra caution while taxiing. It is difficult to judge actual ground speed at night. Pilots can best judge their speed by frequently observing the runway or taxiway close to their aircraft as illuminated by the bottom fuselage light. Taxi slowly for it is possible that unlighted aircraft, vehicles, and/or obstructions are on the taxiways.

TAKEOFF

A night takeoff is accomplished in exactly the same manner as one outlined for daylight operations with the following addition:

Be prepared to transition to complete instrument flight immediately upon leaving the runway.

INFLIGHT PROCEDURES

See section IV of this publication.

LANDING

Night landing procedures are identical to day procedures with the following exceptions:

There is often a tendency to be fast. Be positive about checking angle of attack and airspeed. Determination of altitude and sink rate are difficult at night. This necessitates reference to the vertical velocity indicator. Rates of descent up to 750 feet per minute are acceptable, use mirror when available.

FIELD MIRROR LANDING PRACTICE

PREFLIGHT INSPECTION

A normal preflight inspection will be conducted with specific attention being given to tire condition, nose strut extension, angle of attack probe condition, and windshield cleanliness. Check that the hook bypass switch is in the BYPASS position.

TAKEOFF

The takeoff will be individual using either MIL or MAX power depending on fuel weight, mission, etc.

RADIO PROCEDURES AND PATTERN ENTRY

It is advisable to call Paddles before pattern entry to confirm Charlie Time. Approaches to the field for break will be controlled by the tower and then switched to Paddles for FMLP pattern control. At no time will an aircraft remain in the pattern without a UHF receiver. On each succeeding pass, the following voice report will be made at normal meatball acquisition positions:

Side Number
Type aircraft
Meatball or Clara (no meatball)
Fuel State (nearest 100 lbs.)

PATTERN

The pattern should be a race track pattern with the 180° position approximately 1 ¼ miles abeam at 600 feet above field elevation. The length of the groove should be adjusted to give a wings–level descent on the glide slope of 20–25 seconds (approximately one mile). For maximum gross weight at touchdown, refer to section I, part 4. For a 38,000–pound airplane, an optimum ON SPEED indexer indication results in an airspeed of 141 knots. The turn to downwind leg should be 30° angle of bank and 140–150 knots, climbing to 600 feet above field elevation. Recommended airspeed at the 180° position is 140 to 150 knots. Power will be added to effect a level turn onto final. From the 180° to the 90° position, the airspeed should be corrected for the optimum angle of attack. At approximately the 45° position, the meatball appears on the mirror. A common error is to begin the descent upon first seeing the meatball. Maintain altitude until the meatball is centered on the mirror, then adjust power and angle of attack as necessary to start a rate of descent that will keep the meatball centered. When a Fresnel lens is used, care must be taken to avoid commencing descent until the airplane is aligned with the centerline, since an idiosyncrasy of this lens is to display a false meatball indication when viewed from the approach turn.

APPROACH POWER COMPENSATOR TECHNIQUE

The technique required for an APC approach differs from a manual approach in that all glide slope corrections are made by changing airplane attitude. Since this technique violates the basic rule that altitude is primarily controlled by throttle, practice is required to develop the proper control habits necessary to use APC. For the APC to perform satisfactorily, smooth attitude control is essential. Large abrupt attitude changes result in excessive thrust changes. Close–in corrections are very critical. A large attitude correction for a high–in–close condition causes excessive power reduction and can easily result in a hard landing. If a high–in–close situation develops, the recommended procedure is to stop meatball motion and not attempt to recenter it. A low–in–close condition is difficult to correct with APC and usually results in an over–the–top bolter. It may be necessary to manually override APC in order to safely recover from a low–in–close condition. Throughout the approach the pilot should keep his hand on the throttles in the event it is necessary to manually override the APC.

INTERVAL

The downwind turn should be commenced when the aircraft on the downwind leg is approximately in the 8 o'clock position. The turn should be made with a 30° angle of bank and 140–150 knots, climbing to 500 feet above field elevation.

GLIDE SLOPE

A 2 ¼ to 3° glide slope will be used dependent upon wind conditions. This slope is chosen in order to give the same approximate rate of descent that would be used on the ship.

WAVEOFF TECHNIQUE

Any time the meatball is lost close–in, in the groove, the pilot will initiate his own waveoff. Either MIL or MAX power will be used to effect all waveoffs. Normally, waveoffs will be taken straight ahead, especially when close–in. When using APC, waveoff technique is the same as for manual approaches except that the speedbrake switch should be moved to the closed position, thereby disengaging APC.

NOTE

> If a waveoff is executed by manually overriding the APC, and the APC is not disengaged, the throttles, when released, will retard and attempt to reestablish the optimum approach angle of attack.

BINGO FUEL

No FMLP approach will be commenced with 1500 or less pounds of fuel.

NIGHT FMLP

All provisions which apply to day FMLP also apply to night FMLP plus the following items:

External lights – BRIGHT AND STEADY

Hook bypass switch – BYPASS

When comfortably situated in the pattern, simulated instruments should be flown as much as possible up to the 45° position.

SHORT AIRFIELD FOR TACTICAL SUPPORT (SATS) PROCEDURES

DAY OPERATIONS

GENERAL.

Preflight, start, and poststart checks shall be accomplished in accordance with normal field procedures and the additions noted.

PREFLIGHT

1. Record the expected gross weight of the aircraft for catapult launch on the nose gear door.
2. Ensure that the tension bar retainer clip is installed securely and is in good condition.

START.

1. Start engine sufficiently ahead of time to allow for taxi, catapult launches, and rendezvous before proceeding on the assigned mission.

POSTSTART

1. Set the emergency–jettison armament switch to the proper position prior to taxi.
2. Set trim and flaps as follows:
   a. Rudder: 0
   b. Aileron: 0
   c. Longitudinal: Refer to Pitch Trim Requirements chart
   d. Flaps – DOWN

TAXI

1. Taxiing on advanced airfields presents little difficulty provided attention is given to keeping speed under control.
2. Wet or oily metal runway and taxiway surfaces require especially slow taxi speeds due to a greatly reduced coefficient of friction. Sharp turns cannot be made and the wheels will slide with moderate braking action.

SATS CATAPULT LAUNCHES

Proper positioning on the catapult is not easily accomplished due to surface irregularities in the holdback and arrester area. If the previously launched aircraft utilized afterburner, expect the area aft of the dolly arrester ropes to be wet and slippery. Approach the launch area slowly and be alert for signals from the taxi director.

WARNING

Do not taxi into the dolly arrester ropes area immediately following the launch of another aircraft until the dolly returns and is arrested. Failure of the arrester ropes may occur on dolly rebound.

1. Approximately 80 to 85% rpm is required to taxi up and over the arrester ropes and dolly ramp.

CAUTION

Keep speed under control.

2. As the main wheels roll over the arrester ropes be prepared for immediate braking and power reduction.
3. When the come-ahead signal is given by the taxi director, move ahead cautiously to prevent over-stressing the tension bar.
4. Ground handling crew will extend nose strut upon signal from plane director (prior to tensioning).

CAUTION

If pneumatic pressure approaches 2300 psi with nose strut extended, actuate emergency air brakes to keep pressure below 2300 psi. If the pneumatic pressure exceeds 2300 psi the emergency brakes will not release the pressure in the nose strut. To release this excess pressure the nose strut will have to be deflated and then reinflated. High pressures will subject nose strut to excessive loads during catapulting.

NOTE

Interruption of electrical power, such as cycling both generator switches simultaneously, will cause the nose strut to deflate.

5. After the aircraft is properly positioned and the holdback engaged, the taxi director will signal for the pilot to release brakes while the catapult is tensioned. Advance power to MIL.

CAUTION

Ensure that the brakes are released and power advanced to MIL before tension is taken.

6. After tension is taken and power has been advanced to MIL the taxi director will transfer control of the aircraft to the catapult officer.
7. Recheck ADI and standby attitude indicator, engine instruments, trim indicators, and flap setting. Grip throttle and catapult handgrip firmly.
8. If launch is to be made at MAX A/B, nod to catapult officer after completing MIL power checks. When catapult officer signals, select MIN A/B and ease throttles forward to MAX A/B. Recheck engine instruments.

CAUTION

Failure to ease throttles from MIN A/B to MAX A/B may cause premature tension bar failure.

NOTE

Below 44,000 pounds gross weight, use MIL power. At 44,000 pounds gross weight or above, use MAX (AB) power.

9. When ready for launch, place head against head rest and give an exaggerated left hand salute to the catapult officer while maintaining aft stick with the right hand, observe the green cutoff light, and wait. Launch will occur approximately 3 to 5 seconds after the catapult officer gives the launch signal.

CAUTION

After receiving the signal for full-power turnup, do not allow your hands to appear above the canopy rails unless you intend to salute as a launch signal. Unusual hand movements, such as lowering a helmet visor, will probably result in a premature launch.

TECHNIQUE

Maintain aft stick until start of rotation. Upon perceiving the change in the cutoff light from green to amber, be prepared for rotation, as this is the signal that denotes catapult engine cutoff. Allow the aircraft to rotate to a liftoff attitude (10 – 12° on the ADI) while maintaining positive control of rotation rate by easing stick forward. Do not exceed 21 units AOA. The pilot must avoid gross control movement as the aircraft becomes airborne but should be prepared to make any attitude changes required. When safely airborne, retract gear and flaps as appropriate. Avoid turns until airspeed is well above takeoff speed.

WARNING

- Do not apply brakes during launch.
- Do not change power setting during launch.
- Avoid over–rotation. Over–rotation can result in stabilator contact with matting, excessive distance to clear obstacles, deceleration/stall.
- Rotation rates increase noticeably for CG positions aft of 32% MAC.

AIRCRAFT OR CATAPULT MALFUNCTION

1. If, after established at MIL power, the pilot determines that the aircraft is down, he so indicates to the launching officer by shaking his head from side to side. Never raise a hand into the catapult officer's view in order to give a "thumbs down" signal. Simultaneously broadcast "Suspend" to the tower. When the catapult officer observes the "No–Go" signal, he will immediately give a suspend signal.
2. If bridle shed or bridle failure occurs after holdback release, the pilot will note a sudden loss of acceleration; the dolly will continue to accelerate and move ahead of the aircraft. Wait until the dolly can be seen ahead of the aircraft, then maneuver to the side of the runway to avoid contact with the rebounding dolly. If safe abort or takeoff is not possible and ejection speed has been attained – EJECT.
3. A cold shot can result from inadequate catapult engine acceleration, early catapult cutoff, failure of the capstan brake to release completely, or failure of the dolly jaws. If an abort is attempted and the dolly moves ahead of the aircraft, maneuver to the side of the runway to avoid contact with the rebounding dolly. If safe abort or takeoff is not possible and ejection speed has been attained – EJECT.

LANDING PATTERN

Approach the break point either individually or in echelon, parade formation, at 250 to 300 knots. A 17 to 20 second break interval will provide a 35 to 40 second touchdown interval. Have the landing checklist completed, be at optimum AOA/approach speed by the 180° position.

APPROACH

Plan for and execute an optimum AOA, on–speed approach. Pay particular attention to maintaining the proper airspeed and correct lineup.

WAVEOFF

To execute a waveoff, immediately add full power, and maintain optimum attitude. Make all waveoffs straight ahead until clear of the landing area.

ARRESTED LANDING

The aircraft should be on runway centerline at touchdown. Aircraft alignment should be straight down the runway with no drift. Upon touchdown, maintain the throttle at the approach position. When arrestment is assured, retard the throttle to idle. Allow the aircraft to roll back to permit the hook to disengage from the pendant. When directed by the taxi director, apply both brakes to stop the rollback, raise the hook and flaps. If further rollback is directed, release brakes and allow the aircraft to be pulled back until a brake signal is given. Then apply brakes judiciously to prevent the aircraft from tipping or rocking back.

CAUTION

Use extreme caution when taxiing on a wet SATS runway.

BOLTER

Bolters are easily accomplished. Simultaneously apply full power and retract the arresting gear hook. Smoothly rotate the aircraft to a liftoff attitude and fly away.

WARNING

If landing on a runway with a SATS catapult installed, care must be taken to prevent engagement of the dolly arrester ropes with the aircraft's tailhook. Structural damage to the aircraft and catapult will result.

NIGHT OPERATIONS

GENERAL

This section covers only that portion of night operations significantly different from day operations.

POSTSTART AND TAXI

It is prudent to perform the poststart and taxi phase with the aircraft exterior lights and rotating beacon operating if allowed by local regulations and combat conditions. Wing lights should be on BRT/STDY.

CATAPULT LAUNCHES

Immediately prior to taxi, check master exterior light switch OFF, external light switches BRT and STDY. Rely upon and follow closely the directions of the plane director. Upon receiving the signal from the plane director, release brakes, advance power to MIL and check all instruments as tension is applied. When given the turnup signal by the catapult officer, apply power, and check all instruments. When satisfied that the aircraft is ready for launch, so signify by placing the master exterior light switch in the ON position; fuselage light OFF. Be prepared to establish a wings level climbing attitude on instruments. An initial attitude of 10°–12° noseup is recommended. Retract gear when above 300 feet altitude, retract flaps at no lower than 500 feet. When climbing through 2500 feet, adjust lights and radio as briefed.

PART 4 CARRIER-BASED PROCEDURES

PREFLIGHT

GENERAL

The CVA/CVS NATOPS Manual and the applicable Aircraft Launching Bulletins shall be read by all flight crewmembers prior to carrier qualification. A normal shore–based preflight inspection should be accomplished with particular attention given the landing gear, tires, hook, and underside of the fuselage for possible launching pendant or arresting cable damage. In the cockpit, particular attention should be given to the pilot's scope to ensure that the retaining devices have been installed. Tiedowns shall not be removed from the airplanes unless the emergency brake air pressure gage indicates at least 1500 psi. The pneumatic brake shall be used for stopping the aircraft anytime it is being moved while the engines are not running.

TAXIING

Any signal from the plane director above the waist is intended for the pilot. Any signal below the waist is intended for deck handling personnel. Taxiing aboard ship is much the same as on the land with the exception of additional power requirements. Nose gear steering is excellent and requires use of minimum power while taxiing. Taxi speed should be kept under control at all times especially on wet decks and approaching the catapult area. Be prepared to use the emergency air brake should normal braking fail. The lower ejection handle guard should be down while taxiing.

LAUNCH

PRIOR TO CATAPULT HOOK-UP

Prior to taxi onto the catapult, pilots and RIO's shall ensure through verbal check–off that the takeoff checklists are complete, the compass controller is in the FREE mode, and the radar horizon is set for back–up attitude control. Errors are introduced into the SLAVED mode of the HSI due to the magnetic influence of the ship. To use the radar gyro horizon as an emergency back–up for attitude control if the ADI fails during launch, the pilot should set the radar horizon to zero. The RIO will position the elevation strobe to 17° down, the pilot should verify his el strobe is 17° down, and have the RIO make adjustments as necessary. Refer to figure 3-6 for temperature, gross weight, flap position, and trim setting considerations. Directional and lateral trim should be set at neutral in all cases, regardless of gross weight, flap position, or power settings.

CAUTION

Catapult launching acceleration can force fuel out of the external tanks through the transfer lines to the fuselage cells at a rate beyond tank venting capability, thus creating a partial vacuum in the external tanks. Therefore, to prevent external tank collapse during a catapult launch, insure that the external transfer switch is in the OFF position before launch.

CATAPULT HOOK-UP

Proper positioning on the catapult is easily accomplished if the entry is made with only enough power to maintain forward motion and the plane director's signals are followed explicitly. All functional checks will be performed before taxiing onto the catapult, if practicable. The best technique for positioning is to approach the catapult track with a minimum amount of power utilizing nose gear steering. The pilot should sight down the catapult track, acquire the plane director and follow his

NOSE STRUT EXTENSION PRESSURE MINIMUMS

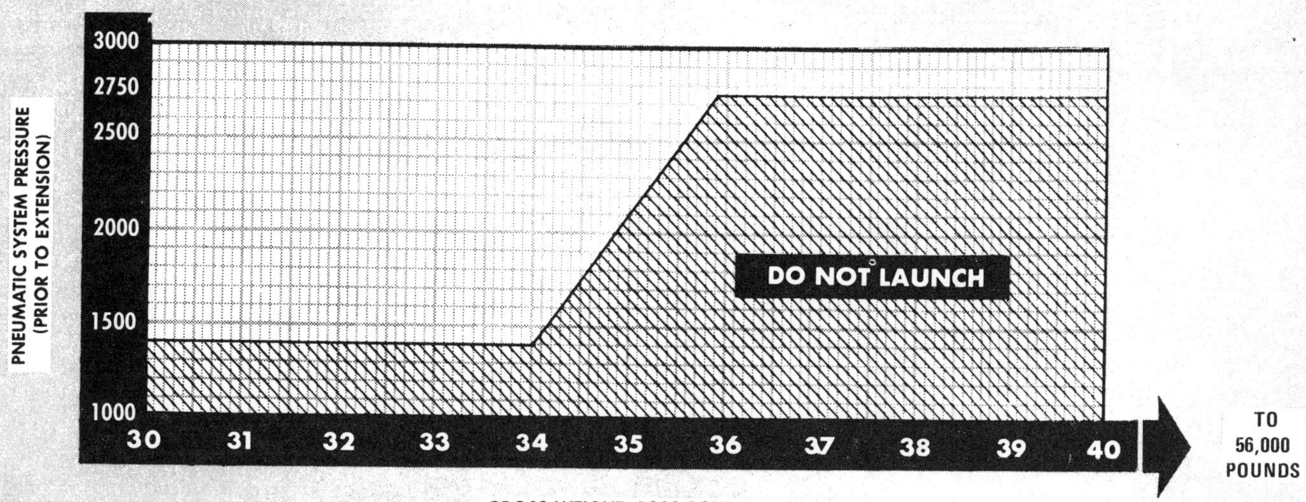

Note

DURING CARRIER QUALIFICATIONS, THE REPEATED NOSE STRUT EXTENSIONS PLACE SUCH A HEAVY LOAD ON THE PNEUMATIC SYSTEM THAT SYSTEM PRESSURE CAN NOT BE BUILT BACK UP TO NORMAL. ALTHOUGH 2750 PSI IS THE NORMAL MINIMUM PRESSURE FOR NOSE STRUT EXTENSION, THIS MINIMUM MAY BE LOWERED AS GROSS WEIGHT DECREASES FOR CARRIER QUALIFICATION PURPOSES. THIS CURVE SPECIFIES THE MINIMUM PNEUMATIC SYSTEM PRESSURE FOR A SPECIFIC GROSS WEIGHT. SATISFACTORY NOSE STRUT EXTENSIONS CAN BE OBTAINED AT THESE MINIMUMS, HOWEVER THERE IS A DECREASE IN ANGLE OF ATTACK ON TAKE OFF.

FDD-1-(48)A

Figure 3-5

signals very closely. The pilot should anticipate an initial hold after the nose wheel drops over the shuttle. After crossing the shuttle, prior to catapult tensioning, the nose strut will be extended (see figure 3-5 for nose strut extension pressure minimums prior to extension). The normal minimum pressure after extension is 1350 psi. On signal of catapult tensioning, release brakes and advance power to MIL, anticipating a handoff signal to the Catapult Officer. After catapult tension, set ADI to zero and recheck standby attitude indicator.

WARNING

Once the nose strut is extended, any interruption of electrical power, such as cycling both generator switches simultaneously, will release the solenoid held pneumatic pressure selector valve. With the pneumatic pressure selector deenergized, the air pressure in the strut extension cylinder will be relieved and the nose strut will deflate from the catapult extended condition.

CAUTION

Do not allow the pneumatic system pressure to exceed 2300 psi with the nose gear extended. If the pneumatic pressure starts to build up above this value, actuate the emergency air brakes as necessary to maintain the pressure below 2300 psi. If the pneumatic pressure exceeds 2300 psi the emergency brakes will not release the pressure in the nose strut. To release this excess pressure the nose strut will have to be deflated and then reinflated. Allowing the pneumatic system pressure to exceed 2300 psi subjects the nose strut to excessive loads during catapulting.

PRELAUNCH

PILOT

1. External transfer switch – OFF
2. Centerline tank – CHECK (IF INSTALLED)
   a. Buddy tank switch – FILL

POWER/FLAP/TRIM SETTINGS

POWER AND FLAP SETTINGS

A/C GROSS WT – LBS	POWER SETTING	AMBIENT AIR TEMPERATURE DEGREES F	FLAP SETTING
BELOW 50,000	MRT MAX A/B	ALL ALL	FULL HALF OR FULL
AT 50,000	MRT MRT MAX A/B	70 OR BELOW ALL ALL	FULL HALF HALF OR FULL
ABOVE 50,000	MRT MAX A/B	ALL ALL	HALF HALF OR FULL

PITCH TRIM SETTING

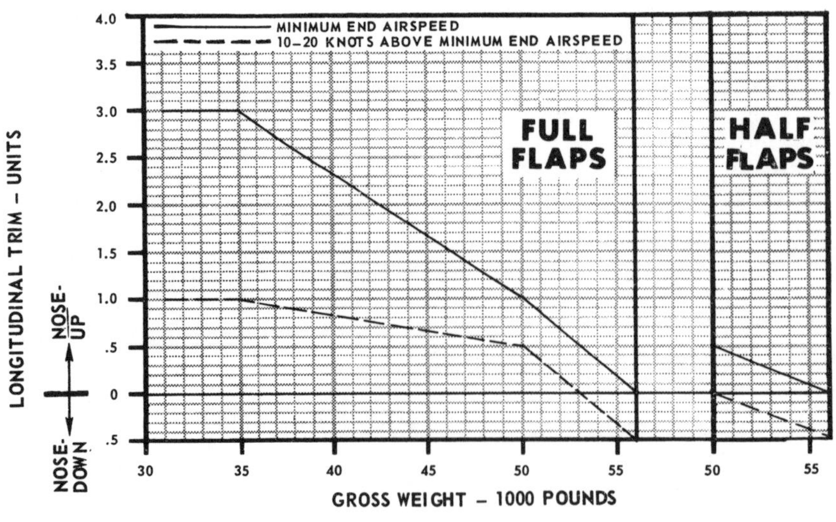

FDD–1–(290)A

Figure 3-6

b. Observe momentary illumination of CTR EXT FUEL light and decreasing sector and counter readings.
c. When CTR EXT FUEL light re-illuminates, Buddy tank switch - STOP FILL
   If the CTR EXT FUEL light fails to illuminate, it may indicate centerline fuel transfer problems and should be investigated prior to flight.

3. Stab aug switches - ENGAGE
4. Engine anti-ice - AS DESIRED
5. Trim - SET (refer to Pitch Trim Requirements chart)
6. Radar horizon - SET
7. Altimeters - CHECK
8. Pneumatic pressure indicators - CHECK
9. After nose extension - RECHECK PNEUMATIC PRESSURE
10. ADI and standby attitude indicator - CHECK AFTER NOSE EXTENSION
11. Defog foot heat handle - AS DESIRED
12. Pitot heat - ON
13. Rain removal - OFF/LOW
14. Bleed air switch - CHECK NORM (some aircraft)
15. Bleed air off light - OFF (some aircraft)
16. Compass - DG MODE
17. Command selector valve - AS DESIRED
18. Lower ejection handle guard - DOWN
19. Complete panel mounted T.O. checklist
20. Engine run-up on signal from catapult officer
21. Flight and engine instruments - CHECK

RIO

1. Compass heading sync - CHALLENGE PILOT
2. Circuit breakers - CHECK IN
3. Equipment - STOWED
4. Command selector valve - INFORM PILOT OF POSITION
5. Lower ejection handle guard - DOWN
6. Takeoff checklist - CHALLENGE PILOT FOR COMPLETION
7. Report circuit breakers in, flaps and ramps - READY FOR TAKEOFF
8. Navigation computer - TGT OR BASE (IMMEDIATELY PRIOR TO LAUNCH)

LAUNCH

WARNING

A verbal report from the pilot to the RIO on stick position, and a visual check of stabilator position in the RIO's mirrors followed by a verbal report to the pilot, must be made before the pilot's salute to the catapult officer.

MILITARY POWER

On catapult tensioning signal, advance throttle to MIL, place the control stick full aft (note fluctuation of PC-1 and PC-2) and check engine instruments and trim settings. Ensure that the head is positioned firmly against the head rest. Use MIL power catapult hand grips or move throttles outboard into the afterburner detent and use as a throttle stop. When satisfied the aircraft is ready, give an exaggerated left hand salute to the catapult officer while maintaining aft stick with the right hand. Control stick positioning during catapult launch is a function of aircraft gross weight and stabilator effectiveness. An increase in gross weight results in an aft CG shift with a resultant decrease in aft stick requirement. For normal-to-heavy gross weights, the control stick shall be placed in the full aft position for initial positioning and then moved forward slightly reducing aft stick approximately one quarter. For carrier qualification weights, the control stick should be positioned full aft and held in this position until rotation off the bow.

WARNING

Holding the control stick in any position that will not place the stabilator in a leading edge down position during a carrier qualification weight launch will impart nose down pitch to the aircraft off the bow from which it may be impossible to recover.

Although the aircraft has no trimmed neutral stick position that will meet the requirements for all gross weight launches, pilot experience is gained rapidly with a minimum number of launches, and stick positioning poses no problem. After launch, establish a 10° to 12° pitch angle on the ADI, cross checking the pressure instruments to ensure a positive rate of climb. If the ADI fails or is unreliable during launch, the standby attitude indicator and radar horizon are available for attitude reference. The altimeter, airspeed, and rate of climb may dip slightly during catapult stroke but will recover shortly after shuttle release.

NOTE

Holding the control stick fully aft during a high gross weight launch will impart a higher than desired airplane rotation rate. Although this overrotation may be stopped with forward stick, it creates an undesirable and unnatural control movement, expecially during night or IFR conditions.

MAXIMUM POWER

When a MAX power launch is scheduled, the following signals will be used:

After completing MIL power checks, the pilot will nod to the catapult officer.

Catapult Officer responds with 5 fingers (open hand held towards pilot)

Pilot selects MAX power, checks instruments and positions himself, then gives an exaggerated left hand salute to the Catapult Officer while maintaining aft stick with the right hand. An optional method of selecting the afterburner may be used by advancing the throttles to minimum AB and assuring an afterburner light-off by noting that the nozzles open slightly. After the catapult fires, advance the throttles to the MAX position.

LAUNCH ABORT PROCEDURES

If, after turn-up on the catapult, (Day Launch), the pilot determines that the aircraft is down, he so indicates by shaking his head from side to side. Never raise the hand into view to give a thumb down or make any motion that might be construed as a salute. After the Catapult Officer observes the no-go signal, he will then cross his forearms over his face. This signal will be followed by the standard release-tension signal. When the bridle has dropped, the Catapult Officer will then step in front of the wing of the aircraft and give the throttle-back signal. Then, and only then, will the power be reduced. If the aircraft is down after the go signal is given, transmit the words "Suspend, Suspend." At night, in addition to transmitting "suspend", turn the aircraft exterior lights OFF.

CARRIER NIGHT FLYING

GENERAL

Night carrier operations will have a much slower tempo than daylight operations and it is the pilot's responsibility to maintain this tempo. The procedures outlined here are different from, or in addition to, normal day carrier operations.

BRIEFING

Before initial night flight operations, all pilots should receive an additional briefing from the following persons.

Flight Deck Officer
Catapult Officer
Arresting Gear Officer
LSO

Individual flight briefings will include all applicable items outlined above, with particular emphasis on weather, and Bingo fuel. The ready room will be lighted for night adaptation during briefings. In addition, pilots may wear night adaptation glasses from the ready room to the flight deck to prevent loss of night vision.

PREFLIGHT

External preflight will be made utilizing the red lensed flashlight. In addition to normal cockpit preflight, insure that external light switches are properly positioned for poststart light check. The general rule of not showing white lights on the flight deck at night should be observed.

POSTSTART

Adjust cockpit light intensity to desired level. When ready for taxi, indicate with appropriate signal.

TAXI

Night deck handling operations are of necessity slower than those used during the day. When a doubt arises as to the meaning of a signal from a taxi director, stop.

CATAPULT HOOK-UP

Maneuvering the aircraft for catapult hook-up at night, is identical to that used in day operation; however, it is difficult to determine your speed or degree of motion over the deck. The pilot must rely upon, and follow closely, the plane director's signals.

CATAPULT LAUNCH

On turnup signal from Catapult Officer, assure throttles in MIL or MAX and check all instruments. When ready to go, place external light master switch ON (dim/bright and steady), fuselage light OFF. After launch, establish a 10° to 12° pitch angle on the ADI, cross checking the pressure instruments to ensure a positive rate of climb. Retract the landing gear. Five hundred feet is considered to be minimum altitude for retraction of flaps. When well established in a climb, switch lights to bright and flashing or as applicable for instrument climb-out. The standby attitude indicator or radar gyro should be used in the event of an ADI malfunction.

CATAPULT ABORT PROCEDURES

The pilot's no-go signal for night launches will be not to turn on his exterior lights. The pilot should also call on land/launch and advise Side No., Cat No., is down. Maintain MIL power until the throttle-back signal is received from the Catapult Officer standing in front of the wing of the aircraft. In the event of a catapult malfunction, the above signals will also apply. If the aircraft is down after the go signal is given, transmit the words "Suspend, Suspend" and turn exterior lights OFF.

ARRESTMENT AND EXIT FROM LANDING AREA

During the approach, all lights shall be on bright and steady. At end of arrestment rollout, turn off external lights and follow director's signals.

INFLIGHT

Refer to Inflight Procedures, section IV, and Performance Data, section XI.

LANDING

CARRIER LANDING PATTERN

A carrier landing pattern (figure 3–7) starts with the level break at 800 feet, 300–350 knots, on the starboard bow of the ship. The break interval will be approximately one–half of the desired ramp interval time. Radio procedures will be in accordance with ship procedures. When established wings–level on the downwind leg, complete the landing checklist, descend to and fly the pattern at 600 feet MSL. The 180° turn is commenced when abeam the LSO platform to arrive at the 45° position at approximately 450 feet MSL. Glide slope/meatball acquisition will occur at approximately 1 to 1 ¼ mile. On rollout to final, slightly overshoot the ship's wake. Optimum time on glide slope is approximately 30 – 35 seconds.

PILOT

1. Landing checklist – COMPLETE
   a. Wheels
   b. Flaps
   c. Hook
   d. Armament
   e. Harness
2. APCS – AS DESIRED
3. Communications antenna selector switch – UPR
4. Lower ejection handle guard – AS DESIRED
5. Command selector valve – AS DESIRED
6. Anti–skid – OFF (some aircraft)
7. Upon touchdown – SPEEDBRAKE SWITCH IN, THROTTLES FULL MILITARY

RIO

1. Pilot's checklist – MONITOR
2. Communication antenna selector switch – UPR
3. Equipment – STOWED
4. Harness – LOCKED
5. Lower ejection handle guard – AS DESIRED
6. Command selector valve – INFORM PILOT OF POSITION
7. Landing checklist – COMPLETE
8. Report – READY FOR LANDING

APPROACH POWER COMPENSATOR TECHNIQUE

Refer to Approach Power Compensator Technique, Field Mirror Landing Practice, this section.

GLIDE SLOPE

The technique of flying the glide slope is the same as FMLP except that more power may be required and line–up will be much harder to maintain. With rough seas and pitching decks, some erratic meatball movements may be encountered. If this is the case, average out the bouncing ball to maintain a smooth and safe rate of descent. In no case overcorrect if the ball moves to a high indication.

ARRESTMENT

NOTE

In the event of a blown tire on landing, do not raise the flaps until the flap area has been inspected.

ARRESTED LANDING AND EXIT FROM LANDING AREA

As the aircraft touches down, advance throttles to MIL. Upon completion of landing rollout, reduce power to IDLE, raise the hook and allow the aircraft to roll aft. Apply brakes on signal. Fold wings and have the RIO report wing fold position. Taxi forward on the come ahead, and keep the engines running until the CUT signal is given by the

CARRIER LANDING PATTERN

TYPICAL

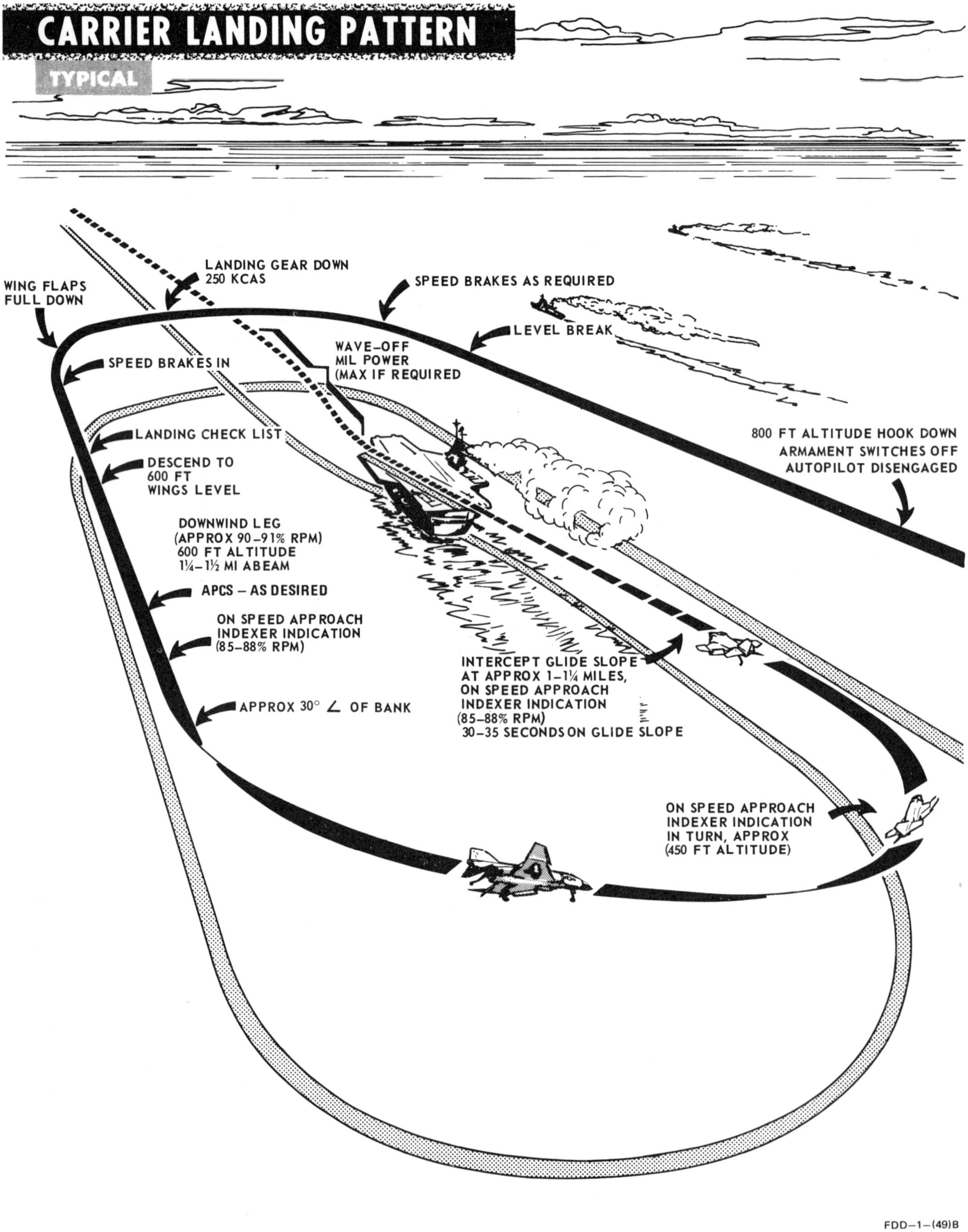

Figure 3-7

plane director. If, at anytime during this phase of operations, one or both brakes fail, utilize the emergency pneumatic brakes, call the tower and drop the arresting hook. Do not leave cockpit until tie-downs have been installed, the number of which will be dictated by the ship.

NOTE

After each arrested landing, inspect the stabilator leading edge for damage from arresting cable.

CARRIER CONTROLLED APPROACHES

GENERAL

Should conflict exist between these procedures and those contained in CVA/CVS NATOPS Manual (and augmenting ships instructions), the latter shall govern. Figure 3-8 illustrates a typical carrier controlled approach. All weather carrier landing system approaches are illustrated in figures 3-9 and 3-10.

HOLDING PHASE

Five minutes before penetration, defogging shall be actuated and maximum comfortable interior temperature will be maintained to prevent possible fogging or icing on the windscreen and canopy.

LET-DOWN PROCEDURES

1. Before descent, check shoulder harness handle locked, set lights as dictated by existing weather, and lower arresting hook.
2. Turn on pitot heat and select engine anti-icing system as appropriate.
3. Accomplish final changes to radio and IFF upon departing Marshal or earlier. After these changes are made, pilot will make no further changes except under emergency conditions.
4. When commencing penetration, initiate a standard descent—250 knots, 4000 feet per minute, speed brakes as desired.
5. Radar and barometric altimeters shall be cross checked continuously when below 10,000 feet.

NOTE

If it becomes necessary to dump fuel during a descent, thrust settings in excess of 85% rpm may be required to ensure rapid inflight dumping.

PLATFORM

At 20 miles passing through 5000 feet, aircraft descent shall be slowed to 2000 feet per minute. At this point, a mandatory unacknowledged voice report will be broadcast by each pilot. The aircraft side number will be given, the word "platform" will be stated

TEN MILE GATE

1. At 10 miles, call "side number" and "ten-mile gate".
2. Commence transition to landing configuration, maintaining 1200 feet.
3. Gear down at 10 miles—flaps down at 195 knots.
4. Complete the landing checklist. Check anti-ice, lights, rain removal, and pitot heat as desired.

SIX MILE GATE

When passing six miles, check exterior lights on, call side number, fuel state, and six-mile gate. For a PAR approach, unless otherwise directed, maintain 1200 feet at approach speed until intercepting the glide path at 2¾ miles. For an ASR approach, a gradual descent of 600 feet shall be commenced departing the six-mile gate. In order that intervals remain constant, ship procedures as to when aircraft are slowed to final approach speed after passing six miles should be followed. Altitude of 600 feet is maintained until the Final Controller calls "commence landing descent", or the meatball is observed to be centered. At 600 feet, aircraft will intercept the center of the glide slope at 1¼ miles on a four-degree slope. If ceilings or visibility preclude visual acquisition of the meatball at 1¼ miles, 500 to 700 feet per minute descent passing the following check points should be continued.

1 mile — 500 feet
3/4 mile — 400 feet
1/2 mile — 300 feet

MEATBALL CONTACT

When ready to continue a visual approach, the pilot reports side number, F-4, meatball or clara (no meatball), fuel state. The LSO will acknowledge and instructions from the Final Controller will cease. Because of this, pilots are cautioned against premature contact reports during night recoveries when visibility permits sighting the ship beyond two to three miles. There is little depth perception even under the most ideal conditions and it is difficult to judge distance from the ship without reference to Tacan. During night VFR conditions, pilots must cross check Tacan DME to ensure that they are actually at 1¼ miles, 600 feet, prior to reporting meatball. The height, dimension of the lens or mirror optical beam at 1¼ miles is over 200 feet and the true center is difficult to distinguish. This, coupled with the relatively short length of the runway lights, will give the pilot the illusion of being on glide slope and high when, in fact, the aircraft may be 50 to 100 feet below the glide slope. An additional

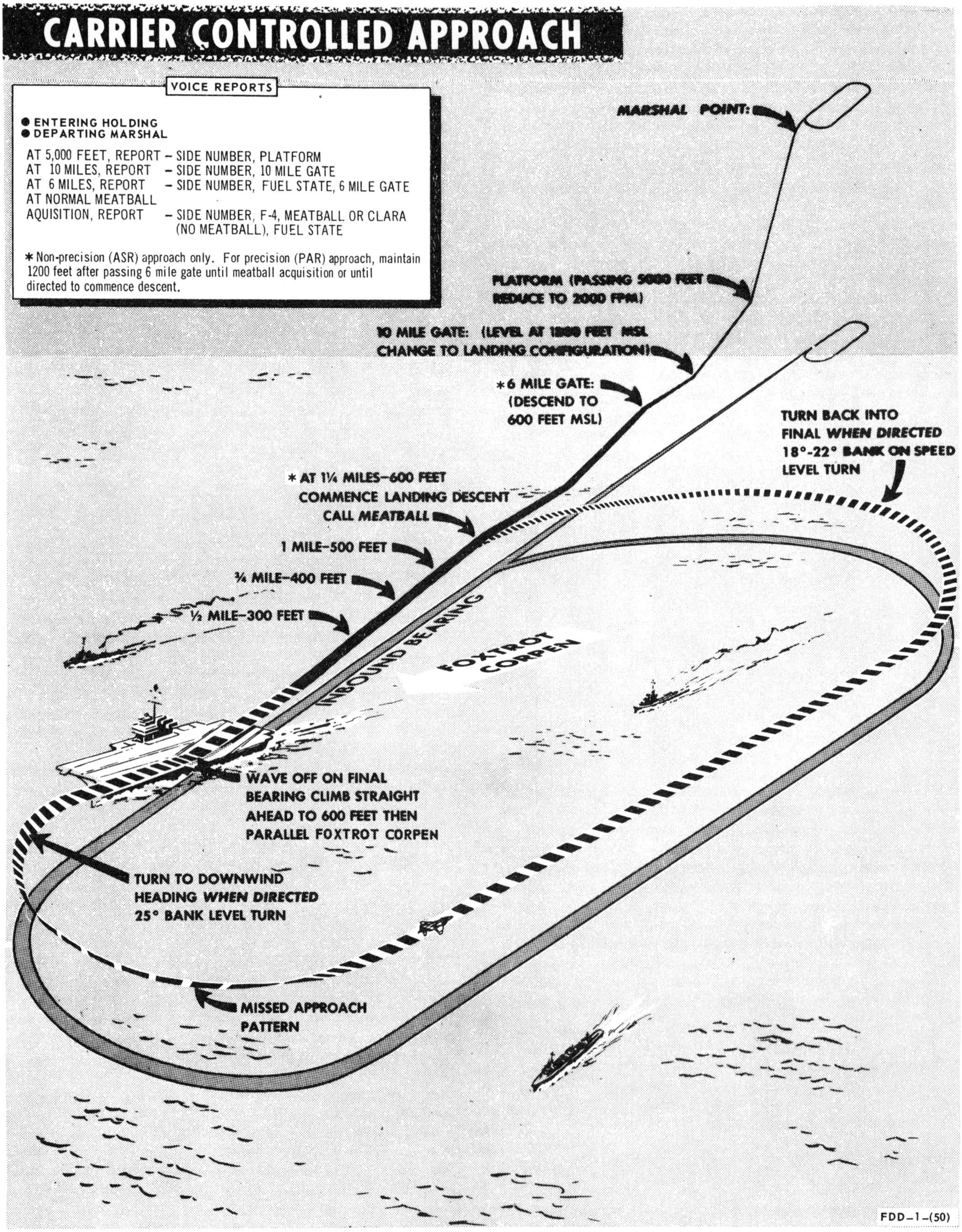
CARRIER CONTROLLED APPROACH
VOICE REPORTS
● ENTERING HOLDING
● DEPARTING MARSHAL
AT 5,000 FEET, REPORT – SIDE NUMBER, PLATFORM
AT 10 MILES, REPORT – SIDE NUMBER, 10 MILE GATE
AT 6 MILES, REPORT – SIDE NUMBER, FUEL STATE, 6 MILE GATE
AT NORMAL MEATBALL AQUISITION, REPORT – SIDE NUMBER, F-4, MEATBALL OR CLARA (NO MEATBALL), FUEL STATE
∗ Non-precision (ASR) approach only. For precision (PAR) approach, maintain 1200 feet after passing 6 mile gate until meatball acquisition or until directed to commence descent.
MARSHAL POINT:
PLATFORM (PASSING 5000 FEET REDUCE TO 2000 FPM)
10 MILE GATE: (LEVEL AT 1200 FEET MSL CHANGE TO LANDING CONFIGURATION)
∗6 MILE GATE: (DESCEND TO 600 FEET MSL)
TURN BACK INTO FINAL WHEN DIRECTED 18°-22° BANK ON SPEED LEVEL TURN
∗ AT 1¼ MILES–600 FEET COMMENCE LANDING DESCENT CALL MEATBALL
1 MILE–500 FEET
¾ MILE–400 FEET
½ MILE–300 FEET
INBOUND BEARING
FOXTROT CORPEN
WAVE OFF ON FINAL BEARING CLIMB STRAIGHT AHEAD TO 600 FEET THEN PARALLEL FOXTROT CORPEN
TURN TO DOWNWIND HEADING WHEN DIRECTED 25° BANK LEVEL TURN
MISSED APPROACH PATTERN
FDD-1-(50)

Figure 3–8

advantage of delaying the meatball report until reaching 1¼ miles — even though the ball is in sight — is that Final Control will continue line-up instructions that can greatly assist the pilot in establishing satisfactory line-up. Use the vertical velocity indicator to set up a rate of descent of 500 to 700 feet per minute.

VOICE PROCEDURE

Detailed pilot/controller voice procedure must be established in accordance with each ship's CCA doctrine.

WAVEOFF AND BOLTER PHASE

WARNING

Due to stabilator ineffectiveness at low airspeeds the aircraft will settle off the angle unless rotation to a takeoff attitude is commenced immediately upon leaving the deck. In order to avoid settling, position the stick well aft during the deck roll.

In the event of a waveoff or bolter, MIL/MAX power should be added as necessary, climb straight ahead to 1200 feet, and maintain 150 knots. When directed by CCA, initiate a level turn to the downwind leg. (If no instructions are received within two minutes or 4 miles distance on Tacan, assume communications failure and initiate the downwind turn to the reciprocal of Base Recovery Course (BRC) and report abeam. If no acknowledgement is received, reenter the final through the 6 mile gate.) A 25-degree bank angle at 150 knots on the upwind turn will establish the aircraft at the desired 1⅜ to 1½ miles abeam on the downwind leg. Aircraft that undershoot or overshoot a proper downwind leg may be vectored back to a proper abeam position. Slow to proper approach speed when approaching the abeam position. This position can be established by using a relative Tacan bearing of 15 degrees aft of the wing at 1¾ to 3 miles on DME when on the downwind heading. Final Control will clear the aircraft to turn back inbound to intercept the final bearing. A level, on speed approach turn of 18 to 20 degrees bank angle from the normal abeam position will allow the aircraft to properly intercept the final inbound bearing at 1¼ to 1½ miles aft of the ship. Traffic spacing ahead may require that the aircraft continue on downwind leg well past the normal abeam position before being directed to turn to final bearing. No attempt should be made to establish visual contact with the ship when executing a CCA until the final approach turn has been executed. When fuel considerations become critical in an extended bolter pattern, 300 to 400 pounds per pass may be saved by raising the landing gear and selecting half flaps. Lower landing gear and full flaps on final. The WHEELS light on the pilot's instrument panel will flash with the gear up and the flaps down.

WARNING

Do not attempt to fly an optimum (on-speed) angle-of-attack with the flaps 1/2 or full down and the landing gear retracted. Gear up airspeeds at optimum (on-speed) angle-of-attack correspond to gear down airspeeds at 22 to 23 units (10 to 12 knots slow). This is especially critical in drooped aileron airplanes because landing configuration stalls occur at 24 units angle-of-attack.

FOULED DECK HOLDING

Detailed procedures for fouled deck holding are contained in the CVA/CVS NATOPS Manual. The best gouge for maximum endurance at any altitude is to fly 9 units on the angle of attack indicator, and utilize minimum bank angles. This pre-supposes that all aircraft in the landing configuration will retract gear and flaps. If necessary, as fuel becomes critical with no Bingo field available, jettison external stores. In extreme emergency situations, a small amount of fuel (approximately 10 pounds per minute) can be saved by securing one engine at sea level with average landing gross weights. If holding must be accomplished with flaps fully extended, the normal endurance time will be decreased by 4.5 minutes per 1000 pounds of fuel for average landing gross weights at sea level. Due to the increase in drag with flaps full down, and decreased efficiency due to the bleed air loss through the BLC system, this decrement increases with altitude. For example, at 25,000 feet the decrement is 9.0 minutes per 1000 pounds of fuel. It is, therefore, recommended that holding with flaps fully extended be accomplished at sea level.

AUTOMATIC CARRIER LANDING SYSTEM APPROACHES

Automatic carrier landing system (ACLS) approaches apply to properly modified data link aircraft utilizing carrier or shorebased SPN-10, SPN-41, SPN-42, or MPN-T1 ACLS radar facilities. Three approach modes are available; however, utilization of specific modes is dependent on the aircraft modifications, which are listed in Limitations, section VIII, Classified Supplement. In Mode 1 approaches, data link/ACLS control signals are coupled to the autopilot after radar lockon and remain coupled until touchdown. Mode 1A approaches differ from Mode 1 approaches in that the data link/ACLS control signals are uncoupled ½ mile (approximately 200 feet above flight deck level) from touchdown. Mode 2 approaches are similar to Mode 1/1A approaches; however, data link/ACLS control signals are not coupled to the autopilot and the pilot flies the approach using the needles presentation.

ACLS MODE 1 AND 1A APPROACHES

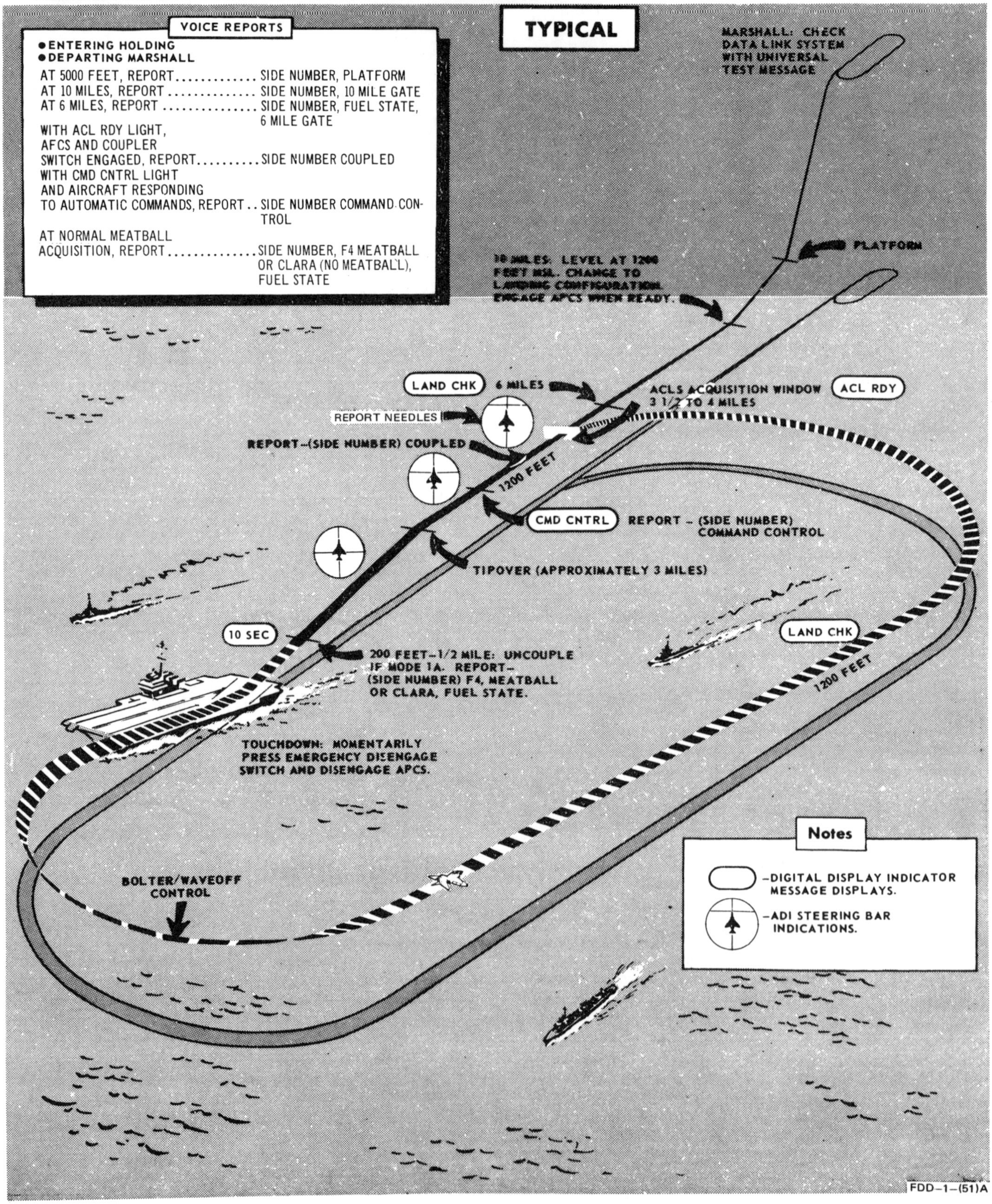

Figure 3-9

ACLS MODE 2 APPROACH

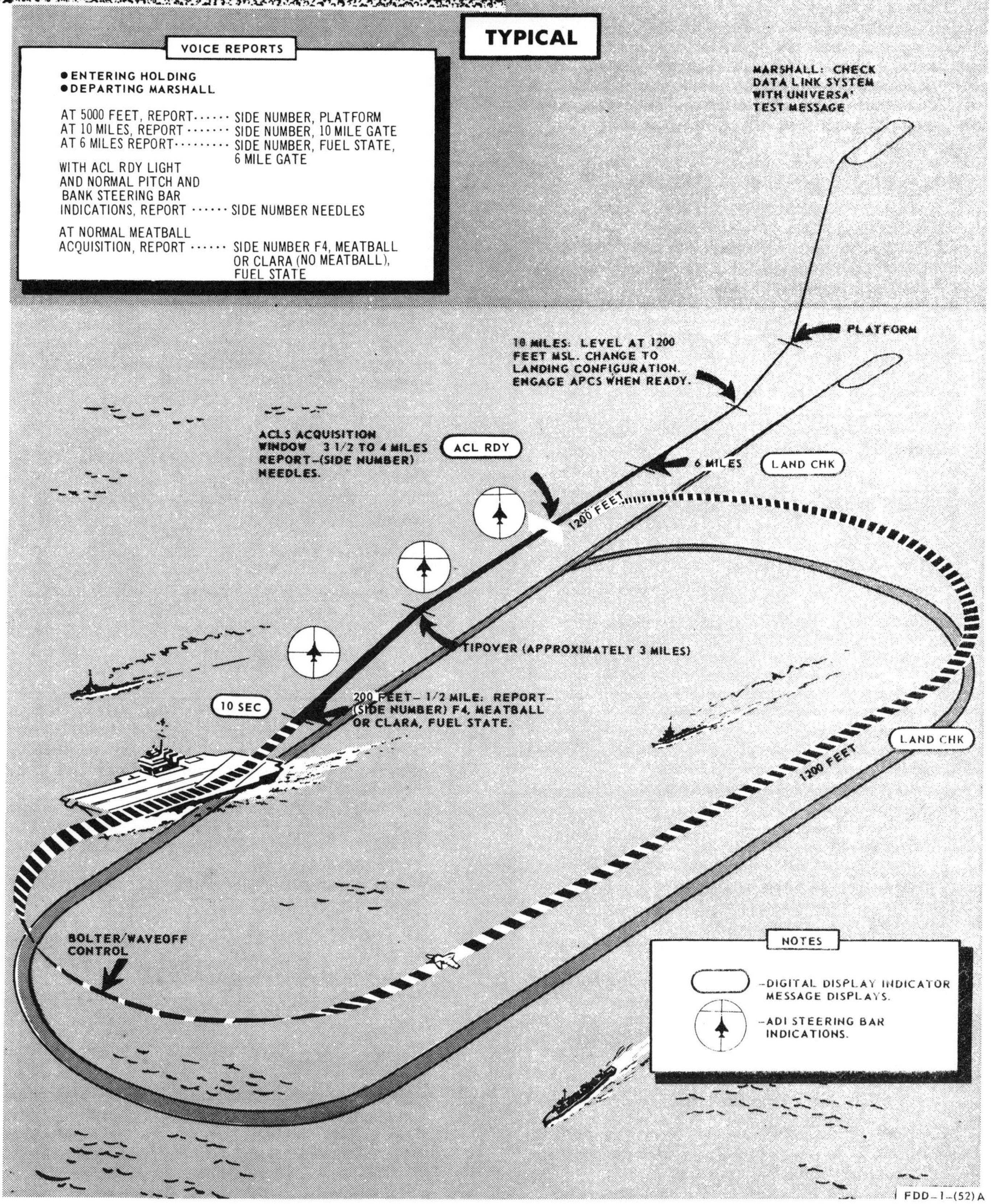

Figure 3–10

ACLS MODE 1 AND 1A APPROACH PROCEDURES

ACLS Mode 1 and 1A approaches are illustrated in figure 3–9. Mode 1 approaches require the Approach Power Compensator System (APCS). The following steps describe typical Mode 1 and 1A approach procedures. Light illumination and extinguishments (other than acknowledge button extinguishments) on the front cockpit digital display indicator (DDI), command display indicator (CDI), telelight panel, and HSI follow each step.

1. Position the following controls as indicated to prepare the aircraft for receiving ACLS signals. (Also refer to section VIII, Classified Supplement.)
   a. Navigation function selector panel.
      (1) Bearing/distance selector switch – TACAN
      (2) Mode selector knob – DL (HSI course deviation indicator is slaved to lubber line).

   NOTE

   On aircraft 157309ar and up, and all others after AFC 470, the AN/ARA–63 Instrument Landing System (ILS) or DL/ILS position on the knob should be utilized until AN/SPN–42 (ACLS) lock–on. At lock–on the DL position should be selected; agreement of the ADI steering bars will verify the AN/SPN–42 is locked on correct aircraft. Shortly after tipover the DL/ILS position should again be selected to verify alignment of AN/SPN–41 with AN/SPN–42 and confirm proper aircraft control. The DL/ILS position should be maintained until the approach is completed. The DL/ILS position does not affect the operation of the AN/SPN–42 system in any way other than the ADI steering bar presention, which will be that of the AN/SPN–41. The pilot may therefore compare the presentations of each system at any time while coupled, but should not attempt switches inside of 1 NM.

   b. DDI ACK button – PRESS WHEN ILLUMINATED
   c. Day–night controls – AS DESIRED
   d. Communications set control
      (1) Data link power switch – ON
      (2) Message selector switch – NORM
      (3) Frequency select dials – AS BRIEFED
   e. Data link BIT switch – NORM

   CDI – TILT
   HSI – DL, TAC (same throughout approach)

2. Before or while in Marshall, receive the universal test message (UTM) to check the data link system for proper operation. Refer to section VIII, Classified Supplement, for UTM procedures and displays.

   CDI – TILT (extinguished during testing)

3. Perform a normal CCA. At 10–mile gate, with aircraft level at 1200 feet MSL or as assigned, change to landing configuration.

   CDI – TILT

4. Engage APCS. While heading inbound, with aircraft in level flight at 1200 feet (or as assigned), stabilize at approach speed.

   NOTE

   To fly mode 1 approaches, the APCS must be utilized. APCS is recommended but not required in mode 1A.

   CAUTION

   In the event of an APCS malfunction, do not attempt to couple the aircraft. Execute a mode 2 (figure 3–10) or a mode 3 (talk–down) approach.

   CDI – TILT

5. Check that corner reflector is extended. Corner reflector extends when gear and hook are lowered. If approach is made with hook up, place hook bypass switch to BY–PASS.

   CDI – TILT

6. Engage AFCS. It is desirable that the aircraft be flown in altitude hold prior to coupling. Altitude hold automatically disengages when aircraft is coupled.

   CDI – TILT

7. At approximately 6 miles, controller inserts aircraft address.

   CDI – TILT (extinguishes)
   DDI – LAND CHK

8. At approximately 4 to 6 miles, with aircraft level at 1200 feet (or as assigned) and on final bearing, controller reports "ACL lockon, report needles." Report position of bank and pitch steering bars (needles) on ADI.
   The ADI pitch and bank steering bars commence providing glide slope line–up indications and the HSI heading marker displays the final bearing (command heading).
9. If the controller concurs he will then transmit "concur, report coupled."

   CDI – CPLR ON (indicates aircraft is receiving autopilot engage/enable signal) or MANUAL (indicates aircraft not receiving autopilot engage enable signal and aircraft cannot be coupled).

   DDI – ACL RDY

   a. A below glide slope indication by the pitch steering bar is normal. An above glide slope indication indicates passage of the glide slope extended and coupling should not be attempted.
   b. If the bank steering bar indicates the aircraft is not on the glide slope centerline extended (lateral error), establish a corrective cut before coupling. This reduces the initial lateral error command.

c. With wings level at assigned altitude (vertical velocity indication less than ±500 feet per minute), engage coupler switch on AFCS control panel.

NOTE

If aircraft is not coupled with wings level, it initially rolls to a wings level attitude upon coupling.

CDI – CPLR ON
DDI – ACL RDY

10. Report (side number) coupled.

CDI – CPLR ON
DDI – ACL RDY

11. Controller sends ACL commands. In event of radio failure, commands are sent 10 seconds after ACLS lockon. Upon observing DDI CMD CNTRL light illumination and aircraft response to commands, report (side number) command control.

CDI – CPLR ON
DDI – ACL RDY (extinguishes)
CMD CNTRL

12. Under certain circumstances the aircraft may become uncoupled. The cockpit indications and recommended pilot action for each instance are as follows:
   a. Intentional disengagement by pilot. Pilot presses the AFCS/ARI emergency disengage switch on control stick and the AFCS and coupler switches disengage. The stab aug and ARI are disengaged as long as the AFCS/ARI emergency disengage switch is pressed.

   CDI – CPLR ON
   DDI – CMD CNTRL
   Telelight panel – AUTO PILOT DISENGAGE
   Windshield – COUPLER OFF (some aircraft)

   (1) If disengagement occurs before intercepting the glide slope, continue the approach in Mode 2/Mode 3 or recouple (pilot's discretion). Recoupling is accomplished by re–engaging the AFCS switch and the coupler switch in that order. If the aircraft is uncoupled for an excessive length of time, large initial corrective commands may result when the aircraft is recoupled. Attempt to center the bank steering bar before recoupling. Never recouple if the ADI pitch steering bar indicates glide slope passage. Continue with a Mode 2 approach.
   (2) If disengagement occurs while on glide slope, continue the approach in Mode 2/Mode 3 or waveoff (pilot's discretion). No recoupling should be attempted at ranges less than 2 miles.
   b. Unintentional disengagement.
   (1) AFCS malfunction. The coupler switch disengages concurrently with the AFCS switch. Cockpit indications and pilot action are the same as with intentional disengagement.

   CDI – CPLR ON
   DDI – CMD CNTRL
   Telelight panel – AUTO PILOT DISENGAGE
   Windshield – COUPLER OFF (some aircraft)

   (2) System malfunction. AFCS and coupler switches disengage. Continue approach in Mode 2 or 3. The aircraft cannot be recoupled.

   CDI – CPLR ON (extinguishes)
   MANUAL (if malfunction due to loss of AFCS engage/enable signal) or TILT (if malfunction due to termination of data link signals).
   DDI – CMD CNTRL
   Telelight panel – AUTO PILOT DISENGAGE
   Windshield – COUPLER OFF (some aircraft)

NOTE

If APCS disengages, it does not uncouple aircraft. Without APCS, a mode 2 or 3 approach should be flown.

   c. System waveoff. A system parameter has been exceeded and the AFCS and coupler switches disengage. If waveoff is received after glide slope interception and in IFR conditions, execute a waveoff. The aircraft cannot be recoupled. Upon placing the speed brake switch in IN to cancel (extinguish) the flashing WAVEOFF lights, the APCS is disengaged.

   CDI – MANUAL
   WAVEOFF (flashing: extinguished when speedbrake switch is positioned to IN)
   DDI – CMD CNTRL
   WAVEOFF (flashing: extinguished when speedbrake switch is positioned to IN)
   Telelight panel – AUTO PILOT DISENGAGE
   Windshield – COUPLER OFF (some aircraft)
   Glareshield/Telelight Panel – APCS OFF (with speedbrake switch IN)

13. At ½ mile in mode 1A approaches, the controller downgrades the approach to mode 2. Although the AFCS and coupler switches automatically disengage, momentary actuation of the AFCS/ARI emergency disengage switch is recommended to ensure that the switches disengage.

CDI – CPLR ON (extinguishes)
MANUAL
DDI – CMD CNTRL
Telelight panel – AUTO PILOT DISENGAGE
Windshield – COUPLER OFF (some aircraft)

14. In Mode 1 approaches, the pilot receives the 10 seconds discrete 12.5 seconds before touchdown.

CDI – CPLR ON
DDI – CMD CNTRL (extinguishes) 10 SEC

15. At touchdown in a Mode 1 approach, momentarily actuate the AFCS/ARI emergency disengage switch to disengage the AFCS and coupler switches.

Actuate speedbrake switch to disengage APCS. Do not hold the AFCS/ARI emergency disengage switch pressed after AFCS disengagement, since on a bolter or a touch and go landing, the aircraft would depart the flight deck with stab aug disengaged.

CDI – CPLR ON
DDI – 10 SEC
Telelight panel – AUTO PILOT DISENGAGE
Windshield – COUPLER OFF (some aircraft)
Glareshield/Telelight Panel – APCS OFF

NOTE

Although the AFCS switch automatically disengages at touchdown on some aircraft, the AFCS/ARI emergency disengage switch shall be momentarily pressed to ensure switch disengagement.

16. If the aircraft bolters, the SPN–10/42 computer clears the data link discretes and terminates data link transmissions to the aircraft approximately 7 seconds after touchdown.

CDI – CPLR ON (extinguishes)
TILT
DDI – 10 SEC (extinguishes)
Telelight panel – AUTO PILOT DISENGAGE
Windshield – COUPLER OFF (some aircraft)
Glareshield/Telelight Panel – APCS OFF

ACLS MODE 2 APPROACH PROCEDURES

An ACLS Mode 2 approach is illustrated in figure 3–10. In a Mode 2 approach, the ADI pitch and bank steering bars present vertical and lateral glide slope errors respectively in relation to the aircraft. The pilot flies the aircraft toward the steering bars as in an ILS approach. Light illumination and extinguishments are noted following the procedural steps as in ACLS Mode 1 and 1A Approach Procedures.

1. Perform steps 1 thru 5 in the ACLS Mode 1 and 1A Approach Procedures. APCS engagement is optional and at the pilots discretion.

CDI – TILT (extinguishes during testing)
HSI – DL, TAC (same throughout approach)

2. At approximately 6 miles, controller inserts aircraft address. Pilot rechecks landing checklist.

CDI – TILT (extinguishes)
MANUAL
DDI – LAND CHK

3. ACLS radar lockon. Pilot observes fly–up signal on ADI pitch steering bar and the final bearing on both the ADI bank steering bar and the HSI heading marker. Report (side number) needles. (a fly–down needle indicates erroneous information if further aft of the carrier than 3 miles at 1200 feet or indicates glide slope passage if within 3 miles at 1200 feet. If glide slope passage has occurred, attempts to intercept it are at pilot's discretion.) Fly aircraft level as the pitch steering bar moves down from the top of the ADI. When the pitch steering bar approaches the ADI miniature wings, pitch the aircraft over onto the glide slope. Fly glide slope by keeping ADI steering bars centered.

CDI – MANUAL
DDI –ACL RDY or CMD CNTRL (refer to following steps a and b)

a. For approaches that were begun in Mode 2, the DDI ACL RDY light illuminates and remains illuminated throughout the approach.
b. If a Mode 1/1A approach was initiated and the pilot uncoupled and continued in Mode 2, the CMD CNTRL light remains illuminated throughout the approach unless the controller recycles his console.

4. At ½ mile to touchdown or sooner (pilot's discretion), transfer to the standard visual approach and call the meatball.

CDI – MANUAL
DDI – ACL RDY or CMD CNTRL (Refer to previous remarks)

NOTE

On aircraft 157309ar and up, and all others after AFC 470, the AN/ARA–63 ILS should be utilized provided the AN/SPN–41 installation is certified because it does not require AN/SPN–42 (ACLS) lock–on.

PART 5 FUNCTIONAL CHECKFLIGHT PROCEDURES

GENERAL

The functional checkflight will be performed after the completion of the calendar maintenance requirements using the applicable Functional Checkflight Checklist. This section contains a detailed description of the checkflight requirements, sequenced in the order in which they will be performed. The checkflight personnel will familiarize themselves with these requirements prior to the flight. NATOPS procedures will apply during the entire checkflight. Only those pilots designated in writing by the Squadron Commanding Officer shall perform squadron checkflights. Checkflight procedures will be in accordance with the current edition of OPNAVINST 4790.2. For ready reference, excerpts from OPNAVINST 4790.2 are quoted below.

> At the discretion of the Commanding Officer, checkflights may be flown in combination with operational flights, provided the operational portion is not conducted until the checkflight requirements have been satisfied and the results have been entered on the checkflight checklist. The general purpose code assigned to a combination check and operational flight will be the one that describes the primary purpose of the flight.
>
> Pilots and crew members who perform checkflights are qualified in accordance with OPNAVINST 3710.7 Series and the applicable NATOPS manual, and are provided a thorough briefing by the Maintenance Officer or his designated representative (normally the QA Officer). This briefing should describe the requirements for that particular flight, the expected results, and corrective emergency action to be taken if required.
>
> Checkflights are conducted with the minimum crew necessary to ensure proper operation of all required equipment.
>
> Checkflights must be of sufficient duration to perform the prescribed checks and to determine whether any additional maintenance work is required.
>
> Checkflights shall be conducted in accordance with the criteria established by OPNAVINST 3710.7 Series (NATOPS).
>
> Checkflight forms must be properly completed and returned to the Maintenance Department.

Checkflights are required to determine whether the airframe, power plant, accessories, and items of equipment are functioning in accordance with predetermined requirements. Depending upon the maintenance performed, the functional checkflight will be either a complete or a partial checkflight. If a complete checkflight is to be flown, all the items contained in the Functional Checkflight Procedures must be accomplished. If a partial checkflight is to be flown due to engine change, flight control rigging, etc., only those items that directly relate to the equipment being checked need be accomplished. Therefore, some items contained in the Functional Checkflight Procedures are coded. This coding is intended to assist the FCF crewmembers in determining which items pertain to the various conditions requiring checkflights. Items coded (E) pertain to engine/fuel control maintenance as outlined in OPNAVINST 4790.2. Items coded (FC) pertain to flight control/rigging maintenance as outlined in OPNAVINST 4790.2. The uncoded items in conjunction with the coded items constitute a complete Functional Checkflight, requirements for which are outlined in OPNAVINST 4790.2. Coding shall appear adjacent to a paragraph title or a step. If it appears adjacent to a paragraph title, all steps following that paragraph title will apply. If the coding appears adjacent to a step, only that step and its subordinates will apply.

CHECKFLIGHT PROCEDURES (PILOT)

PREFLIGHT

NOTE

Due to expanded checks required by the Checkflight Procedures, external intercommunication shall be used by the ground crew.

1. Exterior preflight
   The aircraft exterior preflight will be conducted in accordance with section III of this manual. Particular attention shall be made to check for loose or improperly installed panels in those areas where maintenance has been performed.
2. Interior preflight
   Internal inspection and proper switch positions will be accomplished in accordance with section III of this manual.

(E) STARTING ENGINES

Start engine in accordance with section III of this manual. Check normal indications for the following:

1. Fuel flow
   Fuel flow is 225 to 800 PPH at light–off, up to 2400 PPH during the engine transition to idle.
2. Light–off (within 15 seconds)
   Ignition should occur within 15 seconds after fuel flow or oil pressure is indicated.
3. EGT within limits
4. Nozzle movement
   Afterburner nozzles should be monitored during start. Nozzle positions to full open at approximately 30% RPM and ⅞ to ¾ open at idle.
5. Idle RPM 65 ±1%
6. Boost pump pressure
   Pressure indicating 30 ±5 psi at idle.
7. Idle fuel flow 800 – 1400 PPH

8. Idle EGT 220° - 420°
9. Idle oil pressure 12 psi minimum
10. PC-2 pressure (R eng) 2750 - 3250 PSI
11. Utility pressure (R eng) 2550 - 3000 PSI
12. Extend refuel probe (R eng)
    This will ensure that the probe will extend on the reduced utility pressure.
13. (FC)Left Spoiler Check
    With right engine only operating and PC-1 pressure zero, slowly deflect control stick approximately 1 inch to left. Have ground crew verify that spoiler does not fully deflect and returns to a flush condition when the stick is returned to neutral.
14. PC-1 pressure (both eng) 2750 - 3250 psi
15. Utility pressure (both eng) 2750 - 3250 psi
16. Bus tie light out within 18 seconds.
    a. When the right generator is turned on, the RH GEN OUT light goes out. The BUS TIE light comes on momentarily, then goes out. When turning on the left generator, the LH GEN OUT light goes out and the BUS TIE light comes on and may stay on up to a maximum of 18 seconds, then goes out. The same indication occurs when the right generator is cycled.
    b. With both generators on the line, turn off right generator and check boost pump pressure gages. Turn right generator on and then turn the left generator off and again check boost pump pressure gages. If reverse phasing is present, the boost pumps will run in reverse and the boost pump pressure gages will indicate substantially lower than normal. Select left generator on.
17. SPC light
    STATIC CORR OFF light must go out and remain extinguished after placing SPC switch in the RESET CORR position. After reset, altimeter should indicate within ±40 feet of the before reset indication. Altimeter oscillations of any magnitude are unacceptable.
18. Wing fold
    Operate the wing fold system. Check that the lights on the telelight panel operate and extension or retraction occurs in 12 to 16 seconds.

BEFORE TAXIING

1. (FC) Aileron trim 10-15 seconds
   Move aileron trim left. Have ground crew confirm left spoiler up and right aileron down. Mark time and move aileron trim to the full right position. Maximum time to travel from full left to full right is 15 seconds. Insure stick moves from left to right, and have ground crew confirm that left spoiler going down and right spoiler up after passing neutral position. Move trim back to neutral and have ground crew confirm spoilers and ailerons neutral. Check control stick neutral position ±¼ inch.
   a. Lateral control system checkout
      (1) Mechanical system checks.
      (a) With ailerons trimmed neutral; slowly trim the stick to the left until the left spoiler just breaks the stops. Have ground crew check that the right aileron is deflected down ¼ to 1 inch.
      (b) Repeat this check on the opposite side to insure that when the right spoiler breaks the stop, the left aileron is down ¼ to 1 inch.

CAUTION

If this check fails, the lateral controls must be rerigged before flight, as large yaw angles and possible post-stall gyration can occur at high angles-of-attack since it will be impossible to neutralize the lateral controls.

      (2) Operation system checks.
      (a) With STAB AUG off, trim the lateral system as close to neutral as possible. Spoilers should be flush with the wing and ailerons aligned approximately with the trailing edge flaps.
      (b) Engage STAB AUG while observing both ailerons for movement. Ailerons should not move more than ¼ inch.
2. Tail hook/corner reflector (aircraft 155529ag and up and all others after AFC 388).
   Extension of hook should be 5 seconds maximum. Reflector door opens, reflector extends and CORNER REFLECTOR OUT warning light remains out. Retraction of hook should be 10 ±3 seconds. Reflector retracts and reflector door closes.
3. Hook bypass switch
   Place hook by-pass switch in BY-PASS position: reflector door opens, reflector extends, and CORNER REFLECTOR OUT warning light remains out. The hook remains up and locked. Place hook by-pass switch in NORMAL position; reflector retracts and reflector door closes.
4. Speed brakes
   Should extend to full deflection in 2.5 ±0.5 seconds and retract in 2.5 ±0.5 seconds on the ground.
5. RAT
   Pneumatic pressure should not drop below 2200 ±200 psi on extension and 1200 +400, -200 psi on retraction (when immediately following extension).
6. Flaps/aileron droop/BLC/ARI
   a. Full flaps
      (1) T.E. flaps extend within 8 ±1 seconds.
      (2) Ailerons droop within 3 seconds.
      (3) Check for presence of BLC at trailing edge.
      (4) Check for presence of BLC at leading edge.
      (5) Place engine bleed air switch OFF and ensure leading edge BLC is shut off (On aircraft 155903ap, and 157274ap and up, and all others after AFC 440). On aircraft after AFC 550, ensure BLEED AIR OFF light illuminates and then goes out after bleed air switch is returned to NORM (guard down).
      (6) Place engine bleed air switch NORM.
      (7) Place YAW STAB AUG switch OFF and pull ARI circuit breaker. Move stick full left and check left spoiler up and right aileron down, rudder neutral. Place YAW STAB AUG switch ON and check for 5° left rudder, reset ARI circuit breaker and check for additional 10° left rudder. Repeat this procedure with full right stick.
   b. 1/2 Flap

(1) Check for absence of BLC at trailing edge.
(2) Conduct left and right ARI check as checked with full flaps.
(3) Extend to full flaps.
(4) Flap switch up; retract within 6 ±1 seconds.
(5) Ailerons return to neutral (after initial movement) within 6 seconds.

7. Anti–ice
   Stabilize at 80% engine RPM or higher. Place anti–icing switch to ON and check for illumination of L ANTI–ICE ON and R ANTI–ICE ON lights. Observe approximate 10°C rise in EGT and slight rise in fuel flow. When switch is turned off, check drop in fuel flow, EGT, and that the lights go out.

TAXIING

1. Nose wheel steering
   Engage the nose wheel steering after aircraft begins to roll and ascertain that nose wheel responds to steering command signals in both direction and magnitude.
2. Air brakes
   When a reasonably clear area is reached, move air brake lever slowly to on. Ascertain that both brakes are applied.

(E) BEFORE TAKEOFF

1. MIL power check
   Check engines individually at MIL thrust and note any variation from the following limitations:
2. Throttle check
   Force required to move each throttle from IDLE to MIL requires approximately 5 pounds.
3. EGT
   Exhaust gas temperature at MIL power for the –8 engine is 625° ±10°C.
   Exhaust gas temperature at MIL power for the –10 engine is 660° ±8°C.
4. Fuel flow
   Fuel flow for each engine at MIL power is 7500 to 9000 pounds per hour (approx).
5. Oil pressure
   Engine oil pressure at MIL power for MIL–L–23699 oil is 45 to 70 psi.
6. RPM
   RPM at MIL power for the –8 engine is 100% ±0.5%. RPM at MIL power for the –10 engine refer to Military Power Operating Limits chart in NAVAIR 01–245FDD–1B.

NOTE

Due to T2 cutback, the –10 engine will not reach 100% rpm if CIT is below approximately 45°C (113°F). Nevertheless, engine performance (EGT vs RPM relationship) can be determined during ground run–up by using the Military Power Operating Limits chart and assuming that CIT equals OAT when the aircraft is in a static condition. Theoretically an inflight engine performance check cannot be made since there is no means available for the crewmembers to determine inflight CIT. However, if the engine RPM vs EGT relationship was within tolerance during the ground MIL power check, it can be assumed that RPM is correct. Therefore, a check of inflight EGT can be made by entering the Military Power Operation Limits chart with RPM and obtaining a EGT readout. The actual engine EGT should fall within the range of EGT obtained from the chart.

7. Nozzle position
   Check and record nozzle position, ensure that both openings are approximately the same.

TAKEOFF

Perform a normal afterburner takeoff. Observe and record any variations to following limitations:

1. (E) EGT
   –8 engine – 635° MAX
   –10 engine – 668° MAX
2. (E) Nozzle position – 5/8 to 7/8 open.
3. Gear retraction – 4 to 6 seconds.
4. (FC) Rudder switchover
   Accelerating – 228 to 252 knots
   Decelerating – Approximately 10 to 20 knots below accelerating switchover (Minimum 218 knots)

15,000 FEET

Climb to 15,000 feet and proceed with following checks:

1. Cabin pressure (8000 feet)
   Check that both cockpit pressure altimeters read 8000 feet.
2. (FC) Flight control system
   a. Turn off stab aug. Trim aircraft for hands off flight and check position of spoiler and ailerons. Reengage each stab aug axis separately and check for transits.
   b. STABILATOR – Establish 350 knots and check for normal longitudinal feel. Establish a 2 G pull up and release the stick. Oscillations should dampen in ½ cycle. Trim full nose up, full nose down and check for 15 to 20 pounds of stick force. Return trim to neutral.
   c. AILERONS – Trim for hands off flight. Check ailerons and spoilers in trimmed position. Roll aircraft into a 60° bank and release the stick. Roll should stop and the stick should center. Repeat check in the opposite direction.
   d. RUDDER – With wings level trim aircraft so the yaw string is centered (if installed), or trim aircraft so the needle and ball on the ADI are centered. Displace rudder pedal to develop yaw. Release pedal, yaw should cease in ½ cycle. Repeat check in the opposite direction. Symmetrical flight should occur in 5 seconds following yaw input. Trim right and left ensuring rudder trims in proper direction.
3. Compass system
   a. Standby, ADI, HSI – Heading accuracy (mode selector knob in SLAVED mode) within ±5° of known magnetic heading. No more than 2 ½°

difference can exist between ADI and HSI headings. Standby compass headings should take into account corrections as noted on correction cards located in cockpit.

b. ADF (PRI, AUX) Accuracy is within ±5° of heading toward or away from the transmitter. Bearings off wing tip are accurate with ±20°. Needle oscillations must not exceed ±3°.

c. Tacan – Accuracy of all bearings will be within ±1°; however, reading accuracy prompts an arbitrary tolerance of ±2°. Distance accuracy will be ±0.2 miles plus 0.1% of total distance to station. Course bar pointer and bearing pointer have no discernible difference in bearings. Erratic course bar movement is unacceptable except when close–in to station.

d. Compass system control – No more than 2° difference exists between the AJB–7 and ASN–70 systems output.

4. (E) Engine idle power
   Reduce engine power to idle and observe the following limitations:
   a. EGT – 100° to 300°C.
   b. RPM – minimum 65% at 0.55 to 0.84 Mach.
5. (E) Engines military power
   Advance the throttles to military power and observe the following limitations:
   a. EGT
      (1) –8 engine – 625° ±10°C
      (2) –10 engine – RPM vs EGT (Refer to Military Power Operating Limits chart in NAVAIR 01–245FDD–1B.)
   b. RPM – 100 ±0.5% (–8 only)
   c. Oil pressure – 40 to 70 psi.
6. (E) Engines (AB power)
   Advance the throttles to maximum power and observe the following limitations:
   a. EGT
      (1) –8 engine – 625° ±10°C
      (2) –10 engine – RPM vs EGT (Refer to Military Power Operating Limits chart in NAVAIR 01–245FDD–1B.)
   b. RPM – 100 ±1.0% (–8 only).
7. Cabin Vent
   Perform a functional check of cabin vent by pulling up on emergency vent handle. Observe following action:
   a. All air conditioning and pressurization air to cockpit and pressure suit is shut off.
   b. Cabin pressure regulator and safety (dump) valve is opened and cockpit becomes depressurized.
   c. Cabin and aircraft altimeter will read approximately same within 5 to 6 seconds after handle is pulled.
   d. RAm air shutoff valve is opened and atmospheric air is allowed to enter cockpit through a port located just forward of pilot's feet.
8. Wing fuel dump
   Perform a check of wing fuel dump system, observing through rear view mirror that fuel actually dumps. RIO confirms dump with a visual check.

40,000 FEET

Climb to 40,000 feet and perform following checks:

1. (E) Engines (idle power)
   Reduce engine power to idle, and observe following limitations:
   a. EGT – 100° to 300°C
   b. RPM – 80% (approx)
2. (E) Engines (MIL power)
   a. EGT
      (1) –8 engine – 532° to 537°C at 91%; 619° to 635°C at 100.5%
      (2) –10 engine – RPM vs EGT (Refer to Military Power Operating Limits chart in NAVAIR 01–245FDD–1B)
   b. RPM – 91% to 100.5% (–8 only)
3. (E) Engines (AB power)
   a. EGT
      (1) –8 engine – 625° ±10°C.
      (2) –10 engine – RPM vs EGT (Refer to Military Power Operating limits chart in NAVAIR 01–245FDD–1B.)
   b. RPM – Military rpm ±0.5%
4. (E)/(FC) VMAX
   Maximum power Vmax is attained at 40,000 feet. Special instrumentation is required to check performance precisely; however, a check for minimum max–speed is felt to be a reasonable compromise. By using performance data contained in section XI and subtracting 0.12 TMN from the Vmax given and a further correction for temperature (0.02 IMN/°C) from standard day, an acceptable figure is reached. An aircraft not meeting this minimum requirement may have malfunctioning bellmouth/ramp schedules for low–thrust engines and must be given a complete ground check. Thrust surging, directional oscillations and cyclic ramps are unacceptable characteristics.

DESCENDING TO 20,000 FEET

1. Gyro horizon
   Residual bank angle not to exceed ±2° after erection period. Following 360° turn at a bank angle of 30° or more, the bank angle at roll–out will be no more than ±2° from indication before beginning turn. Horizon is adjustable through minimum angle of 5° climb and 10° dive. A 360° roll will cause no precession in bank or pitch. Erratic motion of sphere is unacceptable. Cross check standby attitude gyro with ADI gyro; the maximum difference is ±1%.
2. Turn and bank indicator
   A single needle width turn results in turn rates of 165°–195° per minute.
3. Vertical velocity indicator
   Must not exhibit erratic motion (except in transonic region where static pressure fluctuation normally occurs) and zeros in level flight.
4. Accelerometer
   All needles return to within 1 ±0.2 G in 1 G flight or on ground after reset button is actuated.

5. Anti-G valve
   Commence air flow to G-suit connection at 1.5 G. Flow rate may vary. At 2 G a suit pressure of near 0 to 1.2 psi may be obtained. The suit pressure may vary between 2.9 and 4.3 psi at 4 G. (Qualitative check only.)
6. Altimeter
   The following lists allowable fluctuation in altitude with normal static pressure compensator operation.
   a. Mach range 0 to 0.91 - Stable (25 feet random variation.)
   b. Accelerating through 0.91 to 0.93 - Two altitude breaks of ±200 feet. Continuous oscillations should not exceed ±50 feet.
   c. Acceleration through 0.93 to 1.05 - Two altitude breaks of ±1000 feet. Continuous oscillations should not exceed ±50 feet.
   d. 1.05 to Vmax - Stable (25 feet random variations).

NOTE

Static pressure compensator must continue to function through flight envelope. Occasional drop-outs (warning light ON) are unacceptable.

7. True airspeed indicator
   The computation range is 150-1500 knots. Readings on ground of 108-150 knots are normal and will vary considerably. In-flight check requires free air temperature information.
8. Air refueling probe (300 knots or 0.90 mach, whichever is less)
   Extend the air refueling probe and check READY light on. Check that reverse fuel transfer does not occur. (Rapid sector decrease.) Retract probe.
9. (FC) Lateral control check
   At approximately 25,000 feet, disengage the stab aug (roll aug) and have the RIO visually check the ailerons for neutral trim. When neutral trim has been established, check the control stick for neutral position. Leave the roll aug disengaged. Slowly trim to establish a 3 to 5 G pull out. Hands off stick and visually check aileron position. Accelerate to 350 knots, increase the angle of attack to 19 units and execute rudder rolls left and right (ailerons neutral). Large erratic yaw oscillations shall not occur. Establish a turn at 300 knots and 19 units AOA and gradually increase AOA until wing rock occurs, approximately 23 to 25 units. If any adverse characteristics are exhibited establish 5-10 units AOA. If the aircraft was not retrimmed, then hands off flight should again show the ailerons neutral.
10. Automatic/manual temperature control
    Cockpit temperature is fully controllable over a range of −20°F to 100°F on AUTO setting. Manual selection may result in inlet temperature of −130°F to well above 220°F depending on flight conditions and techniques in using switch. For example, temperature limiter will limit the inlet air to 220°F ±10°F when active; however, the unit is active only when manual switch is held in HOT position and does not react to higher temperatures that may occur as a result of speed and power changes. Pressure suit vent air control must be fully off to obtain control of the manual system.
11. Defog foot heat control
    Defog air temperature is controllable from 85°F to 200°F, depending upon cabin temperature setting at time defog air is selected. Moving the defog lever through 50% travel position results in an increase in defog air temperature of approximately 100°F. In full defog position, 80% of available air is directed to windshield. In FOOT HEAT position only 10% of the air is directed to windshield.
12. Horizontal situation indicator (HSI/TACAN)
    The following functions will be checked for accuracy:
    (1) DME - Accuracy is ±0.2 miles plus 0.1% of total distance from station; however, reading and flying accuracy will permit an accuracy check of no better than ±1 mile.
    (2) Course indication - The course bar centers within ±1° of known course to station. Erratic course bar movement is unacceptable. Course pointer and bearing pointer have no discernible difference in bearing in TACAN mode. Course selected in window and course pointer have no discernible difference in readings. Erratic movement of the bearing pointer is unacceptable, except for ±3° jitter in ADF mode.
13. Tacan
    Volume control is sufficient to eliminate audible station identification signal. Lock-on normally occurs within 60 seconds following channel selection on T/R mode. This time will vary with bearing and distance from station.
14. Attitude director indicator (ADI)
    Vertical displacement pointer (VDP) moves opposite to direction of bank and centers with no discernible deviation when on selected heading.
15. Antenna functional check
16. Gunsight functional check

20,000 FEET

1. Automatic flight control system
   Engagement of any mode or function of AFCS does not produce transients of sufficient magnitude to cause the aircraft to jump in pitch and/or roll. Performance of the system is unacceptable if an engagement jump occurs.
   a. Stability augmentation mode - Stab aug damps yaw inputs (rudder kicks) within ½ cycle to zero amplitude, and symmetrical flight (ball centered) will occur within 5 seconds following the yaw input. Transients in pitch and roll damp within ½ cycle; however, a slight residual pitch rate is considered normal because the stick longitudinal centering band may exceed the authority of the damper (±2° stabilator travel) where excessive friction exists. Lateral inputs damp within ½ cycle. Random or cyclic transients about any axis are unacceptable.
   b. Control stick steering (CSS) - Control stick longitudinal breakout force will not exceed 4.0 pounds forward or 3.0 pounds aft, while lateral breakout force will not exceed 4.0 pounds. Self-induced lateral, longitudinal or directional transients, random or cyclic, are unacceptable.

Control stick steering in excess of 70° ±3° in pitch and bank will cause AFCS to disengage. While in a 0° to 70° bank, a reduction of stick force within these bank limits is unacceptable. Lateral return to level bank will not exceed 5° ±2°. Heading hold error will not exceed ±10° of the actual heading when return-to-level function becomes operative.

c. Heading hold cutout – While operating in AFC mode, depress and release nose wheel steering button to activate heading hold cutout option. Minute heading changes between 0° and 5° angle of bank can then be made. Again depress and release nose wheel steering button and insure that AFCS heading hold function returns to normal operation.

d. Altitude hold – Altitude hold function maintains altitude within ±50 feet or 0.3% of reference altitude, whichever is greater, up to 0.9 Mach subsonic and above 1.0 Mach supersonic. While accelerating through transonic region, Mach jump may produce transients which will result in noticeably abrupt altitude hold corrections. This condition is considered normal as long as corrections do not become sharply erratic, and once this region is traversed aircraft settles down on reference altitude, or a new reference altitude and holds it. A change in altitude greater than 200 feet will cause reference altitude to slip by the amount that transient altitude limit (200 feet) is exceeded. Altitude hold function will become disengaged if control stick forces exceed 4.0 pounds maximum forward, or 2.75 pounds maximum aft for a period exceeding 0.5 seconds. When performing altitude hold function check, do not make initial engagement while in a climb which exceeds 1000 feet per minute. Engaging altitude hold mode in climbs greater than 1000 feet per minute may result in a reference altitude other than engage altitude.

e. Emergency disengage – Disengaging the AFCS with emergency disengage switch will produce slight transients, but will not result in a G jump of more than ±0.3 G.

f. G-Limiter – Autopilot disengage at 4 ±¼ G (positive) and −1 ±¼ G (negative).

g. Automatic pitch trim

(1) Assume flight conditions of 300 KCAS.

(2) With AFCS switch off, manually trim aircraft for straight and level flight (make a mental note of stabilator trim position indicator).

(3) Engage AFCS.

(4) While holding control stick for straight and level flight, manually mistrim 2 units nose up.

(5) Pull stab feel trim circuit breaker (pilot's right console) as trim indicator reaches 2 units out of trim.

(6) In 2 to 15 seconds, A/P PITCH TRIM warning light shall illuminate.

(7) Depress stab feel trim circuit breaker. The stabilator trim position indicator shall show a retrim toward the predetermined straight and level position.

(8) The autopilot pitch trim indicator light shall go out when indicator is between 1½ and ¾ units from predetermined straight and level position.

(9) Disengage AFCS, transient shall not exceed 0.3 G.

(10) Repeat steps (2) thru (8) only in nose down direction.

10,000 FEET TO LANDING

Between 10,000 feet and landing, perform following checks:

1. Angle of Attack
   a. Basic limitations and pictorial display of system will be found in section I of this manual. The following are additional tolerances.
   b. The airspeed will be within ±4 knots except as noted in section I of this manual.
   c. Pedal shaker activates at 21.3 units angle of attack before AFC 388, and 20.6 units after AFC 388.
2. Flaps – Flap Blow Up Switch Test
   Decrease airspeed below 205 knots and lower flaps. Slowly increase airspeed; the flaps will blow up at 230 to 240 knots. Slowly decrease airspeed; flaps will extend approximately (providing flap switch is down) 10 to 20 knots below noted retraction airspeed (minimum 210 knots). Flaps extend and retract within 8 ±1 second and 6 ±1 second, respectively.
3. Ram air turbine
   a. Reduce airspeed to 250 knots.
   b. Secure tacan, IFF, radar altimeter and STAB AUG; (RIO) secure tacan, radar.
   c. Extend RAT and check rotation.
   d. Secure right engine and relight.
   e. Secure both generators.
   f. Secure left engine and relight.
   g. Check normal operation of the following:
      (1) ICS
      (2) UHF (front and rear)
      (3) CADC RESET
      (4) Wing fuel transfer
      (5) Boost pump press 5–15 psi.
      (6) Landing gear/flap position indicators.
      (7) Longitudinal trim
      (8) ADI/HSI
      (9) Fuel quantity, feed tank check switch and fuel flow indicators.
      (10) RPM, EGT, nozzle indicators
      (11) Oil/pneumatic pressure indicators
      (12) Caution light reset
      (13) Warning lights bright
      (14) Instrument flood lights bright
      (15) AOA indicator
      (16) PC-1/PC-2 pressure gauges.
      (17) Check for electrical output at speeds down to 170 knots
   h. Right and left generator – ON.
   i. Retract RAT and accelerate
4. (E) APCS functional check

WARNING

Do not engage the APCS unless all system components are installed and the landing gear is down.

a. Ensure speed brake switch is in STOP position, the APCS engine selector switch is in BOTH position throttle friction lever is full aft, and emergency speed brake switch is in MAN position (some aircraft).
b. Engage APCS at approximate approach airspeeds and between 75% and MIL power rpm.
c. Move control stick forward and observe that throttles move aft and engine rpm decreases. Perform check with APCS air temp switch in HOT, COLD and NORM positions.
d. Move control stick aft and observe that throttles move forward and engine rpm increases. Perform check with APCS air temp switch in HOT, COLD and NORM positions.
e. Place air temp switch in applicable temperature position.
f. Neutralize controls, grasp throttles and override APCS. 20 to 40 pounds of force is required to override the APCS. Relax throttle pressure and observe that throttles return to their former position. APCS should be operating normally.
g. Select L position on APCS engine selector switch. Place right throttle between 77% and 82% rpm.
h. Move control stick fore and aft and observe that left engine throttle moves in opposite direction from control stick.
i. Neutralize controls and select R position of APCS engine selector switch. Place left throttle between 77% and 82% rpm. Move control stick fore and aft and observe that left engine throttle moves in opposite direction from control stick.
j. Neutralize controls and select BOTH position on APCS engine selector switch. Place speed brake switch to the IN position and observe that APCS engage switch moves to STBY and the APCS OFF and Master Caution lights illuminate. Throttles remain in the same position as when APCS was engaged.

5. Radar altimeter
   a. Drop-out altitude - Drop-out altitude is 5000 feet over land or water.
   b. Low level warning - Accuracy is ± 5 feet of the selected altitude.
   c. Drop-out angles - Occurs at no less than 30° bank angle and 50° pitch angle.
   d. Accuracy - Will be ± 5 feet below 100 feet and $\pm 5\%$ of terrain clearance between 100 and 500 feet.
6. Speed brakes
   (Some aircraft) Extend speed brakes and move emer speed brake switch to RETRACT. Speed brakes will be hydraulically closed and warning light extinguished.
7. Landing gear
   Landing gear extends in 5 to 7 seconds.

LANDING

1. (E) APCS disengagement
   Prior to landing, engage APCS and check for disengagement on touch down.
2. Drag chute
   Force to deploy chute, approximately 50 pounds.
3. Anti-Skid (some aircraft)
   Between 90 and 100 knots on the runway, engage nose gear steering and depress the brakes until the anti-skid cycles. Ensure tires do not skid and the anti-skid system does not cause the aircraft to yaw on the runway.
4. Brakes
   Check for spongy, draggy, or pulling brakes.
5. Rain removal
   During taxi with one or both engine(s) running, at or below 88% and flaps at 1/2 or down actuate the rain removal switch and check that air is coming from the rain removal ports in front of the windshield. Turn the switch off immediately and ensure the air flow decreases within a maximum of 6 seconds and stops within 17 seconds. If it does not stop, pull the EMERG VENT handle to prevent possible damage to the windshield.
6. Fuel level low
   Check that the FUEL LEVEL LOW light illuminates at 1880 ± 200 pounds. Check feed tank level at 1350 ± 200 pounds.
7. (E) Right engine shutdown.
   With right engine stabilized at idle, place right throttle off. Check engine for smooth coast down.
8. (FC) Right spoiler check
   With left engine operating and PC-2 pressure zero, slowly deflect control stick approximately 1 inch to right. Have ground crew verify that spoiler does not fully deflect and returns to a flush condition when the stick is returned to neutral.
9. (E) Left engine
   With left engine stabilized at idle, place left throttle off, check engine for smooth coast down.

AFTER FLIGHT

Pilot

1. Before leaving aircraft, complete procedures in accordance with section III of this manual.
2. All discrepancies or maintenance action requirements shall be properly posted on appropriate forms.

Ground Crew

Perform postflight inspection in accordance with applicable publications and as follows when applicable.

1. Engines for audible bearing roughness; shrouds to turbine rotor seals for interference after shutdown; turbine wheels for contact with adjacent surfaces.
2. Evidence of structural failure in areas where major structural modification or repair was accomplished prior to test flight.
3. New or newly overhauled engine checks.
   a. CSD and engine oil filter elements for metal particles.
   b. Engine low and high pressure fuel filter elements for metal particles or foreign material.
4. Remove emergency generator contact brush access cover.
   a. Rotate RAT blades clockwise and observe

armature for rotation. Failure of armature to rotate indicates probable quill shaft failure.

CHECKFLIGHT PROCEDURES (RIO)

TAXIING

1. AMCS
   a. Built In Test
      Perform Built–In–Test Function checks.

15,000 FEET

After climb to 15,000 feet, proceed with following checks.

1. ADF
   Cross check ADF operation with front cockpit, allowable difference between instrument is $\pm 1°$.
2. Tacan
   Cross check tacan operation with front cockpit, there should be no discernible difference between cockpit indicators.
3. BDHI
   Cross check BDHI with HSI in front cockpit, allowable difference between instrument is $\pm 1°$.
4. Navigation computer
   Inflight check of NAV computer is basically a functional check because of inherent inaccuracy of the necessary high altitude wind information. Normal procedure is to set in your home field as a base, a tacan station as a target and to utilize Weather Bureau altitude winds. System readouts are crosschecked with tacan information during flight and unless an apparent error of appreciable magnitude is obvious, additional detailed checking is not performed. If closer checking is dictated by particular system performance, a leg of at least 12 minutes duration is flown (departing and exact fix) and all applicable parameters are noted at the departure and destination points; i.e., heading, TAS, computer settings and readouts. Resultant information is subsequently checked against the below tolerances.
   a. Present position latitude – ± 1 mile for each 12 minutes of elapsed time after start of problem or $\pm 1\%$ of the distance travelled, whichever is greater.
   b. Present position longitude – Same tolerance as latitude.
   c. Aircraft ground track angle (relative to true heading) – $\pm 2°$ from ground speed of 150 to 1700 knots.
   d. Aircraft bearing (relative to true heading) – Bearing error shall not exceed an amount which is defined by a line passing through the present position and a point 5 miles to either side of the target or base (as selected), or $\pm 2°$ deviation from a line passing through the target or base, whichever tolerance is larger.
   e. Great circle distance (range) – The range error shall not exceed 5 miles $\pm 1\%$ of the theoretical range.

40,000 FEET

After climb to 40,000 feet, proceed with the following checks.

1. Airspeed and mach indicator
   Cross check the airspeed and mach indicator with the front cockpit, allowable airspeed difference at 1.4 Mach is ± 14 knots and allowable Mach difference at 1.4 Mach is ± 0.04 Mach.

DESCENDING TO 20,000 FEET

1. AMCS operational check.

20,000 FEET

1. Attitude gyro
   Cross check remote attitude gyro with ADI in front cockpit. Allowable error between instruments is 1°.
2. Altimeter
   Cross check altimeter with front cockpit, allowable difference between instruments is 30 feet.

10,000 FEET TO LANDING

No check at this altitude.

SECTION IV
FLIGHT CHARACTERISTICS

TABLE OF CONTENTS

FLIGHT CHARACTERISTICS

ANGLE OF ATTACK

Flight characteristics of swept-wing high performance aircraft like the F-4 are predictable and repeatable. The most significant factors influencing flight characteristics are angle of attack and static margin. Angle of attack is defined as that angle formed by the chord line of the wing and the aircraft flight path (relative wind). Static margin is defined and discussed under Longitudinal Stability, this section.

LOW ANGLE OF ATTACK MANEUVERING

Induced drag is at a minimum at approximately 5 units angle of attack (nearly zero G); therefore, acceleration characteristics are exceptional. To achieve maximum performance acceleration from subsonic Mach numbers to supersonic flight, a 5 unit angle of attack pushover will provide minimum drag and allow gravity to enhance airplane acceleration. This technique provides the minimum time, fuel, and distance to accelerate from subsonic Mach numbers to the optimum supersonic climb schedule. When confronted with a recovery from a condition of low airspeed and high pitch attitude, the angle of attack indicator becomes the primary recovery instrument. A smooth pushover to 5 units angle of attack will unload the airplane and reduce the stall speed to nearly zero. Recovery can be accomplished safely at any speed which will provide stabilator effectiveness (ability to control angle of attack). Smooth control of angle of attack is a necessity, and no attempt to control bank angle or yaw should be made. High pitch angles with rapidly decreasing airspeed will result in loss of stabilator effectiveness and subsequent loss of control of angle of attack.

MEDIUM ANGLE OF ATTACK MANEUVERING

Maneuvering at angles of attack from 5 to 15 units will produce normal airplane response to control movement.

HIGH ANGLE OF ATTACK MANEUVERING

Above 15 units angle of attack, airplane response and flight characteristics begin to exhibit the changes expected in swept-wing high performance aircraft. The primary flight characteristics exhibited at high angles of attack are adverse yaw (yaw due to roll) and dihedral effect (roll due to yaw).

Adverse Yaw

Attempts to roll the airplane with lateral stick deflections (ailerons and spoilers) will result in yaw opposite to the direction of the intended turn. This adverse yaw becomes more severe at high angles of attack. In the high angle of attack flight regime, aileron inputs provide very low roll rates. At very high angles of attack (near stall), aileron inputs cause increased adverse yaw and roll opposite to that intended. The natural tendency to raise the wing with aileron must be avoided. Aileron deflection at the point of departure from controlled flight will increase the probability of spin entry. At the first indication of adverse yaw, the ailerons must be neutralized.

NOTE

With roll stab aug engaged, roll rates not commanded by lateral stick deflection will cause aileron deflection against the roll. At high angles of attack this will cause pro-spin adverse yaw in opposition to rudder induced roll and will increase the probability of departure from controlled flight. Roll aug should be selected OFF for air combat maneuvering.

Dihedral Effect

Attempts to yaw the airplane with rudder will produce roll in the same direction as yaw. This dihedral effect becomes more pronounced at high angles of attack. The use of rudder inputs to produce yaw and in turn generate roll, will provide the highest attainable roll rates at high angles of attack. Above 15 units angle of attack, desired roll should be generated primarily through use of the rudder. The rudder must be used judiciously, however, since excessive rudder inputs will induce excessive yaw.

LANDING

The optimum approach (ON SPEED) indicated angle of attack with the landing gear down is 19.0 units, and stall warning (rudder pedal shaker) is set at 21.3 units. On aircraft 155529ag and up, and all others after AFC 388, the optimum approach angle of attack is set at 18.3 units, and stall warning indication is set at 20.6 units. This change in optimum approach (ON SPEED) angle of attack compensates for error induced by extending the corner reflector. The angle of attack reference changes with nose gear position in the approach speed range. The indicated angle of attack increases approximately 3 units when the nose gear is extended. This change in indicated angle of attack with nose gear position is due to a change in the airflow pattern over the fuselage mounted angle of attack probe with the nose gear retracted. All references to indicated angle of attack should take this factor into consideration. Optimum approach angle of attack is adequate for all allowable gross weight and flap configurations. No adjustment is required for gusting crosswinds, runway, or weather conditions.

MAXIMUM PERFORMANCE MANEUVERING

The three factors that determine maximum performance maneuvering capability are structural limitations, stabilator effectiveness, and aerodynamic limitations. Structural limitations are outlined in section I, part 4. The limit in stabilator effectiveness occurs at high altitudes and supersonic speeds where full aft stick can be attained without reaching aerodynamic or structural limits. Aerodynamic limitations (stall) are primarily a function of angle of attack. In this area of flight, maximum performance turns are achieved by maintaining 19 to 20 units angle of attack while utilizing afterburner as required. The adverse yaw produced by the use of aileron during high angle of attack maximum performance maneuvering has been discussed and is of paramount importance in Air Combat Maneuvering. If a high angle of attack must be maintained and a roll is necessary, rudder must be used to produce roll due to yaw as previously discussed. During maximum performance maneuvering, higher roll rates may be achieved by: momentarily unloading the airplane (reducing angle of attack to between 5 and 10 units), utilizing aileron to roll to the desired bank angle, then neutralizing aileron and reestablishing the required angle of attack. At the first indication of departure from controlled flight, controls must be neutralized to preclude aggravating the out of control condition. Maximum performance maneuvering must be fully understood and demonstrated by a qualified instructor pilot prior to attempting any practice in this area of flight.

STALLS

CRUISE/COMBAT CONFIGURATION

Normal Stalls

Normal (1 G) stalls are preceded by a wide band of buffet. First noticeable buffet occurs at 12 to 14 units angle of attack and usually increases from moderate to heavy buffet immediately prior to stall or departure. The rudder pedal shaker is activated at 21.3 units on F-4B airplanes and 20.6 units on F-4N airplanes; (20.6 units on aircraft 155529ag and up and all others after AFC 388); however, it may not be recognizable due to heavy buffet. Wing rock, if encountered, will commence at approximately 23 units and variations in bank angle of up to 30° from wings level can be expected near the stall. The angle of attack at stall varies with loading and is normally above 25 units. The stall is characterized by a slight nose rise and/or yawing motion in either direction. Recovery from the stall is easily and immediately effected when angle of attack is reduced by positioning the stick forward, maintaining neutral ailerons and making judicious use of rudder to avoid inducing excessive yaw.

Accelerated Stalls

Only general stall characteristics are discussed herein; specific characteristics vary with airspeed, Mach number, loading, center of gravity position, G level, aircraft attitude, and control techniques. Accelerated stalls are preceded by moderate buffet which increases to heavy buffet immediately prior to the stall. Wing rock is unpredictable, but generally starts at about 22 to 25 units and progresses to a high frequency, large amplitude roll oscillation. The amplitude of the roll oscillations will be less with a heavy wing loaded aircraft. The angle of attack at stall varies considerably with loading, but is above 25 units for all loadings. Rapidly entered accelerated stalls may occur at lower indicated angles of attack. Increasing the rate of aft stick displacement increases the magnitude and rate of yaw and roll oscillations at the stall. Applying and holding full aft stick, even with ailerons and rudder neutral, can result in a spin. Prompt neutralization of controls will effect recovery from an accelerated stall. Oscillations in roll and yaw, which may be present during recovery, should be allowed to damp themselves out and should not be countered with ailerons or rudder.

Inverted Stalls

An inverted (negative angle of attack) stall can only be obtained with abrupt application of full forward stick in vertical maneuvers or an inverted climb of greater than 20° nose-up. Light to moderate buffet occurs at the stall and there are no distinct yaw or roll tendencies. Recovery from the inverted stall is effected by relaxing the forward stick pressure and maintaining an angle of attack between 5 and 10 units until recovered from the unusual attitude.

LANDING CONFIGURATION STALLS

NOTE

Do not practice landing configuration stalls above 10,000 feet. The effectiveness of the BLC system and engine bleed pressures decrease with altitude. Use of BLC above 10,000 feet may cause the systems using bleed air to become inoperative.

STALL SPEEDS

ALL CONFIGURATIONS

AIRCRAFT THRU 155528ag BEFORE AFC 388

ALTITUDES
10,000 FEET AND BELOW
POWER-ON STALL SPEEDS

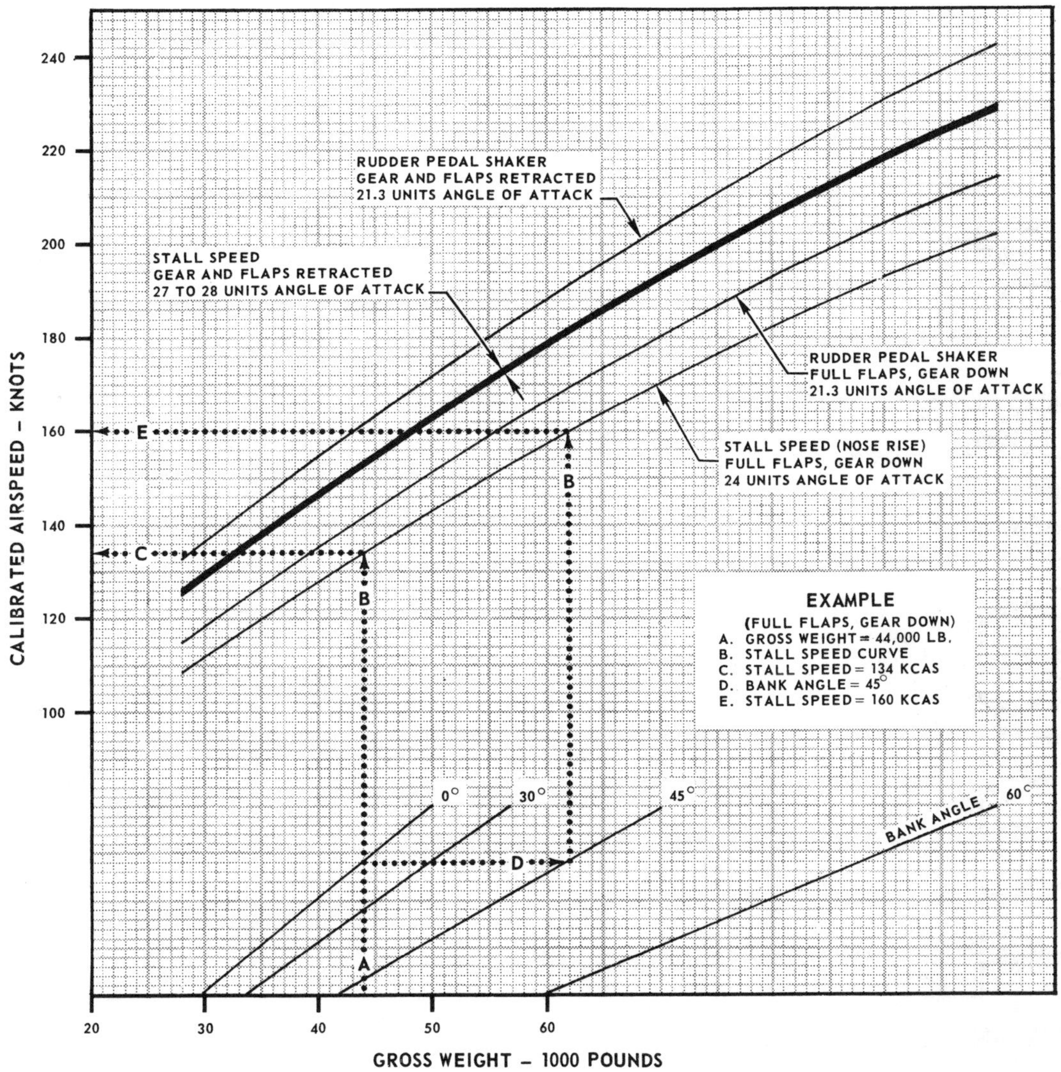

FDD-1-(255)

Figure 4-1

STALL SPEEDS

ALL CONFIGURATIONS

ALTITUDES
10,000 FEET AND BELOW
POWER–ON STALL SPEEDS

AIRCRAFT 155529ag AND UP, AND
ALL OTHERS AFTER AFC 388

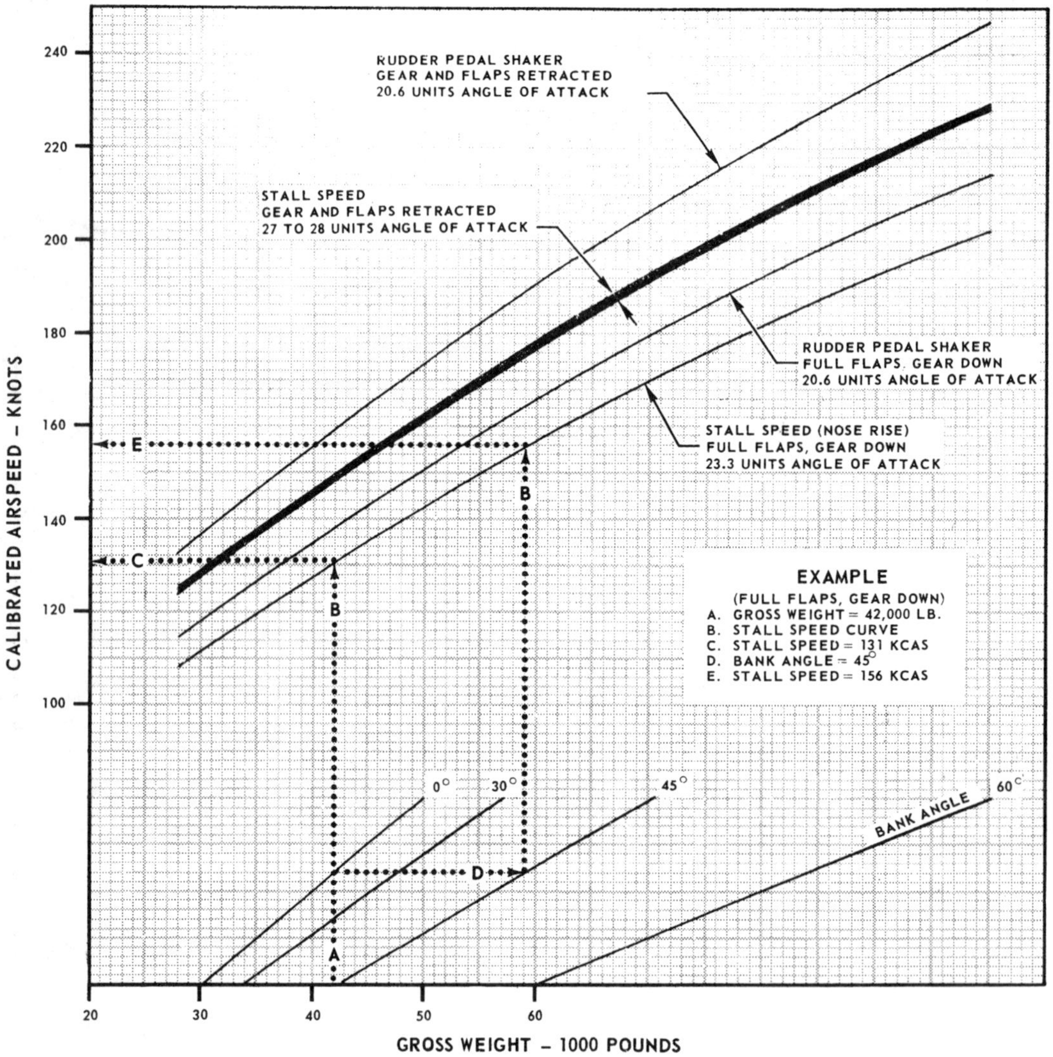

FDD-1-(259)

Figure 4–2

With the gear and flaps down the stall occurs at approximately 24 units angle of attack, and is characterized by simultaneous nose rise and apparent loss of lateral stability. With the flaps down and the landing gear up (bolter configuration), the stall angle of attack will be approximately 3 units lower because of the different airflow pattern over the fuselage mounted angle of attack probe with the nose gear retracted. There is virtually no increase in general airframe buffet during stall approaches, and very little aerodynamic warning (stick force lightening and wing rock is apparent prior to the stall). Stall warning is in the form of a rudder pedal shaker set at 21.3 units regardless of gear or flap position(20.6 units on aircraft 155529ag and up and all others after AFC 388). With the gear and flaps down, the pedal shaker actuates approximately 8 knots prior to the stall. In the bolter configuration, the pedal shaker does not provide stall warning; the only stall warning being stick forces lightening and slight wing rock just prior to the stall. The rate of increase of angle of attack above the stall is difficult to control and is a function of pitch rate and center of gravity position (aft CG most critical). Recovery from the stall is effected by placing the stick forward to reduce the angle of attack to below stall and increasing the throttles to MIL thrust. Recovery attitude is about 30° nose down. Stalls with one–half flaps are similar to full–flap stalls.

LOSS OF CONTROL

A loss of control or departure from controlled flight (departure) is best described as random motions about any or all axes. Departure characteristics are highly dependent on airspeed, Mach number, G level, type of entry, and loading. In addition to the stall warning discussed under Normal and Accelerated Stalls, a build–up in sideforces (tendency to move the pilot to the side of the cockpit) will be encountered immediately prior to a departure. The sideforces may not be detectable in a high speed, high G condition where wing rock will be the most positive indication of impending departure. If angle of attack is not reduced to below stall, departures can be expected to develop into spins. The angle of attack at departure is highly dependent upon loading. Clear or Sparrows–only airplanes will depart at slightly greater than 30 units, while heavy wing loaded air–to–ground configurations may depart as low as 27 to 28 units. Departures are best prevented by proper control of angle of attack. Although aileron deflection may aggravate the situation, excessive angle of attack is the primary cause of departure. Ailerons or excessive rudder deflection at departure increase the probability of spin entry following departure. Departures are characterized by a yawing motion with roll in the direction of yaw. The yawing motion at departure will be more violent when encountered during a high speed, high G condition than during a low speed, low G condition. At the first indication of departure, or when a nose high, rapidly decaying airspeed situation is encountered, attempt to reduce the angle of attack by moving the stick smoothly forward, simultaneously neutralizing ailerons and rudder. The throttles should be retarded to idle to reduce the probability of engine flame out unless, in the pilot's opinion, the altitude is so low that thrust will be required to minimize altitude loss during the recovery. The stick should be smoothly, yet positively, moved forward; not jammed forward. Forward stick should be applied until negative G is felt or until full forward stick is reached. An aircraft unloading (zero or negative G) is the most positive indication of recovery. The majority of recoveries will be effected before the stick reaches the forward stop. If recovery is not apparent after the application of full forward stick, deploy the drag chute without hesitation. Large roll and yaw oscillations may be present during recoveries from departures. Angle of attack indications will be erroneous during these oscillations and should not be used as a departure recovery indication. Applying full forward stick and neutralizing ailerons and rudder is the most effective means of damping the oscillations and should be maintained until the oscillations cease. An out of control situation may be reentered if stick movement off the forward stop is commenced prior to aircraft unloading and the oscillations ceasing. A series of rolls at 15 to 20 units angle of attack may be encountered with full forward stick; however, the unloading will not be present. Do not attempt to fly out of this condition; rather maintain full forward stick until negative G is felt. Do not confuse the rolls with a spin. Maintain 5 to 10 units until airspeed is sufficient for dive pull out (approximately 200 knots). Use angle of attack to minimize altitude loss and do not exceed 19 units during dive pull out. Recovery action upon departure from controlled flight is as follows:

1. Smoothly apply forward stick to reduce angle of attack (full forward if necessary), simultaneously neutralizing aileron and rudder, and reducing power to idle (altitude permitting).
2. If positive indications of recovery are not obvious after application of full forward stick, deploy the drag chute while maintaining full forward stick, and neutral ailerons and rudder.
3. Do not exceed 19 units during dive pull out.

NOTE

Engine flameouts may occur during departure; however, engine relights can be obtained with the throttles at idle even during a developed spin. Disengage the AFCS if in use.

If the angle of attack has been reduced to below the stall, the aircraft will not spin. The drag chute should produce recovery shortly after deployment, and will reduce the oscillations encountered during recovery. It is not necessary to jettison the chute since it will separate as airspeed builds up. The altitude loss following departure is dependent upon nose attitude at recovery, which is usually very nose low. Altitude loss pulling out of a 90° dive, initiated at 200 knots and utilizing 19 units, is approximately 5,000 feet. The out of control recovery procedure is summarized in figure 4–3.

SPINS

Spins have been entered from level flight stalls, accelerated turns, vertical climbs, 60° dive pullouts, and inverted climbs. Departure and spin characteristics were investigated with a clean aircraft, various heavy wing loadings, and with asymmetric loads. The angle of attack indicator is the primary instrument for verifying the type of spin (upright or inverted). During upright spins, the angle of attack indicator will be pegged at 30 units and during inverted spins will indicate zero (0) units. The

OUT-OF-CONTROL RECOVERY

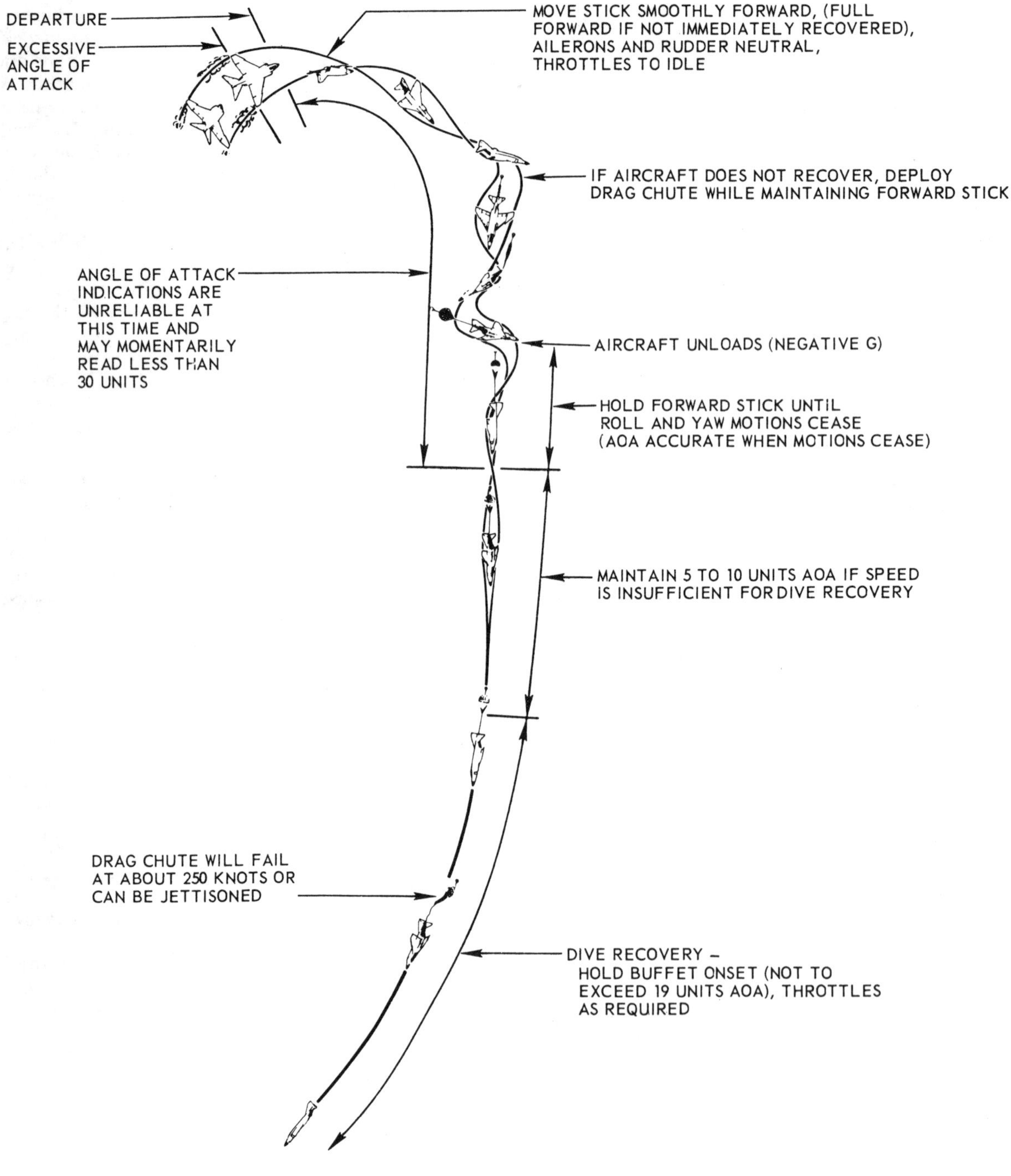

FDD-1-(53)

Figure 4-3

SPIN RECOVERY

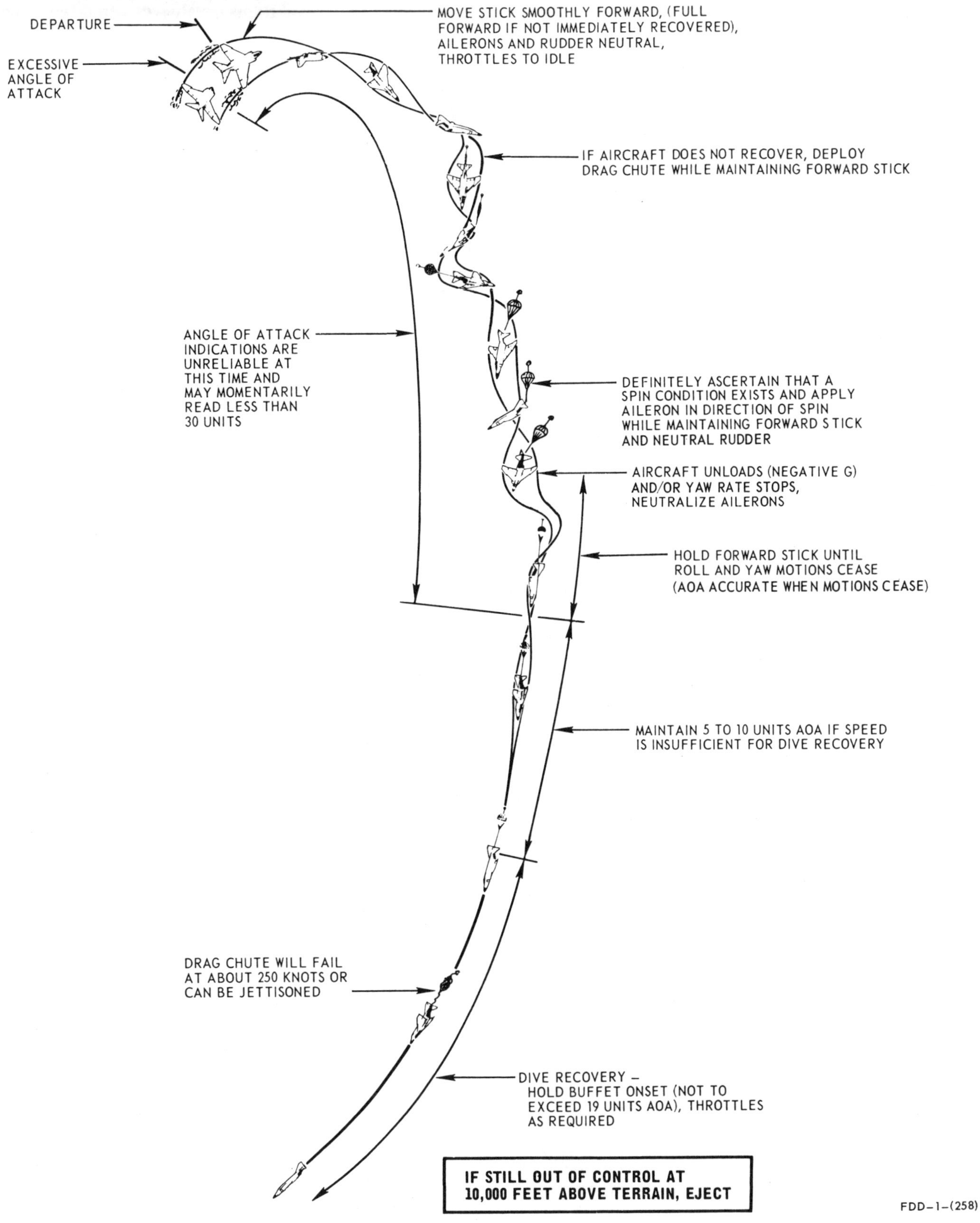

Figure 4–4

direction of spin can easily be determined from visual cues if ground reference is available; however, the direction of spin should be verified by referencing the turn needle (not the ball). The turn needle will always be pegged in the direction of the spin.

NOTE

Angle of attack may momentarily indicate less than 30 units (off the peg) during a spin; however, a sustained yawing motion in one direction verifies the spin condition.

The RIO should relay airspeed and altitude information to the pilot continuously during uncontrolled flight. If the airspeed is increasing through 200 knots, the aircraft is not spinning. Fly the airplane.

UPRIGHT SPINS

Steep Oscillatory Mode

The upright spin is oscillatory in pitch, roll, and yaw. The airplane pitch attitude may vary from slightly above the horizon to 90° nose down, and large roll angle excursions will be encountered. Yaw rate in the spin may vary between 10° and 80° per second, while airspeed will vary between 80 and 150 knots. The altitude lost during an upright spin is approximately 2000 feet per turn, and the spin turn rates average about 5 to 6 seconds per turn. Initially, spin oscillations may produce slightly uncomfortable accelerations in the cockpit; however, the oscillations should not be confusing. If recovery from the departure has not been effected after deploying the drag chute, and the airplane has been determined to be in an upright spin, POSITIVELY determine the spin direction and apply full aileron in the direction of the spin. Recovery from most spins will occur within two turns for a symmetrically loaded aircraft. The upright spin recovery procedure is:

1. Positively determine spin direction.
2. Maintain full forward stick and neutral rudder, and apply full aileron in the direction of the spin. (Right turn needle deflection, right spin, right ailerons.)
3. When the aircraft unloads (negative G) and/or yaw rate stops, neutralize the ailerons and fly out of the unusual attitude.
4. Do not exceed 19 units during dive pull out.
5. If still out of control by 10,000 feet above the terrain – eject.

The most positive indication of recovery from a spin is the aircraft unloading. Incidents have been encountered, however, where the yaw rate stopped and the aircraft entered 15 to 20 unit angle of attack rolls. If this occurs, the ailerons should be neutralized when the yaw rate stops and full forward stick maintained until the rolls cease and the aircraft unloads. Large excursions in roll and yaw may be encountered during recovery; do not mistake these excursions for a spin direction reversal. If the aircraft's nose remains on one spot on the ground or horizon, the aircraft is rolling, not spinning. Spin direction reversals are rare using the recommended recovery procedure; however, if reversal occurs, again positively determine the spin direction and reapply the upright spin recovery procedure. An out of control situation will be reentered if aft stick pressure is applied prior to aircraft unloading and the oscillations ceasing. Maintain 5 to 10 units until airspeed is sufficient for dive pullout (approximately 200 knots). Total altitude loss from a departure that develops into a spin until level flight is achieved can be as little as 10,000 feet; however, it will be closer to 15,000 feet if too much time is consumed determining spin direction. If the pilot considers that there is insufficient altitude for recovery the crew should eject immediately. Figure 4-4 summarizes the upright spin recovery procedure.

Flat Mode

There have been isolated cases of the airplane exhibiting an upright flat spin mode. The flat spin can develop within one or two turns after departure from controlled flight or after several turns of an upright steep oscillatory spin. Oscillations in pitch and roll are not apparent in the flat spin. The spin turn rate is 3 to 4 seconds per turn, and the altitude lost per turn is 1,000 to 1,500 feet. There is no known technique for recovery from a flat spin. Tests indicate that a very high angle of attack, well in excess of 30 units, is required for flat spin entry. Proper use of controls at departure will preclude entering a flat spin.

INVERTED SPINS

The aircraft is highly resistant to an inverted spin entry and tests indicate that pro spin controls are necessary to maintain an inverted spin. The inverted spin is characterized by zero (0) units indicated angle of attack and negative G and is less oscillatory than the upright steep oscillatory spin. Spin direction can be determined visually by the yawing motion of the aircraft or more positively by the deflection of the turn needle which is always pegged in the direction of the spin. Air speed will vary up to 150 knots. During an inverted spin roll rate is opposite yaw rate and can cause pilots to misinterpret spin direction. If recovery from uncontrolled flight is not effected by utilizing the out of control recovery procedure, and the airplane has been determined to be in an inverted spin, apply the following:

1. Positively determine the spin direction.
2. Full rudder against the spin (opposite the turn needle deflection).
3. Stick and ailerons neutral.
4. When the yaw rate stops, neutralize all controls and fly out of the unusual attitude.

The RIO should relay airspeed and altitude information to the pilot continuously during uncontrolled flight. This is particularly important in an apparent inverted spin, since airspeed may be the only recognizable difference between such a spin and a high speed inverted spiral. Rudder deflection at negative angles of attack causes roll opposite to rudder. After recovery from an inverted spin, continued rudder deflection will probably cause entry into an opposite inverted spiral. Angle of attack and turn needle may cause the pilot to believe he is still in the spin as his interpretation of visual cues may be unreliable. The best indication the aircraft is in a spiral will be increasing airspeed and, possibly, high negative G forces (depending on stabilator trim and position). If airspeed is increasing through 200 knots, the aircraft is not spinning.

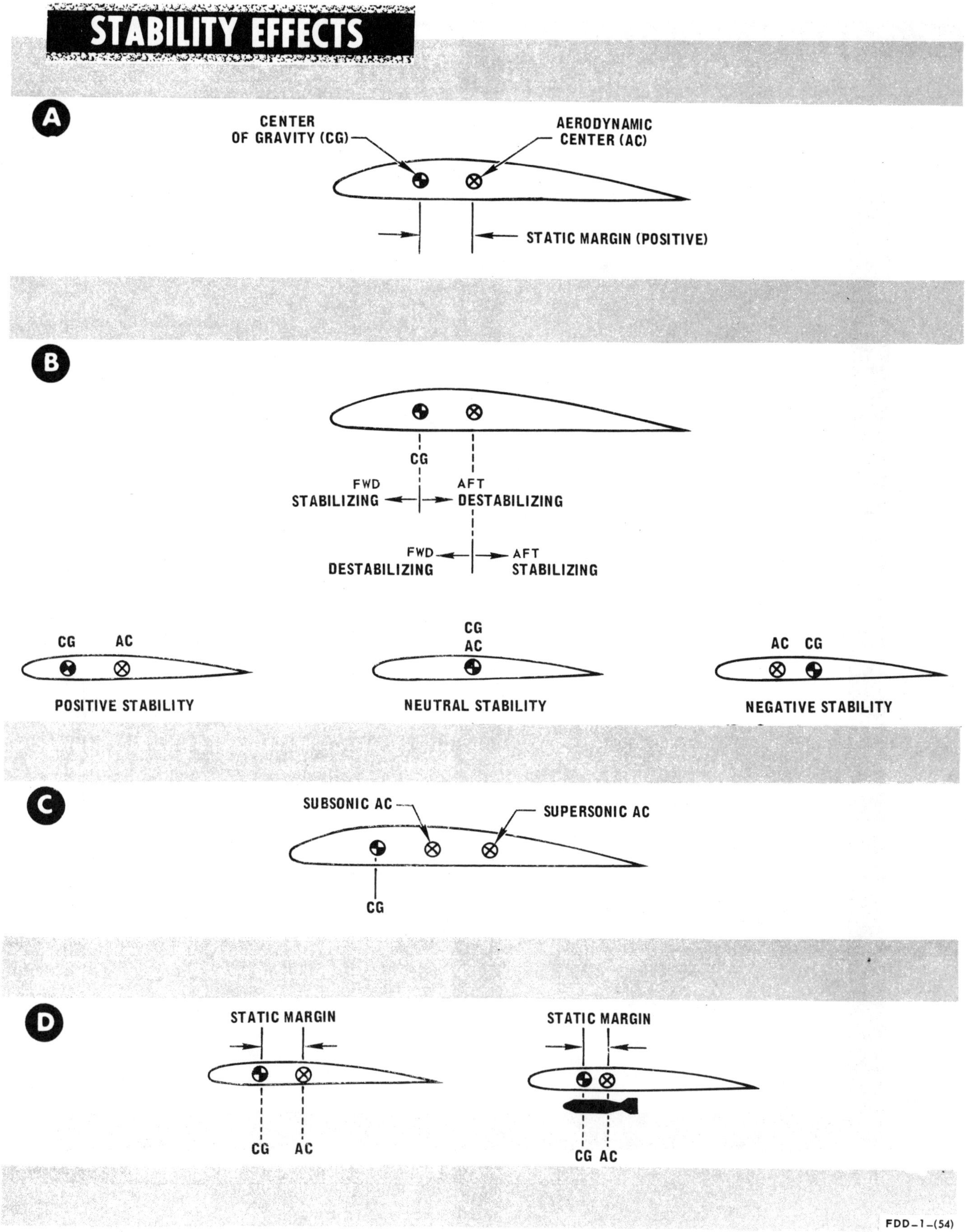
STABILITY EFFECTS
A
CENTER
OF GRAVITY (CG)
AERODYNAMIC
CENTER (AC)
STATIC MARGIN (POSITIVE)
B
CG
FWD
STABILIZING
AFT
DESTABILIZING
FWD
DESTABILIZING
AFT
STABILIZING
CG
AC
POSITIVE STABILITY
CG
AC
NEUTRAL STABILITY
AC
CG
NEGATIVE STABILITY
C
SUBSONIC AC
SUPERSONIC AC
CG
D
STATIC MARGIN
CG
AC
STATIC MARGIN
CG AC
FDD-1-(54)

Figure 4–5

C G TRAVEL DUE TO FUEL CONSUMPTION

APPROXIMATE

AIRPLANES 153071z THRU 153087aa

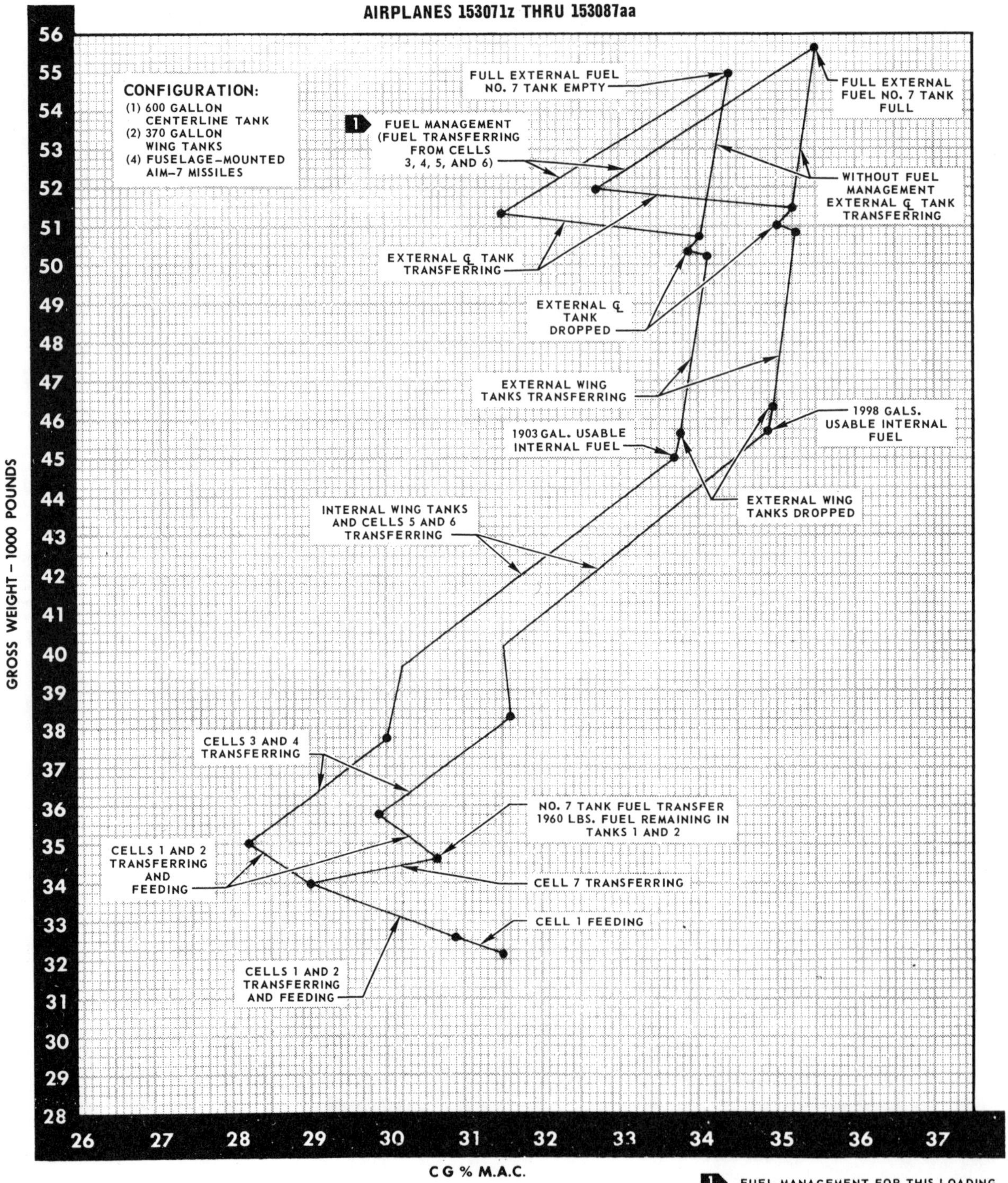

1 FUEL MANAGEMENT FOR THIS LOADING CONFIGURATION WITH NO. 7 TANK EMPTY IS NOT REQUIRED. SHOWN FOR INFORMATION ONLY.

FDD-1-(55)

Figure 4-6

C G TRAVEL DUE TO FUEL CONSUMPTION

APPROXIMATE

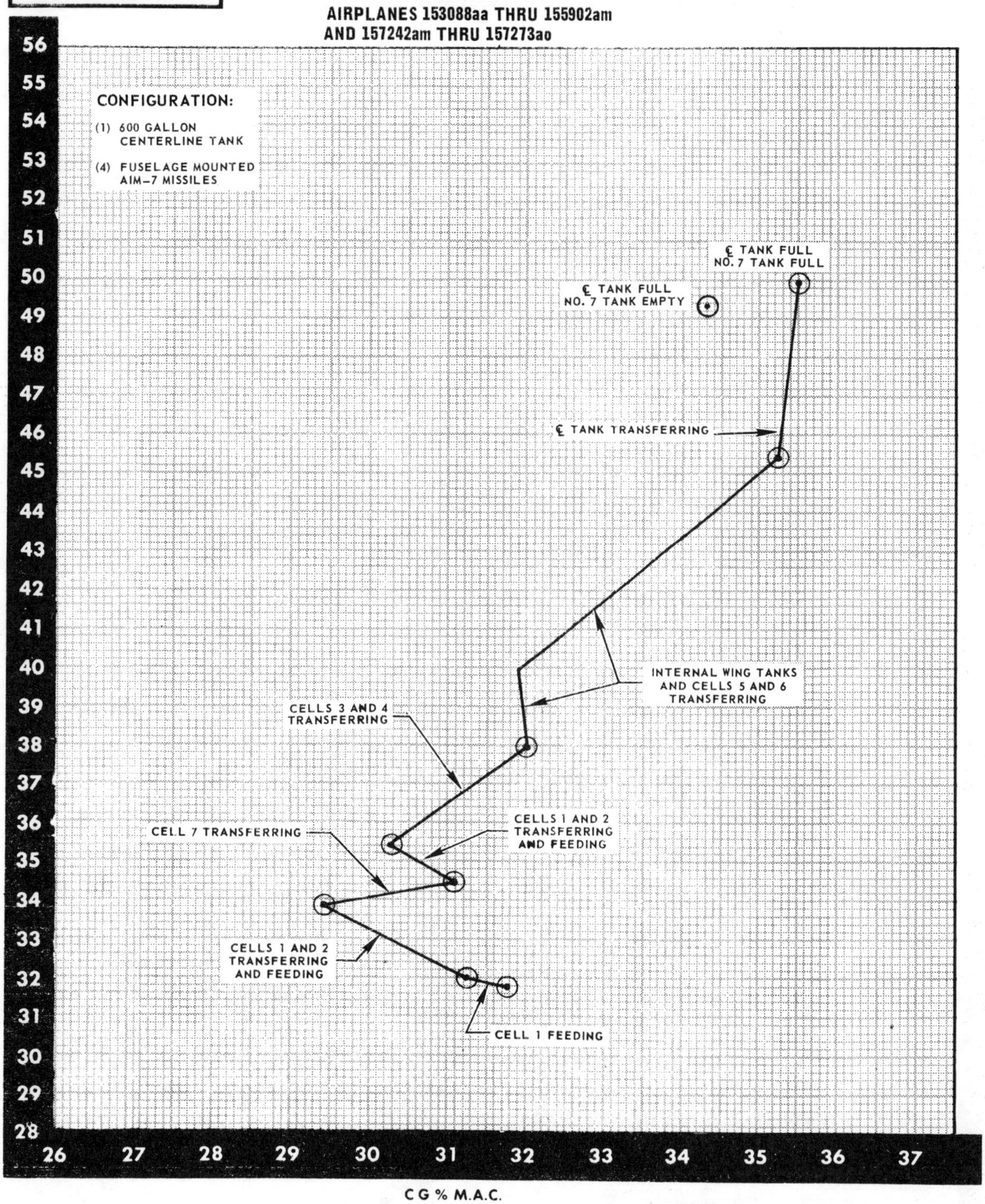

FDD-1-(56)

Figure 4-7

C G TRAVEL DUE TO FUEL CONSUMPTION

APPROXIMATE

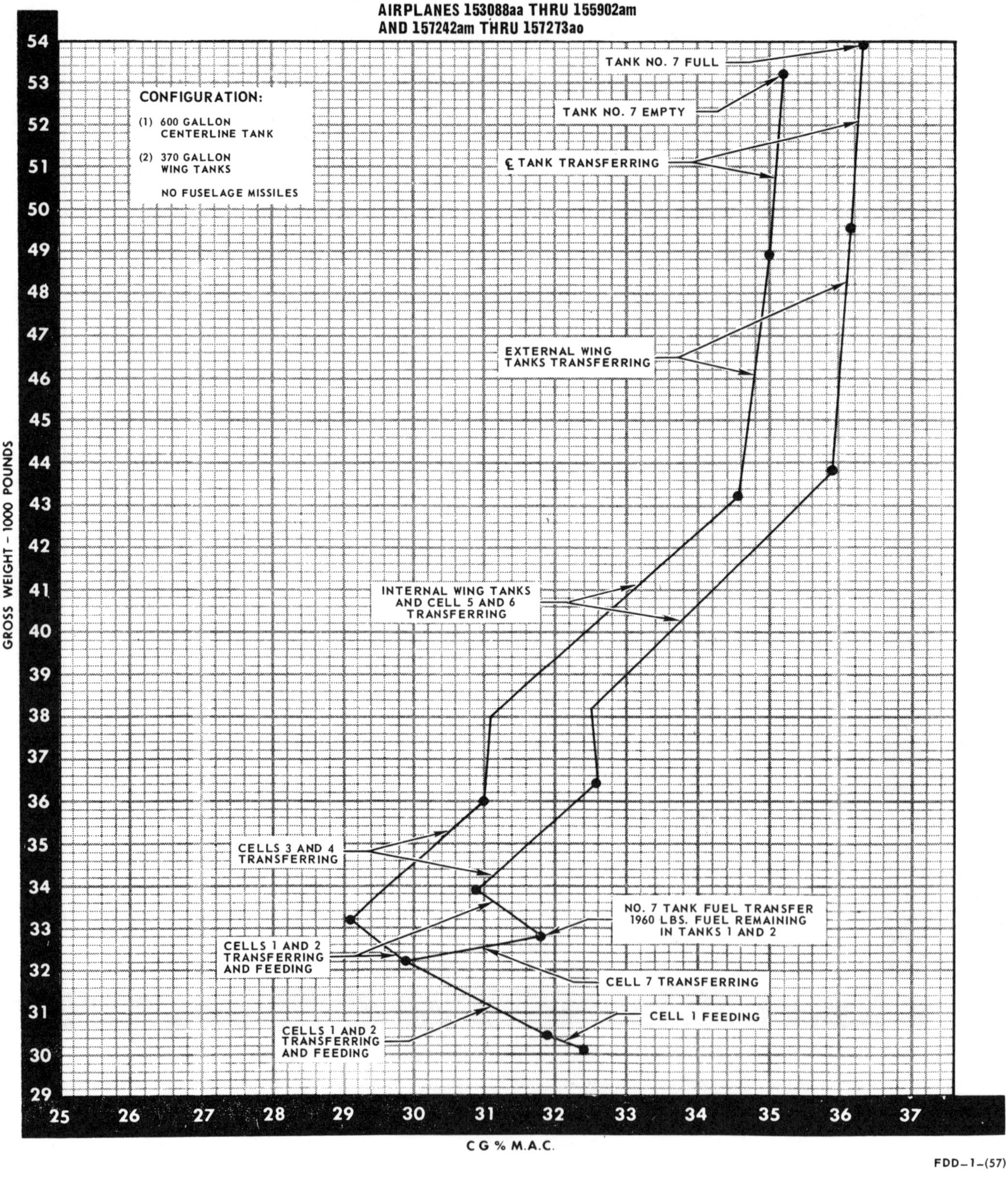

Figure 4–8

C G TRAVEL DUE TO FUEL CONSUMPTION

APPROXIMATE

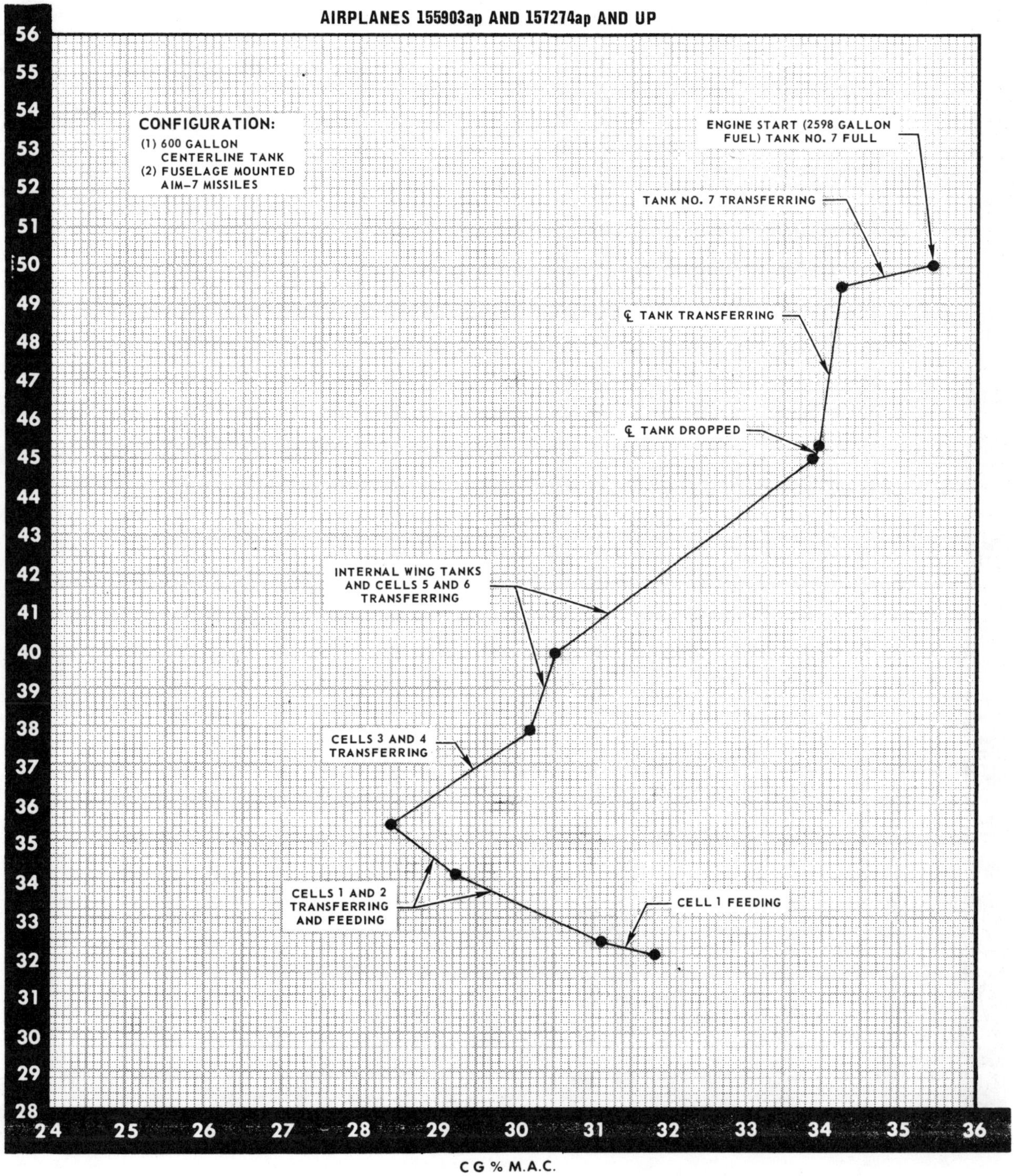

Figure 4-9

C G TRAVEL DUE TO FUEL CONSUMPTION

APPROXIMATE

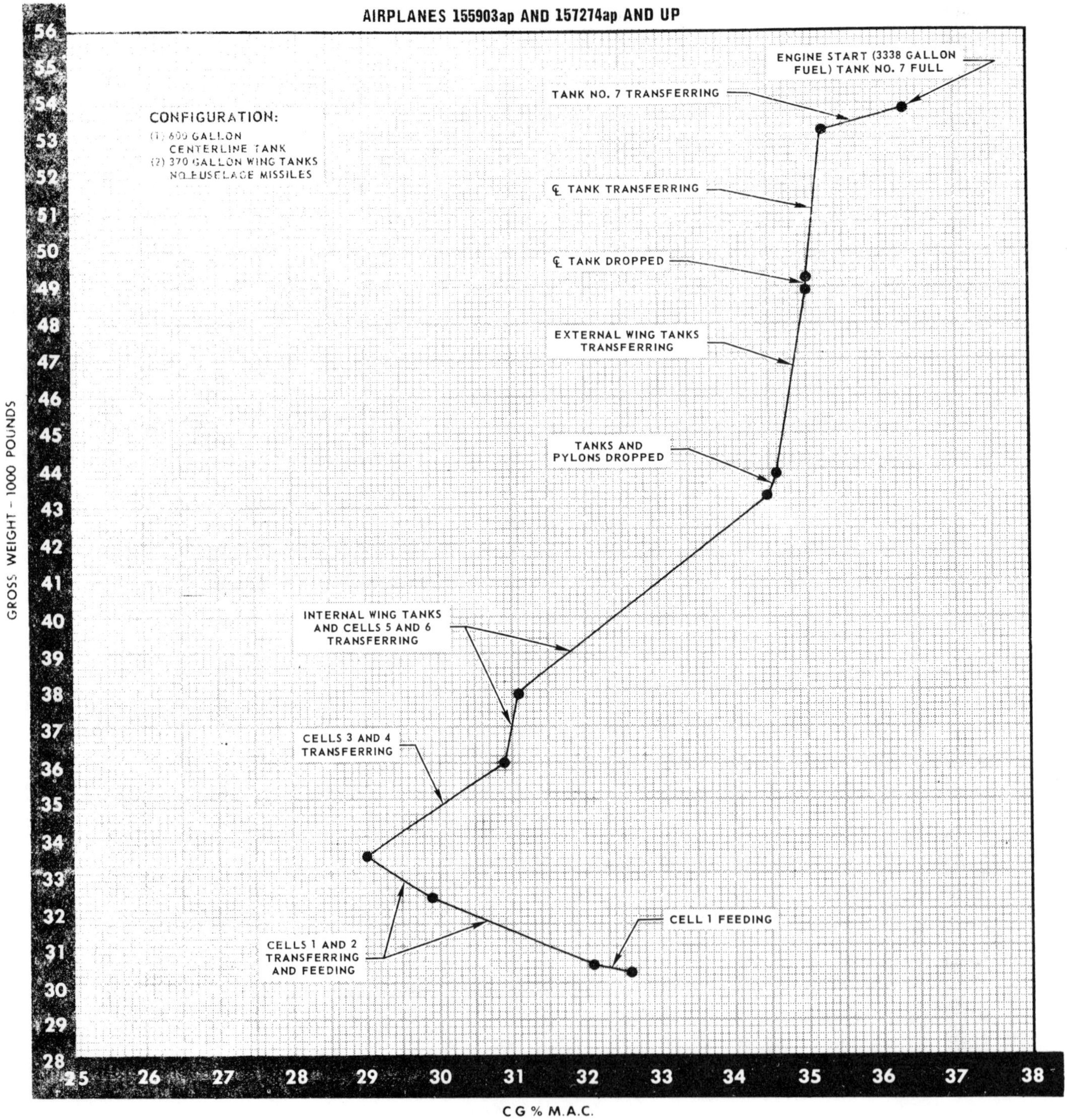

FDD-1-(59)

Figure 4-10

ENGINE EFFECTS

Engine flameout (one or both) may occur during departure. The probability of engine flameout is greatest at MIL or MAX thrust, and least at IDLE thrust. The best indication of a flameout is a MASTER CAUTION light and one or both GEN OUT lights. Engine relights can be accomplished at idle throttle setting during a spin. Normal operation of the flight controls will deteriorate if a relight is not accomplished. The RAT will not be effective while in a spin, but will be an immediate source of electrical power following spin recovery.

ASYMMETRIC LOAD EFFECTS

Airplane departure and spin characteristics were tested to an asymmetric moment of 117,000 inch-pounds. An asymmetrically loaded aircraft will depart at a slightly lower angle of attack (approximately 25 to 27 units) and will always depart in the direction opposite the heavy wing. Asymmetric load spin characteristics are more oscillatory about all axes than spins with symmetric loads. With large asymmetric loads it may be impossible to prevent a spin if a departure occurs. The same procedures for recovery should be utilized for asymmetric load departures and spins as were presented for symmetric loads. Asymmetrically loaded configurations will require more turns for recovery than symmetrically loaded aircraft.

CAUTION

Angle of attack should be closely monitored when maneuvering with an asymmetric load because of the possibility of entering a spin immediately following departure from controlled flight.

INITIAL CG TRAVEL

NORMAL FUEL SEQUENCING

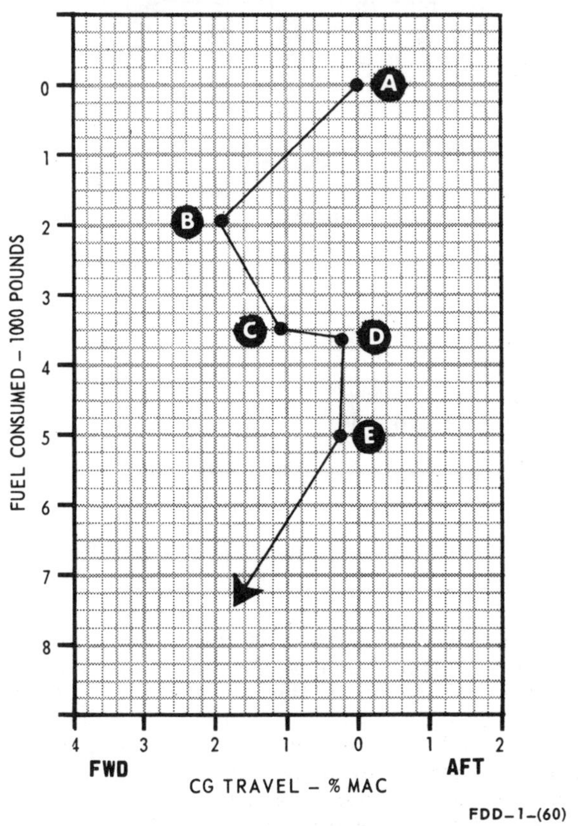

Figure 4–11

STABILITY AND CONTROL

In discussing stability and control, it must be realized that a large variation exists throughout the flight envelope. Stability varies with Mach number and also CG location. Control effectiveness is also affected by Mach number, but just as much, if not more so, by Q (dynamic pressure).

LONGITUDINAL STABILITY

The F–4 forward CG limit is based on airplane strength and longitudinal control effectiveness considerations. The aft CG limit is based on stability. For the clean airplane the aft limit is 36.0% MAC (Mean Aerodynamic Chord). This CG location provides an acceptable static margin. Refer to CG Limitations in part 4 of section I. Static margin is a physical measurement of longitudinal stability and is normally stated as the distance in % MAC between the CG and the aerodynamic center (AC), or nominal point of lift (A, figure 4–5). As the CG moves aft toward the AC, stability is reduced (B, figure 4–5). CG position alone does not define stability, but rather it is defined by the relative positions of the CG and the AC. The movement of the AC is a major contributor to changes in stability. An actual shift in the AC can be felt during an accelerated transonic speed reduction where the AC moves from its aft supersonic location to a forward subsonic location, noticeably changing (reducing) the stability (C, figure 4–5). Attempting to maintain a constant stick force through this deceleration will produce a pitch-up or load factor increase when the AC shifts. At high angles of attack the AC moves forward when the wing tip stalls, and again, a reduction in stick forces occurs. When a store or tank is loaded under the wing, a change in the AC position results. The amount of AC movement is a function of the geometric characteristics of the store/tank (D, figure 4–5). Centerline stores have essentially no aerodynamic effect on stability. The aerodynamic effect of any wing-mounted store/tank is always destabilizing. The static margin for a given configuration will change throughout the flight. Figures 4–6 thru 4–10 illustrate CG travel due to fuel consumption for specific configurations. Figure 4–11 illustrates the trend of CG travel during the normal fuel transfer sequence with two external wing tanks installed. During ground operation, the CG moves forward as fuselage fuel is transferred forward and consumed (A to B). When external fuel transfer is commenced, the CG starts moving aft, and continues to move aft slowly throughout the climb (B to C). During cruise, the CG remains near the aft starting point as external fuel maintains full fuselage cells (C to D). When the external tanks are empty and turned OFF (point D), the CG again moves forward as internal (wing and fuselage) fuel is consumed. In order to preclude

LONGITUDINAL STABILITY

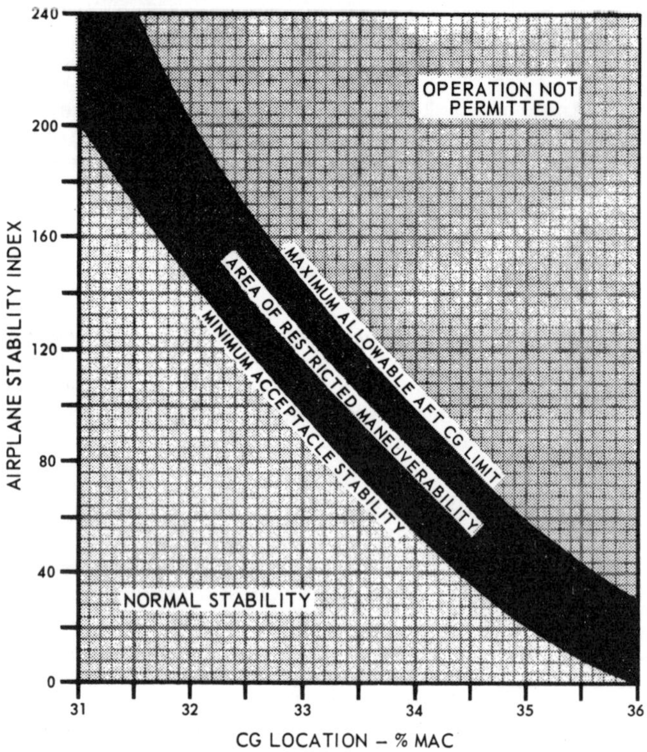

Note

Within the amber area, normal flight only (angle of attack for less than buffet onset) permitted. Only slow deliberate control movements allowed.

FDD-1-(61)

Figure 4–12

flight with unacceptably low static margins, a stability index system has been established. Aerodynamic effects of stores, tanks, and associated suspension equipment are assigned unit stability numbers. For any given configuration, the stability index is the sum of the stability numbers of the tanks/stores and suspension equipment. Once the airplane stability index has been computed and the airplane CG location for the given configuration has been determined, figure 4–12 may be used to plot the point which represents these two numbers. If this point falls within the red area, flight is not permitted due to the excessive pitch attitude control caused by inadequate stability. If the point falls within the amber area, flight is permitted; however, smooth coordinated control inputs are required and the airplane is restricted to non–maneuvering flight (AOA below buffet onset). Within the amber area longitudinal stick forces are very light and the possibility of overstressing the airplane or entering an accelerated stall prohibits air combat maneuvering. Pitch–up (where the nose of the airplane pitches up during maneuvering flight without additional back stick force) can also be encountered, again resulting in possible overstress and stall. High speed, low altitude flight should also be avoided in this area, since extremely sensitive pitch attitude control may lead to a dangerous pilot induced oscillation (PIO). If the point falls within the green area maneuvering flight is permitted but maneuvering within 1 to 2% MAC of the amber area must be undertaken with a reasonable degree of caution. The unsafe maneuvering characteristics present in the amber area are still present near the aft limit in the green area, but of a sufficiently reduced magnitude to allow safe operation. For flight in the green area, within approximately 1 to 2% MAC of the aft limit, every effort should be made to make smooth control inputs during all maneuvering. As the CG moves forward, the longitudinal stability characteristics improve and result in good flying qualities. The airplane exhibits satisfactory longitudinal stability characteristics for non–maneuvering flight in the amber and green area. The aft CG limits of figure 4–12 are based on operation with PITCH AUG engaged. Loss of PITCH AUG results in an apparent loss of longitudinal stability of at least 1% MAC. In this condition, maneuvering flight at the PITCH AUG ON limits should not be attempted. Individual pilot proficiency and operational necessity dictate whether to continue a mission upon failure of the PITCH AUG mode of the AFCS when near the aft CG limits for PITCH AUG ON operation.

LONGITUDINAL STABILITY IMPROVEMENT

Several approaches may be used to improve the longitudinal stability of the airplane. The primary method is by control of airplane CG through fuel management and external stores loadings. The fuel in the number 7 fuselage cell is a major contributing factor toward an aft CG condition. For a clean configuration airplane, a full number 7 cell shifts the CG aft 1.33% MAC. Flight with the number 7 empty thus provides increased longitudinal stability and allows more external stores to be carried prior to reaching the aft CG limits. Selective loading of external stores can significantly affect airplane CG. All stores on stations 2 and 8 result in a forward CG shift, whereas the majority of stores on stations 1 and 9 result in an aft CG shift. Stores on the centerline station normally result in a slight aft CG shift. Maximum utilization of airplane stations 2 and 8 for external stores can thus result in more forward CG conditions. Fuselage–mounted Sparrow III missiles can also be utilized to adjust CG location. Listed below are the approximate CG effects due to loading fuselage–mounted AIM–7E missiles:

CONFIGURATION	CG SHIFT
2 Missiles Forward (Stations 4 & 6 only)	Forward 1.48% MAC
2 Missiles Aft (Stations 3 & 7)	Aft 0.82% MAC
4 Missiles (Stations 3, 4, 6, & 7)	Forward 0.66% MAC

Fuselage fuel distribution is the primary in–flight CG position control. Burning down fuselage fuel provides improved stability as the aft tanks empty first. As previously shown (A to B in figure 4–11), holding external and internal wing fuel causes a favorable rate of forward CG movement. Obviously, sometime during the flight external and internal wing fuel must be transferred and consumed with a resultant aft CG shift. Holding internal wing fuel, as opposed to simultaneous internal wing and fuselage transfer, will provide for a slightly more rapid rate of forward CG movement from the point of full internal fuel. Caution must be exercised during any non–standard fuel transfer sequence to preclude fuel starvation due to not turning external tanks off. In order to transfer internal wing fuel, the external tanks must be turned off and internal wing transfer selected. Automatic fuel transfer should not be depended upon to correct for improper fuel switchology. Since the CG shifts due to fuel transfer and consumption, the fuel quantity gage (figure 4–13) can be used as an approximate indication of CG condition. A sector (fuel quantity of fuselage cells 1 thru 6) indication of 8500 pounds indicates an aft CG condition as all fuselage cells are full. At 6000 pounds on the sector, the CG is approximately mid–range of its possible travel (fuel cells 5 and 6 nearly empty and fuel distributed in cells 1 thru 4). The CG continues to move forward as the sector indication drops to approximately 2000 pounds. The most forward CG shift (due to fuel consumption) occurs immediately after fuel from cell 7 has been consumed and cells 1 and 2 are full.

SUBSONIC REGION

Takeoff Configuration

Takeoff performance is based on airplane gross weight and CG position. Liftoff speed is a function of gross weight and is essentially the speed at which the wings develop sufficient lift to raise the weight of the airplane. Nosewheel liftoff speed is a function of CG, stabilator position, and gross weight. Nosewheel liftoff occurs when the noseup aerodynamic moment exceeds the nosedown weight moment. The stabilator cannot be stalled in the takeoff maneuver. Therefore, full aft stick takeoff technique (full leading edge down stabilator) provides the lowest nosewheel liftoff speed and the shortest takeoff distance. For full aft stick, the CG position will determine when the rotation begins; a lower speed than liftoff with an aft CG, and a higher speed than liftoff with a forward CG. The rate of rotation is a function of airplane acceleration (rate of buildup of down tail lift) and CG. Therefore, because of the rapid buildup of pitch–rate at aft CG's, full aft or rapid stick displacements can cause over–rotation following the nosewheel liftoff phase. In computing CG for takeoff, allowance must be made for the forward shift of CG during ground operation. The CG will move forward approximately 1% for every 1000 pounds of internal fuel used. In the forward CG range, nosewheel liftoff speed is increased approximately 4 – 5 knots for every percent of forward CG movement. After nosewheel liftoff, desired pitch attitude is maintained by using whatever control movement is required.

C G–FUEL REMAINING RELATIONSHIP

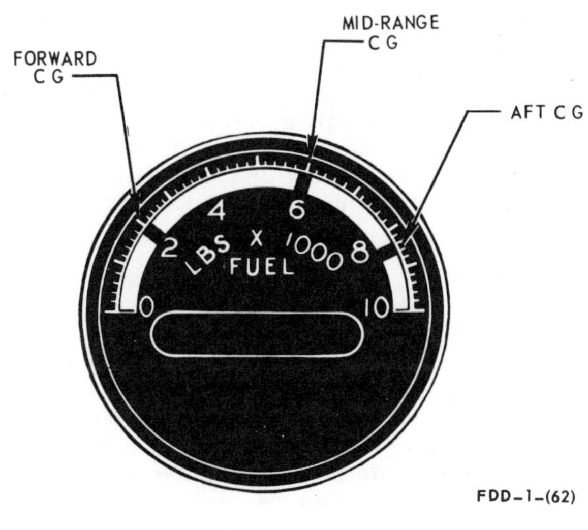

FDD-1-(62)

Figure 4–13

Landing Configuration

In this configuration, the airplane exhibits positive longitudinal static stability except for an area about 10 knots before stall where a mild stick force lightening occurs. This is followed by a regaining of static stability after the stall so that if back pressure is released, the airplane recovers by itself. In the speed range between 130 – 180 knots, where most landing configuration flying is done, the airplane demonstrates almost neutral stick force stability up to about 150 knots and mild stick force stability above this speed. This is due to control system friction and rather weak stick centering at this low Q. Stabilator effectiveness is reduced with full flaps due to an aft center–of–pressure shift and a change in the down–wash pattern over the tail. However, adequate effectiveness still remains for all known configurations. Since ground effect also decreases stabilator effectiveness, the aft stick–stop may be bumped during flare–out from a high sink rate landing. Stabilator effectiveness is not sufficient to hold the nose up after landing since the center of rotation is now about the main gear instead of the CG. Lateral and directional control response in the landing configuration is good; however, adverse yaw generated by high roll rates produces a decrease in commanded roll due to strong dihedral effect. This strong dihedral effect can be utilized in the landing configuration to provide roll with rudder deflection. Judicious use of rudder in the landing configuration at approach angles of attack can provide desired roll response. The ARI (Aileron–Rudder Interconnect) feeds in rudder automatically to counteract yaw so that when large amounts of aileron are being used, the turns will be coordinated. Except for unusually asymmetrical external loadings or very rough, gusty air, only small lateral control motions are required for landing. The approach to stall is characterized by a decrease in lateral stability which becomes evident by a mild wing–rock (5° – 10°) which gets progressively worse as speed is reduced.

Clean Configuration

Lateral and directional control response is good in the clean configuration and the airplane exhibits good pilot feel. Rate of roll is quite high in this area and directional stability is strong enough so that ball–centered turns are made without the use of rudder. During abrupt aileron rolls, where some adverse yaw is experienced, the yaw damper is effective in keeping the ball centered.

TRANSONIC REGION

High Altitude

In the transonic region, longitudinal static stability becomes more positive and stabilator effectiveness somewhat reduced resulting in slightly higher maneuvering stick force gradient. Transition from transonic to subsonic speeds while holding G on the airplane results in a mild to moderate nose rise. If corrective action is not taken, this could place the airplane in buffet at the higher altitudes or cause a significant load factor increase at the lower altitudes. This is characteristic of most swept-wing airplanes and is a result of going from an area of higher stability and lowered stabilator effectiveness to an area of lowered static stability and higher stabilator effectiveness. Speed brakes increase the nose rise tendency during transition from transonic to subsonic speeds. Lateral and directional control in the transonic region is about the same as that experienced in the subsonic region except that roll rate capability is higher. Roll rates are highest in the transonic regions; however, in both transonic and subsonic regions, roll rate resulting from full aileron is much too great for any practical use.

Low Altitude

Transonic flight at low altitude presents low stick force gradient and high stabilator effectiveness which results in an area of high sensitivity and possible over-control. Even though the inherent dynamic stability of the airplane is positive, it may be possible to create a short period longitudinal oscillation if the pilot's response becomes out of phase with the airplane motion, thereby inducing negative damping. Such a condition is commonly known as pilot-induced oscillation (PIO). Since the dampers will decrease the stabilator response to rapid stick inputs, the possibility of inducing PIO is minimized with dampers on. It is recommended that the stability augmentation be used when flying at high speeds and low altitudes. The standard and most effective recovery technique from a pilot-induced oscillation is to release the controls. An out-of-trim condition is conducive to PIO, and releasing controls while stick forces are present, because of an out-of-trim condition, could amplify the oscillation. Therefore, it becomes advisable to trim out longitudinal stick forces during a rapid afterburner acceleration at low altitudes. It should be noted, however, that if longitudinal forces are trimmed out while maneuvering, an out-of-trim condition will be present when returning to wings-level flight necessitating a push-force to hold altitude. If the altitude of a mission is such that it would not be desirable to release the flight controls, recovery from a PIO can be accomplished by making the arm and body as rigid as possible, even bracing the left hand against the canopy and either holding the stick in the approximate trim position or by applying slight positive G-loading. In addition, afterburner shutdown at high indicated airspeeds can produce a pitch transient. Abrupt pitch inputs could cause an oscillation to begin; therefore, all corrections should be performed smoothly. Always lock the shoulder straps when flying under conditions of high speed and low altitude. The body, from the lap belt up, could become the forcing function during an inadvertent pitch input if the shoulder straps are unlocked.

SUPERSONIC REGION

High Altitude

Longitudinal static stability gets more positive as Mach number is increased in the supersonic region. Stabilator effectiveness decreases somewhat, so maneuvering stick forces become higher but do not exceed 10–12 pounds per G. Maneuvering capability is limited by stabilator effectiveness at the higher Mach numbers at higher altitudes; for example, full aft stick at Mach 2 at 50,000 feet will produce about 3.5 G. No abnormal lateral or directional control problems exist during supersonic flight. Directional stability remains strong and rate of roll, although decreasing with Mach number, remains quite adequate out to limit Mach numbers.

EFFECT OF EXTERNAL STORES

The effect on longitudinal stability and control is discussed under Longitudinal Stability. Inertial effects are evidenced during abrupt pitch maneuvers at very high angles of attack. The most noticeable effect is the increase in inertia about the longitudinal axes. This shows up as simply reducing roll rate and rolling acceleration. In other words, it takes longer to build up a given roll rate, but once a roll rate is established, it takes longer to stop it. Most restrictions on high-speed flight with external stores are based on structural considerations.

FLIGHT WITH ASYMMETRIC LOADING

Takeoff

Emergency takeoff and landing with asymmetric loadings can be made with an asymmetric load equivalent to one full external wing tank loaded on a clean aircraft. The recommended technique is essentially the same as that used for a crosswind. A strong turning moment into the heavy wing will exist during takeoff roll. This characteristic becomes more prominent during rapid acceleration. Nose gear steering and rudder (when rudder becomes effective) should be used for directional control. As the aircraft breaks ground, there is a tendency to roll into the heavy wing. Therefore, trimming ailerons prior to takeoff roll is desirable. The actual lift-off should not be abrupt. Establish an attitude and allow the aircraft to fly itself off the ground using rudder and aileron as required.

Maneuvering

Caution must be exercised due to a rapid build-up of asymmetric forces during maneuvering. Roll tendencies increase with load factor whereas control to counter this roll is a function of airspeed. Loss of lateral (roll) control can occur at angles of attack well below buffet and stall. Use of excessive lateral control (aileron) will produce adverse yaw. Control can be regained only when an airspeed is reached which will provide sufficient aerodynamic control to overcome the rolling force of the asymmetric load. The rolling moment produced by failure of one internal wing tank to transfer will be essentially undetectable in 1 G flight. At higher load factors this rolling moment will become significant. Every asymmetric condition has airspeed/load factor combinations beyond which control cannot be maintained. Control can be regained only by an increase in airspeed and/or reduction of load factor.

Landing

When a significant asymmetric condition exists, determine the minimum control airspeed with one–half flaps and gear down (see Minimum Control Airspeed, this section). Make an extended straight–in approach with one–half flaps at minimum control airspeed +10 knots. Directional trim is important, particularly to eliminate any sideslip from the side opposite the heavy wing. For large asymmetric loadings lateral trim may not be sufficient to trim the airplane for wings level flight. However, for those conditions lateral stick forces do not exceed 5 pounds and do not degrade the approach characteristics. If at any time lateral control becomes marginal, use rudder to obtain the roll response required and increase airspeed to increase lateral control effectiveness. During the approach avoid abrupt or accelerated maneuvers, particularly an abrupt landing flare since this may cause a strong roll into the heavy wing. Single and double engine emergency landings can be accomplished with asymmetric loadings up to the limits of section V. Single–engine approaches are slightly worse than two–engine approaches because of asymmetric thrust characteristics. In a left turn with the right engine at IDLE and the asymmetric load on the right wing, full MIL thrust and nearly full left stick is required for a 25° banked turn. Similarly, with the engine operating on the side opposite the asymmetric load, increases in thrust on final result in yaw and roll into the heavy wing which must be countered with additional lateral stick opposite the asymmetric load. This characteristic is especially noticeable during single–engine AB wave–offs; however, the lateral control effectiveness increases rapidly with increasing airspeed. Approaches and landings can be accomplished in crosswinds up to 10 knots from the side opposite the heavy wing. To maintain maximum available lateral control in crosswind approaches, the technique of establishing a crab angle into the wind, rather than the wing down technique, is recommended. If possible, a runway should be selected such that the crosswind is from the same side as the asymmetric load. The airplane should be landed on the down wind side of the runway because the advantages of the crosswind from the heavy wing side are reversed during landing roll–out. The airplane tends to weathercock into the wind during normal landing roll–outs especially with the drag chute deployed. The vertical tail, the drag chute and the higher drag on the upwind wing (the asymmetric load) all tend to turn the airplane into the wind. This tendency is more pronounced than any tendency to turn away from the heavy wing because of differential braking conditions. The weathercock into the wind can be satisfactorily countered with rudder and aileron opposite to the wind direction and the nose wheel steering as a last resort. For carrier operations the same approach techniques apply.

NOTE

Refer to section I, part 4, and section V for asymmetric load limitations.

MINIMUM CONTROL AIRSPEED

Flying characteristics may be degraded under abnormal conditions of flight controls or configuration, to the point that an airspeed higher than normal approach speed is required for adequate control during landing approach. Minimum control airspeed is that airspeed below which the airplane can no longer be adequately controlled to a landing under existing conditions. This speed will vary as a function of airplane condition, turbulence, temperature, field elevation, landing winds, and pilot experience. Minimum control airspeed should be determined at a safe altitude (normally 5000 – 10,000 feet) through investigation of the flying characteristics by applying landing approach type roll and pitch inputs while gradually slowing the airplane. When aircraft reaction to flight path type corrections reaches a minimum acceptable response rate, minimum control airspeed has been reached and the airplane must not be flown slower than this airspeed. Airspeed should normally be used as a reference due to possible errors in angle of attack due to airplane yaw. Caution must be exercised so as not to take this investigation to the point of losing control of the airplane. When minimum control airspeed is determined, a decision must then be made whether this airspeed is slow enough to accomplish a successful landing at available landing facilities. If a landing can be made, an additional ten knots should always be added to minimum control airspeed to provide a safety margin for the landing approach. If landing cannot be safely accomplished, a controlled ejection should be made.

HEAVY GROSS WEIGHT OPERATION

Takeoff

Consideration must be given to gross weight and CG effect on lift–off and nosewheel rotation speeds as discussed under Subsonic Region (Takeoff Configuration), this section. Full aft stick takeoff is required to achieve handbook takeoff performance data.

Maneuvering

Aft CG conditions are usually encountered during heavy gross weight operation. Also refer to Longitudinal Stability, this section. For any given indicated airspeed the airplane will be at a higher angle of attack. This condition reduces the margin of angle of attack prior to stall. A significant change in flight characteristics occurs during air refueling due to the immediate increase in gross weight; mission planning must consider this change in airplane flight characteristics.

Landing

Every reasonable technique must be pursued to attempt to reduce gross weight prior to an emergency landing. Excess weight reduces excess thrust, thereby narrowing the margin for recovery from any subsequent problem. Execute a wider–than–normal or straight–in landing pattern, and avoid abrupt or accelerated maneuvers. Utilize available power (including afterburner) as required to preclude excessive sink rates.

ZOOM CLIMB

A zoom climb can be performed by accelerating to a high energy condition and then slowly rotating to a pitch attitude higher than normal climb. Pitch angles in excess of 60° detract from the zoom climb capability and produce more uncomfortable recovery conditions. During a zoom climb to altitudes above 65,000 feet, the EGT must be monitored. Afterburner blow–out will usually occur around 67,000 to 70,000 feet. When the afterburners blow out, the throttles should be taken out of the afterburner range to preclude unexpected or hard lightoffs during descent. Above 70,000 feet, the engines will have to be shut down if they tend to over–speed or over–temp. Engine windmill speed at altitudes above 70,000 feet are high enough to maintain some cockpit pressurization and normal electrical power. Stabilator effectiveness will decrease noticeably above 50,000 feet and an increased amount of aft stick will be required to hold a given pitch attitude. Zoom climb recovery can be initiated at any time during the zoom maneuver by relaxing back pressure on the control stick and flying the airplane over the top at a G loading which will prevent stall. Maintaining a constant value of angle of attack between 5 and 10 units will properly decrease G with decreasing airspeed during the recovery while still maintaining a safe positive G loading on the aircraft. Negative G recoveries are not recommended due to aircraft and physiological limitations and lack of pilot ability to detect impending stall. Two basic methods of recovering from the zoom climb are possible. A wings–level recovery can be effected by smoothly decreasing angle of attack to the minimum positive G value and holding this until the aircraft is diving. An inverted recovery can be effected by controlling angle of attack while rolling the aircraft to inverted and then increasing angle of attack to produce the maximum G loading on the aircraft. A comparison of the two techniques show that the positive G loading on the aircraft assists the recovery trajectory in the inverted case whereas it detracts from the recovery trajectory in the wings–level case. The resulting flatter trajectory of the wings–level recovery produces a lower minimum airspeed and higher maximum altitude over the top in addition to a longer overall recovery time. Although the inverted recovery is superior from the standpoint of speed, altitude, and exposure time, it exhibits certain risks due to pilot capabilities to properly control angle of attack during the rolling maneuvers. All zoom climb recoveries demand smooth coordinated control action. The angle of attack indication is the primary recovery aid regardless of recovery method. As speed decreases, the stabilator required to develop a given pitch command increases. Higher than normal stick displacements and rates will be necessary to command or hold angle of attack at very low speeds. Inadvertent pitch inputs due to abrupt roll action or pilot's inattention to required pitch control can quickly put the aircraft in a stalled condition. Zoom climb recoveries initiated from indicated airspeeds in excess of 250 knots can be made inverted or wings–level. For the wings–level recovery, smoothly reduce angle of attack to 5 units and hold this value until the aircraft is in a recovery dive, and speed has increased through 250 knots. Attempts to hasten the recovery by pushing over to a value below 5 units angle of attack will produce negative G on the aircraft and possible stall. Precise roll attitude is not important during the recovery. Any aileron used to correct or maintain roll attitude should be smooth and coordinated. For this inverted recovery, smoothly reduce angle of attack to 5 units and holding this value, smoothly roll the aircraft to inverted. Increase and hold angle of attack at 10 units to produce maximum safe G loading on the aircraft. When the aircraft is in an inverted recovery dive, the roll to wings–level must again be accomplished with smooth slow control action while holding angle of attack between 5–10 units. As before, angle of attack should be maintained in the recovery dive until airspeed builds up to 250 knots. Zoom climb recoveries initiated at airspeeds less than 250 knots should be accomplished with pilot's sole attention devoted to proper control of angle of attack between 5–10 units. Roll attitude should be completely ignored with aileron and rudder held generally neutral to maintain coordinated flight. In the event a pilot becomes confused or disoriented during any recovery, he should immediately concentrate only on angle of attack and ignore all other parameters. If angle of attack is maintained between 5–10 units, the aircraft will recover safely to a nose–down accelerating condition, regardless of roll attitude.

FORMATION FLIGHT

PARADE FORMATION

The following formation description is a recommended guideline for pilots flying the aircraft in a basic parade position. This position is flown by the wingman placing the ramp hinge on the lead aircraft in line with the lead pilot's head, and placing his eye level directly abreast of the fuselage seam just aft of the wings on the lead aircraft. This triangulation should position the wingman at approximately a 6–foot stepdown with approximately 6 feet wing tip clearance. See figure 4–14.

NOTE

When flying formation, do not use flaps at an airspeed greater than 200 knots CAS, since the variable actuation between flap airspeed pressure switches might cause unintentional retraction of one aircraft's flaps while the flaps of the other aircraft remain down.

PARADE FORMATION

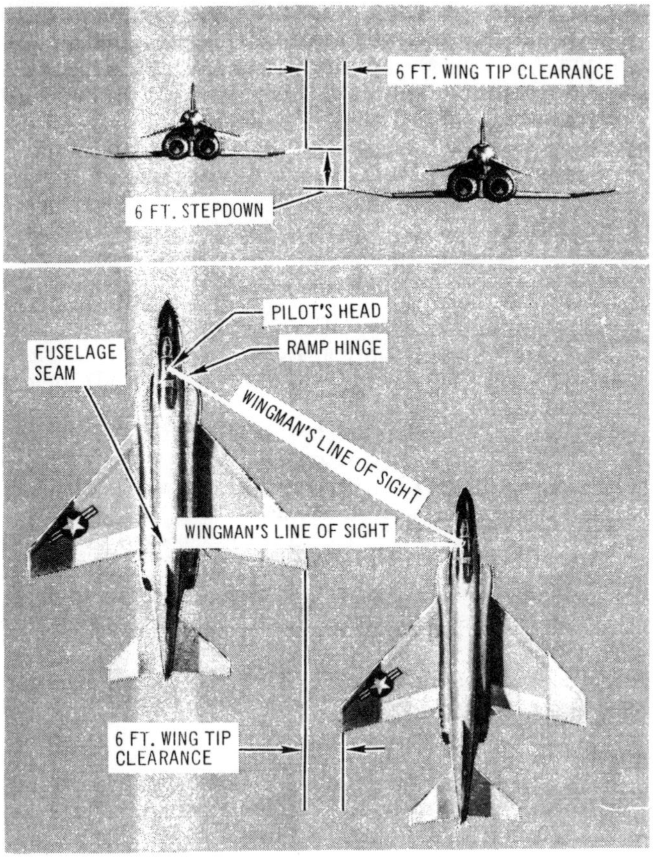

FREE CRUISE FORMATION

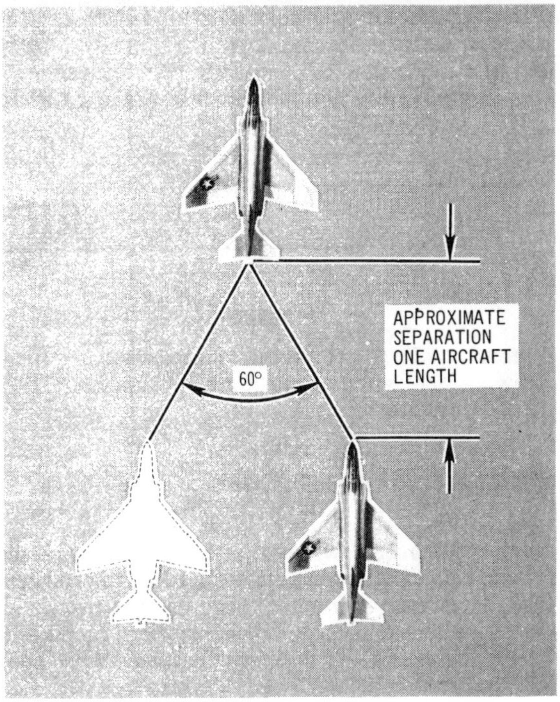

INSTRUMENT WING FORMATION

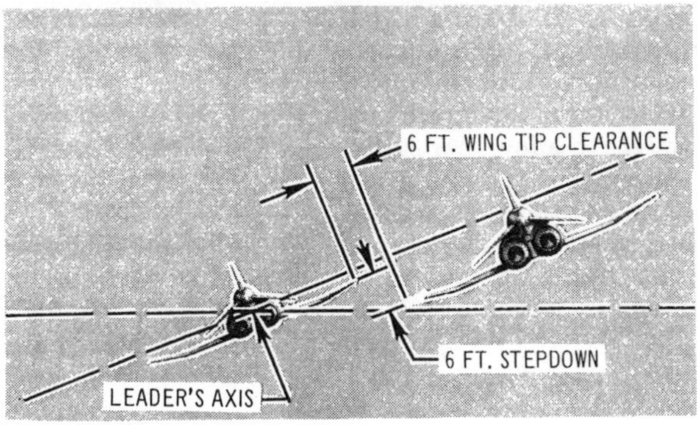

FDD-1-(63)

Figure 4-14

FREE CRUISE FORMATION

Free cruise formation allows for increased flight maneuverability and lookout coverage. The free cruise position is approximately one aircraft length nose-to-tail clearance, slightly stepped down (to avoid the leaders jet wash) and within a 60° cone aft of the leader. Power settings should be constant while maneuvering within the 60° cone.

INSTRUMENT WING FORMATION

The position for instrument wing is identical to the parade position. All turns are performed on the axis of the leader. It is important to maintain this position to avoid possible vertigo in actual instrument conditions.

AIR REFUELING

NOTE

Before air refueling operations, each crewmember will become familiar with NATOPS Air Refueling Manual.

PROBE SWITCH OPERATION

The air refuel probe switch has three positions, marked RETRACT, EXTEND, and REFUEL. Placing the switch to EXTEND merely extends the air refueling probe. However, with the switch in this position, fuselage cells 1 and 3 can accept fuel if/when space is available. Selecting REFUEL conditions the fuel system by venting all tanks to atmosphere, shutting off tank pressurization air, opening the refueling level control valves in cells 3 and 5 and the wing tanks, and by opening the fuel shutoff valves to the external tanks (providing the refuel selection switch is in the ALL TANKS). Immediately after selecting the REFUEL position, monitor the sector portion of the fuel quantity gage. If the fuel quantity shown on the sector depletes more rapidly than the fuel quantity shown on the counter, it is a positive indication that the defuel valve has failed open and that fuselage fuel is transferring into the internal wing tanks. If the wing tanks are sufficiently empty, the fuselage fuel may be depleted completely. Should this indication occur, move the refuel probe switch to the EXTEND position, and allow fuel from the tanker to replenish the fuselage fuel tanks to approximately 6,000 pounds to guard against flameout. Then move the switch back to the REFUEL position and continue the air refueling. It is also possible to deplete the fuselage fuel supply when attempting wet plug-ins. Since all tanks are vented, no internal wing or external fuel will transfer. If an excessive amount of time is used in attempting a plug-in, the fuselage cells may deplete to a low fuel state. In this event, place the air refuel probe switch to EXTEND. This will permit the internal wing and external tanks to transfer to the engine feed tank. Once the engine feed tank is full, the switch may again be placed to REFUEL and air refueling resumed.

BEFORE PLUG-IN

The air refueling checklist should be completed prior to plug in.

1. Radar function switch - STBY
2. Missile power switch - OFF (STBY with AIM-7D/E)
3. Arming switches - OFF/SAFE
4. Internal wing dump switch - NORMAL
5. Refuel probe switch - REFUEL (EXTEND for dry plug-ins)
6. Refuel selection switch - AS DESIRED
7. Check probe fully extended.
8. Check REFUEL READY light illuminated.
9. Visor recommended down.

NOTE

A failed open defuel valve is indicated by a rapid depletion of fuel after selecting the REFUEL position.

For night air refueling -

10. Exterior lights - STEADY BRIGHT
11. Air refuel probe light - ON
12. Tanker lights - AS DESIRED

REFUELING TECHNIQUE

WARNING

When receiving fuel from KC-135 tanker aircraft, the maximum total fuel on board shall not exceed 12,400 pounds of JP-4 or 13,200 pounds of JP-5. Failure to observe these limitations may result in overpressurization and rupture of the fuel tanks.

NOTE

The following procedures, as applied to tanker operation, refer only to single drogue refuelers.

Refueling altitudes and airspeeds are dictated by receiver and/or tanker characteristics and operational needs, consistent with the tanker's performance and refueling capabilities. This, generally, covers a practical spectrum from the deck to 35,000 feet and 190 to 300 KCAS.

Approach

Once cleared to commence an approach, refueling checklists completed, assume a position 10 to 15 feet in trail of the drogue with the refueling probe in line in both the horizontal and vertical reference planes. Trim the aircraft in this stabilized approach position and insure that the tanker's (amber) ready light is illuminated before attempting an approach. Select a reference point on the tanker as a primary alignment guide during the approach phase; secondarily, rely on peripheral vision of the drogues and hose and supplementary remarks by the RIO. Increase power to establish an optimum 3 to 5 knot closure rate on the drogue. It must be emphasized that an excessive closure rate will cause a violent hose whip following contact and/or increase the danger of structural damage to the aircraft in the event of misalignment; whereas, too slow a closure rate results in the pilot fencing with the drogue as it oscillates in close proximity to the aircraft's nose. During the final phase of the approach, the drogue has a tendency to move slightly upward and to the right as it passes the nose of the receiver aircraft due to the aircraft-drogue airstream interaction. Small corrections in the approach phase are acceptable; however, if alignment is off in the final phase, it is best to immediately retire to the initial approach position and commence another approach, compensating for previous misalignment by adjusting the reference point selected on the tanker. Small lateral corrections with a "shoulder probe" are made with the rudder, and vertical corrections with the stabilator. Avoid any corrections about the longitudinal axis since they cause probe displacement in both the lateral and vertical reference planes.

Missed Approach

If the receiver probe passes forward of the drogue basket without making contact, a missed approach should be initiated immediately. Also, if the probe impinges on the canopy lined rim of the basket and tips it, a missed approach should be initiated. Realization of this situation can be readily ascertained through the RIO. A missed approach is executed by reducing power and backing to the rear at an opening rate commensurate with the optimum 3 to 5 knot closure rate made on an approach. By continuing an approach past the basket, a pilot might hook his probe over the hose and/or permit the drogue to contact the receiver aircraft fuselage. Either of the two aforementioned hazards require more skill to calmly unravel the hose and drogue without causing further damage than to make another approach. If the initial approach position is well in line with the drogue, the chance of hooking the hose is diminished when last minute corrections are kept to a minimum. After executing a missed approach, analyze previous misalignment problems and apply positive corrections to preclude a hazardous tendency to blindly stab at the drogue.

Contact

When the receiver probe engages the basket, it will seat itself into the drogue coupling and a slight ripple will be evident in the refueling hose. Here again the RIO can readily inform the pilot by calling contact. The tanker's drogue and hose must be pushed forward 3 to 5 feet by the receiver probe before fuel transfer can be effected. This advanced position is evident by the tanker's (amber) ready light going out and the (green) fuel transfer light coming on. The tanker's (green) fuel transfer light will illuminate only when the receiver aircraft has the REFUEL position selected and the tanker has all required switches selected and other conditions met. While plugged-in, merely fly a close tail chase formation on the tanker. Although this tucked-in condition restricts the tanker's maneuverability, gradual changes involving heading, altitude and/or airspeed may be made. A sharp lookout doctrine must be maintained due to the precise flying imposed on both the tanker and receiver pilots. In this respect, the tanker can be assisted by other aircraft in the formation. The receiver RIO can also assist in maintaining a visual lookout since the receiver radar is in the STBY position.

NOTE

With gross weight above 48,000 pounds, when tanking in the 190-210 knot region, high power settings and one-half flaps may be required. If tobogganing is used with one-half flaps, there is a possibility of flap blow-up.

Disengagement

Disengagement from a successful contact is accomplished by reducing power and backing out at a 3 to 5-knot separation rate. Care should be taken to maintain the same relative alignment on the tanker as upon engagement. The receiver probe will separate from the drogue coupling when the hose reaches full extension. When clear of the drogue, place the refueling probe switch in the RETRACT position. Ensure that the IFR PROBE UNLOCK light is extinguished on the telelight panel before resuming normal flight operations. At night, turn off the probe floodlight.

BUDDY TANK SYSTEM

Before Takeoff

1. Buddy tank - SERVICED
   a. Filler cap - SECURE
   b. Guillotine cartridge - INSTALLED
2. Hose jettison switch - OFF
3. Power switch - OFF
4. Transfer switch - OFF
5. Hose control switch - RET
6. Light switch - AS REQUIRED
   For daylight refueling, place the light switch to BRT; for night refueling, place the light switch to DIM.
7. Buddy fill switch - STOP FILL

Air Refueling

1. Power switch - ON
2. Hose control switch - EXT
   If the airstream driven turbine is not at governed speed when the hose control switch is placed to EXT, the drogue may momentarily delay ejection into the airstream.
3. Buddy fill switch - AS DESIRED
   To transfer buddy tank fuel only, the buddy fill switch must be in the STOP FILL position. To transfer buddy tank fuel plus airplane (tanker) fuel, the buddy fill switch must be held in the FILL position.
4. Drogue position indicator - EXT

When the hose and drogue are fully extended, the amber READY light on the tail cone will illuminate, and the drogue position indicator will display EXT.

5. Fuel transfer switch – TRANS
6. Drogue position indicator – TRA
   After engagement of the drogue and the hose has been retracted a minimum of two feet, the green TRANSFER light on the tail cone will illuminate and the drogue position indicator will read TRA.
7. Gallons delivered indicator – MONITOR
   Periodically check the total gallons transferred. Do not transfer excessive amounts of internal fuel.
8. Fuel transfer switch – OFF
   Upon completion of refueling, place the buddy fuel transfer switch to OFF.
9. Hose control switch – RET
10. Drogue position indicator – RET
    When the hose and drogue is completely retracted, the drogue position indicator will read RET.
11. Power switch – OFF

BUDDY TANK EMERGENCIES

Hose and Drogue Jettisoning

A violently whipping hose and drogue, or the inability to retract the hose for any reason, may require hose and drogue jettisoning. To jettison the hose and drogue, proceed as follows:

1. Hose jettison switch – CUT

CAUTION

Do not change the position of the hose jettison switch after being placed to CUT. If the switch is positioned to NORMAL after jettisoning, the buddy tank electrical system will become energized.

NOTE

The guillotine may not fire immediately, since there is an electrical holding relay that will not close until the hose-reel mechanism is locked. Once the hose-reel mechanism locks, the guillotine hose cutter will fire.

Buddy Tank Jettisoning

The buddy tank may be jettisoned individually from the centerline station or it may be jettisoned along with all external stores. Refer to external Stores Jettison Chart, section V.

TOW TARGET PROCEDURES

TDU–22B TOW TARGET

No special procedures are required for this passive type target.

TDU–22A/B TOW TARGET

FLARE INGITION

1. Set the proper frequency on the UHF.

NOTE

It is desirable to use cold mic when transmitting a flare tone. When hot mic is used, a tone will be heard on the ICS as soon as tone is selected.

2. Select tone on the tone generator box.
3. Depress the mic button for at least 10 seconds, then release.
4. Turn off tone generator and listen for firing results.
5. If another flare is to be fired, wait at least 15 seconds between tone generator signals to allow the stepping relay in the target to actuate.

RMU–8/A REEL LAUNCHER

CAUTION

Failure to adhere to the following preflight and post-start checks and flight procedures during any flight with an RMU-8/A reel launcher installed may result in turbine overspeed and failure with extensive damage to the aircraft.

PREFLIGHT CHECK

After making sure that the proper towline/target combination is installed, place the master power switch OFF, remove the tow reel access panels and check the following:

1. Target secure.
2. Launcher boom latched.
3. Towline threaded through the cable cutter and that a cartridge is installed in the cutter and the cutter assembly lockwired.
4. Pneumatic bottle charged to 2000 psi minimum, filler cap secure.
5. Towline length limit switch assembly for desired

settings. (LAUNCH STOP: 100, IN STOP: 3,000 OUTSTOP: 36,700.)

6. Spool locks secure and lockwired.
7. Level-wind drive and adapter secure, and three cotter pins installed.
8. Towline spool winding, evenly wound with no snarls and insulated leader installed.
9. Oil level.
10. Remove power unit blade guard if used. Note the condition of the spinner, and spinner nut lockwired.
11. Check that the propeller blades have no unacceptable nicks or scratches, and that they are feathered. Check the brakes on.
12. Replace and secure the access panels.
13. Remove safety flag, placing the towline cutter switch in the spring loaded ARM position.

NOTE

For flight missions not involving target towing, if a target is not installed, accomplish all preflight items except 1, 3 and 8. In this case the TARGET OUT light burns continuously when power is applied to the system. All other preflight and post start items must be complied with in any case.

POST-START CHECK

1. Check all reel launcher circuit breakers IN.
2. Check all press-to-test indicator lights on the control panel.
3. Master power switch ON.
4. Check all the indicator lights OFF.
5. Pull RMU-8/A DC PWR circuit breaker and push IN.
6. Check EMERG PWR light ON.
7. Cycle master power switch OFF and ON.
8. Check EMERG PWR light OFF.
9. Reset towline length indicator.
10. Leave master power switch ON during flight.

TAKEOFF AND CLIMB

1. Perform normal takeoff.
2. Do not exceed 0.90 IMN, 500 knots, or the tension limit, whichever is less, with the target in the stowed position. See figure 4-15.

LAUNCHING PROCEDURES

NOTE

If Reel-Out Procedures are initiated first, Launch and Reel-Out procedures are automatically combined.

1. Maximum launch speed is 0.90 IMN, 500 knots, or the tension limit, whichever is less. Optimum launch at 250 to 300 knots, sea level to 40,000 feet altitude.
2. In wings-level flight, actuate the target launch switch to LAUNCH. When the target launch switch is actuated, the two reel brakes are released, the launching boom lowers, and the target leaves the saddle. The LAUNCHER DOWN and TARGET OUT light will then illuminate, and the target begins to deploy at approximately 400 feet per minute. When the launch cycle is completed (preset launch stop setting of approximately 100 feet of towline out), the launcher retracts, the power brakes are applied, and the LAUNCHER DOWN light goes out.

REEL-OUT PROCEDURES

1. Maximum reel-out speed is 0.75 IMN, 300 knots, or the tension limit, whichever is less. See figure 4-15.
2. Actuate the tow reel control switch to REEL-OUT. When the switch is acutated to REEL-OUT, the power brakes release, and the target begins to deploy, at an increasing speed, to a maximum of 5000 feet per minute. At the preselected automatic stop setting, the towline speed decreases and the power unit brakes are applied to stop the reel-out.

CAUTION

- To avoid towline damage due to sheave scuffing, turns should be limited to 20° bank angles while the tow reel is operating.
- To avoid towline damage due to engine exhaust with the tow reel stopped, 45° bank angle turns are desirable in basic engine. If AB is required for turns, use of outboard AB is desirable. If turns are required with both engines in AB, use of 60° bank angles is desirable.
- In the event a normal 10 percent overspeed is observed (in excess of 5500 FPM) and a normal automatic stop is not observed, actuate the stop switch and reduce airspeed to minimum feasible. If overspeed continues, immediately actuate stop and cut switch. If the overspeed persists, jettison the RMU-8/A reel launcher.

NOTE

- If the tow reel should overspeed by 10 percent, a normal stop will occur.
- If the tow reel should overspeed by 20 percent the power unit and tow reel brakes are immediately applied and the towline cutter is acutated. The TOW CUT light will then illuminate.
- Following a manual STOP command after launch, and prior to automatic stop, REEL-IN or REEL-OUT procedures can be initiated. However, following the automatic stop function, only REEL-IN procedures can be initiated.

TARGET PRESENTATION

1. The target can be offered for tracking or live firings within the performance limits of the target.
2. Towline length may be altered; however, its length may never exceed the automatic stop setting (plus a stopping distance of 1,500 to 2,500 feet).
   a. If a stop has been commanded prior to reaching the automatic stop setting, the automatic stop length may be reached by selecting REEL-OUT. If

RMU-8/A LIMITATIONS

TOWING LIMITATIONS

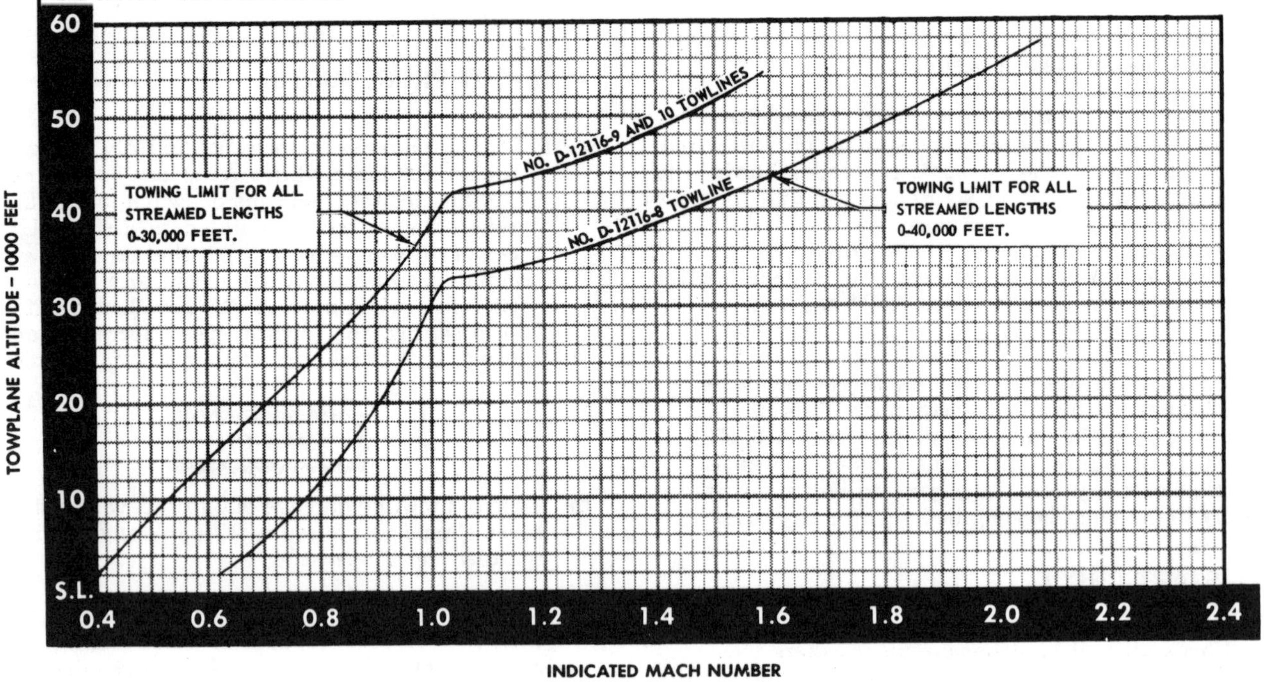

TENSION LIMITATIONS

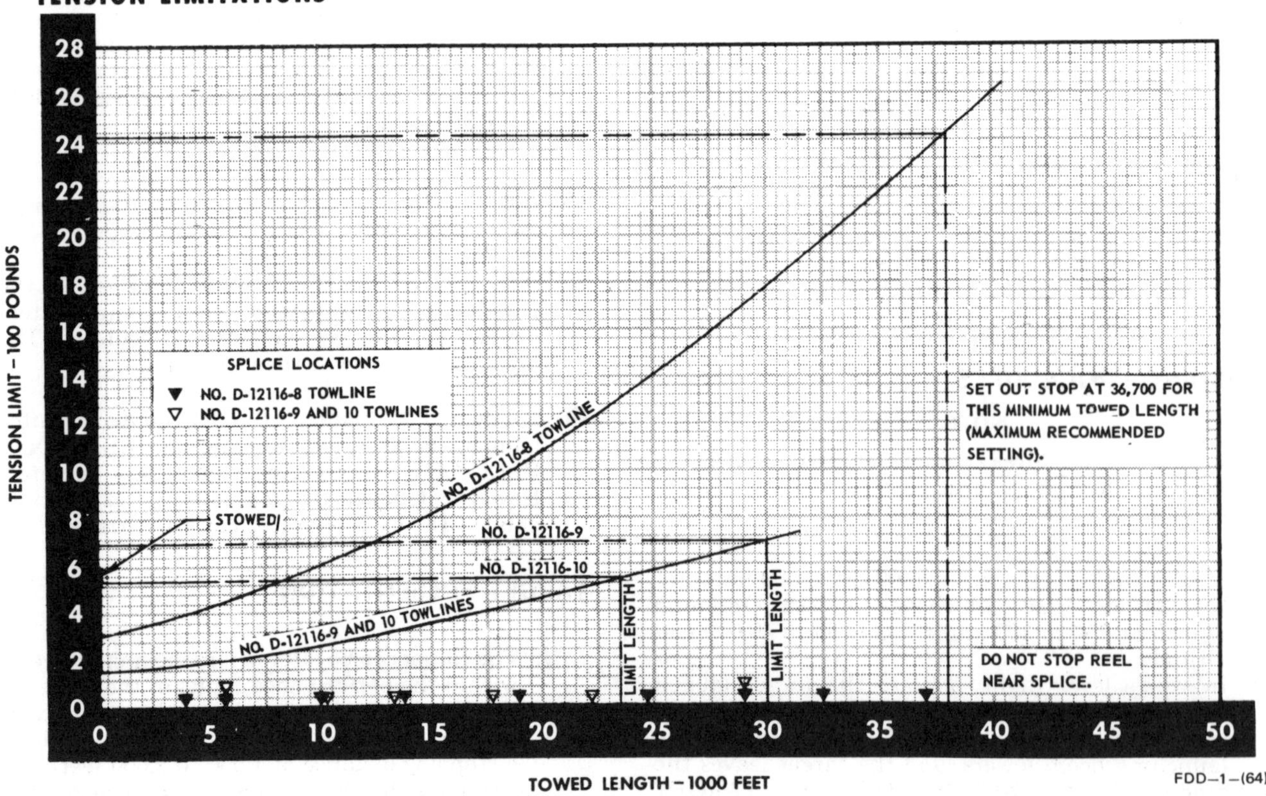

FDD–1–(64)A

Figure 4–15

a length less than the automatic stop length is desired, the tow reel control switch should be actuated to STOP.

NOTE

When commanding a stop prior to automatic stop, anticipate a stopping distance of 1500 to 2500 feet maximum, depending on reel-out speed.

b. If a length less than the automatic stop length is desired, actuate the tow reel control switch to REEL-IN.

NOTE

When commanding a stop during reel-in, anticipate a stopping distance of 1500 to 2500 feet maximum, depending on reel-in speed.

REEL-IN PROCEDURES

1. Maximum reel-in speed is 0.75 IMN, 300 knots, or the tension limit, whichever is less. Anticipate a tension increase due to 0.1 ΔM reel-in speed. See figure 4-15.
2. Actuate the tow reel control switch to REEL-IN. When the switch is actuated to REEL-IN, the SAFETY-ARM light illuminates, the power unit brakes are released. Reel-in speed increases to a maximum of 5000 feet per minute until the automatic stop setting is reached at approximately 3000 feet. The power unit brakes are applied and the target stops 500 to 2000 feet aft of the tow plane.

CAUTION

It is desirable to stabilize aircraft speed for cruise or descent prior to initiating reel-in procedure in order to reduce chance of towline failure.

NOTE

- Following reel-in command, some towline may reel-out until sufficient power is developed to reel-in the towline.
- Following a stop command prior to automatic stop, reel-in or reel-out can be initiated.
- Following an automatic stop, only recovery or reel-out can be initiated.

RECOVERY PROCEDURES

1. Maximum recovery airspeed is 0.85 IMN, 350 knots, or the tension limit, whichever is less. Optimum recovery occurs at 250 knots, 15,000 feet altitude, 180 feet per minute.
2. Actuate the target launch switch to RECOVER. The SAFETY-ARM light goes out, the brakes are released, the launcher extends, and the LAUNCHER DOWN light illuminates.

NOTE

- Reel-in speed can be varied approximately 1000 feet per minute, by actuating the tow reel control switch to INCREASE SPEED and/or FEATHER.
- Manual speed control is required to recover the target.
- Increase speed is restricted to 0.5° of pitch per actuation. The feather switch operates while held until feather pitch position is reached.
- For best recovery, final reel-in speed should be established before target is within 200 feet of aircraft.

3. When the target seats in the saddle, the TARGET OUT light extinguishes. The launcher then retracts to stow the target, and the LAUNCHER DOWN light extinguishes. The power unit blades feather and the brake is applied. Automatic stop length limits reset.

CAUTION

- Abort recovery if tension or speed fluctuation occurs. To abort, actuate stop switch, hold reel-out switch until reel-out occurs, reactuate stop switch at 500 feet. Repeat recovery procedure.
- When the recovery switch is actuated after stopping from reel-out mode, the increase speed switch must be actuated to energize the feather circuit.

EMERGENCY OPERATION

WARNING

If it becomes necessary for the crew to eject from the aircraft during towing operations, actuate tow reel stop switch to STOP & CUT prior to ejection.

NOTE

The RMU-8/A is a jettisonable store in all flight configurations up to 0.95 Mach.

LOW AIR LIGHT ILLUMINATED

If based ashore -

1. Initiate reel-in and/or recovery procedure.

If based aboard ship -

1. During launch or recovery procedure, actuate the tow reel control switch to STOP & CUT.
2. During reel-in or reel-out procedure, complete the mission. When reel-in procedure is complete, actuate the tow reel control switch to STOP & CUT.

TOW CUT LIGHT ILLUMINATED

No action required.

EMERGENCY POWER LIGHT ILLUMINATED

Wait until towline speed instrument indicates that the tow reel has stopped. Turn the master power switch to OFF. Determine the cause of the airplane power failure and follow the electrical system emergency operation procedure. Place the master power switch to ON. If the power trouble is corrected, continue operations. Anticipate a towline length instrument error due to use of emergency power.

NOTE

Actuation of the emergency power circuit may be caused by aircraft generator voltage transients.

SAFETY ARM LIGHT FAILS TO ILLUMINATE

Actuate the tow reel stop switch to STOP when the towline length instrument indicates 03500. Initiate Recovery Procedure.

TOWLINE LENGTH INDICATOR ERROR

Use the launch stop, reel–out automatic stop, and reel–in automatic stop settings to determine target position.

TOWLINE FAILURE

Actuate the tow reel stop switch to STOP & CUT.

TARGET OUT LIGHT PRIOR TO LAUNCH

If towline tension meter indicates zero, do not launch target. If normal tension is indicated, Reel–Out Procedure is required to launch the target.

LAUNCHER LIGHT DOWN ILLUMINATED

1. If the target is at the launch stop position, complete the mission.
2. After target recovery, and if based ashore with chase airplane, check to assure that target is inside clamps. If the target is inside the clamps, actuate the tow reel control switch to INCREASE SPEED to increase towline tension. Do not exceed towline tension limits. Then place the master power switch to OFF. If the target is outside the clamp, actuate the tow reel stop switch to STOP & CUT.
3. After target recovery, and if based aboard ship, or if a chase airplane is not available, actuate STOP, REEL–OUT, and STOP & CUT at 3 second intervals.

NOTE

Field landings are permitted with the launcher down, with or without target. Minor damage to the target will occur.

PREMATURE FLARE IGNITION (TARGET STOWED)

Actuate reel control switch to REEL–OUT. If aircraft is on the ground, actuate tow reel stop switch to STOP & CUT and taxi clear of burning flare and target.

AQM–37A MISSILE TARGET SYSTEM

PREFLIGHT CHECK

1. Ascertain from ground crewman that three target and launcher safety pins are installed.
2. Ascertain from ground crewman that centerline stores rack safety pin is removed.
3. Centerline station safe switch – SAFE.
4. External stores emergency release button – NORMAL
5. Bomb control switch – OFF
6. Master arm switch – SAFE.
7. Target function selector switch – OFF.

POST–START CHECK

1. Launcher no–voltage check performed by ground crewman.
2. Ascertain that target safety pin removed by ground crewman.
3. Ascertain that the launcher safety pins (2) are removed by ground crewman.

TARGET LAUNCH PROCEDURE

During Climb–out to Launch Altitude –

1. Target function selector switch – SELECT.
2. Target PWR ON light – Illuminated.
3. Target READY light – Illuminated (approximately 5 minutes after PWR ON light illuminates).

NOTE

- If the PWR ON light does not illuminate, place the target function selector switch to the OFF position and execute mission abort procedures.
- The target READY light indicates only that power has been applied to the target for at least 5 minutes. If the target READY light does not illuminate, check lamp illumination with the press–to–test feature. Target launch may be executed without a READY light at the pilot's discretion.

At Launch Altitude –

4. Launch altitude – CHECK
5. Launch speed – CHECK
6. Launch heading – CHECK
7. Centerline station safe switch – SAFE
8. Bomb control switch – OFF
9. Centerline station safe switch – READY

10. Bomb control switch – DIRECT
11. Master arm switch – SAFE
12. Bomb release button – Depress until launch is confirmed (PWR ON and READY lights extinguish).

NOTE

During the actual target launch operation, a slight thump will be felt as the simultaneous action of launcher extension and target/aircraft separation occur. Immediately following this separation, it is considered standard practice for the pilot to initiate a hard–to–port turn. The pilot's signal for initiating this turn will be when the PWR ON and READY lights on the missile firing panel are seen to extinguish, thus indicating that target launch has occurred.

13. Place all switches to OFF and SAFE.

GROUND CHECKS (AFTER FLIGHT)

1. Centerline stores rack safety pin – INSTALLED.

EMERGENCY OPERATION

PWR ON LIGHT FAILS TO ILLUMINATE

In Flight –

1. Target function selector switch – OFF
2. Bomb control switch – OFF
3. Centerline station safe switch – SAFE
4. Master arm switch – SAFE

After Landing (Land or Carrier Based Operations) –

5. Target safety pin – INSTALLED
6. Launcher safety pins (2) – INSTALLED
7. Centerline stores rack safety pin – INSTALLED

JETTISON PROCEDURES

Master Jettison (All External Stores Including Target, Launcher, and Adapter)

1. External stores emergency release button – DEPRESS

Centerline Station Jettison (Jettisons Target Only)

1. Target function selector switch – JETTISON

SECTION V
EMERGENCY PROCEDURES

TABLE OF CONTENTS

GENERAL

This section contains procedures to be followed to correct an emergency condition. These procedures will ensure maximum safety for the crew and/or aircraft until a safe landing or other appropriate action is accomplished. Multiple emergencies, adverse weather and other peculiar conditions may require modification of these procedures. The mandatory items (ALL CAPITAL LETTERS) contained in the various emergency procedures cover the most adverse conditions. The nature and severity of the encountered emergency will dictate the necessity for complying with the mandatory items in their entirety. It is essential, therefore, that aircrews determine the correct course of action by use of common sense and sound judgement. As soon as possible, the pilot should notify the RIO, flight/flight leader, and tower of any existing emergency and of the intended action. When an emergency occurs, three basic rules are established which apply to airborne emergencies. They should be thoroughly understood by all aircrews:

1. Maintain aircraft control.
2. Analyze the situation and take proper action.
3. Land as soon as practicable.

When an airborne emergency occurs and the flight conditions permit, the pilot or RIO should record and/or broadcast all available information such as airspeed, altitude, power settings, instrument readings and fluctuations, warning lights illuminated, loss of thrust and unusual noises. Flight leaders, wingmen, other pilots, or any ground station receiving such information should copy it and forward it as soon as possible to the cognizant activity. Wingmen should also record their observations of vapor, smoke, flames or other phenomena. Whenever possible, an effort should be made to escort an aircraft with a declared emergency until it is safely landed. This escort will observe the distressed aircraft for any external indications or symptoms of the problem to provide assistance or advice that may be required and to assist in a SAR effort if required. Throughout section V all reference to AOA will be either ON SPEED, minimum control airspeed, or 17 units as applicable to the emergency.

WARNING

In the troubleshooting of a system discrepancy or in the accomplishment of an emergency procedure, the operation of a system control (such as flap lever, throttle, flight control, electrical control switch, etc.) is usually required. Due to the nature of some failures, and/or the occurrence of successive malfunctions, some control operations may occasionally result in undesirable aircraft responses, such as unexpected roll or pitch motions, smoke, unstable engine operation, etc. Often the most prudent action to take to eliminate such an undesirable response is to immediately return the operated control to its former setting. The pilot must be mentally conditioned to take that action promptly when appropriate.

AIR CONDITIONING/PRESSURIZATION

COCKPIT OVERPRESSURIZATION

1. Emergency vent knob - PULL

WARNING

Do not attempt to open the canopy by the normal method if the cockpit is overpressurized since this may cause the canopy to lift and separate from the aircraft and fall on the banana links resulting in an inadvertent ejection.

If vent knob does not relieve pressure –

2. Cockpit heat and vent circuit breakers – PULL (L9, M9 before AFC 388 – M9, N9 after AFC 388, No. 2 panel)

COCKPIT TEMPERATURE FAILURE

If the cockpit temperature remains full hot despite efforts to moderate the incoming air, the following corrective measures should be accomplished in sequence. Proceeding with the succeeding step only if the preceding effort fails to provide a tolerable environment. Even though corrective measures provide satisfactory results, land as soon as practicable.

1. Temp control auto–manual switch – MANUAL COLD (Hold for 5 seconds)
2. Cockpit heat and vent circuit breakers – PULL (L9, M9 before AFC 388, M9, N9 after AFC 388, No. 2 panel).
3. Emergency vent knob – PULL (if desired)
4. Descend and reduce airspeed below flap blowup speed and lower flaps full down to utilize max BLC air.
5. Bleed air sw – OFF (some aircraft)
6. Jettison aft canopy.

EQUIPMENT COOLING TURBINE FAILURE

A malfunction of the equipment cooling turbine is evidenced by a high pitched whine and/or vibration forward and outboard of the pilot's left foot. Delay in securing the turbine can lead to disintegration and possible pilot injury. The turbine may be secured by the steps listed below. This, in turn, shuts off equipment air conditioning and turns on emergency ram air cooling, but does not affect fuel tank pressurization and transfer, anti–G suits, etc.

To secure the equipment cooling turbine before AFC 440 –

1. Ram air turbine – EXTEND
2. Generators – OFF
   Turn off both generators to secure the cooling turbine.
3. Equipment cooling circuit breakers – PULL (L8, N8 before AFC 388; M8, N8 after AFC 388; No. 2 panel).

After circuit breakers are pulled –

4. Generators – ON
5. Ram air turbine – RETRACT

To secure the equipment cooling turbine after AFC 440 –

1. Bleed air switch – OFF
2. Equipment cooling circuit breakers – PULL (L8, N8 before AFC 388; M8, N8 after AFC 388; No. 2 Panel)

After circuit breakers are pulled –

3. Bleed air switch – NORM

AUXILIARY AIR DOOR(S)

GENERAL

The auxiliary air door warning light(s) illuminate any time the auxiliary air doors are out of phase with the landing gear handle (landing gear up, aux air door(s) open, or landing gear down, aux air door(s) closed). Neither one of these situations constitutes a safety of flight condition if the problem is recognized and the appropriate action is taken. In the event of either a malfunction or failure, the appropriate corrective procedures must be utilized.

AUX AIR DOOR MALFUNCTION (GEAR UP, DOOR(S) OPEN)

1. Reduce power and slow to 250 knots or below.
2. Landing gear – CYCLE

If light(s) remain on –

3. Close aux air door(s) by pulling the applicable circuit breaker(s).
   a. Aircraft before AFC 388 (3D and 4D No. 1 panel)
   b. Aircraft after AFC 388 (15K and 16K No. 1 panel)

CAUTION

If the arresting hook is lowered or the hook bypass switch is in the bypass position, pulling the right aux air door circuit breaker will cause the corner reflector to extend. Airspeed should be reduced to 250 knots prior to lowering hook or selecting hook bypass.

If light(s) still remain on –

4. Maintain no more than cruise power setting.
   A cruise power setting will avoid engine compartment overheating.

CAUTION

Extended engine operation at high power settings will result in engine compartment and aft fuselage overheating.

5. Reset aux air door circuit breaker(s) prior to lowering landing gear.

AUX AIR DOOR FAILURE (GEAR DOWN, DOOR(S) CLOSED)

If the auxiliary air doors fail to open when the landing gear is lowered, there is a possibility that the engines may automatically accelerate up to 100% rpm. A utility hydraulic system failure or double generator failure will render the variable bypass bellmouth and auxiliary air doors inoperative. Operation of an engine with an open variable bypass bellmouth and closed auxiliary air door will allow engine compartment secondary air to recirculate to the engine compressor inlet. During low altitude or ground operation, the temperature of the recirculating air may be high enough to initiate T2 reset through normal detection by the compressor inlet temperature sensor. As T2 reset occurs, it increases the engine idle speed to maintain proper engine airflow and thrust under high temperature conditions and can cause the idle speed to increase up to 100% rpm. The auto–accelerated engine can be shut down, if on the ground, by placing the throttle to OFF. If a false reset occurs while airborne, a near normal landing can be made by modulating the exhaust nozzles of the affected engine(s). If aux air doors fail to open when the landing gear is lowered:

1. Landing gear – CYCLE

AUTO–ACCELERATION OF ONE ENGINE

1. Throttle of affected engine – IDLE
2. Make an ON SPEED approach
   Modulate throttle of unaffected engine for desired thrust. Under no conditions will the combined thrust of the auto–accelerated engine in idle, and the unaffected engine in idle be in excess of that required to make an optimum ON SPEED approach.
3. At touchdown, affected engine – SHUTDOWN

AUTO–ACCELERATION OF BOTH ENGINES

1. Throttle of either engine – IDLE
2. Modulate the exhaust nozzle of the remaining engine for desired thrust.
3. Make an ON SPEED approach.
4. At touchdown, right engine – SHUTDOWN
   Shutting down the right engine at touchdown will leave you in a more favorable position (more systems available) in the event of a bus tie failure.

BLEED AIR SYSTEM

BLEED AIR DUCT FAILURE

Severe damage to the aircraft may result from a bleed air duct failure due to the high temperature produced by the bleed air system. The extremely hot air leaking from a failed duct may ignite flammables in the immediate vicinity of the duct failure. The following symptoms may be indicative of a bleed air system failure: A mild audible thump or bang on the airframe, leading edge barber pole flap indication throughout all flap positions, complete or partial loss of cockpit pressurization; loss of pylons, missiles or other external stores; generator failure, popping of circuit breakers and illumination of fire/overheat warning lights, bleed air overheat warning lights (aircraft 155903ap, and 157274ap and up, and all others after AFC 439) with the MASTER CAUTION lights and/or several other warning/indicator lights, erratic fuel quantity indications; mild stick transients; stiffness of throttles; hydraulic failure; smoke emitting from the intake duct louvers, fuel fumes in the cockpit; high fuel flow/erratic response to throttle movement (indicative of main fuel hose rupture).

CAUTION

Early analysis of a bleed air duct failure is required in order to prevent serious damage to or possible loss of the aircraft.

If several of the preceding symptoms occur in close sequence –

Aircraft before AFC 440 –

1. Reduce power to lowest practicable setting.
2. Check for indications of fire.

If circumstances permit –

3. Lower flaps.
4. Rain removal – ON
5. Land as soon as practicable.

Aircraft after AFC 440 –

1. Bleed air switch – OFF
   With the engine bleed air switch off, operation of the following equipment will be lost.
   a. External fuel and internal wing fuel transfer
   b. Leading edge BLC
   c. Cockpit air conditioning and pressurization
   d. Rain removal air
   e. Defog/Foot heat
   f. Equipment cooling air
   g. Normal pneumatic compressor charging
   h. SPC inputs to all systems
   i. Automatic altitude reporting

NOTE

With the engine bleed air switch in OFF, 17th stage air is not available for pneumatic compressor charging; however, the pneumatic compressor will compress ram air (the rate of charging is dependent on altitude and airspeed).

2. Secure radar and CNI equipment, unless safety of flight/operational necessity dictates otherwise.

If bleed air light(s) go out –

3. Check bleed air overheat detection system for continuity.

If bleed air light(s) stay on –

4. Lower flaps, check for light(s) out.

CAUTION

If fuel considerations preclude flaps down bingo, raise flaps. If bleed air overheat light(s) reilluminate, the individual situation (evidence of fire/overheat damage, etc) must dictate whether to attempt flaps up bingo with light(s) illuminated.

5. Rain removal – ON
6. Check for indications of fire.

If indications of fire or trailing edge bleed air duct failure exists –

7. Affected engine – THROTTLE OFF, MASTER SWITCH OFF

NOTE

If indications of fire exist, or if bleed air duct failure can be isolated to one engine bay, all trailing edge bleed air to the affected side can be shut off by shutting down that engine.

8. Land as soon as practicable.
9. Refer to Landing With Bleed Air Switch Off (No Leading Edge BLC)

BLEED AIR CHECK VALVE FAILURE

No indication of a bleed air check valve failure will be noted in flight until the throttle is retarded and then readvanced on the engine with the failed bleed air check valve. If the throttle has been retarded and then readvanced, either rpm will hang–up or minor compressor stalls and flameout may occur at approximately 85% rpm. If a flameout occurs, a restart can be made, but rpm will probably not go above 65%, EGT will rise to approximately 625° and the nozzle will go full open. In either case, the engine can be regained as follows:

1. Normal operating engine – IDLE
   Idling the normal operating engine will equalize the pressure in the bleed air line.
2. Accelerate the engine with the failed bleed air check valve.
3. Accelerate normal engine.
   The normal operating engine should not be accelerated to or operated at, a rpm greater than that of the affected engine for the remainder of the flight.

BOUNDARY LAYER CONTROL

BLC MALFUNCTION

A boundary layer control system malfunction is indicated by the illumination of the BLC MALFUNCTION warning light. The only type of malfunctions indicated by the light are: a leading edge BLC valve stuck open when the flaps are up, or a trailing edge BLC valve stuck open when the flaps are in any position other than full down. If the BLC MALFUNCTION light flickers or illuminates when flaps are raised, follow the procedures for BLC Malfunction.

1. Flaps – DOWN
2. Maintain airspeed consistent with gross weight and flap blow–up speed.
3. Land as soon as practicable.

CAUTION

- Operating at normal power setting in excess of 30 seconds with the BLC MALFUNCTION light illuminated and flaps up or 1/2 may result in damage to the warning circuit wiring which will extinguish the warning light. Continued flight with the flaps up can then result in additional

APPROACH SPEEDS

– GROSS WEIGHT 36,000 POUNDS –

Note

ADJUST THE APPROACH SPEED 2 KNOTS FOR EACH 1000 POUNDS ABOVE OR BELOW 36,000 POUNDS

NORMAL FLAP OPERATION

– WITH AILERON DROOP –

– BEFORE AFC 388 –

FLAP POSITION	INDICATED ANGLE OF ATTACK	APPROACH SPEED
DOWN	19	138
1/2	19	148
1/2	17	157
UP	17	178
UP	19	165

– AFTER AFC 388 –

FLAP POSITION	INDICATED ANGLE OF ATTACK	APPROACH SPEED	
		COR REFL OUT	COR REFL IN
DOWN	18.3	138	141
1/2	18.3	149	151
1/2	17	155	157
UP	17	175	178
UP	18.3	166	169

EMERGENCY FLAP OPERATION

– WITHOUT AILERON DROOP –

BEFORE AFC 534 OR AFTER AFC 534 WITH THE EMERGENCY AILERON DROOP SWITCH IN DISABLE

1/2	19	154
1/2	17	164

– WITH AILERON DROOP –

AFTER AFC 534 WITH EMERGENCY AILERON DROOP SWITCH IN NORMAL AND ALSO AFTER AFC 388

1/2	18.3	149	151
1/2	17	155	157

– WITHOUT AILERON DROOP –
AFTER AFC 388

1/2	18.3	156	158
1/2	17	164	167

NORMAL FLAP OPERATION

WITH CB PULLED TO DEFEAT ASYMMETRIC DROOP

– BEFORE AFC 388 –

CIRCUIT BREAKER J12 NO.1 PANEL PULLED

DOWN	19	146
1/2	19	154

– AFTER AFC 388 –

CIRCUIT BREAKER H15 NO.1 PANEL PULLED

DOWN	18.3	145	148
1/2	18.3	156	158

IF A BLC LOSS OCCURS IN CONJUNCTION WITH ANY FLAP CONFIGURATION DEPICTED ABOVE, THE AIRSPEED FOR ANY GIVEN AOA WILL INCREASE AS FOLLOWS:

7 KNOTS – LEADING EDGE BLC ON ONE SIDE ONLY.
7 KNOTS – TRAILING EDGE BLC ON ONE SIDE ONLY.
18 KNOTS – LEADING AND TRAILING EDGE BLC ON ONE SIDE ONLY.

FDD–1–(316)A

Figure 5–1

electrical, hydraulic, and structural damage to the wing from overheating. Hot BLC air can melt the insulation off the wiring to the flap up limit switch allowing the bare wire to short to ground. This short results in the same indication as a flaps down condition to the WHEELS light circuit, thereby illuminating the WHEELS light. In view of the above, a steady or flashing WHEELS light with the landing gear handle and flaps up, or the gear handle DN and the flaps up or 1/2, should be treated as a BLC malfunction.

- If the BLC valve fails internally, a BLC malfunction is possible with no BLC MALFUNCTION light.

If fuel state precludes flaps down bingo –

1. Flaps – DOWN
2. BLC MALFUNCTION light and EGT – MONITOR
3. Flaps – 1/2

If BLC MALFUNCTION light comes on with 1/2 flaps –

4. Engine on affected side – SHUT DOWN
   The affected side is the side which registers little or no drop in EGT when raising the flaps to 1/2. The EGT on the unaffected side will decrease 10° to 15°C.
5. Flaps – UP
6. Follow Single–Engine Bingo procedure.

CAUTION

Single–engine bingo fuel required is higher than for two–engine bingo. Bingo charts must be referenced for the particular situation.

If BLC MALFUNCTION light remains out with 1/2 flaps –

4. Bleed air switch – OFF
   With the bleed air switch OFF the following systems will be lost: cockpit air conditioning and pressurization, rain removal, equipment air conditioning, external fuel and internal wing fuel transfer, pneumatic system charging, leading edge BLC, SPC and automatic altitude reporting.
5. Flaps – UP
6. Follow Landing With Bleed Air Switch Off procedure.

BLC FAILURES

A boundary layer control (BLC) system failure affects the handling characteristics and approach speeds of the airplane. The (BLC) system failure usually does not affect the complete BLC system, but rather a portion of the system, and will probably be one of the following variations:

a. Trailing edge BLC inoperative on one side.
   This condition is characterized by a moderate roll when the flaps reach full down. Trim requirements vary only slightly as speed is reduced.
b. Leading edge BLC inoperative on one side.
   This condition is characterized by little or no lateral trim requirements when flaps are first lowered, followed by increased trim requirements as airspeed is reduced.
c. Leading and trailing edge BLC inoperative on the same side.
   With this condition, the initial flap down lateral trim requirements of the trailing edge is first apparent, followed by increased trim as airspeed decreases. Full lateral trim will be required well above approach airspeed.

The BLC failure will probably occur prior to or in the transition to flaps down during a landing approach, with the result being an asymmetric BLC condition. The asymmetric BLC condition has been found to be safe and easily controllable even with both leading edge and trailing edge BLC inoperative on the same side. There is no reason to raise flaps to avoid an asymmetrical BLC condition. In general, for these types of BLC failure, an on speed angle of attack gives satisfactory approach control. BLC loss on both wings presents a different problem in that lateral control may be lost at on speed AOA. For these conditions, refer to Landing With Bleed Air Switch Off (No Leading Edge BLC), this section.

1. Retrim airplane.
2. Fly on speed angle of attack.
3. Land as soon as practicable.
   Refer to Approach Speed chart, this section.

CANOPY

CANOPY MALFUNCTION

If the front or aft CANOPY UNLOCK light remains on after an attempted closure, the canopy actuator shear pin may have failed, and the following procedures will apply:

1. Do not actuate the normal canopy control selector, raise or lower the ejection seat, or allow external canopy control buttons to be operated.
2. Call for an external visual inspection of the ejection seat and canopy rigging, checking especially for foreign objects lodged between the ejection seat and canopy.

If no discrepancy is found or external inspection impossible –

3. Attempt to operate manual canopy release handle to full aft position. Apply approximately 20 pounds force.

If canopy unlocks –

4. Push canopy open and evacuate with extreme caution while staying clear of canopy control lever.

If canopy does not unlock –

5. Actuate normal canopy control lever. Further attempts to relock the canopy are not recommended.

CANOPY UNLOCK LIGHT ON INFLIGHT

1. Power – REDUCE
2. Begin descent below 25,000 feet
3. Tighten mask and lower visor
4. Emergency vent knob – PULL
5. Ejection seat – LOWER
6. Normal canopy control levers – CHECK CLOSED
7. Land as soon as practicable

CHARTS

The charts and illustrations in this section contain procedures and information which complements the text in the rest of the section. These charts should be used as necessary, in conjunction with the text, in order to analyze and judiciously perform the emergency procedures in an expeditious manner.

EJECTION

GENERAL

Escape from the airplane in flight should be made with the ejection seat (figures 5-2 thru 5-6). For most conditions requiring ejection, it is the pilot's responsibility to give the Eject command and then remain with the airplane until the RIO has ejected. In the event of a controlled ejection where the time required to leave the aircraft is not of an immediate nature, it is recommended that individual ejection be performed. A rear cockpit non-sequenced ejection (command selector valve handle in the closed position) followed by a front cockpit ejection has several advantages over pilot/RIO initiated sequenced ejection.

a. Individual ejection eliminates the possibility of a sequencing system failure.
b. If the rear system fails to function properly under individual ejection, the pilot is available to assist the RIO by using the Ejection Seat Failure procedures.
c. Since each crewmember initiates his own ejection, he can properly position his body in preparation for the ejection shock and is aware of exactly when the ejection will commence.

Should the front canopy or both canopies be lost, the front canopy interlock block will also be lost. If ejection is then initiated from the front seat, this could expose the rear crewman to the front seats' rocket blast, and if conditions are right, a collision between seats could result. Should loss of the front canopy or both canopies occur, the rear crewman should be ordered to eject individually, allowing the front crewman to eject once the rear seat leaves, or have the rear crewman select command ejection by actuation of the command selector valve, and eject both seats by automatic sequencing. The automatic sequence when initiated from the aft seat with the front canopy missing is not affected by the fact that the front interlock block is gone. With loss of the rear canopy only, normal sequenced ejection can be initiated from either cockpit. If at any time during an emergency, especially with loss of intercom, the RIO believes that the condition of the aircraft has reached or passed extremes, he must use his own judgement in ejecting. Should a situation arise where it would be desirable for the RIO to eject singly, the command selector valve handle in the rear cockpit must be in the vertical (closed) position. RIO initiated ejection of the pilot shall be limited to emergency/combat situations, when so directed by the pilot or in the event of pilot incapacitation. For catapult shots, the pilot may elect to have the RIO initiate crew ejection under certain emergency conditions such as obvious insufficient flying speed or other situations wherein pilots tasks may impede the pilots timely initiation of crew ejections. In exercising this option, the pilot shall consider the experience level of his RIO, the degree of training/proficiency and meticulously brief on ejection signals (ICS and visual) and the exact circumstances under which the RIO will eject the crew. It is vital that all pilots continuously keep the RIO informed during normal flight as well as in emergency conditions. The following signals are used by the pilot to order the RIO to eject.

a. DAY or NIGHT – WITH INTERCOM – WITH/WITHOUT EJECT LIGHT.
   Pilot actuates EJECT light and verifies light with an EJECT transmission. Without eject light, pilot transmits "eject".
b. DAY – NO INTERCOM – WITH EJECT LIGHT
   Pilot actuates EJECT light and verifies light by repeatedly striking left side of his canopy.
c. NIGHT – NO INTERCOM – WITH EJECT LIGHT
   Pilot actuates EJECT light and verifies light by moving his flashlight in a vertical motion over his left shoulder.
d. DAY – NO INTERCOM – NO EJECT LIGHT
   Pilot will strike the left side of his canopy repeatedly.
e. NIGHT – NO INTERCOM – NO EJECT LIGHT
   Pilot will move his flashlight in a vertical motion over his left shoulder.
g. DAY or NIGHT – NO RIO RESPONSE
   Initiate normal ejection sequence.

The study and analysis of escape techniques by means of the ejection seat reveals that:

a. Ejection at airspeeds ranging from stall speed to 400 knots results in relatively minor forces being exerted on the body, thus reducing injury hazard.
b. Appreciable forces are exerted on the body when ejection is performed at airspeeds of 400 to 600 knots rendering escape more hazardous.
c. At speeds above 600 knots, ejection is extremely hazardous because of excessive forces on the body.

When circumstances permit, slow the airplane prior to ejection to reduce the forces exerted on the body. The emergency restraint release handle should never be actuated before ejection for the following reasons:

a. Actuating the emergency restraint release handle creates a hazard to survival during uncontrollable flight, since negative G forces may prevent the crew from assuming the correct ejection position. A full understanding of the particular situation must be established between crewmembers so that there is no erroneous or time consuming activity.

b. Actuating the emergency restraint release handle creates a hazard to survival if the pilot decides that he has insufficient altitude for ejection and is required to proceed with a forced landing. Once the emergency restraint release handle has been pulled, the lap belt shoulder harnessing is released and cannot be refastened in flight.

c. Actuating the emergency harness release handle prior to ejection causes the occupant to separate from the seat immediately after ejection, and severe shock loads will be imposed on the body.

When it is necessary to perform the actual ejection, the final judgement of which ejection handle to use is left to the individual crewmember. There are circumstances which dictate the use of the lower ejection handle. The circumstances are: high G forces on the aircraft, bulky flight equipment worn by the crewmember, and narrow

MINIMUM EJECTION ALTITUDE vs. SINK RATE

AIRPLANES THRU 155902am BEFORE ACC 176 (SKYSAIL CHUTE)

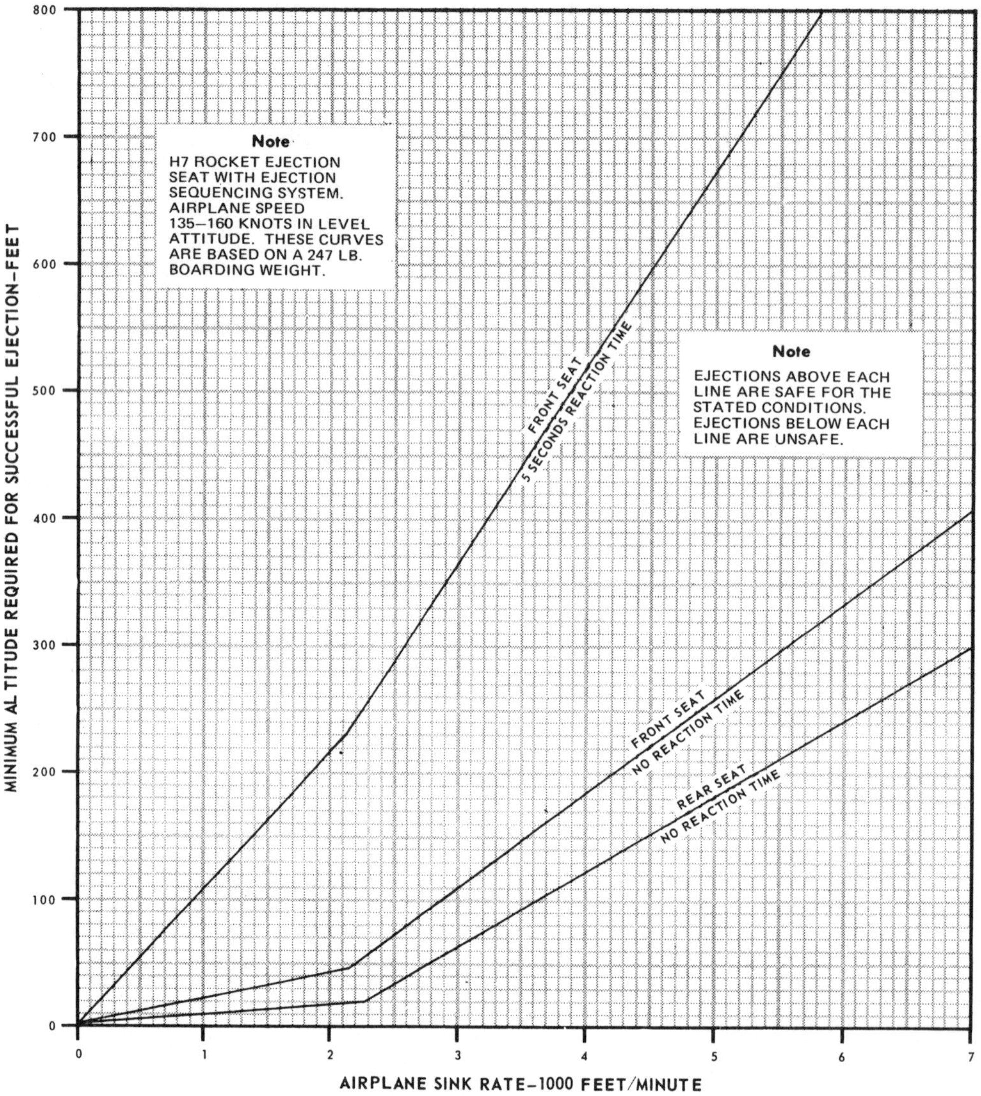

FDD–1–(73)B

Figure 5–2

clearances between the crewmember's helmet and the canopy due to seat adjustment or crewmember size. These factors could make actuation of the face curtain handle difficult if not sometimes impossible.

LOW ALTITUDE EJECTION

Low altitude ejection must be based on the minimum speed, minimum altitude and sink rate limitations of the ejection system (figures 5–2 thru 5–5). Figures 5–2 and 5–3 show minimum ejection altitude for a given sink rate, and figures 5–4 and 5–5 show minimum ejection altitude for a given airspeed and dive angle, such as encountered in a dive bombing run. Although these figures indicate minimum ejection altitudes based on seat capability and a representative pilot reaction time, the ultimate decision as to which altitude to eject must be made by the pilot. The minimum ejection altitude charts are based on a 247 pound boarding weight which is defined to include the crewman, his clothing, and personnel equipment, excluding his parachute and seat pan survival kit. Ejection at low altitudes is facilitated by pulling the nose of the airplane above the horizon (zoom up maneuver). This maneuver affects the trajectory of the ejection seat providing a greater increase in altitude than if ejection is performed in a level flight attitude. This gain in altitude increases time available for seat separation and deployment of the personnel parachute. Ejection should not be delayed when the aircraft is in a descending attitude and cannot be leveled out. Assuming wings level and no aircraft sink rate, the ejection seats provide safe escape within the following parameters:

a. Ground level (zero altitude) – zero airspeed. (canopy must be closed)
b. Ground level and up – 400 knots maximum (based on human factors) – 500 knots or M equal 0.92 maximum, whichever is less (based on seat limitations)

At airspeeds greater than 400 knots, appreciable forces are exerted on the body which makes escape more hazardous.

HIGH ALTITUDE EJECTION

For a high altitude ejection, the basic low level ejection procedure is applicable. Furthermore, the zoom up maneuver is still useful to slow the airplane to a safer ejection speed or provide more time and glide distance as long as an immediate ejection is not mandatory. If the aircraft is descending uncontrolled as a result of a mid–air collision, control failure, spin, or any other reason, the pilot and RIO will abandon the aircraft at a minimum altitude of 10,000 feet above the terrain if possible. If the pilot has decided to abandon the aircraft while still in controlled flight at altitude, the pilot and RIO will abandon the aircraft at a minimum altitude of 10,000 feet above the terrain with the aircraft headed to sea or toward an unpopulated area.

EJECTION HANDLE SELECTION

Due to its greater accessibility and shorter travel when compared to the face curtain, the lower ejection handle should be used during situations requiring an expeditious ejection. Some of these situations are insufficient flying speed from catapult, ramp strike, parting of cross deck pendants during carrier arrestment, low altitude, uncontrolled flight, and under high G during spin or ACM maneuvers.

SURVIVAL KIT DEPLOYMENT

During ground or low level ejection, the crewmember may not have time to deploy or remove the survival kit. However, if time permits, the kit should be deployed and the life preserver inflated during the parachute descent. The following options are available to the crewmember during parachute descent.

a. Land with survival kit intact.
b. Deploy survival kit prior to ground impact or water entry. If water contact is anticipated, ensure that both lap belts remain attached until after water entry and that the upper half is snug against the body prior to contact.
c. Remove survival kit and hold it by carrying strap to be dropped just prior to ground contact.

SURVIVAL KIT JETTISONING

If ground contact is anticipated among trees, rugged terrain, high tension wires, etc., it may be to the crew members advantage to remove the survival kit, hold it by the carrying strap and drop it just prior to impact. To jettison or hand carry survival kit, proceed as follows:

1. Open face visor or disconnect oxygen mask.
2. Composite disconnect release knob – PULL
3. Release left lap belt release fitting.
4. Take hold of survival kit lifting handle with right hand.
5. Release right lap belt release fitting.

EJECTION SEAT FAILURE (BAILOUT)

If the canopy has been jettisoned but the ejection seat fails, proceed as follows:

1. Reduce speed to 200 – 250 knots.
2. Emergency restraint release handle – PULL
   Pull up on the emergency restraint release handle on the right side of the seat bucket to disconnect the parachute harness and the leg restraint harness from the seat. This handle also fires the cartridge actuated guillotine and severs the link line between the drogue chute and the personnel parachute.
3. Full nose down trim, full rudder trim, and opposite aileron trim as required to hold wings level.

MINIMUM EJECTION ALTITUDE VS. SINK RATE

**AIRPLANES 157242an AND UP
AND ALL OTHERS AFTER ACC 176**

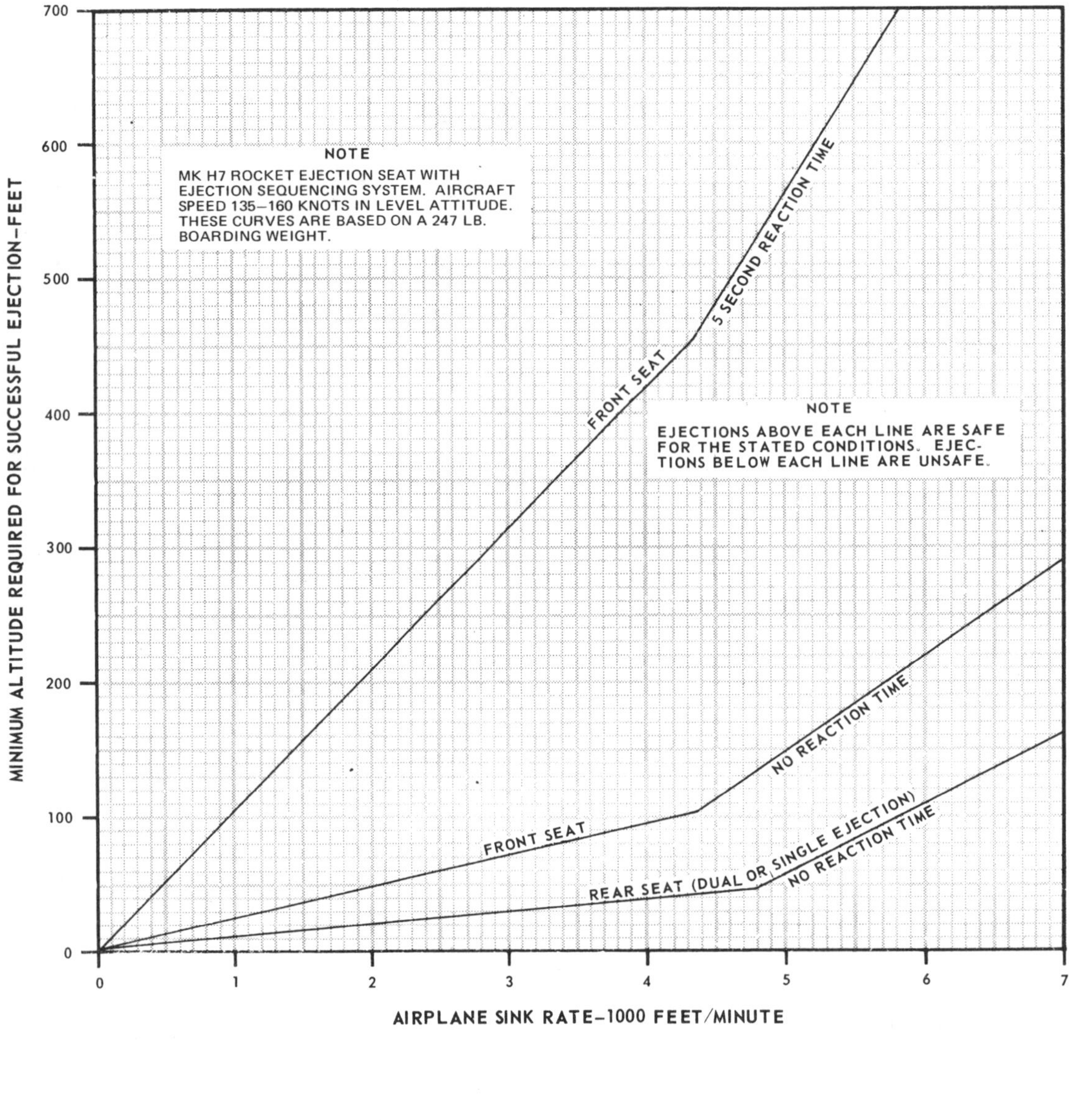

Figure 5–3

MINIMUM EJECTION ALTITUDE vs. AIRSPEED and DIVE ANGLE

AIRPLANES THRU 155902am
BEFORE ACC 176 (SKYSAIL CHUTE)

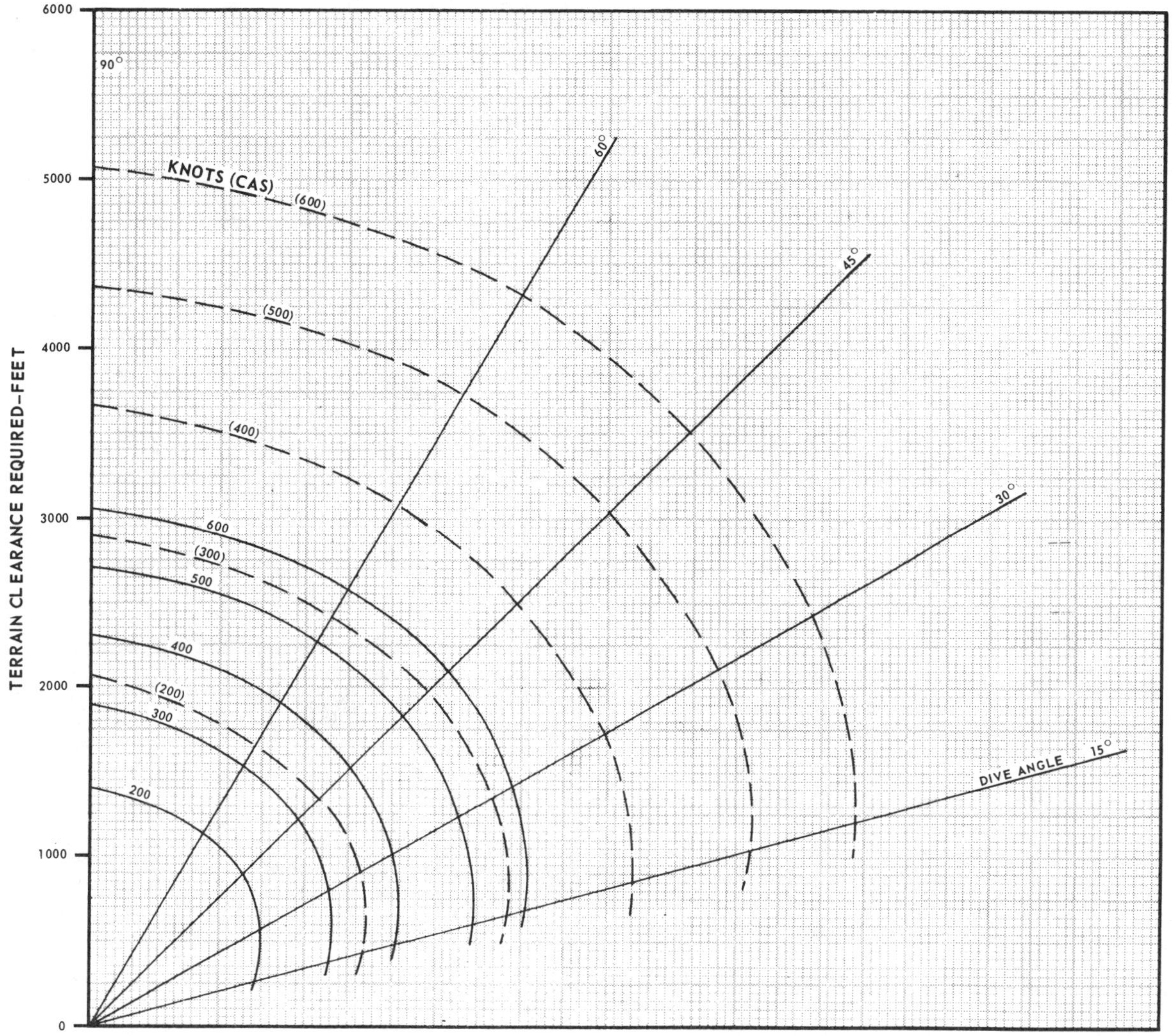

Note

THESE CURVES ARE BASED ON A 247 LB. BOARDING WEIGHT. THE SOLID CURVES INDICATE MINIMUM TERRAIN CLEARANCE WITH NO CREWMEMBER REACTION TIME. THE DASHED CURVES INDICATE MINIMUM TERRAIN CLEARANCE WITH A TWO (2) SECOND CREWMEMBER REACTION TIME. THE CURVES ARE BASED ON WINGS LEVEL BANK ATTITUDE AND APPROPRIATE ANGLE OF ATTACK. TIME REQUIRED FOR THE SEQUENCING SYSTEM TO EJECT BOTH CANOPIES AND SEATS IS INCLUDED. THE CURVES DO NOT INCLUDE A CORRECTION FOR BAROMETRIC ALTIMETER LAG; FOR PROPER VALUES REFER TO PERFORMANCE DATA.

FDD-1-(76)A

Figure 5-4

MINIMUM EJECTION ALTITUDE vs. AIRSPEED AND DIVE ANGLE

AIRCRAFT 157242an AND UP AND ALL OTHERS AFTER ACC 176 (28 FT. CHUTE)

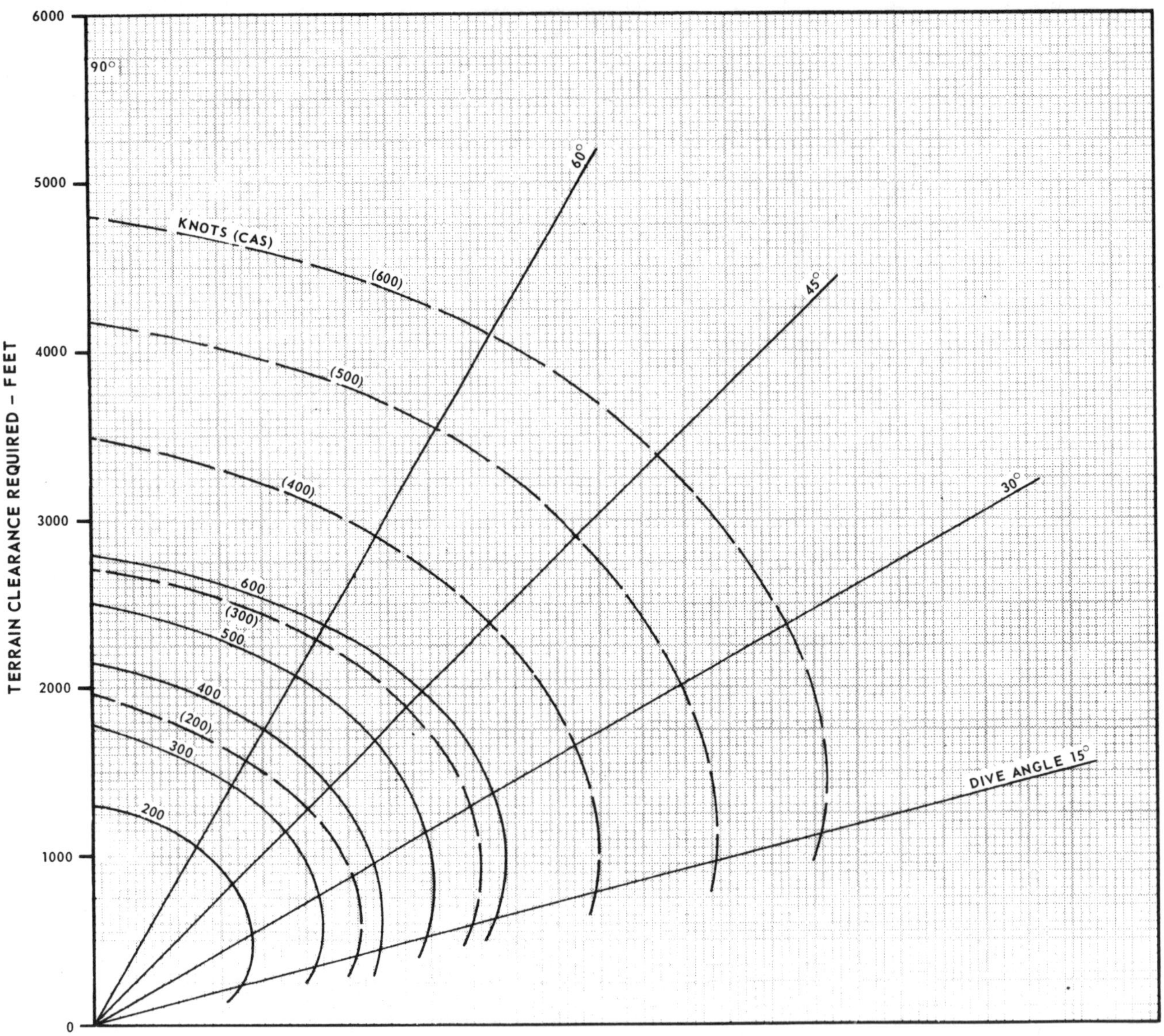

Note

These curves are based on a 247 lb. boarding weight. The solid curves indicate minimum terrain clearance with no crew member reaction time. The dashed curves indicate minimum terrain clearance with a two (2) second crew member reaction time. The curves are based on wings level bank attitude and appropriate angle of attack. Time required for the sequencing system to eject both canopies and both seats is included. It is assumed that the pilot initiates the sequencing system and continues pulling either ejection handle to fire the front seat as soon as the front canopy and interlock block are clear. If the pilot does not continue pulling, or the sequence is initiated by the rear crew member, relying on the front seat mounted time delay initiator to fire the front seat, an additional 250 feet altitude is required for a 90° dive angle at 600 knots, with proportionately less additional altitude required as dive angle and speed decrease. The curves do not include a correction for barometric altimeter lag; for proper values refer to part 1 of applicable index.

FDD-1-(280)A

Figure 5-5

BEFORE EJECTION SEQUENCE

TYPICAL BOTH COCKPITS

If time and conditions permit -

- IFF EMERGENCY
- MAKE RADIO DISTRESS CALL
- SLOW AIRPLANE AS MUCH AS POSSIBLE.
- STOW ALL LOOSE EQUIPMENT
- PULL EMERGENCY VENTILATING KNOB TO MINIMIZE DECOMPRESSION EFFECTS WHEN JETTISONING THE CANOPY.

Note

- THE LOWER EJECTION HANDLE SHOULD BE USED WHEN IT IS NECESSARY TO RETAIN CONTROL OF THE AIRCRAFT AND WHEN IT IS IMPRACTICABLE TO REACH THE FACE CURTAIN HANDLE.

WARNING

- IF CREWMEMBER ELECTS TO USE LOWER EJECTION HANDLE AFTER PULLING FACE CURTAIN HANDLE, THE FACE CURTAIN HANDLE SHOULD BE RETAINED WITH ONE HAND TO PREVENT THE POSSIBILITY OF ENTANGLEMENT WITH DROGUE GUN WHEN IT FIRES.

1. ALERT RIO

2. ASSUME PROPER EJECTION POSITION

ADJUST SEAT POSITION SO TOP OF HELMET IS BELOW FACE CURTAIN HANDLES. BRACE THIGHS ON SEAT CUSHION, LEGS EXTENDED. SIT ERECT, BUTTOCKS BACK, SPINE STRAIGHT, HEAD BACK AGAINST HEADREST, AND CHIN IN.

3. FACE CURTAIN HANDLE-PULL

REACH OVERHEAD, WITH PALMS AFT KEEPING ELBOWS TOGETHER, GRASP FACE CURTAIN HANDLE, PULL FACE CURTAIN AND MAINTAIN DOWNWARD FORCE UNTIL STOP IS ENCOUNTERED. WHEN CANOPY JETTISONS, CONTINUE PULLING FACE CURTAIN UNTIL FULL TRAVEL IS REACHED.

3. LOWER EJECTION HANDLE-PULL

GRASP THE LOWER EJECTION HANDLE, USING A TWO-HANDED GRIP WITH THE THUMB AND AT LEAST TWO FINGERS. OF EACH HAND. PULL UP ON LOWER HANDLE UNTIL STOP IS ENCOUNTERED. WHEN CANOPY JETTISONS, CONTINUE PULLING UP ON LOWER EJECTION HANDLE UNTIL FULL TRAVEL IS REACHED.

WARNING

DURING THE SINGLE EJECTION FROM THE REAR COCKPIT, THE SEAT CATAPULT WILL NOT FIRE AUTOMATICALLY AS IN DUAL EJECTION, AND THE CREWMAN MUST CONTINUE PULL ON THE EJECTION HANDLE AFTER CANOPY REMOVAL TO FIRE THE SEAT GUN. ON AIRCRAFT 155903ap AND UP AND ALL OTHERS AFTER AFC 482, THE SEQUENCING SYSTEM AUTOMATICALLY FIRES THE SEAT AND NO EXTRA PULL IS REQUIRED.

FDD-1-(77-1)B

Figure 5-6 (Sheet 1 of 2)

IF CANOPY FAILS TO JETTISON

Note

IF CANOPY FAILS TO JETTISON RELEASE TENSION ON FACE CURTAIN HANDLE, AND WHILE HOLDING HANDLE WITH ONE HAND, PERFORM THE APPROPRIATE FOLLOWING PROCEDURE. WHEN CANOPY JETTISONS, AGAIN GRASP EJECTION HANDLE WITH BOTH HANDS, AND PULL UNTIL FULL TRAVEL IS REACHED.

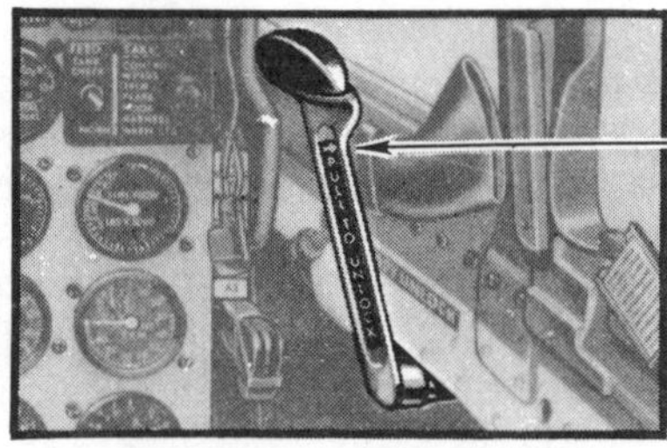

4. Normal canopy control handle – OPEN
5. Manual canopy unlock handle – PULL

If canopy still fails to jettison -

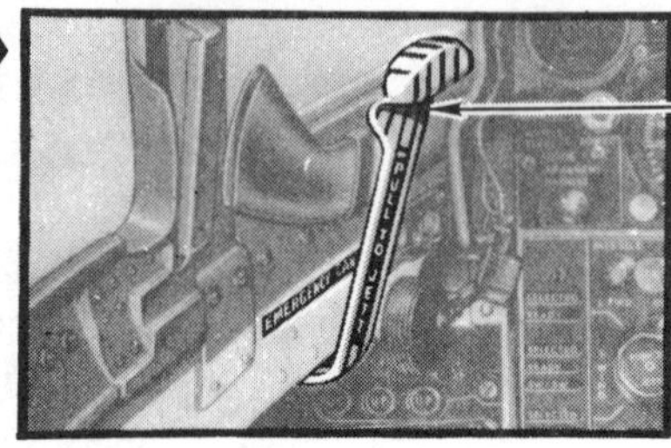

6. Emergency canopy jettison handle – PULL

If canopy still fails to jettison -

7. Put negative G on aircraft

When canopy separates -

8. Face curtain/lower ejection handle – PULL

FDD–1–(77–2)B

Figure 5-6 (Sheet 2 of 2)

AFTER EJECTION SEQUENCE

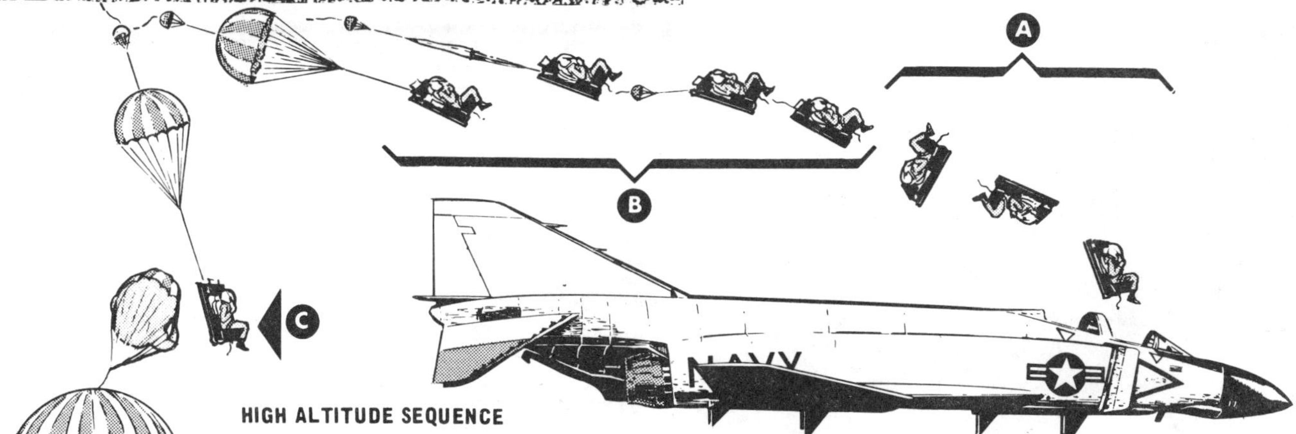

HIGH ALTITUDE SEQUENCE

PULL FACE CURTAIN TO INITIATE THE EJECTION SEQUENCE, CANOPY JETTISONS AND PULLS INTERLOCK BLOCK ALLOWING CATAPULT GUN TO FIRE.

A SEAT IS PROPELLED UP GUIDE RAIL. OCCUPANT'S LEGS ARE RESTRAINED, EMERGENCY OXYGEN IS ACTUATED, TIME RELEASE MECHANISM AND DROGUE GUN ARE TRIPPED, AND EMERGENCY IFF IS ACTUATED. THE ROCKET PACK FIRES NEAR THE END OF THE CATAPULT GUN STROKE.

B DROGUE GUN FIRES APPROXIMATELY 0.75 SECOND AFTER EJECTION, DEPLOYS CONTROLLER DROGUE, WHICH IN TURN, DEPLOYS STABILIZER DROGUE. SEAT IS STABILIZED AND DECELERATED BY DROGUE CHUTES.

C SEAT AND OCCUPANT DESCEND RAPIDLY THRU UPPER ATMOSPHERE. WHEN AN ALTITUDE BETWEEN 14,500 FT. AND 11,500 FT. IS REACHED, THE BAROSTAT RELEASES THE ESCAPEMENT MECHANISM, WHICH IN TURN, ACTUATES TO RELEASE THE OCCUPANT'S HARNESSING, LEG RESTRAINT LINES, AND CHUTE RESTRAINT STRAPS. THE DROGUE CHUTES PULL THE LINK LINE TO DEPLOY THE PERSONNEL PARACHUTE.

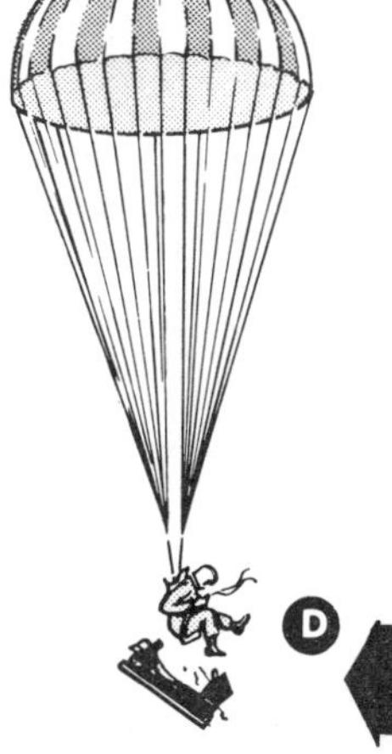

13,000 ± 1500 FEET... IF NECESSARY PROCEED WITH........

D OCCUPANT IS HELD TO SEAT BY STICKER CLIPS UNTIL OPENING SHOCK OF PARACHUTE SNAPS SEAT FROM HIM PERMITTING NORMAL DESCENT.

LOW ALTITUDE SEQUENCE

SAME AS CORRESPONDING STEPS **A-B** IN HIGH ALTITUDE SEQUENCE EXCEPT

C APPROXIMATELY 2 1/4 SECONDS AFTER EJECTION, THE ESCAPEMENT MECHANISM ACTUATES TO RELEASE THE OCCUPANT'S HARNESSING, LEG RESTRAINT LINES, AND CHUTE RESTRAINT STRAPS. THE DROGUE CHUTES PULL THE LINK LINE TO DEPLOY THE PERSONNEL PARACHUTE.

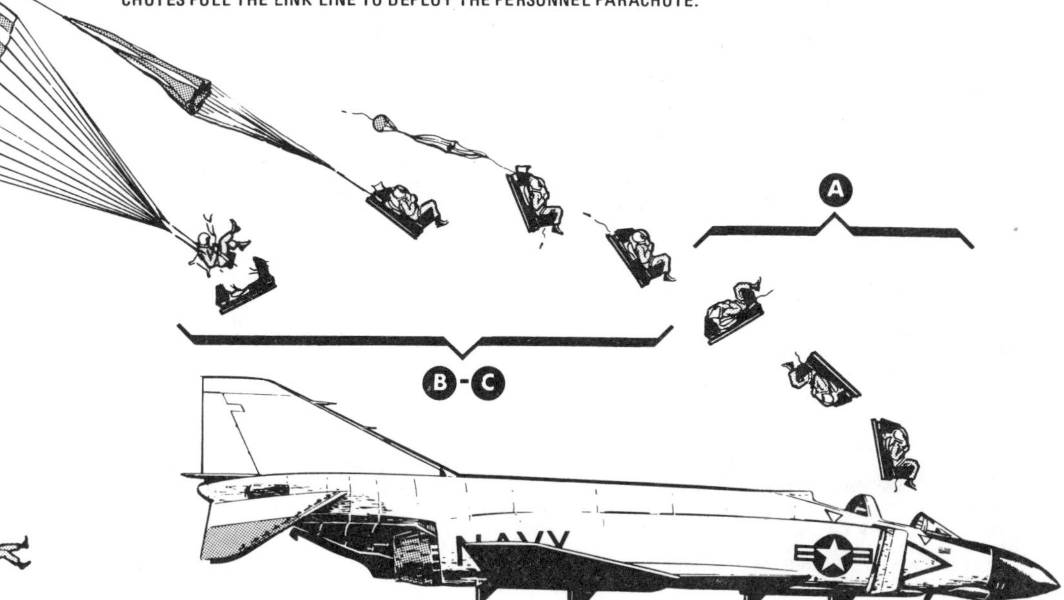

FDD-1-(78-1)A

Figure 5-7 (Sheet 1 of 2)

SURVIVAL KIT DEPLOYMENT

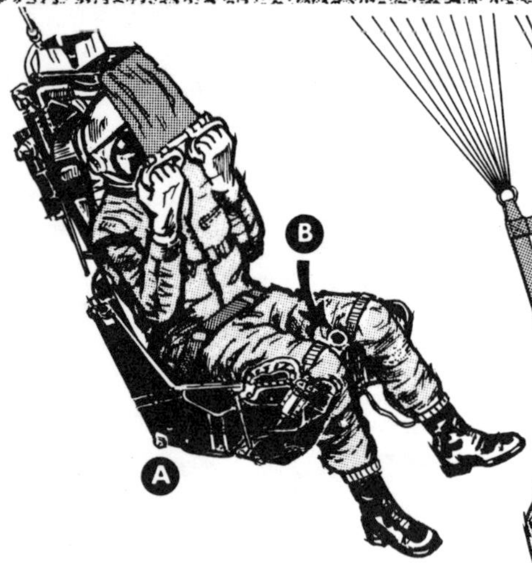

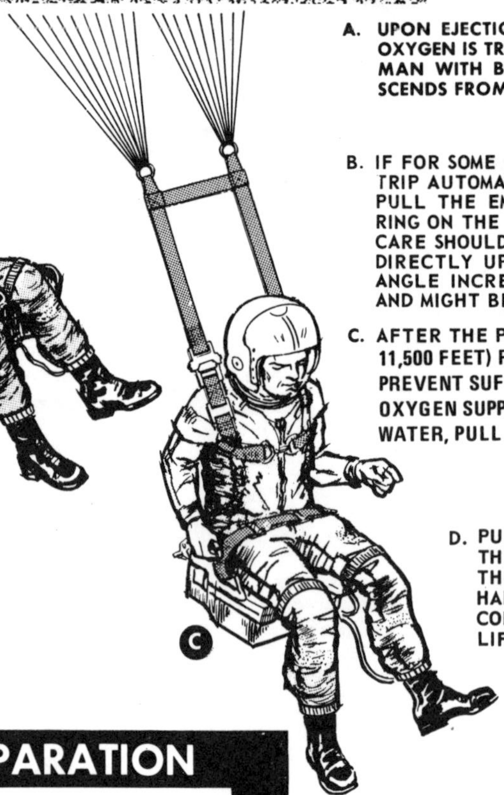

A. UPON EJECTION, SURVIVAL KIT EMERGENCY OXYGEN IS TRIPPED. THIS PROVIDES THE CREWMAN WITH BREATHING OXYGEN AS HE DESCENDS FROM HIGH ALTITUDES.

B. IF FOR SOME REASON THE OXYGEN FAILS TO TRIP AUTOMATICALLY, THE CREWMAN MUST PULL THE EMERGENCY OXYGEN ACTUATOR RING ON THE FRONT OF THE SURVIVAL KIT. CARE SHOULD BE TAKEN TO PULL THE RING DIRECTLY UPWARD, AS PULLING IT AT AN ANGLE INCREASES THE FORCE REQUIRED AND MIGHT BREAK THE RING LANYARD.

C. AFTER THE PARACHUTE HAS OPENED (ABOUT 11,500 FEET) RELEASE OXYGEN MASK TO PREVENT SUFFOCATION WHEN EMERGENCY OXYGEN SUPPLY IS DEPLETED. IF OVER WATER, PULL UP ON KIT RELEASE HANDLE.

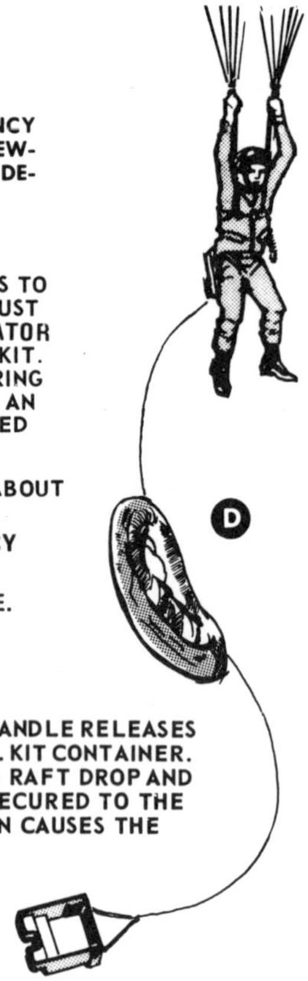

D. PULLING THE KIT RELEASE HANDLE RELEASES THE BOTTOM OF THE SURVIVAL KIT CONTAINER. THE CONTAINER BOTTOM AND RAFT DROP AND HANG FROM A 26 FOOT LINE SECURED TO THE CONTAINER TOP. THIS ACTION CAUSES THE LIFE RAFT TO INFLATE.

MANUAL SEPARATION

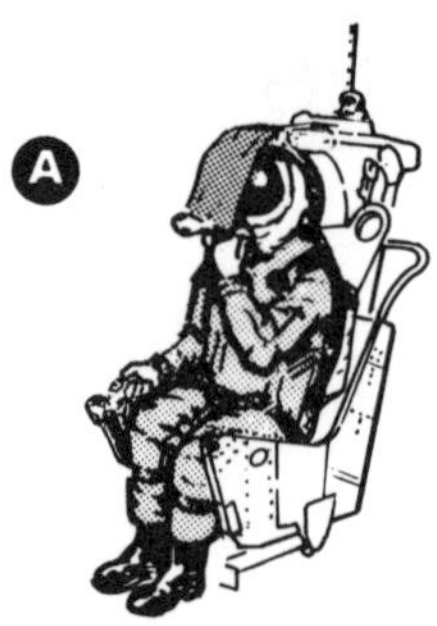

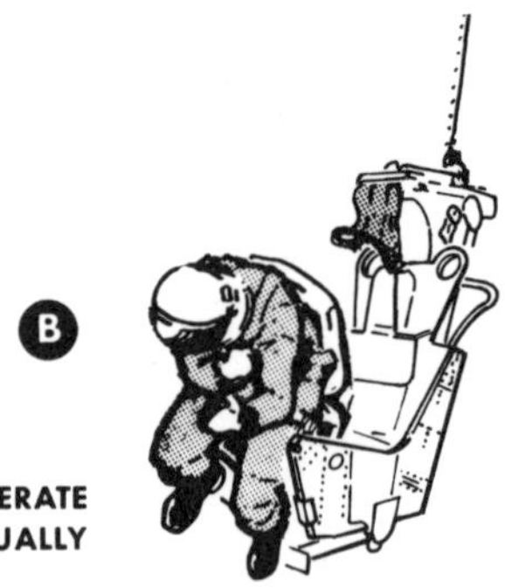

SHOULD THE TIME RELEASE MECHANISM FAIL TO OPERATE AUTOMATICALLY, THE OCCUPANT WOULD MANUALLY SEPARATE FROM THE SEAT AS FOLLOWS:

A. ACTUATE EMERGENCY HARNESS RELEASE HANDLE ON RIGHT SIDE OF SEAT TO ITS FULL AFT POSITION. THIS ACTION WILL RELEASE RESTRAINT HARNESS, LEG RESTRAINT CORD, AND A CARTRIDGE ACTUATED GUILLOTINE WILL SEVER THE LINK LINE BETWEEN THE PERSONNEL CHUTE AND DROGUE CHUTE. THE OCCUPANT IS NOW HELD IN SEAT ONLY BY STICKER CLIPS.

B. PUSH FREE OF STICKER CLIPS AND CLEAR OF SEAT.

C. PULL PARACHUTE RIPCORD D RING (LOCATED ON LEFT SHOULDER) AND MAKE A NORMAL PARACHUTE DESCENT TO THE GROUND.

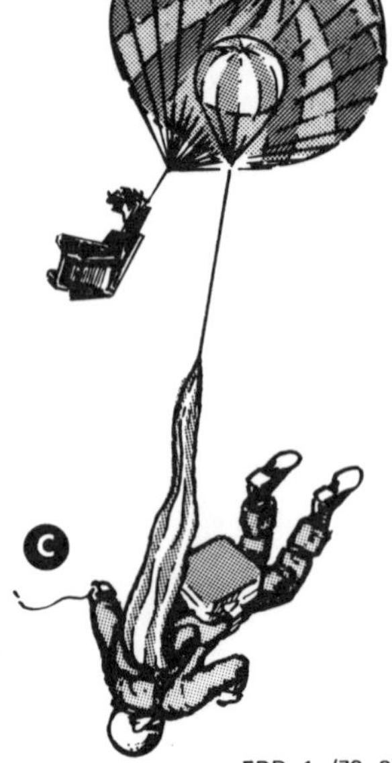

WARNING

AIRCREWS USING THE MANUAL METHOD OF SEAT SEPARATION AND PARACHUTE DEPLOYMENT SHOULD IMMEDIATELY CHECK FOR PARACHUTE ACTUATION AND BE PREPARED TO FORCIBLY DEPLOY THE CHUTE BY HAND AFTER "D" RING ACTUATION

Note

AIRCREWS SHOULD RIGIDLY HOLD LEFT FRONT RISER WITH LEFT HAND AND PULL RIPCORD "D" RING WITH RIGHT HAND IN ORDER TO INSURE FULL TRAVEL OF THE LANYARD

FDD–1–(78–2)A

Figure 5–7 (Sheet 2 of 2)

4. Release stick
5. When clear of airplane and at 11,500 feet or below, parachute D ring – PULL

WARNING

Aircrews using the manual method of seat separation and parchute deployment should immediately check for parachute actuation and be prepared to forcibly deploy the chute by hand after D–ring actuation.

NOTE

Aircrews should rigidly hold left front riser with left hand and pull ripcord D–ring with right hand to ensure full travel of the lanyard.

ELECTRICAL

GENERAL

If an electrical system failure occurs, various components of the combined aircraft systems will be inoperative. Refer to figure 5–8 for equipment that will be lost/available with one or both generators inoperative.

NOTE

- With stability augmentation engaged, small transients in yaw and roll will be experienced whenever power to the system is cutoff and then reapplied. To prevent these transients, disengage the stab aug by the emergency disengage switch before turning on or cycling either of the generator switches.
- If the right engine fails on final and the bus tie remains open, afterburner ignition will not be available. However, if afterburner thrust is required, afterburner light–offs are generally obtainable through turbine torching, by jam accelerating the left engine at 90% rpm or above.

SINGLE GENERATOR FAILURE OR INTERMITTENT GENERATOR OPERATION

In aircraft 155785ai and up and all others after AFC 388 or AFC 535, a left generator failure accompanied by a failure of the bus tie will not be noted by the illumination of the LH GEN OUT and BUS TIE OPEN lights if the pilots instrument lights knob (flight instrument lights knob after AFC 536) is in the OFF position. Failure can be detected initially by noting the OFF flag appearing on the ADI, at which time the instrument lights should be turned on and the generator lights monitored for verification of the failure. If the pilots instrument lights are not in the OFF position when a right generator failure accompanied by a bus tie failure occurs, the RH GEN OUT and BUS TIE OPEN lights will not illuminate. The failure may be detected by loss of all external lights, loss of Com–Nav command lights, appearance of OFF flags in oxygen gages, loss of the front and rear cockpit utility flood lights, or loss of indexer lights with gear down. Warning lights power will be available at all times on the RAT, regardless of the position of the pilots instrument lights knob (flight instrument lights knob after AFC 536).

With or without bus tie failure –

1. Determine malfunctioning/failed generator

Without bus tie open; or bus tie open condition permitting warning lights illumination –

a. Check for illumination of generator warning lights and bus tie open light.

With bus tie open condition preventing warning lights illumination –

Day VFR –

a. With flight instrument lights (instrument lights before AFC 536) OFF
   (1) Initial indications will be:
      OFF flag on ADI
      Loss of intercom
b. Flight instrument lights (instrument lights before AFC 536) – ON

Night or IFR –

a. With flight instrument lights (instrument lights before AFC 536) ON
   (1) Initial indications will be:
      Loss of Com–Nav command lights
      OFF flag in oxygen gage
      Loss of utility flood lights
      Loss of indexer lights (gear down)
      Loss of all external lights
b. Flight instrument lights (instrument lights before AFC 536) – OFF

2. Failed generator switch – CYCLE
3. Check generator warning light – OUT
4. If light remains illuminated, failed generator switch – OFF
5. Monitor engine oil pressure and nozzle operation – SECURE ENGINE IF NECESSARY

EMERGENCY POWER DISTRIBUTION

RH GEN OUT - BUS TIE OPEN

INOPERATIVE BUSES

RIGHT MAIN 115/200 VAC
RIGHT MAIN 28 VAC
RIGHT MAIN 14 VAC
28/14 VAC WARNING LIGHTS (DIM)
– IF SELECTED

INOPERATIVE EQUIPMENT

AFT CKPT FLOOD LT (DIM)
AFT CKPT UTILITY LT
AFTERBURNER IGNITION
4 AILERON DROOP
AILERON FEEL TRIM
AIRSPEED PITOT HEATER
1 ALTIMETER VIBRATOR
19 AN/AWW–1 BOMB FUZING
ANGLE OF ROLL LIGHT
ANTI–COLLISION LT
ANTI–ICE
8 ANTI–SKID
AOA INDEXER LIGHTS
APN–141 ELECTRONIC ALTM
APPROACH LIGHT
AUTO THROTTLE
BELLMOUTH PITOT HEATER

INOPERATIVE EQUIPMENT (CONT)

BUDDY TANK HYD PUMP
℄ GUN POD
CKPT HEAT & VENT
COM–NAV COMMAND LTS
20 CONV WPN BOMB FUZING
33 25 EMER AIL DROOP SWITCH
12 ENG BLEED AIR LEAKAGE DETECTION
EQPT COOLING
FLIGHT REFUEL PROBE LT
FUSLG AND ANTI–COLLISION LTS
FWD CKPT RED CSL FLOOD LT (MED)
FWD CKPT RED INSTR FLOOD LT (DIM)
FWD CKPT UTILITY LT
17 GROUND COOLING FANS
NO. 4 FUEL XFR PUMP
11 NO. 7 TANK FUEL XFR PUMP
OXYGEN GAGE
RADAR ANNUNCIATOR LIGHTS (DIM)
– IF SELECTED
RADAR MISSILES
RIGHT FUEL BOOST PUMP
SIDEWINDER
18 STRIKE CAMERA
TAXI LT
UTILITY FLOOD LTS
UTILITY HYD PRESS INDICATOR
WARNING LTS (ALL/DIM) – IF SELECTED
WARNING LIGHTS TEST (ALL/DIM)
– IF SELECTED
WHITE FLOOD LTS (FWD CKPT)
WING AND TAIL LTS (BRT)
WING AND TAIL LTS (DIM)

LH GEN OUT - BUS TIE OPEN

INOPERATIVE BUSES

LEFT MAIN 115/200 VAC
LEFT MAIN 28 VAC
IGNITION 115 VAC
ESSENTIAL 115/200 VAC
ESSENTIAL 28 VAC
ESSENTIAL 28 VDC
28/14 VAC WARNING LTS(BRT)
– IF SELECTED

INOPERATIVE EQUIPMENT

ADF
AFT CKPT EQPT LT
AFT CKPT FLOOD LT (BRT)
AFT CKPT INST LT
AILERON FEEL TRIM
AILERON POSITION INDICATOR
AILERON RUDDER INTERCONNECT
28 ALQ–51/100 COUNTERMEASURES SET
27 ALQ–91 COUNTERMEASURES SET
29 AN/APQ–126 COUNTERMEASURES SET
21 ALR–45
AN/AJB–7
2 ALT ENCODER
ANGLE–OF–ATTACK XMTR HEATER
22 APR–25 RADAR HOMING AND WARNING
5 APX–76 INTERROGATOR SET
AUTOMATIC FUEL XFR
AUTOPILOT
BUDDY TANK HOSE JETTISON
CADC
CENTERLINE STORE SAFETY
5 DATA LINK
32 23 EMER AIL DROOP SWITCH
7 EMER REFUEL PROBE
3 ENG OIL LEVEL AND CHECK
EXTERNAL STORES EMER JETT
EXTERNAL FUEL XFR
FIRE DETECTOR
FIRE DETECTOR CHECK SWITCH
FUEL FLOW INDICATOR
FUEL LOW WARNING LT
FUEL PRESSURE INDICATOR
FUEL QUANTITY
FLIGHT DIRECTOR GP
26 FWD CKPT FLT INSTR LT
FWD CKPT INST LT
FWD CKPT RED CSL FLOOD LT (BRT)
FWD CKPT RED INSTR FLOOD LT (BRT)
FWD CSL LT

INOPERATIVE EQUIPMENT (CONT)

GVR–10 (VERT FLT REFERENCE)
IFF
INBOARD WG JETTISON
16 INSTRUMENT LANDING SYSTEM (AN/ARA–63)
INTERCOM
INTERNAL WING FUEL XFR
10 KY–28
L & R ENG RAMP CONTROL
L & R ENG TURBINE OUTLET TEMP
L & R MAIN IGNITION
LANDING GEAR AND FLAP POS IND
LEFT ENGINE FUEL SHUTOFF
LH FUEL BOOST PUMP (HI OR LOW SPEED)
MAIN FUEL CONTROL (RT ENG FUEL SHUTOFF)
MASTER CAUTION LT RESET
MISSILE JETTISON
NAV COMPUTER
NO. 6 FUEL XFR PUMP
NOSEWHEEL STEERING
NOZZLE POSITION INDICATOR
OIL PRESSURE INDICATORS
35 OPTICAL SIGHT
OUTBD WG JETTISON
PC1 & PC2 HYD PRESSURE INDICATORS
PNEUMATIC PRESS IND
PNEUMATIC SYSTEM CONT
RADAR
15 RADAR SCOPE CAMERA
REFUEL PROBE
5 R AND L ENG BELLMOUTH CONT
RUDDER FEEL TRIM
RUDDER POSITION INDICATOR
SCOPE LTS
9 SEAT ADJUST
6 SMOKE ABATEMENT SYSTEM
STAB FEEL TRIM
TACAN
UHF COMM
UTILITY RECEPTACLE (AC)
14 VTAS
WARNING LTS (BRT) – IF SELECTED
WINDSHIELD TEMP SENSOR

FDD–1–(68–1)E

Figure 5–8 (Sheet 1 of 3)

RAT OPERATING - LH GEN OUT - BUS TIE OPEN

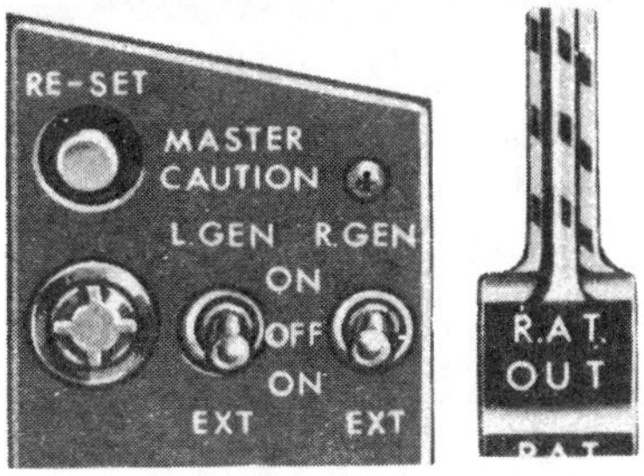

OPERATIVE BUSES

ESSENTIAL 115/200 VAC
ESSENTIAL 28 VAC
ESSENTIAL 28 VDC
IGNITION 115 VAC
RIGHT MAIN 115/200 VAC
RIGHT MAIN 28 VAC
RIGHT MAIN 14 VAC
RIGHT MAIN 28 VDC
LEFT MAIN 28 VDC
ARMAMENT 28 VDC
RADAR 28 VDC
28/14 WARNING LTS – BRT & DIM

OPERATIVE EQUIPMENT

AFT CKPT FLOOD LTS (BRT)
AFT CKPT FLOOD LTS (DIM)
AFT CKPT INSTR LTS
AFT CKPT UTILITY LT
AFTERBURNER IGNITION
4 AILERON DROOP
AILERON FEEL TRIM
AILERON POSITION INDICATOR
AIRSPEED PITOT HEATER
AJB–7

OPERATIVE EQUIPMENT (CONT)

1 ALTIMETER VIBRATOR
AN/ARW–73
19 AN/AWW–1 BOMB FUZING
AOA INDEXER LTS
AOA INDICATOR
AOA PROBE HEATER
ANTI–COLLISION LT
ANTI–ICE
8 ANTI–SKID
5 AN/ALE–29A CHAFF DISPENSER
31 ALE–39 CHAFF DISPENSER
APN–141 ELECTRONIC ALT
APN–154 RADAR BEACON
APPROACH LT
ARMAMENT BUS CONTROL
ARR HOOK CONTROL
AUTOMATIC FUEL XFR
AUTO THROTTLE
AUX RECEIVER
BELLMOUTH PITOT HEATER
BOMB & ROCKET REL CONTROL
BOMB & ROCKET REL POWER
BUDDY TANK FUEL DUMP
BUDDY TANK HOSE JETTISON
BUDDY TANK HYD PUMP
CADC
CKPT HEAT & VENT
℄ GUN POD
20 CONV WPN BOMB FUZING
CTR STORE POWER
CTR STORE SAFETY
33 **24** EMER AIL DROOP SWITCH
7 EMER REFUEL PROBE
12 ENG BLEED AIR LEAKAGE DETECTION
13 ENG BLEED AIR SHUTOFF VALVE
3 ENG OIL LEVEL AND CHECK
EQUIPMENT COOLING
EXT LT CONTROL
EXTERNAL STORES EMER JETT
EXTERNAL FUEL XFR
FIRE AND OVERHEAT DETECTOR
FIRE DETECTOR CHECK SWITCH
FLAPS
FLIGHT DIRECTOR GP
FLIGHT REFUEL PROBE LT
FUEL FLOW INDICATOR
FUEL LOW WARNING LT
FUEL PRESSURE INDICATOR
FUEL QUANTITY
FSLG AND ANTI–COLLISION LTS
26 FWD CKPT FLT INSTR LTS
FWD CKPT INSTR LTS
FWD CKPT RED CSL FLOOD LTS (MED)
FWD CKPT RED CSL FLOOD LTS (BRT)
FWD CKPT RED INSTR FLOOD LTS (DIM)

OPERATIVE EQUIPMENT (CONT)

FWD CKPT RED INSTR FLOOD LTS (BRT)
FWD CKPT UTILITY LT
IFF
INBOARD WG JETTISON
INTERCOM
INTERNAL WING FUEL DUMP
INTERNAL WING FUEL XFR
10 KY–28
LANDING GEAR
LANDING GEAR AND FLAP POS IND
LEFT ENGINE FUEL SHUTOFF
L & R ENGINE TURBINE OUT TEMP
L & R MAIN IGNITION
LH FUEL BOOSTER PUMP (**34** LOW SPEED)
MAIN FUEL CONTROL (R ENG FUEL SHUT-OFF)
MASTER CAUTION LT RESET
MISSILE ARM
MISSILE FIRING
MISSILE JETTISON
NO. 4 FUEL XFR PUMP
11 NO. 7 TANK FUEL XFR PUMP
NOZZLE POSITION IND
OIL PRESSURE IND
OXYGEN GAGE
OUTBOARD WG JETTISON
PC1 & PC2 HYD PRESSURE INDICATORS
PNEUMATIC PRESSURE INDICATORS
PNEUMATIC SYSTEM CONT
R & L AUX DOORS
R & L ENG BELLMOUTH CONT
RADAR MISSILE FAIRING
RAIN REMOVAL
REFUEL PROBE
RIGHT FUEL BOOST PUMP
RUDDER FEEL TRIM
RUDDER POSITION INDICATOR
RUDDER TRIM
SIDEWINDER
6 SMOKE ABATEMENT SYSTEM
SPEED BRAKE
STABILATOR FEEL TRIM
STABILATOR POSITION IND
STATIC ACCEL
STORES RELEASE CONT
TAXI LT
UHF COMM
UTILITY FLOOD LT
UTILITY HYD PRESS IND
UTILITY RECEPTACLE (D–C)
WARNING LTS
WARNING LTS TEST SWITCH
WHITE FLOOD LTS (FWD CKPT)
WINDSHIELD TEMP SENSOR
WING AND TAIL LTS (BRT)
WING AND TAIL LTS (DIM)
WINGFOLD CONTROL

RAT OPERATING - LH GEN OUT - RH GEN OUT

OPERATIVE BUSES

ESSENTIAL 115/200 VAC
ESSENTIAL 28 VAC
ESSENTIAL 28 VDC
IGNITION 115 VAC
28/14 VAC WARNING LTS – BRT

OPERATIVE EQUIPMENT

AFT CKPT FLOOD LTS (BRT)
AFT CKPT INSTR LTS
AILERON POSITION IND
AN/AJB–7
AOA INDICATOR
AUTOMATIC FUEL XFR

OPERATIVE EQUIPMENT (CONT)

BUDDY TANK HOSE JETTISON
CADC
CTR STORE SAFETY
33 **23** EMER AIL DROOP SWITCH
7 EMER REFUEL PROBE
EXT STORES EMER JETT
EXT FUEL XFR
FIRE AND OVERHEAT DETECTOR
FLT DIRECTOR GP
FUEL FLOW INDICATOR
FUEL LOW WARNING LT
FUEL PRESSURE INDICATOR
FUEL QUANTITY
26 FWD CKPT FLT INSTR LTS
FWD CKPT INSTR LTS
FWD CKPT RED CSL FLOOD LTS (BRT)
FWD CKPT RED INST FLOOD LT (BRT)
IFF
INBOARD WG JETTISON
INTERCOM
INTERNAL WING FUEL TRANSFER
10 KY–28
LANDING GEAR AND FLAP POSN IND
LEFT BOOST PUMP EMER (LOW SPEED)

OPERATIVE EQUIPMENT (CONT)

L & R ENG TURBINE OUT TEMP IND
L & R MAIN IGNITION
LEFT ENG FUEL SHUTOFF
MAIN FUEL CONTROL (R ENG FUEL SHUTOFF)
MASTER CAUTION LT RESET
MISSILE JETTISON
NOZZLE POSITION IND
OIL PRESS INDICATORS
OUTBOARD WING JETTISON
PC1 & PC2 HYD PRESS INDICATORS
PNEUMATIC PRESS INDICATOR
PNEUMATIC SYSTEM CONTROL
REFUEL PROBE
5 R & L ENG BELLMOUTH CONT
RUDDER FEEL TRIM
RUDDER POSITION INDICATOR
6 SMOKE ABATEMENT SYSTEM
STAB FEEL TRIM
UHF COMM
WARNING LTS (BRT)
WINDSHIELD TEMP SENSOR

FDD–1–(68–2)D

Figure 5–8 (Sheet 2 of 3)

RAT OPERATING - LH GEN OUT - RH GEN OUT

INOPERATIVE BUSES

RIGHT MAIN 200/115 VAC
RIGHT MAIN 28 VAC
RIGHT MAIN 14 VAC
LEFT MAIN 115/200 VAC
LEFT MAIN 28 VAC
RIGHT MAIN 28 VDC
LEFT MAIN 28 VDC
ARMAMENT 28 VDC
RADAR 28 VDC

INOPERATIVE EQUIPMENT

ADF
AFT CKPT FLOOD LT (DIM)
AFT CKPT UTILITY FLOOD LT
AFT CKPT WARNING LT TEST
AFTERBURNER IGNITION
4 AILERON DROOP
AILERON FEEL TRIM
AIRSPEED PITOT HEATERS
1 ALTIMETER VIBRATORS
2 ALTITUDE ENCODER (AIMS)
AN/AJB-7 (BOMBING CAPACITY)
5 AN/ALE-29 (CHAFF DISPENSER)
31 ALE-39 CHAFF DISPENSER
28 AN/ALQ-51/100 COUNTERMEASURES SET
27 AN/ALQ-91 ELECTRONIC COUNTER MEASURES

INOPERATIVE EQUIPMENT (CONT)

21 AN/ALR-45 RADAR RECEIVER
AN/APN-141 RADAR ALTIMETER
AN/APN-154 RADAR BEACON
22 AN/APR-25 RADAR RECEIVER
29 AN/ALQ-126 COUNTERMEASURES SET
5 AN/APX-76 INTERROGATOR SET
AN/ARW-73
AN/ASN-39A NAV COMPUTER
5 AN/ASW-25 DATA LINK
AN/AWG-10 RADAR SET
AN/AWW-1 BOMB FUZING
AOA INDEXER LTS
AOA PROBE AND TRANSMITTER HEATERS
ANGLE OF ROLL LT
ANTI-COLLISION LT
8 ANTI-SKID
APPROACH LT
ARI
ARRESTING HOOK CONT (RETRACT)
AUTOPILOT
AUTO THROTTLE (APCS)
AUX AIR DOORS
AUX RECEIVER
30 BELLMOUTH
BELLMOUTH PITOT HEATERS
BOMBS, ROCKETS AND BULLPUP RELEASE
CATAPULT EXTENSION
℄ GUN POD
CKPT HEAT TEMP CONTROL
CORNER REFLECTOR
33 25 EMER AILERON DROOP SWITCH
ENG ANTI-ICE
12 ENG BLEED AIR LEAKAGE DETECTION
13 ENG BLEED AIR SHUTOFF VALVE
EQPT COOLING AND PRESSURIZATION
FIRE DETECTOR CHECK SWITCH
FLAPS
FLT REFUEL PROBE LT
FSLG LT
FWD CKPT RED CONSOLE FLOOD LT (DIM-MED)
FWD CKPT RED INSTR FLOOD LT (DIM)
FWD CKPT UTILITY FLOOD LT
FWD CKPT WHITE FLOOD LT

INOPERATIVE EQUIPMENT (CONT)

GVR-10 VERT FLT REFERENCE
17 GROUND COOLING FANS
16 INSTR LANDING SYSTEM (AN/ARA-63)
INTRL WING FUEL DUMP
LANDING GEAR
L FUEL BOOST PUMP (HIGH SPEED)
MISSILE ARMING
MISSILE FAIRING DOORS
MISSILE FIRING
NO. 4 FUEL CELL XFMR PUMP
NO. 6 FUEL CELL XFMR PUMP
11 NO. 7 FUEL CELL XFMR PUMP
NOSE WHEEL STEERING
35 OPTICAL SIGHT
OXYGEN GAGE
15 RADAR SCOPE CAMERA
RADAR MISSILES
RAIN REMOVAL
R FUEL BOOST PUMP
RUDDER TRIM
SEAT ADJUST
SIDEWINDER
SPEED BRAKES
STAB AUG
STABILATOR POSITION INDICATOR
STALL WARNING VIBRATOR
STATISTICAL ACCELERATOR
STORES RELEASE
18 STRIKE CAMERA
TACAN
TAXI LT
UTILITY HYD PRESSURE INDICATOR
UTILITY RCPT (DC)
VARIABLE RAMP SCHEDULING
14 VTAS
WARNING LT (DIM)
WARNING LTS TEST SWITCH
WING FOLD
WING POSITION AND JOIN-UP LT

LEGEND

1 F-4J AIRCRAFT 153071z THRU 153779ab
2 F-4J AIRCRAFT 153780ac AND UP, AND ALL OTHERS AFTER AFC 388
3 F-4J AIRCRAFT 153076aa AND UP
4 F-4J AIRCRAFT 154786ag AND UP
5 F-4J 155529ag AND UP, AND 153768ab THRU 155528ag AFTER AFC 388
6 F-4J AIRCRAFT 155867ak AND UP, AND ALL OTHERS AFTER AFC 373
7 F-4J 155867ak AND UP, AND ALL OTHERS AFTER AFC 370
8 F-4J AIRCRAFT 157242an AND UP
9 F-4J AIRCRAFT 157242an AND UP, AND ALL OTHERS AFTER AFC 486
10 F-4J AIRCRAFT 155529ag AND UP, AND 153768ab THRU 155528ag AFTER AFC 331
11 F-4J AIRCRAFT 155903ap, AND 157274ap AND UP
12 F-4J AIRCRAFT 155903ap, AND 157274ap AND UP, AND ALL OTHERS AFTER AFC 439
13 F-4J 157274ap AND UP, AND ALL OTHERS AFTER AFC 440
14 F-4J 158355at AND UP, AND 153768ab THRU 158354as AFTER AFC 506
15 F-4J 153768ab THRU 158354as BEFORE AFC 506
16 F-4J 157309ar AND UP, AND ALL OTHERS AFTER AFC 470
17 F-4J 158355at AND UP
18 AFTER AFC 518
19 153071z, 153531ac, 153768ab THRU 157309ar,
20 158346as AND UP
21 AFTER AFC 524
22 F-4J 155529ag AND UP, AND 153768ab THRU 155528ag AFTER AFC 388 AND BEFORE AFC 524
23 F-4J 153768ab THRU 153838ad AND 153840ae THRU 154785af AFTER AFC 534
24 AFTER AFC 534
25 F-4J 154786ag THRU 158379au AFTER AFC 534
26 F-4J 153768ab THRU 153838ad AND 153840ae THRU 158379au AFTER AFC 536
27 155570ah AND UP
28 155570ah AND UP BEFORE AFC 541
29 AFTER AFC 541
30 153071z THRU 153088aa; 153768ab THRU 155528ag BEFORE AFC 388
31 153768ab THRU 158379au AFTER AFC 574.
32 AILERONS WILL DROOP WITH SWITCH IN DISABLED POSITION AND FLAPS LOWERED PNEUMATICALLY.
33 AILERONS WILL NOT DROOP WITH SWITCH IN DISABLED POSITION AND FLAPS LOWERED PNEUMATICALLY.
34 HIGH SPEED AND LOW SPEED OPERATIVE 153072z THRU 153088aa; 153768ab THRU 154785af BEFORE AFC 388.
35 153768ab THRU 158379au AFTER AFC 576 PT1.

FDD-1-(68-3)E

Figure 5-8 (Sheet 3 of 3)

NOTE

On airplanes after PPC 62, engine oil loss will be indicated by the illumination of the applicable engine oil low warning light.

During night or IFR conditions prior to penetration and approach –

6 Extend RAT and place console flood lights switch to BRIGHT.

NOTE

Multiple emergencies, adverse weather, and other peculiar conditions may require relighting of the engine. In this case, a relight should be initiated to ensure minimum time on the engine.

7. Land as soon as practicable.

If BUS TIE OPEN light is also illuminated –

8. Set throttles for cruise power or below.
   Throttles should be set for cruise power or below to prevent engine flame–out caused by power interruption to fuel boost pumps when operating generator is cycled.
9. Operating generator switch – CYCLE (OFF then ON)
10. Check bus tie warning light – OUT
11. If light remains illuminated, turn off all electrical equipment not essential to flight.

NOTE

During night or IFR flying, if the right generator remains inoperative with the bus tie open, there will not be power available to illuminate the warning lights while the instrument lights knob (flight instrument lights knob after AFC 536) in the forward cockpit is in any position except OFF. To retain power for the warning lights, turn the instrument lights knob OFF and place the emergency instrument panel flood lights switch to BRT.

To regain use of essential bus and while retaining use of right generator in aircraft 153072z thru 154785af before AFC 388, with left generator out and bus tie open –

1. Both generators – OFF

CAUTION

Failure to turn off both generators before extending RAT will result in RAT burning out.

2. Ram air turbine – EXTEND
3. Left boost pump normal control circuit breaker – PULL (E11, No. 1 panel)
4. GVR–10 circuit breaker – PULL (A16, No. 1 panel)
5. Attitude reference knob – STBY

NOTE

In order for the ADI, HSI and BDHI to receive reliable heading and attitude information on RAT power, the source of these signals must be switched from the GVR–10 to the AJB–7.

6. Right generator switch – ON

CAUTION

With the BUS TIE OPEN light illuminated, cycling of the emergency generator should be done with caution. The original short/fault probably still exists and could be further aggravated by application of emergency power. Should evidence be found that a short/fault exists on the essential buses, power can only be removed by RAT retraction.

To regain use of essential bus while retaining use of right generator in aircraft 153071z, 154786ag and up, and all others after AFC 388, with the left generator out and bus tie open –

1. Ram air turbine – EXTEND

DOUBLE GENERATOR FAILURE

NOTE

The flaps will retract to a trail position if lowered prior to the double generator failure.

On airplanes 153071z thru 155874ah before AFC 388 or AFC 535, a double generator failure is determined by the illumination of all the generator indicator lights. However, typical double failure in airplanes 153071z thru 155784ah prior to AFC 388 or AFC 535 is caused by loss of dc control voltage from the permanent magnet generators. The generator indicator lights will not illuminate since the control voltage is utilized to illuminate the lights. Therefore, detection of double generator failure without illumination of the indicator lights must be determined by other means such as the appearance of OFF flags on the instrument panels and the sudden quiet caused by loss of electrical control power to the air conditioning system. The generator indicator lights cannot be restored by RAT power. On airplanes 155785ai and up, and 153071z thru 155874ah after AFC 388 or AFC 535, the generator indicator lights receive power at all times from the warning lights bus and the lights will not be illuminated after a double generator failure. The lights can be restored, however, when RAT power is applied to the warning lights bus. As above, the double generator failure is determined by the appearance of OFF flags and the quiet caused by shutdown of the air conditioning system. The indicator lights will be restored on RAT power.

1. Ram air turbine – EXTEND
   Reduce airspeed to RAT extension speed and extend RAT.
2. All unessential electrical switches – OFF

3. Generator control switches – CYCLE (OFF then ON)
4. Check generator warning light(s) – OUT
5. If generator warning light(s) go out – RETRACT RAT
6. Judiciously continue mission, be prepared to land as soon as practicable.
7. If both generator warning lights remain illuminated or reilluminate, generator control switches – OFF
8. Console floodlights – BRIGHT
   During night or IFR conditions, turn console floodlights to BRT as it is the only source of console illumination on RAT power.
9. Landing gear – BLOW DOWN FOR LANDING
   If the landing gear is down and locked before loss of electrical power, do not blow the gear down.
10. Flaps – BLOW DOWN (for carrier landing or if utility hydraulic failure is experienced) – UP (for field landings with utility hydraulic pressure available)

CAUTION

Do not blow down the flaps unless making a carrier landing or a utility hydraulic failure has been experienced. Emergency pneumatic extension of the flaps may result in a utility hydraulic failure.

11. Land as soon as practicable.

COMPLETE ELECTRICAL FAILURE

In event of a double generator failure and a ram air turbine failure,neither external fuel nor internal wing fuel wiii transfer (regardless of fuel control panel switch position prior to the failure). Useable fuel at the time of the complete electrical failure will be indicated by the tape on the fuel quantity gauge. Therefore, the pilot must time his remaining fuel to adjust gross weight for landing. The only available method for internal fuselage fuel transfer will be the hydraulic transfer pumps. In the event of a utility hydraulic failure, approximately 1500 pounds of fuselage fuel will not be available. The utility hydraulic gauge will be inoperative; therefore, monitoring of the utility pressure must be accomplished by rudder feel. In any event, regardless of possible multiple failures, the main consideration must be to land as soon as possible giving due regard to available field arresting gear limitations.

WARNING

Normal afterburner ignition and fuel boost pressures will not be available. Attempted use of afterburner may cause a double engine flameout.

If facilities permit a no-flap landing –

1. Land as soon as possible
2. Landing gear – BLOW DOWN FOR LANDING

NOTE

A visual check for landing gear down and locked indication should be made if possible.

3. Flaps – LEAVE UP FOR LANDING
4. Plan for straight-in arrested landing
5. Fly minimum control airspeed for final approach and landing

NOTE

Refer to Approach Speeds chart for approximate approach airspeeds with SPC off.

If facilities do not permit a no-flap landing –

1. Land as soon as possible
2. Speed less than 250 knots
3. Landing gear – BLOW DOWN FOR LANDING
4. Straight and level at safe altitude
5. Speed 200 knots
6. Flaps – BLOW DOWN

NOTE

On airplanes thru 154785af before AFC 534 and on 154786ag and up, the ailerons will not droop when the flaps are blown down. On airplanes thru 154785af after AFC 534, the ailerons will droop regardless of the emergency aileron droop switch position.

WARNING

Blowing the flaps down may cause utility hydraulic failure. On airplanes thru 154785af after AFC 534, a subsequent PC hydraulic failure will cause a split aileron droop condition resulting in an immediate roll into the dead wing from which recovery is impossible.

7. Perform controllability check
8. Plan for straight-in arrested landing
9. Fly minimum control airspeed for final approach and landing.

NOTE

- Refer to Approach Speeds chart for approximate approach airspeeds with SPC off.
- If divert field is not available, a 1/2 flap carrier landing should be attempted only under the most optimum conditions of estimated gross weight, arresting gear limitations, wind over the deck and weather.

If conditions for carrier landing are not optimum –

10. EJECT

ENGINE

GENERAL

Jet engine failures in most cases will be caused by improper fuel scheduling due to malfunction of the fuel control system or incorrect techniques used during certain critical flight conditions. Engine instruments often provide indications of fuel control system failures before the engine actually stops. If engine failure is due to a malfunction of the fuel control system or improper operating techniques, an air start can usually be accomplished, providing time and altitude permit. If engine failure can be attributed to some obvious mechanical failure within the engine proper, do not attempt to restart the engine.

SINGLE–ENGINE FLIGHT CHARACTERISTICS

Single–engine flight characteristics are essentially the same as normal flight characteristics due to proximity of the thrust lines to the center of the airplane. With one engine inoperative, slight rudder deflection is required to prevent yaw toward the failed engine. Thus, good control is assured in the single–engine range. Minimum single–engine control speed varies with gross weight, flap setting, and landing gear position. The aircraft design is such that no one system (flight control, pneumatic, electrical, etc), is dependent on a specific engine. Thus, loss of an engine will not result in a loss of a complete system.

ENGINE FAILURE DURING TAKEOFF

If an engine fails before leaving the ground, the continuation of takeoff is dependent on length of remaining runway, gross weight, airspeed, field elevation and ambient temperature.

NOTE

- During takeoff using Military power, where takeoff will not be aborted, immediately advance both engines to maximum power and follow Engine Failure During Flight procedures, this section, as soon as possible.
- If a single–engine failure occurs using maximum power and takeoff will not be aborted, retard "dead" engine throttle from afterburning range after safely airborne and follow Engine Failure During Flight procedures, this section, as soon as possible.
- On aircraft 155903ap and up, and all others after AFC 440, with flaps down, an engine failure causes a subsequent loss of trailing edge BLC on the same side as the failed engine. Due to asymmetric lift (wing adjacent to failed engine drops), mild trim control is necessary to correct this condition.

If decision to stop is made –

1. ABORT TAKEOFF
   Refer to Aborted Takeoff this section.

If takeoff is continued –

If an engine fails immediately after takeoff, lateral and directional control of the airplane can be maintained if airspeed remains above stalling speed. However, the ability to maintain altitude or to climb depends upon airplane gross weight and airspeed. If level flight cannot be maintained:

1. BOTH ENGINES - MAX (AFTERBURNER)
2. EXTERNAL STORES - JETTISON (IF NECESSARY)
   If altitude cannot be maintained, jettison the external stores.
3. Landing gear - UP
4. Wing flaps - UP
5. Failed engine - SHUTDOWN
6. Non–mechanical failure - INITIATE AIRSTART
7. Land as soon as practicable.

AFTERBURNER FAILURE DURING TAKEOFF

If the afterburner(s) fail during takeoff, the resulting loss of thrust is significant. Takeoff need not be aborted if remaining runway is compatible with power available. After failure, the variable area exhause nozzle will continue to function as directed by exhaust gas temperature. In this circumstance, the nozzle moves as a function of temperature limiting only. In the event of an afterburner failure, proceed as follows:

1. Throttle of failed afterburner - MIL
2. If the exhaust nozzle is operating properly, relight when desired.

ENGINE FAILURE IN FLIGHT

1. Non–mechanical failure – INITIATE AIRSTART
2. Mechanical failure – SHUTDOWN FAILED ENGINE
3. Land as soon as practicable.

DOUBLE ENGINE FAILURE IN FLIGHT

With double engine failure, flight below glide speed will result in degraded flight control response due to insufficient hydraulic pressure from the windmilling engines.

1. RAM AIR TURBINE – EXTEND
   Extend the ram air turbine to provide engine ignition and also to operate the left fuel boost pump at low speed. This will supply enough fuel to either engine for an airstart.
2. EITHER THROTTLE – OFF
3. OTHER ENGINE – INITIATE AIRSTART
   To provide maximum fuel flow for an airstart, retain the throttle of the remaining engine in the OFF position.
4. If no start within 30 seconds, throttle – OFF
5. Remaining engine – ATTEMPT AIRSTART
6. If neither engine can be started – NOTIFY RIO AND EJECT

AIRSTARTS

In general, airstart capability is increased by higher airspeeds and lower altitudes; however, airstarts can be made over a wide range of airspeeds and altitudes.

NOTE

If one or both engines flameout, do not delay the airstart. If no engine mechanical failure is immediately evident, depress and hold the ignition button(s) to restart the engine(s) before excessive rpm is lost.

1. Engine master switch(es) – CHECK ON
2. IGNITION BUTTON – DEPRESS (MORE THAN 12% RPM OPTIMUM)
3. THROTTLES – IDLE
4. EGT and rpm – MONITOR

NOTE

Be sure to give the engine sufficient time for a relight to occur before placing the throttle off and deciding that the engine is not going to start.

5. If any of the following conditions occur, retard the throttle to OFF.
   a. Lightoff does not occur within 30 seconds after ignition.
   b. The engine does not continue to accelerate after lightoff.
   c. The EGT exceeds maximum limitations.
   d. The oil pressure does not attain 12 psi minimum at idle.
6. Wait 30 seconds before initiating another restart.

RUNAWAY ENGINE/STUCK THROTTLE

There is no provision made on the main fuel control for stabilized engine rpm in the event the throttle linkage becomes disconnected from the fuel control. If a disconnect occurs, vibration may cause the fuel control to hunt or assume any position between idle and maximum power. In the event of a runaway engine while on the ground, shutdown the engine by turning the engine master switch OFF. For an inflight runaway engine (throttle disconnected from fuel control), the approach power compensator system (APCS) may be used to control engine rpm. If the throttle is stuck, the APCS cannot control the engine rpm. If the disconnected throttle linkage is not binding and rpm is stabilized between 73 and 99 percent, the use of the engine may be regained for a normal landing utilizing APCS. In the event that APCS disengages or control of rpm is lost, follow single–engine landing procedures this section.

If throttle linkage disconnected –

1. Plan for straight–in approach

At minimum 5000 feet AGL –

2. Speed brake switch – STOP
   APCS will not engage with speed brake switch in any position other than STOP.
3. APCS switch – STBY THEN ENGAGE
   With APCS engaged, the engine will operate between 73 and 99% rpm, depending on AOA.
4. Investigate handling characteristics

If APCS operates satisfactorily –

5. APCS – STBY
6. Landing gear – DOWN
7. Flaps – 1/2
8. APCS – ENGAGE
9. Fly ON–SPEED AOA

At touchdown –

10. Affected engine master switch – OFF
    The APCS is automatically disengaged on touchdown and the fuel control may assume any position between IDLE and MIL.

If throttle stuck or APCS will not control rpm –

1. Affected engine – SECURE BEFORE ENTERING PATTERN
   Should both throttles stick, pilot action will depend on thrust available versus thrust required for flight.

WARNING

With one engine at or near military power, the pilot's task is considerable. The decision to attempt a landing should be made only when conditions are optimum.

2. Follow Single Engine Landing procedure, this section.

VARIABLE AREA INLET RAMP FAILURE

RAMPS RETRACTED AT SPEEDS ABOVE 1.5 MACH

If the inlet ramps fail to move toward the extended (minimum duct area) position while accelerating between 1.5 and 1.8 Mach reduce airspeed to below 1.5 Mach and continue the mission. Engine performance and operating characteristics with the ramps failed to the retracted (maximum duct area) position are normal below 1.5 Mach.

CAUTION

A gradual failure of the inlet ramp to the extended position at any power setting from idle to max AB at airspeeds from 400 knots to landing approach speed does not cause unstable engine operation. However, a sudden failure of the inlet ramp to the extended position at high power settings and low airspeeds will disrupt inlet flow and may cause compressor stalls. In addition, compressor stalls may also occur at airspeeds above 1.7 Mach with the inlet ramps in the retracted position.

RAMPS EXTENDED BELOW 1.5 MACH

If the inlet ramps fail to the fully extended position, reduce engine power to below 80% rpm and descend to below 18,000 feet altitude. Engine operation will be unaffected below 18,000 feet altitude but a substantial loss of thrust will occur at all power settings. Jam accelerations, afterburner operation and airstarts may be performed without overtemperature conditions or compressor stalls below 18,000 feet altitude. It may be possible to retract a ramp which has failed in the EXTENDED position by cycling the applicable ramp control circuit breaker (J13 or J19 No. 1 panel before AFC 388, D12 or D13 No. 1 panel after AFC 388).

1. Applicable circuit breaker(s) – CYCLE (D12, D13, No. 1 panel)

If ramp(s) remain extended –

2. Reduce power below 80% and descent below 18,000 feet
3. Use power settings below 94%

CAUTION

Compressor stall and flameout may occur at power settings above 80% rpm at 18,000 feet altitude and above with the inlet ramps in the extended position.

Range and waveoff performance are degraded. Power settings above 94% rpm produce increased fuel flows without increasing engine thrust output. If the inlet ramps fail to the extended position, maintain the highest altitude below 18,000 feet at which the maximum range Mach number recommended for existing gross weight and configuration can be maintained with 94% rpm or less.

NOTE

With the inlet ramps extended, the reduction in maximum range varies from 5% at 10,000 feet to 18% at 30,000 feet. Single–engine range is reduced by 10% at all attainable altitudes.

LANDING

If both engines are operating, full–flap landings (both field and carrier) can be safely made with the inlet ramp fully extended. Normal power settings must be increased 1% to 2% rpm to maintain an on speed approach. Safe waveoffs can be performed with Military power at gross weights up to 33,000 pounds. At higher gross weights, afterburner may be required for a late waveoff.

SINGLE ENGINE LANDING

Single–engine carrier landings should not be attempted with the inlet ramps fully extended. Thrust necessary to maintain approach angle–of–attack and rate of descent would require a throttle setting between MIL and MIN AB. Waveoff performance, at all gross weights, is marginal under these conditions. Approximately 100 feet of altitude and 15 seconds are required to stop a normal rate of descent after MAX AB has been attained. Field landings at minimum gross weight can be made with 1/2 flaps. However, afterburner will be required occasionally to maintain normal approach angle–of–attack and rate of descent.

COMPRESSOR STALL

A compressor stall is an aerodynamic disruption of airflow through the compressor, and is caused by subjecting the compressor to a pressure ratio above its capabilities at the existing conditions. The compressor capability may be reduced by FOD, corrosion, misrigged or malfunctioning IGV's, or the compressor may be subject to abnormal operating conditions as a result of a malfunction of the ramp or bellmouth system. Compressor stalls may be self clearing, may cause the engine to flameout, or may result in a steady state, fully developed stall. The first case requires no immediate action. In the second case, the flameout clears the stall and an airstart is required. The third case requires recognition and corrective action to restore thrust and prevent engine damage by overtemperature. The stall can be recognized by the simultaneous existence of high EGT, low rpm, low fuel flow, open nozzle, loss of thrust, and lack of engine response to throttle. Compressor stalls may be accompanied by muffled bangs. The most positive stall clearing procedure is to shut down the engine and perform an airstart. Retard throttle to idle immediately to prevent engine over-temp. A throttle chop to IDLE may clear the stall if a significant fuel flow reduction from the stalled condition is achieved. In the event of a compressor stall, proceed as follows:

1. Throttle(s) – IDLE
2. Ignition – HOLD DEPRESSED
3. Throttle affected engine – OFF
4. Throttle – IDLE
5. RPM, EGT, and fuel flow – MONITOR

NOTE

If stall is cleared but desired thrust cannot be obtained because of repeated stalling, the engine may be operated at any obtainable rpm, as long as EGT is within limits.

EXHAUST NOZZLE FAILED OPEN

If an exhaust nozzle fails to the full open position, a significant loss of thrust will be noted; however, it is not necessary that the engine be shut down. Continued engine operation with a fully open nozzle will not damage the engine. The majority of exhaust nozzle failures result from engine oil starvation, but, because the CSD oil standpipe is above that of the nozzle control, oil starvation will usually be first indicated by a generator warning light. If the exhaust nozzle fails in the open position, military thrust on the effected engine will be reduced approximately 70%. Maximum afterburner thrust will be reduced approximately 20%. Retarding the throttle from maximum afterburner may cause afterburner blowout. Once afterburner is terminated, relight may require one to three minutes. In the event of a failed open nozzle:

1. Throttle (engine indicating failed open nozzle) – IDLE
2. Do not advance thrust on affected engine unless necessary.
3. Monitor oil pressure.
4. Land as soon as practicable.

BOTH EXHAUST NOZZLES FAILED OPEN

If both exhaust nozzles fail to the full open position, the thrust available above 80% rpm will be approximately equal to the thrust available during normal single-engine operation. Afterburner lightoff above 15,000 feet is marginal; however, afterburner lightoff probability increases with a decrease in altitude.

1. Follow single engine landing procedures.

EXHAUST NOZZLE FAILED CLOSED IN AFTERBURNER

Upon initiating afterburner, a rapid increase in exhaust temperature and a drop in rpm indicates that the exhaust nozzle has not opened. If an over-temperature condition exists:

1. Immediately move throttle to MIL or below.

CAUTION

Do not attempt to relight afterburner. Damage to engine and airframe structure could result.

AFTERBURNER BLOWOUT IN FLIGHT

In the event of an afterburner blowout or loss of afterburning, proceed as follows:

1. Throttle of failed afterburner – MIL
   The failed engine afterburner throttle should be moved inboard immediately to terminate fuel flow to the afterburner nozzles.
2. If no obvious overheat is discernible – RELIGHT AFTERBURNER
   If cockpit indications of resumed afterburner are normal, continue afterburner operation.

OIL SYSTEM FAILURE

An oil system failure of either engine is recognized by a drop in oil pressure or a complete loss of pressure. Since the constant speed drive unit which drives the generator is supplied with oil under pressure by the engine oil system, a GEN OUT light, followed by sluggish exhaust nozzle action, are early indications of impending engine oil starvation. The engine oil pressure gage should be monitored closely after a generator failure.On aircraft after PPC 62, L and R ENG OIL LOW lights are provided to give an indication of impending oil starvation. In general, it is advisable to shut the engine down as early as possible after a loss of oil supply is indicated, to minimize the possibility of damage to the engine and the constant speed drive unit. The engine will operate satisfactorily at military power for period of one minute with an interrupted oil supply. However, continuous operation, at any engine speed, with the oil supply interrupted will result in bearing failure and eventual engine seizure. The rate at which a bearing will fail, measured from the moment the oil supply is interrupted, cannot be accurately predicted. Such rate depends upon the condition of the bearing before oil starvation, temperature of the bearing and loads on the bearing. Malfunctions of the oil system are indicated by a shift (high or low) from normal operating pressure, sometimes followed by a rapid increase in vibration. A slow pressure increase may be caused by partial clogging of one or more oil jets; while a rapid increase may be caused by complete blockage of an oil line. Conversely, a slow pressure decrease may be caused by an oil leak; while a sudden decrease is probably caused by a ruptured oil line or a sheared oil or scavenge pump shaft. Vibration may increase progressively until it is moderate to severe before the pilot notices it. At this time, complete bearing failure and engine seizure is imminent. Limited experience has shown that the engine may operate for 4–5 minutes at 80 to 90% rpm before a complete failure occurs. In view of the above, the following operating procedures are recommended:

1. If a minimum oil pressure of 35 psi (MIL–L–23699 oil) at Military rpm cannot be maintained throttle – IDLE
2. If a minimum of 12 psi at idle rpm cannot be maintained or if the oil pressure change is accompanied by vibrations, engine – SHUTDOWN

CAUTION

Increasing vibration is an indication of bearing failure. Severe vibration indicates that engine seizure will occur within a few seconds. Chop the throttle to OFF to prevent major engine, and possible aircraft, damage.

3. If engine shutdown is not feasible – REDUCE THRUST
   Shut down the engine when partial power is not required from the affected engine. Where mission or flight requirements demand partial power from the affected engine, set engine speed at 86–89%.
4. Avoid abrupt maneuvers causing high positive G or negative G forces.
5. Avoid unnecessary or large throttle movements.

NOTE

To keep bearing temperature and loads at a minimum, do not use high thrust settings.

6. In either of the above instances – LAND AS SOON AS PRACTICABLE USING SINGLE–ENGINE PROCEDURE
   Use Single–Engine Landing procedure, this section.

ENGINE OIL LEVEL LOW LIGHT(S) ILLUMINATED

If operational requirements permit –

1. Affected engine – SHUT DOWN

NOTE

If no other system failure indications are apparent, the engine may be restarted for landing.

If shutdown not feasible –

1. Affected engine oil pressure – MONITOR
2. Affected engine GEN OUT light – MONITOR
3. Affected engine nozzle – MONITOR
4. Obtain visual inspection by another aircraft, if available.
5. Avoid high positive or negative G.
6. Land as soon as practicable.

FIRE

GENERAL

The pilot's first indication of a fire will be noted by the steady illumination of the FIRE/OVERHT warning light. However, a momentary illumination of the FIRE or OVERHT warning light should not be completely ignored. A momentary illumination should be followed by a check of the fire test circuit to determine if a fire actually exists and has burned through the fire detector wiring.

ENGINE FIRE DURING START

1. THROTTLES – OFF
2. ENGINE MASTER SWITCHES – OFF
3. GENERATOR CONTROL SWITCHES – EXT ON
4. CONTINUE TO CRANK ENGINE
5. Leave airplane as soon as possible.

ENGINE FIRE OR OVERHEAT DURING TAKEOFF

If FIRE or OVERHT warning light illuminates during takeoff roll, it is preferable to abort immediately if sufficient runway is available to stop safely. Refer to Critical Field Length charts, section XI, part 2.

If decision to stop is made –

1. ABORT TAKEOFF
   Refer to Aborted Takeoff, this section.

If takeoff is continued –

1. NORMAL OPERATING ENGINE – MAX (AFTERBURNER)
2. EXTERNAL STORES – JETTISON (IF NECESSARY)
3. THROTTLE (ENGINE INDICATING FIRE OR OVERHEAT) – IDLE
4. Landing gear – UP
5. Wing flaps – UP
6. Climb to safe ejection altitude and investigate.
   After reaching safe ejection altitude, proceed with Engine Fire or Overheat During Flight procedures.

ENGINE FIRE OR OVERHEAT IN FLIGHT

1. THROTTLE (ENGINE INDICATING FIRE OR OVERHEAT) – IDLE
2. If warning light goes out – CHECK FIRE DETECTION SYSTEM
   Depress fire test button to determine that the fire detecting elements are not burned through.
3. If fire detection system check is normal, land as soon as practicable; do not advance thrust on the affected engine unless absolutely necessary.

CAUTION

Advancing thrust on the affected engine after the throttle has been retarded and the warning light has been extinguished may cause fire or overheat damage, and/or possibly burn through the fire detecting elements.

If light remains on and/or fire is confirmed –

4. Throttle – OFF
   Shut down engine if the FIRE/OVERHT warning light remains illuminated after reducing thrust; if the fire detection system has been found to be inoperative; or fire is apparent.
5. Engine master switch – OFF
6. If fire persists – NOTIFY RIO & EJECT
7. If fire ceases – LAND AS SOON AS PRACTICABLE.

AFT SECTION FUEL LEAK/FIRE

In-flight fires may be caused by fuel or hydraulic leaks in the aft fuselage/empennage sections without the engine or overheat warning circuits being activated. Hence, secondary indications must be used to detect this condition. Such fires often relate to maneuvering flight conditions and afterburner use. Discovery of any fluid leaking in the aft sections prior to a fire being ignited is critical. Any indication of abnormal (i.e., prolonged or continuous) fuel streaming from the fuselage fuel vent, empennage, or engine nozzle areas must be considered a potential aft section fire situation. Indications that an aft section fire has already occurred include: fuel streaming, noises similar to compressor stalls, utility hydraulic failure, pronounced lateral control stick transients, loss of rudder control, loss of the rudder position indicator, or the popping of any rudder-associated circuit breakers.

If fuel streaming occurs or fire is evident –

1. Disengage afterburner. Do not re-engage.
2. Cease fuel transfer.
3. Reduce speed to below 300 knots.
4. Air refuel switch – REFUEL
5. Do not allow subsequent fuel transfer to exceed 3500 pounds fuselage fuel
6. Prepare for ejection.

WARNING

With an aft section fire, simultaneous loss of PC-1 and PC-2 pressure to the stabilator is imminent. This will cause a violent noseup pitching condition and loss of flight control.

7. Land as soon as possible.
8. Hook bypass switch – NORM
   If the hook bypass switch is in BYPASS when the right aux air door circuit breaker is pulled, the corner reflector will extend and, if above 250 knots, may depart the aircraft causing a utility hydraulic failure and/or engine FOD.
9. Aux air door circuit breakers – PULL PRIOR TO LOWERING LANDING GEAR (3D, 4D, before AFC 388; 15K, 16K, after AFC 388; No. 1 panel)
10. Field arrested landing not recommended
    Sparks and heat from a lowered arresting hook may ignite the streaming fuel.
11. Throttles – OFF AFTER AIRCRAFT COMES TO A COMPLETE STOP
12. Engine master switches – OFF

FIRE AFTER CATAPULT LAUNCH

An external fire after catapult launch may not be indicated by cockpit fire or overheat warning lights. The aircrew's first warning that such a fire exists may come from a source outside the aircraft, i.e., another aircraft, ship's tower, LSO, etc. A centerline tank fire, an aft section fire, or an engine fire could be similar in appearance to outside observers and difficult to distinguish immediately from within the cockpit. If advised of an external fire after catapult launch, immediately jettison all external stores using the external stores emergency release button. This eliminates explosive or inflammable stores, reduces A/C gross weight, and optimizes the climb to a safe altitude. The effect of jettisoning a partially full centerline tank will be minimized. If afterburner was selected for launch, select military power as soon as maximum thrust is not essential.

If fire persists and no cockpit indications are present –

1. Follow Aft Section Fire procedures

If cockpit indications of fire are present –

1. Follow Engine Fire Or Overheat procedures as appropriate

If a fire exists after catapult launch –

2. External stores – JETTISON

NOTE

On airplane 158355at and up, the three position jettison select switch on the air-to-air panel shall be placed in the ALL or AIR-GRD+CL position prior to launch. In the AIR-GRD position the centerline tank will not jettison upon activation of the external stores emergency release button.

WARNING

Collision of a partially full centerline tank with the aircraft is possible and may cause reduced control effectiveness, abrupt-but momentary-loss of control or even total loss of control. Should control be lost and not regained immediately, eject.

3. Throttles – MIL (if max thrust not required)

If fire persists –

4. Continue climbing to safe ejection altitude and follow Aft Section Fire or Engine Fire/Overheat procedures as indications dictate.

ENGINE FIRE DURING SHUTDOWN

1. THROTTLES – OFF
2. MASTER SWITCHES – OFF
3. Generator switches – EXT ON
4. Engine start switch – LEFT or RIGHT (engine indicating fire)
5. Leave airplane as soon as possible.

ELECTRICAL FIRE

1. Generator switches – OFF
2. Ram air turbine – EXTEND
3. All electrical switches – OFF
4. When fire subsides, generator control switches – ON
5. If fire still persists when generator switches are turned ON, turn generator switches OFF
   The RAT will supply power to the essential buses only.
6. Land as soon as practicable.
7. If no fire is apparent when generator switches are turned ON, individually reposition only the mission essential equipment first.
8. If malfunctioning item is found, turn electric power switch OFF and pull applicable circuit breaker.

ELIMINATION OF SMOKE AND FUMES

1. Emergency vent knob – PULL
2. Initiate an immediate descent to 25,000 feet or below

If smoke or fumes persist –

3. Command selector valve handle – AS DESIRED
4. Rear ejection seat – FULL DOWN
5. Aft canopy – JETTISON

If smoke or fumes still persist –

6. Put the aircraft in a nose high attitude
   A nose high attitude precludes front canopy to rear cockpit collision.
7. Front canopy – JETTISON

WARNING

The aft cockpit occupant must eject first with the front canopy off. A possibility of seat to seat collision exists if ejection is sequenced from the front cockpit with the front canopy off.

FLIGHT CONTROLS

GENERAL

Upon initial detection of any abnormal flight control movement, immediately depress the paddle switch to determine if the stab aug ARI or AFCS was causing the abnormality.

RUNAWAY STABILATOR TRIM

If stabilator trim appears to be running away, it is possible under certain conditions to lessen the situation. Runaway stabilator trim can be alleviated by engaging the AFCS, providing:

a The stab feel trim circuit breaker has been pulled immediately upon detection of runaway trim.

b The runaway trim has not exceeded 2 ½ units nose down.

If the above conditions are met:

1. Reduce airspeed – BELOW 300 KNOTS
2. AFCS – ENGAGE

NOTE

- The AUTO PILOT PITCH TRIM light may illuminate when the AFCS is engaged.
- When the AFCS is used to alleviate a runaway trim condition and excessive out of trim forces are present (full nose down runaway trim), the AFCS may not be able to hold the engagement pitch attitude and pitch oscillations could result. If this occurs, disengage the AFCS.

3. Plan to land as soon as practicable.
4. Prior to landing disengage AFCS.
   When in the landing configuration (gear and flaps extended) and at 180 to 190 knots, grasp stick firmly and disengage the AFCS. Depending upon severity of the malfunction, the airplane may or may not be in trim; if out of trim, the forces should not be too high and the airplane can be landed with the out–of–trim condition, or the AFCS can be reengaged and the landing made with control stick steering.
5. If landing is made with AFCS engaged disengage immediately after touchdown.
   Immediately after touchdown disengage AFCS to prevent damage to AFCS components.

Flight below 350 KCAS with stabilator trim full nose up or down does not require extremely high stick forces.

If runaway stabilator trim has exceeded the limits at which the AFCS can be engaged –

1. Reduce airspeed – BELOW 300 KNOTS
2. Land as soon as practicable

STABILATOR FEEL TRIM FAILURE

PARTIAL BELLOWS FAILURE

Partial bellows failure is recognized by a mild nose down stick force proportional to the airspeed unless the failure occurs during maneuvering flight at which time it may not be noticeable. Reduction of stick centering and pitch stability will result. If this failure occurs:

1. Reduce airspeed – 250–300 KNOTS
2. Retrim airplane.
3. Avoid abrupt fore and aft stick movements.
4. Land as soon as practicable.

COMPLETE BELLOWS FAILURE

A complete bellows failure is recognized by a nose down feel force heavier than with partial bellows failure, with a maximum nose down feel force of 5 pounds/G. This force cannot be trimmed out. If a complete bellows failure occurs:

1. Reduce airspeed to 250–300 KNOTS
2. Set trim for your airspeed
3. Avoid abrupt fore and aft stick movements.
4. Pitot heat – ON
5. Land as soon as practicable.

AILERON RUDDER INTERCONNECT (ARI) SYSTEM DISENGAGEMENT

The ARI system can be temporarily disengaged by depressing the AFCS/ARI emergency disengage switch; this will disengage the ARI only as long as it is held depressed. To permanently disengage the ARI system, the ARI circuit breaker on the left utility panel must be pulled and the yaw stab aug switch must be disengaged. Pulling the circuit breaker only, and keeping the stab aug engaged still provides 5° of ARI rudder authority. To permanently disengage the ARI while retaining complete stab aug, pull the rudder feel trim circuit breaker on circuit breaker panel No. 1. For an ARI system malfunction, proceed as follows:

1. To temporarily disengage system, depress AFCS/ARI emergency disengage switch. Depressing the emergency disengage switch will disengage the ARI system only as long as the switch is held depressed.
2. ARI circuit breaker, front cockpit – PULL
3. To permanently disengage system, pull rudder feel trim circuit breaker (G15 before AFC 388 – B13 after AFC 388, No. 1 panel).

HARDOVER RUDDER

A hard over rudder that cannot be corrected by disengaging the AFCS, yaw stab aug, and rudder trim is in all probability caused by a mechanical malfunction. In this event, the only possible corrective action is an intentional failing of the utility system. If the rudder still fails to streamline, the rudder is being held hard over by a mechanical jam from some unknown object. If a hardover rudder occurs:

1. Emergency quick release lever – DEPRESS
2. Maintain 250 knots and 5,000 feet altitude minimum.
3. Yaw stab aug – DISENGAGE
4. Rudder trim circuit breaker – PULL

If hardover rudder persists, fail utility system –

5. Gear – DOWN
6. Flap circuit breaker – IN
7. Flaps – BLOW DOWN

NOTE

It may take as long as 8 minutes before the utility system eventually fails.

If hardover rudder still persists –

8. Perform controllability check.

If minimum control speed/gross weight combinations do not exceed available arresting gear limits –

9. Land as soon as practicable.

If control speed/gross weight combinations exceed available arresting gear limits –

10. Eject

FUEL

FUEL BOOST PUMP FAILURE

The possibility of a simultaneous mechanical failure of both boost pumps is highly remote; however, double failure may occur as a result of an electrical malfunction. Provisions are made to supply fuel to the engines by gravity flow in the event of a double boost pump failure. This will allow engine operation at a reduced power setting. If both boost pumps fail above 20,000 feet and/or at a high power setting, flameout or an unstable rpm indication on one or both engines may occur.

Double boost pump failure –

1. If both engines have flamed out, execute procedure for double engine failure.
2. If an airstart has been accomplished, or the engines have not flamed out, adjust engine thrust or descend until a stable rpm can be maintained.

NOTE

Afterburner operation is not recommended. Military power operation is unrestricted at any airspeed from sea level to 20,000 feet.

3. Land as soon as practicable.

Single boost pump failure –

1. Adjust throttle to maintain a minimum of 5 psi boost pump pressure, if practicable.

NOTE

Afterburners may have to be modulated or shut off, depending on airspeed and altitude, to maintain 5 psi boost pressure.

FUEL TRANSFER FAILURES

INTERNAL WING FUEL FAILS TO TRANSFER

Failure of internal wing fuel to transfer can be caused by either the wing tanks failing to pressurize, or the wing transfer valves failing to open. If internal wing fuel is not transferring, check the following switch and circuit breaker positions:

1. External transfer switch – OFF (some aircraft)
2. Internal wing transfer switch – NORMAL (some aircraft)
3. Fuel transfer selector knob – INT WING (some aircraft)
4. Refuel probe switch – RETRACT
5. Wing transfer control circuit breaker – IN (H4 before AFC 388 – J13 after AFC 388, No. 1 panel)

To ensure wing tank pressurization –

6. Wing transfer pressure switch – EMERG
   Continue in level flight for 30 seconds.

NOTE

If fuel transfers when the wing transfer pressure switch is placed in EMERG, leave the switch in this position for remainder of flight.

Pressurization has been obtained and fuel can be transferred.

CENTERLINE FUEL FAILS TO TRANSFER

Failure of the centerline fuel to transfer can be caused by either the defueling shutoff valve failing to the open position, the refueling valve failing to the closed position, the tank shutoff valve failing to the closed position, or the tank failing to become pressurized. If the centerline fuel fails to transfer, check the following switch and circuit breaker positions:

1. Buddy tank switch – STOP FILL
2. External transfer switch – CENTER (some aircraft)
3. Fuel transfer selector knob – CTR (some aircraft)
4. External wing fuel control circuit breaker – IN (H3 before AFC 388, J12 after AFC 388, No. 1 panel)
5. L fuel valve power circuit breaker – IN (G4 before AFC 388, J15 after AFC 388, No. 1 panel)
6. Refuel probe switch – RETRACT

To ensure tank pressurization –

7. Wing transfer pressure switch – EMERG
   Continue in level flight for 30 seconds. If fuel transfers when the wing transfer pressure switch is placed in EMERG, leave the switch in this position for remainder of flight. Pressurization has been obtained and fuel can be transferred.
8. Buddy tank switch – CYCLE
9. Refuel probe switch – CYCLE
10. Apply alternating symmetrical positive and negative G up to the acceleration limits of the store

EXTERNAL WING FUEL FAILS TO TRANSFER

Failure of the external wing fuel to transfer can be caused by either the external wing tank shutoff valve failing to the closed position or the tanks failing to become pressurized. If the external wing tanks fail to transfer, check the following switch and circuit breaker positions:

1. External wing transfer switch – OUTBD (some aircraft)
2. Fuel transfer selector knob – OUTBD (some aircraft)
3. External wing fuel control circuit breaker – IN (H3 before AFC 388, J12 after AFC 388, No. 1 panel)
4. Refuel probe switch – RETRACT
5. Fuel valve power circuit breaker – IN (G4 before AFC 388, J15 after AFC 388, No. 1 panel)

To ensure tank pressurization –

6. Wing transfer pressure switch – EMERG

NOTE

- If fuel transfers when the wing transfer pressure switch is placed in EMERG, leave the switch in this position for remainder of flight. Pressurization has been obtained and fuel can be transferred.
- If the above procedures fail to produce fuel transfer from either centerline or external wing tanks, it may be possible to effect transfer by using high power settings for a period of time and/or climbing to an altitude of 35,000 feet or higher.

REVERSE TRANSFER OF FUSELAGE FUEL

Reverse transfer is caused by a failed open wing fuel level control valve and/or a failed open defuel valve. Reverse transfer normally does not occur until the fuel tanks are depressurized. The fuel sector and counter should be monitored after lowering the landing gear or placing the refuel probe switch to REFUEL. If external wing tank fuel is available, and transferring, the counter may slowly increase. Boost pump pressure forces fuel from the engine fuel manifold through the open defuel valve to the internal wing tanks. If reverse fuel transfer occurs, proceed as follows:

1. Buddy tank switch – STOP FILL
2. Ram air turbine – EXTEND
3. Generators – OFF
4. Wing transfer control circuit breaker – IN (H4 before AFC 388, J13 after AFC 388, No. 1 panel)
5. Boost pump control circuit breakers – PULL (A11, D8, E11, before AFC 388; A6, H6, H7, after AFC 388; No. 1 panel)
   With boost pump circuit breakers pulled, it is recommended that flight be restricted to below 20,000 feet.
6. Generators – ON
7. Ram air turbine – RETRACT

FUEL STREAMING FROM UNDERSIDE AFTER CENTERLINE TANK JETTISON

If the centerline tank is jettisoned with an open defuel valve, cockpit indications are identical to Reverse Transfer of Fuselage Fuel. Fuel loss on the sector and counter will be 500 to 700 pounds per minute and will not cease when a low fuel state is reached. The fuel streaming may be confirmed by a wingman or the RIO in his external rear view mirror.

1. Buddy tank switch – STOP FILL
2. Wing transfer pressure switch – NORMAL
3. Refuel probe circuit breaker – PULL (G5; J14 after AFC 388; J5 after AFC 545; No. 1 panel)
4. Refuel probe switch – REFUEL

If fuel continues to stream –

5. Boost pump control circuit breakers – PULL (A11, D8, E11, before AFC 388; A6, H6, H7, after AFC 388; No. 1 panel)
   Pulling the boost pump control circuit breakers will minimize fuel loss. With the circuit breakers pulled, it is recommended that flight be restricted to below 20,000 feet.
6. Hook bypass switch – NORM
   If the hook bypass switch is in BYPASS when the right aux air door circuit breaker is pulled, the corner reflector will extend and, if above 250 knots, may depart the aircraft causing a utility hydraulic failure and/or engine FOD.
7. Aux air door circuit breakers – PULL PRIOR TO LOWERING LANDING GEAR (3D, 4D, before AFC 388; 15K, 16K, after AFC 388; No. 1 panel)
8. Field arrested landing not recommended
   Sparks and heat from a lowered arresting hook may ignite the streaming fuel.
9. Throttles – OFF AFTER AIRCRAFT COMES TO A COMPLETE STOP
10. Engine master switches – OFF

FAILED OPEN DEFUEL VALVE DURING AIR REFUELING

When the refuel probe switch is placed to REFUEL and there is an indication of a failed open defuel valve, proceed as follows:

1. Buddy tank switch – STOP FILL
2. Refuel probe switch – EXTEND
3. External transfer switch – OFF (some aircraft)
4. Fuel transfer selector knob – STOP (some aircraft)
5. Refuel selector – INT ONLY
6. Boost pump control circuit breakers – PULL
   a. L.H. boost pump norm cont – E11 before AFC 388, H7 after AFC 388, No. 1 panel
   b. L.H. boost pump EMER cont – A11 before AFC 388, H6 after AFC 388, No. 1 panel
   c. R.H. fuel boost pump control – D8 before AFC 388 A6 after AFC 388, No. 1 panel

NOTE

With boost pump circuit breakers pulled, it is recommended that air refueling be accomplished at altitudes of less than 20,000 feet.

7. Refuel probe switch – REFUEL
8. Commence refueling

When approximately 6000 pounds is indicated on the sector –

9. Boost pump control circuit breakers – RESET
10. Refuel selector – ALL TANKS
11. Continue refueling.

FUSELAGE FUEL DUMP

Transfer of Fuselage Fuel to Internal Wings for Dumping

To allow expeditious reduction to landing weight, excess fuel can be transferred from fuselage cells 1 and 2 to the internal wings allowing it to be dumped using normal dump procedures. Reverse transfer can be accomplished by the following procedures:

1. Wing transfer control circuit breaker – PULL (H4 before AFC 388, J13 after AFC 388, No. 1 panel)
2. Buddy fill switch – FILL
3. Internal wing dump switch – DUMP.

NOTE

- Fuel transfers at approximately 1000 lbs per minute and dumps at approximately 680 lbs. per minute.
- Care must be exercised when using reverse transfer, for it is possible to deplete the fuel supply in fuselage cells 1 and 2.
- With centerline tank aboard, some fuel is transferred into the centerline tank using reverse fuel transfer.

After fuel is transferred/dumped –

4. Buddy fill switch – STOP FILL
5. Internal wing dump switch – NORMAL

WING FUEL LEAK

1. Afterburner – DISENGAGE, DO NOT REENGAGE
2. Speed – REDUCE TO BELOW 300 KNOTS
3. Air refuel switch – REFUEL
4. Land as soon as practicable
5. Arrested landing recommended
   With a serious wing fuel leak, fuel dripping in the vicinity of the main landing gear constitutes a strong potential for fire if hot brakes occur during a landing rollout.

AIR REFUELING FUSELAGE TANKS ONLY

If the internal wing tanks become damaged and cannot hold fuel, emergency refueling of the fuselage tanks only can be accomplished as follows:

1. External transfer switch – OFF (some aircraft)
2. Fuel transfer selector knob – STOP (some aircraft)
3. Refuel selector switch – INT ONLY
4. Fuel valve power circuit breaker – PULL (G4 before AFC 388, J15 after AFC 388, No. 1 panel)
5. Refuel probe switch – REFUEL
6. Commence refueling.

NOTE

Do not attempt to refuel external tanks. Damage to internal wing tanks may prevent external wing tanks from transferring. The centerline tank cannot be refueling using the above procedure.

HYDRAULIC

GENERAL

The loss of a hydraulic pump in power control systems No. 1, No. 2 or in the utility hydraulic system, is indicated by the illumination of the CHECK HYD. GAGES warning light. This single light serves all three systems, and the pilot should check the hydraulic gages to determine which system has malfunctioned.

SINGLE POWER CONTROL SYSTEM FAILURE

A hydraulic pump failure of either PC–1 or PC–2 presents no immediate problem since the utility system is capable of assuming the full demand of either system. In the event of a single PC system failure:

1. Land as soon as practicable.

NOTE

- The AFCS, if engaged, should be immediately disengaged since the stab aug system will be lost and erratic AFCS operation may occur due to the loss of AFCS stabilator authority. No indication of a stab aug failure will be noted since the PITCH AUG OFF light will only illuminate upon switch disengagement, and in case of a PC–1 failure the switch will remain engaged. With a PC–1 failure, disengage pitch stab aug to avoid erratic pitch oscillations caused by fluctuating hydraulic pressure.
- If the CHK HYD GAGES indicator light illuminates and remains illuminated, monitor the hydraulic system gages for the remainder of the flight since warning of a second hydraulic system failure will not be given.

SINGLE POWER CONTROL AND UTILITY SYSTEM FAILURE

In the event of simultaneous loss of the utility system and one of the power control systems, the operable aileron and spoiler provides adequate lateral control for an emergency landing:

1. Reduce speed to below 500 knots as soon as possible and discontinue mission.

WARNING

Do not allow airspeed to drop below 220 KCAS.

2. Jettison any asymmetric load.
3. Proceed to a suitable divert field.
4. Emergency aileron droop switch – DISABLE (some aircraft)

WARNING

If emergency aileron droop switch is not placed to DISABLE or, on aircraft prior to 154786ag, essential 28 volt dc bus power is not available, a split aileron condition will occur if flaps are lowered pneumatically.

5. Extend gear and flaps pneumatically at a minimum altitude of 5,000 feet.
6. Investigate landing approach characteristics before descending for a landing.
7. Enter a wide and extended downwind leg.
8. Plan early line–up for straight–in approach (select runway to minimize crosswind, if possible). Full lateral stick displacement may be necessary to provide adequate roll response.
9. Make field arrested landing – fly minimum control airspeed for final approach and landing

CAUTION

In the event the arresting gear is missed and the decision is made to go around, the airplane should not be rotated until the airspeed is at least that corresponding to the angle of attack used during the approach.

If field landing cannot be made –

10. Eject

NOTE

The minimum control airspeed is greater than carrier arresting gear limits.

DOUBLE POWER CONTROL SYSTEM FAILURE

The pilot should, upon initial detection of hydraulic power loss, note trend of failure as to whether the gages show a definite steady drop, or gage fluctuations. With a steady drop indication, hydraulic power will probably not recover. In the event of complete power control hydraulic failure, the aircraft will become uncontrollable.

1. If hydraulic pressure does not recover – NOTIFY RIO AND EJECT
2. If hydraulic pressure recovers – LAND AS SOON AS PRACTICABLE

UTILITY HYDRAULIC SYSTEM FAILURE

Failure of the utility hydraulic system will prevent/degrade hydraulic operation of the following essential items:

a. Auxiliary Air Doors
   There are no alternate or emergency provisions for opening the auxiliary air doors. Refer to Auxiliary Air Door Malfunction, for procedures to follow in the event of an auxiliary air door failure.
b. Flaps/Drooped Ailerons
   Emergency pneumatic operation is provided. Refer to Flap Emergency Extension, this section.
c. Fuel Transfer Pumps (hydraulic)
   There are no alternate or emergency provisions for hydraulic transfer pump operation.
d. Landing Gear
   Emergency pneumatic operation of the landing gear is provided. Refer to Landing Gear Emergency Lowering, this section.
e. Nose Gear Steering
   There are no emergency provisions for nose gear steering.
f. Rudder Power Control System
   Limited manual control of the rudder is available; however, pedal forces will be much higher than normal.
g. Variable Bypass Bellmouth
   There are no alternate or emergency provisions for operation of the variable bypass bellmouth. Refer to Auxiliary Air Door Malfunction for side effects of simultaneous variable bypass bellmouth and auxiliary air door failures.
h. Variable engine Intake Ramps
   There are no alternate or emergency provisions for operation of the variable engine intake ramps.
i. Wheel Brakes
   Emergency pneumatic and hydraulic operation of the wheel brakes is provided. Refer to Wheel Brake Emergency Operation, this section.

If a utility hydraulic system failure occurs, proceed as follows:

NOTE

If the CHK HYD GAGES indicator light illuminates and remains illuminated, monitor the hydraulic system gages for the remainder of the flight, since warning of a second hydraulic system failure will not be given. With a utility hydraulic failure, disengage roll and yaw stab aug to avoid control surface transients caused by fluctuating utility hydraulic pressure.

1. Land as soon as practicable.
2. Landing gear – BLOW DOWN WHEN READY FOR LANDING
   If the landing gear is down and locked, a subsequent utility hydraulic or electrical failure does not require that the landing gear be blown down.

For field arrested landing –

3. Flaps – UP
4. Engines – SHUT DOWN AFTER ARRESTMENT

For carrier landing or field arresting gear not available –

3. Flaps – BLOW DOWN WHEN READY FOR LANDING
4. Engines – SHUT DOWN AFTER ARRESTMENT

DIRECTIONAL CONTROL WITH UTILITY HYDRAULIC SYSTEM FAILURE

Without utility hydraulic pressure nose gear steering is lost and rudder control reverts to manual operation. Only small rudder deflections are available at landing speeds because of the high pedal forces required to overcome the airloads. Therefore, differential braking becomes the primary heading control during landing and is accomplished through the use of the brake hydraulic accumulator pressure (several applications normally are available), or through application of the emergency brakes (pneumatic) in conjunction with manual braking (which requires high pedal forces and large pedal deflections) for differential control. Use of accumulator pressure provides braking action and "feel" identical to the normal utility system but the number of applications are limited. For this reason the pilot should hold steady pedal pressure to obtain braking action, once pedal force has been applied, even though differential pedal force may be required. If accumulator pressure becomes depleted through consecutive brake applications or a system malfunction, manual brakes must provide any differential braking required to maintain heading while using emergency pneumatic brakes to stop the aircraft. It should be reemphasized that the normal brake system accumulator is capable of providing sufficient brake power to stop the aircraft while providing heading control through differential braking. Ailerons and spoilers provide some directional control and will be more effective at higher speeds. The drag chute should be used with caution in a strong crosswind if the utility hydraulic system has failed. A deployed drag chute in a strong crosswind will increase substantially the requirement for differential braking. This is especially applicable when the hydraulic accumulator pressure has been depleted and the emergency pneumatic brakes are employed to brake the aircraft, which leaves only the manual brake, with associated high pedal forces and deflections, for heading control.

LANDING GEAR UNSAFE

An unsafe gear indication does not necessarily constitute an emergency. The unsafe indication could be caused by a malfunction within the indicating system, or the result of incorrect gear lowering procedure coupled with a low pressure condition of the utility hydraulic system. Upon initial detection of unsafe gear indication, proceed as follows:

1. Airspeed – BELOW 250 KNOTS
2. If utility hydraulic pressure is normal, recycle landing gear.
3. Landing gear position indicators – CHECK

If unsafe condition still exists –

4. Landing gear circuit breaker – CYCLE
5. Landing gear handle – UP
6. Apply negative G to airplane.
7. While under negative G, place gear handle down. Negative G will help if unsafe gear is caused by high break-out forces.
8. If unsafe condition still exists, utilize landing gear emergency lowering procedure.

LANDING GEAR EMERGENCY LOWERING

If normal gear operation fails, the gear can be lowered by utilizing the following procedures:

CAUTION

If, after a utility hydraulic failure, the pneumatic pressure begins to drop and fuel permits, extend the landing gear before the pressure drops below 2000 psi.

1. Airspeed – BELOW 250 KNOTS
2. Landing gear handle – DOWN
3. Landing gear circuit breaker – PULL (do not reset)
4. Landing gear handle – PULL AFT AND HOLD
   Pull handle full aft, (full limit of travel) and hold in full aft position until gear indicates down and locked.

CAUTION

- Hold handle in full aft position until gear indicates down and locked, and then leave the landing gear handle in the full aft position. Returning the handle to its normal position allows compressed air from the gear down side of the actuating cylinder to be vented overboard.
- If the landing gear is inadvertently extended in flight by emergency pneumatic pressure, it must be left in the extended position until post flight servicing. If retraction in flight is attempted, rupture of the utility reservoir will probably occur with subsequent loss of the utility hydraulic system.

NOTE

It is possible to actuate the landing gear emergency system by pulling the landing gear control handle aft while the handle is in any position from UP thru DOWN. If the handle cannot be pulled aft while in the down position, slowly raise the handle while continuing to pull aft. Once the handle moves aft, hold the handle in the full aft position until the landing gear indicates down and locked; then continue to hold back pressure on the handle and return it to the full down position.

5. Landing gear position indicators – CHECK

CAUTION

To prevent drop tank collapse during high altitude descent with wheels down, place wing transfer pressure switch to EMERG. Prior to landing, return wing transfer pressure switch to NORMAL.

If landing gear is still unsafe –

6. Yaw airplane to assist locking main gear, or bounce airplane on main gear to assist lowering/locking the nose gear.
7. If one main gear is still unsafe and the utility system pressure is within limits, refer to Landing Gear Emergency Retraction this section.
8. If gear is still unsafe and the utility system pressure is not within limits, refer to Landing Gear Malfunction – Emergency Landing Guide, figure 5-9.

NOTE

Any pneumatic extension of the landing gear shall be logged in the yellow sheet (OPNAV Form 3760-2).

LANDING GEAR EMERGENCY RETRACTION

If gear retraction is desired after an attempted Landing Gear Emergency Lowering, and the utility system pressure is within limits, retract the gear using the following procedures:

NOTE

Unless operational necessity dictates otherwise, this procedure should be used only when an unsafe gear condition exists after an attempted

LANDING GEAR MALFUNCTION-EMERGENCY LANDING GUIDE

FINAL CONFIGURATION	CARRIER LANDING		FIELD LANDING ARRESTING GEAR AVAILABLE		FIELD LANDING NO ARRESTING GEAR	
		Notes		Notes		Notes
ALL GEAR UP	EJECT	2,4,5,12	NO ARRESTED LANDING	3, 5, 7, 8, 12, 14, 15	LAND	3, 5, 7, 8, 12, 14, 15
NOSE GEAR UP	LAND	4,12	NO ARRESTED LANDING	5, 7, 8, 12, 14, 15, 16	LAND	5, 7, 8, 12, 14, 15, 16
STUB NOSE GEAR	LAND	4	NO ARRESTED LANDING	1, 5, 7, 8, 13, 14, 15, 16	LAND	1, 5, 7, 8, 14, 15, 16
ONE MAIN GEAR UP	LAND	4, 5	**ARRESTED LANDING**	1, 5, 6, 14, 15	LAND	1, 5, 7, 8, 10, 11, 14
ONE OR BOTH STUB MAIN GEAR	LAND	4	NO ARRESTED LANDING	1, 5, 7, 8, 10, 11, 14	LAND	1, 5, 7, 8, 10, 11, 14
ONE MAIN GEAR UP NOSE GEAR UP	EJECT	1,2,4,5,12	NO ARRESTED LANDING	1, 3, 5, 7, 8, 10 11, 12, 14, 16	LAND	1, 3, 5, 7, 8, 10 11, 12, 14, 16
BOTH MAIN GEAR UP	LAND	3,4,5	NO ARRESTED LANDING	5, 7, 8, 9, 11, 14, 16	LAND	5, 7, 8, 11, 14, 16

Notes

1. If External tanks installed and landing gear can be actuated, emergency retract all gear. Refer to All Gear Up Configuration.
2. Option to land if wing tanks installed.
3. Option to eject if wing tanks not installed.
4. Hook down barricade engagement in accordance with current aircraft barricade engagement recovery bulletin.
5. Retain External tanks.
6. Foam runway beyond arresting gear if practicable.
7. Foam runway if practicable.
8. Deploy drag chute at touchdown in accordance with crosswind landing criteria.
9. Option of midfield arrestment if 4000 feet of runway plus overrun available.
10. Land off–center to gear down side/good tire.
11. Keep engines operating in order to retain nose wheel steering and/or power boost brakes.
12. Angle of attack will indicate 3 to 4 units low with the nose gear up.
13. Remove all field arresting cables (for field landing with stub nose gear).
14. Land with fuel weight as low as practicable.
15. Both engines, secure after touchdown.
16. Hold nose off the runway until stabilator effectiveness is lost.

- Multiple emergencies, adverse weather and other peculiar conditions may require modifications to these procedures.
- If available, an LSO is recommended for all field arrested landings.
- If a gear up landing is made with external tanks aboard, depressurize the external tanks by first pulling the landing gear circuit breaker, and then place the landing gear handle down.
- A field arrested landing should not be attempted during any gear malfunction involving a stub main gear due to the probability of severing the arresting cables with the stub.
- During some emergencies it is desirable to remove certain crossdeck pendants or field arresting cables. For additional information see applicable aircraft recovery bulletins.
- Jettison all external ordnance including missiles.

FDD-1-(66)B

Figure 5-9

Landing Gear Emergency Lowering. It should not be used if the emergency gear was inadvertently actuated and all three gears are down and indicating locked.

1. Return the emergency gear handle to the normal position.
2. Wait a minimum of 1 minute.
3. Landing gear handle – UP
4. Landing gear circuit breaker – RESET

CAUTION

The landing gear circuit breaker must not be reset until the emergency handle is returned to normal, maintained in that position for a minimum of 1 minute, and the landing gear handle placed UP. Only then may the circuit breaker be safely reset.

5. After gear is retracted, refer to Landing Gear Malfunction – Emergency Landing Guide, figure 5-9.

FLAP EMERGENCY EXTENSION

The flaps may be pneumatically extended to the 1/2 position if normal extension fails. Emergency extension of the flaps with good utility hydraulic pressure will eventually cause utility hydraulic system failure with loss of nosewheel steering and normal wheel brakes. Emergency flap extension should normally be used only if there is an existing utility hydraulic system failure. For carrier landings or field landings where runway length and/or arresting gear limits dictate, the flaps should be pneumatically extended even if there is no failure of the utility hydraulic system. Any pneumatic extension of the flaps shall be logged on the yellow sheet (OPNAV Form 3760.2).

CAUTION

If the flaps are inadvertently extended by emergency pneumatic pressure, they must be left in the extended position. Rupture of the utility hydraulic reservoir and loss of the utility hydraulic system will occur if retraction is attempted.

On aircraft before AFC 534, the ailerons will not droop when the flaps are extended by the emergency system. On aircraft after AFC 534, with the emergency aileron droop switch in NORMAL, the ailerons will droop when the flaps are extended by the emergency system. The ailerons will not droop with the switch in DISABLE. Emergency aileron droop should only be used for carrier landings with insufficient wind-over-the-deck to permit landing without aileron droop or for field landings where runway length and/or arresting gear limitations do not permit a landing without aileron droop. If the ailerons are not drooped when the flaps are extended by the emergency system, the trailing edge flaps will indicate barberpole because the trailing edge flap indicating circuit is wired through the aileron droop down limit switch. Therefore, there is no positive indication that the trailing edge flaps have reached 1/2, however, the barberpole indication may be used as an indirect indication of trailing edge flap extension. The leading edge flap indications are not affected by aileron droop position and should read DN. Pneumatic pressure from the emergency air bottle is the only means for maintaining the flaps in the 1/2 position after the flaps are blown down. A subsequent massive air leak on one side may cause an asymmetric flap retraction as airloads push the flaps up. Approach speeds with various flap/drooped aileron configurations are tabulated in the Approach Speeds chart, this section. The flaps can be extended using the following procedure.

1. Airspeed – 200 KNOTS
2. Flap circuit breaker – PULL (do not reset)
   Failure to pull the flap circuit breaker can result in an asymmetrical flap condition with some flaps extended and others retracted.

For most cases, if flaps are required –

3. Emergency aileron droop switch – DISABLE (some aircraft)

WARNING

If a PC hydraulic failure has occurred, or occurs after emergency extension of the flaps, and the emergency aileron droop switch is not in DISABLE or, on aircraft prior to 154786ag, regardless of switch position, essential 28 volt dc bus power is not available, a split aileron condition will occur which will cause an immediate uncontrollable roll into the dead wing.

4. Emergency flap handle – PULL FULL AFT
   Pull the emergency flap handle full aft to limit of travel. Leave the handle full aft. Returning the handle to the forward position allows the compressed air from the flap down side of the actuator to be vented. Emergency extension of the flaps may cause a momentary roll which can be easily countered by normal control application.
5. Flap position indicators – CHECK (T.E. barberpole)

If an asymmetric leading edge flap occurs –

6. Emergency flap handle – PUSH FORWARD
   Push the emergency flap handle forward to vent air and allow the flaps to retract under air loads. An asymmetric trailing edge flap is easily controllable and approach speed with an asymmetric trailing edge flap is lower than with flaps up.

If aileron droop is required -

3. PC-1 and PC-2 hydraulic pressure - CHECK NORMAL
4. Emergency aileron droop switch - NORMAL (some aircraft)

WARNING

If a PC hydraulic failure has occurred, or occurs after emergency extension of the flaps, and the emergency aileron droop switch is in NORMAL or, on aircraft prior to 154786ag, regardless of switch position, essential 28 volt dc bus power is not available, a split aileron condition will occur which will cause an immediate uncontrollable roll into the dead wing.

5. Emergency flap handle - PULL FULL AFT
   Pull the emergency flap handle full aft to limit of travel. Leave the handle full aft. Returning the handle to the forward position allows the compressed air from the flap down side of the actuator to be vented. Emergency extension of the flaps may cause a momentary roll which can be easily countered by normal control application.
6. Flap position indicators - CHECK (T.E. 1/2)

If asymmetric droop occurs -

7. Emergency aileron droop switch - DISABLE
   The ailerons will return to the undrooped position
8. Determine approach speed and notify carrier.
   Refer to Approach Speed chart, this section

If asymmetric trailing edge flaps occur -

7. Determine minimum control airspeed and notify carrier.
   At a safe altitude, slow and determine minimum control airspeed for an approach. An asymmetric trailing edge flap is easily controllable and approach speed with an asymmetric trailing edge flap is lower than with flaps up.

If an asymmetric leading edge flap occurs -

7. Emergency flap handle - PUSH FWD
   Push the emergency flap handle forward to vent air and allow the flaps to retract under air loads.
8. Determine minimum approach speed and notify carrier.
   Refer to Approach Speed chart, this section.

WHEEL BRAKE EMERGENCY OPERATION

If a utility hydraulic system failure or loss of brake action occurs, the airplane can be stopped by using the emergency brake system. However, if arresting gear is available, plan to make a short field arrested landing.

1. Hydraulic wheel brakes - APPLY
   Depress brakes and keep a constantly increasing brake pressure, do not pump the brakes. There may be brake application available from the emergency hydraulic accumulator to provide a limited amount of differential braking.

NOTE

With no utility hydraulic system pressure available, the manual hydraulic brakes are still capable of furnishing flow and pressure to accomplish a limited amount of differential braking. Manual braking utilized with the emergency brake system becomes the primary method of maintaining directional control of the airplane. The number of such applications is limited, and higher pedal travel and higher brake pedal forces will be necessary.

2. If unable to stop airplane, pull emergency brake handle.
   The emergency brake system meters air pressure in proportion to applied pilot effort but does not provide differential braking. There will be a time lag between pulling the emergency brake handle and the application of pneumatic pressure to the wheel cylinders.
3. Manual braking (heavy pedal forces) - AS REQUIRED
   Manual braking may be required for directional control as asymmetrical braking is prevalent during use of the emergency braking handle. The asymmetrical braking could be due to runway crown or crosswinds as well as unequal brake torque.
4. Do not taxi with emergency brakes

SPEED BRAKE EMERGENCY OPERATION

Three basic failures, and their combinations, can affect the speed brakes. They are: switch failures, electrical failure, and utility hydraulic system failure. In the event of a utility hydraulic system failure with the speed brakes extended, the speed brakes will be forced by air loads to a low drag trail position, regardless of switch positions. If an electrical failure occurs, the speed brakes automatically retract to a fully closed position. In the event the throttle-mounted switch fails:

1. Emergency speed brake switch - RETRACT (before AFC 534)

If switch fails or after AFC 534 –

2. Speed brake circuit breaker – PULL

AIR REFUEL PROBE EMERGENCY OPERATION

On aircraft 155867ak and up, and all others after AFC 370, if the air refueling probe cannot be extended normally, it can be extended with pneumatic pressure.

1. Refuel probe circuit breaker – PULL (G5; J14 after AFC 388; J5 after AFC 545; No. 1 panel)
2. Refuel probe switch – REFUEL
3. Emergency refuel probe switch – EMER EXT

After refueling is accomplished –

4. Refuel probe switch – EXTEND

CAUTION

To prevent possible damage to the utility reservoir and/or loss of utility pressure, the emergency refuel probe switch must be left in EMER EXT.

LANDING EMERGENCIES

LANDING GEAR MALFUNCTION

Refer to Landing Gear Malfunction – Emergency Landing Guide, figure 5-9.

FIELD ARRESTING GEAR

The various types of field arresting gear in use are the anchor chain cable, water squeezer, and the morest type. All of these types require engagement of the arresting hook in a cable pendant rigged across the runway. Location of the pendant in relation to the runway will classify the gear as follows:

Midfield Gear – Located near the half way point of the runway. Usually requires prior notification in order to rig for arrestment in the direction desired.
Abort Gear – Located 1500 to 2500 feet short of the upwind end of the duty runway and usually will be rigged for immediate use.
Overrun Gear – Located shortly past the upwind end of the duty runway, and is usually rigged for immediate use.

All pilots should be aware of the type, its location, and compatibility of the arresting gear in use with the aircraft; and the policy of the local air station with regard to which gear is rigged for use, and when. The approximate maximum permissible engaging speed, gross weight, and off-center engagement distance for field arrestment of aircraft are listed in figure 5-10.

WARNING

Under no circumstances should a pilot's decision to abort a takeoff be delayed because of knowledge that an emergency arresting gear is available at the end of the runway. Decision to abort should be based on the usual parameters of remaining runway and distance required for stopping, using wheel brakes. The arresting gear will then serve as an assist to stop the aricraft from rolling off the runway onto unprepared surfaces.

- If off center, just prior to engaging the arresting gear, do not attempt to go for the center of the runway. Continue straight ahead parallel to the centerline.

As various modifications to the basic type of arresting gear are used, exact speeds will vary accordingly. Certain aircraft service changes may also affect engaging speed and weight limitations. Severe damage to the aircraft is usually sustained if an engagement into the chain gear is made in the wrong direction.

In view of the existing emergency runway conditions (i.e., weather, time, fuel remaining and other considerations), it may be impractical or impossible to adhere strictly to the following general recommendations. In an emergency situation, first determine the extent of the emergency by whatever means are possible (instruments, other aircraft, LSO, RDO, tower or other ground personnel). Next determine the most advantageous arresting gear available and the type of arrestment to be made under the conditions which prevail. Whenever deliberate field arrestment is intended, notify control tower personnel as much in advance as possible and state estimated landing time in minutes. If gear is not rigged, it will probably require 10 to 20 minutes to prepare it for use. If foaming of the runway or area of arrestment is required or desired it should be requested by the pilot at this time. In general the arresting gear is engaged on the centerline at as slow a speed as possible. Burn down to 1500 pounds of fuel or less. While burning down, make practice passes to accurately locate the arresting gear. Engagement should be made with feet off the brakes, shoulder harness locked, and with the aircraft in a 3—point attitude. When speed has been reduced to approximately 20 knots, braking should be applied to prevent the aircraft with idle power from pulling the gear through to a two—block position. In the event of brake malfunction the aircraft engines should be shut down.

FIELD ARRESTMENT GEAR DATA

ARRESTING GEAR	AIRCRAFT GROSS WEIGHT/1000 POUNDS									MAXIMUM OFF-CENTER ENGAGEMENT FEET
	31	34	38	40	42	46	50	54	56	
	MAXIMUM ENGAGING SPEED (KNOTS)									
M2	107	103	98	96	94	91	88	85	83	20
E-14-1	160	160	160	160	160	160	160	145	140	50
E-27	156	153	149	147	145	141	136	132	130	35
E-15 (200′ SPAN)	180	180	180	180	180	178	176	173	172	35
E-15 (300′ SPAN)	180	180	180	180	180	180	180	179	177	50
M-21	150	150	145	145	140	135	135	135	125	10
E-28	180	180	180	180	178	175	163	160	160	40
E-5 (STD CHAIN)	150	150	150	150	147	142	138	134	132	*
E-5-1 (STD CHAIN)	159	159	153	150	147	142	138	134	132	*
E-5 (HVY CHAIN)	150	150	150	150	150	150	150	150	150	*
E-5-1 (HVY CHAIN)	165	165	165	165	165	165	165	165	162	*
BAK-6	160	160	156	154	153	150	146	142	140	15
BAK-9	187	184	180	177	173	165	158	151	147	30
BAK-12	190	190	190	190	190	187	177	169	165	50

* Off-center engagement may not exceed 25% of the runway span.

Note

Maximum engaging speed limited by arresting gear.

FDD-1-(80)B

Figure 5-10

SHORT FIELD ARRESTMENT

If at any time prior to landing, it is known that a directional control problem exists or a minimum rollout is desired, a short field arrestment should be made and the assistence of an LSO requested. He should be stationed near the touchdown point and equipped with a radio. Inform the LSO of the desired touchdown point. A constant glide slope approach to touchdown is permitted (mirror or Fresnel Lens Landing aid utilized) with touchdown on centerline at or just prior to the arresting wire with the hook extended. The hook should be lowered while airborne and a positive hook-down check should be made (observe light in the hook control handle). If midfield gear or morest type is available, it should be used. If neither are available, use abort gear. Use an approach speed commensurate with the emergency experienced. Landing approach power will be maintained until arrestment is assured or a waveoff is taken. Be prepared for a waveoff if the gear is missed. After engaging the gear, retard the throttle to IDLE or secure engine and abandon aircraft, depending on existing conditions.

1. Notify tower and request LSO assistance.
   Notify tower of intended action, request LSO assistance (if required) and supply required information; e.g., type of emergency, estimated landing time, gross weight, etc.
2. Reduce gross weight.
   Dump and/or burn fuel to reduce gross weight to lowest practicable.
3. Fly pattern as dictated by emergency.
4. Arresting hook - DOWN.
5. Fly final approach on LSO or for touchdown just short of wire. If LSO is available, fly final approach and touchdown as directed. If LSO is not available, fly final approach touchdown on centerline and just short of the crossdeck pendant.
6. Maintain landing approach power until arrestment is assured.
7. Longitudinal control neutral until engagement.
8. Engage wire with feet off brakes.
9. Throttles IDLE or OFF at engagement as desired.

NOTE

Placing longitudinal control to neutral reduces the possibility and limits the severity of stabilator slaps.

LONG FIELD ARRESTMENT

The long field arrestment is used when a stopping problem exists with insufficient runway remaining (i.e., aborted take-offs, icy or wet runways, loss of brakes after touchdown, etc.). Lower the hook, allowing sufficient time for it to extend fully prior to engagement. Do not lower the hook too early and weaken the hook point. Line up the aircraft on the runway centerline. Inform the control tower of your intentions to engage the arresting gear so that aircraft landing behind you may be waved off. If no directional control problem exists (crosswind, brakes out, etc.), secure the engine.

1. ENGINES - IDLE
2. ARRESTING HOOK - DOWN
   Lower the tail hook at least 1000 feet ahead of the wire; 5 seconds are required for full extension.
3. DRAG CHUTE - DEPLOY
4. Apply optimum wheel braking.
5. Aim for center of wire.
   Steer to engage the wire near the center and at a 90 degree angle.
6. Release brakes 100 feet ahead of wire.
7. Control stick - FULL AFT

BARRICADE ENGAGEMENT

1. Jettison all missiles.
2. Lower hook if possible.
   Lowering the arresting hook will assist the barricade in stopping the aircraft and will help to keep the aircraft on the deck at barricade entry.
3. Fly a normal on speed approach, on centerline and on meatball. Anticipate loss of meatball for a short period of time late in approach due to barricade stanchions obscuring the meatball.

EXPENDING HUNG ORDNANCE

Before making an arrested landing with hung ordnance on a MER/TER, the following should be accomplished in an effort to expend ordnance.

1. Reselect station on which ordnance is hung and again depress bomb button.
2. Rock wings and/or pull positive G load.

LANDING WITH HUNG ORDNANCE

For emergency landing conditions only, the following asymmetric moments due to external stores loading are permitted:

Field landing (minimum sink-rate)	379,000 inch-pounds
twin or single-engine	2870 lb. total on stations 1/9 or 4650 lb. total on stations 2/8
Field arrested landings and FMLP	212,000 inch-pounds
twin or single-engine	1600 lb. total on stations 1/9 or 2610 lb. total on stations 2/8
Carrier landing, twin-engine	212,000 inch-pounds.
	1600 lb. total on stations 1/9 or 2610 lb. total on stations 2/8

Carrier landings, single engine	60,000 inch–pounds
	450 lb. total on stations 1/9 or 740 lb. total on stations 2/8

The above asymmetrics are defined as the asymmetric load at the outboard wing stations (1 or 9) x 132.5 plus the asymmetric load at the inboard wing stations (2 or 8) x 81.5.

CAUTION

- The hung ordnance restrictions of the TACTICAL Manual, NAVAIR 01–245FDB–1T, appendix A, External Stores Limitations chart, apply and take precedence over the above limits.
- If a landing must be made with an asymmetric load, refer to Flight With Asymmetric Loadings section IV for handling characteristics.

BLOWN TIRE

The situation may occur when a landing with a blown tire must be made, or a tire may rupture during the landing ground roll. A blown tire and high speed require immediate corrective action to keep the airplane aligned with the runway, therefore:

LANDING WITH A KNOWN BLOWN TIRE

1. Plan to make a short field arrestment (if available)

If short field arresting gear is not available –

2. Make a normal on speed approach.
3. Anti–skid switch – OFF
4. Land on side of runway opposite blown tire.
5. Touchdown with weight on good tire.
6. Use nose gear steering to maintain directional control.
   If nose gear steering is inoperative, use of aerodynamic steering should be considered.
7. Drag chute – DEPLOY
8. Use light opposite braking to slow aircraft.

CAUTION

Avoid braking on the wheel with the blown tire. Heavy braking could cause a flat spot on the wheel which could prevent further wheel rotation and make aircraft control more difficult.

9. Do not retract flaps.
   The wing flap seals may have been damaged by pieces of broken tire and retracting the wing flaps will increase the damage.
10. If fire equipment is available, throttles – OFF
    If possible, do not shut down engines until adequate fire fighting equipment is available.

WARNING

The damaged wheel may either be on fire or very hot, and fuel drained overboard during engine shutdown could contact the hot wheel and cause a fire.

BLOWN TIRE DURING LANDING ROLLOUT

1. NOSE GEAR STEERING – ENGAGE (RUDDER CENTERED)
2. Anti–skid switch – OFF
3. HOOK – DOWN FOR FIELD ARRESTMENT (IF ARRESTING GEAR AVAILABLE)
4. Use light opposite braking to slow aircraft.

CAUTION

Avoid braking on the wheel with the blown tire. Heavy braking could cause a flat spot on the wheel which could prevent further wheel rotation and make aircraft control more difficult.

If arresting gear is not available –

5. Drag chute – DEPLOY

After coming to stop –

6. Do not retract flaps.
   The wing flap seals may have been damaged by pieces of broken tire and retracting the wing flaps will increase damage.
7. If fire equipment is available, throttles – OFF
   If possible, do not shut down engines until adequate fire fighting equipment is available.

WARNING

The damaged wheel may either be on fire or very hot, and fuel drained overboard during engine shut down could contact the hot wheel and cause a fire.

CAUTION

If nose wheel tires are blown on landing, secure both engines as soon as possible to preclude portions of the tire entering either or both intake ducts.

NOTE

The airplane can be safely taxied or towed with a flat tire on either main or nose gear.

SINGLE–ENGINE FAILURE ON FINAL

At the first indication of engine failure, the throttle of the operating engine must be advanced to obtain full military or afterburner thrust and the flaps raised to the one–half position as soon as sink–rate is arrested or under control. Any unnecessary delay in applying power to the good engine will result in excessive sink rates and/or airspeed bleed–off that you may not be able to overcome (even with full afterburner thrust) prior to ground impact. The excessive sink rate condition will develop if the angle of attack is allowed to increase beyond 20 units. If the angle of attack reaches 22 units, even the use of full afterburner thrust will not stop the sink rate prior to ground impact, unless the angle of attack is again reduced to 20 units or below. Raising the flaps to one–half while flying full flap on–speed will reduce drag and minimize power loss to the BLC system, but an altitude loss of 100 to 300 feet can be expected. Do not attempt to immediately level the aircraft; accept a continued rate of descent and maintain aircraft control with smooth coordinated attitude corrections, being careful not to exceed 20 units angle of attack. External store drag is negligible at low airspeeds and will have little effect on aircraft performance, the aircrafts gross weight will dictate the thrust required to continue the approach or execute a go–around. At normal landing gross weights, once the situation is stabilized; i.e., flaps at one–half and the sink rate stopped or under control, afterburner thrust will no longer be required. Very little yaw rate is induced by military thrust on one engine; however, a slight asymmetrical control problem is created by use of afterburner thrust on one engine. These yaw rates must be controlled with the rudder, since excessive aileron inputs could induce additional adverse control conditions. You can normally continue the approach, and land the aircraft on the runway, from a situation created by a single engine failure on final.

1. THROTTLES – MIL/AFTERBURNER

NOTE

If the right engine fails on final and the bus tie remains open, afterburner ignition will not be available. However, if afterburner thrust is required, afterburner light–offs are generally obtainable through turbine torching, by jam accelerating the left engine at 90% rpm or above.

2. FLAPS – RAISE TO 1/2
3. LAND OR WAVEOFF

If decision to waveoff is made:

4. Initiate waveoff – CLIMB TO SAFE ALTITUDE
5. Landing gear – UP
6. Non–mechanical failure – INITIATE AIRSTART
7. Mechanical failure – SHUTDOWN FAILED ENGINE
8. Land as soon as practicable.

SINGLE–ENGINE LANDING

A single–engine landing is basically the same as a normal landing except that the pattern is expanded to avoid steep turns, final approach speeds are increased for better lateral control, and 1/2 flaps are used in lieu of full flaps. Single–engine carrier landings are to be considered as emergency situations and, as such, diverting to a land base should be a first consideration. For single–engine carrier landings, all stores including racks (MER, TER, etc.) should be jettisoned. Sparrows may be retained provided that weight and atmospheric temperature restrictions are adhered to.

1. Turn off non–essential electrical equipment.
2. Reduce gross weight to 36,000 pounds or less.
   Reduce gross weight to 36,000 pounds or less at atmospheric temperatures below 80°F with MK 7 Mod 2 or 3 arresting gear. Above 80°F reduce weight linearly to 32,000 pounds at 100°F. The maximum gross weight for MK 7 Mod 1 arresting gear is 32,000 pounds. Recommended approach speeds (1/2 flaps) versus weight are presented in the Approach Speed chart, this section.
3. Variable area inlet ramp on operating engine – CHECK FULL OPEN

NOTE

If the inlet ramp on the operating engine is in the closed position, the afterburner may be required to make a safe approach.

4. Cycle afterburner of operating engine to ensure rapid lightoff for possible afterburner operation.
5. Ram air turbine – EXTEND
6. Wing flap switch – 1/2

NOTE

- Full flaps shall not be used during a single–engine approach and landing, since the engine bleed air that would be utilized for the trailing edge BLC system deprives the engine of fully rated thrust. Also, the full flaps configuration increases drag appreciably over the 1/2 flaps configuration. In the 1/2 flaps configuration, approximately 25% of the available thrust of the good engine is diverted to maintain leading edge BLC.
- If the generator on the operating engine is lost, the flaps solenoid operated selector valves revert to a full trail position, and the flaps are blown up by the airstream.
- The ARI system is inoperative with flaps retracted.

7. Landing gear handle – DOWN
8. Field Landing – Fly 17 units angle of attack until landing is assured.
   Carrier Landing – ON–SPEED APPROACH
   Do not exceed 20 units angle of attack as the aircraft waveoff performance will be degraded.
9. For waveoff, place throttle of operating engine to MIL or MAX as required.
10. For carrier landing, place throttle of operating

engine to MAX AB upon main gear touchdown.

NOTE

If the right engine is shutdown, and the bus tie is open, afterburner ignition will not be available. However, if afterburner thrust is required, afterburner light-offs are generally obtainable through turbine torching, by jam accelerating the left engine at 90% rpm or above.

UTILITY SYSTEM AND ENGINE FAILURE

Handling qualities are critically degraded when operating near approach speeds with a combination of single engine and utility hydraulic system failure.

WARNING

- Carrier landing shall not be attempted.
- If the combination of weather, landing facilities and pilot experience is less than ideal, consideration should be given to a controlled ejection.

If the decision is made to land, plan for a straight-in no-flap approach. In order to reduce the roll due to the asymmetric thrust required in the dirty configuration, flaps should not be lowered. Although the single engine thrust line is near the aircraft centerline, the effects of losing powered rudder, rudder trim, yaw and roll augmentation and, possibly, roll control surfaces on one wing, with single engine asymmetric thrust, seriously degrades control of the aircraft. Maximum application of manual rudder in combination with lateral control deflection is required to counter roll due to yaw. Banking into the good engine will reduce the rudder required. Higher than normal airspeeds must be maintained to provide adequate lateral control. The yaw induced by asymmetric thrust causes an AOA indication error, therefore, airspeed rather than AOA should be used on final. Extend the gear at a minimum altitude of 5000 feet AGL, perform a controllability check and investigate the landing approach characteristics. The airspeed for final approach is the minimum control airspeed +10 knots, The emergency pattern should be flown so that all turns are away from the dead engine. Care should be taken to avoid excessive sink rates as acceleration of the good engine to reduce the sink rate could result in loss of control. Touchdown should be planned to be in the first third of the runway. Plan to lower the hook after landing to make a midfield or long field arrestment.

WARNING

- Any throttle movements should be made slowly to minimize yaw/roll transients. Minimum practical power should be used at all times to minimize yaw and roll due to yaw. Minimum control airspeed is a direct function of throttle setting.
- If afterburner is required at any time during the approach there may not be sufficient roll control to compensate for the asymmetric thruts.
- Power must be added slowly and airspeed increased to insure adequate control during any attempted wave-off. Wave-off capability in afterburner may not exist below 230 knots due to inadequate lateral and directional control. Afterburner should not be selected below 230 knots unless absolutely necessary to prevent catastrophe and the pilot must be aware that this could cause loss of control.

1. Reduce airspeed to 250 knots.
2. External stores - JETTISON
3. Non-essential electrical equipment - OFF
4. Reduce gross weight to 36,000 pounds or less.
5. Ram air turbine - EXTEND
6. Landing gear - BLOW DOWN AT MINIMUM 5000 FEET AGL
7. Flaps - UP
8. Perform controllability check and investigate landing approach characteristics. Approach speed is 230 knots until wings level on final then fly minimum control speed +10 knots.
9. Plan a straight-in approach with all turns away from the dead engine.
10. Make a long field arrested landing.

UTILITY SYSTEM, SINGLE PC AND ENGINE FAILURE

Aircraft handling characteristics are of the same nature as with utility and engine failure. However, lateral control is more critical due to the loss of aileron and spoiler on one wing.

WARNING

For any of the above combination failures, maintain a minimum maneuvering airspeed of 230 knots. Yaw and roll encountered with asymmetric thrust and heavy weight significantly degrade handling characteristics when operating below 230 knots. Avoid rapid roll rates, rapid or large thrust changes and turns toward the inoperative wing/dead engine. If turns are required minimum bank angles should be used.

LANDING WITH BLEED AIR SWITCH OFF (NO LEADING EDGE BLC)

1. Investigate landing approach characteristics before descending for a landing.
2. Jettison any asymmetric load (as required).
3. Enter a wide and extended downwind leg.
4. Plan for early line-up and straight in approach.
5. For field or carrier landings fly 17.0 units AOA for approach and touchdown.

WARNING

- Actual stall occurs before the rudder pedal shaker is activated in the no leading edge BLC configuration. The aircraft stalls at approximately 20 units AOA and is characterized by a nose rise.
- In the single engine, half flap, leading edge BLC off configuration, the pilot–task during a carrier landing and approach is considerable. The decision to recover the aircraft aboard ship should be made only when conditions are optimum. Suitability of a divert field should be considered.

ARRESTING HOOK MALFUNCTION

If the hook fails to extend upon handle actuation –

1. Arresting hook control circuit breaker – PULL (H7 before AFC 388; K7 after AFC 388, No. 2 panel)

CONTROLLABILITY CHECK

During any inflight emergency, when structural damage or any other failure is known or suspected that may adversely affect aircraft handling characteristics, a controllability check should be performed as follows:

1. Proceed to a safe altitude (min 5000 feet AGL).
2. Reduce gross weight to minimum practicable.
3. Establish landing configuration required by type of emergency.
4. Slow aircraft to determine minimum control speed, but no slower than 17 units AOA.

If adequate control is available –

5. Plan for straight–in approach.
6. Fly final approach no slower than minimum control speed +10 knots.

APPROACH POWER COMPENSATOR SYSTEM DISENGAGEMENT

If a malfunction occurs in the system, APCS disengagement should be attempted in the following sequence:

1. Emergency speed brake switch – MANUAL (aircraft thru 155866aj before AFC 392).
2. Speedbrake circuit breaker – IN (aircraft thru 155866aj before AFC 392).
3. Speed brake switch – IN
   On aircraft 155867ak and up and all others after AFC 392, the APCS may be disengaged at any time by moving the speedbrake switch to IN, regardless of the position of the emergency speedbrake switch or the speedbrake circuit breaker.

If this fails –

4. APCS power switch – OFF or STBY.

If this fails –

5. Auto throttle circuit breakers – PULL
   (J3, K4, K5 before AFC 388 – M3, K3, L3 after AFC 388, No. 2 panel)

This should ensure positive disengagement, but if it still remains engaged –

6. Manually override throttles.
   The pilot can manually override the throttles by exerting a 20 to 25 pound force per throttle. In this case, the throttles must be held in the desired position.

If time does not permit the accomplishment of the preceding steps –

1. RAT – OUT
2. Generators – OFF
3. Climb to a safe altitude
4. Auto throttle circuit breakers – PULL (J3, K4, K5 before AFC 388; M3, K3, L3 after AFC 388, No. 2 panel)
5. Generators – ON
6. RAT – IN

NOTE

When using the above procedures, discretion must be used as the flaps will blow up.

NO–FLAP LANDING

A no–flap landing pattern is basically the same as for a normal landing, except that it should be expanded to avoid steep turns. In the no–flap landing configuration, stabilator effectiveness in increased and aileron/spoiler effectiveness is decreased relative to that experienced in normal flap landing configurations. For field landings, no–flap approach AOA is reduced to 17 units to ensure adequate roll control throughout the pattern. For carrier approaches, on speed AOA must be flown on final to remain within shipboard arresting gear engagement limitations. Careful control of nose attitude is required and large pitch control inputs should be avoided. For field landings, care must be exercised so as not to become airborne again during landing rollout.

Ashore -

1. Fly 17 units AOA

Afloat -

1. Fly on-speed AOA

ASYMMETRIC AILERON DROOP (AIRCRAFT 154786ag AND UP)

Effective airplanes 154786ag and up, aileron droop is provided by electro-mechanical actuators when either 1/2 or full flaps are selected. If one aileron fails to droop, a pronounced aircraft roll will occur. The following action allows either a half or full flap landing without aileron droop. Refer to Approach Speed chart, this section.

1. Retract flaps
2. Pull aileron droop actuator circuit breakers (J12 before AFC 388 - H15 after AFC 388 No. 1 panel)
3. Extend flaps (ailerons will not droop)
4. Fly on speed angle-of-attack

HOT BRAKE PROCEDURES

Hot brake procedures are contained in BUAER/BUWEPS INST 13420.1. In view of the varied climatic conditions, field conditions, and safety devices available, specific procedures must be covered in local squadron/field SOP.

LANDING WITH BOTH ENGINES INOPERATIVE

Landing with both engines inoperative will not be attempted.

GLIDE DISTANCE

The aircraft will glide approximately 6 nautical miles for every 5000 feet of altitude. The recommended glide airspeed is 215 knots CAS. Below 50,000 feet, 215 knots CAS will provide near maximum glide distance and will allow the windmilling engines to maintain power control hydraulic pressure within safe limits.

PRECAUTIONARY EMERGENCY APPROACH

The standard precautionary emergency approach for the aircraft is the straight-in GCA/CCA approach modified to accomodate single-engine, half-flap, or no-flap approach speeds as power available dictates. The precautionary emergency approach depicted in figure 5-11 will be used for field landings if one engine has failed and the remaining engine has suffered a malfunction that results in only partial power. This procedure may be used day or night, provided ceiling and visibility are such that visual contact can be maintained with the field. Although the approach depicted is the classic overhead entry to a left hand pattern, the precautionary approach may be initiated from any check point, using either a left or right hand pattern. The pilot must select a check point in either the straight-in or overhead approach at which the decision to continue the approach or eject must be made. Sink rate, power available, configuration, and position relative to the runway or obstructions must be considered. The flaps may have to be blown down due to airspeed being above flap blow-up speed. The check point selected must be early enough to permit safe ejection and in such a position so as not to compromise or endanger the safety of populated areas, military installations or other aircraft. In no case should this check point be lower than 1000 feet AGL for the straight-in approach, or 3000 feet AGL for the overhead approach. The pilot should plan the approach to utilize available field arresting gear. If the success of the approach and landing appear to be marginal to the pilot, consideration should be given to heading the aircraft into a clear area and ejecting instead of attempting the approach.

FORCED LANDING

WARNING

All forced landings on land shall be made with the landing gear extended, regardless of terrain. A greater injury hazard is present whenever emergency landings are made with the landing gear retracted. Increased airspeed or nose high angle of impact during landings with landing gear retracted is common practice and contributes greatly to pilot injury and damage to the airplane. This nose high attitude causes the airplane to slap the ground on impact, subjecting the pilot to possible spinal injury. Less airplane damage will result with the gear extended.

It is recommended that a landing on unprepared terrain not be attempted with this airplane; the crew should eject. If a forced landing is unavoidable, proceed as follows:

1. Ram air turbine - EXTEND
2. If time and conditions permit, dump or burn excess fuel.
3. Notify RIO of existing emergency and intended action.
4. Shoulder harness - LOCK
5. Canopies - JETTISON (forward canopy first)
   The aft canopy should be jettisoned last to preclude the possibility of the forward canopy entering the aft cockpit when jettisoned.
6. Armament - JETTISON
7. Landing gear - DOWN AND LOCKED
8. Wing flaps - DOWN
9. External tanks - RETAIN IF EMPTY AND UNPRESSURIZED
   Empty external tanks should be retained to absorb the shock of landing.

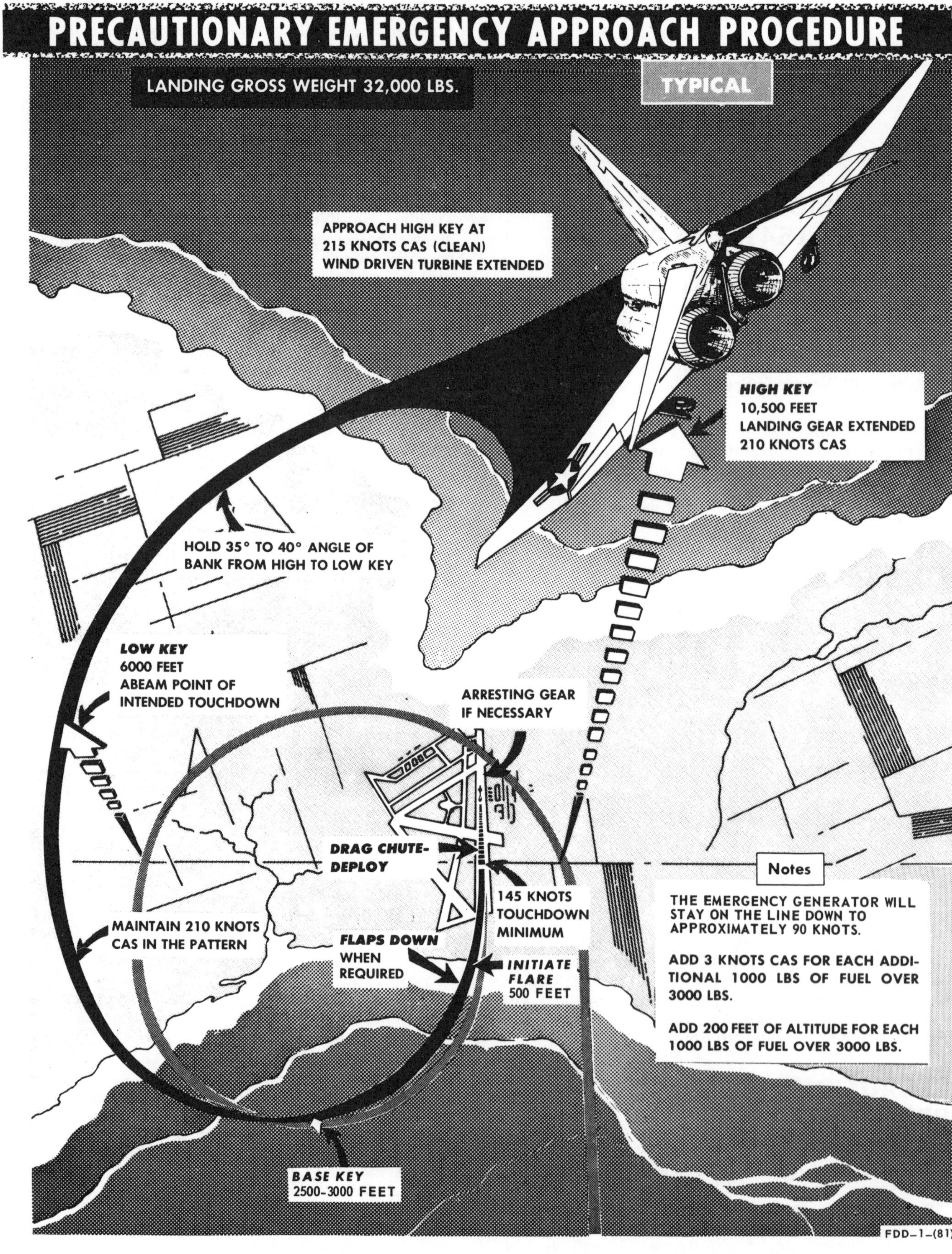
PRECAUTIONARY EMERGENCY APPROACH PROCEDURE
LANDING GROSS WEIGHT 32,000 LBS.
TYPICAL
APPROACH HIGH KEY AT
215 KNOTS CAS (CLEAN)
WIND DRIVEN TURBINE EXTENDED
HIGH KEY
10,500 FEET
LANDING GEAR EXTENDED
210 KNOTS CAS
HOLD 35° TO 40° ANGLE OF
BANK FROM HIGH TO LOW KEY
LOW KEY
6000 FEET
ABEAM POINT OF
INTENDED TOUCHDOWN
ARRESTING GEAR
IF NECESSARY
DRAG CHUTE-
DEPLOY
145 KNOTS
TOUCHDOWN
MINIMUM
MAINTAIN 210 KNOTS
CAS IN THE PATTERN
FLAPS DOWN
WHEN
REQUIRED
INITIATE
FLARE
500 FEET
BASE KEY
2500-3000 FEET
Notes
THE EMERGENCY GENERATOR WILL STAY ON THE LINE DOWN TO APPROXIMATELY 90 KNOTS.
ADD 3 KNOTS CAS FOR EACH ADDITIONAL 1000 LBS OF FUEL OVER 3000 LBS.
ADD 200 FEET OF ALTITUDE FOR EACH 1000 LBS OF FUEL OVER 3000 LBS.
FDD-1-(81)

Figure 5-11

NOTE

If gear is not lowered, external tanks can be unpressurized by pulling the landing gear circuit breaker and placing the landing gear handle down.

10. Make normal approach.
11. Drag chute - DEPLOY AFTER TOUCHDOWN
12. Engines - SHUTDOWN
    a. Throttles - OFF
    b. Engine master switches - OFF
13. As soon as stopped - EVACUATE AIRPLANE

WHEEL BRAKE ANTI-SKID FAILURE (AIRCRAFT 157242an, AND UP)

In the event of brake loss with normal utility hydraulic pressure and the anti-skid switch ON:

1. Wheel brakes - RELEASE

CAUTION

The brakes should be released before the anti-skid system is disengaged to prevent the possibility of skidding. When the anti-skid system is disengaged manually or due to a malfunction, the brakes will immediately be applied proportional to pedal displacement.

NOTE

The anti-skid system will not protect against skids below approximately 10-20 knots. A malfunctioning anti-skid system which relieves brake pressure below this speed range might be interpreted as wheel brake failure. Assuming a normal utility hydraulic pressure, brakes can be regained by deactivating the anti-skid system and reapplying the brakes.

2. Anti-skid switch - OFF.

NOTE

The anti-skid system can be temporarily disengaged by depressing the emergency quick release lever on the control stick. This lever, however, must be held depressed to keep the system disengaged.

3. Wheel brakes - REAPPLY
   If braking returns, leave anti-skid switch OFF; it not, activate the emergency brake system.

CAUTION

If normal braking is available and anti-skid protection is lost, it is possible to lock a wheel(s) and produce a skid which could result in a blown tire(s). The runway condition and aircraft speed will dictate the amount of braking that can be applied without skidding a wheel.

EMERGENCY EGRESS

Due to forced landing, ditching or landing emergency such as barricade engagement or runway overrun, rapid egress is essential. The most rapid method of egress is by divestment of both the seat survival kit and the parachute. On land, if the aircraft is burning, the extra time required to egress retaining the seat kit could cause serious injury or death. After safely egressing, if conditions permit, return to the aircraft and retrieve the survival kit. If ditching, the parachute should be divested and the seat kit retained to allow for underwater breathing and use of the raft/survival equipment.

WARNING

During emergency egress from aircraft with the H-7 rocket assist ejection seat, actuation of the emergency harness release handle allows the personnel parachute to slide down and wedge against the seat survival kit; thereby, causing serious difficulty in egressing with the seat survival kit attached.

To evacuate cockpit without survival kit and parachute -

1. Canopy - OPEN
   a. Normal
   b. Canopy emergency release handle
   c. Manual unlock handle

NOTE

If circumstances dictate, open canopy after remaining steps are performed.

2. Rotate lower ejection handle guard - UP
3. Leg restraint release handle - PULL AFT
4. Arch back to apply tension to lap belt, and release lap belt fittings.
5. Oxygen mask/face plate - RELEASE/OPEN
6. Composite disconnect - RELEASE
7. Shoulder harness fittings - RELEASE

NOTE

To evacuate cockpit with survival kit only, refer to ditching chart, this section.

DITCHING CHART

WARNING
THE AIRCRAFT SHOULD BE DITCHED ONLY WHEN ALL OTHER ATTEMPTS OF EGRESS HAVE FAILED.

CREW MEMBER	DUTIES BEFORE IMPACT	POSITION	DUTIES AFTER IMPACT	EQUIPMENT	EXIT
PILOT	1. RIO–ALERT. 2. Make radio distress call. 3. IFF–EMERGENCY 4. External stores–JETTISON. 5. Landing Gear–UP. 6. Wing Flaps–DOWN. 7. Arresting Hook–DOWN. 8. Leg restraint release handle–PULL AFT. Release leg restraint lines before ditching to expedite egress from the cockpit. 9. Visor–DOWN. 10. Oxygen mask or face visor–TIGHTEN or SEAL. 11. Canopy–JETTISON (forward canopy first). **Note** In the event of a ditching following such situations as catapult shots or bolters, attempt to jettison the canopy prior to hitting the water. Once the aircraft is immersed under water the canopy might not jettison until water introduced by the folloding doors equalize pressures between the inside and outside of the canopy, an event which would occur at a considerable water depth. 12. Lower seat, assume position for ditching. 13. Shoulder Hurness–LOCK. 14. Fly parallel to swell pattern. 15. Attempt, to touch down along wave crest. 16. When hook contacts water–SHUT DOWN ENGINES.	1. In Seat. 2. Feet on rudder pedals knees flexed.	1. Lower ejection handle Guard–UP 2. Shoulder harness release fittings– RELEASE 3. Emergency restraint release handle – PULL. **WARNING** In airplanes 153071z thru 153832ad before AFC 307, if handle is allowed to return to the down position, the leg restraint lines will be re-locked. 4. Stand straight up without twisting to release survival kit sticker clips from the seat. **Note** The Bail–Out Bottle will actuate when the crewmember stands up. In the event of ditching and sinking in water when immediate escape is impossible, it is possible to survive under water with oxygen equipment or full pressure suit until escape can be made. 5. Abandon aircraft. **WARNING** Should aircraft be abandoned under water, exhale while ascending to the surface to prevent bursting of lungs due to pressure differential between lungs and outside of body. 6. Inflate life vest. 7. See that RIO is clear. 8. Inflate life raft and secure Emergency equipment. 9. Proceed away from aircraft and tie rafts together.	1. One man raft and Emergency equipment. 2. Life vest 3. Flash light.	Over canopy sill.
RADAR INTERCEPT OFFICER	1. Acknowledge pilots ditching order. 2. Radar equipment–STOW. 3. Leg restraint release handle–PULL AFT. Release leg restraint lines before ditching to expedite egress from the cockpit. 4. Visor–DOWN. 5. Oxygen mask or face visor–TIGHTEN or SEAL. 6. Canopy–JETTISON (Aft canopy last). **Note** In the event of a ditching following such situations as catapult shots or bolters, attempt to jettison the canopy prior to hitting the water. Once the aircraft is immersed under water the canopy might not jettison until water introduced by the flocding doors equalize pressures between the inside and outside of the canopy, an event which would occur at a considerable water depth. 7. Lower seat, assume position for ditching. 8. Shoulder Harness–LOCK.	1. In Seat. 2. Feet on deck knees flexed.	7. Same as for Pilot Except: See that Pilot is clear.	1. One man raft and Emergency equipment. 2. Life vest. 3. Flash light.	Over canopy sill.

FDD–1–(82)A

Figure 5–12

DITCHING

Ditching the airplane should be the pilot's last choice. All survival equipment is carried by the crewmember, thus ejection is advisable. However, if altitude and situation demand ditching, the procedures set forth on the ditching chart (figure 5-12) should be observed.

NOTE

In the event of ditching and sinking in water when immediate escape is impossible, it is possible for the crewmember to survive under water with oxygen equipment until escape can be made. The oxygen regulator is a suitable underwater breathing device since the regulator is always on 100% oxygen. It is essential that the mask be tightly strapped in place.

OPERATIONAL EMERGENCIES

PILOT/RIO EMERGENCY COMMUNICATIONS

Provided that the aircraft intercom system is in working order, it is assumed that any communication required between the pilot and RIO concerning any emergency will be carried out on the intercom. If the intercom is inoperative for any reason, the following procedures will be utilized:

1. Check mic and earphone plugs.
2. Check upper block connections.
3. Use Emergency ICS and Emergency Radio positions in conjunction with the override switch.
4. Try intercommunication with UHF transceiver.
5. Check all circuit breakers.
6. If conditions are suitable, remove mask and shout.

PILOT/RIO ATTENTION SIGNALS

The following may be used as pilot/RIO attention signals under emergency conditions with no method of communicating.

1. The pilot will attract the RIO's attention by a rapid rocking of the wings.
2. The RIO will attract the pilot's attention by slamming home his radar scope to the stow position, or actuate the hand control to the limits of the antenna.
3. Acknowledgement of the attention signals will be a thumbs-up and future communications will be conducted by visual signals.

HEFOE signals (figure 5-14) may be utilized by the pilot and RIO. As the RIO is always flying with the left side of the curtain on his hood open, the pilot will be signaling over his left shoulder and looking in his left mirror for the return signals. The same signals will apply at night except that a flashlight must be held up to outline the fingers. If the RIO's upper block is unplugged as indicated by the 4-finger signal, the pilot will maintain a cockpit altitude of 10,000 feet or below for the duration of the flight. If the RIO desires an immediate landing, he will give a thumbs-down signal to the pilot.

DOWNED PLANE PROCEDURES

DECLARATION OF AN EMERGENCY

When flying without a wingman or section leader, it is critically important that the pilot advise someone of his trouble and location. Even a deferred emergency can develop into a first rate emergency. The initial radio contact should be preceded with the word PAN when the situation requires urgent action, but is not an actual distress; the word MAYDAY should be used when threatened by serious or imminent danger and immediate assistance is required. If a serious emergency has arisen, shift immediately to EMERGENCY IFF; set up SIF Mode 3, Code 77; place UHF to GUARD; and broadcast MAYDAY. The following information should be relayed to a ground station immediately:

1. PAN or MAYDAY (depending upon situation)
2. Identification
3. Model aircraft
4. Position
5. Situation
6. Intentions

Single Aircraft

If the situation permits, prior to ejection or crash landing:

1. Switch IFF to EMERGENCY.
2. Transmit MAYDAY over guard channel.

Conditions existing following the ejection or crash landing will dictate whether to remain near the scene of the crash or attempt to find assistance.

Section

If one member of a section goes down, the other member should:

1. Establish contact with a ground station, preferably a GCI site or RADAR control agency. Switch IFF to EMERGENCY and UHF to GUARD.

2. Make every effort to follow the other aircraft or crew during descent. It is of primary importance to keep the crew in sight at all times, while on the ground or in the water. Note as accurately as possible, bearings, distances from known prominent landmarks or navigational aids, in order to direct rescue planes or boats to the scene.
3. Establish a RESCAP.
4. Maintain sufficient altitude to assure radio contact with the rescue facility.
5. Leave the area with sufficient fuel to POSITIVELY ensure return to base or alternate field.

Division

Everything mentioned earlier holds true if there are more than two members to the flight. Some additional procedures can be followed which will ensure a greater likelihood of a successful rescue. The other member of the section in which the downed crew has been flying, should:

1. Follow the aircraft/crew and circle them at low altitude, making every effort to keep the downed crew in sight.

Other members of the flight:

1. Remain at altitude.
2. Alert appropriate facilities.
3. Relay communications.
4. Conserve fuel.

NAV–COMM EMERGENCY PROCEDURES

These procedures deal with communication emergencies. Other types of emergencies where navigation and communication aids are available should be handled according to the individual circumstances under which they arise and as the factors involved indicate. An aircraft with running lights flashing usually indicates that an emergency condition exists.

LOST AIRCRAFT (WITHOUT NAVIGATION AIDS)

The pilot will have navigated to best position by dead reckoning. The following procedures will apply.

With Radio Receiver

1. Fly a minimum of two triangular patterns to the right with 1–minute legs. Repeat pattern at 20–minute intervals.
2. Conserve fuel and maintain altitude.
3. Squawk appropriate Mode and be alert for aircraft vectored to join.

Without Radio Receiver

1. Fly a minimum of two triangular patterns to the left with 1–minute legs. Repeat pattern at 20–minute intervals.
2. Conserve fuel and facilitate radar pickup by maintaining highest feasible altitude consistent with situation.
3. Squawk appropriate Mode and be alert for aircraft vectored to join.
4. After joining, inform healthy aircraft of all emergency conditions by appropriate hand signals in order to prevent separation during penetration/letdown.

NO RADIO AIRCRAFT (WITH NAVIGATION AIDS)

1. Proceed to alternate marshal.
2. Energize I/P function at least once each minute.
3. Commence penetration/letdown at EAC as briefed.
4. Be alert for aircraft vectored to join.
5. If immediate assistance is required, energize emergency IFF.

PENETRATION/LETDOWN NAV–COMM EMERGENCIES

Even though communication aids have failed, if navigation equipment is still available:

1. Continue approach.
   Regardless of weather, any jet aircraft having passed Platform must continue its approach.
2. If no contact has been made after 2 minutes past individual expected ramp time, conduct lost aircraft (without navigation aids, and without radio receiver) procedures.

If all communication and navigation equipment is lost, and last known weather at the ship was 800 feet with two miles visibility or better:

1. Continue approach by dead reckoning.
2. Maintain dead reckoning until 2 minutes past individual expected ramp time.
3. Conduct lost aircraft (without navigation aids, and without radio receiver) procedures.

If last known weather at ship was below 800 feet with two miles visibility or better:

1. Level off.
2. Conserve fuel.
3. Execute a one–half standard rate timed turn to a heading of 90 degrees to the right of previous penetration heading.
4. Maintain new heading for two minutes.
5. Conduct lost aircraft (without navigation aids, and without radio receiver) procedures.

LANDING WITHOUT UHF COMM (ESCORTING WINGMAN)

Exact procedures to be followed in the event one aircraft in a flight experiences communications failures must be covered in detail on every pre–flight brief. In general, the following procedures are recommended for the approach and landing:

1. Escorting aircraft determine the escorted (no radio) aircrafts fuel state through hand signals.
2. Escorting aircraft determine escorted aircrafts UHF guard and auxiliary receiver status.
3. Escorting aircraft establishes flight for a straight–in approach, lowers gear and flaps through hand signals, and passes the lead to the escorted aircraft after clearance to land is received.
4. Passing of the lead should ideally be done at about 1 mile from touchdown. At this time the flight should be lined up with the runway with the proper rate of descent established.

SECTION CARRIER CONTROLLED APPROACH

Should a section approach become necessary because of radio or instrument failure:

1. Place wingman on right side prior to commencing descent.
2. Reduce speed to 145 knots during last part of final approach so as to be approximately on speed when meatball is sighted.
3. Indicate meatball to wingman (blink external lights at night) to indicate carrier in sight.
4. Wingman will continue approach and land.
5. Leader will make definite turn to port and parallel final bearing in order to be in position should wingman bolter.
6. Following the wingman trap/bolter, the leader will execute the normal CCA waveoff procedure and be vectored in for an additional section approach or final landing.

CARRIER EMERGENCY NAV–COMM SIGNALS

Refer to CVA/CVS NATOPS Manual.

OUT-OF-CONTROL/SPIN RECOVERY

OUT–OF–CONTROL RECOVERY

1. Smoothly apply forward stick to reduce angle of attack (full forward if necessary), simultaneously neutralizing aileron and rudder, and reducing power to idle (altitude permitting).
2. If positive indications of recovery are not obvious after full forward stick application, deploy the drag chute while maintaining full forward stick, and neutralize ailerons and rudder.
3. Do not exceed 19 units AOA during dive pull–out.

NOTE

Engine flameouts may occur during departure; however, engine relights can be obtained with the throttles at idle even during a developed spin. Disengage the AFCS if in use.

SPIN RECOVERY

1. Positively determine spin direction.
2. Maintain full forward stick and neutral rudder and apply full aileron in the direction of the spin (right turn needle deflection, right spin, right aileron).
3. When the aircraft unloads (negative G) and/or yaw rate stops, neutralize the ailerons and fly out of the unusual attitude.
4. Do not exceed 19 units AOA during dive pull–out.
5. If still out of control by 10,000 feet above the terrain — EJECT.

INVERTED SPIN RECOVERY

1. Positively determine spin direction.
2. Full rudder against the spin (opposite the turn needle deflection).
3. Stick and ailerons neutral.
4. When the yaw rate stops, neutralize all controls and fly out of the unusual attitude.

TAKEOFF EMERGENCIES

ABORTED TAKEOFF

Where an aircraft takeoff must be aborted, a roll–in type engagement of all arresting gear is recommended to prevent overrun. The aircraft is cleared up to the maximum takeoff gross weight specified in figure 5–10. Additionally, the data may be used for light weight aborted takeoff, where applicable.

NOTE

The taxi light may be of use in locating arresting/abort gear at night.

ABORT–GEAR ENGAGEMENT

An abort gear long arrestment is used when a stopping problem exists with insufficient runway remaining; i.e., aborted takeoff, icy or wet runways, or loss of brakes after touchdown. Lower the hook allowing sufficient time for it to extend fully (normally 1000 feet prior to reaching the arresting gear). Do not lower the hook too early, the hook scraping on the runway will weaken the hook point. Line up the aircraft on the runway centerline and inform the control tower of your intention to engage the arresting gear, so that aircraft landing behind you may be waved off. After engagement, when no directional control problems exist, secure the engines.

1. ENGINES – IDLE
2. ARRESTING HOOK – DOWN
   Lower the tail hook at least 1000 feet ahead of the wire; 5 seconds is required for full extention.
3. DRAG CHUTE – DEPLOY
4. CONTROL STICK – FULL AFT
5. Aim for center of wire.
   Steer to engage the wire near the center and at a 90° angle.
6. Release brakes 100 feet ahead of wire.

BLOWN TIRE ON TAKEOFF

A situation may occur when the main wheel tire(s) or nose wheel tire(s) blow on takeoff roll. If the nose wheel tire(s) blow on takeoff, it is likely that one or both engines will receive FOD.

If decision to stop is made –

1. ABORT TAKEOFF
   Refer to Aborted Takeoff, this section.
2. LEAVE FLAPS IN POSITION SET FOR TAKEOFF.
   Leave flaps in their takeoff position to prevent additional damage to wing flap seals in the event they have been damaged by pieces of broken tire.

If takeoff is continued –

1. LEAVE LANDING GEAR DOWN, AND FLAPS IN POSITION SET FOR TAKEOFF.
   Leave the landing gear extended to preclude fouling the blown tire(s) in the wheel well(s). Leave flaps in their takeoff position to prevent additional damage to wing flap seals in the event they have been damaged by pieces of broken tire.
2. Monitor engine instruments.
   If any abnormal indications, such as RPM, EGT, or engine vibrations are noted, it is possible that FOD is present.
3. Plan to make a short field arrested landing.
   Refer to Short Field Arrestment, this section.

FLAP RETRACTION AND/OR BLC LOSS DURING FIELD AND CATAPULT TAKEOFFS

Inadvertent flap retractions and/or BLC loss during field and catapult takeoffs may occur. The above failures can be caused by any one or combination of the following: loss of electrical power, utility hydraulic failure, or failure of the flap blow–up switch.

FIELD TAKEOFF

If takeoff is continued –

1. ENGINES – MAX AB
2. External stores – JETTISON (IF NECESSARY)

CAUTION

Avoid tendency of over rotation beyond desired 10° to 12° takeoff attitude.

If decision to abort is made –

1. Employ aborted takeoff procedures, this section.

CARRIER TAKEOFF

Crewmember decisions will be based on when the failure occurs during the takeoff phase. If the failure occurs during the catapult stroke or shortly thereafter –

1. EJECT IMMEDIATELY

If the failure occurs at a time after the cat shot such that airspeed, altitude, and sink rate will allow the aircraft to be flown straight and level at less than 27 units AOA (above stall speed for gross weight), proceed as follows:

1. Engines – MAX AB
2. External stores – JETTISON
3. Flaps – BLOW DOWN
4. Attempt to maintain level flight or climb
5. Eject if above steps fail to effect minimum flight control and recovery.

ENGINE FAILURE DURING CATAPULT LAUNCH

1. Both engines – MAX AB
2. External stores – JETTISON
3. Landing gear – UP
4. Flaps – 1/2 ABOVE 180 KNOTS
5. Flaps – UP ABOVE 200 KNOTS
6. Failed engine – SHUTDOWN

If non–mechanical failure –

7. Failed engine – AIRSTART
8. Land as soon as practicable

EMERGENCY VISUAL COMMUNICATIONS

MALFUNCTION AND EMERGENCIES

TYPE OF EMERGENCY	SIGNAL		RESPONSE	
	DAY	NIGHT	DAY	NIGHT
I am in trouble, I want to land immediately.	Arm bent across forehead, followed by landing motion with open palm.	Circular motion of flashlight shined at other aircraft.	Carry out squadron doctrine of escort for disabled planes. Assume lead if indicated, and return to base or nearest suitable field.	
Are you having difficulty?	Point to pilot and give series of thumb down movements.	Flash a series of dots using exterior lights.	Thumb up, I am all right. Thumb down, I am having trouble.	Lights off once then on steady, I am all right. Lights flashing, I am having trouble
Initiation of "HEFOE" code.	Clench fist held to top of canopy.	Hold flashlight to top of canopy with steady light pointed at wingman, then flash exterior lights in accordance with code listed below.	Repeat signal to show acknowledgement.	

"HEFOE" SIGNALS

TYPE OF EMERGENCY	SIGNAL		RESPONSE	
	DAY	NIGHT	DAY	NIGHT
Hydraulic trouble.	One finger extended upward.	One flash of exterior lights.	Nod of head (I understand)	Series of Flashes (I understand)
Electrical trouble.	Two fingers extended upward.	Two flashes of exterior lights.	Nod of head (I understand)	Series of Flashes (I understand)
Fuel trouble.	Three fingers extended upward.	Three flashes of exterior lights.	Nod of head (I understand)	Series of Flashes (I understand)
Oxygen trouble.	Four fingers extended upward.	Four flashes of exterior lights.	Nod of head (I understand)	Series of Flashes (I understand)
Engine trouble.	Five fingers extended upward.	Five flashes of exterior lights.	Nod of head (I understand)	Series of Flashes (I understand)

FDD-1-(71)

Figure 5-13

AIRSPEED INDICATOR FAILURE

FLIGHT CONDITION		INDICATED AOA-UNITS
CATAPULT		
Transition From Catapult (Gear Down)		21.0
(Gear Up)		18.0
MILITARY POWER CLIMB		
Drag Index = 0	Sea level	5.5
	combat ceiling	7.6
Drag Index = 120	Sea level	7.6
	combat ceiling	8.5
MAXIMUM POWER CLIMB		
All Drag Indexes	Sea level	5.0
	combat ceiling	9.2
CRUISE AT ALTITUDES BELOW 20,000 FT. (all gross weights)		
Drag Index = 0		5.8
Drag Index = 130		8.0
CRUISE AT OPTIMUM ALTITUDE		
Drag Index = 0		6.9
Drag Index = 130		8.0
ENDURANCE AT OPTIMUM ALTITUDE		
Drag Index = 0		7.9
Drag Index = 130		9.3
DESCENTS (low to medium gross weight)		
250 KCAS, idle power		8.5
300 KCAS, 80% rpm		6.5
GEAR AND FLAPS EXTENSION		
Safe Gear Extension (with flaps up)		9.0
Safe Flap Extension (with gear down and corner reflector retracted)		13.0
(with gear down and corner reflector extended)		12.3
STALL WARNING (pedal shaker) before AFC 388		21.3
after AFC 388		20.6
STALL		
Gear down, Flaps Down, (with corner reflector retracted)		24.0
Gear Down, Flaps Down (with corner reflector extended)		23.3
Gear Down, 1/2 Flaps (with corner reflector retracted)		24.5
Gear Down, 1/2 Flaps (with corner reflector extended)		23.8
Gear Down, Flaps Up (with corner reflector retracted)		30.0
Gear Down, Flaps Up (with corner reflector extended)		29.3
Gear Up, Flaps Down		21.5
Gear Up, 1/2 Flaps		22.0
Gear Up, Flaps Up		27.0
APPROACH		
CCA/GCA Pattern (200 KCAS, gear up, half flaps)		9.5
Final ON SPEED approach (gear down, all engine/flap configurations) before AFC 388		19.0
after AFC 388		18.3

Notes

- Due to the basic inaccuracy of setting up flight conditions (other than landing approach) by reference to the angle of attack indicator, the information included in this table should be used only in an emergency situation.
- The ranges shown for angle of attack versus drag index, while not entirely linear, may be interpolated linearly for practical purposes.

FDD-1-(70)A

Figure 5-14

WARNING INDICATOR LIGHTS

LIGHT	CAUSE	CORRECTIVE ACTION	REMARKS
WHEELS	FLAPS DOWN-GEAR UP	LOWER LANDING GEAR OR RAISE FLAPS TO EXTINGUISH LIGHT	AN ILLUMINATED WHEELS LIGHT WITH THE FLAPS SWITCH UP AND THE GEAR HANDLE UP SHOULD BE TREATED AS A BLC MALFUNCTION.
FIRE OVERHT	FIRE OR OVERHEAT CONDITION EXISTS IN INDICATED ENGINE COMPARTMENT	CARRY OUT PRESCRIBED EMERGENCY PROCEDURE AS NECESSARY	LIGHT INDICATES TEMPERATURE IN EXCESS OF 765° F IN AFFECTED ENGINE COMPARTMENT.
MASTER CAUTION	SOME SYSTEM HAS A CAUTION CONDITION	CHECK TELELIGHT PANEL.	LIGHT ILLUMINATES WITH LIGHTS ON THE TELELIGHT PANEL EXCEPT: ● L/R/CTR EXT FUEL. ● REFUEL READY ● SPEEDBRAKES OUT.
L. ENG OIL LOW R. ENG OIL LOW	CORRESPONDING ENGINE OIL CAPACITY IS LOW	MONITOR OIL PRESSURE– INFO ONLY	THESE LIGHTS INOPERATIVE WITH J–79–8 ENGINES BEFORE PPC 62.
L WING PIN UNLOCK R WING PIN UNLOCK	WING(S) IS UNLOCKED.	WING PIN HANDLE–LOCKED.	MAY BE NECESSARY TO RECYCLE THE WING FOLD SYSTEM.
L ANTI-ICE ON R ANTI-ICE ON	SWITCH ON – NORMAL INDICATION.	INFO ONLY.	SWITCH OFF: ● IF LIGHT GOES OUT, ACCELERATE AND DISREGARD INDICATION. ● IF LIGHT STAYS ON, REMAIN AT REDUCED SPEED.
	SWITCH OFF – ERRONEOUS INDICATION	REDUCE SPEED.	
IFF			HIS LIGHT IS INOPERATIVE UNTIL MODE 4 IS MADE OPERATIVE
L AUX AIR DOOR R AUX AIR DOOR	DOOR(S) OUT OF PHASE WITH GEAR HANDLE.	CARRY OUT PRESCRIBED EMERGENCY PROCEDURES AS NECESSARY.	DISREGARD MOMENTARY LIGHT.
ALT ENCODER OUT	UNRELIABLE SIGNAL OR NO SIGNAL FROM ALTITUDE ENCODER UNIT	NONE–INFO ONLY	IF LIGHT STAYS ON: PERFORM THIS FUNCTION THROUGH VOICE COMMUNICATION.
IFR PROBE UNLOCKED	NORMAL INDICATION WITH PROBE EXTENDED.	INFO ONLY.	IF LIGHT STAYS ON: ● REFUEL PROBE SWITCH – EXTEND. ● REMAIN BELOW PUBLISHED AIRSPEED LIMITATIONS.
	PROBE IS NOT FULLY RETRACTED.	REFUEL PROBE SWITCH – EXTENDED, THEN RETRACT.	
CHK FUEL FILTERS	ONE OR BOTH FUEL FILTERS ARE PARTIALLY CLOGGED.	NONE – INFO ONLY	LOG ON YELLOW SHEET (OPNAV 3760-2). MONITOR ENGINE OPERATION
REFUEL READY	FUSELAGE PRESSURIZATION AND VACUUM RELIEF VALVE OPEN	NONE – INFO ONLY	INDICATES THAT FUSELAGE TANKS ARE PROPERLY VENTED FOR REFUELING.
CABIN TURB OVERSPEED	TURBINE SUBJECTED TO EXTREME PRESSURE AND/OR TEMPERATURE.	● REDUCE THRUST. ● REDUCE SPEED.	IF LIGHT STAYS ON: ● EMERGENCY VENT KNOB–PULL
W'SHIELD TEMP HIGH	WINDSHIELD IS OVERHEATED.	● RAIN REMOVAL SWITCH – OFF. ● EMER. VENT KNOB – PULL (IF NECESSARY)	INTERMITTENT ILLUMINATION DURING HIGH MACH FLIGHT, RAIN REMOVAL SYSTEM OFF, IS NORMAL.
CORNER REF L OUT	CORNER REFLECTOR EXTENDED	DO NOT EXCEED 250 KCAS	CORNER REFLECTOR EXTENDED AND THE NOSE LANDING GEAR IS UP
L EXT FUEL CTR EXT FUEL R EXT FUEL	● TANK IS EMPTY. ● FUEL FLOW HAS STOPPED. ● TANK IS FULL (DURING AIR REFUELING) ● MERS INSTALLED AND AUTOMATIC FUEL TRANSFER CIRCUIT ENERGIZED.	NONE – INFO ONLY	INTERMITTENT ILLUMINATION DURING FUEL TRANSFER IS NORMAL.
PRIMARY GYRO OFF	PRIMARY ATTITUDE REFERENCE IS UNRELIABLE	SELECT STBY ON THE REFERENCE SYSTEM SELECTOR KNOB	MALFUNCTION HAS OCCURRED IN THE AN/ASN-70 VERTICAL FLIGHT REFERENCE SET USE AN/AJB-7 ATTITUDE REFERENCE

FDD–1–(67–1)

Figure 5–15 (Sheet 1 of 2)

LIGHT	CAUSE	CORRECTIVE ACTION	REMARKS
RADAR LIQ COOL OVERHT	LIQUID COOLANT TEMP ABOVE 145 ± 5°F	NONE	IF LIGHT ILLUMINATES SELECT A RADAR LOW HEAT MODE
RADAR-CNI COOL OFF	EQUIPMENT COOLING TURBINE SHUTDOWN.	● AIRSPEED – REDUCE. ● WAIT 15 SECONDS. ● RESET – PUSH	IF LIGHT STAYS ON: ● REMAIN AT REDUCED AIR SPEED TO PREVENT EQUIPMENT DAMAGE. ● IFF & RADAR – STBY ● TACAN – OFF
AUTO PILOT PITCH TRIM	PITCH TRIM CIRCUIT IS NOT FUNCTIONING	● STICK – GRASP FIRMLY. ● AUTOPILOT – DISENGAGE.	DISREGARD A MOMENTARY LIGHT.
BLC MALFUNCTION	● FLAPS ARE UP AND AT LEAST ONE BLC VALVE IS STUCK OPEN. ● FLAPS ARE 1/2 DOWN AND TRAILING EDGE BLC VALVE IS OPEN.	● REDUCE SPEED BELOW – 200 KTS CAS ● FLAPS FULL DOWN	CONTINUED FLIGHT WITH FLAPS UP AND LIGHT ON MAY RESULT IN ELECTRICAL HYDRAULIC AND STRUCTURAL DAMAGE TO AIRCRAFT.
ENG INLET TEMP HIGH	INLET TEMPERATURE IS ABOVE 121°C.	BELOW 30,000 FEET – REDUCE SPEED.	LIGHT ON OPERATION ABOVE 30,000 FEET: SEE ENGINE LIMITATIONS.
AUTO PILOT DISENGAGE	AUTOPILOT HAS DISENGAGED.	AUTOPILOT – RE-ENGAGE IF PRACTICAL.	NONE
OXYGEN LOW	SUPPLY DEPLETED TO 1 LITER.	DESCEND TO SAFE ALTITUDE.	REFER TO OXYGEN DURATION TABLE.
SPEED BRAKE OUT	SPEED BRAKES ARE NOT CLOSED.	NONE – INFO ONLY	MASTER CAUTION WILL NOT ILLUMINATE.
PITCH AUG OFF	PITCH STAB AUG IS DISENGAGED.	NONE – INFO ONLY	ILLUMINATES ANY TIME BUSES ARE ENERGIZED AND PITCH STAB AUG IS DISENGAGED.
FUEL LEVEL LOW	1880 ± 200 POUNDS (JP–5) TOTAL FUEL REMAINING IN CELLS 1 & 2	NONE – INFO ONLY	CHECK ALL FUEL TRANSFERRED TO THE FUSELAGE.
TANK 7 FUEL	FUSELAGE CELL 7 TRANSFER VALVE HAS FAILED TO OPEN WHEN FUEL LEVEL LOW LIGHT ILLUMINATES.	INFO ONLY	FUSELAGE CELL 7 TANK WILL NOT BE AVAILABLE AND THE FUEL QUANTITY COUNTER WILL READ APPROX . 700 POUNDS HIGH.
TANK 7 EMPTY	FUSELAGE CELL 7 IS EMPTY	INFO ONLY	LIGHT CAN BE EXTINGUISHED BY DEPRESSING THE MASTER CAUTION RESET BUTTON.
APCS OFF	APCS IS DISENGAGED.	NONE – INFO ONLY	ILLUMINATES ANYTIME BUSES ARE ENERGIZED AND APCS BECOMES DISENGAGED.
CHK HYD GAGES	PRESSURE BELOW 1500 ± 100 PSI IN ANY SYSTEM.	ANALYZE SITUATION.	CARRY OUT PRESCRIBED EMERGENCY PROCEDURES.
STATIC CORR OFF	AIR DATA COMPUTER MALFUNCTION.	● AUTOPILOT – DISENGAGED. ● STATIC COMPENSATOR SWITCH – RESET.	IF LIGHT STAYS ON: ● STATIC COMPENSATOR SWITCH – CORR OFF. ● USE INSTRUMENT POSITION CORRECTION DATA AS NECESSARY.
CANOPY UNLOCKED	FORWARD, AFT OR BOTH CANOPIES ARE UNLOCKED.	CARRY OUT PRESCRIBED EMERGENCY PROCEDURE AS NECESSARY.	INSURE ADEQUATE PNEUMATIC PRESSURE FOR PROPER CANOPY OPERATION.
RADAR COOL DIV CLOSED	AIR DIVERTER CLOSED	NONE	IF LIGHT ILLUMINATES DURING FLIGHT: SELECT A RADAR LOW HEAT MODE
ANTI–SKID INOPERATIVE	ANTI-SKID IS NOT ENGAGED OR HAS MALFUNCTIONED.	ANTI-SKID SWITCH – ON.	LIGHT ON AND SWITCH ON: CARRY OUT PRESCRIBED EMERGENCY PROCEDURE. DISREGARD MOMENTARY LIGHT
FUS BLEED AIR OVERHT	TEMPERATURE IN EXCESS OF 410°F EXISTS IN FORWARD FUSELAGE AREA.	● REDUCE POWER ● BLEED AIR SWITCH – OFF	● MASTER CAUTION ILLUMINATES ● CONTINUED OPERATION WITH LIGHT(S) ON MAY RESULT IN EQUIPMENT DAMAGE ● **SEE BLEED AIR OFF LIGHT.**
ENG BLEED AIR OVERHT	TEMPERATURE IN EXCESS OF 575°F EXISTS IN KEEL WEB AREA.		
WING BLEED AIR OVERHT	TEMPERATURE IN EXCESS OF 410°F EXISTS IN WING L. E.		
BLEED AIR OFF	BLEED AIR SHUTOFF VALVE IS CLOSED.	NONE – INFO ONLY	● LOSS OF L.E. BLC, FUEL PRESSURIZATION, PNEUMATIC SYSTEM CHARGING, CABIN AND EQUIPMENT REFRIGERATION AND SPC. ● **LANDING SPEED INCREASES 7 KNOTS**
COUPLER OFF	DATA LINK COUPLER HAS DISENGAGED	ASSUME MANUAL CONTROL OF AIRCRAFT	INDICATES UNRELIABLE DATA LINK CONTROL SIGNALS OR INTENTIONAL DISENGAGEMENT

FDD–1–(67–2)B

Figure 5–15 (Sheet 2 of 2)

EXTERNAL STORES JETTISON CHART

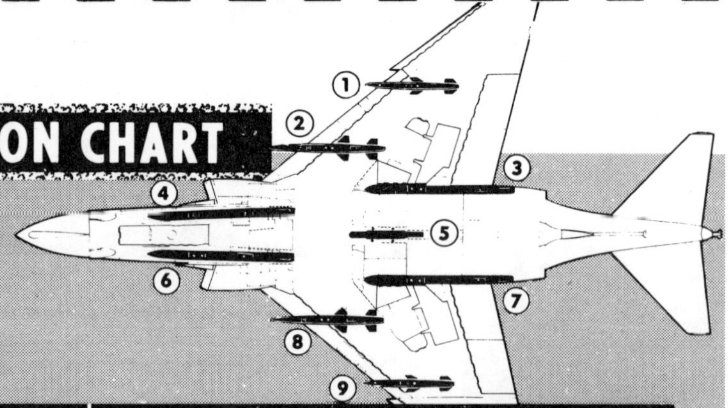

* WEIGHT MUST BE OFF GEAR
† MISSILES MOUNTED ON FUSELAGE STATIONS 4 AND 6 CANNOT BE JETTISONED IF A 600-GALLON TANK, BUDDY TANK OR MER IS INSTALLED ON CENTERLINE STATION 5.
†† ARMAMENT SAFETY OVERRIDE SWITCH BYPASSES THE GEAR HANDLE SWITCH AND ALLOWS RELEASE WITH GEAR HANDLE DOWN.

EXTERNAL STATION	GEAR HDLE. POSITION	METHOD OF RELEASE
1 thru 9	UP or DOWN*	EXTERNAL STORES EMERG RELEASE BUTTON - **DEPRESS** [2]
1 & 9	UP or DOWN	EXTERNAL (WING) TANKS JETTISON SW. - **JETT**
2 & 8	UP or DOWN	MISSILE JETTISON SELECTOR SW. - **APPROPRIATE STA (L. WING, R. WING)** MISSILE JETTISON SELECTOR SW. - **PUSH**
3, 4 †, 6 † & 7	UP or DOWN	MISSILE JETTISON SELECTOR SW. - **APPROPRIATE STATION (L. FWD, R. FWD L. AFT, R. AFT)** MISSILE JETTISON SW. - **PUSH**
2,3,4 †,6 †,7 & 8	UP or DOWN	MISSILE JETTISON SELECTOR SW. - **ALL** [1] MISSILE JETTISON SELECTOR SW. - **PUSH**
5	UP ††	CENTERLINE STATION SAFE SW. - **READY** BOMB CONTROL SW. - **DIRECT** WEAPONS SW. - **CONV OFF-NUCL ON** BOMB / ℄ STORE RELEASE BUTTON - **DEPRESS**

[1] THE **ALL** POSITION OF THE MISSILE JETTISON SELECTOR SW IS ELIMINATED ON **F-4J** AIRCRAFT 153900af AND UP, AND ALL OTHERS AFTER AFC 346. HOWEVER, AFTER AFC 346 EITHER THE FORMER ALL POSITION OR THE R AFT POSITION CAN BE USED TO JETTISON THE RIGHT AFT FUSELAGE MISSILE, BUT ON 153900af AND UP ONLY THE R AFT POSITION CAN BE USED.

[2] ON F-4J 158355at AND UP, AND ALL OTHERS AFTER AFC 506, THE CHART BELOW LISTS STORES JETTISONED BY THE EXTERNAL STORES RELEASE BUTTON FOR EACH POSITION OF THE JETTISON SELECT SW.

JETTISON SELECT SWITCH POSITION	EXTERNAL STORES STATIONS					
	LEFT OUTBOARD	LEFT INBOARD	CENTER-LINE	RIGHT INBOARD	RIGHT OUTBOARD	FUSELAGE MISSILES
ALL	X	X	X	X	X	X
AIR-GRD	X	O	⊙	O	X	
AIR-GRD+CL	X	O	X	O	X	

X - DENOTES JETTISONING OF STORE AND ITS ASSOCIATED JETTISONABLE ARMAMENT ATTACHMENT (PYLONS, MERS, ETC.). THE WORD STORE DENOTES NON-WEAPON ITEMS, SUCH AS FUEL TANKS, AS WELL AS WEAPONS.
⊙ - STORE RETAINED ON CENTERLINE.

O - JETTISONING FROM INBOARD WING STATIONS DEPENDS ON LOADING AS SHOWN BELOW.

LOADING	LEFT INBOARD	RIGHT INBOARD	JETTISON
1	PYLON ONLY	PYLON ONLY	YES
2	TER	TER	YES
3	SIDEWINDER OR SPARROW	SIDEWINDER OR SPARROW	NO
4	SIDEWINDER OR SPARROW	EMPTY	NO
5	EMPTY	SIDEWINDER OR SPARROW	NO
6	SIDEWINDER OR SPARROW	SIDEWINDER/TER OR TER	ONLY TER SUSPENDED STORES JETTISONED
7	SIDEWINDER/TER OR TER	SIDEWINDER OR SPARROW	ONLY TER SUSPENDED STORES JETTISONED
8	SIDEWINDER/TER	SIDEWINDER/TER	ONLY TER SUSPENDED STORES JETTISONED

FDD-1-(69)

Figure 5-16

SECTION VI
ALL WEATHER OPERATION

TABLE OF CONTENTS

SIMULATED INSTRUMENT PROCEDURES

GENERAL

Instrument flight is primarily a problem of time and distance navigation wherein all, or part, of the flight is conducted under instrument conditions. To complete a successful instrument flight, crewmembers must be properly prepared and have conducted the necessary planning. All pilots will be current in latest instrument flight rules and regulations published by higher authority and, when operating aircraft under instrument flight conditions, will be guided by the current OPNAV INSTRUCTION 3710.7 (General Flight and Operating Instructions for Naval Aircraft) and Federal Air Regulations.

SAFETY PRECAUTIONS

It is the responsibility of the chase pilot to ensure that the flight is clear of other aircraft at all times.

The instrument pilot will not go hooded on departure until reaching a minimum of 2000 feet above the terrain.

At a minimum of 500 feet above terrain, the instrument pilot will go contact on any hooded penetration or ground controlled approach.

The chase pilot will conduct communications checks with the instrument pilot and receive an acknowledgement at 1–minute intervals below FL 240 and at 3–minute intervals above FL 240, if not under positive control.

If loss of radio contact occurs, the instrument pilot will immediately go contact and remain VFR until radio contact is reestablished. If necessary, the chase pilot will pass to the right and pull ahead/in front to attract the instrument pilot's attention to go contact. Lighting afterburner when passing the instrument pilot will usually get his attention.

Radio contact will be positively established immediately before and after any channel or frequency change.

Unless under positive control, the instrument pilot will call the indicated altitude at each 5000–foot interval during descent and at level–off.

HOODED RADAR INTERCEPTS

a. All hooded intercept flights will be pre–briefed prior to launch.
b. Positive UHF communication must exist prior to "going hooded" and commencing intercepts. All aircraft in the flight will have UHF set on T/R + G. The chase and target aircraft will have the tactical frequency with comm command in one cockpit. The other cockpit will have Guard set in the comm channel, allowing the pilots to take comm command and transmit on guard, if the need arises.
c. If the chase or target pilot does not have a "Tally Ho" prior to 5 miles, the hooded pilot will go contact.
d. If the hooded aircraft does not have a "Judy" by 10 miles, the target aircraft will call his altitude and the intercept pilot will go contact, and remain so, for the completion of the run, even if a "Judy" is subsequently obtained.
e. Minimum altitude above all terrain will be 2000 feet.
f. Target aircraft will not "jink" in altitude.

CHASE PLANE PROCEDURES

The chase pilot's duties on instrument flights are to act as lookout and to be a flight monitor. The best position for this is a loose tactical wing position where airspeed, attitude, and altitude may be monitored while maintaining a good lookout. During GCA approaches, the chase will fly a position as directed by GCA. This position is normally about four or five o'clock from the GCA aircraft, 500 feet away, and slightly stepped up.

CHASE PLANE RADIO PROCEDURE

The chase aircraft will set up its radios in the following manner: the frequency in use with comm command in one cockpit, and the guard position set in the comm channel of the other cockpit. The instrument airplane shall monitor the TR + G position. If the chase pilot suspects radio failure or cannot "burn through" transmissions by GCA or other controlling agencies, he can take command and transmit instructions to the instrument pilot on guard channel.

ACTUAL INSTRUMENT PROCEDURES

INSTRUMENT FLIGHT

This is an all-weather airplane designed to perform operational missions in all extremes of weather. Rapid acceleration rates and high pitch angles during climb, of necessity dictate some modification of standard instrument procedures.

INSTRUMENT FLIGHT PLANNING

On instrument flights, delays in departure and descent and low climb rates to altitude are often required in high density control areas. These factors make fuel consumption and flight endurance critical. All instrument flights should be carefully planned and consideration given to the additional time and fuel which may be required. A complete weather briefing for all pilots on the flights will be obtained and the appropriate flight plan will be filed.

BEFORE STARTING ENGINES

When practical, an ATC clearance should be obtained from the tower before starting engines. When operating on external power, all CNI equipment except the UHF and intercom are limited to 10 minutes of accumulated operation in a 1-hour span.

BEFORE TAKEOFF

It is essential that the instrument and navigation equipment be thoroughly checked prior to takeoff. If a climb through precipitation or clouds is anticipated, place the pitot heat switch to ON and the engine anti-ice switch to DE-ICE. At idle rpm, the operation of the engine de-icers can be checked by noting a slight rise (approx. 10°C) in EGT and a slight increase in fuel flow. The ADI wings symbol should be set with the wings 1° above the horizon and the compass should be aligned with the runway heading. The stab aug should be engaged.

INSTRUMENT TAKEOFF

Same as normal Takeoff.

INSTRUMENT CLIMB

The simplified climb technique described in part 3 of section III can be used with minimum sacrifice in fuel consumption and climb rates. Turns should be kept to a minimum during climb due to the difficulty of determining bank angles and rates of turn while at high pitch angles. Upon reaching clear air, turn off the engine anti-icing and the pitot heat. Follow the clearance exactly as given. If unable to comply with the clearance, it is mandatory that ATC be advised immediately. Climb speed will conform to local procedures, but should be a comfortable airspeed with transition to the published climb schedule accomplished at a comfortable altitude above the terrain.

PENETRATION PROCEDURES

Three to five minutes prior to making a descent, the cabin temperature control should be set at the maximum comfortable level and the defog/footheat lever to the DEFOG position. Contact approach control 10 minutes prior to EAT or as directed by ARTC, and conform to the provision of section 2, Flight Planning Document. Three minutes prior to entering holding, adjust power to arrive at the holding fix with maximum endurance airspeed (265 knots CAS maximum). Prior to descent, the pilot will check missed approach procedures and will obtain the latest weather information at the destination and at the alternate if required. Refer to Descent/Instrument Penetration procedures, section III.

PENETRATION WITH GEAR AND FLAPS EXTENDED

Under certain conditions, it may be necessary to penetrate with the gear and flaps extended. If such is the case, advise approach control that the approach will be executed with a non-standard approach speed. Prior to commencing the approach, slow to below 250 knots and lower the landing gear and flaps. If external tanks are carried, place the wing transfer switch to EMERG before extension of the landing gear to avoid de-pressurizing the fuel tanks. Commencing penetration, reduce the pitch attitude to maintain 195 knots. This attitude will seem extremely nose low. Make throttle adjustments as necessary to maintain a 3500 fpm rate of descent. Initiate a round-out in order to reach GCA pickup or tacan gate altitude at a speed of 150-160 knots. If the descent has been made with the wing transfer in the EMERG position, return the switch to NORMAL prior to landing.

RADAR CONTROLLED PENETRATION

The approaches are basically the same as previously described with the following additions. The controlling activity will normally ask for turns or specific IFF squawks for positive identification. The controlling activity will advise turns or headings which will produce the desired flight path. They will also advise as to distance from the destination and direct a descent to lower minimum altitudes as traffic and terrain permit.

GCA (PAR) APPROACHES

This airplane handles exceptionally well in the GCA pattern. It is very stable directionally and is very responsive to minor corrections about all axes. When directed, descend to GCA pickup altitude and transition to landing configuration. Slow to 150–160 knots which will require approximately 88–92%. Trim required will be approximately 3 units nose-up. When the pilot is told to commence a normal rate of descent for his aircraft, he should retard the power to approximately 82–84%. Allow the airplane to slow to 145 knots CAS minimum (section). If alone, adjust the nose to maintain 140 knots CAS or a donut on the angle-of-attack indexer, whichever is greater. While holding the attitude constant, make smooth but positive power adjustments to maintain a desired rate of descent of approximately 600–800 feet per minute or as directed. For heading corrections after starting descent, recommended bank angle is 10°. Up to 3° heading corrections may be made by using rudder alone. When the controller announces "minimums have been reached", the pilot will look up. If the runway is not in sight, he will immediately execute a missed approach. If the pilot has the runway in sight, he will adjust power to establish optimum angle-of-attack/airspeed and complete the landing.

GCA Box Pattern

To enter the GCA pattern from other than a penetration, contact Approach Control and proceed as follows. The downwind leg is flown at 230–250 knots with gear and flaps retracted. The base leg is flown at 150 knots with gear and flaps down. After completing the turn on final and slowing to approach speed, the normal GCA procedures apply. If entering the GCA pattern after a touch-and-go landing, aircraft will comply with Approach Control instructions.

TURBULENT AIR AND THUNDERSTORM OPERATION

Intentional flight through thunderstorms should be avoided, unless the urgency of the mission precludes a deviation from course, because of the high probability of damage to the airframe by impact ice, hail, and lightning. The radar provides an excellent means of navigating between or around storm cells and the airplane is capable of climbing over the top of small and moderately developed thunderstorms.

PENETRATION

If necessary to penetrate, the basic structure of the airplane is capable of withstanding the accelerations and gust loadings associated with the largest thunderstorms. The airplane is exceptionally stable and comparatively easy to control in the severe turbulence; however, the effects of turbulence become noticeably more abrupt and uncomfortable at airspeeds above optimum cruise and below 35,000 feet. The airplane is not displaced significantly from the intended flight path and desired heading. Altitude, airspeed, and attitude can be maintained with reasonable accuracy.

Penetration Airspeeds

The optimum thunderstorm penetration speeds, based on pilot comfort, controllability, and engine considerations are between optimum cruise and 280 knots below 35,000 feet, and no less than 300 knots above 35,000 feet. The afterburner should be utilized if necessary to maintain airspeed.

APPROACHING THE STORM

If the storm cannot be seen, it may be located by radar. Adjust power to establish the recommended approach speed. Note stabilator trim position. Place the pitot heat switch to ON, the engine anti-ice switch to DE-ICE, and the autopilot OFF. The seat should be lowered in order to view the instruments and to minimize the buffeting due to turbulence. Do not try to top thunderstorms at subsonic speeds above 40,000 feet; the stall margin of both the airframe and engines becomes critical in this region. Flight through a thunderstorm at the proper airspeed and attitude is much more advantageous than floundering into the storm at a dangerously slow airspeed while attempting to reach the top. If the penetration is made at night, the daylight floodlights should be ON, and the console and instrument lights should be full bright.

IN THE STORM

Maintain a normal instrument scan with added emphasis on the attitude gyro (ADI). Attempt to maintain a constant pitch attitude and, if necessary, accept moderate altitude and airspeed fluctuations. In heavy precipitation, a reduction in engine rpm may be necessary due to the increased thrust resulting from water ingestion. If compressor stalls or engine stagnation develops, attempt to regain normal engine operation by momentarily retarding the throttle to IDLE then advance to the operating range. If the stall persists, shut down the engine and attempt to relight. If the engine remains stagnated at reduced power and the EGT is within limits, maintain reduced power until clear of the thunderstorm. While in the storm, the longitudinal feel trim, angle-of-attack, total temperature, windshield overheat, static pressure correction, and cabin pressurization systems may experience some abnormalities due to rain, ice, or hail damage. No difficulty should be encountered in maintaining control of the airplane; however, the rapid illumination of numerous warning lights may be somewhat distracting to the pilot if he is not prepared.

Longitudinal Feel Trim Failure

Longitudinal feel trim system failures are caused by ice blockage of the ram air bellows venturi located in the vertical fin. The failure is characterized by either a complete or intermittent failure of the feel system. The failure can also be verified by noting that the pointer on the stab trim indicator has moved from the previously noted position.

Angle-of-Attack System Failure

The angle-of-attack system may become temporarily inaccurate due to probe icing, or it may permanently fail due to structural damage of the probe from ice or hail. Icing of the probe is usually characterized by a zero angle of attack indication which will return to normal in clear air. Structural damage may cause erroneous readings or fail the system completely.

Total Temperature System Failure

The total temperature sensor may fail because of water or ice damage. This failure is characterized by a flashing or steady ENG INLET TEMP HI light erroneous true airspeed indications, and possible cycling of the engine intake ramps.

Windshield Overheat Sensor Failure

The windshield overheat sensor may fail in heavy rain or icing conditions. The failure is characterized by a steady W'SHIELD TEMP HIGH light.

SPC Failure

The SPC system will monitor off during thunderstorm penetrations between speeds of 0.93 to 1.0 Mach. Immediately prior to SPC disengagement, the altimeter, airspeed indicator, and vertical velocity indicator will be highly erratic. Attempts to reset the SPC in precipitation will usually be unsuccessful; however, the SPC can be reset for normal operation upon reaching clear air.

Cabin Pressurization System Failure

Cabin pressurization fluctuation occurs in precipitation above the freezing level. This is evidenced by a significant decrease in available cabin pressurization flow and is caused by ice at the pressure suit heat exchanger.

ICE AND RAIN

The possibility of engine and/or airframe icing is always present when the airplane is operating under instrument conditions. Icing is most likely to occur when takeoffs must be made into low clouds with temperature at or near freezing. Normal flight operations are carried on above the serious icing levels, and the airplane's high performance capabilities will usually enable the pilot to move out of the dangerous areas quickly. When an icing condition is encountered, immediate action should be taken to avoid further accumulation by changing altitude and/or course and increasing the rate of climb or airspeed. When icing conditions are anticipated, actuate the engine anti-ice switch to DE-ICE and the pitot heat switch to ON.

WINDSHIELD RAIN REMOVAL

The following precautions must be observed when contemplating the use of the windshield rain removal system.

1. Do not operate on a dry windshield.
2. Turn system OFF immediately if W'SHIELD TEMP HIGH indicator light illuminates.
3. Do not operate above Mach 1.0.

LONGITUDINAL FEEL TRIM

When flying through areas of precipitation, partial or complete failure of the longitudinal control artificial feel system may result due to ice and/or water blockage of the bellows ram air line. If this condition occurs, excessive stick force will be required to maintain the desired airplane attitude. Since sudden longitudinal trim changes may occur several minutes after flying through freezing precipitation (especially during descent to altitudes below the freezing level), the application of corrective longitudinal trim when a blocked bellows inlet is suspected is not recommended.

AIR DATA COMPUTER

The Air data computer may malfunction during flight through ice and/or because of impact forces imposed by water and ice on the ADC total temperature sensor. A momentarily flashing ENG INLET TEMP HIGH warning light usually indicates that the sensor probe has been blocked or shorted by ice accumulation.

HYDROPLANING

Operations on wet or flooded runways may produce three conditions under which tire traction may be reduced to an insignificant value.

1. Dynamic hydroplaning
2. Viscous hydroplaning
3. Reverted rubber skids

DYNAMIC HYDROPLANING

As the tire velocity is increased, the hydrodynamic pressure acting on the leading portion of the tire footprint will increase to a value sufficient to support the vertical load acting on the tire. The speed at which this occurs is called total hydroplaning speed, and can be computed using the empirically derived relation VH 9 × p , where VH is the total hydroplaning speed in knots and p is the tire inflation pressure in pounds per square inch. Any increase in ground speed above this critical value lifts the tire completely off the pavement, leaving it supported by the fluid alone. Since the fluid cushion is incapable of sustaining any appreciable shear forces, braking and sideforce coefficients become almost non-existent. Total hydroplaning speed of the main landing gear tires at 400 psi is 180 knots, the nose gear tire inflated to 150 psi is 110 knots.

VISCOUS HYDROPLANING

Viscous hydroplaning occurs due to the inability of the tire to penetrate the very thin fluid film found under damp runway conditions. This condition is aggravated when more viscous fluids such as oil, or road dust and water mixed are present, and is improved in the presence of a coarse textured runway surface. Viscous hydroplaning occurs at medium to high speed with rolling or skidding tires, and the speed at which it occurs is not dependent on tire pressure.

REVERTED RUBBER SKIDS

Reverted rubber skids occur after a locked-wheel skid has started on wet runway. Enough heat may be produced to turn the entrapped water to steam. The steam in turn melts the rubber. The molten rubber forms a seal preventing the escape of water and steam. Thus the tire rides on a cushion of steam which greatly reduces the friction coefficient and may continue to do so to very low speeds.

EXTREME WEATHER PROCEDURES

COLD WEATHER OPERATION

PREFLIGHT

Check that the airplane is free of frost, snow, and ice. These accumulations present a major flight hazard resulting in loss of lift and increased stall speeds. Do not allow ice to be chipped or scraped from the airplane; damage to the airframe may result. Shock struts, actuating cylinders, pitot-static sources, and fuel vents should be inspected for ice and dirt accumulation. In addition to the above checks, the BLC duct tape (if applied) should be removed.

WARM-UP AND GROUND CHECK

If the airplane has been parked in heavy rain when subsequent freezing has been anticipated, a protective pressure-sensitive tape should have been applied to the trailing edge wing/flap junctures to prevent precipitation and, thereby, ice from accumulating in the BLC ducts and valves. If the tape has not been applied and it is suspected that ice has formed in the ducts and valves, the flap actuating check should not be performed until the engines have been running for 12 to 15 minutes at 85% rpm. This permits hot BLC air to thaw any ice which may have accumulated in the BLC valves. After it is felt that the valves have been thawed:

1. Flap switch - CYCLE AND CHECK
   Actuate the flaps to the 1/2 and full down positions. Check that the BLC system is operating. Actuate the flaps to up and monitor the BLC MALFUNCTION light for a malfunction indication.

CAUTION

A BLC MALFUNCTION light illumination with flaps up constitutes a flight hazard. Refer to Boundary Layer Control Malfunction, section V of this manual.

NOTE

A BLC MALFUNCTION light with flaps in any position other than up, indicates that the flap-up limit switches are not returning to their normal open position. By cycling the flaps three or four times to allow circulating hydraulic fluid to warm the actuators, the switches may be freed so that the BLC malfunction indicating system functions normally.

In addition to normal walk-around check that:

1. Shock struts, pitot tube, fuel vents, and actuating cylinders are free of ice or dirt.
2. Fuel drain cocks free of ice and drain condensation and ensure that all pneumatic bottles have been adequately serviced.
3. All exterior covers and BLC duct tape (if applied) removed.
4. Closely inspect the nozzle shroud flaps for any signs of ice deposits. If any ice is present apply heat to the nozzle control feedback housing area for 5 to 10 minutes just prior to engine start.

INTERIOR CHECK

In temperatures below 0°F, difficulty may be experienced when connecting the oxygen mask hose to the T-connector, due to a stiff O-ring in the T-connector. Application of a small amount of heat to the T-connector will alleviate this problem. Also, if the oxygen mask is not fastened, keep it well clear of the face to prevent freezing of the inhalation valves. Due to the bulk of arctic flying clothes, difficulty may be experienced in getting strapped in and in removing the face curtain safety pin. Assistance from the plane captain may be required.

ENGINE START

During engine start operation, press ignition buttons at approximately 6% rpm; however, do not advance the throttles until approximately 10% rpm is reached. Pressing the ignition buttons prior to throttle advance will dry out the igniter plugs, thereby enhancing a successful start. If any abnormal sounds or noises are present during starting, discontinue starting and apply intake duct preheating for 10 to 15 minutes. Immediately after starting the engine at extremely low temperatures, the engine oil pressure indication will become excessive and may peg out at 100+ psi. When this condition occurs, the generators may not connect to the aircraft load when the generator control switch is placed to GEN ON. This is caused by either the limit governor not porting oil to the underspeed switch or the contacts within the underspeed switch not closing. Both of these problems are associated with low temperature, high viscosity oil. Repeated cycling of the generator switch in an attempt to get a generator on the line will not damage the CSD or generator, but may prove futile until the engine/oil warms up (oil pressure drops below 50 psi). Occasionally because of cold oil, the spool valve in the limit governor does not move fully to the open position. If this occurs, it also prevents the generators from connecting to the aircraft load. An increase in oil temperature does not always correct this condition. However, engine shut down and restart often exercises the system sufficiently to free the spool valve. Spool valve operation is not critical as this valve operates once to open at engine start and again to close at engine shut down. An aircraft flight should not be aborted for failure of a generator to come on or stay on the line prior to engine/oil warm-up nor should an abort be declared if an engine shut-down and restart will bring and keep a generator on the line.

TAXIING

Avoid taxiing in deep or rutted snow; frozen brakes will probably result. Increase the interval between taxiing airplanes to insure a safe stopping distance and to prevent icing of the airplane surfaces by the snow and ice melted by the jet blast of the preceding airplane.

BEFORE TAKEOFF CHECK

During the engine runups, an ice-free area should be selected if possible. The engine thrust is noticeably greater at low temperatures and the probability of skidding the airplane is likely. If icing conditions are encountered or expected, place the engine anti-ice switch in the DE-ICE position and the pitot heat ON.

TAKEOFF

When operating from runways which are covered with excessive water, snow or slush, high-speed aborts may result in engine flame-out due to precipitation ingestion. The probability of flame-out is highest when throttles are chopped from afterburner to IDLE at speeds above 100 knots. With a double flame-out, normal braking, anti-skid on aircraft 157242an and up, and nose gear steering will be lost. After takeoff from runways covered with snow or slush, packed snow/slush in the auxiliary air door area may make throttle movement difficult until the snow/slush can be melted. Check applicable takeoff distance charts, part 2 of section XI.

NOTE

If inflight freezing within the longitudinal control system is experienced, excessive stick forces may be required to move the control stick. Normal aiplane control is available but requires higher force inputs. Normal control forces and AFCS operation will return at lower (warmer) altitudes.

LANDING

If snow and ice tires are installed, use brakes intermittently to keep the tire thread from filling and glazing. As soon as practicable after the landing roll, the flaps should be placed in the full UP position. This will shut off the BLC air which otherwise causes the loose snow to swirl and be drawn in through the auxiliary air doors and pass along the engine. If this happens, the snow melts and deposits of ice form shortly after engine shut down. The ice can cause binding of the nozzle feedback housing and possibly result in nozzle failure upon the next engine start.

AFTER LANDING

During operations where the temperature is below freezing with heavy rain, or expected to drop below freezing with heavy rain, the aircraft may be parked with wings spread and flaps in the full down position.

BEFORE LEAVING AIRPLANE

Weather permitting, leave the canopy partially open to allow for air circulation. This will help prevent canopy cracking from differential cooling and decreases the possibility of windshield and canopy frosting.

HOT WEATHER OPERATION

TAXIING

While taxiing in hot weather, the canopies may be opened, if necessary, to augment crew comfort. Do not operate the engines in a sand or dust storm if avoidable. Park the airplane crosswind and shut down the engines to minimize damage from sand or dust.

TAKEOFF

The required takeoff distances are increased by a temperature increase. Check the applicable Takeoff Distance charts, part 2 of section XI.

SECTION VII
COMMUNICATIONS PROCEDURES

TABLE OF CONTENTS

RADIO COMMUNICATIONS

GENERAL

It is the responsibility of the pilot to ensure that all radio and electronic transmissions from the aircraft are in compliance with applicable directives and squadron doctrine. It is mandatory that the pilot and RIO be thoroughly indoctrinated in all communications equipment, methods and procedures including hand signals. Radio communications will be in accordance with procedures set forth in NWIP 41-1A, NWP-41A, ACP 165, JANAP 119 and local fleet/shore instructions.

PILOT/RIO INTERCEPT COMMENTARY

DESCRIPTIVE COMMENTARY

Descriptive commentary is given in a specific order when existing conditions allow; i.e., azimuth, elevation, closing rate. Under certain conditions, it is impossible to give the description in the desired sequence since adhering rigidly to the sequence may interfere with positive control of target movement. The description of target action or position is normally given in a conversational tone of voice. Upon initial contact, the RIO will immediately start giving descriptive commentary. It is apparent that if the RIO gives "CONTACT" followed immediately by range reading, the pilot is aware of the urgency or need of positive action by the RIO. The RIO will also, upon initial contact, give any directive commentary that is necessary to ensure that the target does not exceed the limitations of the set. Descriptive commentary is not required to be particularly accurate at long ranges, but as the fighter approaches visual or attack range, the description must be accurate, and still not interfere with commands to the pilot. Sufficient descriptive commentary should be given to keep the pilot constantly informed of the position of the target in terms of azimuth angle, range, elevation angle, and overtaking speed.

Contact Report

Contact reports will be given to the controlling agency in the following manner:

"CONTACT" followed by AZIMUTH – Degrees followed by LEFT or RIGHT.

Range – Nearest mile

Example: CONTACT, 25 LEFT, 30 MILES.

Target Position Reporting

Position reports will be given in a specific order; i.e., azimuth angle, range, elevation angle, and overtaking speed. The following items will be used:

AZIMUTH – Degrees followed by LEFT or RIGHT

RANGE – In miles (yards may be used when appropriate)

ELEVATION – Degrees followed by above, below, or level

OVERTAKING SPEED – Knots

Examples:

10 RIGHT, 8 MILES, LEVEL, OVERTAKE 300.
20 LEFT, 12 MILES, 5 ABOVE, OVERTAKE 50.

Judy

"JUDY" will be given to the controlling agency when assuming responsibility and control of the intercept.

DIRECTIVE COMMENTARY

Directive commentary is divided into three categories and is used when the situation calls for a change of the aircraft's direction, speed, or elevation. A considerable amount of information as to the urgency may be obtained from the inflection of the RIO's voice as well as speed with which one command follows the other. Voice modulation properly employed will give flexibility to commentary. If the RIO places emphasis in his voice commands, it ensures that the pilot hears each and every command and will also cause the pilot to react accordingly. Directive commentary will at all times take precedence over descriptive commentary. The pilot will also inform the RIO whenever the limitations of the aircraft are reached. Upon achieving a speed change, leveling off, or resuming straight and level flight, the pilot will inform the RIO with commentary such as "SPEED SET" "STEADY AND LEVEL", or "ALTITUDE SET".

Heading Change Commands

"EASY PORT" or "EASY STARBOARD"	15° angle of bank.
"PORT" or "STARBOARD"	30° angle of bank.
"PORT HARD" or "STARBOARD HARD"	45° angle of bank.
"HARD AS POSSIBLE"	Maximum possible turn, maintaining airspeed and altitude.
"EASE"	Roll out slowly toward steady.
"HOLD"	Maintain present bank angle.
"STEADY"	Roll out of turn.
"HARDER"	Increase angle of bank to next higher Increment. Example: If at 30° angle of bank, pilot will increase bank angle to 45°; if at 15° bank, increase to 30°, etc.
"WRAP IT UP"	Maximum possible turn within aircraft acceleration limitations.
"REVERSE"	Immediate identical turn in opposite direction.
"OVERBANK"	More than 90° of bank angle, (can be further amplified by "HARDER").

NOTE

Turns may be given as a specific number of degrees. For example: "PORT HARD 40", etc.

Elevation Commands

"GO DOWN "	Descend a specified number of feet designated by the RIO.
"CLIMB "	Climb as directed by RIO (GATE, BUSTER, or specified number of feet).
"NOSE-UP"/"-DOWN"	Change pitch angle until given "HOLD" by RIO. Leave throttles at same setting.
"DIVE"	Maximum rate descent until given "HOLD DIVE" by RIO or until maximum permissible rate of descent is established.

CONTINUED

Elevation Commands (CONT)

"LEVEL OFF"	Return to level flight.

Speed Commands

"BUSTER"	Full military power.
"GATE"	Maximum power.
"BUSTER ______" or "GATE ______"	Military or maximum power to CAS or indicated Mach number specified.
"SPEED UP ______"	Increase airspeed by amount specified (CAS or IMN).
*"THROTTLE BACK ______"	Decrease speed by amount specified (CAS or IMN).
*"THROTTLE RIGHT BACK"	Decrease airspeed as rapidly as possible until minimum airspeed reached or RIO gives "HOLD SPEED".
"HOLD SPEED"	Maintain present airspeed.

*Speed will normally be reduced as rapidly as possible utilizing speed brakes and/or throttle as appropriate.

Less Frequent Commands

"BREAK STARBOARD" or "BREAK PORT"	Immediate maximum possible turn, within aircraft acceleration limitations, in the direction indicated. An "UP" or "DOWN" elevation command may be given as appropriate.
"COMPASS RECOVERY"	Immediate, hard as possible, turn 30° beyond target's last known heading.

VISUAL COMMUNICATIONS

Communications between aircraft will be conducted visually whenever practicable, provided no sacrifice in operational efficiency is involved. Flight leaders shall ensure that all pilots in the formation receive and acknowledge signals when given. The visual communications section of NWP 41 must be reviewed and practiced by all pilots and RIO's. For ease of reference, visual signals applicable to flight operations are contained in figure 7–1 and deck/ground handling signals are contained in figure 7–2.

VISUAL COMMUNICATIONS

GENERAL CONVERSATION

MEANING	SIGNAL	RESPONSE
Affirmative (I understand).	Thumb up, or nod of head.	
Negative (I do not know).	Thumb down, or turn of head from side to side.	
Question (repeat). Used in conjunction with another signal, this gesture indicates that the signal is interrogatory.	Hand cupped behind ear as if listening.	As appropriate.
Wait.	Hand held up with palm outward.	
Ignore last signal.	Hand waved in an erasing motion in front of face, with palm turned forward.	
Numerals, as indicated.	With forearm in vertical position, employ fingers to indicate desired numerals 1 through 5. With forearm and fingers horizontal, indicate number which, added to 5, gives desired number from 6 through 9. A clenched fist indicates zero.	A nod of the head (I understand). To verify numerals, addressee repeats. If originator nods, interpretation is correct. If originator repeats numerals, addressee should continue to verify them until they are understood.

CONFIGURATION CHANGES

MEANING	SIGNAL	RESPONSE
Lower landing gear.	Rotary movement of hand in cockpit, as if cranking wheels.	Execute.
Lower arresting gear hook.	Leader lowers hook.	Wingman lowers arresting gear hook. Leader indicates wingman's hook is down with thumb up signal.
Extend or retract flaps or speed brakes as appropriate.	Open and close four fingers and thumb.	

FUEL AND ARMAMENT

MEANING	SIGNAL	RESPONSE
How much fuel have you?	Raise fist with thumb extended in a drinking position.	Indicate fuel in hundreds of pounds by finger numbers.
1—Arm or safety missiles as applicable; 2—how much ammo do you have? 3—I am unable to fire.	1—Pistol cocking motion with either hand; 2—followed by question signal; 3—followed by nose-held signal.	1—Execute and return signal; 2—thumb up, over half; down, less than half; 3—nod head (I understand).
1—Arm or safety tanks as applicable; 2—how many tanks do I have? 3—I am unable to drop.	1—Shaking fist; 2—followed by question signal; 3—followed by nose-held signal.	1—Execute and return signal; 2—indicate with appropriate finger numerals; 3—nod head (I understand).

FDC-1-(83-1)

Figure 7-1 (Sheet 1 of 2)

FORMATION

MEANING	SIGNAL	RESPONSE
1) I have completed my takeoff check list and am, in all respects ready for takeoff; 2) I have completed my takeoff check list and am, in all respects, ready for a section takeoff; 3) Takeoff path is clear, I am commencing takeoff.	1) Section takeoff leader raises arm (either) over head; 2) Wingman raises arm over head; 3) Leader lowers arm.	1) Stands by for reply from wingman, holding arm over head until answered; 2) Wingman lowers arm and stands by for immediate section takeoff; 3) Executes section takeoff.
Leader shifting lead to wingman. (Day)	Leader pats self on head points to wingman.	Wingman pats head and assumes lead.
Leader shifting lead to wingman. (Night)	1) Two aircraft – lead aircraft puts external lights on bright and flash. 2) More than two aircraft – leader places flight in echelon and then use two aircraft procedures.	Wingman turns external lights to DIM and STEADY and assumes lead.
	3) With external light failure – leader shines flashlight on hardhat, then shines at Wingman.	Wingman turns external lights to DIM and STEADY and assumes lead. With external light failure – Wingman shines flashlight at leader, then on his hardhat.
Leader shifting lead to division designated by numerals.	Leader pats self on head points to wingman and holds up two or more fingers.	Wingman relays signal; division leader designated assumes lead.
Take cruising formation.	Thumb waved backward over the shoulder.	Execute.
I am leaving formation.	Any pilot blow kiss.	Nod (I understand).
Aircraft pointed out leave formation.	Leader blows kiss and points to aircraft.	Execute.
Directs plane to investigate object or vessel.	Leader beckons wing plane, then points to eye, then to vessel or object.	Wingman indicated blows kiss and executes.
Refers to landing of aircraft, generally used in conjunction with another signal: 1) I am landing; 2) Directs indicated aircraft to land.	Landing motion with open hand: 1) Followed by patting head; 2) Followed by pointing to another aircraft.	1) Execute; 2) Execute.
a) Join up or break up, as appropriate. b) On GCA/CCA final: Leader has runway/ship in sight.	Flashing external lights.	a) Comply. b) Wingman repeats, indicating runway/ship in sight. Ship: Leader waves-off wingman lands. Field: When runway conditions preclude a safe section landing leader will wave-off.

FORMATION SIGNALS MADE BY AIRCRAFT MANEUVER

COMBAT OR FREE CRUISE

MEANING	SIGNAL	RESPONSE
Single aircraft cross under in direction of wing dip.	Single wing dip.	Execute.
Section cross under.	Double wing dip.	Execute.
Close up.	Series of small zooms.	Execute.
Join up; join up on me.	Series of pronounced zooms.	Expedite join-up.

DD-1-(83-2)

Figure 7-1 (Sheet 2 of 2)

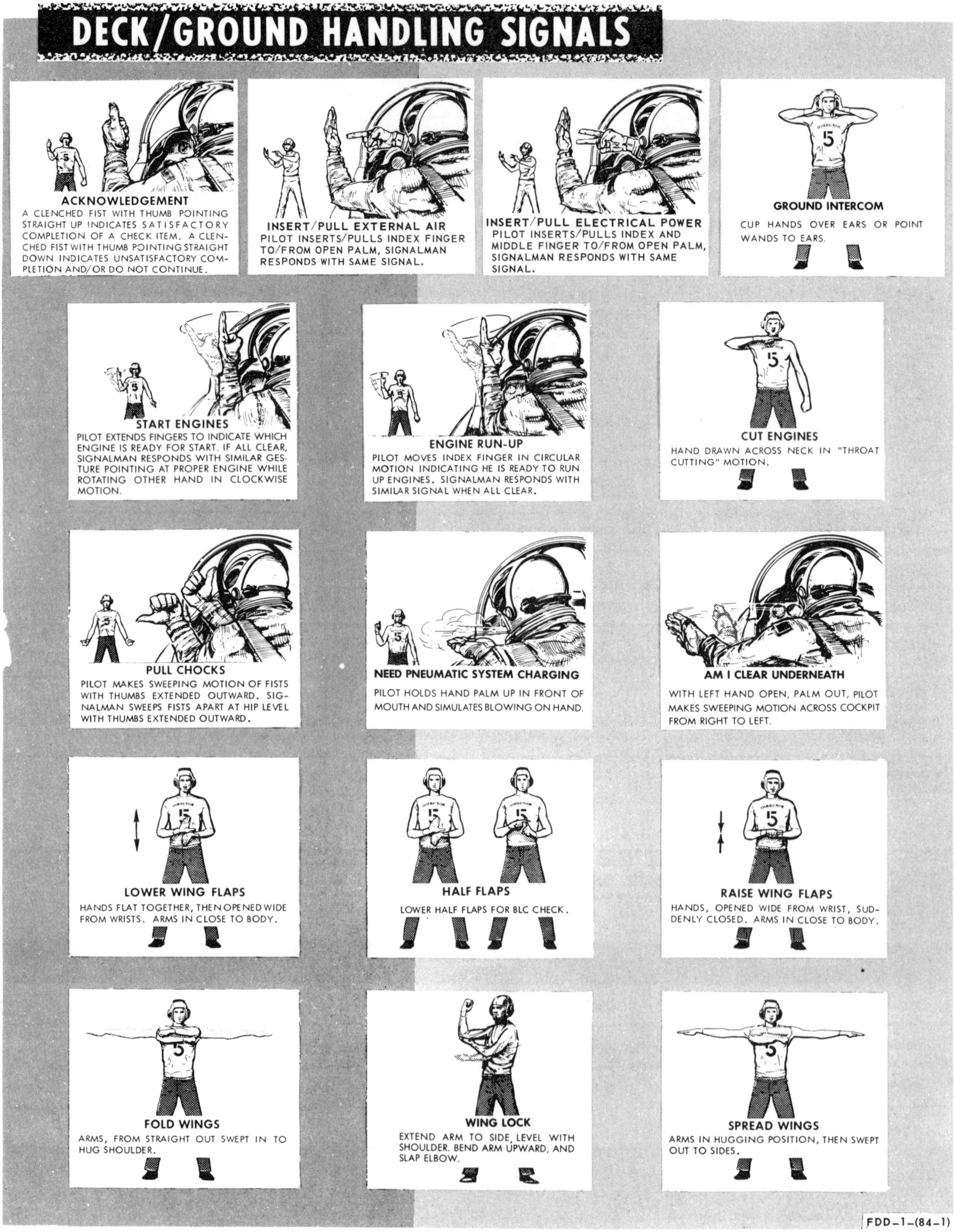

Figure 7-2 (Sheet 1 of 3)

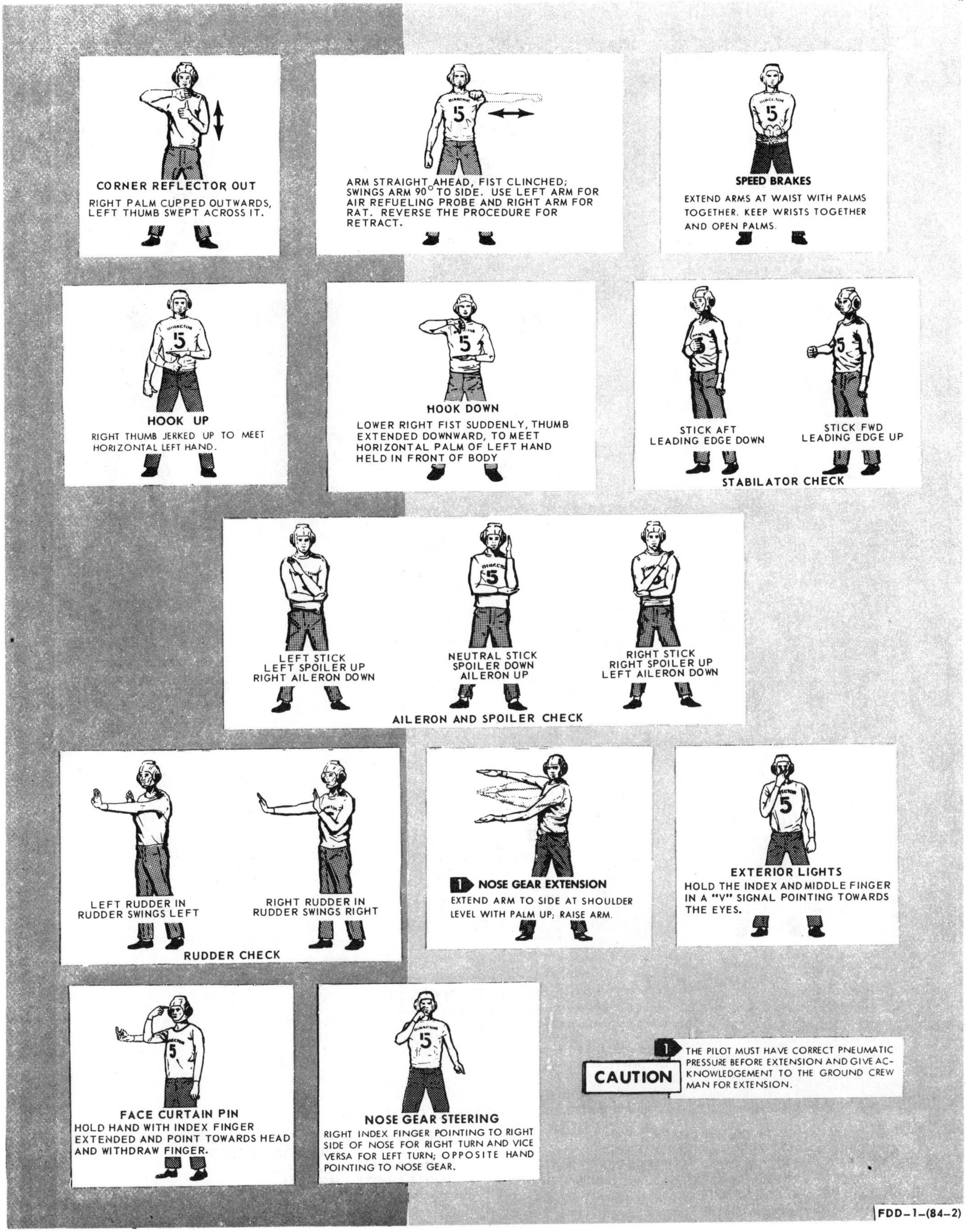

Figure 7-2 (Sheet 2 of 3)

CARRIER FLIGHT DECK PERSONNEL COLOR CODING

RED SHIRTS – ORDNANCE, and CRASH CREW
YELLOW SHIRTS – PRI FLY, PLANE DIRECTORS, CATAPULT OFFICER and ARRESTMENT OFFICER
BLUE SHIRTS – PLANE HANDLERS (Pushers, Chock Men, etc.)
PURPLE SHIRTS – FUEL HANDLING
GREEN SHIRTS – AIRCRAFT MAINTENANCE, CATAPULT CREW, ARRESTMENT CREW
BROWN SHIRTS – PLANE CAPTAINS
WHITE SHIRTS – MEDICAL

FDD–1–(84–3)A

Figure 7–2 (Sheet 3 of 3)

SECTION VIII
WEAPONS SYSTEMS

TABLE OF CONTENTS

BOMBING EQUIPMENT

Refer to Tactical Manual, NAVAIR 01—245FDB—1T.

DATA LINK SYSTEM

Refer to section VIII, Flight Manual Supplement.

DIRECT RADAR SCOPE CAMERA

Refer to section VIII, Flight Manual Supplement.

ELECTRONIC COUNTERMEASURES EQUIPMENT

Refer to Tactical Manual Supplement, NAVAIR 01—245FDB—1TA.

GUNNERY EQUIPMENT

Refer to Tactical Manual, NAVAIR 01—245FDB—1T.

MISSILE CONTROL SYSTEM

Refer to section VIII, Flight Manual Supplement.

STRIKE CAMERA SYSTEM

DESCRIPTION

Aircraft after AFC 518 are equipped with a strike camera system. The system consists of a camera pod LB-30A mounted in the left forward fuselage missile cavity. The camera pod contains a panoramic camera KB-18B and a forward looking motion picture camera KB-19A. The panoramic camera provides continuous film documentation of the strike area throughout an air-to-ground armament delivery. The forward looking camera provides motion picture coverage of the area forward of the aircraft along the boresight line during air-to-air or air-to-ground armament delivery. Both cameras operate simultaneously when energized by automatic or manual control circuits.

KB-18B CAMERA

The KB-18B camera components contained in the pod are the camera body with an accessory magazine LB-29A, and a camera control LB-17A. The camera has a 3-inch focal length f/2.8 lens. The shutter speeds range from approximately 1/100 second to 1/4000 second depending on limits established by a cycle rate switch and an aerial exposure index (AEI) switch. The accessory magazine accommodates 250 feet of film (300 exposures) and produces photographs with a format size of 2.25 by 9.40 inches. Panoramic photography is accomplished by rotating a double dove prism in front of the lens while the film is advanced across a narrow slit at the focal plane of the camera. Film advance is synchronized with prism rotation, projecting the panoramic image of the strike area on the film as the prism scans 180° fore and aft. Side coverage is 20° either side of the airplane vertical axis. The automatic exposure control senses variations in the light source and initiates compensatory aperture adjustments. The camera control contains switches for ground testing the camera and preselecting camera operating functions required for the mission. The controls include the ground test switch, cycle rate switch, overrun dial, and the AEI switch.

Overrun Dial

The overrun dial setting determines the amount of time the camera will run (automatic mode only) after the trigger switch or bomb button is released. Dial settings are calibrated in 2-second increments from 0 to 20 seconds plus an additional setting of 32 seconds.

Cycle Rate Switch

The cycle rate switch permits selection of 1, 2, or 4 frames per second during camera operation. The switch is set in accordance with the planned mission altitude/airspeed combination.

AEI Switch

The AEI switch is set as determined by the type of film in the camera. Film speeds (exposure sensitivity) of 40, 64, 80, or 200 are selectable.

KB-19A CAMERA

The KB-19A camera mounted in the forward fairing of the pod, provides 16mm motion picture coverage of the area forward of the aircraft along the boresight line. The film magazine attached to the camera contains 100 feet of film. An automatic exposure control (AEC) controls the shutter sector opening to assure proper exposure.

Overrun Switch

The overrun switch controls the time the camera continues to operate after trigger or bomb button release and provides the following overrun selections: 0, 5, 10, or 15 seconds. An overrun indicator light marks the film at the release of the trigger or bomb button and continues marking until completion of the overrun period.

Test Switch

The test switch on the rear of the camera is utilized by the ground crew to test operation of the camera.

STRIKE CAMERA CONTROL PANEL

The strike camera control panel, on the left console (figure A-1, appendix A), controls the operating modes of the strike camera system. The panel labeled CAMERAS contains a mode switch and a PAN OPERATE light. The mode switch has three positions: OFF, AUTO, and MANUAL. With the mode switch in AUTO, when the armament circuits are energized and the pilot presses the trigger switch or the bomb button the cameras begin operating. When the trigger switch or bomb button is released the cameras continue operating for the preselected overrun time. MANUAL is a momentary position and the cameras operate as long as the switch is held in MANUAL. The manual mode does not involve the armament systems and enables film coverage of specific terrain without armament delivery. The PAN OPERATE light blinks on for each cycle of operation of the KB-18B camera (i.e., illuminates during each panoramic scan and extinguishes between scans). If a film failure occurs or the film supply is exhausted in the KB-18B camera, the PAN OPERATE ceases to blink. This light may be tested by pressing the light.

NORMAL OPERATION

With the camera mode switch in AUTO, the cameras operate automatically during armament delivery; i.e., with the specific arming circuits energized, both cameras operate when the trigger switch or bomb button is pressed. Furthermore, camera operation continues for the preselected overrun time. The pilot may operate the cameras without expending armament by holding the mode switch in MANUAL.

EMERGENCY OPERATION

There are no special provisions for emergency operation of the strike camera system.

LIMITATIONS

The limits of the basic airplane apply to the strike camera system.

SECTION IX
FLIGHT CREW COORDINATION

TABLE OF CONTENTS

PILOT/RIO RESPONSIBILITIES

GENERAL

The most successful fighter teams are those who work together continuously and know each other's reactions, weaknesses and strengths. The mission commander bears responsibility for mission success, sharing safety of flight responsibilities whether he be pilot or RIO. The pilot is the aircraft commander and is responsible for the safe control of the aircraft throughout its entire mission. The RIO should assist the pilot in every way possible, anticipating rather than awaiting developments. The duties of the Pilot/RIO team are necessarily integrated, and each must support and contribute to the performance of the other. Each crewmember must share the responsibility of every evolution encountered. In other words, the crew concept is a system of checks and balances and, as such, it is a dual obligation equally shared. The interruption of routine functions is usually caused by an emergency; however, routine habit patterns may be broken through complacency. The assistance of the RIO in these instances is of the utmost value, challenging the pilot with the appropriate emergency checklist and monitoring flight performance of the aircraft.

The crew responsibilities are written with the premise that the RIO will handle UHF communications throughout the flight; however, it is recommended that prior to takeoff the pilot initiate at least one transmission to check out his communications.

The importance of each crewmember being completely aware of all his responsibilities has to be continuously stressed. Both crewmembers must realize that successful mission accomplishment and safety depends on flight crew coordination. In this respect, it is stressed that the crew coordination functions, duties and responsibilities listed in this section are not meant to be utilized as in–flight checklists. While frequent reference is made to checklists and certain important procedures are highlighted, the contents of this section are designed to be utilized as a basis for squadron and unit ground training syllabi. So utilized, this section will enhance the successful and safe completion of each unit's mission through intelligent and proper compliance with all NATOPS procedures and other applicable aviation directives.

To further the strengthening of aircrew integrity, it is imperative that the team concept be employed as early in the training evolution as is practicable. The best balance of talent and experience divided into permanent Pilot/RIO teams will develop the coordination required of the truly professional combat team.

SPECIFIC RESPONSIBILITIES

FLIGHT PLANNING

Mission Commander

1. Responsible for all phases of the assigned mission and for the effectiveness of the flight.

Flight Leader

1. Assist the mission commander in preparing required charts, flight logs, navigation computations (including fuel planning, checking weather and NOTAMS), and for completing required flight plans.

Pilot

1. Assist the mission commander/flight leader in preparing required charts, flight logs, navigation computations (including fuel planning, checking

weather and NOTAMS), and for completing required flight plans.
2. Be prepared to assume flight lead; thus have a thorough understanding of the conduct of the flight.

RIO

1. Assist the mission commander/flight leader/pilot in preparing required charts, flight logs, navigation computations (including fuel planning, checking weather and NOTAMS), and for completing required flight plans.
2. Be prepared to assist the pilot in assuming flight lead; thus have a thorough understanding of the conduct of the flight.

BRIEFING

Mission Commander

1. Responsible for briefing all crewmembers on all aspects of the operation and conduct of the flight.
2. Utilize a briefing guide or syllabus card as outlined in part 1 of section III, this manual.

Flight Leader

1. Assist the mission commander in preparing required flight or briefing forms and if directed brief all crewmembers on all aspects of the operation and conduct of the flight.
2. When directed, in addition to using the briefing guide or syllabus card, brief the following:
   a. Crew coordination in each flight evolution.
   b. Command/control of aircraft.
   c. Utilization of each weapons system.
   d. System degradation.

Pilot

1. Assist the mission commander/flight leader in preparing required flight or briefing forms, and, if directed, brief all crewmembers on all aspects of the operation and conduct of the flight.
2. When directed, in addition to using the briefing guide or syllabus card, brief the following:
   a. Crew coordination in each flight evolution.
   b. Command/control of aircraft.
   c. Utilization of each weapons system.
   d. System degradation.

RIO

1. Assist the mission commander/flight leader in preparing required flight or briefing forms, and, if directed, brief all crewmembers on all aspects of the operation and conduct of the flight.
2. When directed, in addition to using the briefing guide or syllabus card, brief the following:
   a. Crew coordination in each flight evolution.
   b. Command of aircraft.
   c. Utilization of each weapons system.
   d. System degradation.

PREFLIGHT

Pilot

1. Check yellow sheet for flight status, configurations, armament loading, and servicing.
2. Review at least the ten previous B sections for discrepancies noted and corrective actions taken.
3. Responsible for accepting the aircraft assigned.
4. Responsible for preflighting the aircraft assigned in accordance with this manual, appropriate NATOPS pocket checklists and TACTICAL manual.
5. Note any preflight discrepancies and ensure they are logged on yellow sheet.
6. Calculate gross weight of aircraft for catapult launch and compare with RIO's figures.

RIO

1. Check yellow sheet for flight status, configuration, armament loading, and servicing.
2. Review at least the ten previous B sections for discrepancies noted and corrective action taken.
3. Conduct a thorough individual preflight of entire aircraft in accordance with this manual, appropriate NATOPS pocket checklist and TACTICAL manual.
4. Check missile security and log missile serial numbers.
5. Note any preflight discrepancies and ensure they are logged on yellow sheet.
6. Calculate gross weight of aircraft for catapult launch and compare with pilot's figures.

BEFORE STARTING ENGINES

Pilot

1. Execute pre–start checks in accordance with this manual and appropriate NATOPS pocket checklist.
2. During intercom system check, report "ordnance switches safe, CW power off, interlocks in".
3. Receive a "clear to start" signal from plane captain ensuring security of aircraft and adjacent area.
4. Inform RIO, "Pre–start checks complete, ready to start".

RIO

1. Execute pre–start checks in accordance with this manual and appropriate NATOPS pocket checklist.
2. Challenge pilot to check ordnance switches safe, CW power off, interlocks in.
3. Respond to ICS check with "intercom status, eject light status, CB in, transformer/rectifier check status and position of command selector valve handle".
4. Ensure intake ramps are clear of any personal flight equipment.
5. Inform pilot, "Pre–start checks complete".

STARTING ENGINES

Pilot

1. Start engines in accordance with this manual and appropriate NATOPS pocket checklist.
2. Remain alert for any emergency signals from ground crew or RIO.

RIO

1. Notify pilot of any emergency signals noted from ground crew or any unusual occurrences you may observe.
2. Report any circuit breakers that do not remain seated.
3. Observe wings, reporting any movement to pilot.
4. Alert pilot of possibility of any foreign object which might be drawn into the compressor.
5. Report any movement of aircraft to the pilot.

BEFORE TAXIING

Pilot

1. After switching to internal power, inform RIO "on internal power, bus tie light out" (missile power switch in standby – in accordance with appropriate directives).
2. Complete all before taxi check in accordance with this manual and appropriate NATOPS pocket checklists.
3. Acknowledge "CW power switch on" when requested (in accordance with appropriate directives).
4. Report time to close and lock front canopy, checking rod alignment marks after closure.
5. Obtain verbal clearance from RIO prior to cycling refueling probe.
6. Inform RIO when "ready to taxi".

RIO

1. Ascertain generator control switches ON and bus tie and generator warning lights out.
2. Complete all before taxi check in accordance with this manual and appropriate NATOPS pocket checklists.
3. Request pilot to turn on CW power (in accordance with appropriate directives).
4. Advise pilot when performing BIT checks (in accordance with appropriate directives).
5. Inform pilot on status of all systems.
6. Check navigation computer operation.
7. Check pilot's radio/navigation equipment.
8. Take command of radio and navigation equipment and report any malfunction to pilot.
9. Report time of close and lock rear canopy, checking rod alignment marks after closure.
10. Scan instruments and circuit breakers, reporting any discrepancies.
11. Challenge pilot on position of UHF antenna switch.
12. Challenge pilot on following items:
    a. Fuel transfer – "STOP TRANSFER/OFF"
    b. Bleed air switch – "NORM"
    c. Tacan/UHF – "channel no/frequency"
13. Advise pilot when "ready for taxi".
14. Call clearance delivery and ground control when directed by pilot.

TAXIING

Pilot

1. Aboard ship remain on HOT MIC position during taxi.
2. After aircraft has started rolling, check brakes, and report to RIO.
3. Remain watchful for obstructions and taxi signals and inform RIO.
4. Check wing clearance.
5. Be alert for calls from ground/tower control concerning taxi information.

RIO

1. Aboard ship remain on HOT MIC position during taxi.
2. Challenge pilot on brake check.
3. Monitor taxi speed.
4. Remain watchful for obstructions and taxi siganls, relaying to pilot.
5. Check wing clearance.
6. Be alert for calls from ground/tower control concerning taxi information.
7. Complete BIT checks (in accordance with appropriate directives).
8. Return radar function switch to STBY/OFF (as required) after completion of BIT checks.

BEFORE TAKEOFF

Pilot

1. Complete pre-takeoff checks in accordance with this manual and appropriate NATOPS pocket checklist.
2. During engine run-up, relay engine instrument readings to RIO.
3. Use challenge and reply system when going over takeoff checklist.
4. Inform RIO of acknowledged aircraft gross weight for catapult launch.
5. Complete takeoff checklist prior to crossing shuttle aboard ship, ensuring compass FREE DG mode, radar horizon set at zero.
6. Report gages normal, controls free, gyro set.
7. Inform RIO when ready for takeoff.

RIO

1. Complete pre-takeoff checks in accordance with this manual and appropriate NATOPS pocket checklist.
2. During engine run-up:
    a. Check variable area inlet ramps fully retracted.

b. Record oil pressure reported by pilot.
3. Use challenge and reply system when going over takeoff checklist.
4. Report command eject handle position.
5. Inform pilot of acknowledged aircraft gross weight for catapult launch.
6. Report visual check of the wing lock indicator pins and leading edge flaps.
7. Complete takeoff checklist prior to crossing shuttle aboard ship.
8. After nose-strut extension, RIO challenge pilot – "reset gyro/check gages normal".
9. Report stabilator position.
10. Maintain hands below canopy sill while on catapult.
11. Report "ready for takeoff, circuit breakers in".

TAKEOFF/DEPARTURE

Pilot

1. Ensure ICS remains in HOT MIC position.
2. Inform RIO when ready for tower clearance.
3. Copy and ensure RIO acknowledges all clearances.
4. Report engine operation normal to RIO after throttles are advanced to takeoff power.
5. Report rolling, saluting or lights as appropriate to RIO.
6. Advise RIO of any unusual occurrences.
7. Report "gear up" and "flaps up" to RIO.
8. Inform RIO when going from gages to visual or vice versa.
9. Inform RIO when changing radio/tacan/IFF.
10. Advise RIO prior to engaging autopilot.
11. Confirm headings and altitudes with RIO.
12. Maintain a lookout doctrine.

RIO

1. Ensure ICS remains in HOT MIC position.
2. Copy and acknowledge all clearances.
3. Challenge pilot for engine operation after throttels are advanced to takeoff power.
4. During section takeoff check wingman's airplane and position.
5. Monitor airspeed and runway remaining.
6. Remind pilot to deploy drag chute and drop hook if takeoff aborted.
7. Challenge pilot if aircraft doesn't have 10° – 12° on rotation with a positive rate of climb.
8. Challenge pilot on After Takeoff checklist.
9. Monitor instruments, maintain a diligent lookout doctrine.
10. Monitor the published or clearance departure procedures and challenge the pilot on any deviations from prescribed heading or altitude.
11. When not monitoring gages, inform pilot.

RENDEZVOUS

Pilot

1. Maintain a good lookout doctrine.
2. Ensure RIO establishes radio contact, as briefed.
3. Advise RIO of any signals received from other aircraft or signals to be passed to other aircraft.

RIO

1. Maintain a good lookout doctrine.
2. Monitor airspeed, angle of bank, altitude.
3. Establish radio contact, as briefed.
4. Confirm all lights BRIGHT/FLASH (if applicable).
5. Utilize radar to provide azimuth, range, closure rates to targets.
6. Monitor closure rate, relative bearing, altitude separation.
7. Advise pilot of any signals received from other aircraft.
8. Be prepared to give hand signals to other aircraft.

CRUISE AND MISSION

Pilot

1. Maintain a good lookout doctrine.
2. Report fuel switchology, transfer and fuel state to RIO periodically.
3. Inform RIO when going from gages to visual or vice versa.
4. Inform RIO when changing radio/tacan/IFF.
5. Advise RIO when engaging AFCS.
6. Be constantly aware of geographical position.
7. Maintain howgozit chart, flight log and frequencies assigned.

RIO

1. Maintain a good lookout doctrine.
2. Maintain a scan pattern.
3. Challenge pilot on fuel switchology, transfer and state periodically.
4. Inform pilot when going from gages to visual or vice versa.
5. Be constantly aware of geographical position and bingo distance for alternate/emergency airfields along route of flight. Be prepared to give immediate steer to home base.
6. Know winds at operating and bingo profile altitude.
7. Copy all clearances, heading/altitude changes.
8. Maintain howgozit chart, flight log and frequencies assigned.
9. Update Nav computer regularly.

INSTRUMENTS

Pilot

1. Transitioning from VFR to IFR conditions, the pilot should shift completely to the gages prior to entering IFR flight and so inform the RIO.
2. Remain on instruments anytime the aircraft is climbing or descending through layers or when in and out of IFR conditions at low altitudes.
3. If experiencing vertigo/disorientation, notify RIO immediately, stay on gages and refrain from making any UHF or IFF changes.
4. If RIO experiences vertigo/disorientation, inform

him of altitude, attitude and airspeed.

5. Inform RIO anytime you break your scan to change an IFF squawk, tacan channel, radio frequency or light configuration.
6. Inform RIO anytime a change in heading or altitude is to be made.

RIO

1. Assume all lookout responsibilities anytime the pilot is flying instruments, monitor the flight instruments and advise the pilot in VFR conditions.
2. If pilot experiences vertigo/disorientation, carefully monitor gages; informing pilot of attitude, altitude and airspeed. Be prepared to advise pilot how to maneuver the aircraft back to a wings–level attitude.
3. Inform pilot if you go off the gages or experience vertigo/disorientation.
4. Inform pilot anytime prior to changing radio frequencies or tacan channel.
5. Challenge pilot if not at assigned altitude or heading.
6. Be particularly aware of attitude, altitude, airspeed and geographical position at all times.
7. Advise the pilot of bank angle and changes in altitude, airspeed or heading when flying formation.
8. Utilize radar as appropriate and ensure radar gyro is indicating properly.
9. Do not conduct BIT checks under actual instrument conditions.

AIR INTERCEPTS/MISSILEX

Pilot

1. Complete with RIO the checklists in accordance with the Confidential Supplement to the NATOPS Manual.
2. Report synchronization of compass to RIO.
3. Maneuver the aircraft as directed by GCI/CIC/RIO, observing normal operating limitations.
4. Report weapon status, data link status, weapon selected and armed, and visual contact on target to RIO on each run.
5. Monitor aircraft positioning from initial vector through breakaway by pigeons or navigational display.
6. Make missile away reports.
7. Report "switches safe" to RIO after each run.
8. Report fuel switches and state to RIO periodically.
9. Call aborting if any unsafe circumstances are observed.
10. Maintain a good lookout doctrine.

RIO

1. Complete with pilot the checklist in accordance with the Confidential Supplement to the NATOPS Manual.
2. Challenge pilot on synchronization of compass.
3. Monitor the maneuvering of the aircraft as directed by GCI/CIC.
4. Direct and coordinate maneuvering the aircraft with the pilot as necessary to complete the intercept.
5. Challenge pilot on weapons status, weapon selected and armed, data link and visual contact with target on each run.
6. Know aircraft positioning from initial vector through breakaway by pigeons or navigational display.
7. Conduct UHF communications from initial vector through breakaway, excluding missile away transmission.
8. Provide pilot with descriptive/directive commentary.
9. Challenge pilot on armament switchology after each run.
10. Challenge pilot on fuel switches and state periodically.
11. Call aborting if any unsafe circumstances are observed.
12. Be especially alert and precise to safely complete a MISSILEX; i.e., differentiate between tractor and target radar return.

TANKING

Pilot

1. Maintain good lookout doctrine.
2. Complete with RIO the air refueling checklist.
3. Monitor sector position of the fuel quantity gage foı a failed–open defuel valve.
4. Report to RIO when fuel transfer is complete.
5. After disengagement, stay behind the drogue until all members of the flight are sighted.
6. Complete with RIO the post–refueling checklist.
7. Report fuel received, fuel remaining, and fuel switchology to the RIO.

RIO

1. Acquire tanker on radar.
2. Keep pilot informed on azimuth, range and overtake during joinup.
3. Maintain good lookout doctrine.
4. Complete with the pilot the air refueling checklist.
5. Challenge pilot for status of defuel valve.
6. Insure amber light visible prior to commencing approach.
7. Monitor closure rate on and "coach" pilot into the drogue if requested.
8. Make required voice reports as briefed.
9. Maintain sharp lookout doctrine while engaged.
10. Monitor heading, airspeed and altitude, and report variances to pilot while engaged.
11. Advise pilot of unusual fuel spillage or unsafe probe/drogue conditions.
12. Complete with pilot the post–refueling checklist.
13. Challenge pilot on fuel received, fuel remaining, and fuel switchology.

AIR COMBAT MANEUVERING

Pilot/RIO

Delineating the entire realm of aircrew responsibility during ACM is beyond the scope of this manual. Careful preplanning and briefing are necessary to ensure adequate crew coordination prior to any ACM mission. As a minimum, each crewmember must have a constant awareness of the rules of engagement, flight safety, fuel state (including Bingo), attitude and minimum prebriefed base altitude; and carry on a continuous supportive commentary.

CONVENTIONAL WEAPONS

Pilot

1. Complete with RIO the combat checklists.
2. Report fuel switches and state to RIO periodically.
3. Confirm with RIO dive angle, airspeed, release altitude and minimum safe altitude for pull-out and ejection.
4. Report visual acquisition of target to RIO.
5. Advise RIO of other than prebriefed release parameters if weather/target dictates.

RIO

1. Complete with pilot the combat checklist.
2. Challenge pilot on fuel state periodically.
3. Confirm with pilot dive angle, airspeed, release altitude and minimum safe altitude for pull-out and ejection.
4. Report visual acquisition of target to pilot.
5. Be prepared to give pilot other than prebriefed release parameters if weather/target dictates.
6. Provide CW commentary of dive angle (steep/shallow), airspeed (fast/slow), altitude, "standby", "mark", and "pull".
7. Be prepared to provide pilot with unusual attitude recovery procedures.

DESCENT/APPROACH

Pilot

1. Select the type approach to be made with respect to existing weather.
2. Complete with RIO the descent/instrument penetration checklist.
3. Challenge RIO on approach plate availability and bingo profile.
4. Monitor UHF communication conducted by the RIO and copy all clearances.
5. When executing a CCA conform to procedures contained in this manual, CVA/CVS NATOPS Manual and augmenting ships instructions.
6. Inform RIO when going from gages to visual or vice versa.
7. Inform RIO when changing IFF/UHF/tacan.
8. Confirm headings and altitudes with RIO.
9. Cross-check with RIO pressure altimeter and radar altimeter below 5000 feet.
10. Advise RIO if fuel dump switch has been activated/deactivated.
11. Advise RIO if experiencing vertigo/disorientation.
12. If RIO experiences vertigo/disorientation, inform him of attitude, altitude and airspeed.

RIO

1. Assist the pilot in his decision to commence the approach with existing weather and the selection of the type approach to be made.
2. Prior to penetration complete with the pilot the descent/instrument penetration checklist.
3. Update Nav computer prior to commencing an approach.
4. Conduct UHF communications being prepared to request, copy and readback clearance, and provide the pilot with any required information.
5. Aid in planning holding/marshall entry and pattern timing.
6. Monitor altitude, attitude, airspeed during holding, penetration and approach.
7. Monitor appropriate approach plate, advising pilot of heading changes and altitudes prescribed on the approach plate.
8. When executing a CCA monitor the procedures of this manual, CVA/CVS NATOPS Manual and augmenting ship's instruction.
9. Utilize radar during CCA prior to landing. Inform pilot if radar horizon does not match aircraft attitude, and place gyro switch out if directed by pilot.
10. When not monitoring the gages inform the pilot.
11. Challenge pilot on deviations from published/assigned course or altitude.
12. Know ship's/field position relative to bingo field.
13. Report to pilot during penetration and/or descents (VFR or IFR) the aircraft descending through each 5000 feet of altitude above 5000 feet and each 1000 feet below 5000 feet, until reaching desired altitude.
14. Challenge pilot on cross-checking pressure altimeter and radar altimeter passing 5000 feet.
15. Monitor the fuel dump.
16. Advise the pilot of experiencing vertigo/disorientation.
17. If pilot experiences vertigo/disorientation, inform him of attitude, altitude and airspeed.
18. Be prepared to talk-down pilot through distance/altitude check points when executing a CCA utilizing radar, tacan, Nav computer and data link.
19. Inform pilot when going contact or achieving a "ball" when on GCA/CCA.
20. For field landing brief pilot on location of arresting gear available in terms of runway remaining.

LANDING

Pilot

1. Complete with RIO landing checklist.
2. Double check pressure altimeter with radar altimeter and inform RIO.
3. Report indexer lights to RIO. (Secondary gear

check)
4. During Mode 1/1A/2 approaches, maintain a running commentary on position of the needles.

RIO

1. Monitor traffic pattern entry, altitude, airspeed as prescribed by local course rules or CVA/CVS NATOPS Manual while maintaining a good lookout doctrine and assisting pilot in determining interval.
2. Complete with pilot landing checklist.
3. Challenge pilot on pressure versus radar altimeter.
4. Look in mirrors to check droop of ailerons; look over ducts and check leading edge flaps down; attempt to check gear handle down, hook handle, indexer lights, and no flashing wheels light.
5. Make UHF transmissions as directed by pilot.
6. Be alert for and report conflicting traffic to the pilot.
7. Call for "power" to pilot on touchdown aboard ship.
8. On roll–out confirm status of drag chute. On section landing, lead RIO calls wingman's drag chute status and be prepared to advise of arresting gear location.
9. Report to pilot airspeed in conjunction with runway remaining.

AFTER LANDING

Pilot

1. Perform post–flight checklist with RIO.
2. Report to RIO when wing fold is actuated.

RIO

1. Perform post–flight checklist with the pilot.
2. When informed wing fold is actuated, report to pilot wing position.
3. Advise pilot of clearances when taxiing.
4. Complete BIT checks remaining.
5. Monitor and report drag chute clearance from obstructions.
6. Advise pilot when ready for shutdown.

DEBRIEFING

Mission Commander

1. Review the entire flight from takeoff to landing.
2. Specifically comment on errors and techniques, and review procedures for correcting/improving them.

Flight Leader

1. Assist the mission commander when requested, and, if directed, debrief all crewmembers on all aspects of the flight from taekoff to landing.
2. When directed comment on errors and techniques, and review procedures for correcting/improving them.
3. When directed cover completely any deviations from standard operating procedures.
4. Complete all required debriefing forms.
5. Utilize Anymouse reporting to pass the word on unusual flight occurrences, if applicable.

Pilot

1. Assist the mission commander/flight leader when requested, and, if directed, debrief all crewmembers on all aspects of the flight from takeoff to landing.
2. When directed comment on errors and techniques, and review procedures for correcting/improving them.
3. When directed cover completely any deviations from standard operating procedures.
4. Complete the yellow sheet and all required debriefing forms.
5. Utilize Anymouse reporting to pass the word on unusual flight occurrences, if applicable.

RIO

1. Assist the mission commander/flight leader when requested, and, if directed, debrief all crewmembers on all aspects of the flight from takeoff to landing.
2. When directed comment on errors and techniques, and review procedures for correcting/improving them.
3. When directed cover completely any diviations from standard operating procedures.
4. Assist pilot in completing the yellow sheet and all required debriefing forms.
5. Utilize Anymouse reporting to pass the word on unusual flight occurrences, if applicable.

SECTION X
NATOPS EVALUATION

TABLE OF CONTENTS

NATOPS EVALUATION PROGRAM

CONCEPT

The standard operating procedures prescribed in this manual represent the optimum method of operating this aircraft. The NATOPS Evaluation is intended to evaluate compliance with NATOPS procedures by observing and grading individuals and units. This evaluation is tailored for compatability with various operational commitments and missions of both Navy and Marine Corps units. The prime objective of the NATOPS Evaluation program is to assist the unit commanding officer in improving unit readiness and safety through constructive comment. Maximum benefit from the NATOPS Program is achieved only through the vigorous support of the program by commanding officers as well as flight crewmembers.

IMPLEMENTATION

The NATOPS Evaluation program shall be carried out in every unit operating naval aircraft. The various categories of flight crewmembers desiring to attain/retain qualification in this aircraft shall be evaluated initially in accordance with OPNAV Instruction 3510.9 series, and at least once during the twelve months following initial and subsequent evaluations. Individual and unit NATOPS Evaluations will be conducted annually; however, instruction in observation of adherence to NATOPS procedures must be on a daily basis within each unit to obtain maximum benefits from the program. The NATOPS Coordinators, Evaluators, and Instructors shall administer the program as outlined in OPNAVINST 3510.9 series. Evaluees who receive a grade of Unqualified on a ground or flight evaluation shall be allowed 30 days in which to complete a reevaluation. A maximum of 60 days may elapse between the date the initial ground evaluation was commenced and the date the flight evaluation is satisfactorily completed. Crewmembers possessing a valid F/RF-4 NATOPS Evaluation Report Form, OPNAV Form 3510-8 are considered qualified in all F/RF-4 model aircraft provided the conditions outlined in section II, Indoctrination, are met.

DEFINITIONS

The following terms, used throughout this section, are defined as to their specific meaning within the NATOPS program.

NATOPS EVALUATION

A periodic evaluation of individual flight crewmember standardization consisting of an open book examination, a closed book examination, an oral examination, and a flight evaluation.

NATOPS REEVALUATION

A partial NATOPS Evaluation administered to a flight crewmember who has been placed in an Unqualified status by receiving an Unqualified grade for any of his ground examinations or the flight evaluations. Only those areas in which an unsatisfactory level was noted need be observed during a reevaluation.

QUALIFIED

Well standardized; evaluee demonstrated highly professional knowledge of and compliance with NATOPS standards and procedures; momentary deviations from or minor omission in non-critical areas are permitted if prompt and timely remedial action is initiated by the evaluee.

CONDITIONALLY QUALIFIED

Satisfactorily standardized; one or more significant deviations from NATOPS standards and procedures, but no errors in critical areas and no errors jeopardizing mission accomplishment or flight safety.

UNQUALIFIED

Not acceptably standardized; evaluee fails to meet minimum standards regarding knowledge of and/or ability to apply NATOPS procedures, one or more significant deviations from NATOPS standards and procedures which could jeopardize mission accomplishment or flight safety.

AREA

A routine of preflight, flight, or postflight.

SUB-AREA

A performance sub-division within an area, which is observed and evaluated during an evaluation flight.

CRITICAL AREA/SUB-AREA

Any area or sub-area which covers items of significant importance to the overall mission requirements, the marginal performance of which would jeopardize safe conduct of the flight.

EMERGENCY

An aircraft component, system failure, or condition which requires instantaneous recognition, analysis, and proper action.

MALFUNCTION

An aircraft component or system failure or condition which requires recognition and analysis, but which permits more deliberate action than that required for an emergency.

GROUND EVALUATION

GENERAL

Prior to commencing the flight evaluation, an evaluee must achieve a minimum grade of Qualified on the open book and closed book examinations. The oral examination is also part of the ground evaluation but may be conducted as part of the flight evaluation. To assure a degree of standardization between units, the NATOPS instructors may use the bank of questions contained in this section in preparing portions of the written examinations.

OPEN BOOK EXAMINATION

The open book examination shall consist of, but not be limited to, the question bank. The purpose of the open book examination portion of the written examination is to evaluate the crewmember's knowledge of appropriate publications and the aircraft.

CLOSED BOOK EXAMINATION

The closed book examination may be taken from, but not limited to, the question bank and shall include questions concerning normal/emergency procedures and aircraft limitations. Questions designated critical will be so marked.

ORAL EXAMINATION

The questions may be taken from this manual and drawn from the experience of the Instructor/Evaluator. Such questions should be direct and positive and should in no way be opinionated.

COT/WST PROCEDURES EVALUATION

A COT may be used to assist in measuring the crewmember's efficiency in the execution of normal operating procedures and his reaction to emergencies and malfunctions. In areas not served by the COT facilities, this may be done by placing the crewmember in an aircraft and administering appropriate questions.

NAMT SYSTEMS CHECK

If desired by the individual squadron, Naval Air Maintenance Trainer facilities may be utilized to evaluate pilot and RIO knowledge of aircraft systems and normal and emergency procedures.

GRADING INSTRUCTIONS

Examination grades shall be computed on a 4.0 scale and converted to an adjective grade of Qualified or Unqualified.

Open Book Examination

To obtain a grade of Qualified, an evaluee must obtain a minimum score of 3.5.

Closed Book Examination

To obtain a grade of Qualified, an Evaluee must obtain a minimum score of 3.3.

Oral Examination and OFT Procedure Check (If Conducted)

A grade of Qualified or Unqualified shall be assigned by the Instructor/Evaluator.

FLIGHT EVALUATION

GENERAL

The flight evaluation may be conducted on any routine syllabus flight with the exception of flights launched for FMLP/CARQUAL or ECCM training. Emergencies will not be simulated.

(*) The number of flights required to complete the flight evaluation should be kept to a minimum; normally one flight. The areas and sub-areas to be observed and graded on a flight evaluation are outlined in the grading criteria with critical areas marked by an asterisk (*). Sub-area grades will be assigned in accordance with the grading criteria. These sub-areas shall be combined to arrive at the overall grade for the flight. Area grades, if desired, shall also be determined in this manner. At the discretion of the squadron or unit commander, the evaluation may be conducted in a WST, OFT, or COT.

OPERATIONAL DEPLOYABLE SQUADRONS

Pilots and RIO's assigned to operational deployable squadrons will normally be checked as a team with the flight evaluation being conducted by the check-crew flying wing. RIO commentary will be transmitted on the GCI/CIC control frequency in use.

TRAINING AND EVALUATION SQUADRONS

Units with training or evaluation missions that are concerned with individual instructor pilot/RIO standardization, rather than team standardization, may conduct the flight evaluation with the check-crew/pilot flying wing or on an individual basis. A pilot may be individually checked with the Instructor/Evaluator conducting the flight evaluation from the rear seat. The RIO may be individually checked by flying with the Instructor/Evaluator as his crewmember.

FLIGHT EVALUATION

The areas and sub-areas in which pilots and RIO's may be observed and graded for adherence to standardized operating procedures are outlined in the following paragraphs.

NOTE

If desired, units with training missions may expand the flight evaluation to include evaluation of standardized training methods and techniques.

(*) The IFR portions of the Flight Evaluation shall be in accordance with the procedures outlined in the NATOPS Instrument Flight Manual.

MISSION PLANNING/BRIEFING

a. Flight Planning (Pilot/RIO)
b. Briefing (Pilot/RIO)
c. (*) Personal Flying Equipment (Pilot/RIO)

PREFLIGHT/LINE OPERATIONS

Inasmuch as preflight/line operations procedures are graded in detail during the ground evaluation, only those areas observed on the flight check will be graded.

a. Aircraft Acceptance (Pilot/RIO)
b. Start
c. Before Taxiing Procedures (Pilot)

TAXI/RUN UP

(*) TAKEOFF/TRANSITION

a. ATC Clearance (Pilot/RIO)
b. Takeoff
c. Transition to Climb Schedule

CLIMB/CRUISE

a. Departure (Pilot)
b. Climb and Level-Off (Pilot)
c. Procedures Enroute (Pilot)

(*) APPROACH/LANDING

a. Radar, ADF (Pilot)
b. Recovery (Pilot)

COMMUNICATIONS

a. R/T Procedures (Pilot/RIO)
b. Visual Signals (Pilot/RIO)
c. IFF Procedures (Pilot)

(*) EMERGENCY/MALFUNCTION PROCEDURES

In this area, the Pilot/RIO will be evaluated only in the case of actual emergencies, unless evaluation is conducted in the COT/WST.

POST FLIGHT PROCEDURES

a. Taxi-in (Pilot)
b. Shutdown (Pilot/RIO)
c. Inspection and Records (Pilot/RIO)
d. Flight Debriefing (Pilot/RIO)

CREW COORDINATION

MISSION EVALUATION

This area includes missions covered in the NATOPS Flight Manual, F-4 Tactical Manual, and NWP/NWIP's for which standardized procedures/techniques have been deployed.

APPLICABLE PUBLICATIONS

The NATOPS Flight Manual contains the standard operations criteria for this aircraft. Publications relating to environmental procedures peculiar to shorebased and shipboard operations and tactical missions are listed below:

F-4 Tactical Manual	ATC/CATCC Manual
NWP's	Local Air Operations Manual
NWIP's	Carrier Air Operations Manual

FLIGHT EVALUATION GRADING CRITERIA

Only those sub-areas provided or required will be graded. The grades assigned for a sub-area shall be determined by comparing the degree of adherence to standard operating procedures with adjectival ratings listed below. Momentary deviations from standard operating procedures should not be considered as unqualifying provided such deviations do not jeopardize flight safety and the evaluee applies prompt corrective action.

FLIGHT EVALUATION GRADE DETERMINATION

The following procedure shall be used in determining the flight evaluation grade: A grade of Unqualified in any critical area/sub-area will result in an overall grade of Unqualified for the flight. Otherwise, flight evaluation (or area) grades shall be determined by assigning the following numerical equivalents to the adjective grade for each sub-area. Only the numerals 0, 2 or 4 will be assigned in sub-area. No interpolation is allowed.

Unqualified 0.0
Conditionally qualified 2.0
Qualified 4.0

To determine the numerical grade for each area and the overall grade for the flight, add all the points assigned to the sub-areas and divide this sum by the number of sub-areas graded. The adjective grade shall then be determined on the basis of the following scale.

0.0 to 2.19 – Unqualified
2.2 to 2.99 – Conditionally Qualified
3.0 to 4.0 – Qualified

EXAMPLE: (add Sub-area numerical equivalents)

4+2+4+2+4 ÷ 5 3.20 Qualified

FINAL GRADE DETERMINATION

The final NATOPS Evaluation grade shall be the same as the grade assigned to the flight evaluation. An evaluee who receives an Unqualified on any ground examination or the flight evaluation shall be placed in an Unqualified status until he achieves a grade of Conditionally Qualified or Qualified on a reevaluation.

RECORDS AND REPORTS

A NATOPS Evaluation Report (OPNAV Form 3510-8) shall be completed for each evaluation and forwarded to the evaluee's commanding officer only. This report shall be filed in the individual flight training record and retained therein for 18 months. In addition, an entry shall be made in the pilot/RIO flight log book under "Qualifications and Achievements" as follows:

Qualification			Date	Signature
NATOPS EVAL.	(Aircraft Model)	(Crew Position)	(Date)	(Authenticating Signature) (Unit which administered Eval.)

FDD-1-(294)

In the case of enlisted crewmembers, an entry shall be made in the Administrative Remarks of his Personnel Record upon satisfactory completion of the NATOPS Evaluation as follows:

(Date) Completed a NATOPS Evaluation in (Aircraft Designation) as (Flight crew position) with an overall grade of (Qualified or Conditionally Qualified).

CRITIQUE

The critique is the terminal point in the NATOPS evaluation and will be given by the Evaluator/Instructor administering the check. Preparation for the critique involves processing, reconstructing data collected, and oral presentation of the NATOPS Evaluation Report. Deviations from standard operating procedures will be covered in detail using all collected data and worksheets as a guide. Upon completion of the critique, the pilot/RIO will receive the completed copy of the NATOPS Evaluation Report for certification and signature. The completed NATOPS Evaluation Report will then be presented to the Unit Commanding Officer.

NATOPS EVALUATION QUESTION BANK

The following bank of questions is intended to assist the unit NATOPS Instructor/Evaluator in the preparation of ground examinations and to provide an abbreviated study guide. The questions from the bank may be combined with locally originated questions in the preparation of ground examinations. The closed book exam will consist of no less than 50 questions nor more than 75 questions. The time limit for the closed book exam is 1 hour and 30 minutes. The requirements for the open book exam are the same as those for the closed book exam, except there is no time limit.

F-4J NATOPS QUESTION BANK

1. An operating procedure, practice, or condition etc., which, if not strictly observed, may damage equipment are indicated in the NATOPS manual by__________.

2. Operating procedures, practices, etc., which will result in personnel injury or loss of life if not carefully followed are indicated in the NATOPS manual by________.

3. The eight circuit breakers in the front cockpit are:

   a. ______

   b. ______

   c. ______

   d. ______

   e. ______

   f. ______

   g. ______

   h. ______

4. List the warning lights in the RIO's cockpit.

   a. Before AFC 506

      (1) ______

      (2) ______

      (3) ______

      (4) ______

      (5) ______

      (6) ______

   b. After AFC 506

      (1) ______

      (2) ______

      (3) ______

      (4) ______

      (5) ______

      (6) ______

      (7) ______

5. The cockpit will begin to pressurize at______feet. At 40,000 feet, the cabin altitude should be approximately________.

6. The rain removal system directs__________over the center windshield panel.

7. To stop the cooling turbine if the CABIN TURB OVERSPEED light illuminates__________________.

QUESTION BANK CONTINUED

8. The only control the RIO has over his ventilated wet suit environment is the volume of airflow. T / F

9. To prevent smoke and fumes from entering the cockpit through the airconditioning system, pull the ______________.

10. Illumination of the RADAR CNI COOL OFF light shall be logged on the yellow sheet. T / F

11. Either the pilot or RIO can reset the CNI cooling reset button. T / F

12. Prior to resetting the CNI cooling reset button, what action must be taken:

    a. ______________________________

    b. ______________________________

13. If the CNI COOL OFF light cannot be extinguished, high speed flight should be avoided. T / F

14. Suggested angle of attack settings are:

    a. Climb (400 KCAS)________units

    b. Max Endurance________units

    c. Stall Warning________units

    d. Safe gear extension________units

    e. Safe flap extension (gear already down)________units

    f. No flap landing________units

15. The arresting hook is lower primarily by________and raised by________________. Once the arresting hook is lowered it is held on deck by________________.

16. The arresting hook up latch is mechanically operated. T / F

17. With a loss of electrical power the arresting hook cannot be extended. T / F

18. After AFC 534, emergency retraction of the speed brakes is accomplished by________________ ________.

19. The two modes of operation of the automatic flight control system are (PILOT ONLY)

    a. ______________________________

    b. ______________________________

20. The stab aug switches can be engaged individually or in any combination. T / F

21. In the AFCS mode of operation, the pitch limits are________degrees, and the roll limits are________ degrees.

22. The autopilot will disengage when acceleration forces exceed plus_____or minus______G.

23. A malfunction in the static pressure compensator will have no affect on autopilot operation. T / F

QUESTION BANK CONTINUED

24. Only_____maximum effort applications of the normal wheel brakes should be anticipated when utility hydraulic pressure is lost.

25. Actuation of the emergency pneumatic braking system will introduce air into the wheel brake hydraulic system. T / F

26. The emergency pneumatic brake system does not provide differential braking. T / F

27. The canopies are capable of being stopped at any intermediate position but should not be stopped in any position other than fully open or fully closed for safety reasons. T / F

28. Before actuating the canopy manual unlock handle, the normal canopy control handle must be placed in the__________position to relieve any opposing pneumatic pressure.

29. The pilot's CANOPY UNLOCKED indicator light will illuminate if the RIO's canopy is jettisoned. T / F

30. The canopy is designed to remain in the full open position up to_____knots and to separate from the airplane at_________knots.

31. Canopy closure should not be attempted with engines running above a stabilized idle rpm. T / F

32. Actuation of the canopy emergency system also actuates the cockpit flooding doors. T / F

33. Under what circumstances is it advisable for the RIO to select command eject?

    a. ______________________________

    b. ______________________________

    c. ______________________________

34. On ejection the radar indicator and the control box stow automatically; however the hand control remains out (before AFC 506). T / F

35. With the STATIC CORR OFF light illuminated actual altitude and airspeed will usually be_________than that indicated by the cockpit instruments.

36. List the five inputs to the ADC:

    a. ______________________________

    b. ______________________________

    c. ______________________________

    d. ______________________________

    e. ______________________________

37. The ADC supplies corrected information to the following instruments and systems.

    a. ______________________________

    b. ______________________________

    c. ______________________________

    d. ______________________________

    e. ______________________________

    f. ______________________________

    g. ______________________________

QUESTION BANK CONTINUED

h. ____________________

i. ____________________

j. ____________________

k. ____________________

m. ____________________

n. ____________________

o. ____________________

38. The static pressure compensator will automatically reset during engine start. T / F

39. When in the CNI mode, the RIO's #1 needle on the BDHI will indicate__________bearing to the__________ and the #2 needle will indicate__________bearing to the__________.

40. a. Who can select the UHF antenna to be utilized. ____________________

    b. Who can select the TACAN antenna to be utilized. ____________________

41. Emergency operation of the intercom system is provided by placing the amplifier selector knob in the B/U position, thereby by-passing the faulty amplifier. T / F

42. If the pilot selects emergency intercom operation, it is necessary for the RIO to also select emergency in order to regain satisfactory intercom communications. T / F

43. The position of the intercom controls will not affect UHF radio operation from either cockpit. T / F

44. The CNI equipment operating on external power without cooling air, is limited to________minutes of accumulated operation in a 1 hour span. (excluding ICS and UHF).

45. The items that are released or disconnected when the emergency harness release handle is actuated are:

    a. ____________________

    b. ____________________

    c. ____________________

    d. ____________________

46. The primary source of all electrical power is two,__________cycle,_____phase_____volt a-c generators.

47. With external electrical power connected, and either generator control switch in the EXT position, the aircraft buses will be energized. T / F

48. If the left generator is inoperative, and the BUS TIE OPEN light is illuminated, all dc buses will be energized. T / F

49. When operating on emergency electrical power, the following items of CNI equipment will be lost:

    a. ____________________

    b. ____________________

50. In airplanes 152965y and up, and all others after AFC 220 the emergency generator will drop off the line at approximately_____knots.

QUESTION BANK CONTINUED

51. Pilot action, upon the illumination of a generator warning light, is to______the generator control switch. If the generator does not come back on the line, secure the______and monitor__________, __________and__________.

52. After a complete electrical failure, the landing gear should be lowered by______________.

53. The BUS TIE OPEN light may illuminate momentarily when one generator drops off the line. T / F

54. The sea level, standard day, static thrust ratings for the J79-10 engine are,__________pounds in MIL, and__________pounds in MAX.

55. The torch igniter plug will operate momentarily each time the throttles are moved to afterburner. T / F

56. The Engine Start switch should be moved to__________when the engine reaches__________rpm.

57. In the event of a generator or exhaust nozzle failure, the pilot should immediately check the corresponding________gage and________light.

58. The ignition duty cycle is________minutes ON,______minutes OFF,______minutes ON and____________ minutes OFF.

59. Operation with the ENG INLET TEMP HIGH warning light illuminated is prohibited below______feet and is limited to__________above__________feet.

60. MIL and MAX thrust are time limited to__________minutes below 35,000 feet, and______hours above 35,000 feet.

61. Due to limited oil distribution to the variable nozzle system during negative "g" and zero "g" flight, the aircraft is limited to the following:

    a.________seconds of negative G flight.

    b.________seconds of zero G flight.

62. The emergency flap system air bottles store enough air for______extension(s) of the wing flaps.

63. If pneumatic pressure is lost when the flaps are blown down, they (will/will not) be blown up by the airstream.

64. When lateral trim corrections are made, the control stick moves in proportion to the amount of trim applied. T / F

65. The stabilator system utilizes a____________to increase stick forces during rapid fore and aft stick movement.

66. To temporarily disengage the ARI:

    1. ______________________________

    To permanently disengage the ARI and retain YAW Stab Aug:

    1. ______________________________

    2. ______________________________

67. The stall warning vibrator is set at:

    a.__________units angle of attack.

68. Disengagement of the yaw stability augmentation mode decreases the ARI authority by 5°. T / F

69. Normal rudder trim range is______±______degrees of rudder deflection.

70. All fuel tanks can be pressure or gravity filled. T / F

QUESTION BANK CONTINUED

71. Two electrically operated fuel boost pumps are located in fuselage fuel cell number__________. In the event of an electrical failure, the left pump, which is a__________speed pump, will operate on low speed when________________.

72. The internal and external wing tanks are vented to the atmosphere when the landing gear is down and the wing transfer pressure switch is in the EMERG position. T / F

73. Two methods available to pressurize the internal fuel cells and the external wing tanks are:

    a. Landing gear handle - ____

    b. Wing transfer pressure switch - ____

74. With an engine running, fuel from the centerline tank will start transferring upon selection of the __________position of the external transfer switch provided that the landing gear handle is in the __________ position or the wing press switch is in the__________position.

75. When operating on RAT electrical power, external tanks may be transferred normally. T / F

76. The internal wing tanks can be pressurized on the ground by moving the wing transfer pressure switch to the__________position. Transfer of internal wing fuel can then be accomplished by moving the internal wing transfer switch to the   position if the external tanks transfer switch is in the OFF position.

77. The internal wing tanks will be pressurized any time the wing transfer pressure switch is placed to EMERG and an engine is running and the probe is retracted. T / F

78 Fuel transfer pumps are located in fuselage cells________and_________. Two are_________and two are ________ operated.

79. 17th stage engine bleed air is used to transfer________and________fuel. Fuel in fuselage cells_________, ________,________and________is transferred by________only. (aircraft thru block 41)

80. With the bleed air switch in the OFF position, what will be lost?

    a. ____

    b. ____

    c. ____

    d. ____

    e. ____

    f. ____

    g. ____

    h. ____

    i. ____

81. Wing fuel may be dumped at any time, regardless of any other transfer switch position, by selecting the DUMP position on the internal wing dump switch with external power or normal internal electrical power applied. T / F

82. Wing fuel will be dumped on the deck if the internal wing dump switch is in the DUMP position and external electrical power is applied to the airplane. T / F

83. The fuel quantity indicator counter indicates fuel quantity in the________ and________. The tape indicates fuel quantity in__________cells__________through__________.

QUESTION BANK CONTINUED

84. To prevent external tank collapse during high altitude descent with the gear down, place the ______________ ______________ in the __________ position.

85. The internal wing dump control operates normally on RAT electrical power. T / F

86. Internal wing fuel can be dumped, regardless of the landing gear handle position when operating on normal electrical power. T / F

87. Illumination of the external fuel warning light is an indication of no flow, rather than low fuel quantity. T / F

88. When the air refueling probe is fully extended, the IFR PROBE UNLOCK warning light will not be illuminated. T / F

89. Fuselage fuel call number seven will transfer either by_________ or ______________________.

90. The FUEL LEVEL LOW warning light will illuminate any time the combined fuel in fuselage cells and____________ reaches_________ pounds.

91. The total internal fuel capacity is approximately___________ pounds.

92, Fuel boost pump pressure limits on preflight check is________ to ________ psi.

93. Wing fuel transfer limits are_______ degrees nose-up, and_______ degrees nose-down.

94. a. List the procedures for emergency probe extension.

(1) ____

(2) ____

(3) ____

(4) ____

b. Why is it not recommended to return the emergency probe switch to the normal position?

95. Internal wing fuel will not transfer unless the external transfer switch is in the___________ position.

96. If both fuel boost pumps fail while operating above________ feet, flameout of both engines may occur.

97. Fuel will be supplied to the engines by gravity feed, if a failure of both fuel boost pumps occur. T / F

98. If one fuel boost pump fails, engine thrust settings should be reduced until a minimum boost pump pressure of__________ psi can be maintained.

99. The power control systems supply hydraulic power to the_________, _________ and_________.

100. There are three independent hydraulic systems in the airplane; they are the______, the______ and the_________ systems.

101. Which hydraulic pumps utilize the same reservoir. ______________________________.

102. The three hydraulic systems operate in a pressure range of_____ PSI. The CHK HYD GAGES, light will illuminate when any system pressure drops below________ ± 100 PSI, and will go out when system pressure recovers to above______ PSI.

103. A PC-1 or a PC-2 failure will limit full travel of the lateral control surfaces and the stabilator at high airspeeds. T / F

104. A CHK HYD GAGES warning light, with all three gages indicating 3000 psi, indicates________________ ________________.

QUESTION BANK CONTINUED

105. The PC-1 and PC-2 hydraulic systems utilize the same accumulators and reservoirs. T / F

106. The rudder will be completely inoperative in the event of a utility hydraulic system failure. T / F

107. There is no safety feature incorporated in the landing gear system to prevent retraction on the ground. T / F

108. Nose gear steering is limited to__________degrees either side of center.

109. Two possible ways to deflate the nose gear strut while airborne are:

a. ______________________________

b. ______________________________

110. The nose gear steering system also functions as a__________.

111. Actuation of the warning lights test switch by the RIO will illuminate the pilot's master caution light. T / F

112. List the data that should be pre-set into the navigation computer.

a. ______________________________

b. ______________________________

c. ______________________________

d. ______________________________

e. ______________________________

113. The OXYGEN LOW warning light will illuminate when the supply is reduced to one liter. T / F

114. The normal pneumatic system pressure range on the cockpit indicator is________to________psi.

115. The systems operated with pneumatic system pressure are:

a. Normal system operation:

1. ______________________________
2. ______________________________
3. ______________________________

b. Emergency system operation: (after AFC 370)

1. ______________________________
2. ______________________________
3. ______________________________
4. ______________________________
5. ______________________________
6. ______________________________

116. If the SPEED BRAKE OUT indicator light illuminates, the MASTER CAUTION light will also illuminate. T / F

117. The wings cannot be folded with the landing gear retracted. T / F

QUESTION BANK CONTINUED

118. The authorized fuels for the airplane are:

a. Ashore - ______________________________

b. Afloat - ______________________________

c. Emergency - ______________________________

119. For ground pressure refueling, with engines off and external electrical power connected, list the fuel control panel switch positions:

a. External transfer switch - ______________________________

b. Wing transfer pressure switch - ______________________________

c. Refuel selection switch - ______________________________

d. Buddy fill switch - ______________________________

e. Refuel probe switch - ______________________________

120. When you are pressure refueling on the ground, engines not running and electrical power connected, the engine master switches should be in the________position.

121. The ground fueling switch is located in the__________wheel well.

122. The main and afterburner fuel controls should be adjusted for the specific density of the fuel grade being used. T / F

123. The maximum airspeed for drag chute deployment is_________knots CAS.

124. List the airspeed limitations for:

a. Landing gear extended - ______________________________

b. Wing flap extended - ______________________________

c. Canopy open - ______________________________

d. RAT extension -________Knots CAS or________Mach, whichever is less.

e. Boarding ladder extended ______________________________

125. The center of gravity limitations for all current permissible gross weights and configurations are between_______% and_______% of the Mean Aerodynamic Chord.

126. The weight limitations for the following conditions are:

a. Field takeoff -________pounds.

b. Field landing (flared) -________pounds.

c. Catapulting -_________pounds.

d. Arresting landing, touch-and-go, and FMLP -________pounds.

e. Single engine approach -_________pounds.

QUESTION BANK CONTINUED

127. What is the maximum airspeed and Mach number for the air refueling probe:

a. Extension ________knots CAS or__________Mach, whichever is less.

b. Fully extended__________knots CAS or____________Mach, whichever is less.

128. The maximum gross weight for a flared field landing is___________pounds.

129. Maximum gross weight for a field mirror landing is___________pounds.

130. List nine prohibited maneuvers:

a. ______________________________

b. ______________________________

c. ______________________________

d. ______________________________

e. ______________________________

f. ______________________________

g. ______________________________

h. ______________________________

i. ______________________________

131. Full pressure suits will be worn on all flights above________feet.

132. The pressure altimeter should be set and checked prior to engaging the static pressure compensator. T / F

133. The limiting crosswind component for drag chute deployment is:

a. Dry runway________Knots.

b. Low RCR________Knots.

134. The FMLP pattern altitude is________above ground level.

135. The pilot's NO-GO signals for catapult abort are:

a. Day - ______________________________

b. Night - ______________________________

136. If a BLC failure is experienced, approach and landing speeds must be increased. T / F

137. The day signal for HEFOE code commencing is______________________________

____

138. The RIO has just given you a HEFOE signal - four fingers - followed by a thumbs down. What pilot action is required?______________________________

139. The HEFOE signals and meanings are:

FINGERS -	ONE	TWO	THREE	FOUR	FIVE
MEANING -	____	____	______	_____	____

QUESTION BANK CONTINUED

140. The RIO will normally eject before the pilot. According to the NATOPS manual, what are the only foreseeable deviations from this?

a. ____________________

b. ____________________

141. The ejection signals given to the RIO if the intercom is inoperative with no eject light are:

a. day ____________________

b. night ____________________
If there is no response to either (a) or (b):

142. You are attempting to eject but the canopy fails to jettison. What would you do to manually jettison the canopy?

a. ____________________

If the canopy still fails to jettison:

b. ____________________

c. ____________________

143. After ejection, the crewmember must manually actuate his oxygen supply. T / F

144. The recommended maximum airspeeds for ejection are:

145. The minimum published altitude for safe ejection is 200 feet. T / F

146. List the procedures required to obtain the use of the essential bus when the left generator failed and the BUS TIE OPEN light is illuminated.

a. ____________________

147. List the NATOPS procedure for engine failure during takeoff, takeoff continued.

a. ____________________

b. ____________________

c. ____________________

d. ____________________

e. ____________________

f. ____________________

g. ____________________

148. List the NATOPS procedure for double engine failure during flight.

a. ____________________

b. ____________________

c. ____________________

d. ____________________

e. ____________________

QUESTION BANK CONTINUED

f. ____

g. ____

149. List the NATOPS air start procedure.

a. ____

b. ____

c. ____

d. ____

e. ____

150. List the NATOPS procedure for engine fire during start.

a. ____

b. ____

c. ____

d. ____

e. ____

151. List the procedures for recovery from post stall gyrations.

a. ____

b. ____

c. ____

152. List the procedures for recovery from an upright spin.

a. ____

b. ____

c. ____

d. ____

e. ____

153. After the gear and flaps are extended, using the emergency systems, their respective control handles should be reset by the pilot to ready the systems for normal operation. T / F

154. The emergency procedure for lowering the landing gear is:

a. ____

b. ____

c. ____

d. ____

e. ____

QUESTION BANK CONTINUED

155. If a hardover rudder occurs, proceed as follows:

a. __________

b. __________

c. __________

d. __________

e. __________

f. __________

g. __________

h. __________

i. __________

If unable -

j. __________

156. The flap position for a single-engine landing is__________ flaps.

157. If a forced landing on unprepared terrain is unavoidable, land with the wheels__________.

158. If it becomes necessary to jettison both canopies, jettison the__________canopy first.

149. If it is necessary to ditch the airplane, ditch with the wheels__________.

160. What steps should be taken in the event the intercom became inoperative:

a. __________

b. __________

c. __________

d. __________

e. __________

161. The pilot/RIO attention signals, under emergency conditions, with no other means of communication are:

a. Pilot - __________

b. RIO - __________

162. If you are lost and have an operative radio receiver, you should fly two triangles to the__________, repeating the pattern at__________minute intervals.

163. If you are lost and have no operative radios, you should fly two triangles to the__________, repeating the pattern every__________minutes.

QUESTION BANK CONTINUED

164. Supply the bank angles for the following directive commentary:

 a. EASY ____________________

 b. PORT (STBD) ____________________

 c. HARD ____________________

 d. HARD AS POSSIBLE ____________________

 e. WRAP IT UP ____________________

165. The proper sequence for information given in description commentary is:

 a. ____________________

 b. ____________________

 c. ____________________

 d. ____________________

166. The pilot will echo all directive commentary given by the RIO. T / F

167. Supply the interpretation of the following directive commentary:

 a. Ease - ____________________

 b. Hold - ____________________

 c. Harder - ____________________

 d. Break - ____________________

 e. Compass recovery - ____________________

 f. Buster - ____________________

 g. Gate - ____________________

168. The term JUDY informs the controlling agency that the aircraft radar is locked on the target. T / F

169. The daylight, emergency stop signal is raised crossed arms with clenched fists. T / F

170. BIT checks should not be conducted during.

 a. ____________________

 b. ____________________

171. In the landing pattern, the RIO will challenge the pilot on the gear, flaps, and hook positions if a report is not received. T / F

172. The RIO will monitor all instrument departures and approaches, and will advise the pilot of any course or minimum altitude deviations. T / F

173. A gradual afterburner shutdown must be made at speeds above ________ knots CAS or __________ Mach, whichever is less, when operating at 40,000 feet.

174. The maximum limiting Mach number for a clean airplane is _______ and begins at ________ feet on a standard day.

175. The maximum allowable airspeed for a clean airplane below 10,000 feet with C.G. forward of 30% MAC is ________ knots CAS.

QUESTION BANK CONTINUED

176. Maximum allowable acceleration limits of the centerline tank are:

   a. Full ________ to ________.

   b. Empty to 3/4 Full _______ to _________.

177. Maximum allowable acceleration limits of the external wing tank are:

   a. Full ________ to _________.

   b. Empty to 3/4 Full _______ to _________.

178. The maximum speed for jettisoning:

   a. Wing tanks - ________ knots CAS below 30,000 feet and ________ knots CAS above 30,000 feet.

   b. Centerline tank - ________ knots CAS below 15,000 feet and _______ knots CAS above 15,000 feet.

179. The maximum speed at which an LAU-17A launcher may be jettisoned is ______ knots CAS or __________ Mach, whichever is less.

180. The wing missile stations may carry either Sparrow III AIM-7D, Sparrow III AIM-7E, Sidewinder AIM-9B, or Sidewinder AIM-9D missiles. T / F

181. When the wing missiles are jettisoned, the missiles remain attached to the pylon and the pylon is jettisoned. T / F

182. Which of the following jettison switch(es) is/are hot anytime electrical power is on the airplane?

   a. External stores emergency release button.

   b. External wing tanks jettison switch.

   c. Missile jettison selector switch.

Note

For questions pertaining to Weapons System, refer to NATOPS Flight Manual Supplement, NAVAIR 01-245FDD-1A.

SECTION XI PERFORMANCE DATA

Part 1	DRAG NUMBERS, AIRSPEED and ALTIMETER POSITION ERROR CORRECTIONS, and STANDARD USAGE CHARTS.	**Part 1 INTRODUCTION**
Part 2	TAKEOFF SPEEDS and DISTANCES.	**Part 2 TAKEOFF**
Part 3	CLIMB SPEEDS, TIMES, FUEL, and DISTANCES.	**Part 3 CLIMB**
Part 4	CRUISE SPEEDS, ALTIMETER, and MACH NUMBER.	**Part 4 RANGE**
Part 5	ENDURANCE TIME, SPEEDS, and ALTITUDES.	**Part 5 ENDURANCE**
Part 6	AIR REFUELING TRANSFER TIME.	**Part 6 AIR REFUELING**
Part 7	DESCENT SPEEDS, TIMES, FUEL, and DISTANCES.	**Part 7 DESCENT**
Part 8	APPROACH SPEEDS and LANDING DISTANCES.	**Part 8 LANDING**
Part 9	COMBAT POWER FUEL FLOW RATES and NAUTICAL MILES PER POUND PER POUND, DIVE RECOVERY, MAXIMUM PERFORMANCE ENVELOPES and ACCELERATION.	**Part 9 COMBAT PERFORMANCE**
Part 10	Sample problems utilizing the collective performance data to plan a complete mission from engine-start to engine-shutdown.	**Part 10 MISSION PLANNING**

FDD-1-(86)A

PART 1 INTRODUCTION

CHARTS

This section is divided into parts (1 through 10) to present performance data in proper sequence for preflight planning. Two concepts of data presentation are utilized to show drag effects on aircraft performance; i.e., specific configuration charts and drag index charts. The drag index concept presents subsonic climb data, nautical miles per pound for cruise endurance, and descents. All other data are presented as a specific-configuration per chart. All performance data is based on flight tests or the contractor's estimate, ICAO standard day conditions and/or provisions to correct for nonstandard temperatures. Where required, separate charts are provided for J79-GE-8 and J79-GE-10 engines using JP-5 Fuel.

Note

The indication of the fuel quantity indicator presents the readings of actual fuel weight remaining. This is accomplished by means of compensator capacitors which provide accurate readings regardless of changes in the dielectric value of the fuel or variations in specific density due to temperature changes. Therefore, adjustment for various fuel densities is not necessary.

ARMAMENT ATTACHMENT ASSOCIATION CHART

The information necessary to determine the total weight of the stores loaded on the airplane, and their effect on the aircraft center-of-gravity is contained in the Armament Attachment Association chart (figure 11-1) and the Station Loading chart (figure 11-2). The Armament Attachment Association chart lists the various attachments (launcher, pylons, racks, and adapters) that are needed to carry an external store on any one particular station.

STATION LOADING

The Station Loading chart (figure 11-2) lists the individual weight, drag number, stability number, stations location and incremental center-of-gravity shift, of the various pylons, adapters, racks and external stores. It also lists the average operating weight with its corresponding center-of-gravity and the basic takeoff gross weight with its corresponding center-of-gravity for various airplanes. The chart does not intend to list the quantity and total gross weight of the external stores that can be carried on each station. However, the takeoff gross weight and approximate takeoff center-of-gravity can be computed by first referring to the Armament Attachment Association chart and determining the various attachments necessary to carry the particular stores that are to be loaded. Next refer to the Station Loading chart to find the individual weights and incremental center-of-gravity shifts of the selected stores and attachments. Once the individual weights have been noted, multiply the individual weights by the quantity to be carried (this figure will be the total external store weight). The external store weight, added to the airplane basic takeoff weight will result in a close approximation of the takeoff gross weight. The takeoff center-of-gravity can be computed by adding the incremental center-of-gravity values for each station that the various pylons, adapters, racks and external stores are intended to be carried on. The summation of the center-of-gravity values, added or subtracted as necessary from the center-of-gravity corresponding to the basic takeoff weight will result in a close approximation of the actual takeoff center-of-gravity.

INCREMENTAL CG SHIFT

Sample Problem

Configuration: 6 LAU-68/A rocket launchers on aircraft stations 1 & 9 (FULL) (3 each station).
6 MK 82 LDGP bombs on aircraft station 5.

Estimated CG of 33.7% MAC (estimated operating weight plus weight of full internal fuel, not including tank no. 7)

A.	LAU-68/A (stations 1 & 9)	(+0.10 x 6)	= +0.60
B.	Suspension equipment (stations 1 & 9)		
	Wing tank pylon	(+0.04 x 2)	= +0.08
	MW adapter	(+0.01 x 2)	= +0.02
	TER	(+0.06 x 2)	= +0.12
C.	MK-82 LDGP (station 5)		
	Forward cluster	(-0.31 x 3)	= -0.93
	Aft cluster	(+0.26 x 3)	= +0.78
D.	Suspension equipment (station 5)		
	MW ℄ adapter	0	0.00
	MER (shifted forward)	(-0.02 x 1)	= -0.02
E.	Incremental CG shift		+0.65
F.	Estimated takeoff CG	(33.7% +0.65)	= 34.35%

DRAG INDEX SYSTEM

The drag number for each externally carried store and its associated suspension equipment is listed. The drag index for a specific configuration may be found by multiplying the number of stores carried by its drag number and adding the drag number of the applicable suspension equipment (if not included). The total drag index may then be used to enter the planning data charts. Charts applicable for all loads and configurations are labeled ALL DRAG INDEXES. Charts labeled INDIVIDUAL DRAG INDEXES contain data for a range of drag numbers; i.e., individual curves/columns for a specific drag number. Supersonic data is not compatible to the drag index system; therefore, each chart is labeled for a specific configuration.

Sample Problem

Configuration: 6 LAU-68/A rocket launchers on aircraft stations 1 & 9 (FULL) (3 each station).
6 MK 82 LDGP bombs on aircraft station 5.

A.	LAU-68/A drag number	2.7 x 6 = 16.2
B.	Wing tank pylon, MW adapter, TER	2(1.1 +0.4 +5.5) = 14.0
C.	MK 82 LDGP drag number	1.1 x 6 = 6.6
D.	Centerline MW adapter, MER	2 + 8 = 10.0
E.	Total drag index	46.8

STABILITY INDEX SYSTEM

With the many possible external loading configurations and their resulting aerodynamic effects, it is possible to load the airplane past the aft CG limit. Adding wing-mounted stores tends to shift the aerodynamic center forward toward the CG of the aircraft, thereby reducing the longitudinal maneuvering stability. To be assured of an acceptable static margin, it is necessary to consider stability effects in conjunction with CG location. Each wing-mounted store and its associated suspension equipment is assigned a unit stability number corresponding to its aerodynamic effect. Each stability index (sum of stability numbers) has a corresponding aft CG limit. After the loading configuration has been determined, compute the airplane stability index. Enter the Aft CG Limits chart (part 4 of section I) with the airplane stability index to obtain maximum allowable aft CG location. The CG location is determined in the normal manner by using a Weight and Balance Clearance Form F in conjunction with the Handbook of Weight and Balance Data, AN-1-1B-40.

Note

- In some cases where the originally desired configuration is not within the allowable envelope, an acceptable static margin may be achieved through rearrangement of wing-mounted stores.
- Tandem-mounted weapons count as a single weapon when computing the airplane stability index.
- Fuselage-mounted stores are not used in determining airplane stability index but they are used in computing takeoff CG location.
- Unit stability numbers are assigned for single mounted and cluster mounted weapons. The cluster mounted unit stability number will be used when two or more weapons are mounted on the same rack, with each weapon being assigned this number.

Sample Problem

Configuration: 6 LAU-68/A rocket launchers on aircraft stations 1 & 9 (FULL) (3 each station).
6 MK 82 LDGP bombs on aircraft station 5.

A.	LAU-68/A (cluster mounted-stations 1 & 9)	5.2 x 6 = 31.2
B.	Wing tank pylon, MW adapter, TER	2(4.3 + 6.6) = 21.8
C.	Stability index	53.0
D.	Aft CG limit based on stability index	35.2% MAC

DRAG DUE TO ASYMMETRIC LOADING

This chart (figure 11-3) provides the drag number that results from trimming out an asymmetric store loading. The drag number is added to the computed drag of the airplane to obtain the drag index. Asymmetric drag varies with Mach number and altitude.

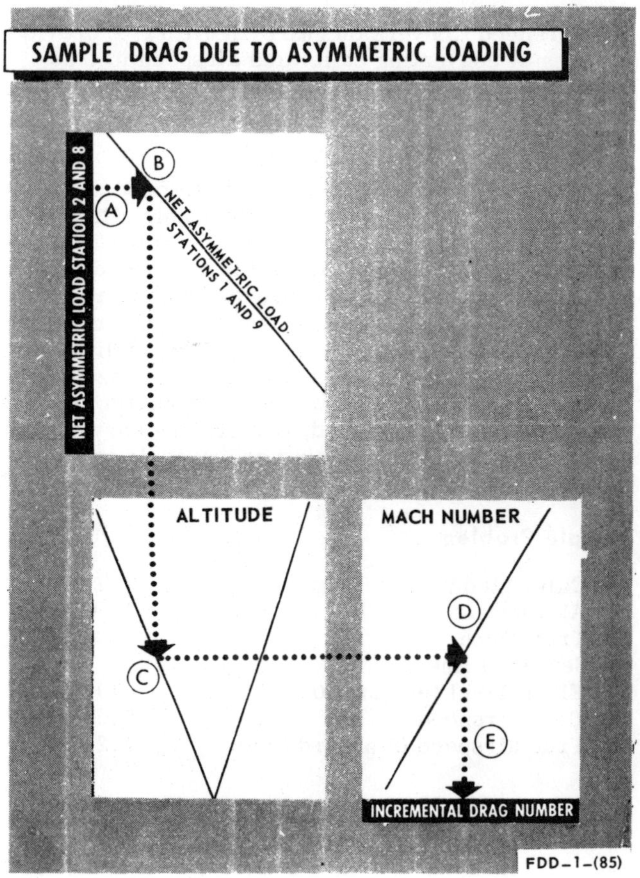

USE

Find the net asymmetric load on stations 2 and 8 (stations 2 and 8 are indicated on the left vertical axis and stations 1 and 9 are indicated by the diagonal parallel lines) by subtracting the lighter from the heavier weight. Attach to this net load the position, RWH (right wing heavy) or LWH (left wing heavy) as appropriate. In the same manner, find the net asymmetric load on stations 1 and 9. Enter the chart with the net asymmetric load for stations 2 and 8 corresponding to the load position. Proceed horizontally to the right to the net asymmetric load on stations 1 and 9 and its position. Proceed vertically downward to the altitude, horizontally to the right to the Mach number, and then vertically downward to obtain the incremental drag number.

Sample Problem

A. Load on station 2 = 1000 Lbs.
   Load on station 8 = 3000 Lbs.
   Net asymmetric load on stations 2 & 8 = 2000 Lbs. RWH
B. Load on station 1 = 2500 Lbs.
   Load on station 9 = 2000 Lbs.
   Net asymmetric load on stations 1 & 9 = 500 Lbs. LWH

C.	Altitude	25,000 Ft
D.	Mach number	0.7
E.	Incremental drag number	5.8

AIRSPEED CONVERSION

The Airspeed Conversion chart (figure 11-5) provides a means of converting calibrated airspeed to true Mach number and true airspeed.

USE

Enter the chart with the calibrated airspeed and proceed vertically to intersect the applicable altitude. From this point, proceed horizontally to the left to read true Mach number. From the calibrated airspeed-altitude intersection, proceed horizontally to the right to intersect the sea level line. From this point descend vertically to intersect the applicable flight level temperature. Then proceed horizontally to the right to read true airspeed. To obtain the Standard Day true airspeed, parallel the curved dash lines from the calibrated airspeed-altitude intersection to the sea level line.

Sample Problem

A.	Calibrated airspeed	330 Kt
B.	Altitude	25,000 Ft
C.	True Mach	0.782
D.	Sea level line	
E.	Flight level temperature	-20°C
F.	True airspeed	486 Kt
G.	True airspeed (Standard Day)	472 Kt

INDICATED AIRSPEED

Indicated airspeed (IAS) is the uncorrected airspeed read directly from the indicator when the ADC is inoperative.

CALIBRATED AIRSPEED

Calibrated airspeed (CAS) is indicated airspeed corrected for static source error. In this airplane, the ADC automatically compensates for this error so that calibrated airspeed may be read directly from the indicator.

EQUIVALENT AIRSPEED

Equivalent airspeed (EAS) is calibrated airspeed corrected for compressibility effect. There is no provision for reading EAS; however, it may be obtained by multiplying the TAS by square root of the density ratio.

TRUE AIRSPEED

True airspeed (TAS) is equivalent airspeed corrected for atmospheric density. Refer to Airspeed Conversion, (figure 11-5).

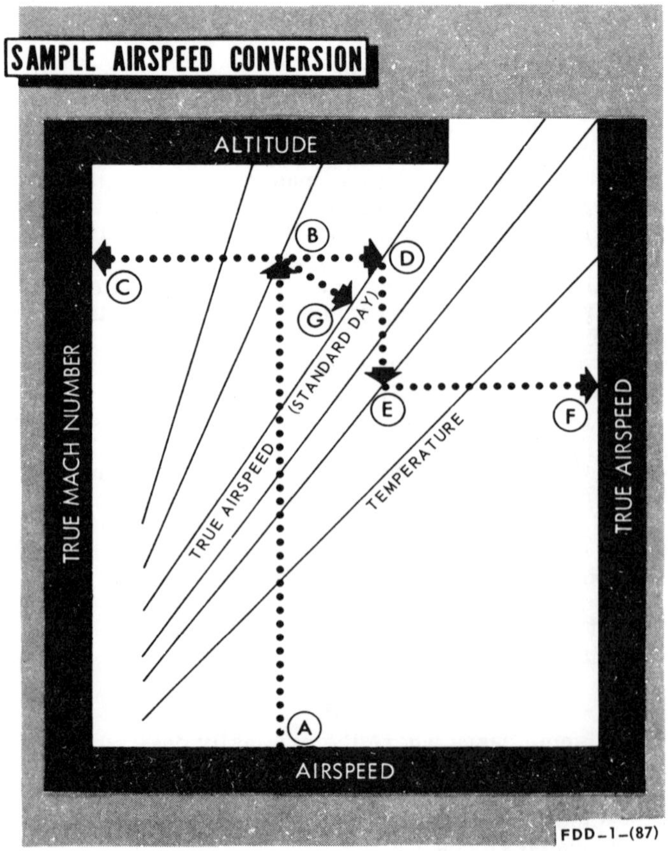

SPC/ALTIMETER TOLERANCE

The SPC/Altimeter Tolerance Check chart (figure 11-6) provides a means of checking the accuracy of the static pressure compensator in flight. The chart is plotted for 9 units angle of attack between 5000 and 20,000 feet. The Δ altitude between the curves represents the allowable tolerance of the system.

USE

With the static pressure compensator on and operating, establish the airplane at 9 units angle of attack between 5000 and 20,000 feet at a constant Mach number. Record the Mach number and altitude. Enter the chart with the Mach number and proceed vertically to intersect both curves. From these intersections, proceed horizontally to the left and record the two corresponding Δ altitudes. Add these to the indicated altitude to obtain the upper and lower allowable limits. Move the static pressure compensator switch (labeled CADC), located on the pilot's left console, to the OFF position. The altimeter must jump. Note and record the indicated altitude. If the indicated altitude, with the static pressure compensator turned off, falls on or between the previously computed limits, reset the static pressure compensator and ensure that the STATIC CORR OFF light is extinguished.

WARNING

If the altimeter does not jump, or the indicated altitude with the static pressure compensator off does not fall within the limits established by the SPC/Altimeter Tolerance Check chart, leave the static pressure compensator off during the remainder of the flight and utilize the Altimeter and Airspeed Position Error Correction charts (STATIC CORRECTION OFF), this section.

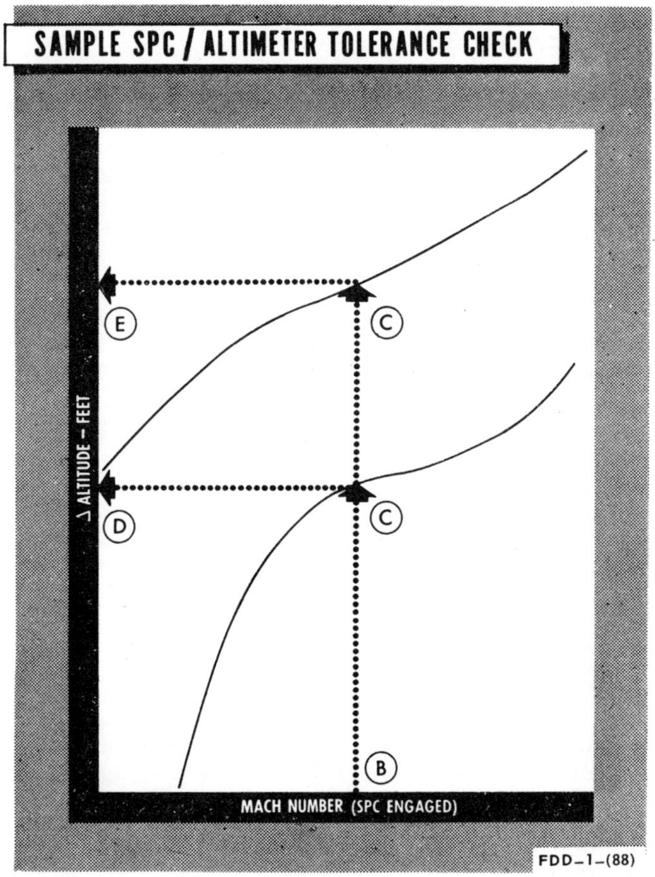

FDD-1-(88)

Sample Problem

A.	Altitude	15,000 Ft
B.	Mach	0.77
C.	Intersect both curves	
D.	Lower Δ altitude	180 Ft
E.	Upper Δ altitude	320 Ft
F.	Lower limit (A +D)	15,180 Ft
G.	Upper limit (A +E)	15,320 Ft

AIRSPEED POSITION ERROR CORRECTION

Under normal conditions, airspeed position error is automatically compensated for by the air data computer system (ADC). However, if a malfunction of the ADC occurs, position error must be applied to the cockpit indication. These charts (figures 11-7 and 11-8) provide a direct-reading conversion from indicated to calibrated airspeed, and from indicated to true Mach number.

SAMPLE AIRSPEED POSITION ERROR CORRECTION

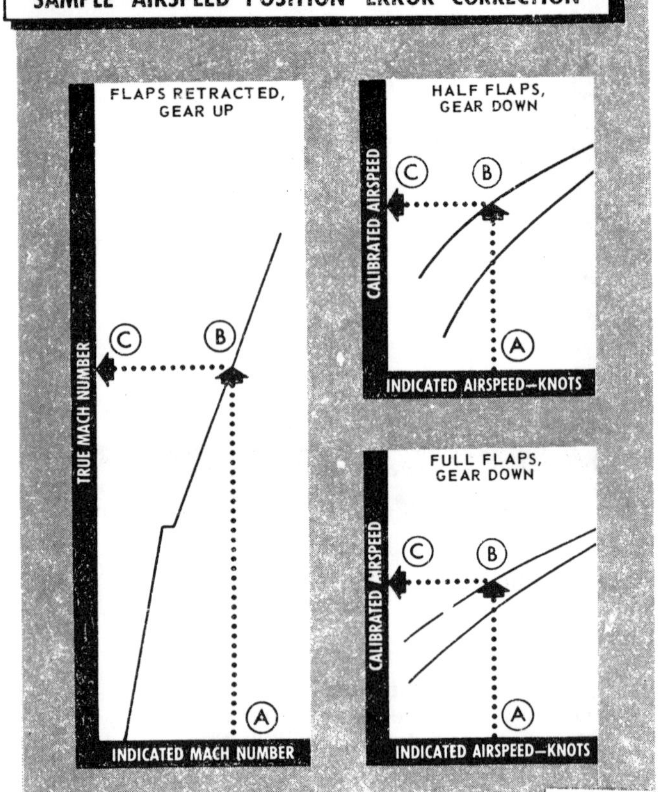

FDD-1-(89)

Sample Problem

Configuration: Gear Down, Full Flaps

A.	Indicated airspeed	160 Kt
B.	Gross weight	40, 000 Lb
C.	Calibrated airspeed	155 Kt

Configuration: Gear and Flaps Up

A.	Indicated Mach number	1.4
B.	Indicated altitude	40, 000 Ft
C.	True Mach number	1.34

ALTIMETER POSITION ERROR CORRECTION

Under normal operating conditions, compensation for the static source position error, as it affects the altimeter, is provided by the ADC. If the ADC fails in flight (STATIC CORR OFF), the altimeter must be corrected by means of the Altimeter Position Error Correction chart (figure 11-9). The chart contains three altitude correction (Δ H) plots; one cruise configuration and two landing configurations. The altitude correction (Δ H) must be added (algebraically) to the assigned altitude to obtain indicated altitude.

Note

Assigned altitude +Δ H = indicated altitude. Fly indicated altitude.

SAMPLE ALTIMETER POSITION ERROR CORRECTION

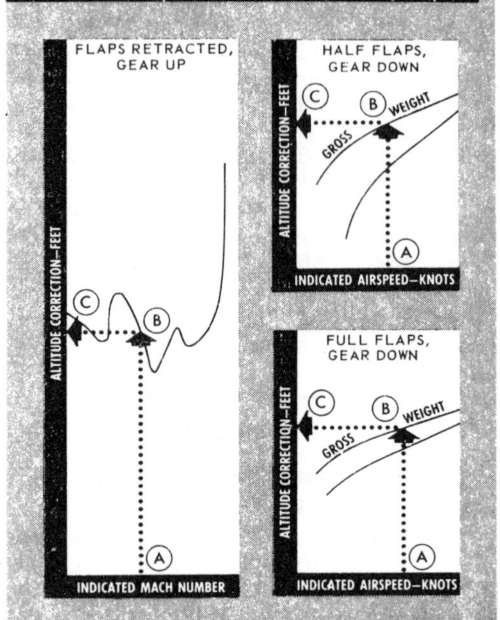

FDD-1-(90)

USE

Enter the cruise plot with the indicated Mach number. Proceed vertically upward to intercept the assigned altitude curve, then horizontally to the left to read the altitude correction(Δ H). Enter either of the landing plots with the indicated airspeed, proceed vertically to the applicable gross weight reflector, then horizontally to the left to read altitude correction (Δ H). Apply Δ H to the assigned altitude to obtain the indicated altitude.

Sample Problem

To maintain 35, 000 feet at 1.4 Mach (cruise plot), proceed as follows:

A.	Indicated Mach number	1.4
B.	Assigned altitude	40, 000 Ft
C.	Altitude correction (Δ H)	+2270 Ft
D.	Fly indicated altitude (B + C)	42, 270 Ft

ALTIMETER LAG CHART

These charts (figures 11-10 and 11-11) provide a means of obtaining the altimeter lag (difference between indicated altitude and actual altitude) resulting from diving flight. Data is provided for dive angles up to 60° and airspeeds up to 600 knots TAS.

USE

Enter the chart with dive airspeed, and project horizontally to the right to intersect the dive angle curve. From this point, project vertically downward to read the resulting altimeter lag. Add the altimeter lag data to desired/required pullout altitude to obtain indicated altitude for pullout.

Sample Problem

Aircraft with SPC operative

A. Dive airspeed (TAS)	400 Kt
B. Dive angle	45°
C. Altimeter lag	92 Ft

WIND COMPONENTS CHART

A standard Wind Components chart (figure 11-12) is included. It is used primarily for breaking a forecast wind down into crosswind and headwind components for takeoff and landing computations. It may, however, be used whenever wind component information is desired. To determine effective wind velocity, add one-half the gust velocity to the steady state velocity; e.g., reported wind 050/20 G 30, effective wind is 050/25.

USE

Reduce the reported wind direction to a relative bearing by determining the difference between the wind direction and the runway heading. Enter the chart with the relative bearing. Move along the relative bearing to intercept the wind speed arc. From this point, descend vertically to read the crosswind component. From the intersection of bearing and wind speed, project horizontally to the left to read headwind component.

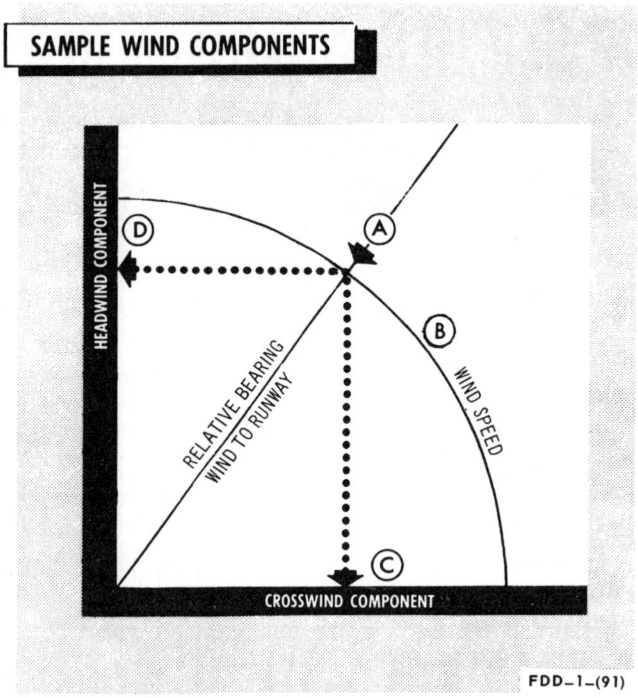

FDD-1-(91)

Sample Problem

Reported wind 050/35; runway heading 030.

A. Relative bearing	20°
B. Intersect windspeed arc	35 Kt
C. Crosswind component	12 Kt
D. Headwind component	33 Kt

ARMAMENT ATTACHMENT ASSOCIATION

STORES TO BE CARRIED	STATIONS AND ARMAMENT ATTACHMENTS								
	1	2	3	4	5	6	7	8	9
370 GAL. EXT. WING TANK	1								1
MK 81, MK 82, MK 83 LDGP BOMB, MK 86, MK 87, MK 88, MK 124 PRACTICE BOMB	2 14	4			10 15			4	2 14
ADSID I (NORMAL) SENSOR	2 14	4						4	2 14
MK 81, MK 82 LDGP BOMB, MK MK 36 DST, MK 40 DST	2 14	4			10 15			4	2 14
MK 77 MOD 4 FIRE BOMB	2 14	4			10 15			4	2 14
MK 20 MOD 2/3, CHAFFEYE	2 14	4			10 15			4	2 4
MK 12 CHEMICAL TANK	14	4						4	14
LAU–10/A, LAU–10A/A OR LAU–10B/A ROCKET POD	14	4			15			4	14
LAU–32A/A, LAU–32B/A, LAU–56/A, LAU–61/A, LAU–61A/A, LAU–68/A, LAU–68B/A, LAU–69/A, LAU–69A/A ROCKET POD	14	4			15			4	14
MK 76, MK 89, MK 105 PRACTICE BOMB	2 3 14	4 8			10 11 15			4 8	2 3 14
600 GAL. EXT. TANK					9				
CBU–24, –29 –49	2 14	4			10 15			4	2 14
NAMAR CAMERA POD		17						17	
CTU–1/A DELIVERY CONTAINER	2 14				10				2 14

STORES TO BE CARRIED	STATIONS AND ARMAMENT ATTACHMENTS								
	1	2	3	4	5	6	7	8	9
SUU–40/44 FLARE DISPENSER	2 14								2 14
D–704 AIR REFUELING STORE					9				
RCPP–105 STARTER POD					9				
AIM–7 MISSILES		5	16	16		16	16	5	
AIM–9 MISSILE		6						6	
MK 24, MK 45 FLARE	2 3 14	4 8			10 11 15			4 8	2 3 14
RMU–8/A REEL LAUNCHER					9				
CNU–169/A FERRY EQUIPMENT STORE		5						5	
AQM–37A MISSILE TARGET					7				
MK 4 MOD 0 GUN POD					13				
LAU–33A/A		6						6	
MK 82 LGB, MK 83 LGB	2 14	4			10 15			4	2 14
LB–30A STRIKE CAMERA			12						

1 WING TANK PYLON

2 WING TANK PYLON, MULTIPLE WEAPONS ADAPTER, MER

3 WING TANK PYLON, MULTIPLE WEAPONS ADAPTER, A/A37B–3 PMBR

4 LAU–17/A WING MISSILE PYLON, MULTIPLE WEAPONS ADAPTER, TER

5 LAU–17/A WING MISSILE PYLON

6 LAU–17/A WING MISSILE PYLON, LAU–7/A MISSILE LAUNCHER

7 AERO 27A RACK WITH LAU–24 LAUNCHER

8 LAU–17/A WING MISSILE PYLON, MULTIPLE WEAPONS ADAPTER, A/A37B–3 PMBR

9 AERO 27A RACK

10 AERO 27A RACK, MULTIPLE WEAPONS ADAPTER AND CENTERLINE MER

11 AERO 27A RACK AND A/A37B–3 PMBR

12 MISSILE STATION 3 WITH AERO 7A LAUNCHER REMOVED.

13 AERO–27A RACK, MULTIPLE WEAPONS ADAPTER AND GUN POD ADAPTERS

14 WING TANK PYLON, MULTIPLE WEAPONS ADAPTER, TER

15 AERO–27A RACK, MULTIPLE WEAPONS ADAPTER AND TER

16 NO ADDITIONAL EQUIPMENT REQUIRED

17 LAU–17/A WING MISSILE PYLON, MULTIPLE WEAPONS ADAPTER

FDD–1–(92)A

Figure 11-1

STATION LOADING

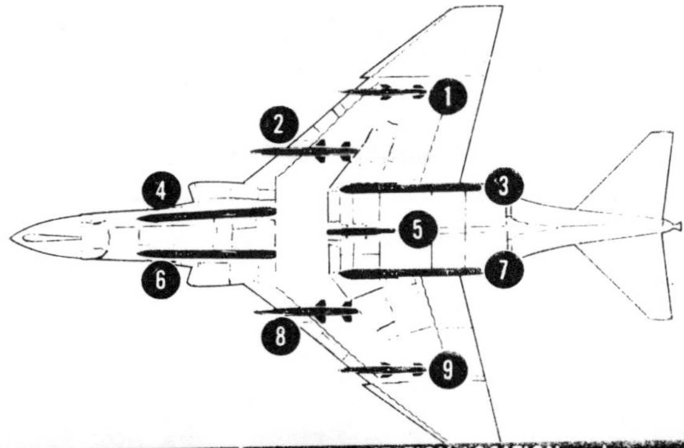

WARNING

For precise external store and attachment information, refer to charts C and E of the Weight and Balance Data Handbook (AN-01-1B-40) for your airplane.

STORES – BOMBS	UNIT WEIGHT (LBS)	UNIT DRAG	STABILITY NUMBER – SINGLE MOUNTED	STABILITY NUMBER – CLUSTER MOUNTED	BOMB RACK or BOMB CLUSTER POSITION		STATION LOCATION and INCREMENTAL C G SHIFT FOR INDIVIDUAL UNIT – 1	2	3	4	5	6	7	8	9
CBU–24/29/49	835	4.2	7.4	9.9	MER	FWD CLUSTER	–.26				–.52				–.26
					MER	AFT CLUSTER	+.57				+.36				+.57
						TER	+.31	–.42			+.05			–.42	+.31
CBU–59/B APAM	750	2.9	7.1	9.4	MER	FWD CLUSTER	–.24				–.49				–.24
					MER	AFT CLUSTER	+.57				+.55				+.57
						TER	+.35	.49			+.08			–.49	+.33
MK 40 DESTRUCTOR (WITH MAU–91 FIN)	1057	3.9	NE	NE	MER	FWD CLUSTER									
					MER	AFT CLUSTER									
						TER									
MK 81 LDGP (WITH CONICAL FIN)	270	0.8	1.8	2.4	MER	FWD CLUSTER	–.10				–.16				–.10
					MER	AFT CLUSTER	+.21				+.10				+.21
						TER	+.10	–.16			+.05			–.16	+.10
MK 81 LDGP (WITH MK 14 FIN OR BSU–33/B)	300	1.2	1.8	2.4	MER	FWD CLUSTER	–.10				–.19				–.10
					MER	AFT CLUSTER	+.23				+.14				+.23
						TER	+.10	–.20			+.05			–.20	+.10
MK 82 LDGP, MK 36 DESTRUCTOR (WITH CONICAL FIN)	531	1.1	2.8	3.7	MER	FWD CLUSTER	–.16				–.31				–.16
					MER	AFT CLUSTER	+.41				+.26				+.41
						TER	+.26	–.36			+.05			–.36	+.26
MK 82 LDGP, MK 36 DESTRUCTOR (WITH MK–15 FIN)	570	2.4	2.8	3.7	MER	FWD CLUSTER	–.16				–.31				–.16
					MER	AFT CLUSTER	+.41				+.26				+.41
						TER	+.26	–.36			+.05			–.36	+.26
MK 82 LGB (KMU–388B) (NON–EXTENDED FIN)	668	2.3	7.8	NA	MER	FWD CLUSTER									
					MER	AFT CLUSTER									
						TER									
MK 82 LGB (MUU–388) (EXTENDED FIN)	682	3.1	7.8	NA	MER	FWD CLUSTER									
					MER	AFT CLUSTER									
						TER									
MK 83 LDGP	985	1.8	4.6	6.1	MER	FWD CLUSTER	–.05				–.62				–.05
					MER	AFT CLUSTER					+.44				
						TER	+.36	–.68			+.10			–.68	+.36
MK 83 LGB	1088	3.6	13.1	NA	MER	FWD CLUSTER									
					MER	AFT CLUSTER									
						TER									
MK 77 MOD 4	520	3.5	14.3	19.1	MER	FWD CLUSTER	–.16				–.21				–.16
					MER	AFT CLUSTER	+.38				+.37				+.38
						TER	+.21	–.36			+.05			–.36	+.21
MK 20 MOD 2/3	475	2.9	NE	NE	MER	FWD CLUSTER	–.16				–.31				–.16
					MER	AFT CLUSTER	+.36				+.21				+.36
						TER	+.15	–.31			+.05			–.31	+.15

STORES – MISSILES		UNIT WEIGHT (LBS)	UNIT DRAG	STABILITY NUMBER – SINGLE MOUNTED	STABILITY NUMBER – CLUSTER MOUNTED	BOMB RACK or BOMB CLUSTER POSITION	STATION LOCATION and INCREMENTAL C G SHIFT FOR INDIVIDUAL UNIT – 1	2	3	4	5	6	7	8	9
SPARROW III	AIM–7D	402	FUSELAGE MOUNTED 1.3 WING MOUNTED 2.6	FUSELAGE MOUNTED NA WING MOUNTED 2.7	NA	——		–.30	+.35	–.68		–.68	+.35	–.30	
	AIM–7E	455			NA	——		–.33	+.39	–.76		–.76	+.39	–.33	
	AIM–7E–2	427			NA	——		–.33	+.39	–.76		–.76	+.39	–.33	

FDD-1-(93-1)B

Figure 11-2 (Sheet 1 of 5)

STATION LOADING (CONTINUED)

Stores		Unit Weight	Unit Drag	Single Mounted	Cluster Mounted	Bomb Rack or Bomb Cluster Position	1	2	3	4	5	6	7	8	9
SIDEWINDER	AIM-9B	157	1.7	1.0	1.4	—		-.10						-.10	
	AIM-9D/G	197	1.7	1.0	1.4	—		-.14						-.14	
	AIM-9H	195	1.7	1.0	1.4	—		-.14						-.14	

A NOSE AND TAIL CONES ON
B NOSE CONE OFF TAIL CONE ON
C NOSE AND TAIL CONES OFF – FULL
D NOSE AND TAIL CONES OFF – EMPTY

STORES / ROCKETS	F FULL / E EMPTY	UNIT WEIGHT	UNIT DRAG A / B	UNIT DRAG C / D	STABILITY NUMBER SINGLE MOUNTED	STABILITY NUMBER CLUSTER MOUNTED	BOMB RACK or BOMB CLUSTER POSITION	1	2	3	4	5	6	7	8	9
LAU-10/A, A/A, B/A	F	533	3.6	11.2	8.0	10.6										
	E	105	10.2	10.0			TER	+.21	-.36			+.05			-.36	+.21
LAU-32A/A	F	174	2.7	6.8	3.9	5.2										
	E	51	6.3	6.2			TER	+.05	-.14			.00			-.14	+.05
LAU-32B/A	F	175	2.7	6.8	3.9	5.2										
	E	53	6.3	6.2			TER	+.05	-.14			.00			-.14	+.05
LAU-33A/A	F	286	F	2.2	NE	NA										
	E	72	E	2.2			TER									
LAU-56/A	F	188	2.7	6.8	3.9	5.2										
	E	60	6.3	6.2			TER	+.05	-.14			.00			-.14	+.05
LAU-60/A	F	434	4.2	13.5	10.1	13.5										
	E	74	12.5	12.0			TER	+.21	-.36			+.05			-.36	+.21
LAU-61/A, A/A	F	502	4.2	13.5	10.1	13.5										
	E	115	12.5	12.0			TER	+.21	-.36			+.05			-.36	+.21
LAU-68/A, B/A	F	200	2.7	6.8	3.9	5.2										
	E	58	6.3	6.2			TER	+.10	-.16			.00			-.16	+.10
LAU-69/A, A/A	F	483	4.2	13.5	10.1	13.5										
	E	95	12.5	12.0			TER	+.21	-.16			+.05			-.16	+.21

STORES / PRACTICE BOMBS & ROCKETS	UNIT WEIGHT (LBS)	UNIT DRAG	STABILITY NUMBER SINGLE MOUNTED	STABILITY NUMBER CLUSTER MOUNTED	BOMB RACK or BOMB CLUSTER POSITION		1	2	3	4	5	6	7	8	9
MK 76	24	0.3	1.0	1.0	PMBR WITH (6) UNITS		+.10	-.16			+.05			-.16	+.10
MK 76	24	0.3	NE	NE	MER	FWD CLUSTER									
					MER	AFT CLUSTER									
					TER										
MK-86 (WATER-SAND FILL)	200	0.8	1.8	2.4	MER	FWD CLUSTER	-.08				-.15				-.08
					MER	AFT CLUSTER	+.17				+.10				+.17
					TER			-.15			+0.0			-.15	
MK-87 (WATER-SAND FILL)	330	1.1	2.8	3.7	MER	FWD CLUSTER	-.12				-.22				-.12
					MER	AFT CLUSTER	+.25				+.15				+.25
					TER			-.24			+.05			-.24	
MK-88 (WATER-SAND FILL)	750	1.8	4.6	6.1	MER	FWD CLUSTER	-.07				-.52				-.07
					MER	AFT CLUSTER					+.36				
					TER			-.56			+.10			-.56	

FDD-1-(93-2)-B

Figure 11-2 (Sheet 2 of 5)

STATION LOADING (CONTINUED)

STORES: PRACTICE BOMBS & ROCKETS	UNIT WEIGHT (LBS)	UNIT DRAG	STABILITY NUMBER: SINGLE MOUNTED	STABILITY NUMBER: CLUSTER MOUNTED	BOMB RACK or BOMB CLUSTER POSITION	1	2	3	4	5	6	7	8	9
MK 89	56	0.2	NE	NE	PMBR WITH (6) UNITS	+.15	−.26			−.05			−.26	+.15
MK 89	56	0.2	NE	NE	MER FWD CLUSTER									
					MER AFT CLUSTER									
					TER									
MK 106	5	0.4	1.0	1.0	PMBR WITH (6) UNITS	+.05	−.05			.00			−.05	+.05
MK 106	5	0.4	NE	NE	MER FWD CLUSTER									
					MER AFT CLUSTER									
					TER									
MK 124	565	2.8	NE	NE	NE									

Station columns 1–9: STATION LOCATION and INCREMENTAL C G SHIFT FOR INDIVIDUAL UNIT

STORES: TANKS, RACKS AND PODS		UNIT WEIGHT (LBS)	UNIT DRAG	STABILITY NUMBER: SINGLE MOUNTED	STABILITY NUMBER: CLUSTER MOUNTED	BOMB RACK or BOMB CLUSTER POSITION	1	2	3	4	5	6	7	8	9
370 GALLON EXTERNAL WING TANKS (INCLUDES PYLON)	MCDONNELL	F 2856	4.8	29.8 [1]	NA		+.13								+.13
		E 340		20.0 [2]			+.04								+.04
	SARGENT FLETCHER	F 2824	6.4	29.8 [1]	NA		+.13								+.13
		E 308		20.0 [2]			+.04								+.04
	ROYAL JET	F NE	6.4	NE	NE										
		E NE													
600 GALLON EXTERNAL CENTERLINE TANK	MCDONNELL	F 4329	9.6	NA	NA						+.21				
		E 249									+.01				
	ROYAL JET	F 4384	9.6	NA	NA						+.21				
		E 304									+.01				
MK 4 GUN POD		F 1390	11.6	NA	NA						+.21				
		E 787									.00				
MK 12 MOD 0 CHEMICAL TANK		F 1000	3.3	8.0	10.6	TER									
		E 350	3.3												
WING TANK PYLON		92	1.1	4.3	NA		+.04								+.04
LAU-17/A GUIDED MISSILE LAUNCHER		150	2.4	6.9	NA			−.07						−.07	
MULTIPLE WEAPONS ADAPTER (OUTBOARD)		24	0.4	NA	NA		+.01								+.01
MULTIPLE WEAPONS ADAPTER (INBOARD)		24	0.3	NA	NA			−.02						−.02	
MULTIPLE WEAPONS TER ℄ ADAPTER		55	2.0	NA	NA						.00				
LAUNCHER LAU-7/A		87	0.4	2.2	NA			−.06						−.06	
MER	℄ STATION	215	8.0	NA	NA	RACK SHIFTED FWD					−.02				
						RACK SHIFTED AFT					+.05				
	WING STATION	225	8.0	7.1	NA	RACK SHIFTED FWD	+.05								+.05
						RACK SHIFTED AFT	+.11								+.11
TER	℄ STATION	95	5.5	NA	NA						+.01				
	WING STATION	95	5.5	6.6	NA		+.06	−.06						−.06	+.06
℄ BOMB RACK AERO-27A		51	NA	NA	NA						.00				
RCPP-105 STARTER POD (FULLY SERVICED)		2016	7.4	NA	NA						+.03				
D-704 AIR REFUEL STORE		F 2773	10.0	NA	NA						+.20				
		E 733	10.0	NA	NA						+.25				
CNU-169/A FERRY MISSION EQUIPMENT STORE		F 435	4.4	18.0	NA										
		E 185													
LB-30A STRIKE CAMERA POD		126	3.0	NA	NA										
ALE-37 DISPENSER		NE	2.8	7.9	10.5										

[1] WING TANK AND PYLON (WITH WEAPONS OR PYLONS INSTALLED ON STATIONS 2 & 8)

[2] WING TANK AND PYLON (WITHOUT WEAPONS OR PYLONS INSTALLED ON STATIONS 2 & 8)

FDD-1-(93-3)B

Figure 11-2 (Sheet 3 of 5)

STATION LOADING (CONTINUED)

STORES — FLARES, SENSORS	UNIT WEIGHT (LBS)	UNIT DRAG	STABILITY NUMBER — SINGLE MOUNTED	STABILITY NUMBER — CLUSTER MOUNTED	BOMB RACK or BOMB CLUSTER POSITION		1	2	3	4	5	6	7	8	9
MK-24, MK 45 FLARE	27	1.0	1.0	1.3	MER	FWD CLUSTER	.00				.00				.00
					MER	AFT CLUSTER	.00				.00				.00
					TER		.00	.00			.00			.00	.00
SUU-40/44 FLARE DISPENSER	F 365	3.6	NE	NE	MER	FWD CLUSTER									
	E 125	3.6			MER	AFT CLUSTER	+.31				+.04				+.31
					TER		+.31	−.16			+.10			−.16	+.31
ADSID I (NORMAL) SENSOR	26	4.5	NE	NE	–										
SUU-40/44 DISPENSER WITH AN/GSQ-117/-117LS/-141 SENSOR	NE	3.6 3.6	NE	NE	–										
NAMAR CAMERA POD	275	3.8	NE	NE	–										
CTU-1/A DELIVERY CONTAINER (EMPTY) *SEE NOTES	230	4.9	NE	NE	MER										
					TER										
ASDC (AIR-SHIP DELIVERY CONTAINER SYSTEM) — DELIVERY CONTAINER	180 Max	3.7	NA	8.3	TER		+.08	−.12						−.12	+.08
ASDC (AIR-SHIP DELIVERY CONTAINER SYSTEM) — WIRE CONTAINER	F85 E41	4.8	9.3	12.3	TER		+.03	−.06						−.06	+.03

FDD-1-(93-4)B

Figure 11-2 (Sheet 4 of 5)

STATION LOADING (CONTINUED)

Notes

- THE DRAG INDEX OF THE CLEAN AIRPLANE IS ZERO.
- INDIVIDUAL STORE DRAG X NUMBER OF STORES TO BE CARRIED + SUSPENSION EQUIPMENT DRAG (IF NOT INCLUDED) = DRAG INDEX
- DRAG NUMBERS FOR SINGLE STORES ARE SLIGHTLY CONSERVATIVE. INTERFERENCE DRAG BETWEEN MULTIPLE STORES HAS BEEN CONSIDERED.
- FUSELAGE-MOUNTED STORES ARE NOT USED IN DETERMINING AIRPLANE STABILITY INDEX BUT THEY ARE USED IN COMPUTING TAKEOFF CG LOCATION.
- TANDEM-MOUNTED WEAPONS COUNT AS A SINGLE WEAPON WHEN COMPUTING THE AIRCRAFT STABILITY INDEX.
- UNIT STABILITY NUMBERS ARE ASSIGNED FOR SINGLE MOUNTED AND CLUSTER MOUNTED WEAPONS. THE CLUSTER MOUNTED UNIT STABILITY NUMBER WILL BE USED WHEN TWO OR MORE WEAPONS ARE MOUNTED ON THE SAME RACK, WITH EACH WEAPON BEING ASSIGNED THIS NUMBER.
- NE = NOT ESTABLISHED
- NA = NOT APPLICABLE
- E = EMPTY, F = FULL

* CTU–1/A CARRIER OPERATIONS NOT ALLOWED.

ESTIMATED OPERATING WEIGHT (Basic airplane plus the weight of oil, unusable fuel, and two crew members)
Airplanes 153071z thru 153087aa .... 29,834 Lbs. 31.6% M.A.C.
153768ab thru 155528ag .... 30,280 Lbs. 32.6% M.A.C.
155529ag and up (Shoehorn equipment included) .... 31,538 Lbs. 31.4% M.A.C.
158355at and up .... 31,785 Lbs. 31.0% M.A.C.

ESTIMATED TAKEOFF GROSS WEIGHT (Estimated operating weight plus weight of full internal fuel including tank no. 7)
Airplanes: 153071z thru 153087aa .... 43,421 Lbs. 35.2% M.A.C. (NOT INCLUDING NO. 7 TANK 42,775 Lbs. 33.9% M.A.C.)
153768ab thru 155528ag .... 43,867 Lbs. 35.9% M.A.C. (NOT INCLUDING NO. 7 TANK 43,221 Lbs. 34.6% M.A.C.)
155529ag and up (Shoehorn equipment included) .... 45,125 Lbs. 35% M.A.C. (NOT INCLUDING NO. 7 TANK 44,479 Lbs. 33.7% M.A.C.)
158355at and up .... 45,372 Lbs. 34.7% M.A.C. (NOT INCLUDING NO. 7 TANK 44,726 Lbs. 33.4% M.A.C.)

NOTES:
1. The Incremental CG shift effects are in terms of % M.A.C. (+ = AFT CG shift). (– = FWD CG shift). These unit stores increments are approximations only, and will vary depending on the individual airplane gross weight and CG.
2. Fuel weight based on JP–5 at 6.8 lb. per gallon.

FDD–1–(93–5)

Figure 11-2 (Sheet 5 of 5)

DRAG DUE TO ASYMMETRIC LOADING

REMARKS
ENGINE(S): (2) J79-GE-8 or -10
ICAO STANDARD DAY

GUIDE

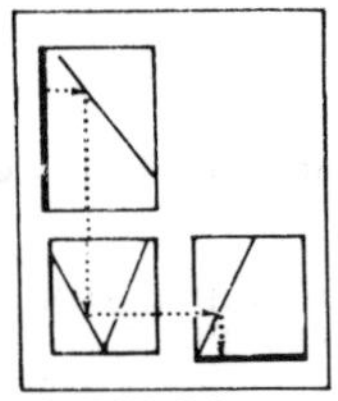

FUEL GRADE: JP-5
FUEL DENSITY: 6.8 LB/GAL

DATE: 1 MAY 1968
DATA BASIS: ESTIMATED (BASED ON FLIGHT TEST)

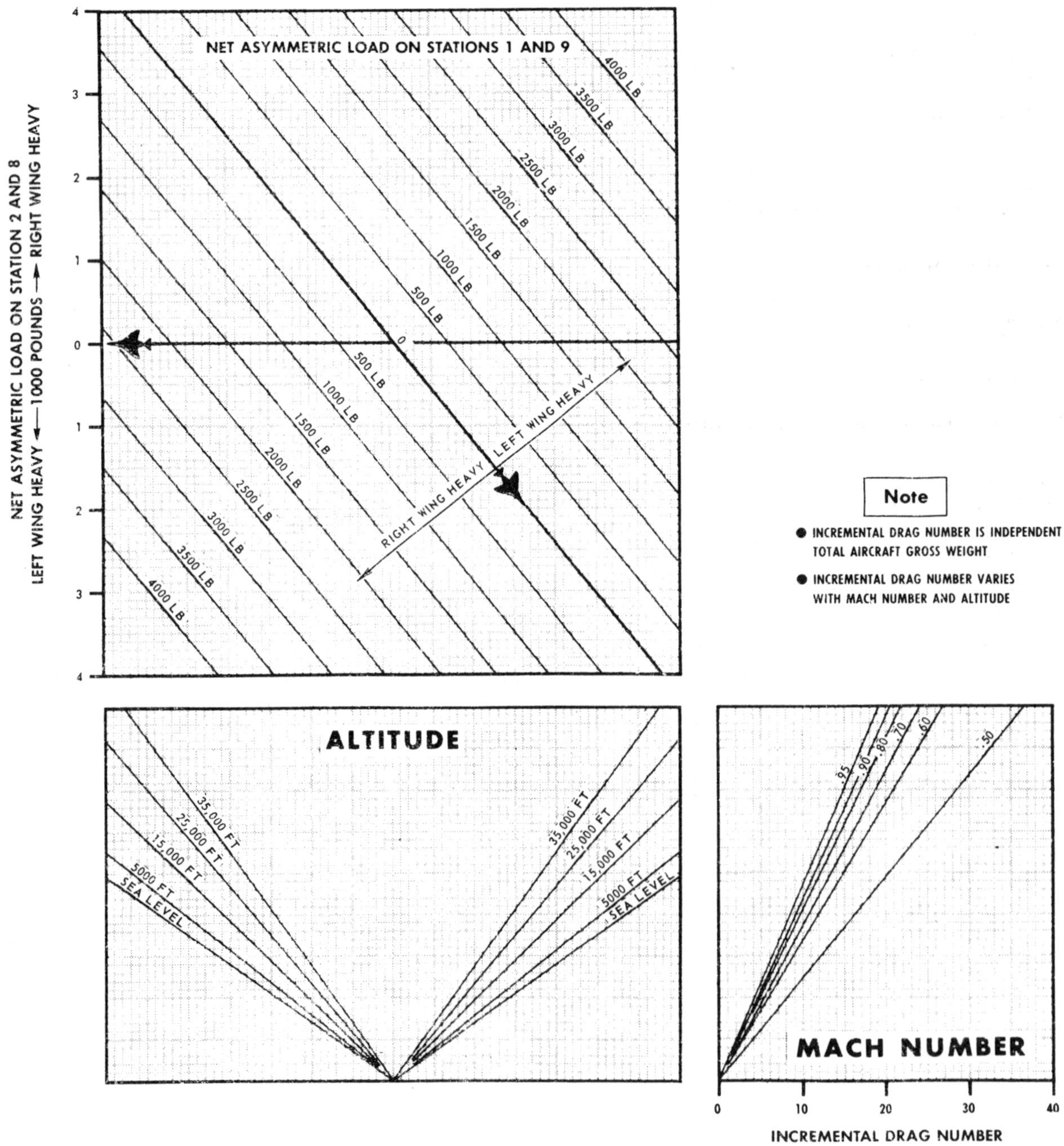

Note

- INCREMENTAL DRAG NUMBER IS INDEPENDENT OF TOTAL AIRCRAFT GROSS WEIGHT
- INCREMENTAL DRAG NUMBER VARIES WITH MACH NUMBER AND ALTITUDE

Figure 11-3

STANDARD ATMOSPHERE TABLE

Standard Sea Level Air:
T = 15°C.
P = 29.921 in. of Hg.

W = .07651 lb/cu. ft. ρ_0 = .002378 slugs/cu. ft.
1" of Hg. = 70.732 lb/sq. ft. = 0.4912 lb/sq. in.
a = 1116 ft./sec.

This table is based on NACA Technical Report No. 218

ALTITUDE FEET	DENSITY RATIO ρ/ρ_0	$\frac{1}{\sqrt{\sigma}}$	TEMPERATURE		SPEED OF SOUND RATIO a/a_0	PRESSURE	
			DEG. C	DEG. F		IN. OF Hg	RATIO P/P_0
0	1.0000	1.0000	15.000	59.000	1.000	29.92	1.0000
1000	.9710	1.0148	13.019	55.434	.997	28.86	.9644
2000	.9428	1.0299	11.038	51.868	.993	27.82	.9298
3000	.9151	1.0454	9.056	48.301	.990	26.81	.8962
4000	.8881	1.0611	7.075	44.735	.986	25.84	.8636
5000	.8616	1.0773	5.094	41.169	.983	24.89	.8320
6000	.8358	1.0938	3.113	37.603	.979	23.98	.8013
7000	.8106	1.1107	1.132	34.037	.976	23.09	.7716
8000	.7859	1.1280	-0.850	30.471	.972	22.22	.7427
9000	.7619	1.1456	-2.831	26.904	.968	21.38	.7147
10,000	.7384	1.1637	-4.812	23.338	.965	20.58	.6876
11,000	.7154	1.1822	-6.793	19.772	.962	19.79	.6614
12,000	.6931	1.2012	-8.774	16.206	.958	19.03	.6359
13,000	.6712	1.2206	-10.756	12.640	.954	18.29	.6112
14,000	.6499	1.2404	-12.737	9.074	.950	17.57	.5873
15,000	.6291	1.2608	-14.718	5.507	.947	16.88	.5642
16,000	.6088	1.2816	-16.699	1.941	.943	16.21	.5418
17,000	.5891	1.3029	-18.680	-1.625	.940	15.56	.5202
18,000	.5698	1.3247	-20.662	-5.191	.936	14.94	.4992
19,000	.5509	1.3473	-22.643	-8.757	.932	14.33	.4790
20,000	.5327	1.3701	-24.624	-12.323	.929	13.75	.4594
21,000	.5148	1.3937	-26.605	-15.890	.925	13.18	.4405
22,000	.4974	1.4179	-28.586	-19.456	.922	12.63	.4222
23,000	.4805	1.4426	-30.568	-23.022	.917	12.10	.4045
24,000	.4640	1.4681	-32.549	-26.588	.914	11.59	.3874
25,000	.4480	1.4940	-34.530	-30.154	.910	11.10	.3709
26,000	.4323	1.5209	-36.511	-33.720	.906	10.62	.3550
27,000	.4171	1.5484	-38.493	-37.287	.903	10.16	.3397
28,000	.4023	1.5768	-40.474	-40.853	.899	9.720	.3248
29,000	.3879	1.6056	-42.455	-44.419	.895	9.293	.3106
30,000	.3740	1.6352	-44.436	-47.985	.891	8.880	.2968
31,000	.3603	1.6659	-46.417	-51.551	.887	8.483	.2834
32,000	.3472	1.6971	-48.399	-55.117	.883	8.101	.2707
33,000	.3343	1.7295	-50.379	-58.684	.879	7.732	.2583
34,000	.3218	1.7628	-52.361	-62.250	.875	7.377	.2465
35,000	.3098	1.7966	-54.342	-65.816	.871	7.036	.2352
36,000	.2962	1.8374	-55.000	-67.000	.870	6.708	.2242
37,000	.2824	1.8818	-55.000	-67.000	.870	6.395	.2137
38,000	.2692	1.9273	-55.000	-67.000	.870	6.096	.2037
39,000	.2566	1.9738	-55.000	-67.000	.870	5.812	.1943
40,000	.2447	2.0215	-55.000	-67.000	.870	5.541	.1852
41,000	.2332	2.0707	-55.000	-67.000	.870	5.283	.1765
42,000	.2224	2.1207	-55.000	-67.000	.870	5.036	.1683
43,000	.2120	2.1719	-55.000	-67.000	.870	4.802	.1605
44,000	.2021	2.2244	-55.000	-67.000	.870	4.578	.1530
45,000	.1926	2.2785	-55.000	-67.000	.870	4.364	.1458
46,000	.1837	2.3332	-55.000	-67.000	.870	4.160	.1391
47,000	.1751	2.3893	-55.000	-67.000	.870	3.966	.1325
48,000	.1669	2.4478	-55.000	-67.000	.870	3.781	.1264
49,000	.1591	2.5071	-55.000	-67.000	.870	3.604	.1205
50,000	.1517	2.5675	-55.000	-67.000	.870	3.436	.1149

FDD-1-(95)

Figure 11-4

AIRSPEED CONVERSION

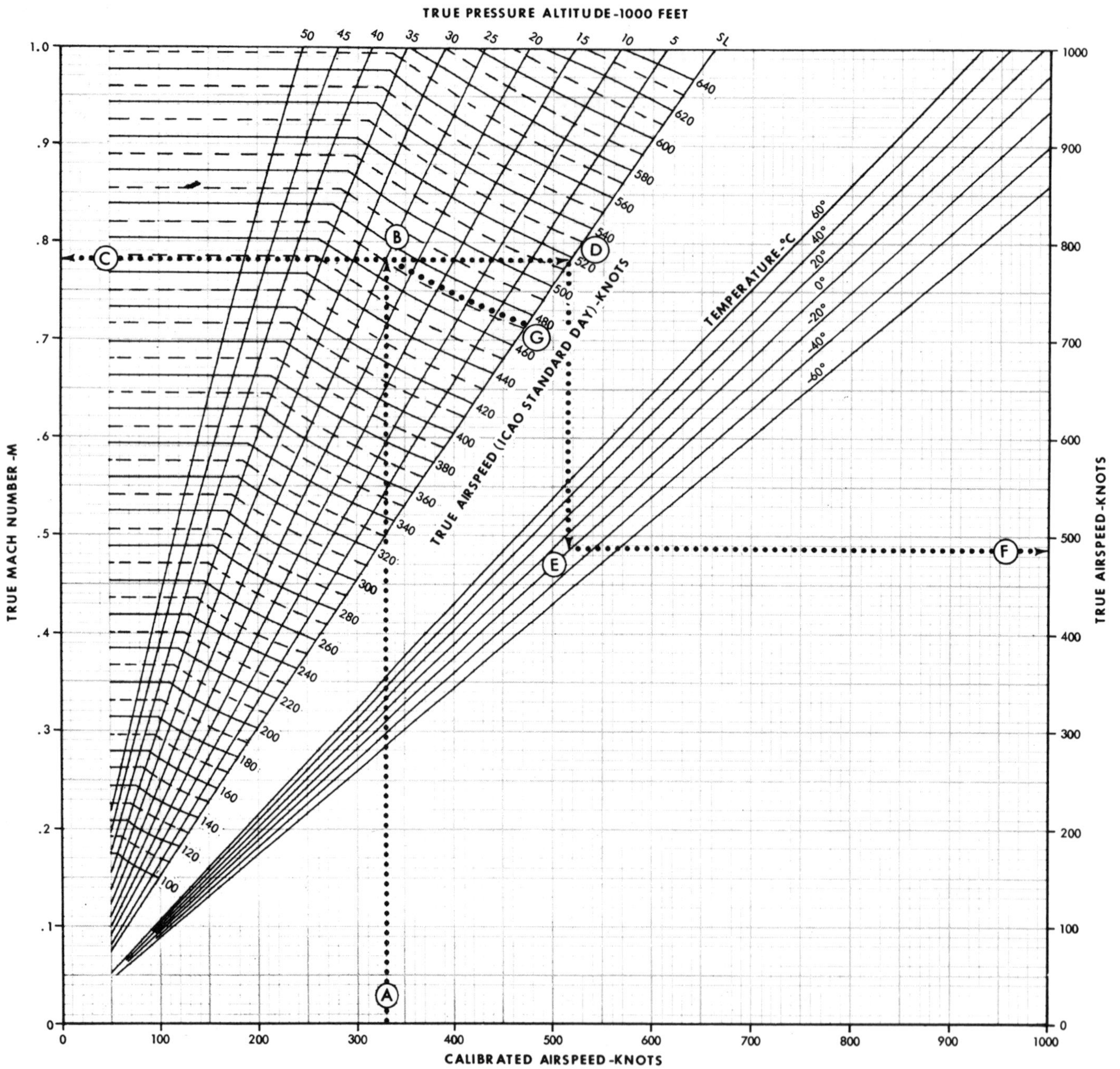

EXAMPLE

A = CAS = 330 KNOTS
B = ALT = 25,000 FT.
C = MACH = .782
D = SEA LEVEL LINE
E = TEMP. = -20°C
F = TAS = 486 KNOTS
G = TAS (STANDARD DAY) = 472 KNOTS

FDD-1-(96-1)

Figure 11-5 (Sheet 1 of 2)

AIRSPEED CONVERSION

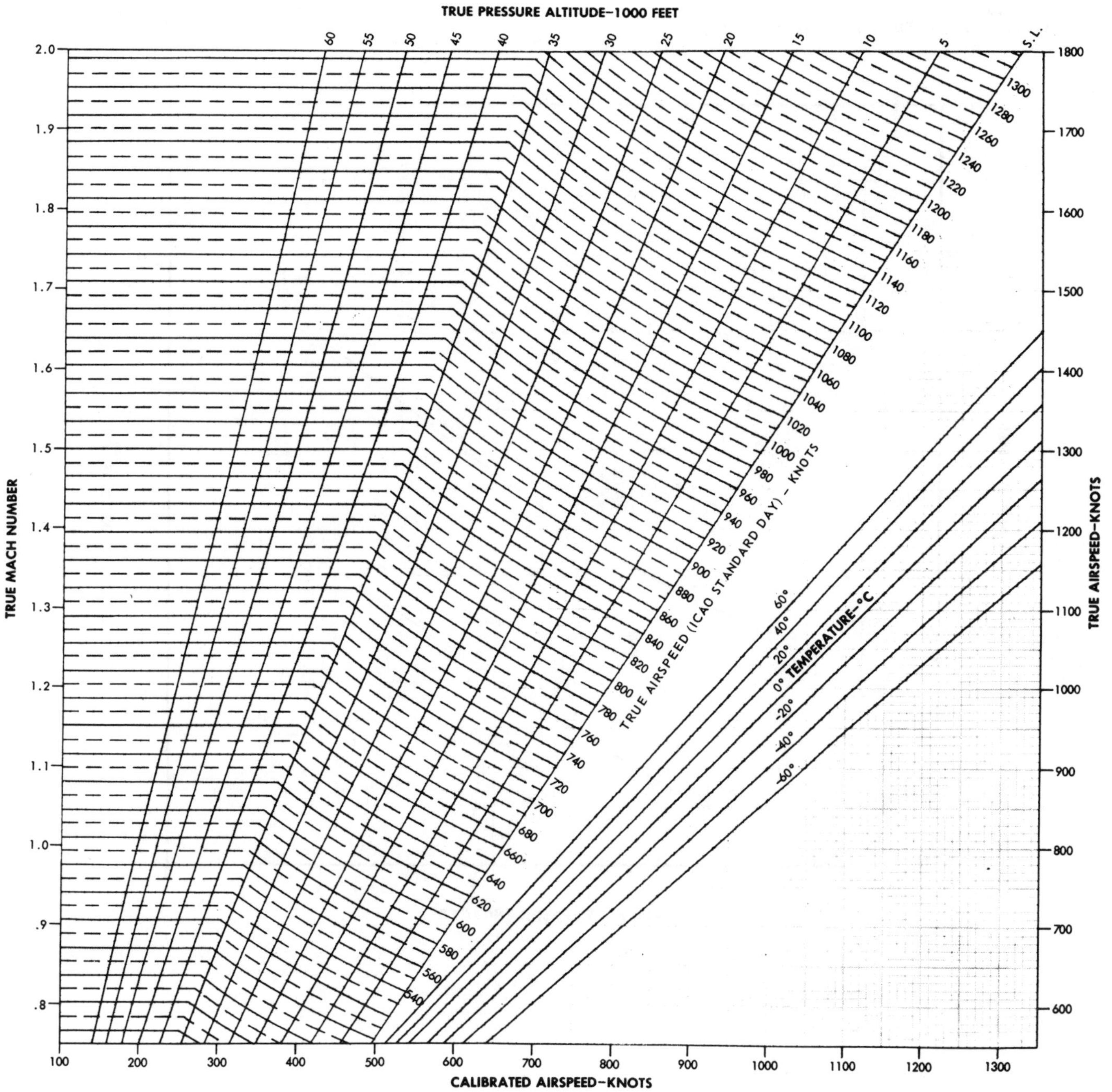

FDD-1-(96-2)

Figure 11-5 (Sheet 2 of 2)

SPC/ALTIMETER TOLERANCE CHECK

5000 TO 20,000 FEET

REMARKS

ENGINE(S): (2) J79-GE-8 OR-10
ICAO STANDARD DAY

GUIDE

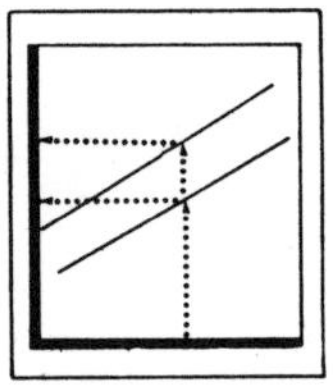

AIRPLANE CONFIGURATION

ALL DRAG INDEXES
FLAPS RETRACTED, GEAR UP

DATE: 15 MAY 1967
DATA BASIS: **ESTIMATED** (BASED ON FLIGHT TEST)

FUEL GRADE: JP-5
FUEL DENSITY: 6.8LB/GAL

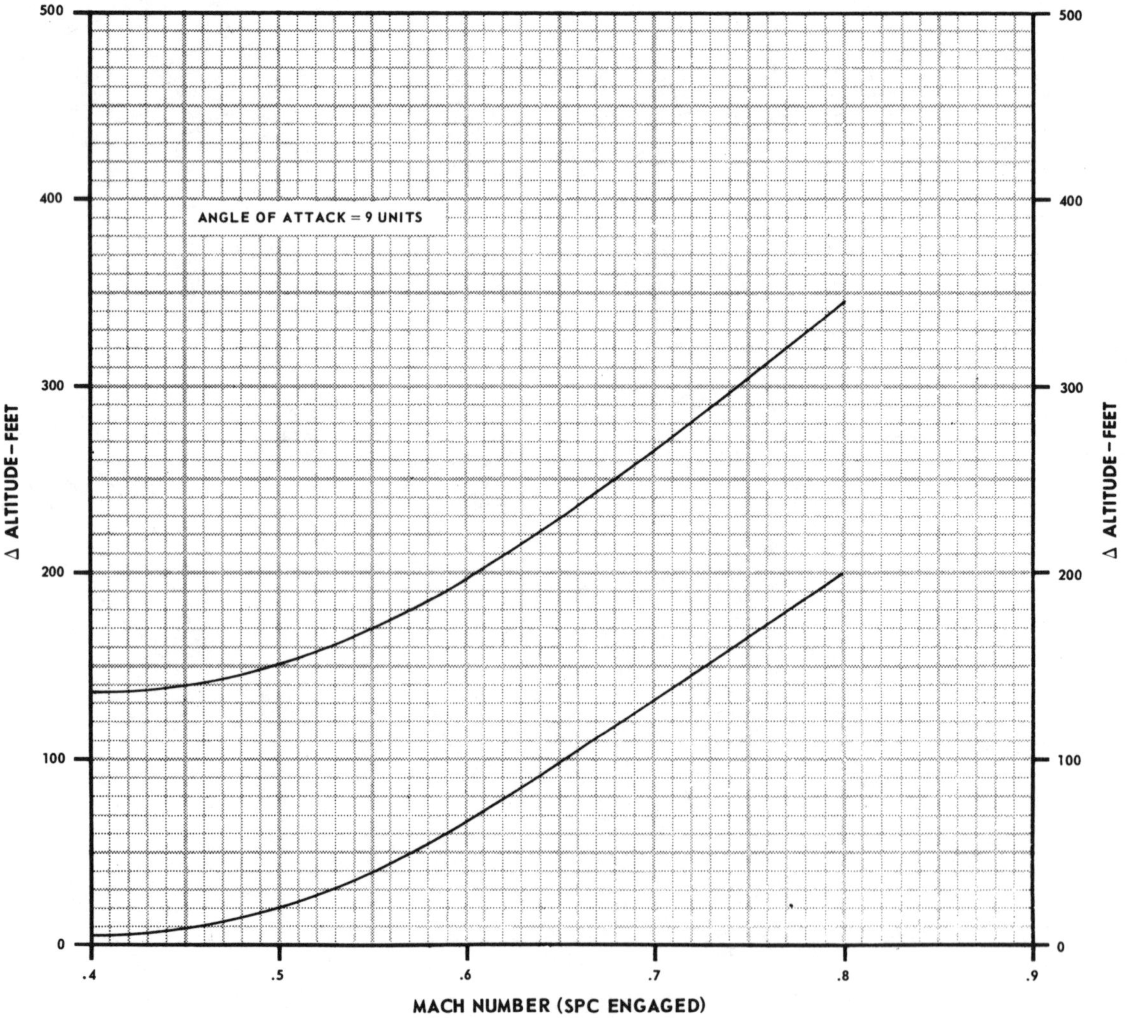

FDD-1-(97)

Figure 11-6

AIRSPEED POSITION ERROR CORRECTION

STATIC CORRECTION OFF

AIRPLANE CONFIGURATION
ALL DRAG INDEXES
FLAPS AND GEAR AS NOTED

REMARKS
ENGINE(S): (2) J79-GE-8 OR-10
ICAO STANDARD DAY

GUIDE

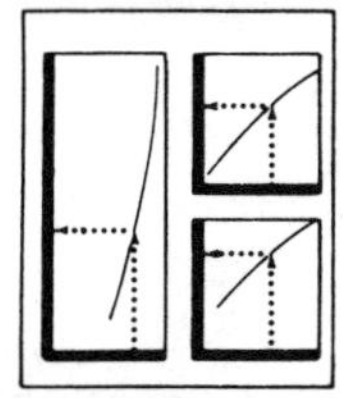

DATE: 15 MAY 1967
DATA BASIS: FLIGHT TEST

FUEL GRADE: JP-5
FUEL DENSITY: 6.8 LB/GAL

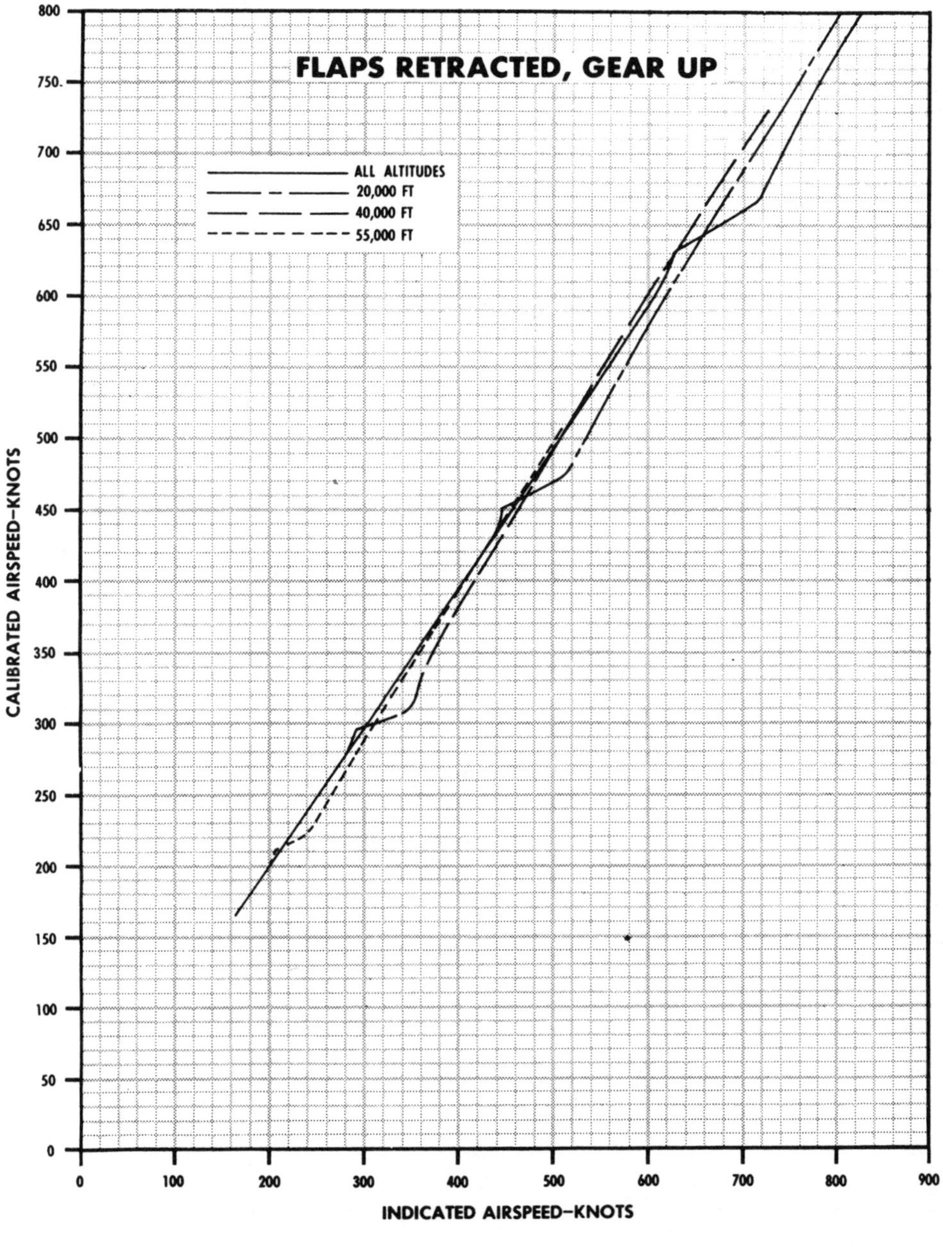

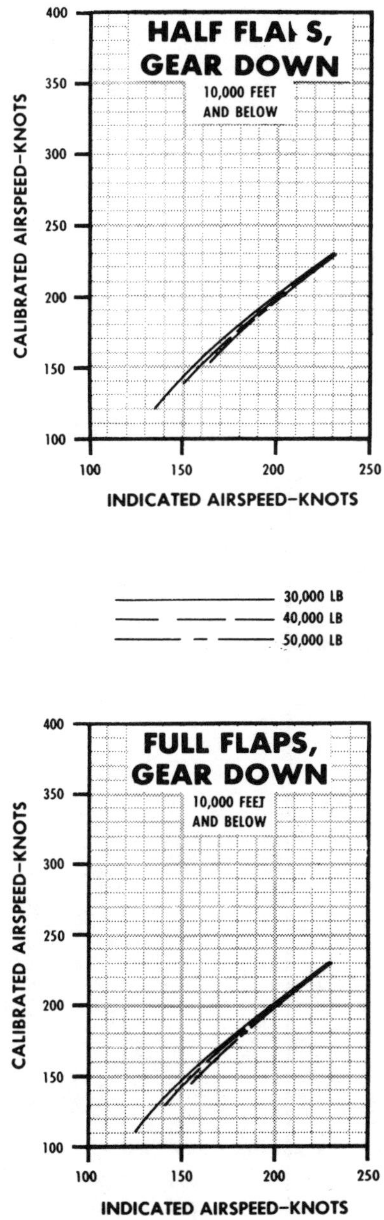

FDD-1-(98)

Figure 11-7

AIRSPEED POSITION ERROR CORRECTION

STATIC CORRECTION OFF

AIRPLANE CONFIGURATION
ALL DRAG INDEXES
FLAPS RETRACTED, GEAR UP

REMARKS
ENGINE(S): (2) J79-GE-8 OR-10
ICAO STANDARD DAY

GUIDE

DATE: 15 MAY 1967
DATA BASIS: FLIGHT TEST

FUEL GRADE: JP-5
FUEL DENSITY: 6.8 LB/GAL

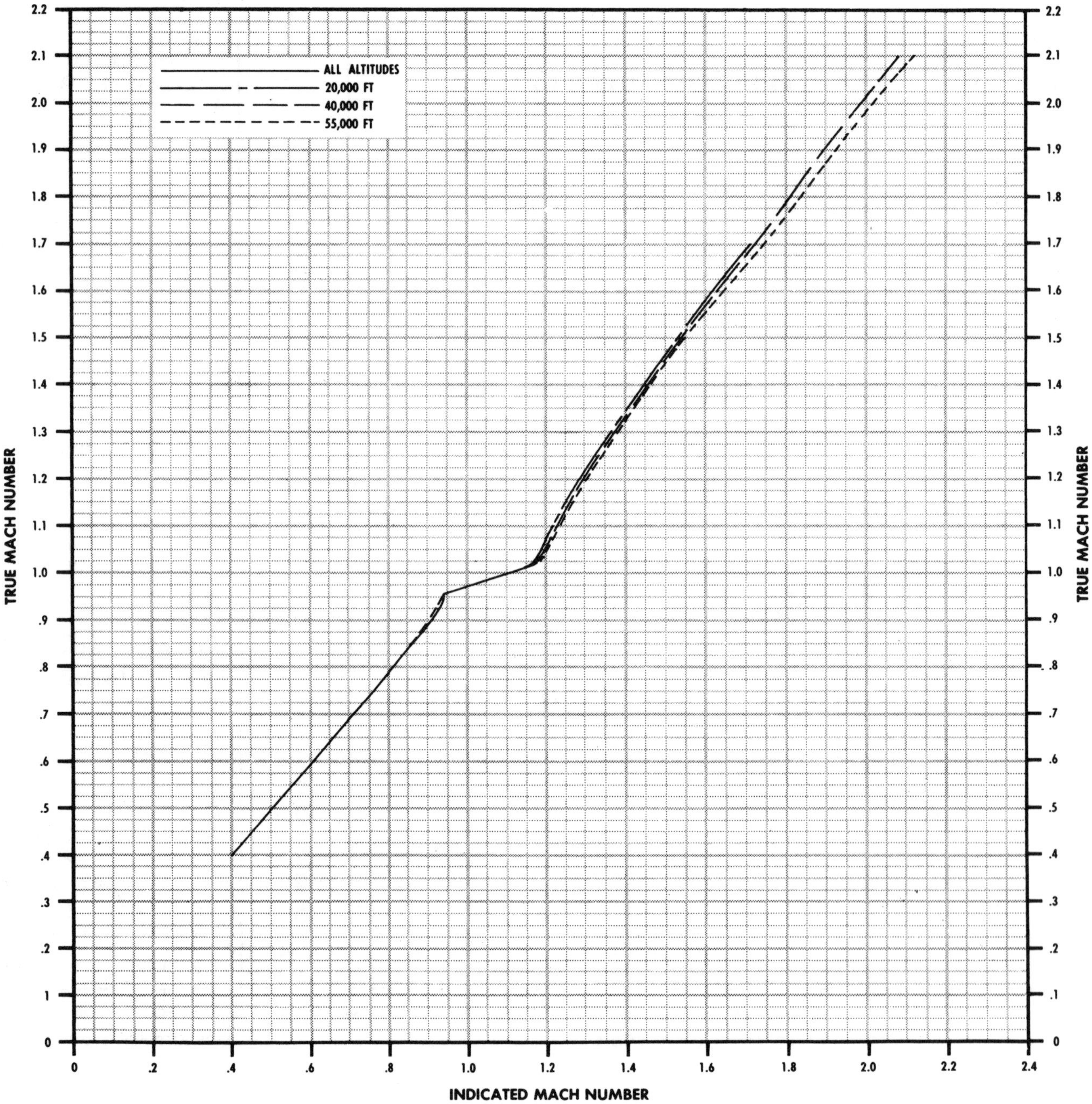

FDD-1-(99)

Figure 11-8

ALTIMETER POSITION ERROR CORRECTION

STATIC CORRECTION OFF

AIRPLANE CONFIGURATION
ALL DRAG INDEXES
FLAPS AND GEAR AS NOTED

REMARKS
ENGINE(S): (2) J79-GE-8 OR-10
ICAO STANDARD DAY

GUIDE

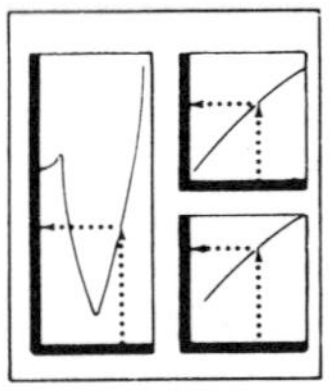

NOTE: ASSIGNED ALTITUDE +ΔH = INDICATED ALTITUDE. FLY INDICATED ALTITUDE.

DATE: 15 MAY 1967
DATA BASIS: FLIGHT TEST

FUEL GRADE: JP-5
FUEL DENSITY: 6.8 LB/GAL

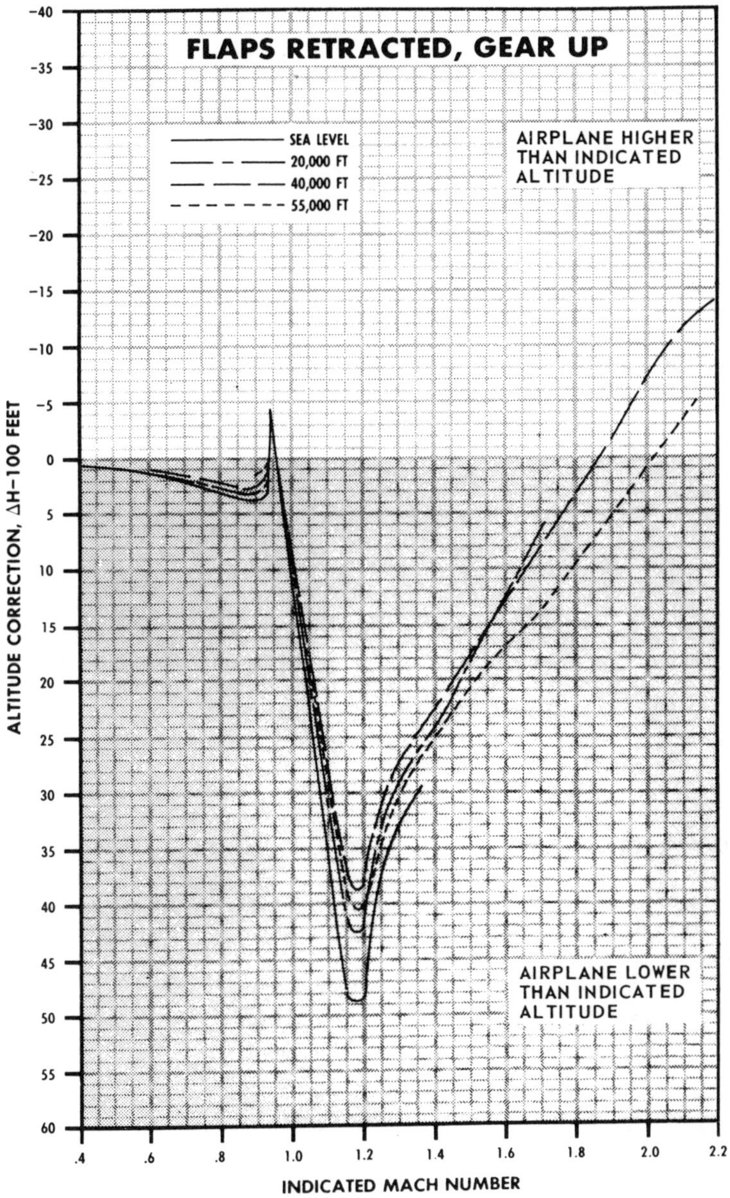

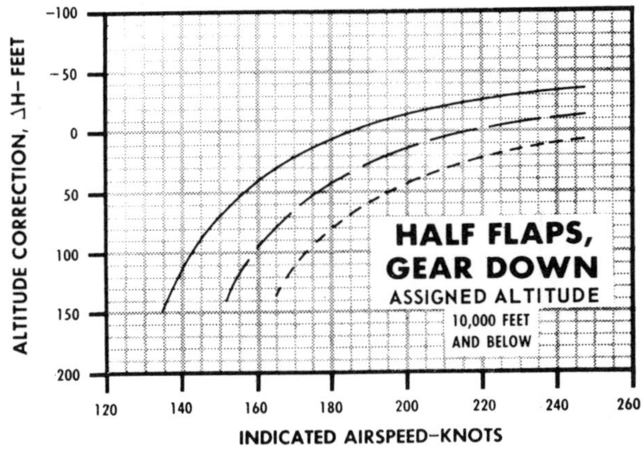

30,000 LB
40,000 LB
50,000 LB

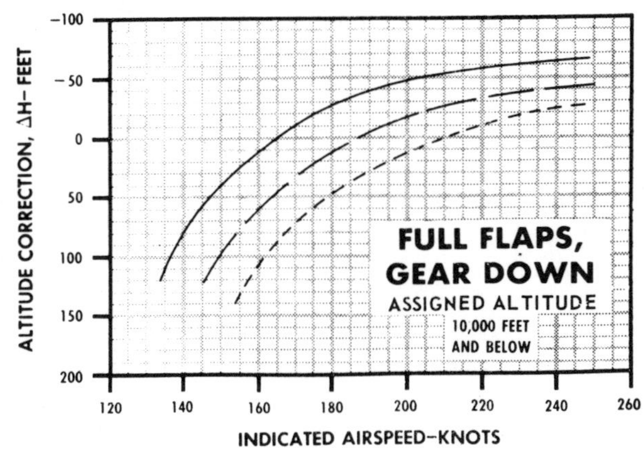

FDD-1-(100)

Figure 11-9

ALTIMETER LAG

AIRPLANE CONFIGURATION
ALL DRAG INDEXES

REMARKS
ENGINE(S): (2) J79-GE-10

GUIDE

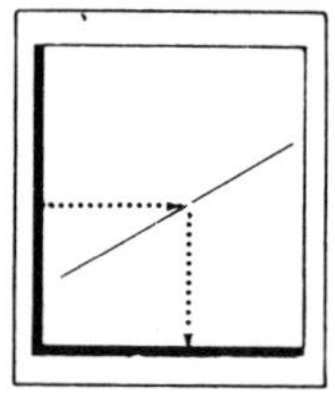

WITH SPC INOPERATIVE

DATE: 1 AUGUST 1969
DATA BASIS: ESTIMATED

FUEL GRADE: JP-5
FUEL DENSITY: 6.8 LB/GAL

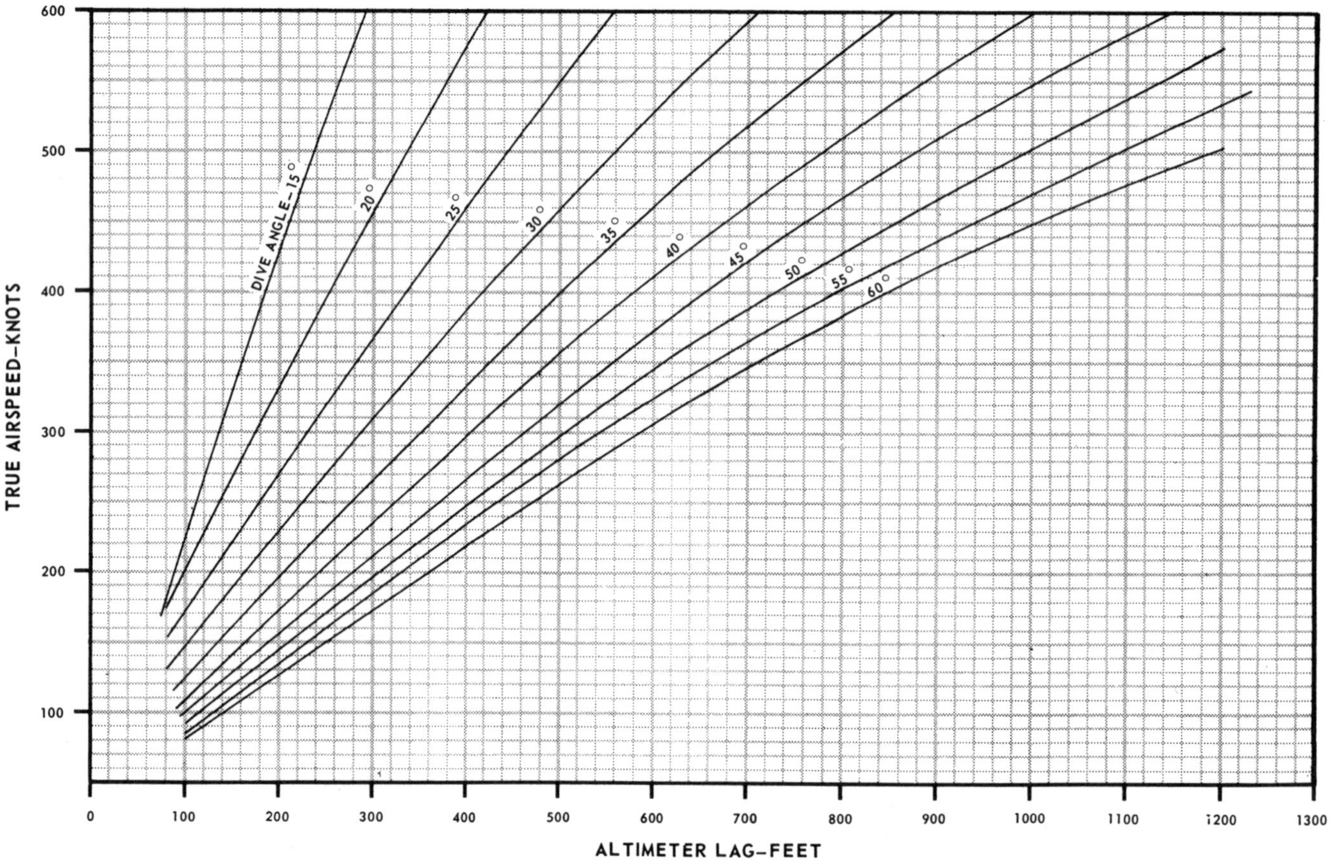

FDD-1-(257)

Figure 11-10

ALTIMETER LAG

AIRPLANE CONFIGURATION
ALL DRAG INDEXES

REMARKS
ENGINE(S): (2) J79-GE-10

WITH SPC OPERATIVE

GUIDE

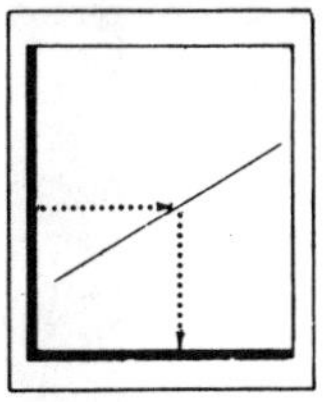

DATE: 15 DECEMBER 1968
DATA BASIS: FLIGHT TEST

FUEL GRADE: JP-5
FUEL DENSITY: 6.8 LB/GAL

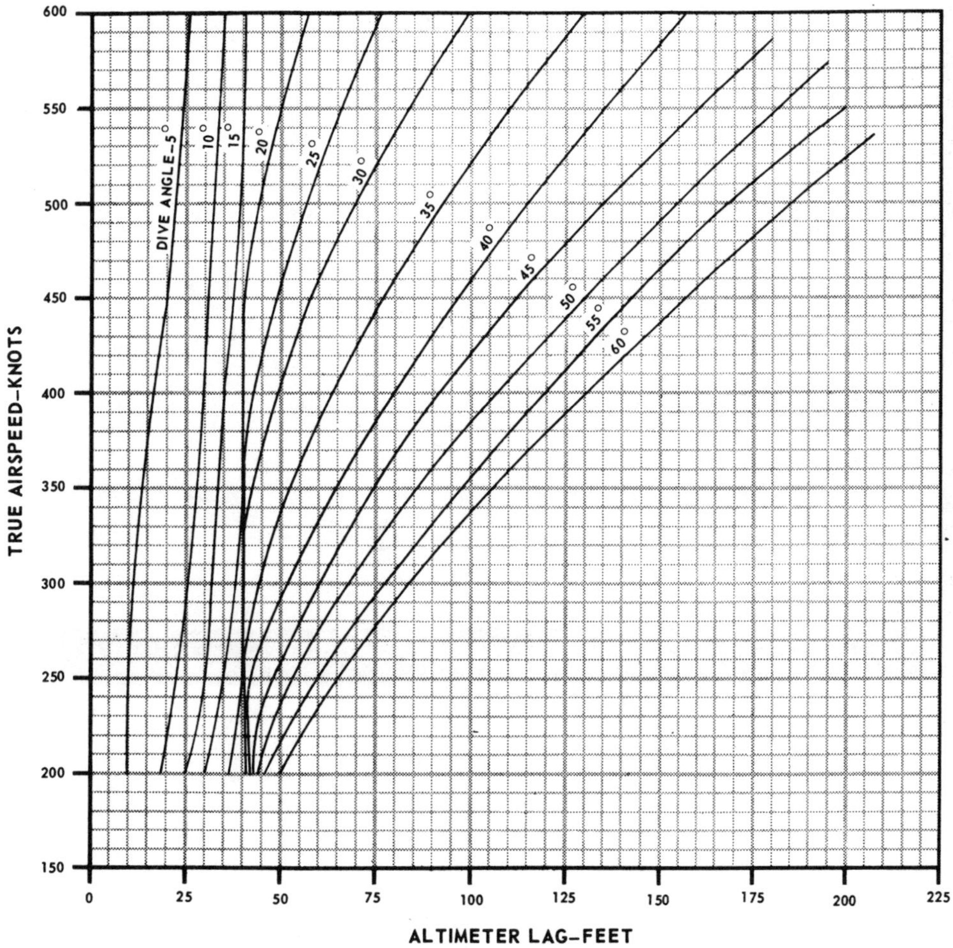

FDD-1-(256)

Figure 11-11

WIND COMPONENTS

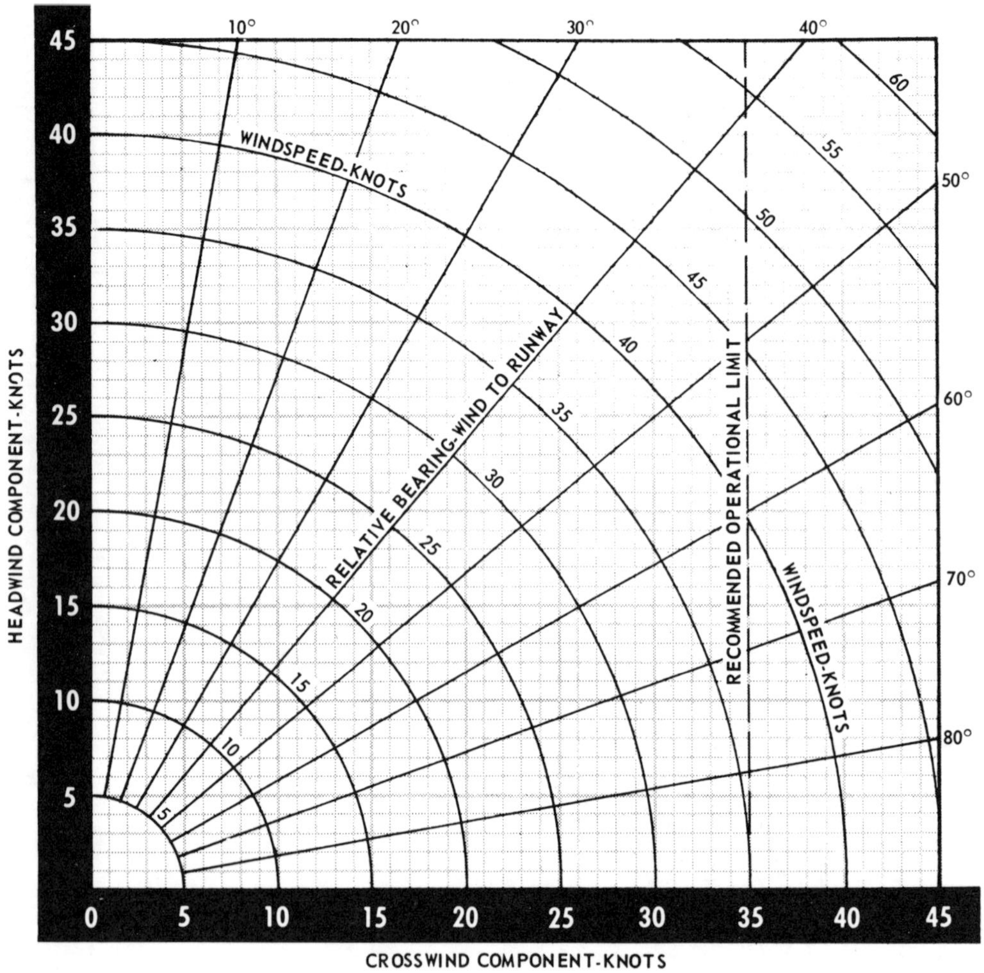

FDD-1-(101)

Figure 11-12

PART 2 TAKEOFF

CHARTS

DENSITY RATIO CHART

This chart (figure 11-13) provides a means of obtaining a single factor (density ratio) that may be used to represent a combination of temperature and pressure altitude. Density ratio must be determined before the takeoff data charts can be utilized.

USE

Enter the chart with the pressure altitude and project horizontally to the right to intersect the appropriate temperature curve. From this intersection, project vertically downward to read density ratio.

Sample Problem

A.	Pressure Altitude	5000 Ft
B.	Temperature	0°C
C.	Density ratio	0.88

MINIMUM GO SPEED CHART

This chart (figure 11-14) provides the means of determining the minimum speed at which the aircraft can experience an engine failure and still take off under existing conditions of temperature, pressure altitude, gross weight, and the runway length remaining. Separate plots are provided for maximum and military thrust conditions. The data is based on an engine failure occurring at the minimum go speed and allows for a 3-second decision period with one engine operating at its initial thrust setting. In the case of a military thrust takeoff, an additional 3-second period is allowed for advancing the operating engine throttle to maximum thrust.

WARNING

Under heavy gross weight/high temperature and/or low RCR factors, it is possible to have a minimum go speed that is higher than the maximum abort speed. Under these conditions, if an engine is lost above the maximum abort speed but below the minimum go speed, the pilot can neither abort nor take off safely on the runway length remaining without considering such factors as reducing gross weight or engaging the overrun end arrestment cable.

USE

Enter the applicable plot with the prevailing density ratio, and project horizontally to the available runway length grid line. Parallel the nearest guide line up or down to intersect the base line. From this point descend vertically to intersect the applicable takeoff gross weight curve, then horizontally to read minimum go speed. If this projected line does not intersect the computed takeoff gross weight curve, then there will be no corresponding minimum go speed. If the gross weight curve lies to the right of the projected line, a single-engine takeoff cannot be made under the combined conditions.

SAMPLE MINIMUM GO SPEED

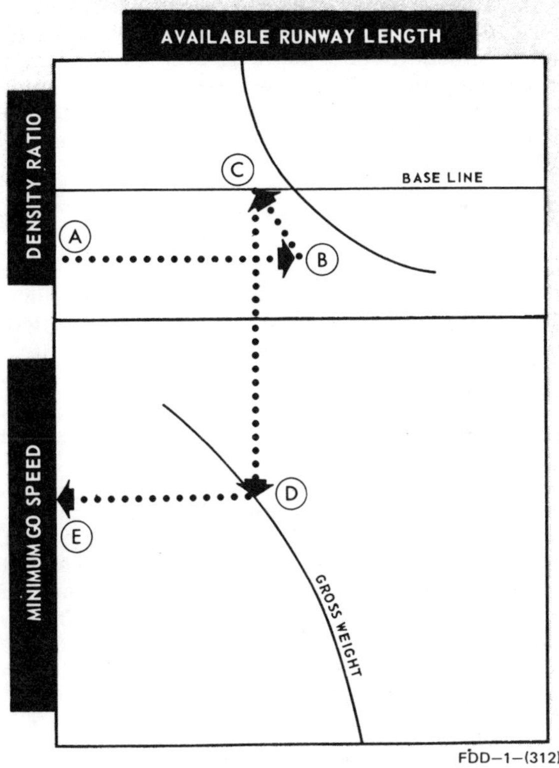

FDD-1-(312)

Sample Problem

Note

This problem assumes maximum thrust on operating engine within 6 seconds after engine failure.

Military Thrust Takeoff

A.	Density ratio	0.95
B.	Available runway length	11,000 Ft
C.	Parallel guide line to base line	
D.	Takeoff gross weight	52,000 Lb
E.	Minimum go speed	156 KCAS

MAXIMUM ABORT SPEED CHARTS

Note

> The maximum abort speed charts do not include the capability of any arrestment gear which may be installed, and take into account only aircraft stopping performance for the given field conditions. However, the capability of the arrestment gear should be considered when computing maximum abort speed. Takeoff may be aborted at the maximum engagement speed for the arrestment gear installed or the maximum abort speed computed from the charts, whichever is higher.

These charts (figures 11-15 and 11-16) provide a means of determining the maximum speed at which an abort may be started and the aircraft stopped within the remaining runway length. Separate charts are provided for maximum and military thrust, and each chart has separate plots to relate drag chute effects. Allowances included in this data are based on 3-second decision period (with both engines operating at the initial thrust setting) and a 5-second period to accomplish abort procedures (throttles to IDLE, wheel brakes applied, and drag chute deployed (if used)).

RCR FACTORS

Runway Condition Reading (RCR) factors are synonymous with runway condition and climatic conditions. If RCR factors are not available (i.e., not provided at local base of operation), use RCR factor 23 for a dry runway, RCR factor 14 for a wet runway, and RCR factor 5 for an icy runway.

USE

Enter applicable plot with the prevailing density ratio, and project horizontally to intersect the available runway length curve. From this point descend vertically to the RCR base line, and parallel nearest guide line down to the forecast RCR factor. From this point, descend further to intersect the computed takeoff gross weight, then horizontally to read the corresponding maximum abort speed.

SAMPLE MAXIMUM ABORT SPEED

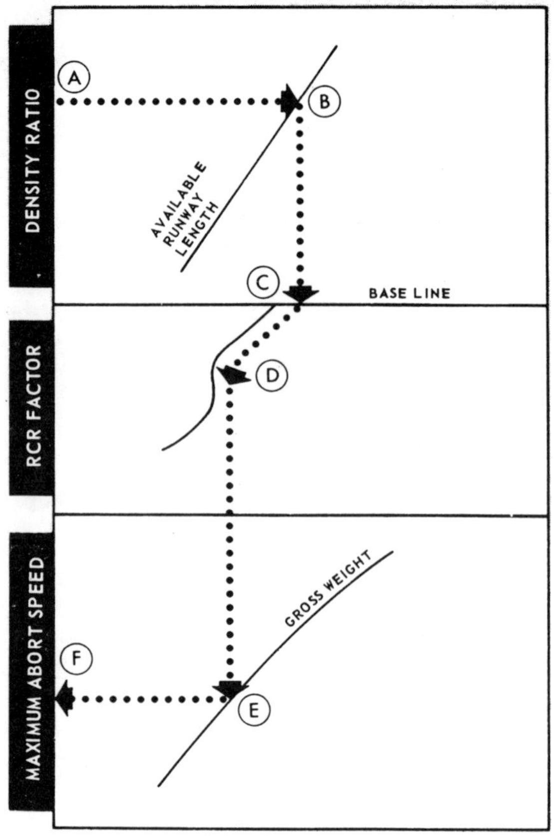

FDD-1-(313)

Sample Problem

Maximum Thrust Takeoff, Without Drag Chute

A.	Density ratio	1.0
B.	Available runway length	8000 Ft
C.	RCR base line	
D.	RCR factor	16
E.	Gross weight	50,000 Lb
F.	Maximum abort speed	69 Kt

TAKEOFF DISTANCE CHARTS

These charts (figures 11-17 and 11-18) are used to determine the no wind ground run distance, wind adjusted ground run and the total distance required to clear a 50-foot obstacle. Separate charts are provided for maximum and military thrust. A table has been provided to show nosewheel lift-off speed with the corresponding aircraft takeoff speed for various gross weight and CG combinations.

USE

Enter the chart with the applicable density ratio, and proceed horizontally to the right and intersect the takeoff weight line. Then descend vertically to read no wind ground run distance. Parallel the appropriate wind guide line (headwind or tailwind) to intersect the takeoff wind velocity. From this point project vertically down to read the ground run adjusted for wind effects. To find the total distance required to clear a 50-foot obstacle, continue downward to the reflector line and project horizontally to the left scale.

Sample Problem

Maximum Thrust

A.	Density Ratio	0.98
B.	Gross weight	46,000 Lb
C.	No wind Ground run distance	2600 Ft
D.	Effective headwind	10 Kt
E.	Ground run (wind corrected)	2400 Ft
F.	Intersect reflector line	
G.	Total distance required to clear 50-foot obstacle	3500 Ft

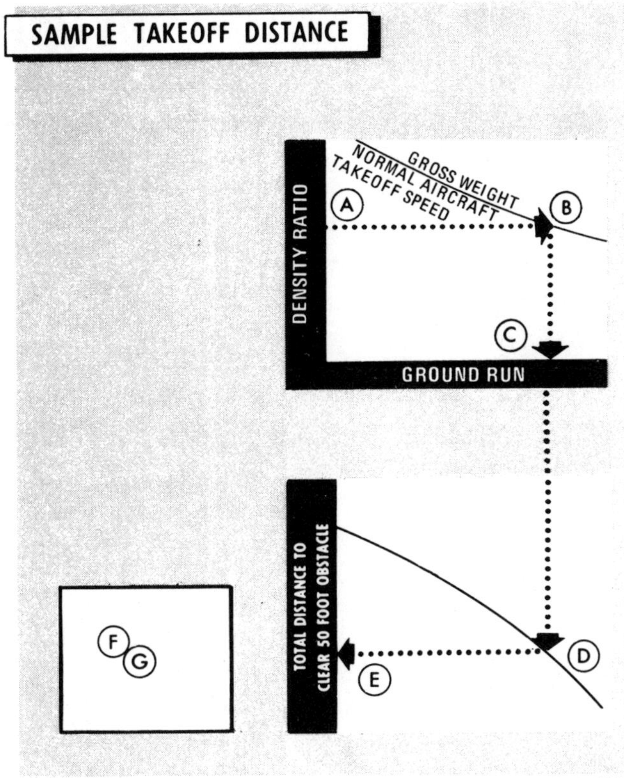

FDD-1-(314)

H.	Nosewheel lift-off speed for CG of 27 MAC (from table)	163 Kt
I.	Takeoff Speed (from table)	176 Kt

DENSITY RATIO

AIRPLANE CONFIGURATION

ALL DRAG INDEXES

GUIDE

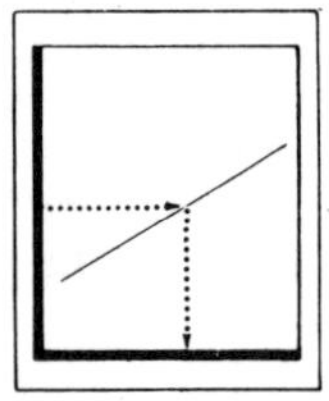

DATE: 1 FEBRUARY 1969

DATA BASIS: ESTIMATED (BASED ON FLIGHT TEST)

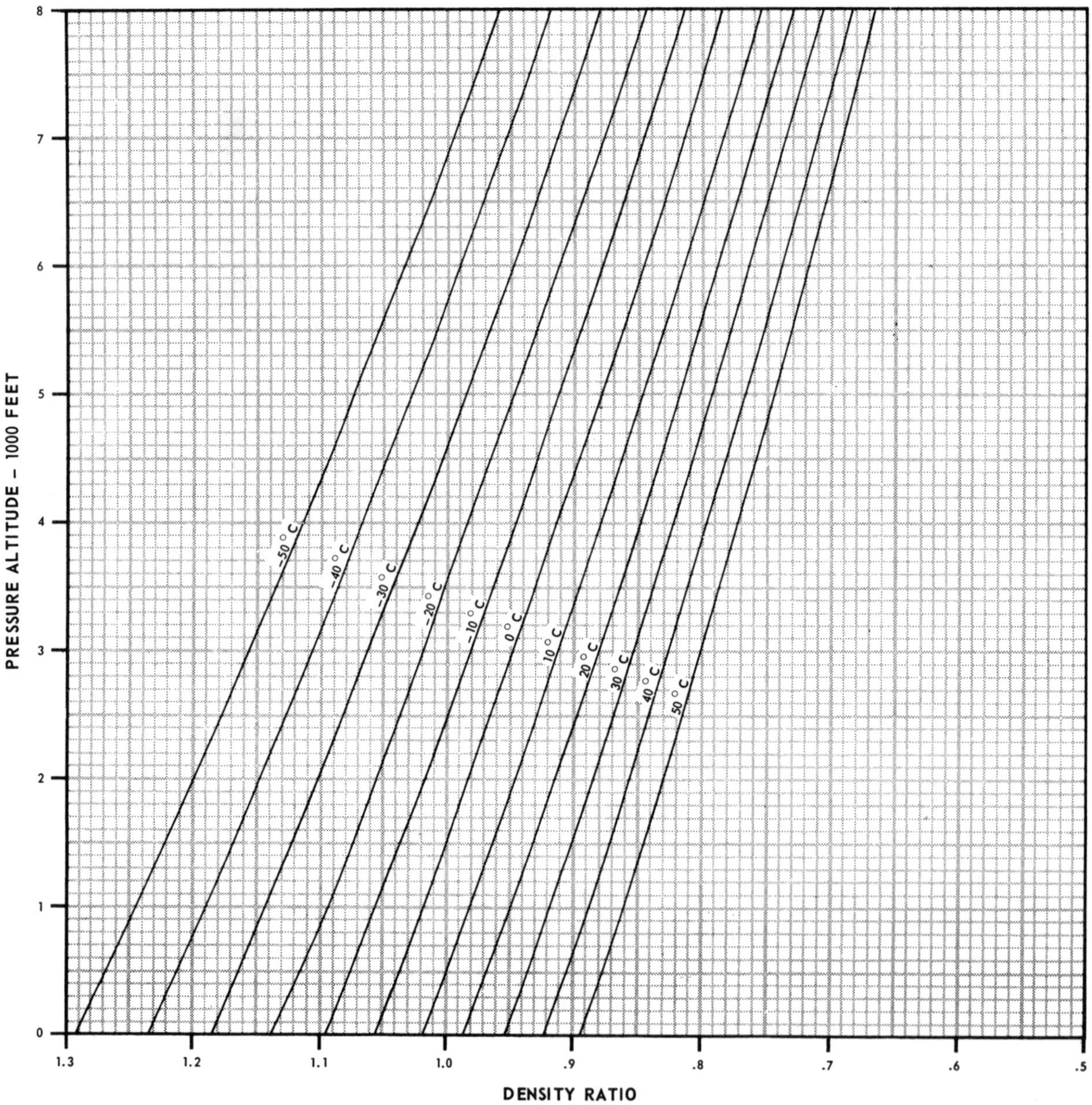

FDD-1-(110)

Figure 11-13

MINIMUM GO SPEED

(WITH SINGLE-ENGINE FAILURE)

AIRPLANE CONFIGURATION
1/2 FLAPS, DOWN
ALL DRAG INDEXES

REMARKS
ENGINE(S): (2) J79-GE-10

NOTES

- SINGLE-ENGINE TAKEOFF, WITH AFTERBURNER IGNITED ON OPERATING ENGINE AFTER FAILURE DURING MILITARY THRUST TAKEOFF.
- SINGLE-ENGINE TAKEOFF/CLIMB-OUT CAPABILITY IS CRITICAL WITH HIGH GROSS WEIGHT AT LOW DENSITY RATIOS.

GUIDE

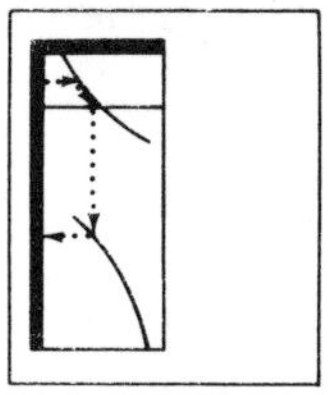

DATE: 1 MAY 1975
DATA BASIS: **FLIGHT TEST**

FUEL GRADE: JP-5
FUEL DENSITY: 6.8 LB/GAL

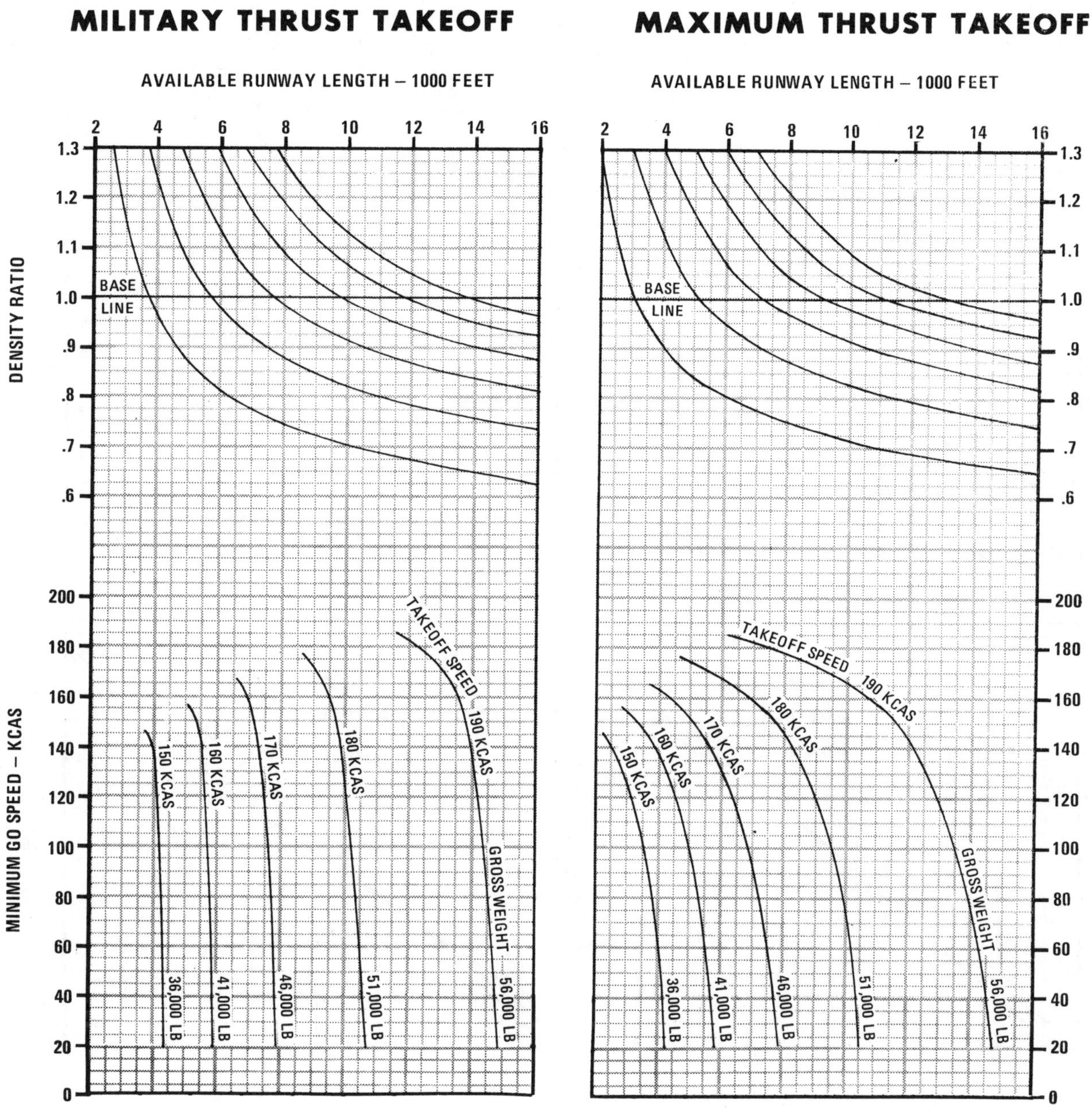

FDD-1-(323)

Figure 11-14

MAXIMUM ABORT SPEED

MAXIMUM THRUST

AIRPLANE CONFIGURATION
ALL DRAG INDEXES
1/2 FLAPS

REMARKS
ENGINE(S): (2) J79-GE-10

DATE: 1 MAY 1975
DATA BASIS: FLIGHT TEST

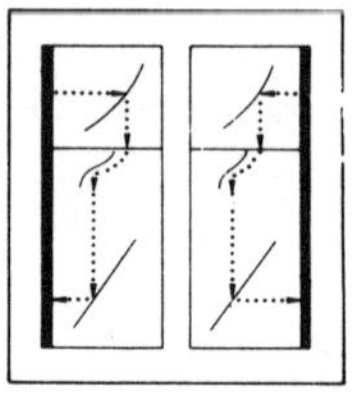

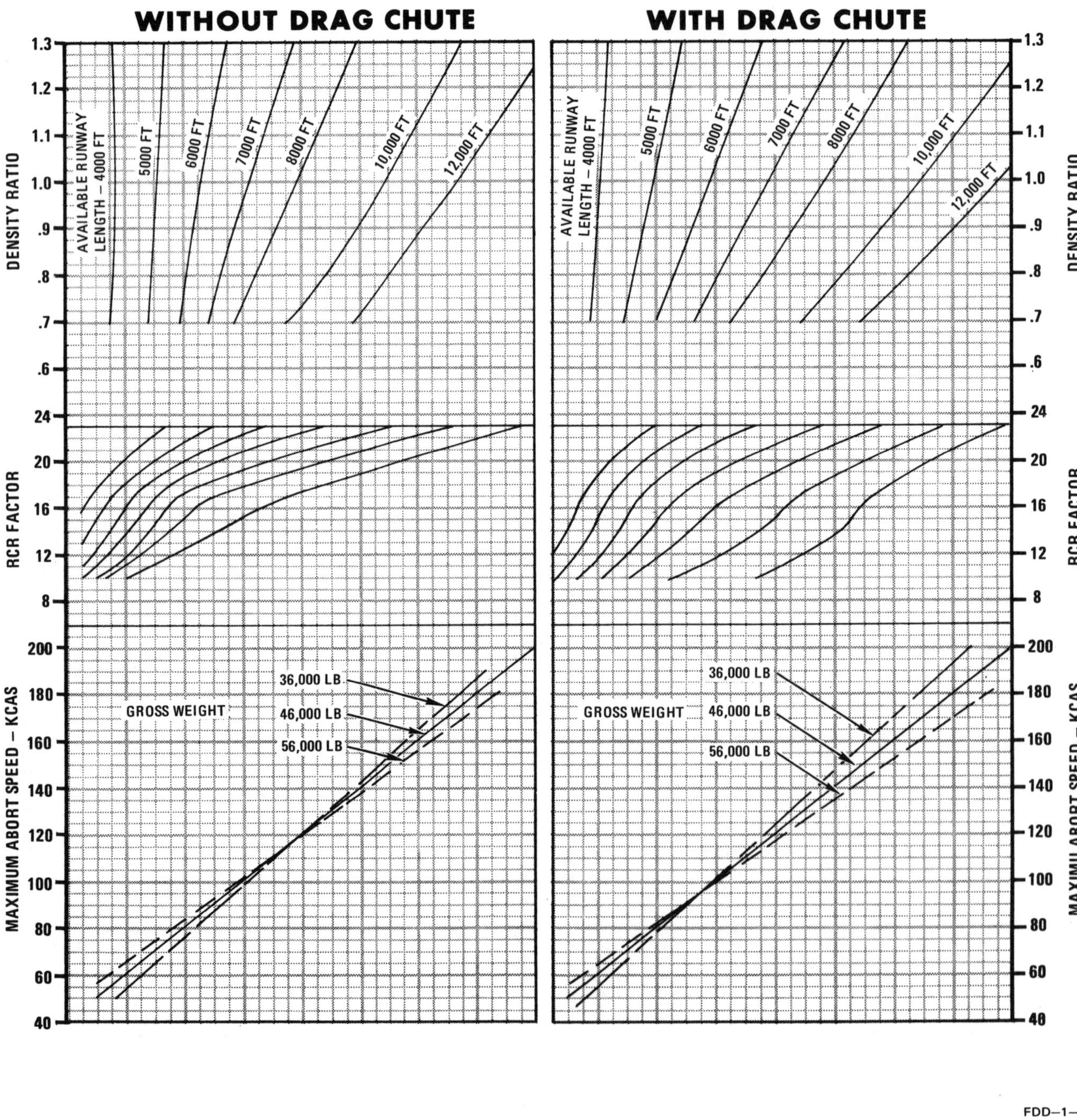

FDD-1-(318)

Figure 11-15

MAXIMUM ABORT SPEED

MILITARY THRUST

AIRPLANE CONFIGURATION
ALL DRAG INDEXES
1/2 FLAPS

REMARKS
ENGINE(S): (2) J79-GE-10

DATE: 1 MAY 1975
DATA BASIS: **FLIGHT TEST**

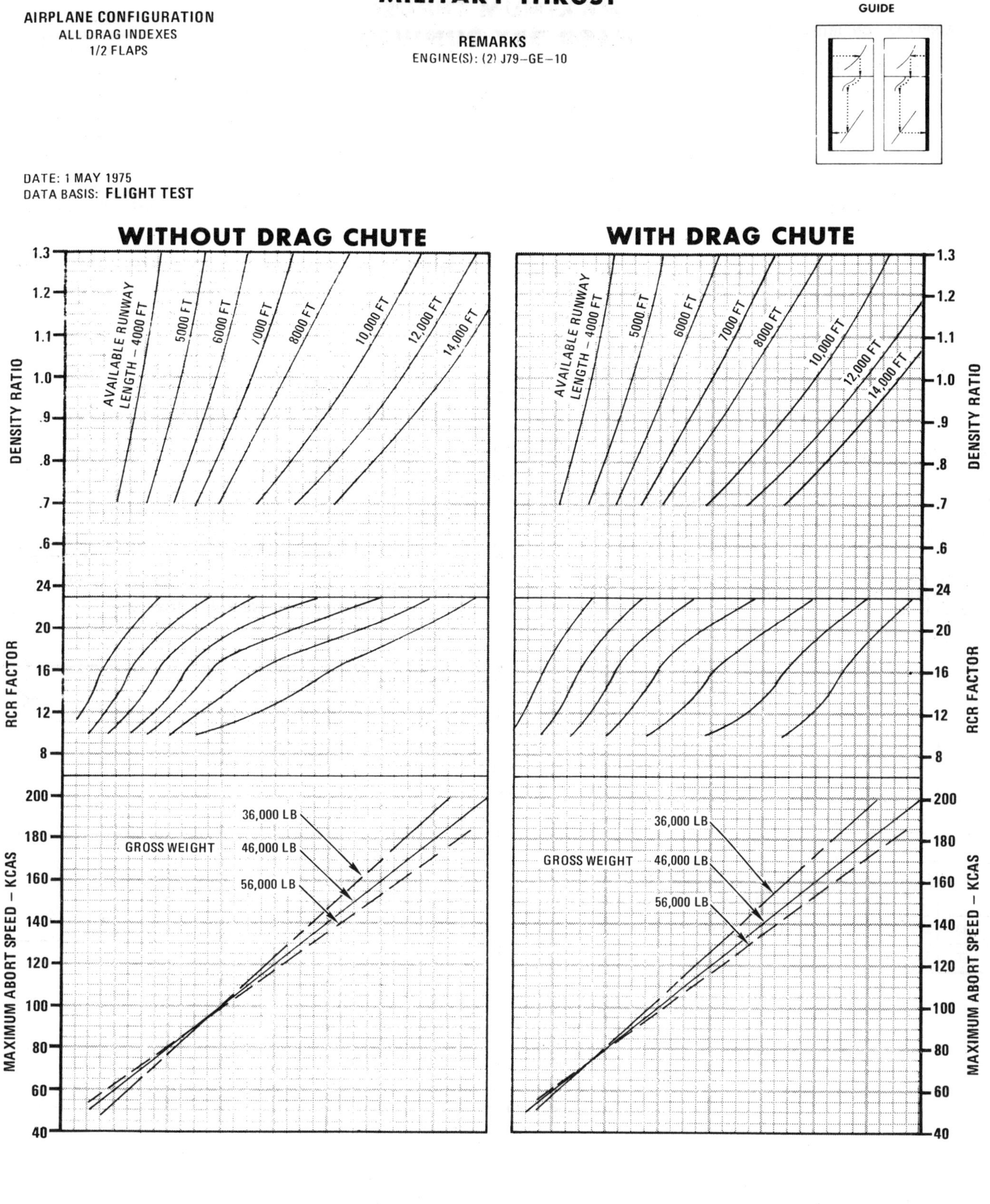

FDD-1-(319)

Figure 11-16

TAKEOFF DISTANCE

MAXIMUM THRUST
HARD DRY RUNWAY

AIRPLANE CONFIGURATION
1/2 FLAPS, GEAR DOWN
ALL DRAG INDEXES

REMARKS
ENGINE(S): (2) J79-GE-10

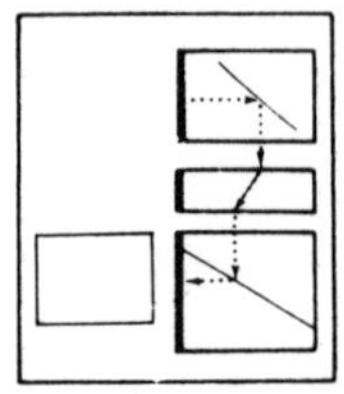

DATE: 1 MAY 1975
DATA BASIS: **FLIGHT TEST**

FUEL GRADE: JP-5
FUEL DENSITY: 6.8 LB/GAL

NOTES

- IF ONE AFTERBURNER FAILS TO LIGHT, TAKEOFF DISTANCE WILL BE INCREASED BY 35%.
- GROSS WEIGHTS SHOWN REFLECT LIFT-OFF VALUES.
- ENGINE START AND TAXI FUEL WEIGHTS ARE FOUND IN PART 3.
- LIFT-OFF CENTER OF GRAVITY SHALL BE CALCULATED USING WEIGHT AND BALANCE HANDBOOK, AN1-1B-40.

CG % MAC	GROSS WEIGHT – 1000 POUNDS									
	36		41		46		51		56	
	NOSEWHEEL LIFT-OFF SPEED / TAKEOFF SPEED									
27	144	158	154	167	163	176	172	184	180	192
29	138	154	147	163	156	171	165	179	173	186
31	131	149	140	157	149	166	157	173	165	183
33	124	145	133	152	141	162	149	172	156	181
35	117	140	126	149	134	160	141	170	148	180

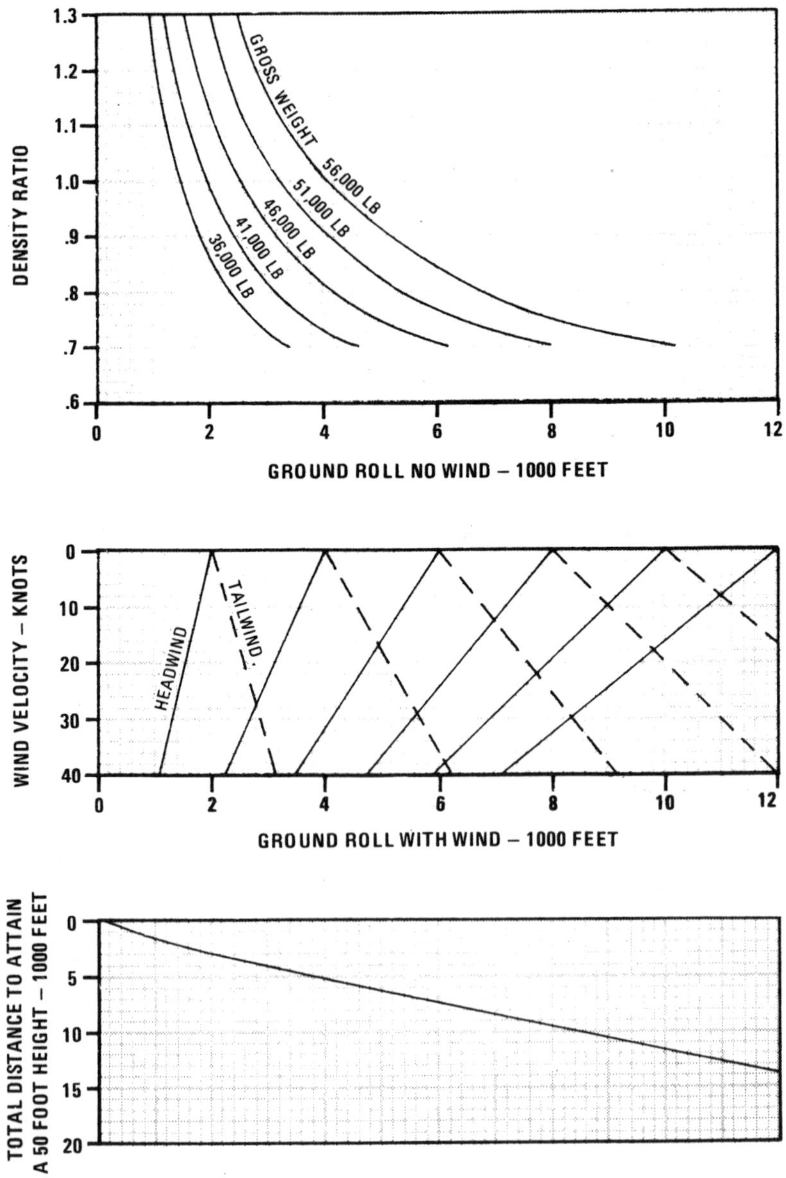

FDD-1-(320)

Figure 11-17

TAKEOFF DISTANCE

MILITARY THRUST
HARD DRY RUNWAY

AIRPLANE CONFIGURATION
1/2 FLAPS, GEAR DOWN
ALL DRAG INDEXES

REMARKS
ENGINE(S): (2) J79-GE-10

GUIDE

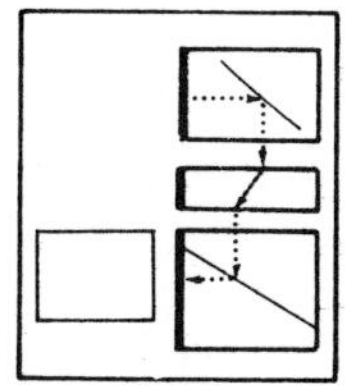

DATE: 1 MAY 1975
DATA BASIS: **FLIGHT TEST**

FUEL GRADE: JP-5
FUEL DENSITY: 6.8 LB/GAL

NOTES

- GROSS WEIGHTS SHOWN REFLECT LIFT-OFF VALUES.
- ENGINE START AND TAXI FUEL WEIGHTS ARE FOUND IN PART 3.
- LIFT-OFF CENTER OF GRAVITY SHALL BE CALCULATED USING WEIGHT AND BALANCE HANDBOOK, AN1-1B-40.

CG % MAC	GROSS WEIGHT – 1000 POUNDS									
	36		41		46		51		56	
	NOSEWHEEL LIFT-OFF SPEED / TAKEOFF SPEED									
27	145	156	155	165	164	173	173	181	181	190
29	139	151	148	159	157	170	166	180	174	189
31	132	146	141	158	150	169	158	179	166	188
33	125	143	134	157	142	167	150	177	157	187
35	118	145	127	156	135	166	142	176	149	186

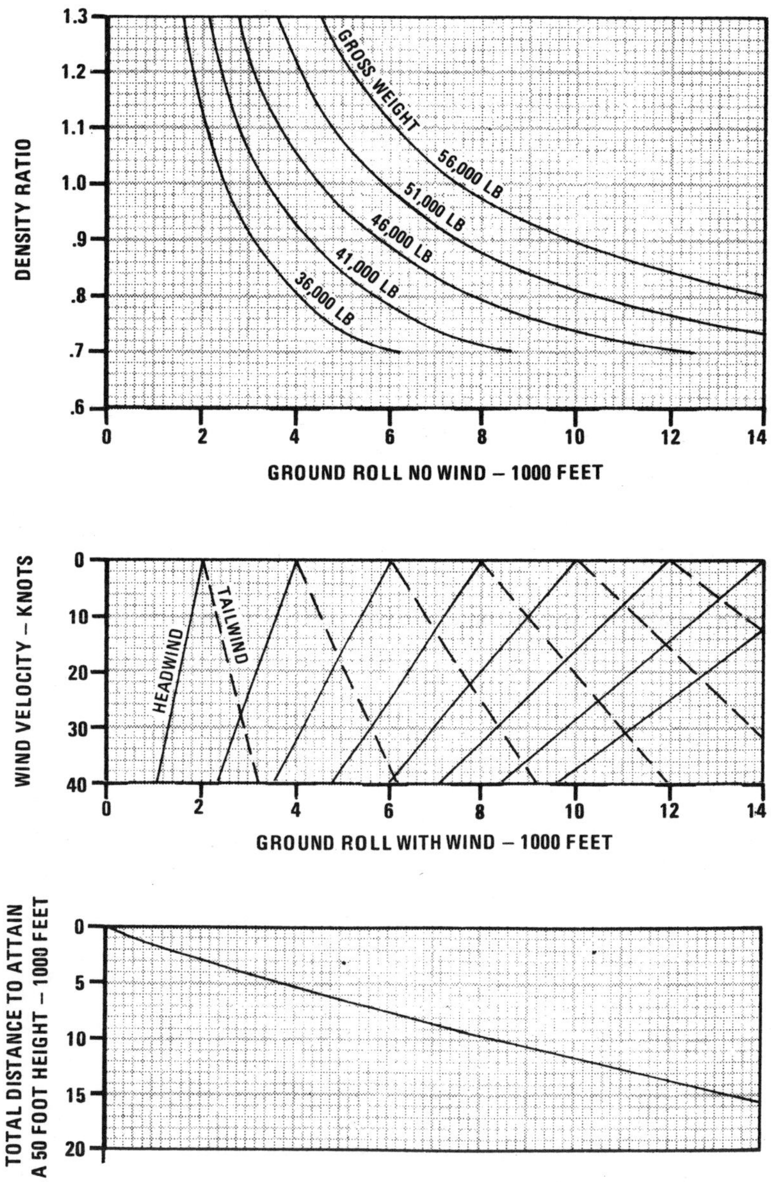

FDD-1-(321)

Figure 11-18

PART 3 CLIMB

CHARTS

CLIMB CHARTS

Two series of charts are presented, one for military and one for maximum thrust climb schedules (figures 11-19 and 11-20). Each series includes charts for determining time, distance covered and fuel used while in the climb, and tables for determining climb indicated airspeed and Mach number. Preclimb requirements are included in a table that presents time, fuel, and distance to intercept the climb schedule after takeoff. Time, fuel, and distance for a simplified military thrust climb are presented in figure 11-21. This data is based on climbing at 350 knots until interception of optimum cruise Mach number, then maintaining cruise Mach to cruise altitude.

USE

Tables

Enter the Climb Speed Schedule tables corresponding to the climb thrust and the computed drag index. Read the column of indicated airspeeds and the Mach numbers to be used during climb. Determine the preclimb fuel, distance, and time to intercept the climb schedule which corresponds to the applicable takeoff and acceleration options.

CHARTS

The method of presenting data on the time, distance, and fuel charts is identical, and the use of all three charts will be undertaken simultaneously here. Enter the charts with the initial climb gross weight. Project horizontally to the right and intersect the assigned cruise altitude, or the optimum cruise altitude for the computed drag index. Project vertically downward to intersect the applicable drag index line, then project horizontally to the left to the temperature base line (corresponds to ICAO Standard Day (°C)). Parallel the applicable guide line (hotter or colder) to intersect a vertical grid line corresponding to the degree of deviation between forecast flight temperature and standard ICAO day temperature. From this point continue horizontally to the left to read the planning data.

Sample Problem

Fuel Required - Military Thrust

A.	Gross weight	50,000 Lb
B.	Cruise altitude	30,000 Ft
C.	Drag index	60
D.	Temperature base line	
E.	Temperature deviation	+5°C
F.	Fuel required	1975 Lb
G.	Time to climb	8 Min
H.	Distance	60 Nautical Miles

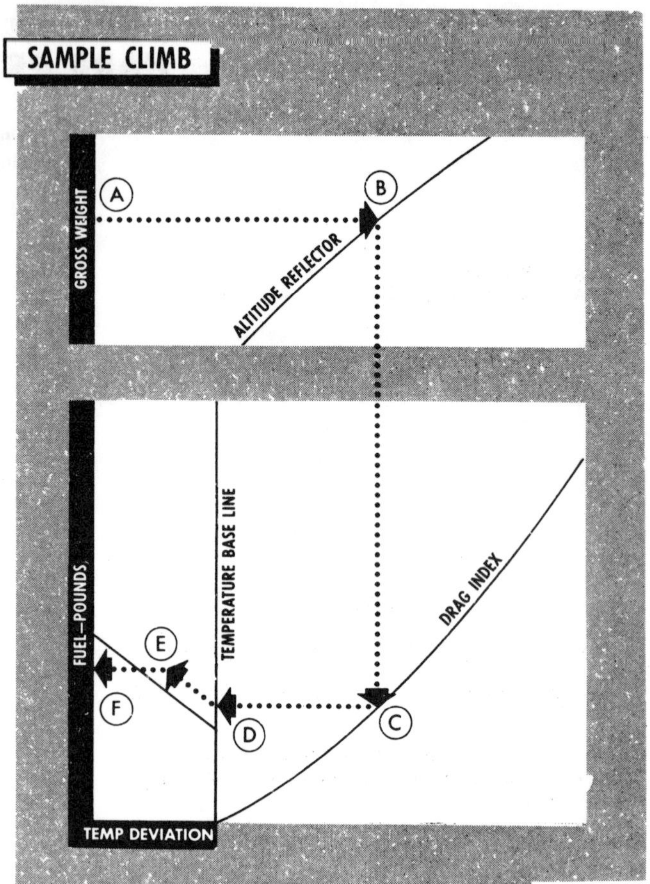

FDD-1-(121)

COMBAT CEILING CHARTS

This chart (figure 11-22) presents the military and maximum thrust combat ceiling for various combinations of gross weight and drag index.

USE

Enter the applicable graph with estimated gross weight at end of climb. Project vertically upward to intersect applicable drag index, then horizontally to the left to the temperature base line (corresponds to ICAO Standard Day (°C)). From this point, parallel the applicable guide line (hotter or colder) to intersect a vertical grid line corresponding to the degree of deviation between altitude at end of climb and standard day temperature. From this point continue horizontally to the left to read combat ceiling.

Sample Problem

Combat ceiling - Maximum Thrust - (2) Engines

A.	Gross weight at end of climb	45,000 Lb
B.	Drag index	40
C.	Temperature base line	
D.	Temperature deviation	+8°C
E.	Combat ceiling	47,200 Ft

CLIMB SPEED SCHEDULE

AIRPLANE CONFIGURATION
INDIVIDUAL DRAG INDEXES

MAXIMUM THRUST

REMARKS
ENGINE(S): (2) J79-GE-10
ICAO STANDARD DAY

DATE: 1 APRIL 1971
DATA BASIS: FLIGHT TEST

FUEL GRADE: JP-5
FUEL DENSITY: 6.8 LB/GAL

	DRAG INDEX													
	0		10		20		30		40		50		60	
ALTITUDE	KCAS	MACH	KCAS	MACH	KCAS	MACH	KCAS	MACH	KCAS	MACH	KCAS	MACH	KCAS	MACH
S. L.	606	.92	602	.91	598	.90	591	.89	583	.88	573	.87	562	.85
5000	564	.92	561	.92	556	.91	550	.90	543	.89	535	.87	525	.86
10000	520	.92	519	.92	517	.92	513	.91	508	.90	499	.89	490	.87
15000	481	.93	479	.93	476	.92	471	.91	466	.90	460	.89	453	.88
20000	442	.94	439	.93	436	.93	432	.92	427	.91	423	.90	418	.89
25000	404	.94	402	.94	398	.93	395	.92	392	.92	388	.91	384	.90
30000	367	.95	364	.94	362	.94	359	.93	357	.93	355	.92	352	.91
35000	332	.95	330	.95	328	.94	326	.94	324	.94	322	.93	321	.93
40000	296	.96	294	.95	292	.94	291	.94	289	.94	288	.93	287	.93
45000	265	.96	263	.95	261	.94	259	.94	258	.94	257	.93	256	.93

	DRAG INDEX															
	70		80		90		100		110		120		130		140	
ALTITUDE	KCAS	MACH	KCAS	MACH	KCAS	MACH	KCAS	MACH	KCAS	MACH	KCAS	MACH	KCAS	MACH	KCAS	MACH
S. L.	547	.83	531	.80	511	.77	492	.74	472	.71	453	.69	439	.66	427	.65
5000	512	.84	498	.82	482	.79	465	.76	448	.74	432	.71	418	.69	408	.67
10000	478	.85	467	.83	453	.81	437	.78	424	.76	408	.73	396	.71	386	.69
15000	444	.86	436	.85	426	.83	416	.81	403	.79	388	.76	376	.74	367	.72
20000	412	.88	404	.86	396	.85	387	.83	379	.81	368	.79	358	.77	349	.75
25000	380	.89	374	.88	368	.87	361	.85	353	.83	346	.82	338	.80	331	.79
30000	349	.91	345	.90	341	.89	337	.88	331	.86	324	.85	318	.83	312	.82
35000	319	.92	318	.92	315	.91	313	.91	310	.90	305	.89	300	.87	293	.86
40000	286	.93	284	.92	283	.92	282	.92	281	.91	279	.91	277	.90	275	.90
45000	255	.93	254	.92	252	.92	251	.92	-	-		-	-	-		-

TAKEOFF ALLOWANCES & ACCELERATION TO CLIMB SPEED

START – 65 LBS / ENG

RUNUP 50 LBS / ENG

TAXI – 21 LB / MIN / ENG

	BRAKE RELEASE TO CLIMB SPEED MAX T. O. MAX ACCEL TO MAX CLIMB SPEED
FUEL – LBS	1225
DIST – N M	5.0
TIME – MIN	1.0

.J–1–(126–1)A

Figure 11-19 (Sheet 1 of 4)

TIME TO CLIMB
MAXIMUM THRUST

AIRPLANE CONFIGURATION
INDIVIDUAL DRAG INDEXES

REMARKS
ENGINE(S): (2) J79-GE-10

GUIDE

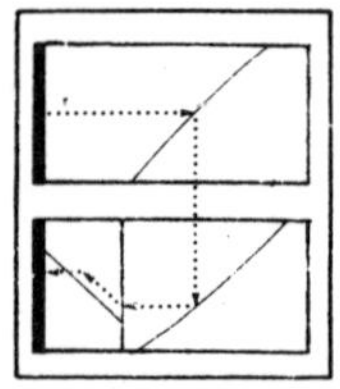

DATE: 1 APRIL 1971
DATA BASIS: FLIGHT TEST

FUEL GRADE: JP-5
FUEL DENSITY: 6.8 LB/GAL

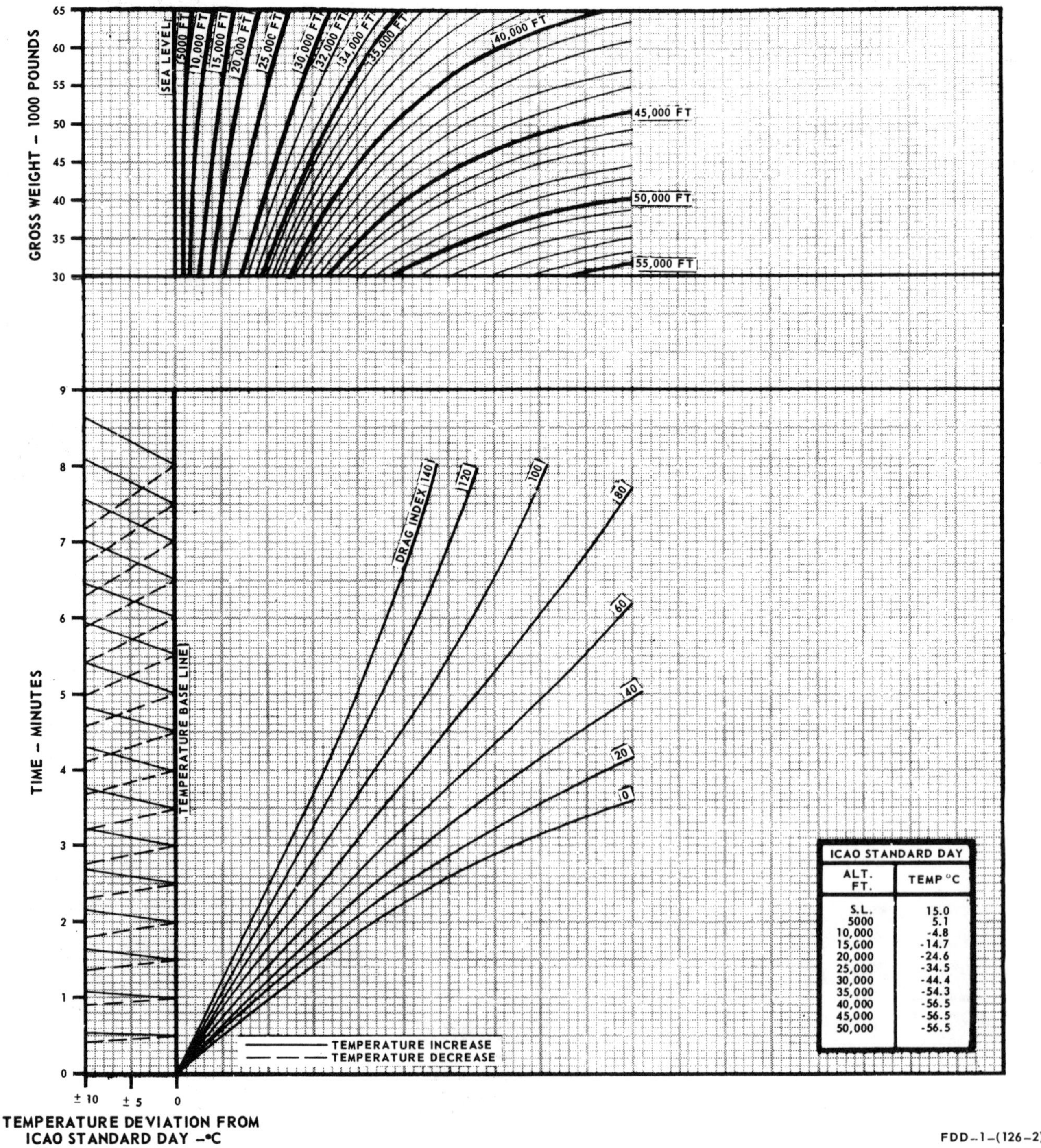

ICAO STANDARD DAY	
ALT. FT.	TEMP °C
S.L.	15.0
5000	5.1
10,000	-4.8
15,000	-14.7
20,000	-24.6
25,000	-34.5
30,000	-44.4
35,000	-54.3
40,000	-56.5
45,000	-56.5
50,000	-56.5

FDD-1-(126-2)A

Figure 11-19 (Sheet 2 of 4)

FUEL REQUIRED TO CLIMB

MAXIMUM THRUST

AIRPLANE CONFIGURATION
INDIVIDUAL DRAG INDEXES

REMARKS
ENGINE(S): (2) J79-GE-10

DATE: 1 APRIL 1971
DATA BASIS: FLIGHT TEST

FUEL GRADE: JP-5
FUEL DENSITY: 6.8 LB GAL

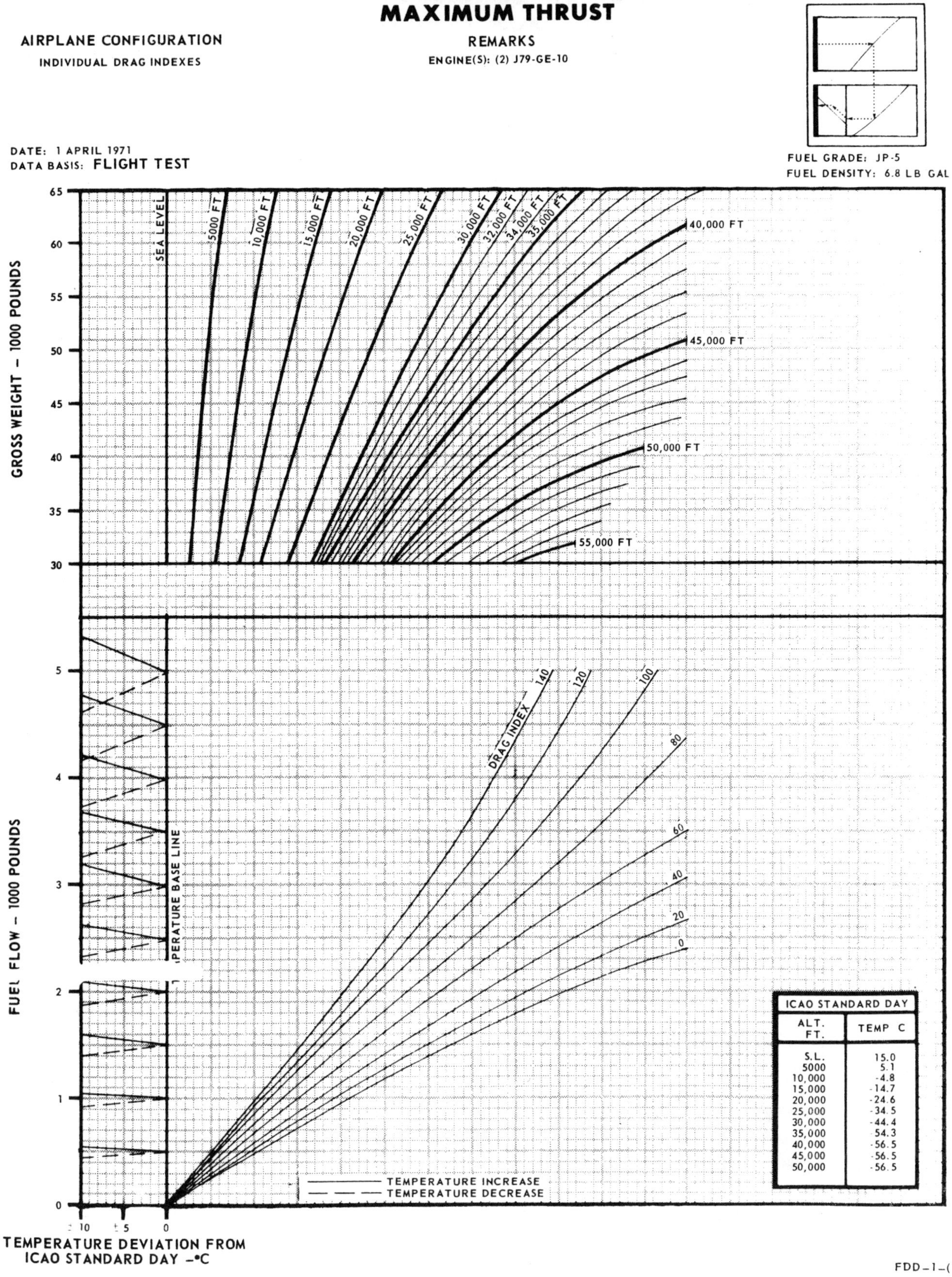

ICAO STANDARD DAY	
ALT. FT.	TEMP C
S.L.	15.0
5000	5.1
10,000	-4.8
15,000	-14.7
20,000	-24.6
25,000	-34.5
30,000	-44.4
35,000	54.3
40,000	-56.5
45,000	-56.5
50,000	-56.5

FDD-1-(126-3)A

Figure 11-19 (Sheet 3 of 4)

DISTANCE REQUIRED TO CLIMB

MAXIMUM THRUST

AIRPLANE CONFIGURATION
INDIVIDUAL DRAG INDEXES

REMARKS
ENGINE(S): (2) J79-GE-10

GUIDE

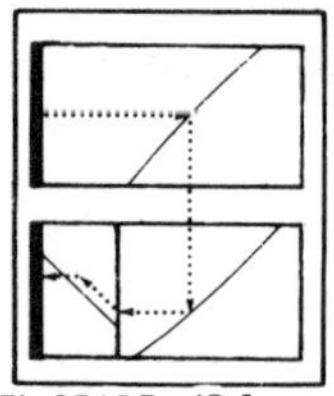

DATE: 1 APRIL 1971
DATA BASIS: FLIGHT TEST

FUEL GRADE: JP-5
FUEL DENSITY: 6.8 LB/GAL

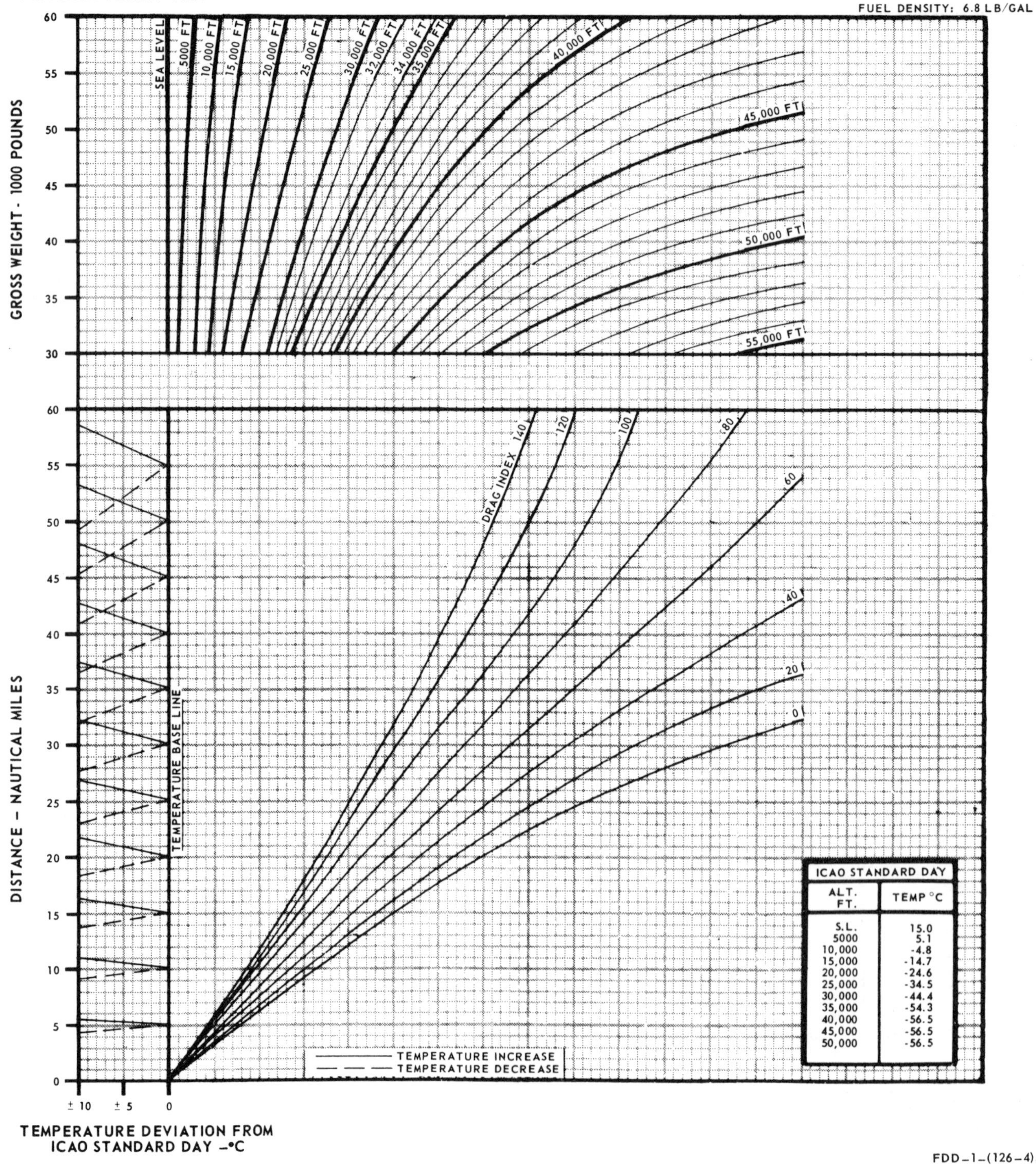

ICAO STANDARD DAY	
ALT. FT.	TEMP °C
S.L.	15.0
5000	5.1
10,000	-4.8
15,000	-14.7
20,000	-24.6
25,000	-34.5
30,000	-44.4
35,000	-54.3
40,000	-56.5
45,000	-56.5
50,000	-56.5

FDD-1-(126-4)A

Figure 11-19 (Sheet 4 of 4)

CLIMB SPEED SCHEDULE

MILITARY THRUST

AIRPLANE CONFIGURATION
INDIVIDUAL DRAG INDEXES

DATE: 1 APRIL 1971
DATA BASIS: FLIGHT TEST

REMARKS
ENGINE(S) (2) J79-GE-10
ICAO STANDARD DAY

FUEL GRADE: JP-5
FUEL DENSITY: 6.8 LB/GAL

ALTITUDE	DRAG INDEX 0		10		20		30		40		50		60	
	KCAS	MACH	KCAS	MACH	KCAS	MACH	KCAS	MACH	KCAS	MACH	KCAS	MACH	KCAS	MACH
S. L.	508	.77	484	.73	459	.69	435	.66	413	.62	396	.60	379	.57
5000	474	.78	454	.74	432	.71	413	.68	395	.65	382	.63	368	.61
10000	443	.79	429	.77	413	.74	398	.71	383	.69	369	.66	356	.64
15000	414	.81	403	.79	392	.77	380	.74	369	.72	357	.70	345	.68
20000	386	.83	374	.81	369	.79	360	.77	350	.75	340	.73	330	.71
25000	358	.84	353	.83	348	.82	342	.81	336	.80	329	.78	320	.76
30000	331	.86	330	.86	328	.86	325	.85	322	.84	317	.83	309	.81
35000	308	.89	306	.89	305	.89	303	.88	300	.87	294	.86	286	.84
40000	278	.90	277	.90	276	.90	-	-	-	-	-	-	-	-

ALTITUDE	DRAG INDEX 70		80		90		100		110		120		130		140	
	KCAS	MACH	KCAS	MACH	KCAS	MACH	KCAS	MACH	KCAS	MACH	KCAS	MACH	KCAS	MACH	KCAS	MACH
S. L.	365	.55	351	.53	340	.51	329	.50	319	.48	311	.47	304	.46	296	.45
5000	353	.58	343	.56	328	.54	318	.52	309	.51	303	.50	297	.49	291	.48
10000	344	.62	330	.59	318	.57	310	.56	301	.54	295	.53	289	.52	284	.51
15000	332	.65	321	.63	310	.61	302	.60	293	.58	287	.57	282	.56	278	.55
20000	319	.69	308	.67	298	.65	289	.63	281	.61	275	.60	270	.59	266	.58
25000	311	.74	300	.72	291	.70	283	.68	276	.66	269	.65	264	.64	260	.63
30000	300	.79	287	.76	274	.73	267	.71	259	.69	254	.68	-	-	-	-
35000	276	.81	265	.78	-	-	-	-	-	-	-	-	-	-	-	-
40000	-	-	-	-	-	-	-	-	-	-	-	-	-	-	-	-

TAKEOFF ALLOWANCES & ACCELERATION TO CLIMB SPEED

START - 65 LBS /ENG

RUNUP 50 LBS /ENG

TAXI - 21 LB /MIN /ENG

	BRAKE RELEASE TO CLIMB SPEED		
	MIL T. O. MIL ACCEL TO MIL CLIMB SPEED	MAX T. O. MIL ACCEL TO MIL CLIMB SPEED	MAX T. O. MAX ACCEL TO MIL CLIMB SPEED
FUEL - LBS	525	725	925
DIST - N M	6.0	5.3	3.0
TIME - MIN	1.7	1.3	.8

FDD-1-(127-1)A

Figure 11-20 (Sheet 1 of 4)

TIME TO CLIMB
MILITARY THRUST

REMARKS
ENGINE(S): (2) J79-GE-10

AIRPLANE CONFIGURATION
INDIVIDUAL DRAG INDEXES

GUIDE

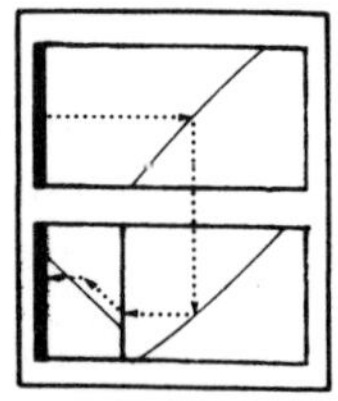

DATE: 1 APRIL 1971
DATA BASIS: FLIGHT TEST

FUEL GRADE: JP-5
FUEL DENSITY: 6.8 LB/GAL

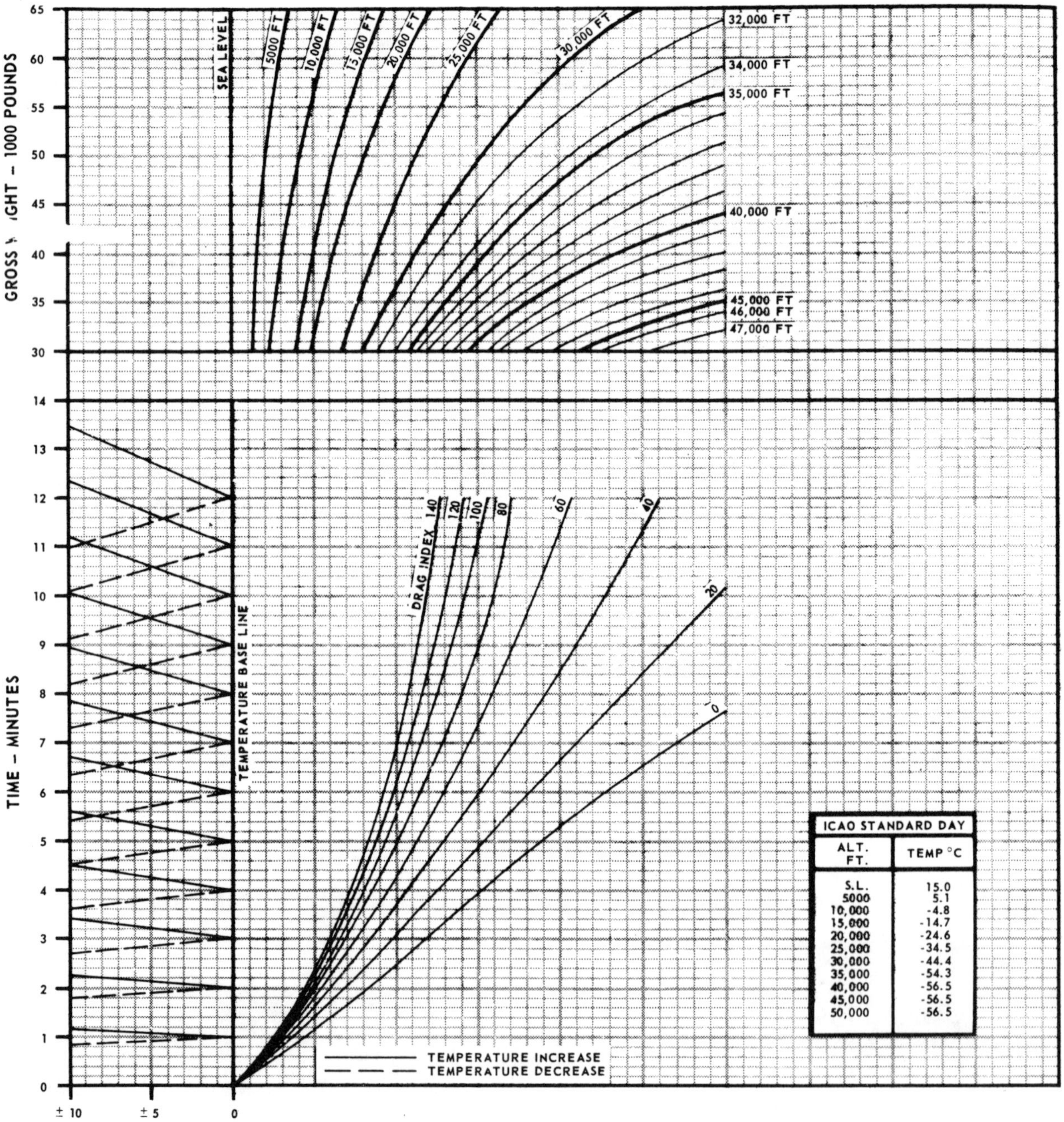

ICAO STANDARD DAY	
ALT. FT.	TEMP °C
S.L.	15.0
5000	5.1
10,000	-4.8
15,000	-14.7
20,000	-24.6
25,000	-34.5
30,000	-44.4
35,000	-54.3
40,000	-56.5
45,000	-56.5
50,000	-56.5

FDD-1-(127-2)A

Figure 11-20 (Sheet 2 of 4)

FUEL REQUIRED TO CLIMB
MILITARY THRUST

AIRPLANE CONFIGURATION
INDIVIDUAL DRAG INDEXES

REMARKS
ENGINE(S): (2) J79-GE-10

GUIDE

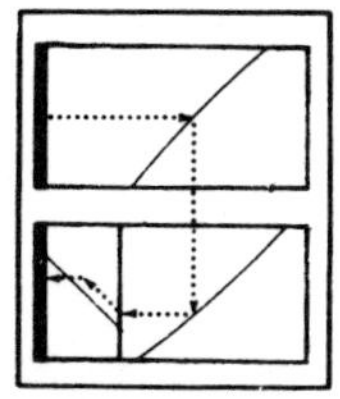

DATE: 1 APRIL 1971
DATA BASIS: FLIGHT TEST

FUEL GRADE: JP-5
FUEL DENSITY: 6.8 LB/GAL

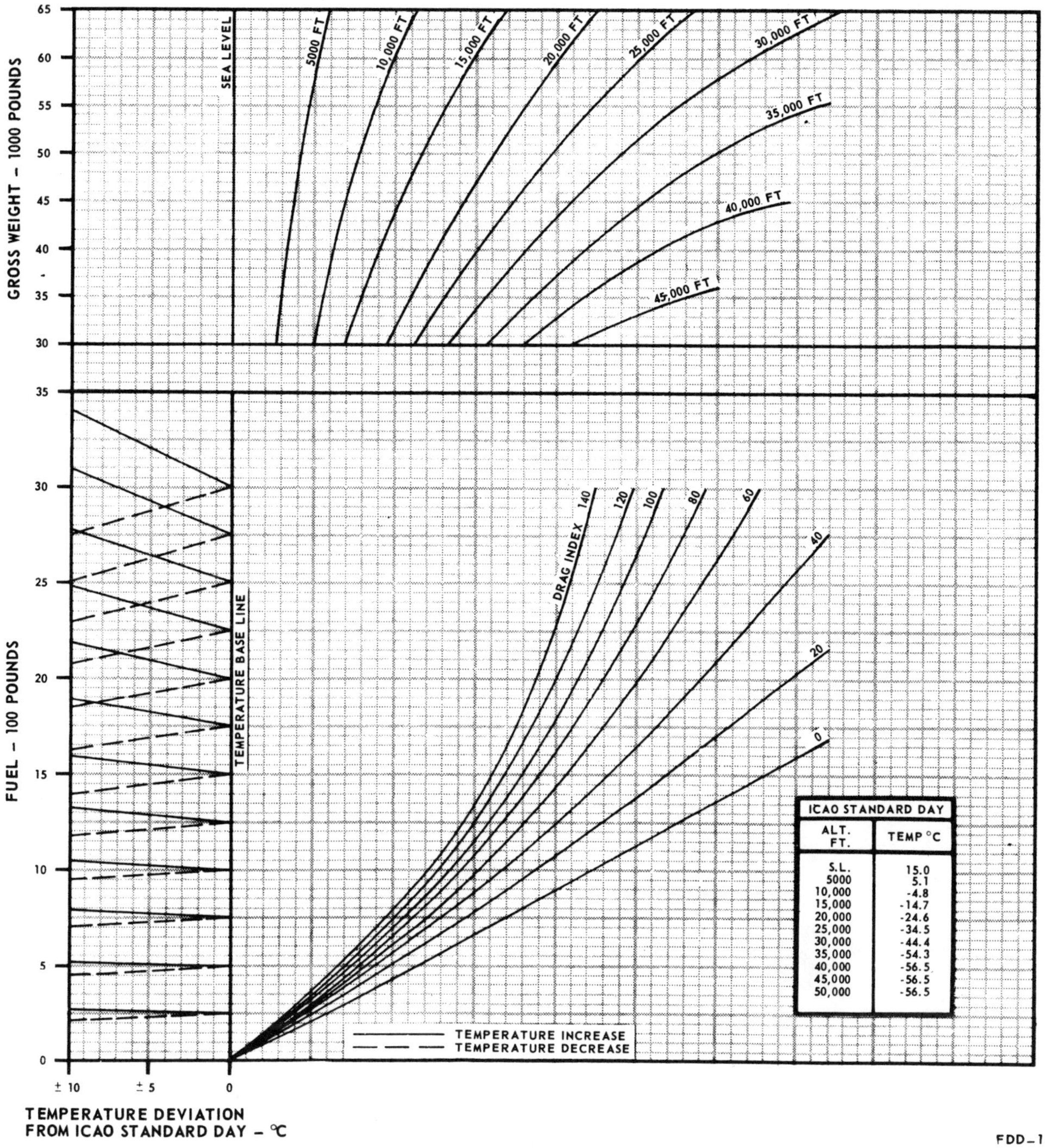

ICAO STANDARD DAY	
ALT. FT.	TEMP °C
S.L.	15.0
5000	5.1
10,000	-4.8
15,000	-14.7
20,000	-24.6
25,000	-34.5
30,000	-44.4
35,000	-54.3
40,000	-56.5
45,000	-56.5
50,000	-56.5

FDD-1-(127-3) A

Figure 11-20 (Sheet 3 of 4)

DISTANCE REQUIRED TO CLIMB

MILITARY THRUST

AIRPLANE CONFIGURATION
INDIVIDUAL DRAG INDEXES

REMARKS
ENGINE(S): (2) J79-GE-10

GUIDE

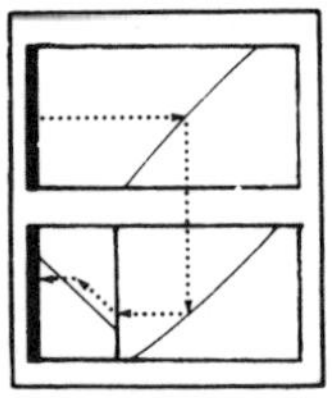

DATE: 1 APRIL 1971
DATA BASIS: FLIGHT TEST

FUEL GRADE: JP-5
FUEL DENSITY: 6.8 LB/GAL

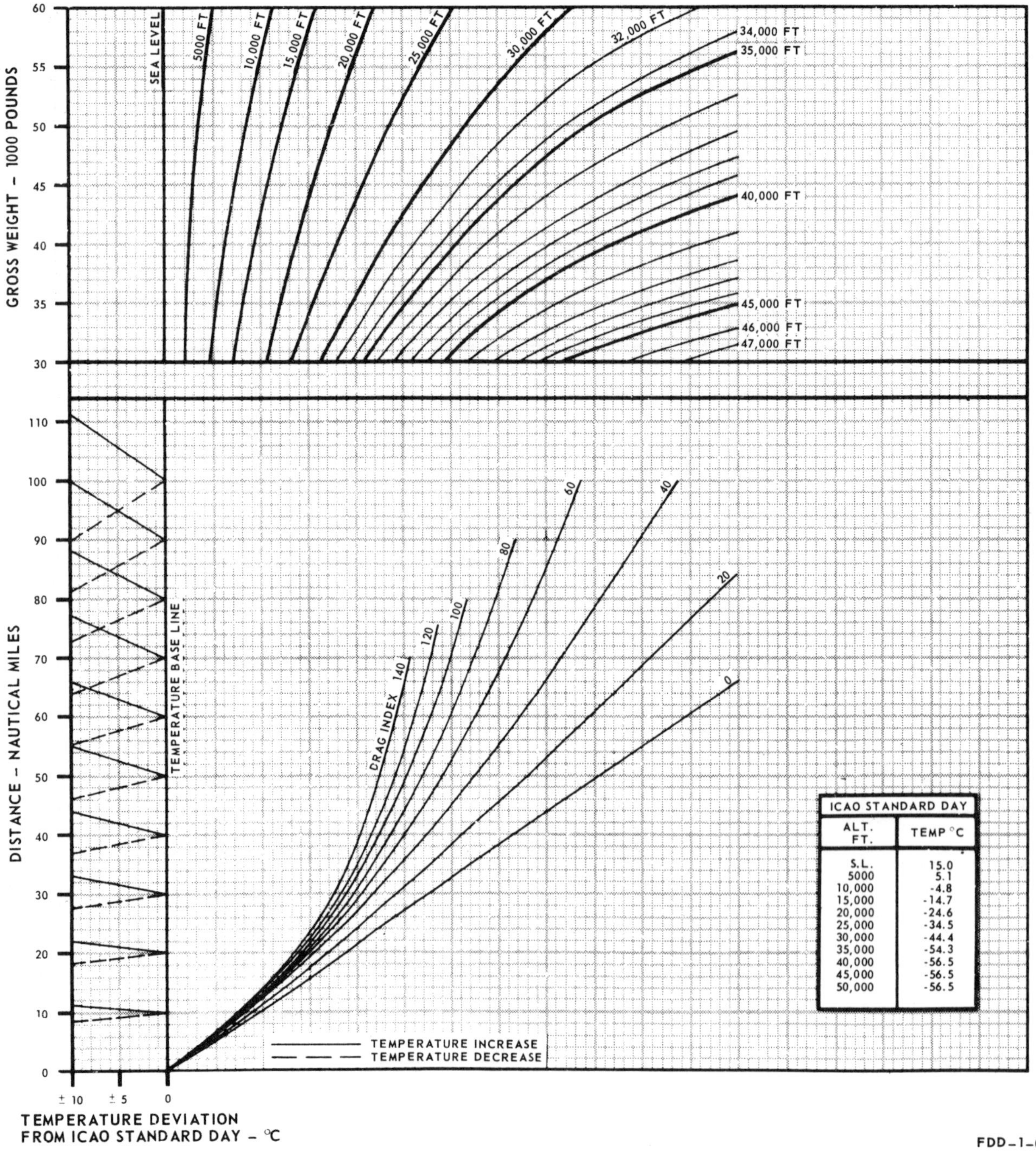

ICAO STANDARD DAY	
ALT. FT.	TEMP °C
S.L.	15.0
5000	5.1
10,000	-4.8
15,000	-14.7
20,000	-24.6
25,000	-34.5
30,000	-44.4
35,000	-54.3
40,000	-56.5
45,000	-56.5
50,000	-56.5

FDD-1-(127-4)A

Figure 11-20 (Sheet 4 of 4)

TIME TO CLIMB

350 KCAS-MILITARY THRUST

AIRPLANE CONFIGURATION

INDIVIDUAL DRAG INDEXES

REMARKS
ENGINE(S); (2) J79-GE-10

NOTE
DATA BASED ON 350-KNOT CLIMB UNTIL INTERCEPTION OF OPTIMUM CRUISE MACH/ TAS, THEN MAINTAIN CRUISE MACH TO CRUISE ALTITUDE. REFER TO PART 4 TO OBTAIN CRUISE ALTITUDES.

GUIDE

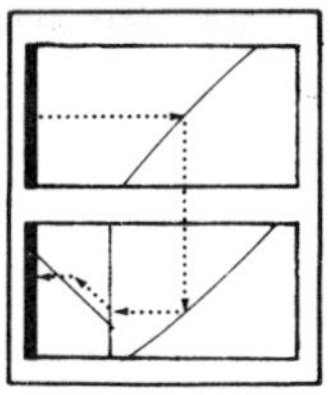

DATE: 1 MARCH 1971
DATA BASIS: FLIGHT TEST

FUEL GRADE: JP-5
FUEL DENSITY: 6.8 LB/GAL

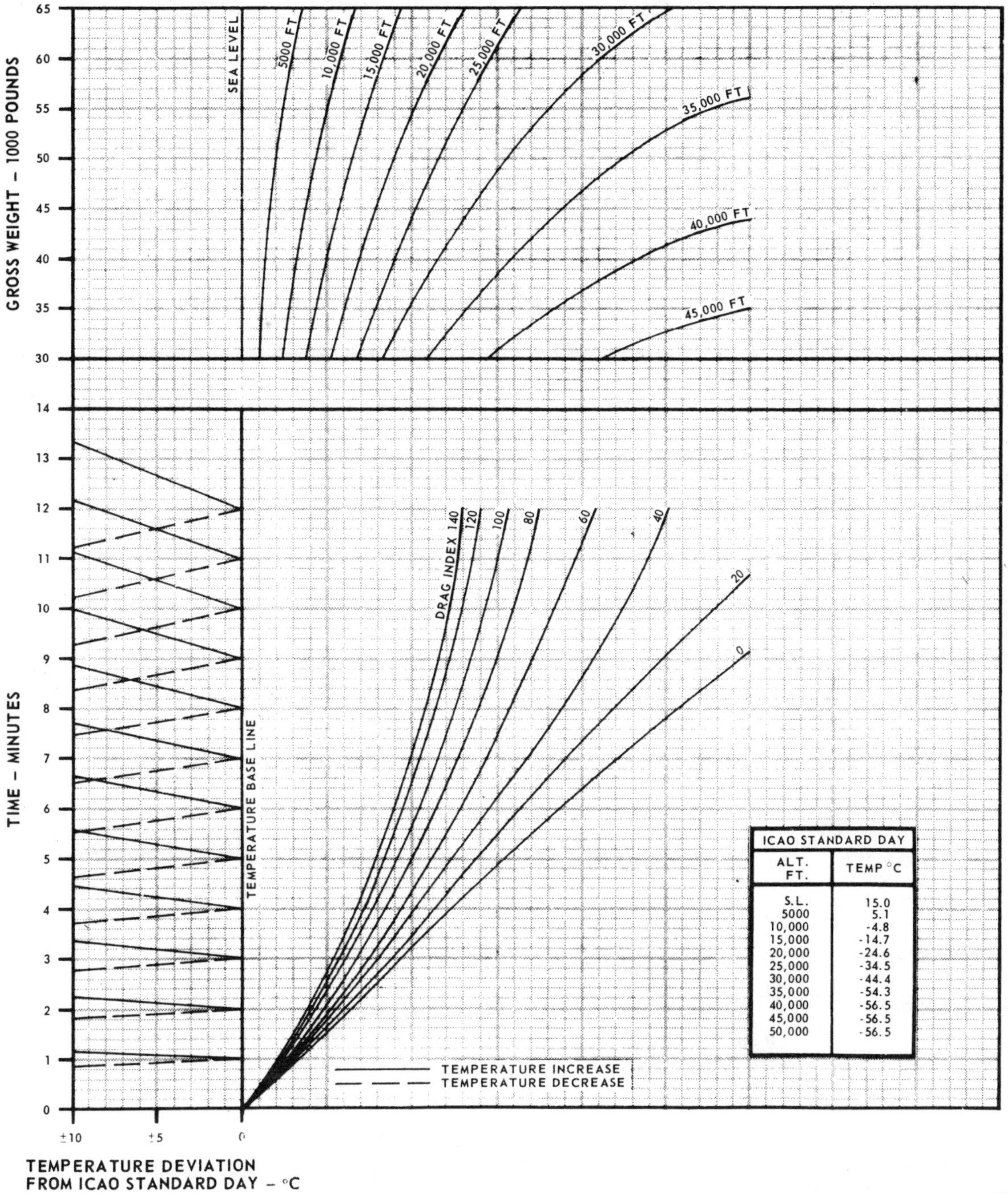

ICAO STANDARD DAY	
ALT. FT.	TEMP °C
S.L.	15.0
5000	5.1
10,000	-4.8
15,000	-14.7
20,000	-24.6
25,000	-34.5
30,000	-44.4
35,000	-54.3
40,000	-56.5
45,000	-56.5
50,000	-56.5

FDD-1-(296-1)

Figure 11-21 (Sheet 1 of 3)

FUEL REQUIRED TO CLIMB
350 KCAS-MILITARY THRUST

AIRPLANE CONFIGURATION

INDIVIDUAL DRAG INDEXES

REMARKS

ENGINE(S): (2) J79-GE-10

NOTE

DATA BASED ON 350-KNOT CLIMB UNTIL INTERCEPTION OF OPTIMUM CRUISE MACH/TAS, THEN MAINTAIN CRUISE MACH TO CRUISE ALTITUDE. REFER TO PART 4 TO OBTAIN CRUISE ALTITUDES.

GUIDE

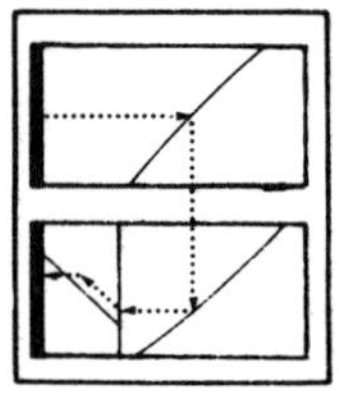

DATE: 1 MARCH 1971
DATA BASIS: FLIGHT TEST

FUEL GRADE: JP-5
FUEL DENSITY: 6.8 LB/GAL

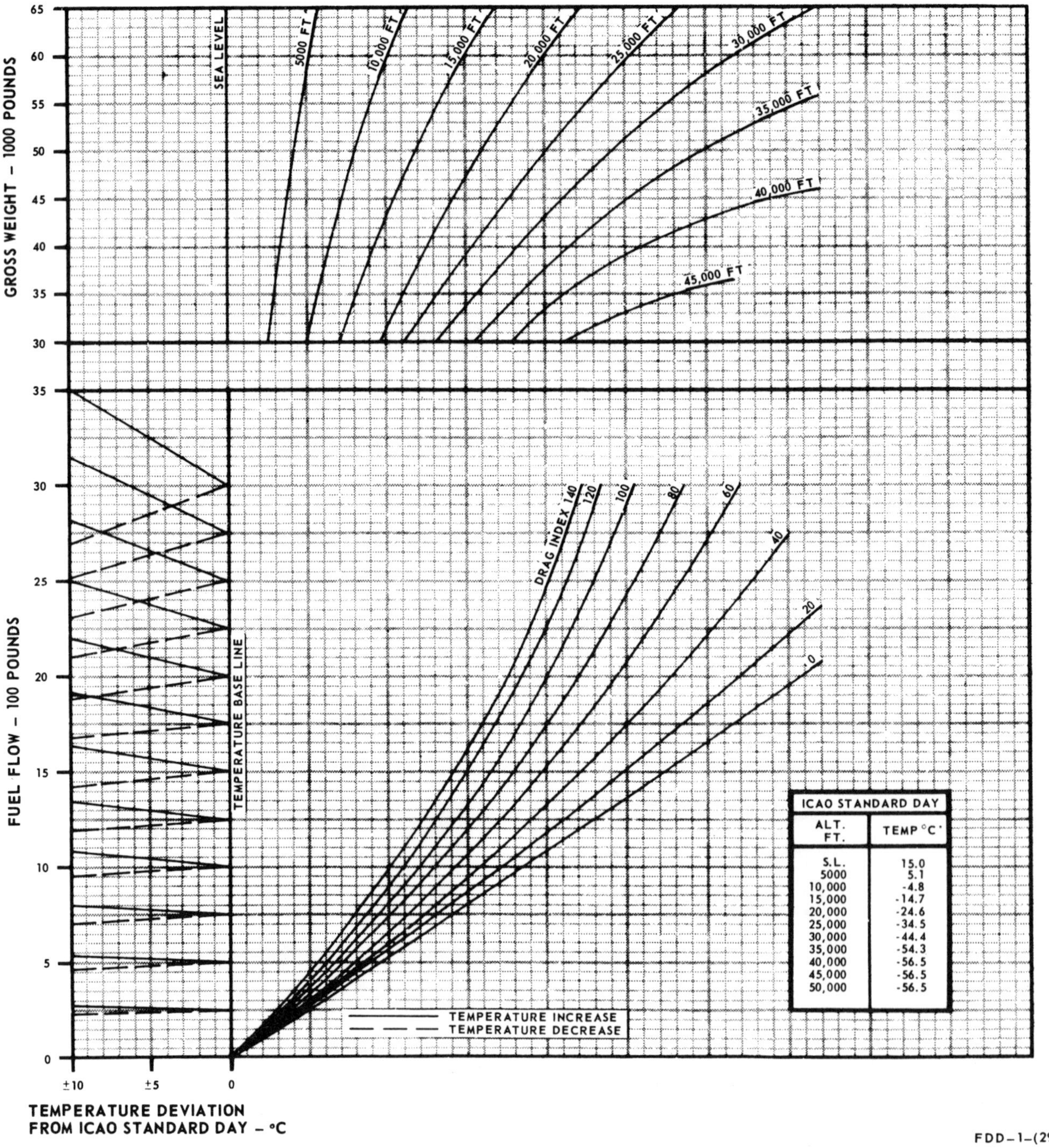

ICAO STANDARD DAY	
ALT. FT.	TEMP °C
S.L.	15.0
5000	5.1
10,000	-4.8
15,000	-14.7
20,000	-24.6
25,000	-34.5
30,000	-44.4
35,000	-54.3
40,000	-56.5
45,000	-56.5
50,000	-56.5

FDD-1-(296-2)

Figure 11-21 (Sheet 2 of 3)

DISTANCE REQUIRED TO CLIMB

350 KCAS-MILITARY THRUST

AIRPLANE CONFIGURATION

INDIVIDUAL DRAG INDEXES

REMARKS

ENGINE(S): (2) J79-GE-10

NOTE

DATA BASED ON 350-KNOT CLIMB UNTIL INTERCEPTION OF OPTIMUM CRUISE MACH/TAS, THEN MAINTAIN CRUISE MACH TO CRUISE ALTITUDE. REFER TO PART 4 TO OBTAIN CRUISE ALTITUDES.

GUIDE

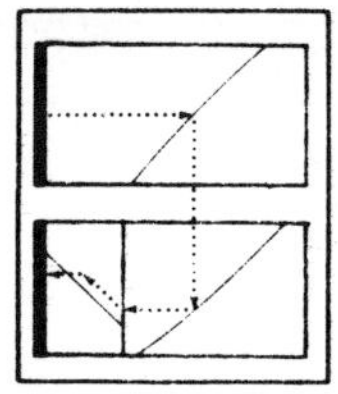

DATE: 1 MARCH 1971
DATA BASIS: **FLIGHT TEST**

FUEL GRADE: JP-5
FUEL DENSITY. 6.8 LB/GAL

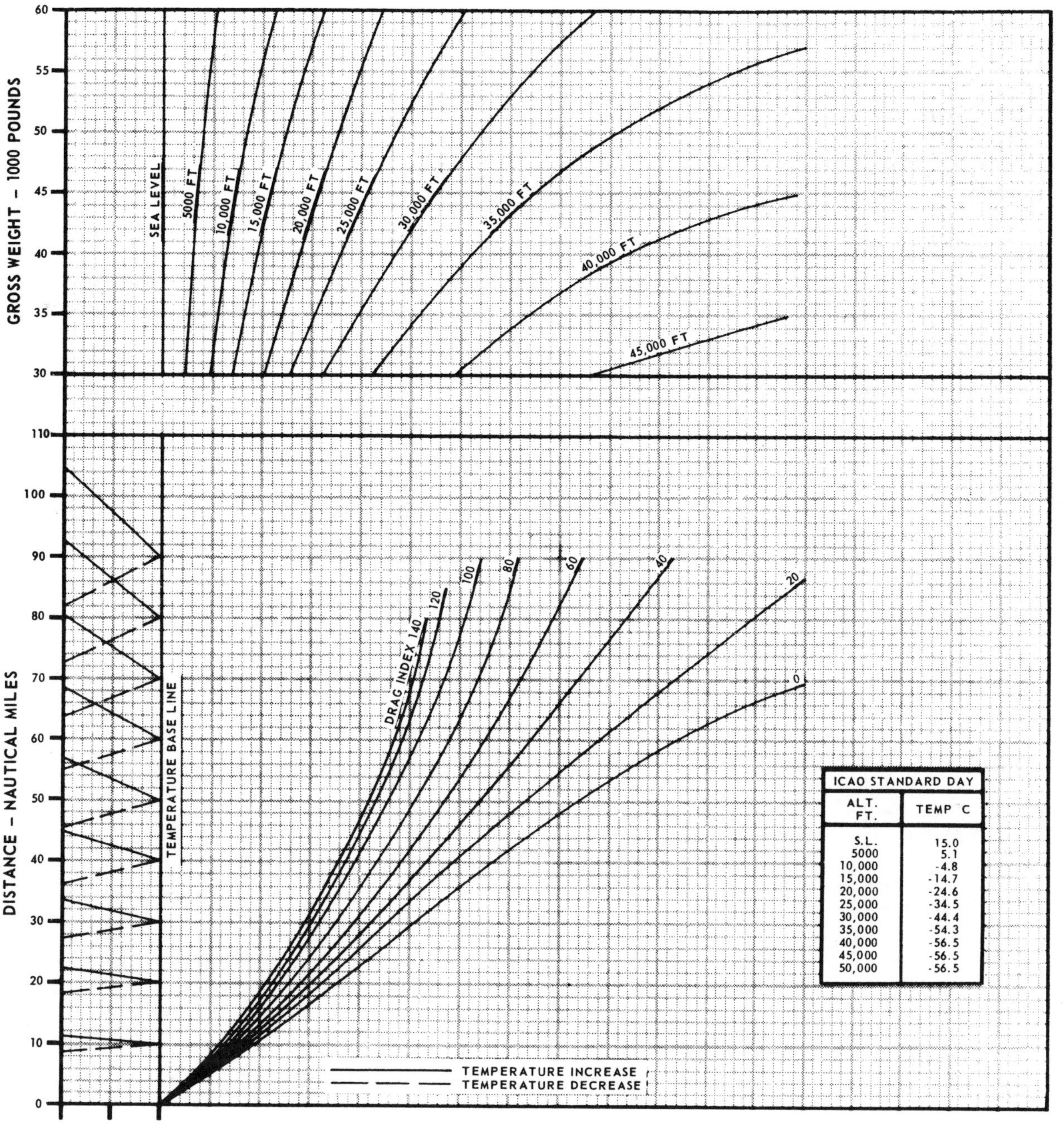

ICAO STANDARD DAY	
ALT. FT.	TEMP °C
S.L.	15.0
5000	5.1
10,000	-4.8
15,000	-14.7
20,000	-24.6
25,000	-34.5
30,000	-44.4
35,000	-54.3
40,000	-56.5
45,000	-56.5
50,000	-56.5

FDD-1-(296-3)

Figure 11-21 (Sheet 3 of 3)

COMBAT CEILING

AIRPLANE CONFIGURATION
INDIVIDUAL DRAG INDEXES

ICAO STANDARD DAY	
ALT. FT.	TEMP C
S.L.	15.0
5000	5.1
10,000	-4.8
15,000	-14.7
20,000	-24.6
25,000	-34.5
30,000	-44.4
35,000	-54.3
40,000	-56.5
45,000	-56.5
50,000	-56.5

DATE: 1 APRIL 1971
DATA BASIS: FLIGHT TEST

REMARKS
ENGINE(S): (2) J79-GE-10
ICAO STANDARD DAY

NOTE

COMBAT CEILING IS THE PRESSURE ALTITUDE AT WHICH THE AIRCRAFT CAN CLIMB AT A MAXIMUM RATE OF 500 FEET PER MINUTE.

GUIDE

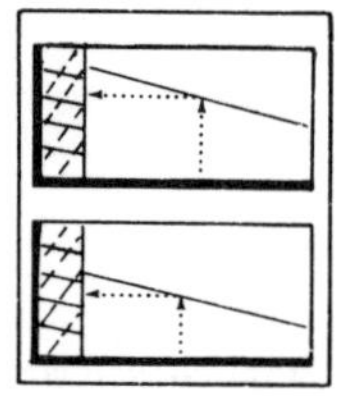

FUEL GRADE: JP-5
FUEL DENSITY: 6.8 LB/GAL

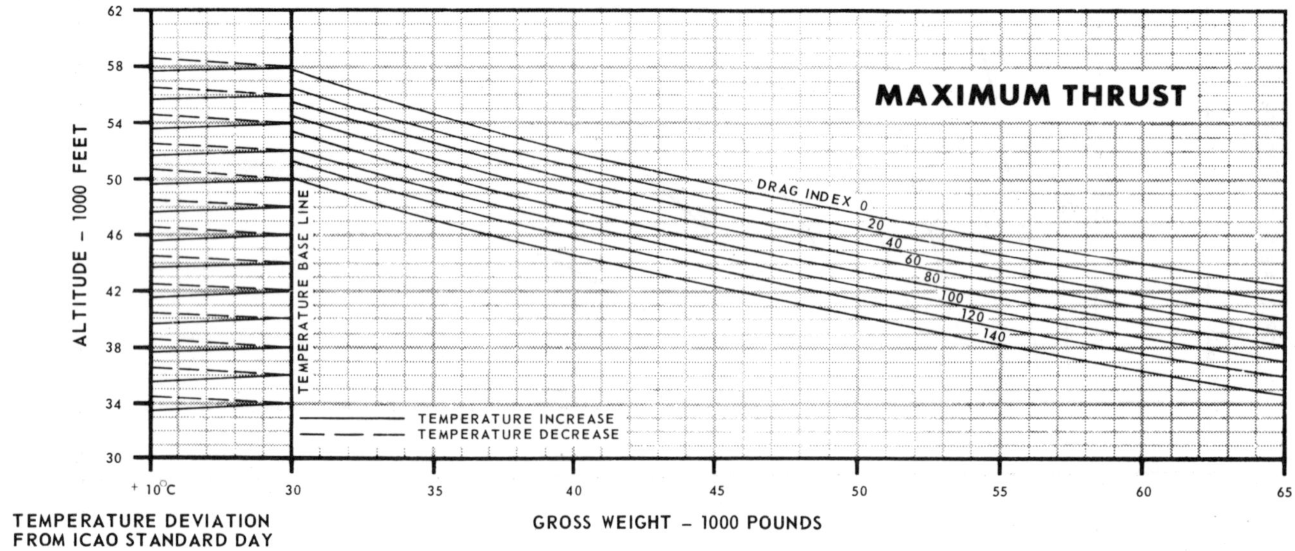

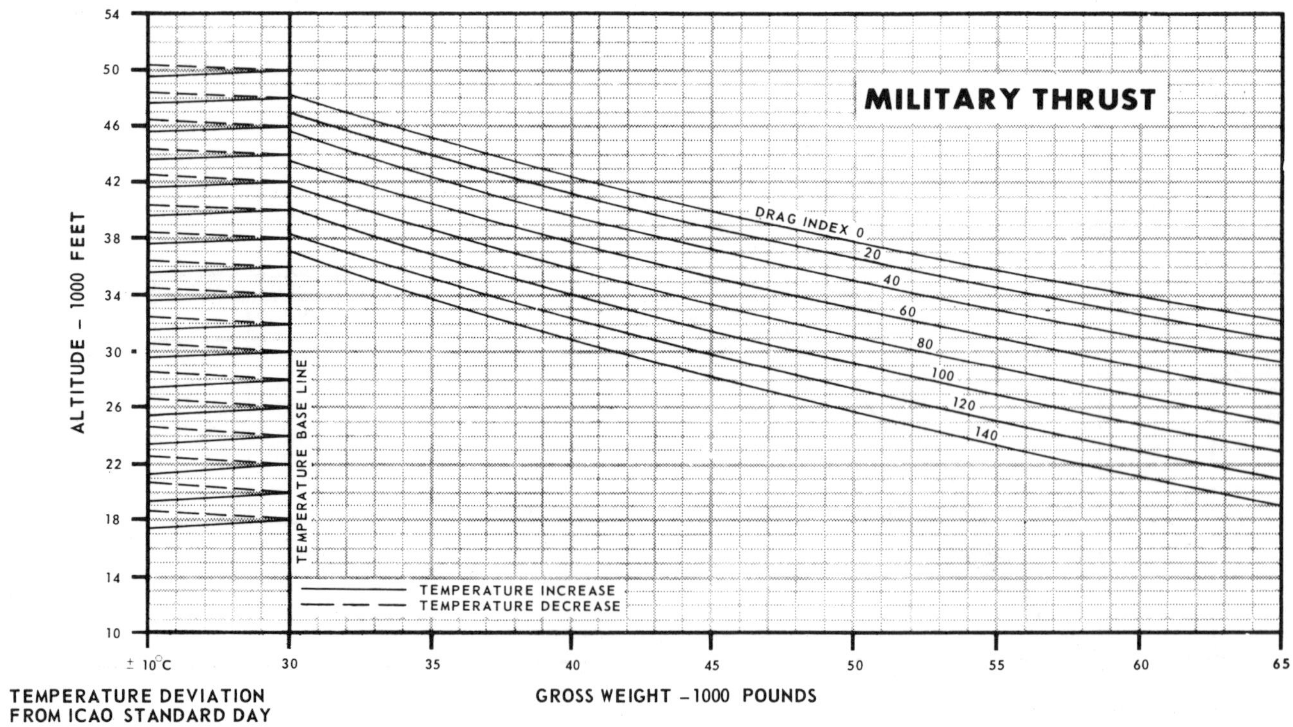

FDD-1-(130)A

Figure 11-22

PART 4 RANGE

CHARTS

RANGEWIND CORRECTION CHART

This chart (figure 11-23) provides a means of correcting computed range (specific or total) for existing wind effects. The presented range factors consider wind speeds up to 150 knots from any relative wind direction for airplane speeds of 200 to 1300 knots (TAS).

USE

Determine the relative wind direction by subtracting the aircraft heading from the forecast wind direction. If the aircraft heading is greater than the forecast wind direction, add 360° to the wind direction and then perform the subtraction. Enter the chart with relative wind direction and proceed vertically to the interpolated wind speed. From this point, project horizontally to intersect the airplane true airspeed and reflect to the lower scale to read the range factor. Multiply computed range by this range factor to find range as affected by wind.

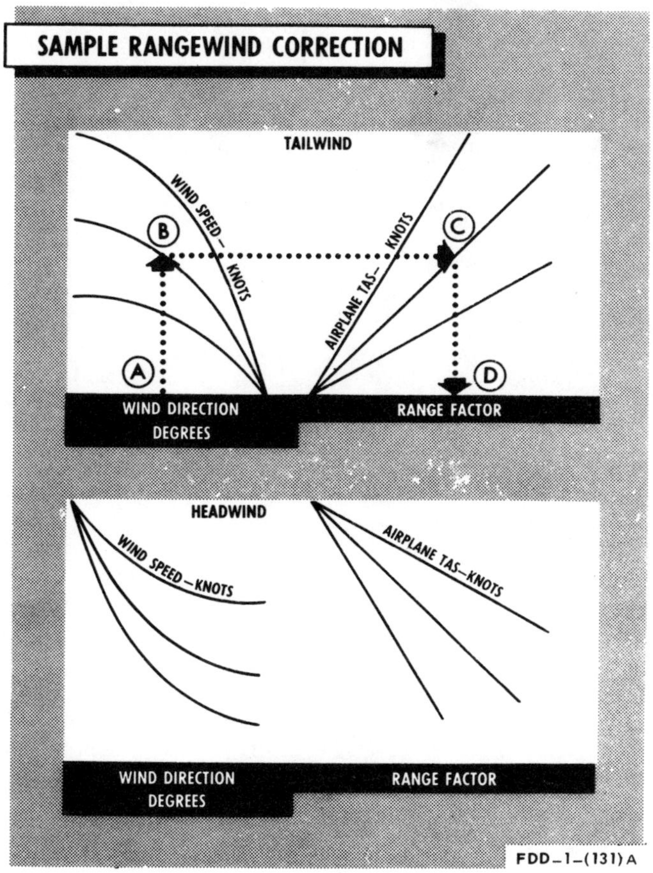

FDD-1-(131)A

Sample Problem

A.	Relative wind direction	150°
B.	Wind speed	125 Kt
C.	Airplane speed (TAS)	400 Kt
D.	Range factor	1.25

OPTIMUM CRUISE SUMMARY

This chart (figure 11-24) presents optimum cruise data for two-engine operation. The chart depicts cruise altitude, specific range (in nautical miles per pound) and cruise Mach number for all gross weights and drag indexes.

USE

Enter the chart with the previously computed drag index, and project vertically to intersect the gross weight curves of all three plots. At the intersection of the appropriate gross weight curves, reflect horizontally to the left and read cruise Mach number, specific range in nautical miles per pound, and cruise altitude.

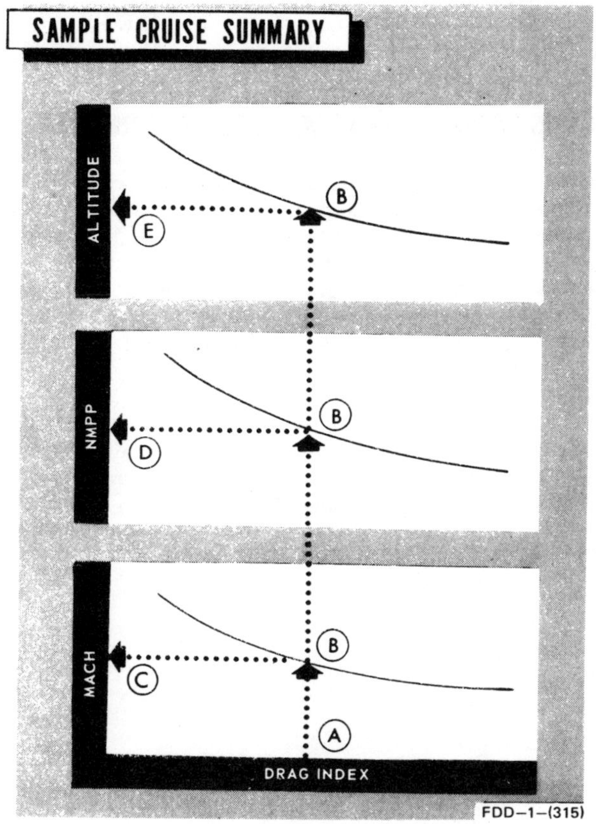

FDD-1-(315)

Sample Problem

A.	Drag index	20
B.	Gross weight	40,000 Lb
C.	Mach number	0.88
D.	Specific range	0.92 NMPP
E.	Cruise altitude	37,900 Ft

LOW ALTITUDE CRUISE TABLES

These charts (figures 11-25 thru 11-27) present total fuel flow values for various combinations of true airspeed and drag index at altitudes of Sea Level 4,000, 8,000, and 12,000 feet. Also included is the resultant V_{max} (maximum attainable TAS) for a particular altitude-drag index combination at a MIL thrust setting). Separate charts are provided for several gross weights. Fuel flow values are tabulated for ICAO Standard Day; however, correction factors are given for non-standard temperatures.

USE

After selecting the applicable chart for gross weights and altitudes, enter with the desired true airspeed and project horizontally to the applicable drag index column. Read total fuel flow for a standard day.

Sample Problem

Gross weight 50,000 Lb, Sea Level (15°C)

A.	Airspeed (TAS)	540 Kt
B.	Drag index	20
C.	Fuel flow	16,829 PPH
D.	Non-standard day temperature	20°
E.	Correction factor	1.008
F.	Fuel flow for non-standard day (E x C)	16,963 PPH

CONSTANT MACH/ALTITUDE CRUISE

These charts (figures 11-28 thru 11-33) present nautical miles per pound and total fuel flow for various combinations of Mach number, gross weight, altitude and drag index. This data is based on cruise at a constant Mach number and a constant altitude. Specifics are presented for 0°C; however, correction factors are provided for temperature deviations.

USE

After selecting the desired cruise Mach, enter the chart with the estimated gross weight at end of climb. Project horizontally to the right to intersect the desired cruise altitude, then vertically downward to intersect the applicable drag index. From this point, project horizontally to both sides of the graph and read nautical miles per pound and total fuel flow for 0°C temperature. If required, correct these values for the actual temperatures.

Sample Problem

A.	Mach number	0.85
B.	Gross weight	40,000 Lb
C.	Altitude	30,000 Ft
D.	Drag index	40
E.	Total fuel flow	7850 PPH
F.	Specific range	0.069 NMPP
G.	Actual temperature	-20°C
H.	Total fuel flow (corrected)	7536 PPH

SAMPLE CONSTANT MACH/ALTITUDE CRUISE

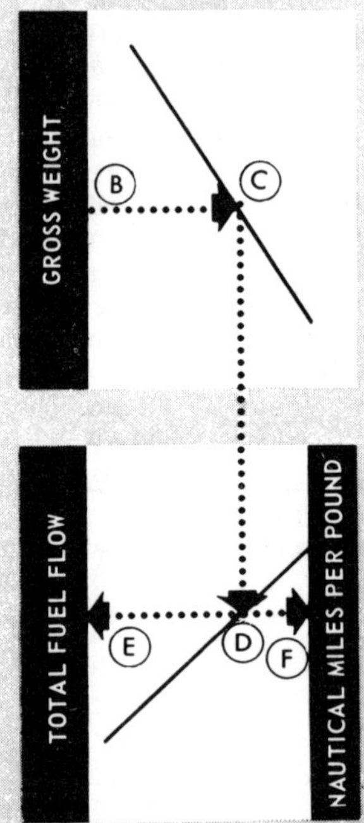

FDD-1-(132)

CONSTANT ALTITUDE CRUISE

These charts (figures 11-34 and 11-35) present the necessary planning data to set up optimum cruise schedules for normal two-engine and single engine operation, at a constant altitude. The recommended procedure is to use an average gross weight for a given leg of the mission. One way to find the average gross weight is to divide the mission into weight segments. With this method, readjust the cruise schedule each time a given amount of fuel is used. Subtract one-half of the fuel weight allotted for the first leg from the initial cruise gross weight. The remainder is the average gross weight for the leg. It is possible to obtain instantaneous data if desired.

USE

Enter the left side of sheet 1 with the average gross weight. Project horizontally to the right to intersect desired cruise altitude, and then vertically downward to the computed drag index, then horizontally to the right to obtain specific range (nautical miles per pound). Repeat these projections on the right side of sheet 1 to obtain optimum cruise Mach number for the desired altitude. Enter sheet 2 with the optimum cruise Mach number. Project horizontally to the right to intersect predicted flight-level temperature, then vertically downward to obtain corresponding true airspeed. Continue this projection vertically downward to intersect the interpolated specific range (obtained from sheet 1), then horizontally to the left to obtain total fuel flow required in pounds per hour.

Sample Problem

(2) Engines

A.	Average gross weight for first leg	45,000 Lb
B.	Cruise altitude	35,000 Ft
C.	Computed drag index	40
D.	Specific range	0.074 NMPP
E.	Gross weight	45,000 Lb
F.	Altitude	35,000 Ft
G.	Drag index	40
H.	True Mach number	0.86
J.	True Mach number	0.86
K.	Temperature at flight altitude	-40°C
L.	True airspeed	510 Kt
M.	Specific range	0.08 NMPP
N.	Total fuel flow	6800 PPH

SAMPLE CONSTANT ALTITUDE CRUISE

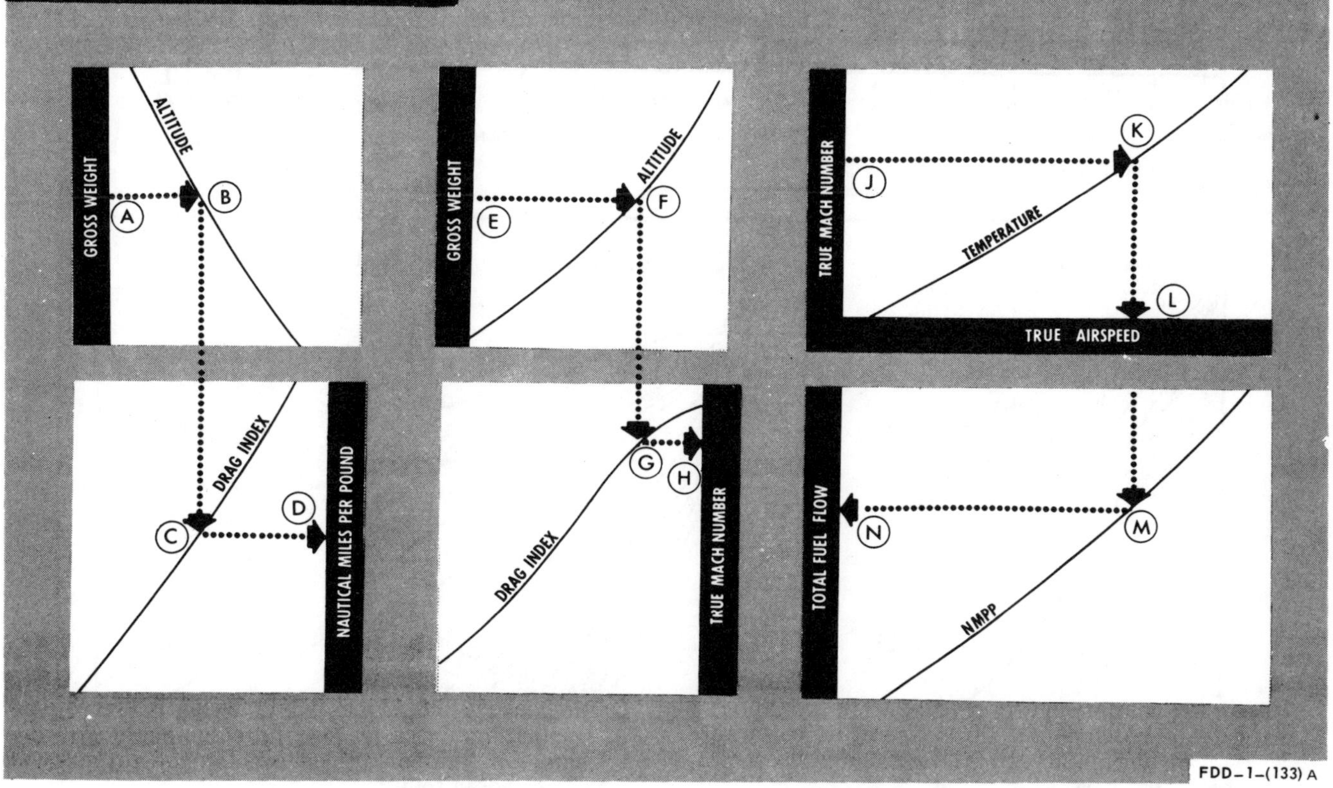

FDD-1-(133) A

RANGEWIND CORRECTION

AIRPLANE CONFIGURATION

ALL CONFIGURATIONS

RELATIVE WIND DIRECTION

NOTE: RELATIVE WIND DIRECTION = ANGULAR DIFFERENCE MEASURED CLOCKWISE, BETWEEN AIRPLANE HEADING AND TRUE WIND DIRECTION

GUIDE

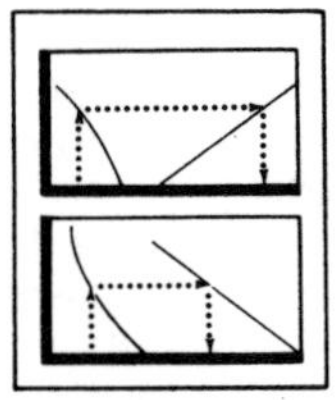

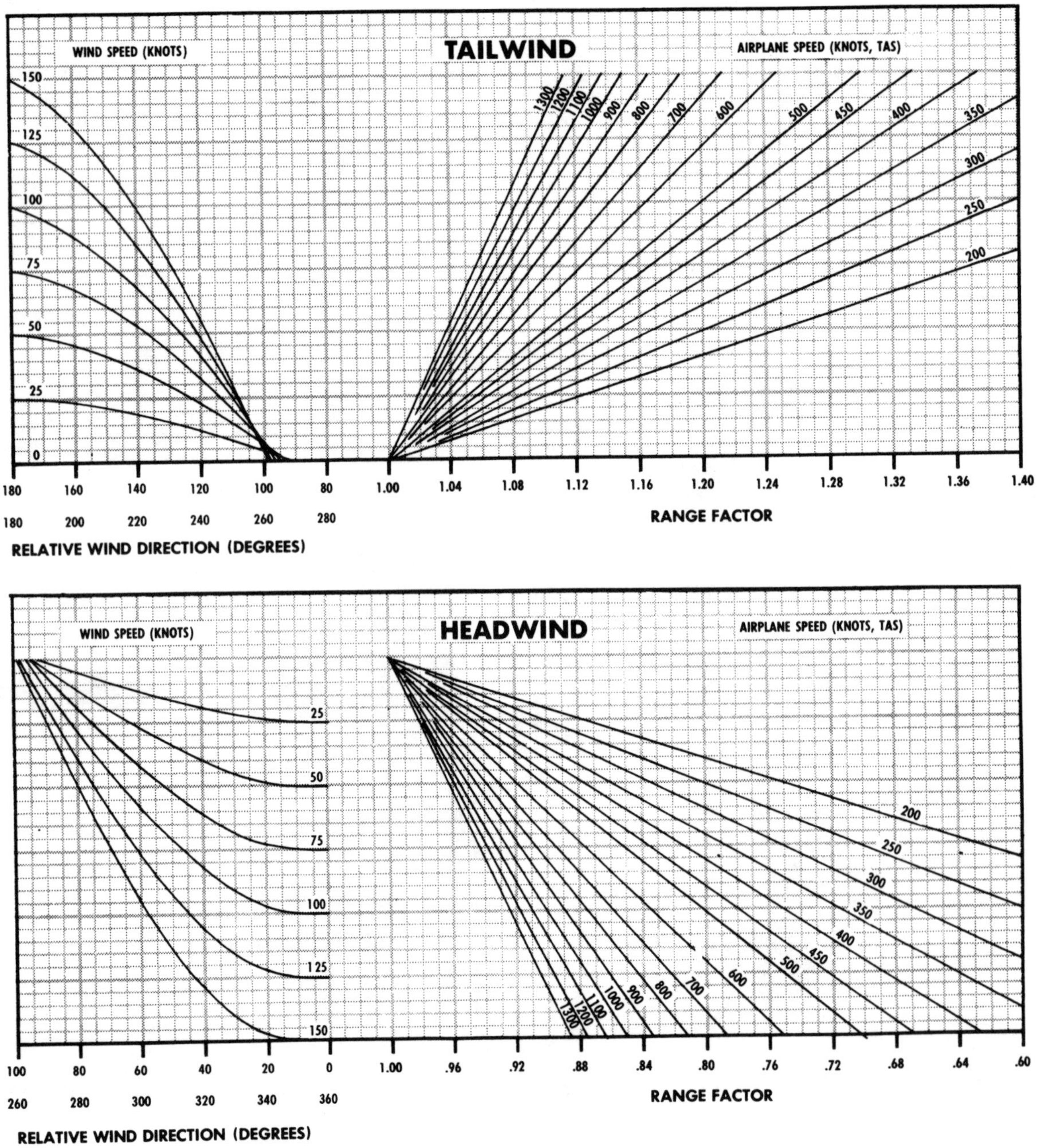

FDD-1-(134)A

Figure 11-23

OPTIMUM CRUISE SUMMARY

AIRPLANE CONFIGURATION
INDIVIDUAL DRAG INDEXES

REMARKS
ENGINE(S): (2) J79-GE-10
ICAO STANDARD DAY

GUIDE

DATE: 1 MAY 1971
DATA BASIS: FLIGHT TEST

FUEL GRADE: JP-5
FUEL DENSITY: 6.8 LB/GAL

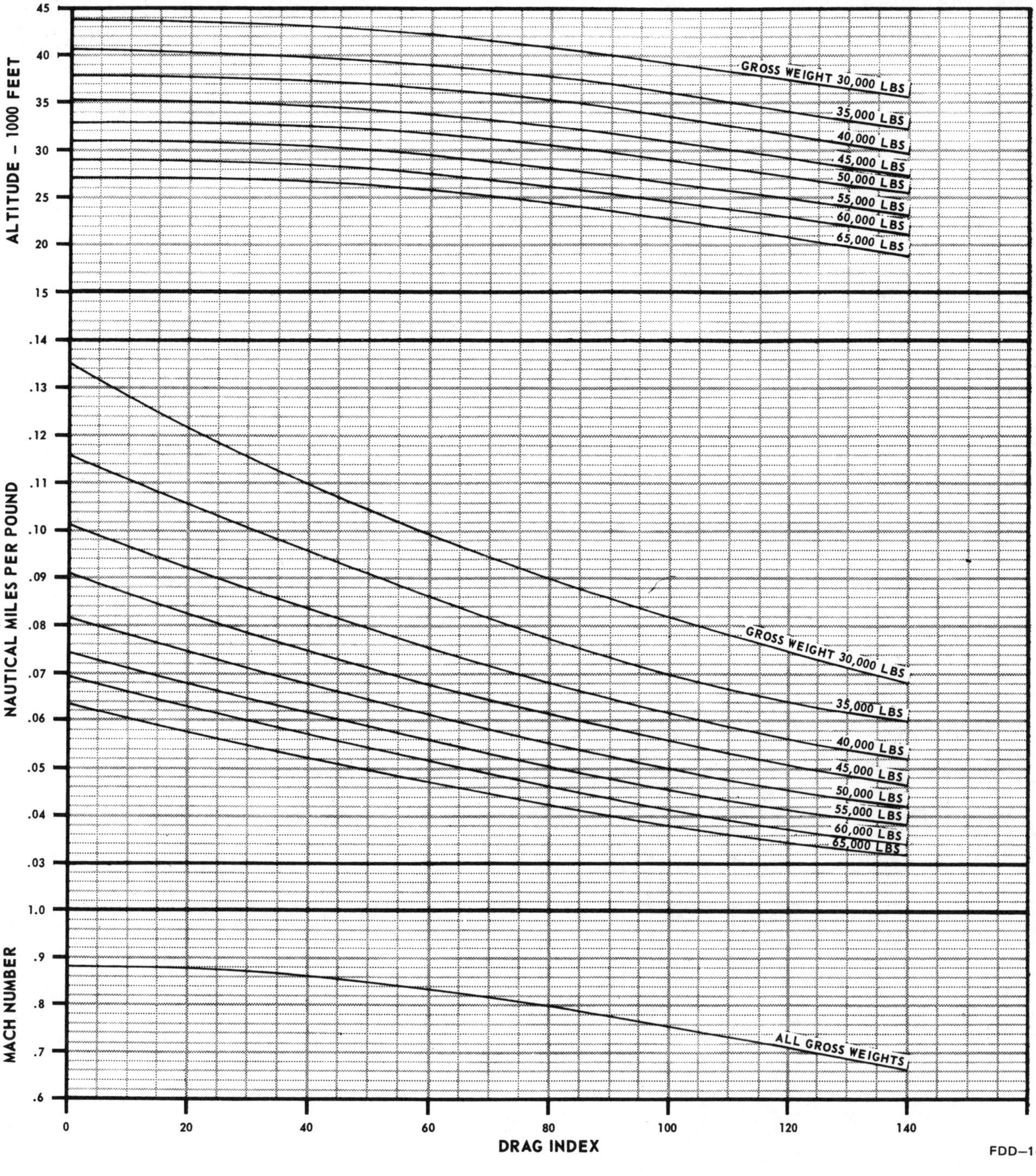

Figure 11-24

LOW ALTITUDE CRUISE

GROSS WEIGHT – 35,000 POUNDS

AIRPLANE CONFIGURATION
INDIVIDUAL DRAG INDEXES

REMARKS
ENGINES: (2) J79–GE–10

DATE: 1 JUNE 1971
DATA BASIS: FLIGHT TEST

FUEL GRADE: JP-5
FUEL DENSITY: 6.8 LB/GAL

	KTAS	DRAG INDEX 0	20	40	60	80	100	120	TEMP. EFFECTS °C	FACTOR
		TOTAL FUEL FLOW-LB/HR								
SEA LEVEL (15°C)	360	7185	7963	8767	9590	10436	11304	12125		
	420	9053	10188	11373	12547	13876	15143	16530	–40	.899
	480	11381	12954	14641	16358	18191	20092	22408	–20	.937
	540	14296	16394	18496	21118	24390			0	.973
	600	20230	23256	27556					20	1.008
	MIL	28200	27894	27643	27099	26604	26085	25421	40	1.042
	VMAX	630.5	614.7	601.9	576.1	553.9	531.9	504.3		
4,000 FEET (7°C)	360	6477	7194	7896	8624	9371	10135	10927		
	420	8067	9093	10132	11194	12274	13492	14813	–40	.913
	480	10096	11477	12859	14469	16352	18238	20149	–20	.949
	540	12743	14599	16475	19405	22153			0	.987
	600	18779	22276	25117					20	1.022
	MIL	25869	25467	25146	24566	24110	23725	23159	40	1.057
	VMAX	631.9	614.3	600.7	575.5	555.2	537.1	509.7		
8,000 FEET (-1°C)	360	5837	6457	7087	7740	8387	9059	9765		
	420	7191	8088	9005	9927	10911	11874	12986	–40	.925
	480	8934	10149	11388	12673	14233	16115	18154	–20	.963
	540	11303	12835	14530	16870	19998	21480		0	1.001
	600	17164	20660						20	1.037
	MIL	23513	23006	22654	22154	21799	21528	21099	40	1.072
	VMAX	633.3	613.6	597.9	574.4	556.1	541.2	514.7		
12,000 FEET (-9°C)	360	5272	5811	6366	6916	7511	8084	8713		
	420	6405	7186	8009	8817	9672	10548	11440	–40	.939
	480	7890	8967	10041	11216	12413	13879	15670	–20	.978
	540	10012	11382	12800	14732	17157	19399		0	1.016
	600	15549	18742						20	1.052
	MIL	21219	20725	20351	19993	19722	19500	19135	40	1.088
	VMAX	634.0	612.3	593.7	572.6	556.2	542.8	517.9		
16,000 FEET (-16°C)	360	4781	5252	5732	6218	6700	7235	7764		
	420	5747	6432	7104	7831	8614	9383	10252	–40	.953
	480	6991	7909	8841	9883	11033	12174	13582	–20	.993
	540	8963	10156	11545	13254	15086	17548		0	1.032
	600	14366	17333						20	1.069
	MIL	19102	18622	18289	18038	17825	17633	17232	40	1.105
	VMAX	634.3	609.0	588.1	569.1	554.5	542.6	518.3		

FDD-1-(138)A

Figure 11-25

LOW ALTITUDE CRUISE

GROSS WEIGHT – 45,000 POUNDS

AIRPLANE CONFIGURATION
INDIVIDUAL DRAG INDEXES

REMARKS
ENGINE(S): (2) J79-GE-10

DATE: 1 JUNE 1971
DATA BASIS: FLIGHT TEST

FUEL GRADE: JP-5
FUEL DENSITY: 6.8 LB/GAL

	KTAS	TOTAL FUEL FLOW-LB/HR, DRAG INDEX 0	20	40	60	80	100	120	TEMP EFFECTS °C	FACTOR
SEA LEVEL (15°C)	360	7672	8475	9277	10102	10953	11788	12678		
	420	9401	10559	11774	12930	14247	15564	16917	-40	.899
	480	11681	13256	14932	16712	18536	20431	22830	-20	.937
	540	14587	16681	18780	21421	24796			0	.973
	600	20463	23565	27573					20	1.008
	MIL	28183	27879	27629	27057	26562	26007	25348	40	1.042
	V_{MAX}	629.5	613.9	601.2	574.2	552.2	528.5	501.2		
4,000 FEET (7°C)	360	6988	7689	8423	9175	9918	10678	11456		
	420	8493	9527	10571	11634	12711	13963	15355	-40	.913
	480	10438	11856	13233	14861	16770	18705	20556	-20	.949
	540	12987	14854	16787	19700	22525			0	.987
	600	19082	22676						20	1.022
	MIL	25838	25443	25120	24521	24071	23649	23088	40	1.057
	V_{MAX}	630.5	613.3	599.6	573.5	553.4	533.5	506.1		
8,000 FEET (-1°C)	360	6365	7002	7639	8292	8969	9635	10333		
	420	7654	8573	9500	10431	11413	12389	13586	-40	.925
	480	9308	10541	11778	13079	14681	16604	18734	-20	.963
	540	11602	13139	14830	17299	20406			0	1.001
	600	17447	21064						20	1.037
	MIL	23471	22975	22608	22114	21761	21481	21038	40	1.072
	V_{MAX}	631.7	612.4	595.8	572.4	554.1	538.3	510.5		
12,000 FEET (-9°C)	360	5829	6384	6934	7532	8100	8709	9317		
	420	6930	7715	8526	9345	10213	11093	11990	-40	.939
	480	8327	9394	10458	11600	12858	14376	16265	-20	.978
	540	10357	11740	13170	15113	17750	19442		0	1.016
	600	15788	19093						20	1.052
	MIL	21177	20696	20317	19956	19686	19464	19061	40	1.088
	V_{MAX}	632.2	611.0	591.7	570.4	554.0	540.6	512.7		
16,000 FEET (-16°C)	360	5420	5900	6395	6876	7420	7929	8494		
	420	6300	6975	7706	8404	9186	9996	10891	-40	.953
	480	7447	8383	9325	10374	11488	12730	14218	-20	.993
	540	9335	10513	11884	13642	15551	17586		0	1.032
	600	14687	17797						20	1.069
	MIL	19042	18586	18252	18004	17786	17596	17122	40	1.105
	V_{MAX}	631.6	606.8	585.3	566.5	552.1	540.3	511.8		

FDD-1-(139)A

Figure 11-26

LOW ALTITUDE CRUISE

GROSS WEIGHT – 55,000 POUNDS

AIRPLANE CONFIGURATION
INDIVIDUAL DRAG INDEXES

REMARKS
ENGINES: (2) J79-GE-10

DATE: 1 JUNE 1971
DATA BASIS: FLIGHT TEST

FUEL GRADE: JP-5
FUEL DENSITY: 6.8 LB/GAL

	KTAS	DRAG INDEX / TOTAL FUEL FLOW LB/HR							TEMP. EFFECTS	
		0	20	40	60	80	100	120	°C	FACTOR
SEA LEVEL (15°C)	360	8196	8999	9823	10672	11528	12361	13341		
	420	9901	11060	12247	13470	14742	16133	17455	−40	.899
	480	12076	13654	15312	17077	18947	20871	23381	−20	.937
	540	14926	17014	19108	21770	25265			0	.973
	600	20702	23882	27592					20	1.008
	MIL	28164	27862	27615	27012	26516	25912	25256	40	1.042
	VMAX	628.4	613.1	600.5	572.1	550.2	524.6	497.4		
4,000 FEET (7°C)	360	7544	8282	9021	9756	10519	11290	12066		
	420	9001	10037	11096	12137	13313	14579	16100	−40	.913
	480	10825	12236	13653	15300	17188	19134	21004	−20	.949
	540	13335	15200	17202	20100	23049			0	.987
	600	19395	23090						20	1.022
	MIL	25807	25419	25067	24468	24022	23547	23004	40	1.057
	VMAX	629.2	612.2	597.3	571.1	551.1	528.7	501.9		
8,000 FEET (-1°C)	360	6976	7608	8263	8942	9603	10278	10994		
	420	8238	9153	10073	11006	11961	13037	14318	−40	.925
	480	9797	11014	12223	13630	15288	17219	19339	−20	.963
	540	11987	13528	15211	17851	20923			0	1.001
	600	17821	21601						20	1.037
	MIL	23425	22939	22549	22064	21714	21391	20962	40	1.072
	VMAX	629.9	611.0	593.1	569.7	551.6	532.7	505.1		
12,000 FEET (-9°C)	360	6592	7142	7743	8324	8912	9528	10136		
	420	7499	8316	9143	9970	10829	11675	12673	−40	.939
	480	8836	9909	10987	12113	13477	15084	17072	−20	.978
	540	10740	12072	13589	15542	18431			0	1.016
	600	16223	19736						20	1.052
	MIL	21121	20649	20260	19907	19642	19382	18968	40	1.088
	VMAX	629.7	608.9	588.4	567.5	551.4	535.1	506.2		
16,000 FEET (-16°C)	360	6385	6866	7410	7910	8459	8988	9590		
	420	6937	7659	8360	9089	9900	10751	11558	−40	.953
	480	8069	9008	9946	11020	12102	13458	15086	−20	.993
	540	9786	10996	12370	14152	16261			0	1.032
	600	15100	18399						20	1.069
	MIL	18965	18541	18204	17959	17734	17452	16972	40	1.105
	VMAX	628.2	604.1	581.7	563.1	548.9	531.4	502.9		

FDD-1-(140)A

Figure 11-27

CONSTANT MACH/ALTITUDE CRUISE

AIRPLANE CONFIGURATION
INDIVIDUAL DRAG INDEXES

REMARKS
ENGINE(S): (2) J79-GE-10

DATE: 1 JULY 1971
DATA BASIS: FLIGHT TEST

GUIDE

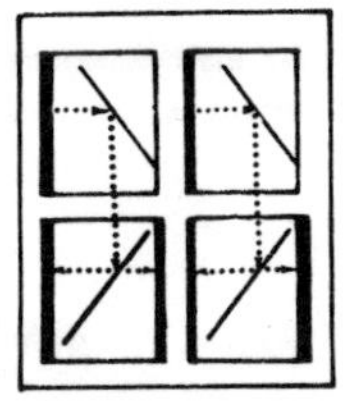

FUEL GRADE: JP-5
FUEL DENSITY: 6.8 LB/GAL

ICAO STANDARD DAY	
ALT. FT.	TEMP °C
S.L.	15.0
5000	5.1
10,000	-4.8
15,000	-14.7
20,000	-24.6
25,000	-34.5
30,000	-44.4
35,000	-54.3
40,000	-56.5
45,000	-56.5
50,000	-56.5

FUEL FLOW CORRECTION FACTORS						
TEMPERATURE °C	-60	-40	-20	0	20	40
FUEL FLOW FACTOR	0.88	0.92	0.96	1.00	1.04	1.07

0.60 MACH

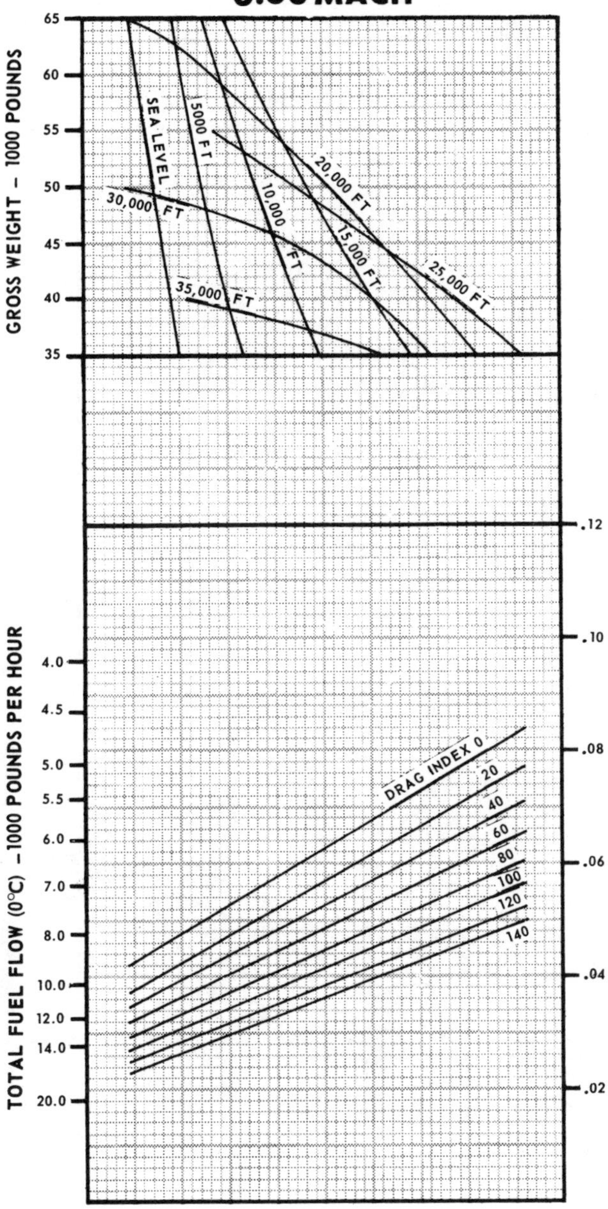

0.70 MACH

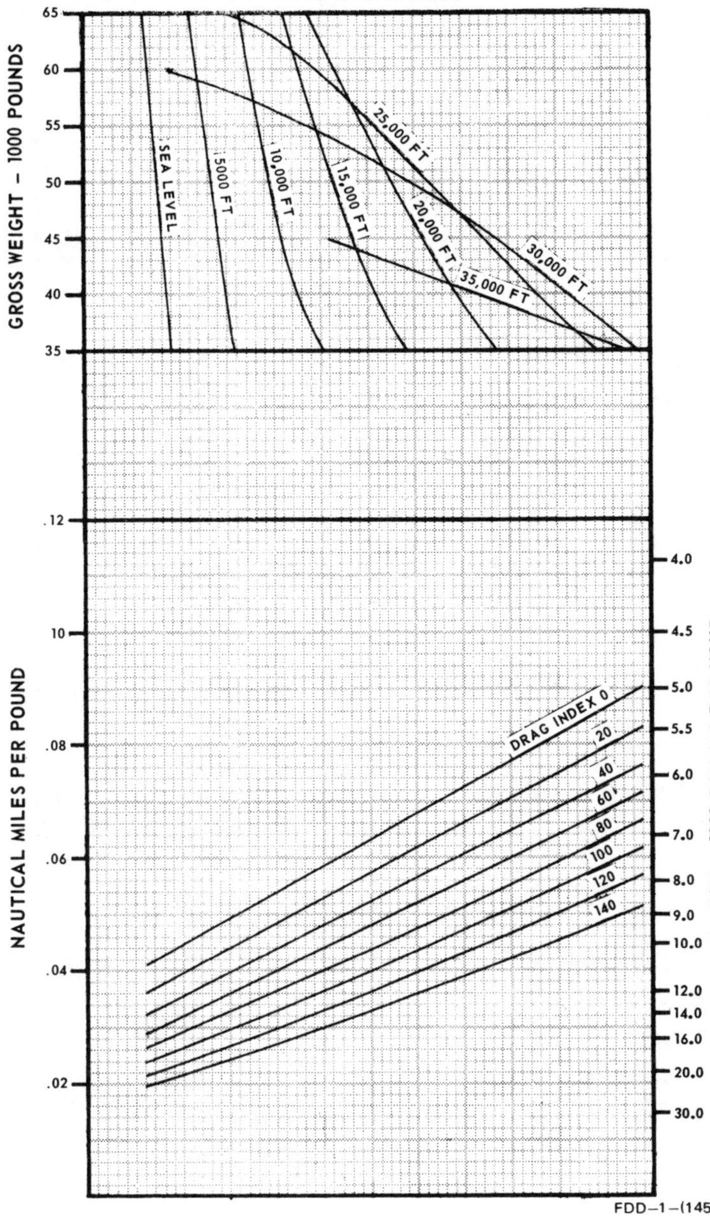

FDD-1-(145)A

Figure 11-28

CONSTANT MACH/ALTITUDE CRUISE

AIRPLANE CONFIGURATION
INDIVIDUAL DRAG INDEXES

REMARKS
ENGINE(S): (2) J79-GE-10

DATE: 1 JULY 1971
DATA BASIS: FLIGHT TEST

GUIDE

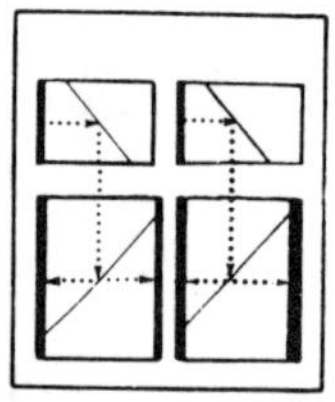

FUEL GRADE: JP-5
FUEL DENSITY: 6.8 LB/GAL

ICAO STANDARD DAY	
ALT. FT.	TEMP °C
S.L.	15.0
5000	5.1
10,000	-4.8
15,000	-14.7
20,000	-24.6
25,000	-34.5
30,000	-44.4
35,000	-54.3
40,000	-56.5
45,000	-56.5
50,000	-56.5

FUEL FLOW CORRECTION FACTORS						
TEMPERATURE °C	-60	-40	-20	0	20	40
FUEL FLOW FACTOR	0.88	0.92	0.96	1.00	1.04	1.07

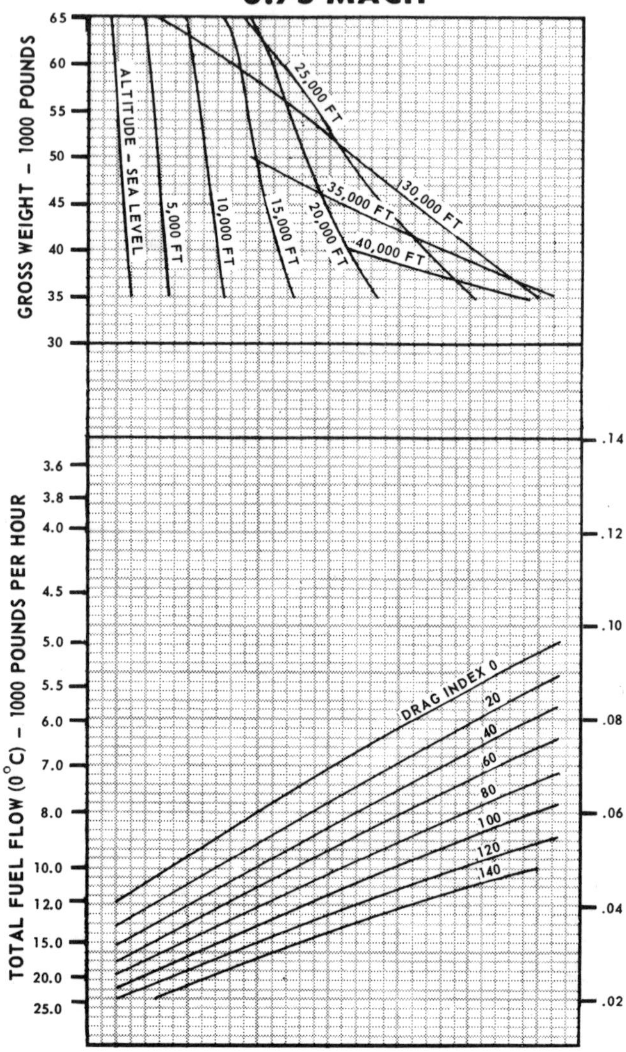

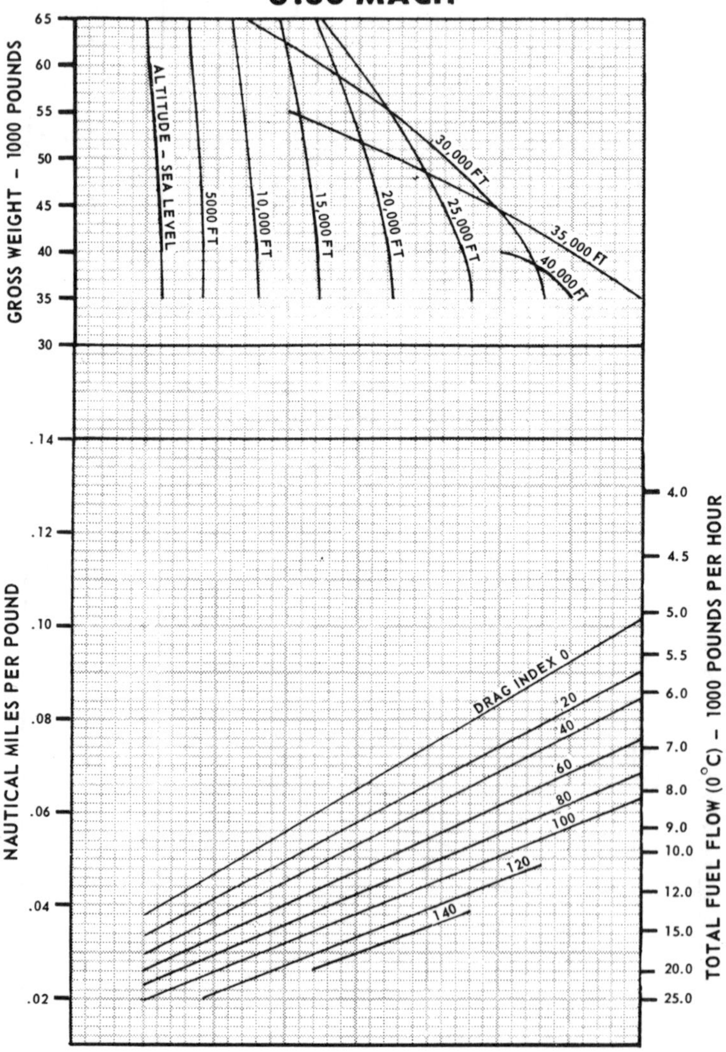

FDD-1-(146)A

Figure 11-29

CONSTANT MACH/ALTITUDE CRUISE

AIRPLANE CONFIGURATION
INDIVIDUAL DRAG INDEXES

REMARKS
ENGINE(S): (2) J79-GE-10

DATE: 1 JULY 1971
DATA BASIS: FLIGHT TEST

GUIDE

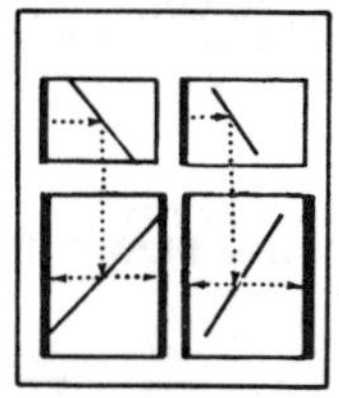

FUEL GRADE: JP-5
FUEL DENSITY: 6.8 LB/GAL

ICAO STANDARD DAY	
ALT. FT.	TEMP °C
S.L.	15.0
5000	5.1
10,000	-4.8
15,000	-14.7
20,000	-24.6
25,000	-34.5
30,000	-44.4
35,000	-54.3
40,000	-56.5
45,000	-56.5
50,000	-56.5

FUEL FLOW CORRECTION FACTORS						
TEMPERATURE °C	-60	-40	-20	0	20	40
FUEL FLOW FACTOR	0.88	0.92	0.96	1.00	1.04	1.07

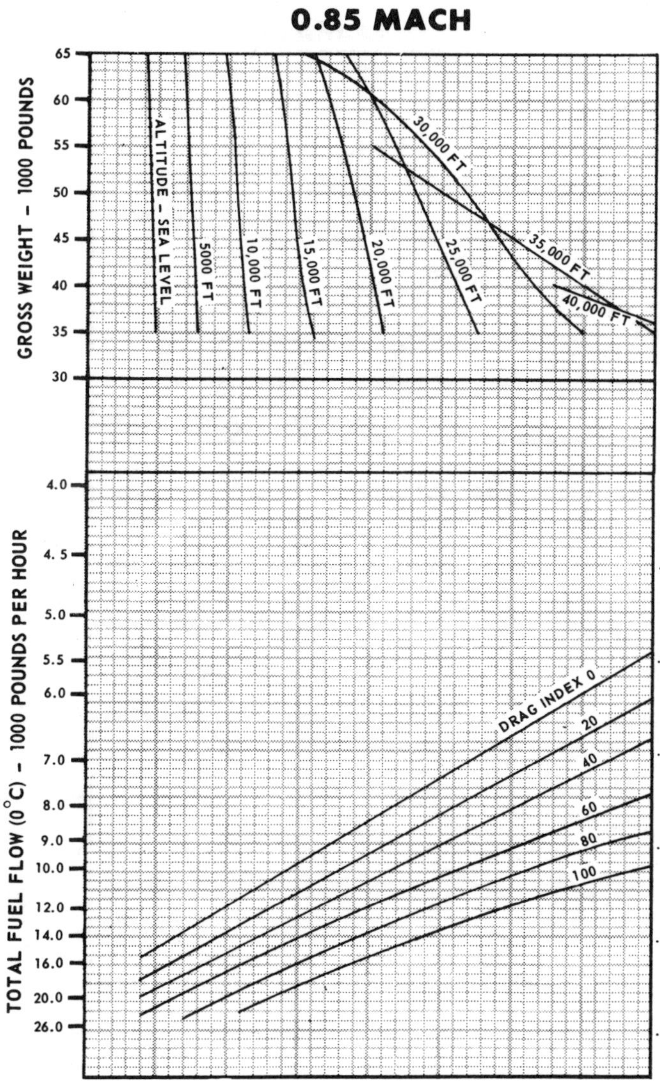

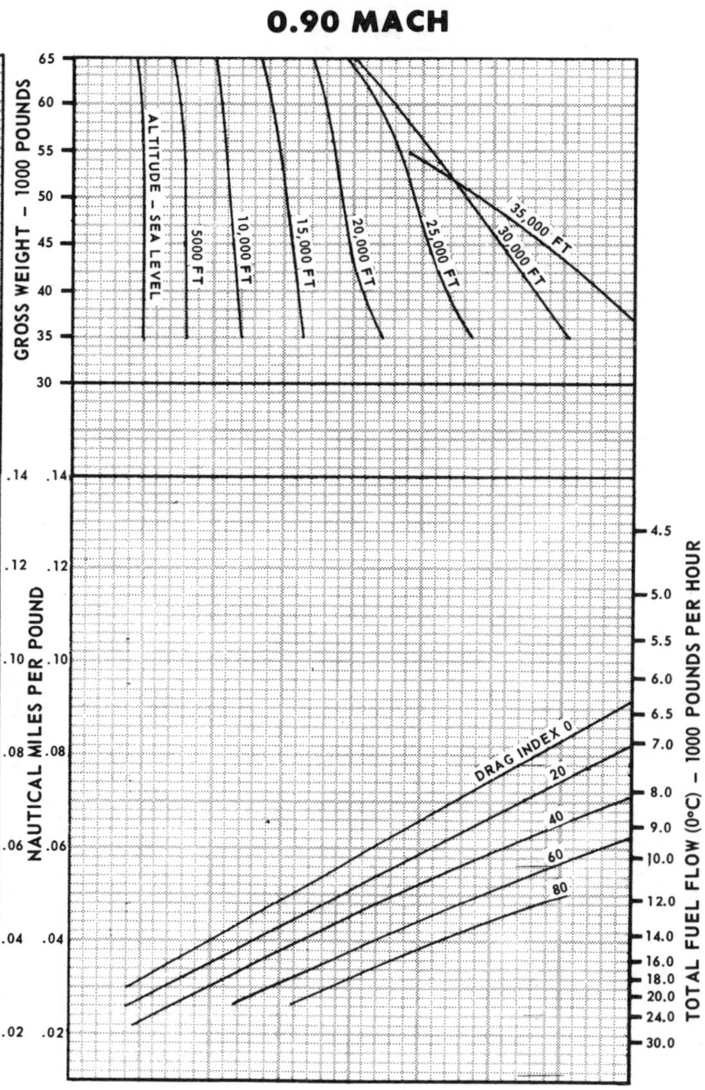

FDD-1-(147)A

Figure 11-30

CONSTANT MACH/ALTITUDE CRUISE

ONE ENGINE OPERATING

AIRPLANE CONFIGURATION
INDIVIDUAL DRAG INDEXES

REMARKS
ENGINE(S): (2) J79-GE-10
INOPERATIVE ENGINE WINDMILLING

GUIDE

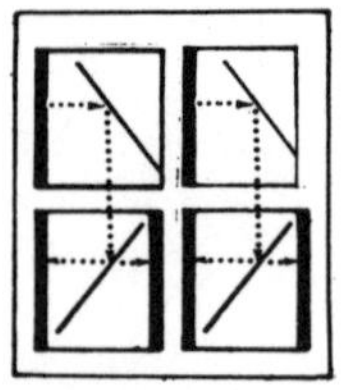

DATE: 1 JULY 1971
DATA BASIS: FLIGHT TEST

NOTE
IF INOPERATIVE ENGINE IS NOT WINDMILLING, INCREASE DRAG BY 53 ADDITIONAL UNITS.

FUEL GRADE: JP-5
FUEL DENSITY: 6.8 LB/GAL

ICAO STANDARD DAY	
ALT. FT.	TEMP °C
S.L.	15.0
5000	5.1
10,000	-4.8
15,000	-14.7
20,000	-24.6
25,000	-34.5
30,000	-44.4
35,000	-54.3
40,000	-56.5
45,000	-56.5
50,000	-56.5

FUEL FLOW CORRECTION FACTORS						
TEMPERATURE °C	-60	-40	-20	0	20	40
FUEL FLOW FACTOR	0.88	0.92	0.96	1.00	1.04	1.07

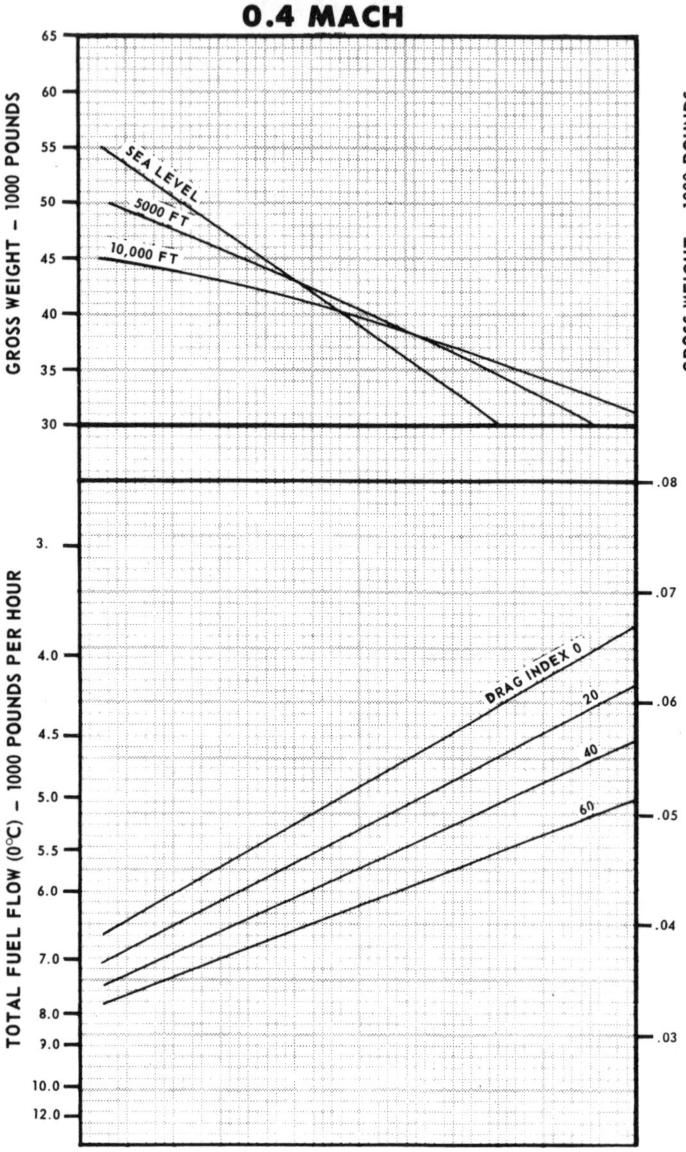

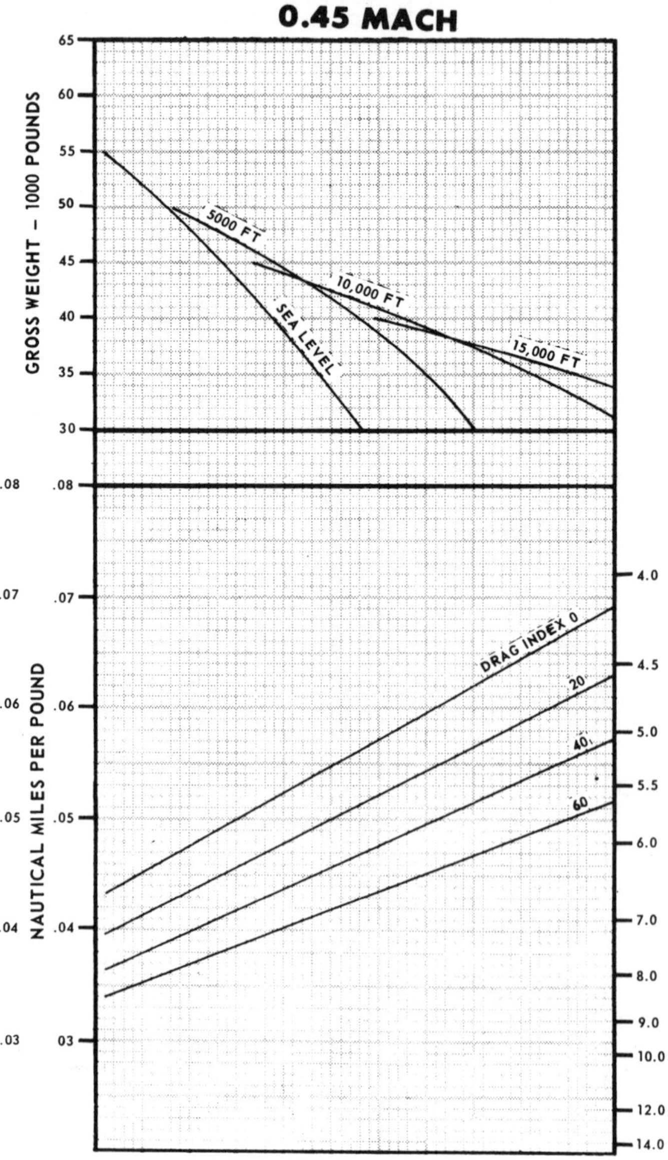

FDD-1-(148)A

Figure 11-31

CONSTANT MACH/ALTITUDE CRUISE
ONE ENGINE OPERATING

AIRPLANE CONFIGURATION
INDIVIDUAL DRAG INDEXES

REMARKS
ENGINE(S): (2) J79-GE-10
INOPERATIVE ENGINE WINDMILLING

GUIDE

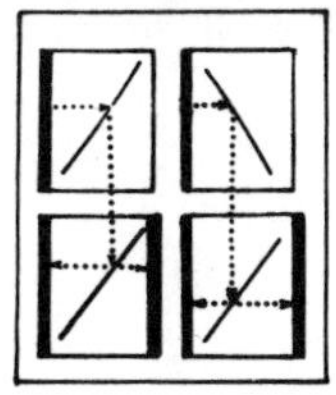

DATE: 1 JULY 1971
DATA BASIS: FLIGHT TEST

NOTE
IF INOPERATIVE ENGINE IS NOT WINDMILLING, INCREASE DRAG BY 53 ADDITIONAL UNITS.

FUEL GRADE: JP-5
FUEL DENSITY: 6.8 LB/GAL

ICAO STANDARD DAY	
ALT. FT.	TEMP °C
S.L.	15.0
5000	5.1
10,000	-4.8
15,000	-14.7
20,000	-24.6
25,000	-34.5
30,000	-44.4
35,000	-54.3
40,000	-56.5
45,000	-56.5
50,000	-56.5

FUEL FLOW CORRECTION FACTORS						
TEMPERATURE °C	-60	-40	-20	0	20	40
FUEL FLOW FACTOR	0.88	0.92	0.96	1.00	1.04	1.07

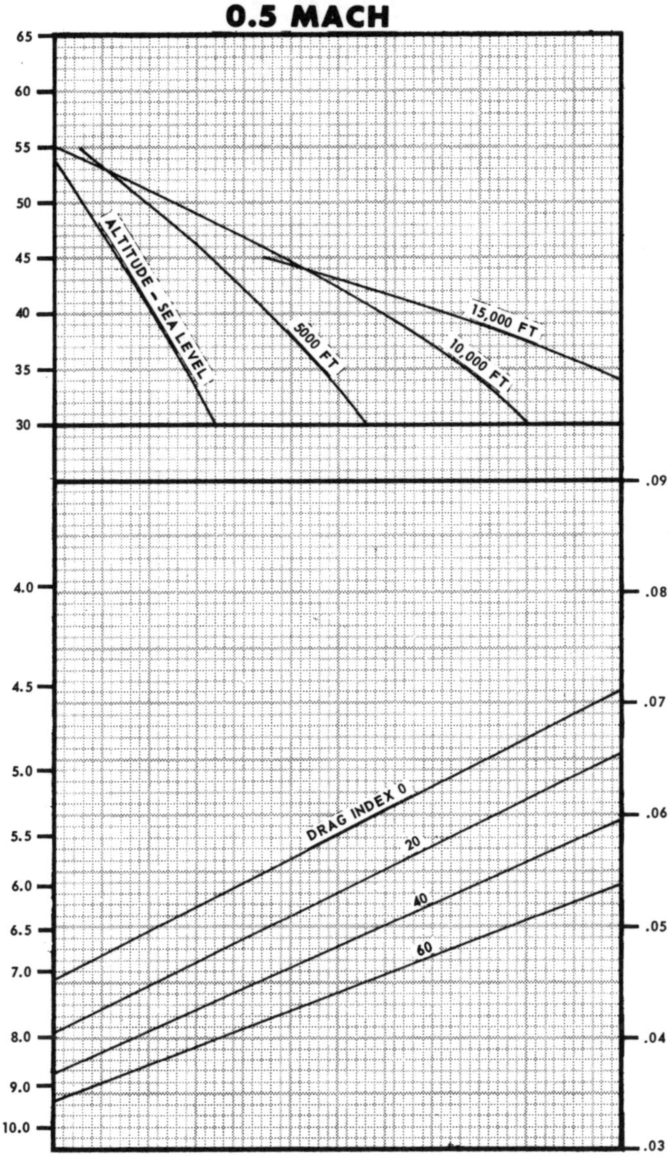

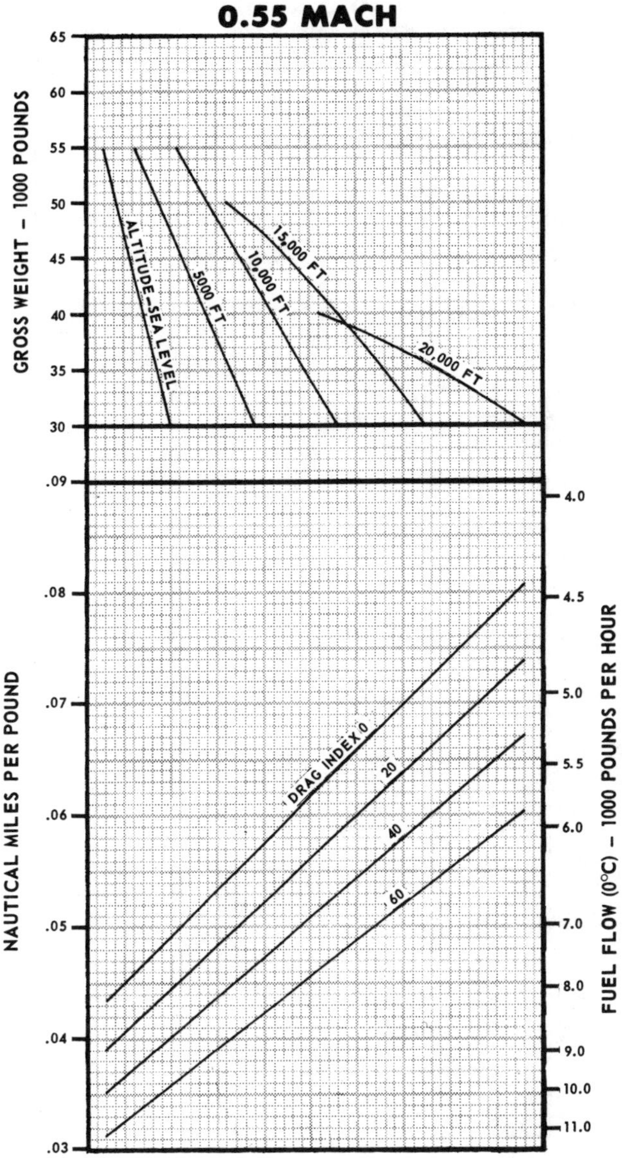

FDD-1-(149)A

Figure 11-32

CONSTANT MACH/ALTITUDE CRUISE

ONE ENGINE OPERATING

AIRPLANE CONFIGURATION
INDIVIDUAL DRAG INDEXES

DATE: 1 JULY 1971
DATA BASIS: FLIGHT TEST

REMARKS
ENGINE(S): (2) J79-GE-10
INOPERATIVE ENGINE WINDMILLING

NOTE
IF INOPERATIVE ENGINE IS NOT WINDMILLING, INCREASE DRAG BY 53 ADDITIONAL UNITS.

GUIDE

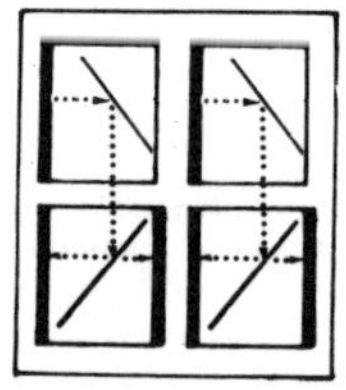

FUEL GRADE: JP-5
FUEL DENSITY: 6.8 LB/GAL

ICAO STANDARD DAY	
ALT. FT.	TEMP °C
S.L.	15.0
5000	5.1
10,000	-4.8
15,000	-14.7
20,000	-24.6
25,000	-34.5
30,000	-44.4
35,000	-54.3
40,000	-56.5
45,000	-56.5
50,000	-56.5

FUEL FLOW CORRECTION FACTORS						
TEMPERATURE °C	-60	-40	-20	0	20	40
FUEL FLOW FACTOR	0.88	0.92	0.96	1.00	1.04	1.07

0.6 MACH

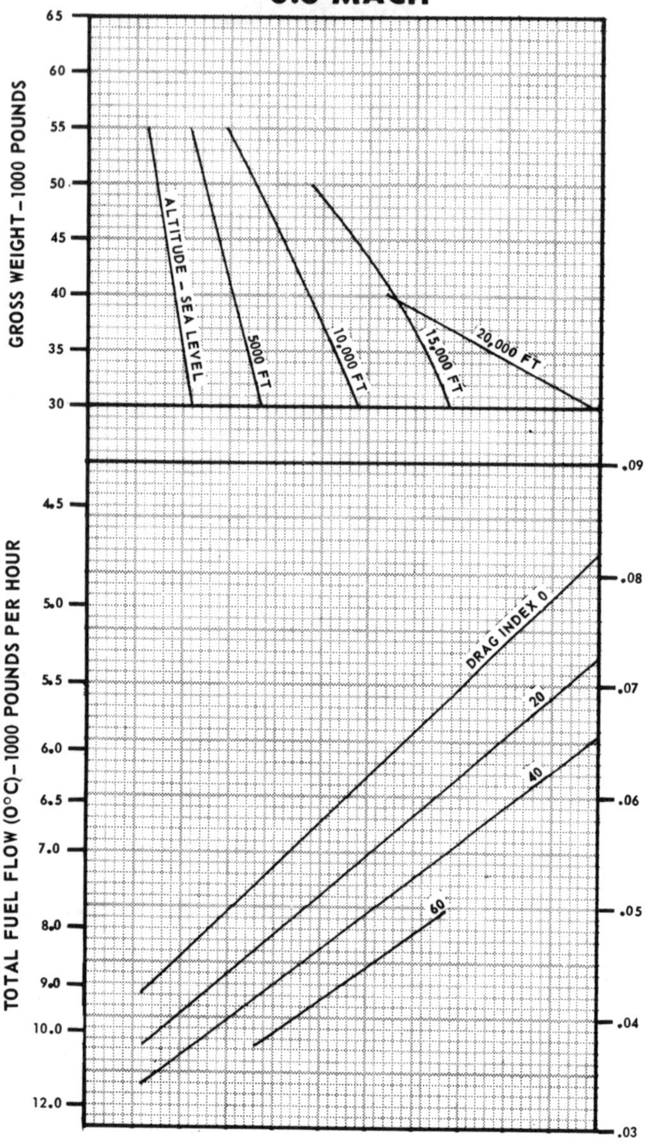

0.65 MACH

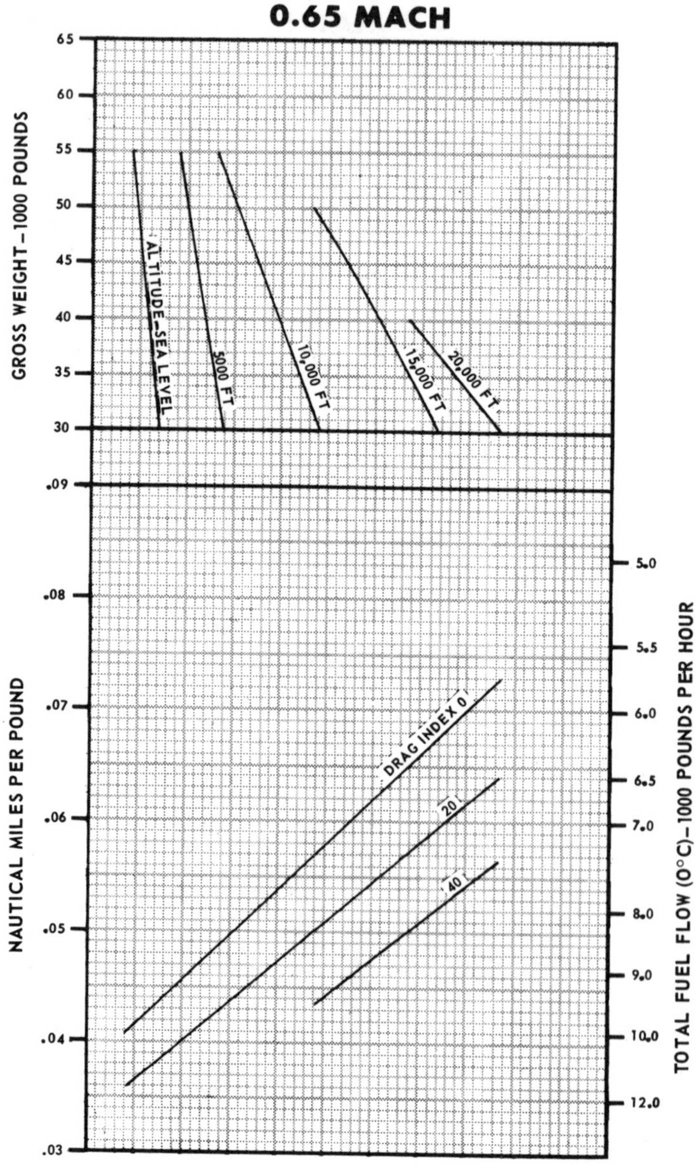

FDD-1-(150)A

Figure 11-33

CONSTANT ALTITUDE CRUISE

LONG RANGE SPEED NAUTICAL MILES PER POUND AND MACH NUMBER

AIRPLANE CONFIGURATION
INDIVIDUAL DRAG INDEXES

REMARKS
ENGINE(S): (2) J79-GE-10

DATE: 1 JUNE 1971
DATA BASIS: FLIGHT TEST

FUEL GRADE: JP-5
FUEL DENSITY: 6.8 LB/GAL

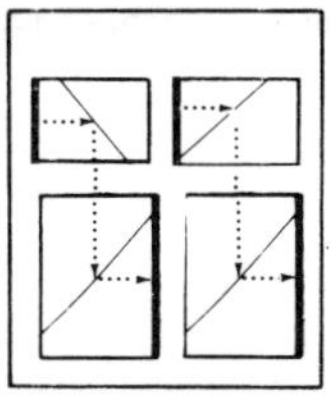

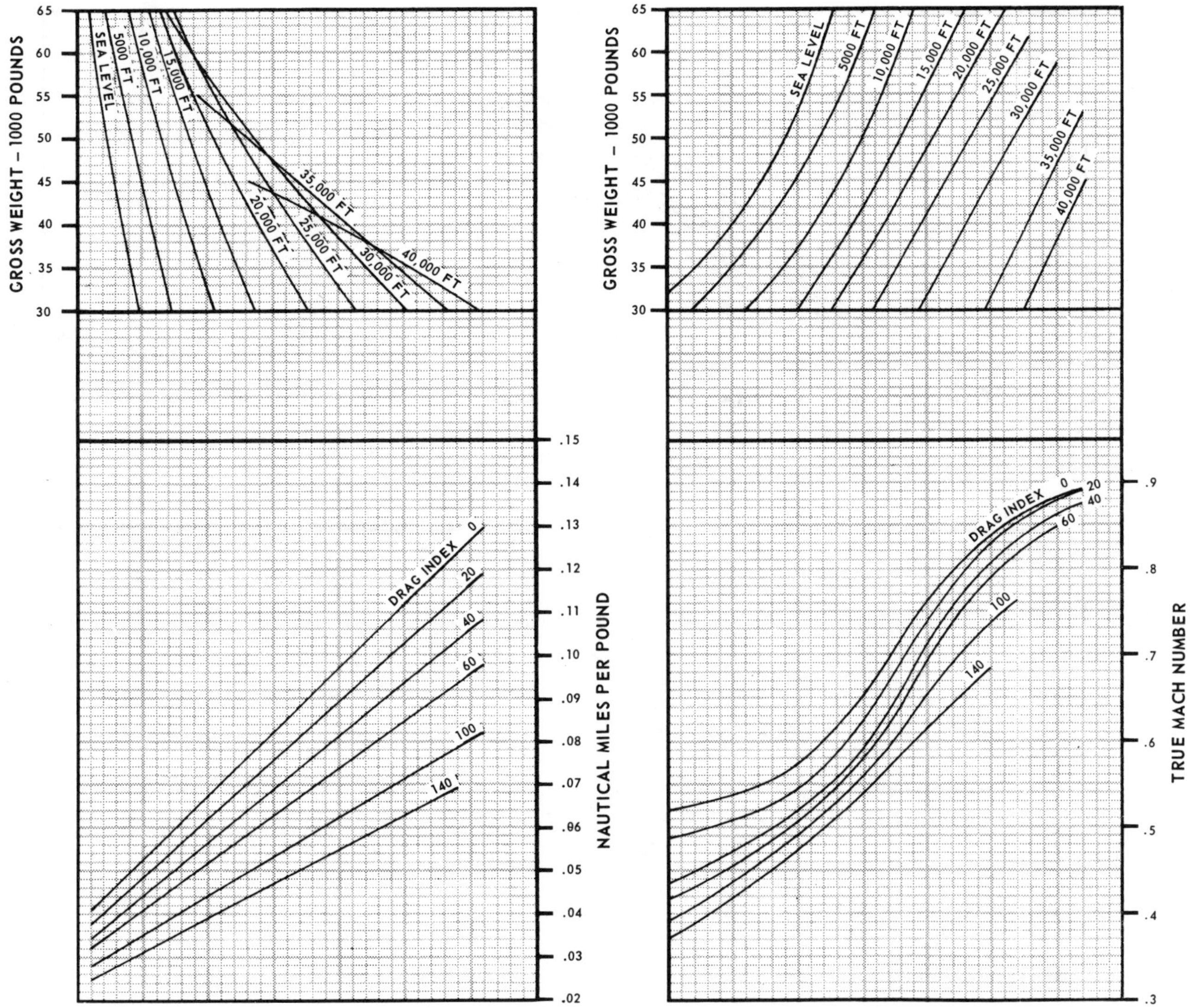

FDD-1(297-1)

Figure 11-34 (Sheet 1 of 2)

CONSTANT ALTITUDE CRUISE

AIRPLANE CONFIGURATION
INDIVIDUAL DRAG INDEXES

LONG RANGE SPEED
TRUE AIRSPEED AND FUEL FLOW

REMARKS
ENGINE(S): (2) J79-GE-10

GUIDE

DATE: 1 MARCH 1968
DATA BASIS: FLIGHT TEST

FUEL GRADE: JP-5
FUEL DENSITY: 6.8 LB/GAL

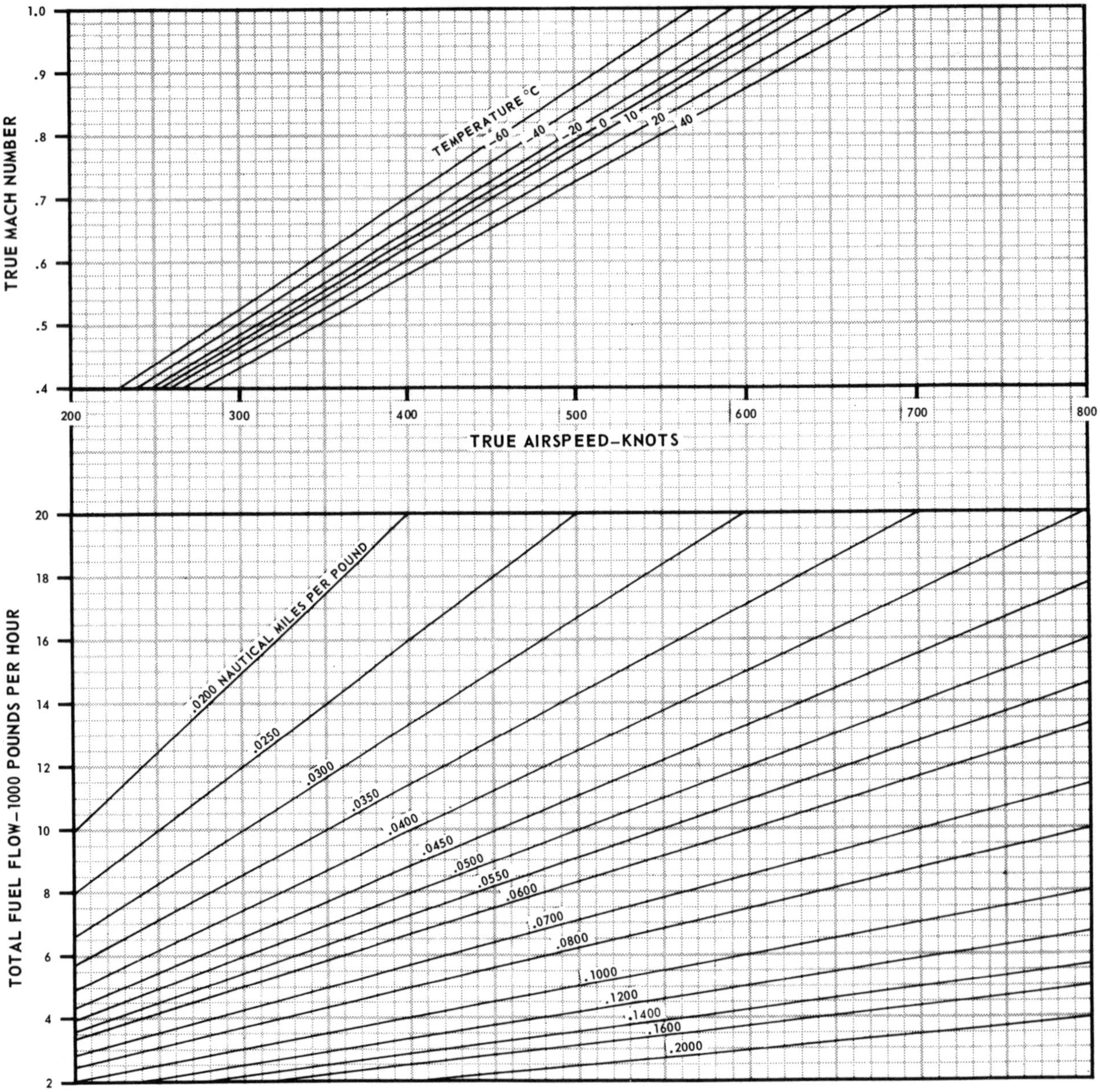

FDD-1-(297-2)

Figure 11-34 (Sheet 2 of 2)

CONSTANT ALTITUDE CRUISE

LONG RANGE SPEED NAUTICAL MILES PER POUND AND MACH NUMBER ONE ENGINE OPERATING

AIRPLANE CONFIGURATION
INDIVIDUAL DRAG INDEXES

REMARKS
ENGINE(S): (2) J79–GE–10
INOPERATIVE ENGINE WINDMILLING

NOTE
IF INOPERATIVE ENGINE IS NOT WINDMILLING, INCREASE DRAG BY 53 ADDITIONAL UNITS.

DATE: 1 JUNE 1971
DATA BASIS: FLIGHT TEST

FUEL GRADE: JP–5
FUEL DENSITY: 6.8 LB/GAL

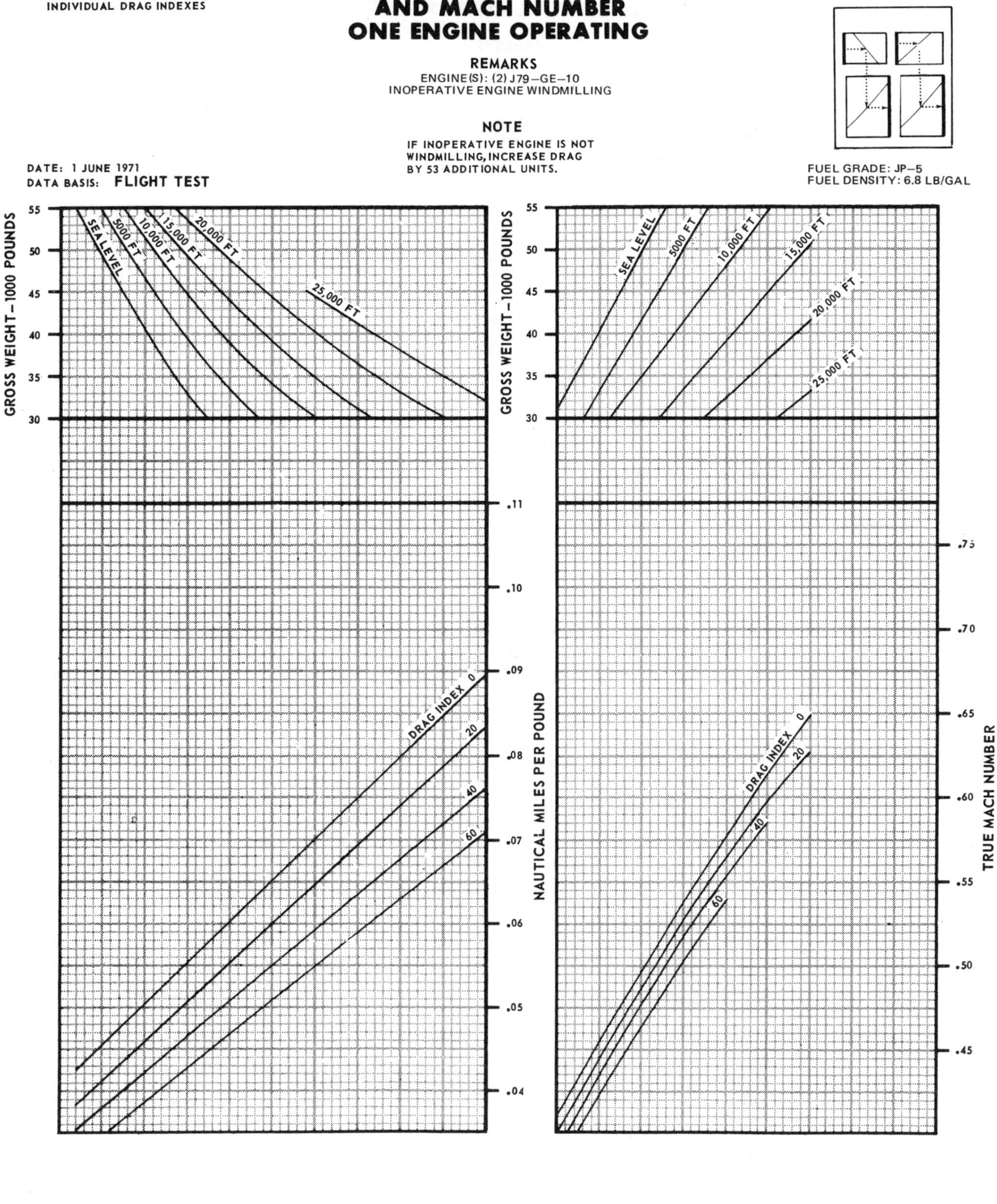

FDD–1–(298–1)

Figure 11-35 (Sheet 1 of 2)

CONSTANT ALTITUDE CRUISE
LONG RANGE SPEED
TRUE AIRSPEED AND FUEL FLOW
ONE ENGINE OPERATING

AIRPLANE CONFIGURATION
INDIVIDUAL DRAG INDEXES

REMARKS
ENGINE(S): (2) J79—GE—10
INOPERATIVE ENGINE WINDMILLING

NOTE
IF INOPERATIVE ENGINE IS NOT WINDMILLING, INCREASE DRAG BY 53 ADDITIONAL UNITS.

GUIDE

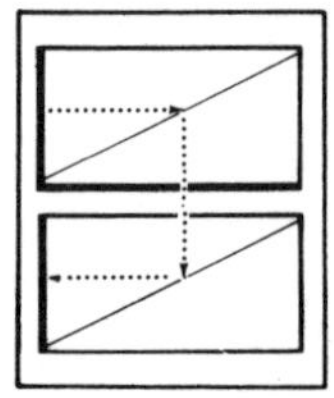

DATE: 15 AUGUST 1968
DATA BASIS: FLIGHT TEST

FUEL GRADE: JP—5
FUEL DENSITY: 6.8 LB/GAL

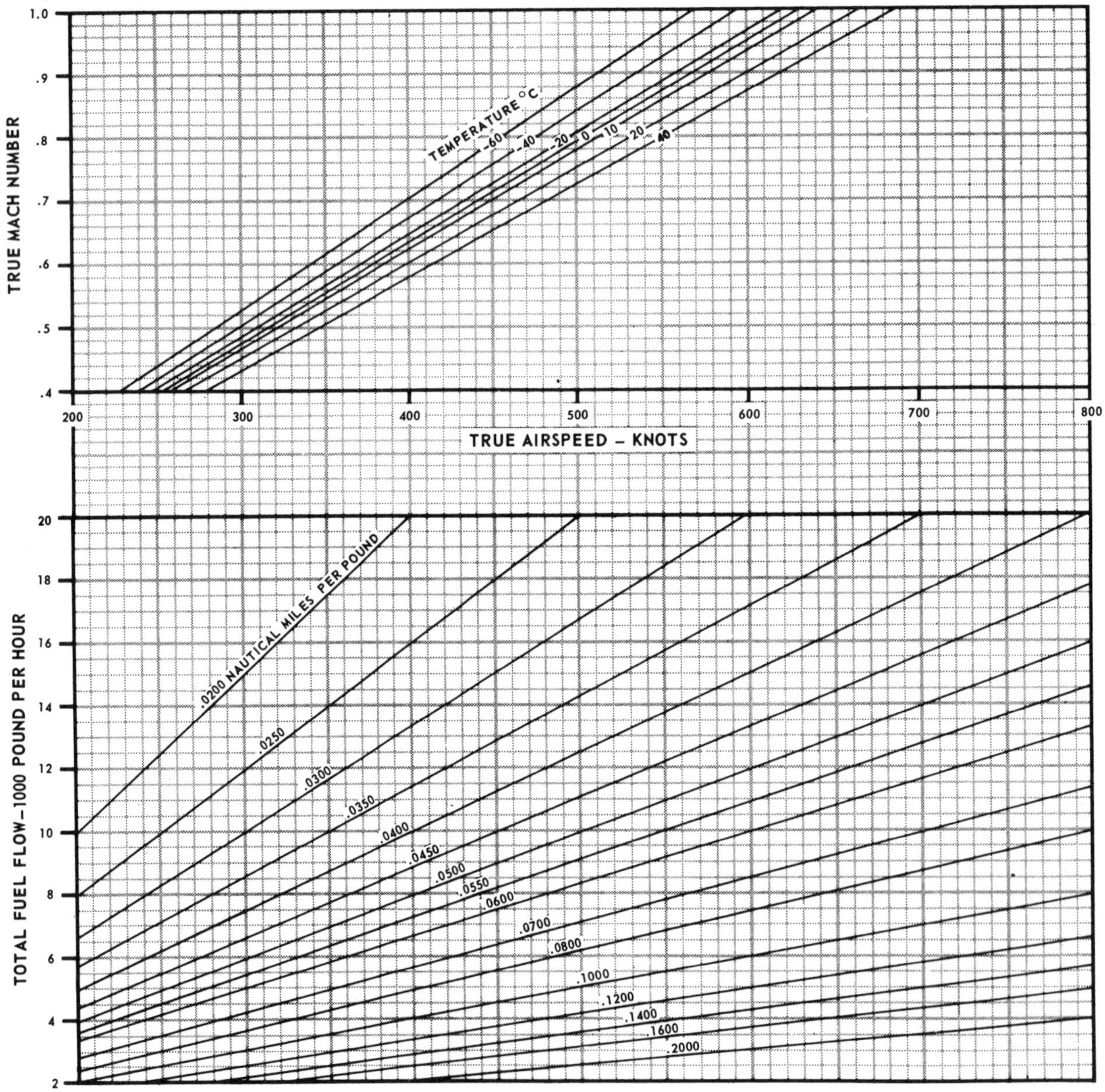

FDD-1-(298-2)

Figure 11-35 (Sheet 2 of 2)

PART 5 ENDURANCE

MAXIMUM ENDURANCE CHARTS

These charts (figure 11-36 thru 11-41) present optimum endurance altitude and maximum endurance specifics (fuel flow and Mach number) for all combinations of effective gross weight and altitude. Separate charts are included for single-engine operation.

USE

Enter the Altitude and Bank Angle chart with the average gross weight. If bank angles are to be considered, follow the gross weight curve until it intersects the bank angle to be used, then horizontally to the right to obtain effective gross weight. (If bank angles are not to be considered, enter the chart at the effective gross weight scale.) From this point proceed horizontally to the right and intersect the computed drag index. Reflect downward and read the optimum endurance altitude. Enter the Mach number plots with the effective gross weight, and proceed horizontally to intersect the optimum endurance altitude. Then descend downward and intersect the computed drag index and horizontally to read true Mach number. From the intersection of endurance altitude and drag index proceed horizontally to the right and intersect the optimum altitude. At this point read endurance airspeed. Enter the Fuel Flow plots with the effective gross weight, proceed horizontally to intersect the optimum endurance altitude. Reflect downward to the computed drag index, and then horizontally to read total fuel flow.

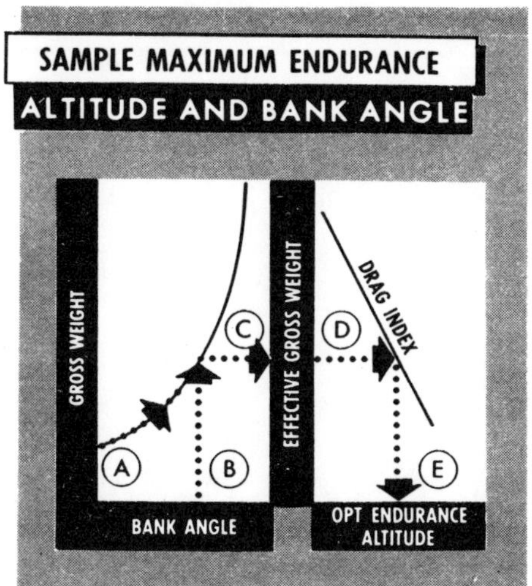

FDD-1-(154)

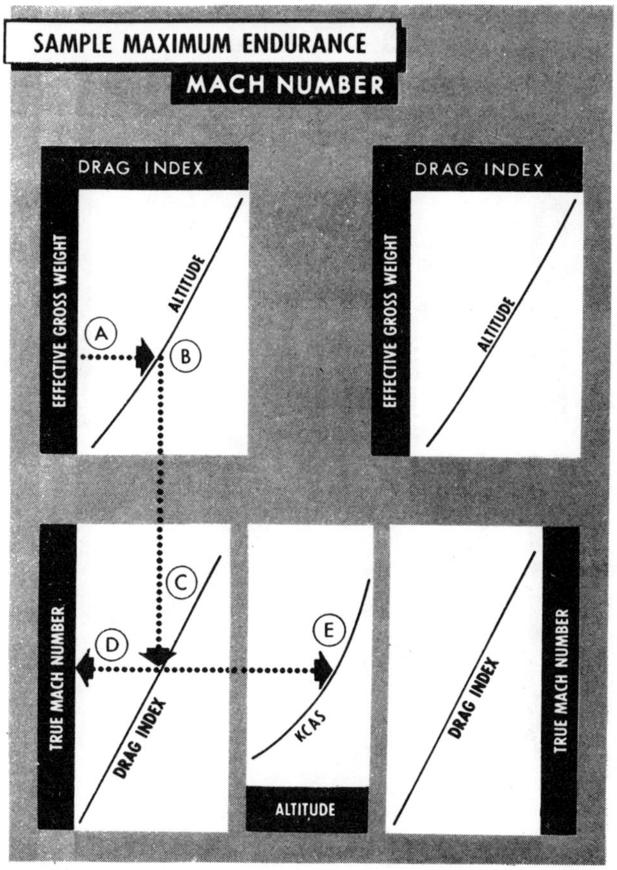

FDD-1-(155)

Sample Problem

Altitude and Bank Angle

A.	Gross weight	45,000 Lb
B.	Bank angle	20°
C.	Effective gross weight	48,000 Lb
D.	Drag index	40
E.	Optimum endurance altitude	25,000 Ft

Mach Number

A.	Effective gross weight	48,000 Lb
B.	Endurance altitude	25,000 Ft
C.	Drag index	40
D.	Mach number	0.64
E.	Airspeed (IAS)	265 Kt

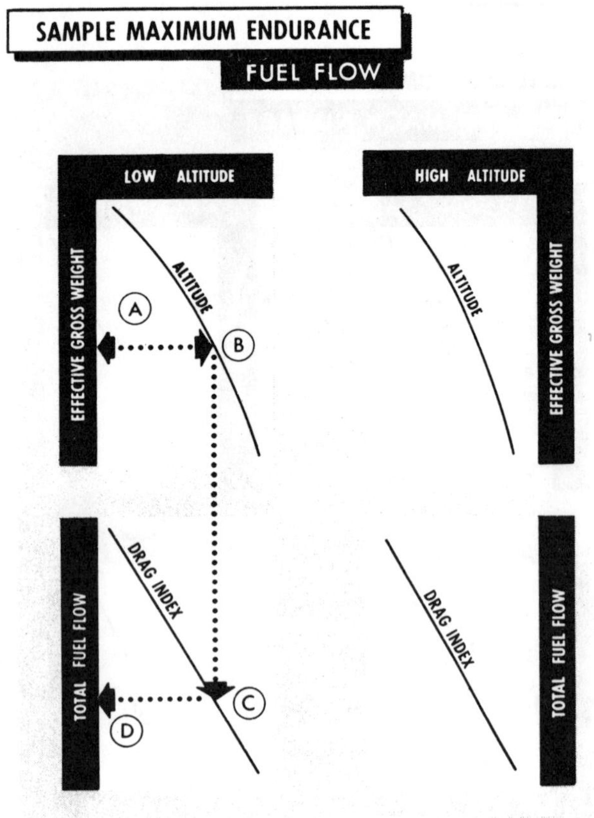

FDD-1-(156)

Fuel Flow

A.	Effective gross weight	48,000 Lb
B.	Endurance altitude	25,000 Ft
C.	Drag index	40
D.	Fuel flow	6200 PPH

MAXIMUM ENDURANCE
ALTITUDE AND BANK ANGLE

AIRPLANE CONFIGURATION
INDIVIDUAL DRAG INDEXES

REMARKS
ENGINE(S): (2) J79-GE-10
ICAO STANDARD DAY

GUIDE

DATE: 1 MAY 1971
DATA BASIS: FLIGHT TEST

FUEL GRADE: JP-5
FUEL DENSITY: 6.8 LB/GAL

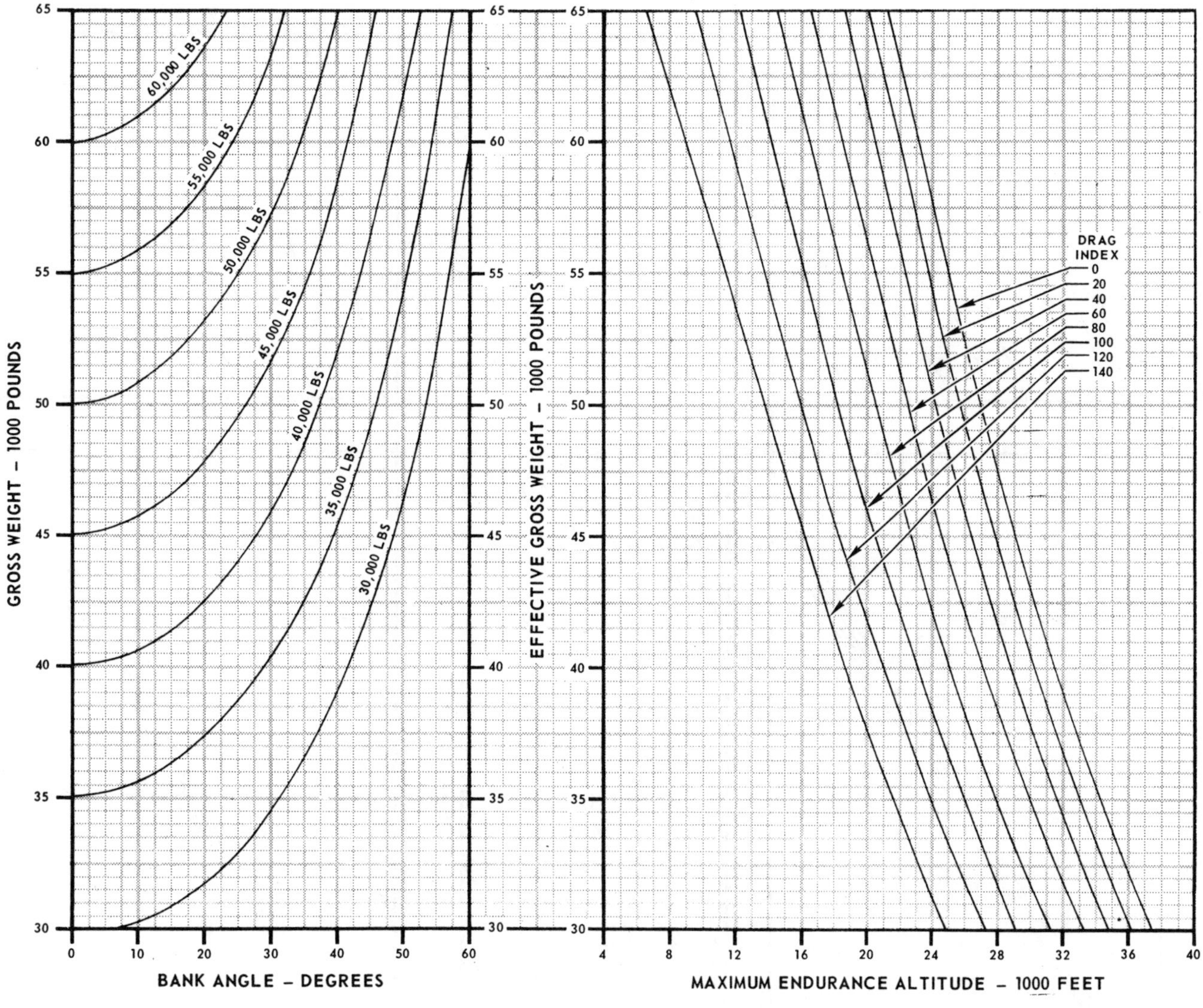

FDD-1-(160) A

Figure 11-36

MAXIMUM ENDURANCE
MACH NUMBER

AIRPLANE CONFIGURATION
INDIVIDUAL DRAG INDEXES

REMARKS
ENGINE(S): (2) J79-GE-10
ICAO STANDARD DAY

DATE: 1 MAY 1971
DATA BASIS: FLIGHT TEST

FUEL GRADE: JP-5
FUEL DENSITY: 6.8 LB/GAL

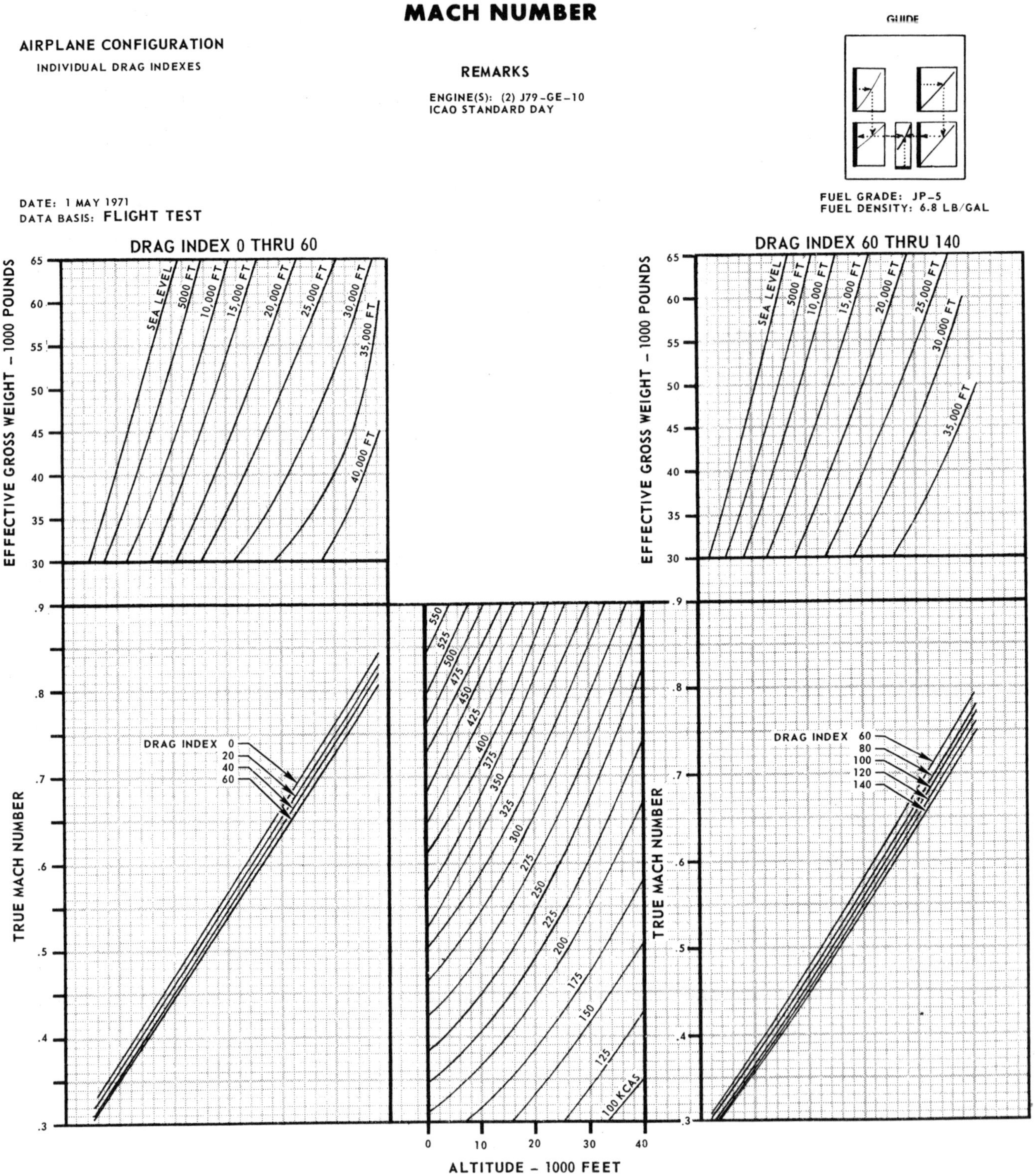

FDD-1-(161)A

Figure 11-37

AIRPLANE CONFIGURATION
INDIVIDUAL DRAG INDEXES
(0–60)

ICAO STANDARD DAY	
ALT. FT.	TEMP °C
S.L.	15.0
5000	5.1
10,000	-4.8
15,000	-14.7
20,000	-24.6
25,000	-34.5
30,000	-44.4
35,000	-54.3
40,000	-56.5
45,000	-56.5
50,000	-56.5

DATE: 1 JUNE 1971
DATA BASIS: FLIGHT TEST

MAXIMUM ENDURANCE
FUEL FLOW

REMARKS

ENGINE(S): (2) J79-GE-10
ICAO STANDARD DAY

NOTE

TOTAL FUEL FLOW IS DIRECTLY PROPORTIONAL TO TEMPERATURE CHANGE, INCREASING OR DECREASING 2% FOR EACH 10°C INCREMENT FROM STANDARD DAY.

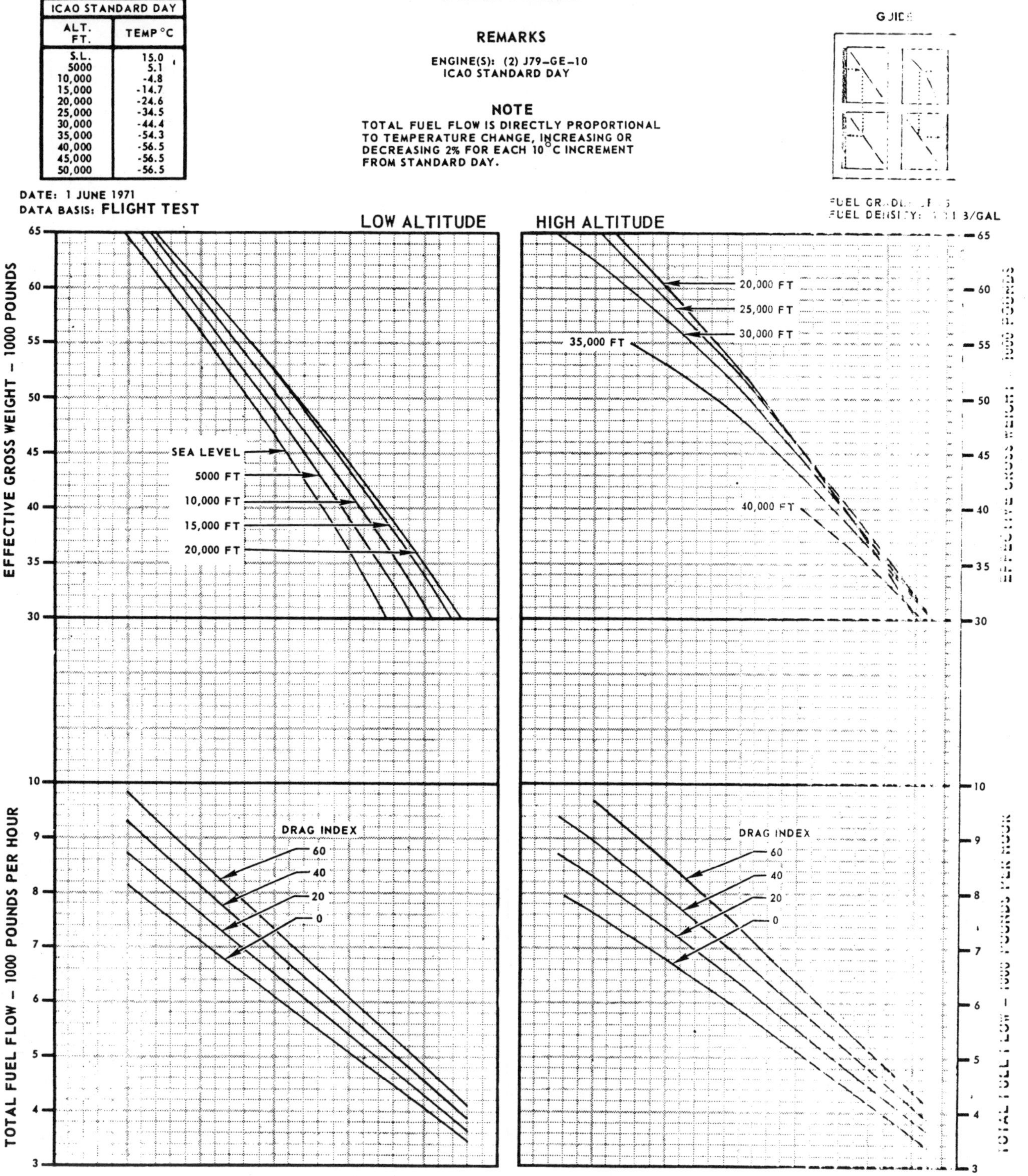

FDD-1-(162)A

Figure 11-38

MAXIMUM ENDURANCE
FUEL FLOW

AIRPLANE CONFIGURATION
INDIVIDUAL DRAG INDEXES
(60–140)

ICAO STANDARD DAY	
ALT. FT.	TEMP °C
S.L.	15.0
5000	5.1
10,000	-4.8
15,000	-14.7
20,000	-24.6
25,000	-34.5
30,000	-44.4
35,000	-54.3
40,000	-56.5
45,000	-56.5
50,000	-56.5

DATE: 1 MAY 1971
DATA BASIS: FLIGHT TEST

REMARKS

ENGINE(S): (2) J79-GE-10
ICAO STANDARD DAY

NOTE

TOTAL FUEL FLOW IS DIRECTLY PROPORTIONAL TO TEMPERATURE CHANGE, INCREASING OR DECREASING 2% FOR EACH 10°C INCREMENT FROM STANDARD DAY.

GUIDE

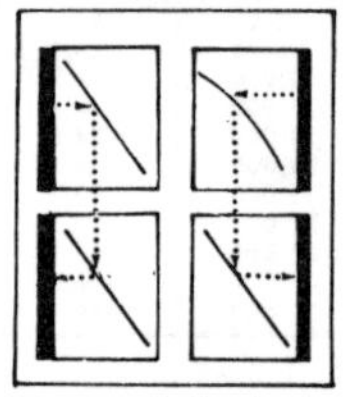

FUEL GRADE: JP-5
FUEL DENSITY: 6.8 LB/GAL

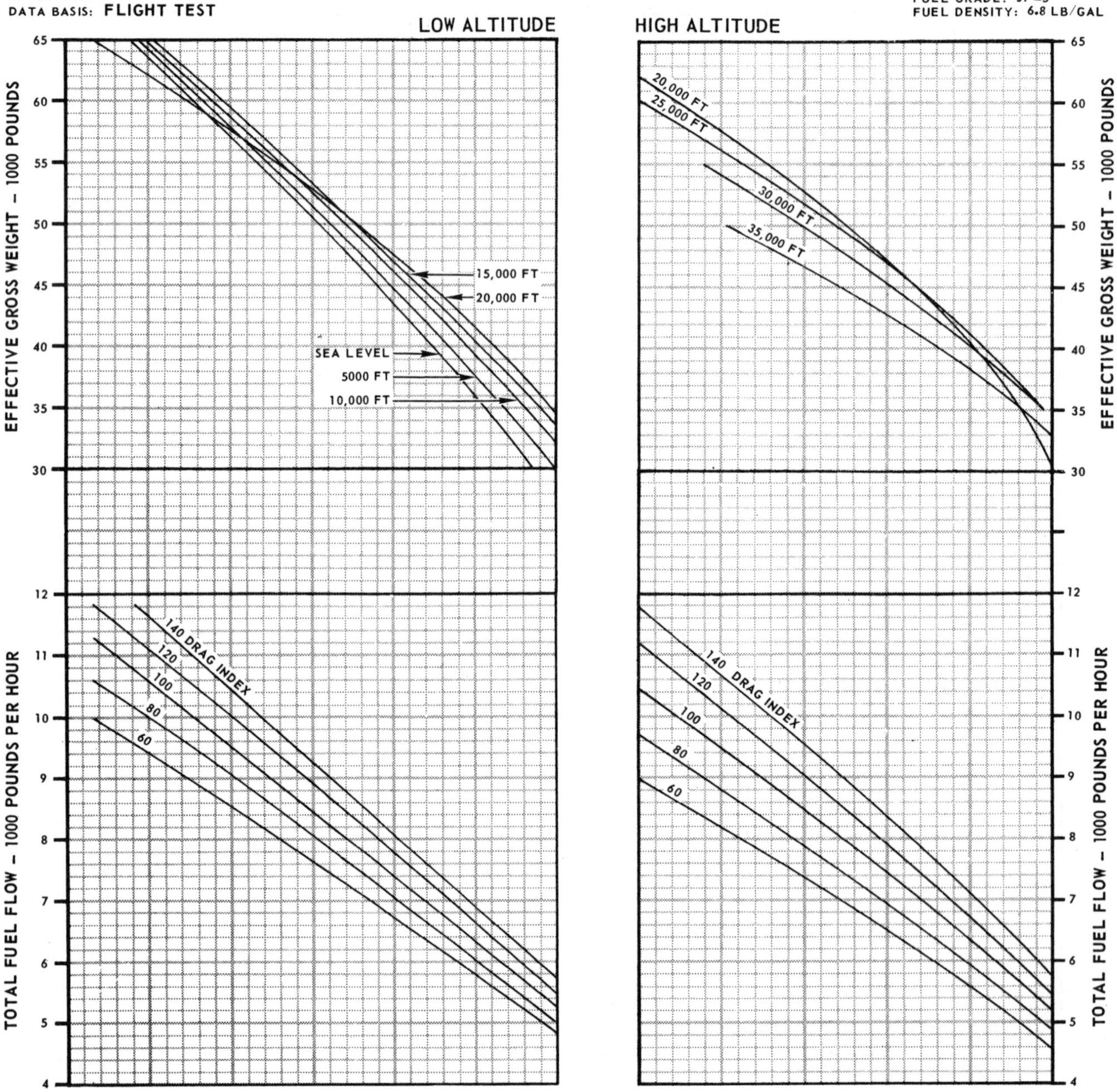

FDD-1-(163)A

Figure 11-39

MAXIMUM ENDURANCE
ALTITUDE AND BANK ANGLE ONE ENGINE OPERATING

AIRPLANE CONFIGURATION
INDIVIDUAL DRAG INDEXES

REMARKS

ENGINE(S): (2) J79-GE-10
ICAO STANDARD DAY
INOPERATIVE ENGINE WINDMILLING

NOTE

IF INOPERATIVE ENGINE IS NOT WINDMILLING, INCREASE DRAG BY 53 ADDITIONAL UNITS.

GUIDE

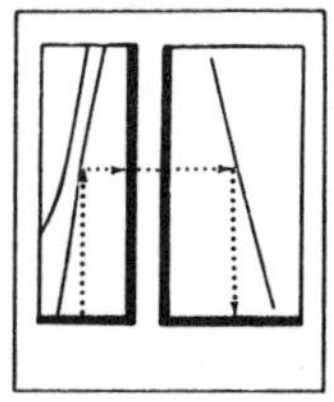

DATE: 1 MAY 1971
DATA BASIS: FLIGHT TEST

FUEL GRADE: JP-5
FUEL DENSITY: 6.8 LB/GAL

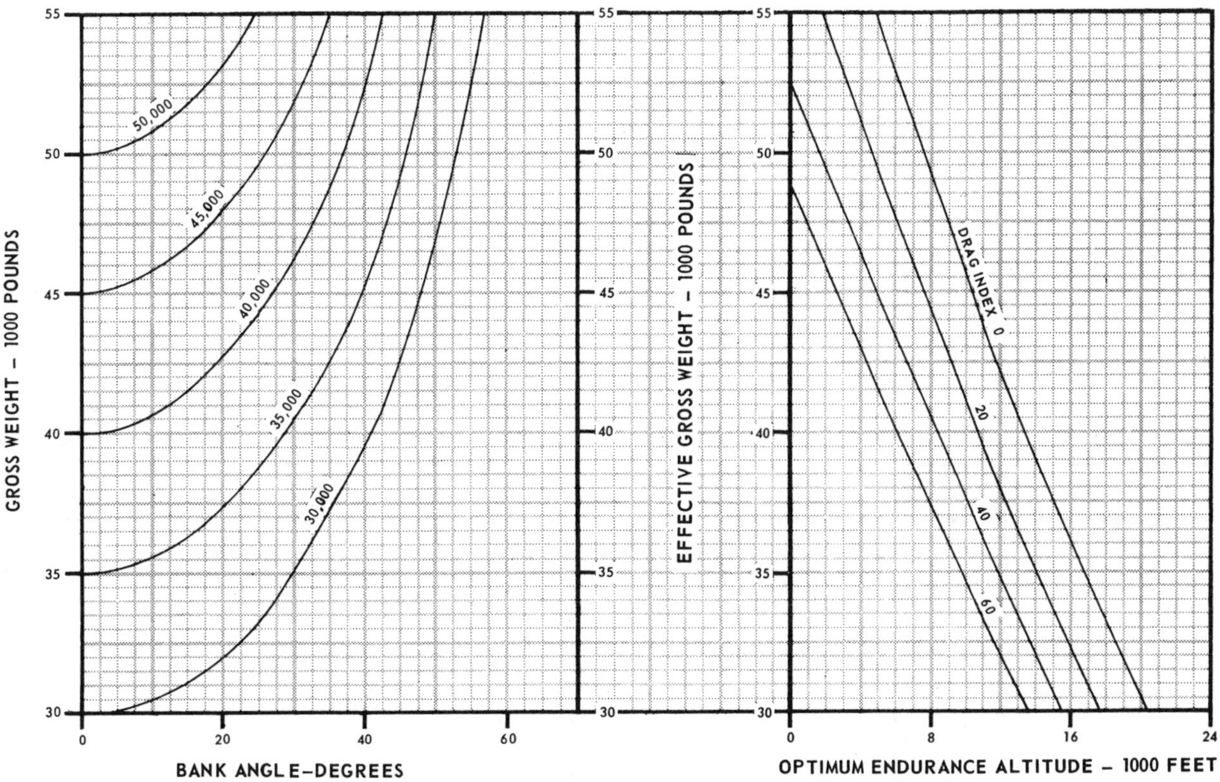

FDD-1-(164)A

Figure 11-40

MAXIMUM ENDURANCE

MACH NUMBER AND FUEL FLOW
ONE ENGINE OPERATING

AIRPLANE CONFIGURATION
INDIVIDUAL DRAG INDEXES

REMARKS

ENGINE(S): (2) J79-GE-10
ICAO STANDARD DAY
INOPERATIVE ENGINE WINDMILLING

NOTE

- IF INOPERATIVE ENGINE IS NOT WINDMILLING, INCREASE DRAG BY 53 ADDITIONAL UNITS.
- TOTAL FUEL FLOW IS DIRECTLY PROPORTIONAL TO TEMPERATURE CHANGE, INCREASING OR DECREASING 2% FOR EACH 10°C INCREMENT FROM STANDARD DAY.

GUIDE

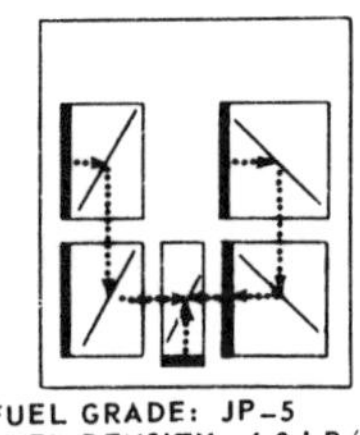

DATE: 1 MAY 1971
DATA BASIS: FLIGHT TEST

FUEL GRADE: JP-5
FUEL DENSITY: 6.8 LB/GAL

ICAO STANDARD DAY	
ALT. FT.	TEMP °C
S.L.	15.0
5000	5.1
10,000	-4.8
15,000	-14.7
20,000	-24.6
25,000	-34.5
30,000	-44.4
35,000	-54.3
40,000	-56.5
45,000	-56.5
50,000	-56.5

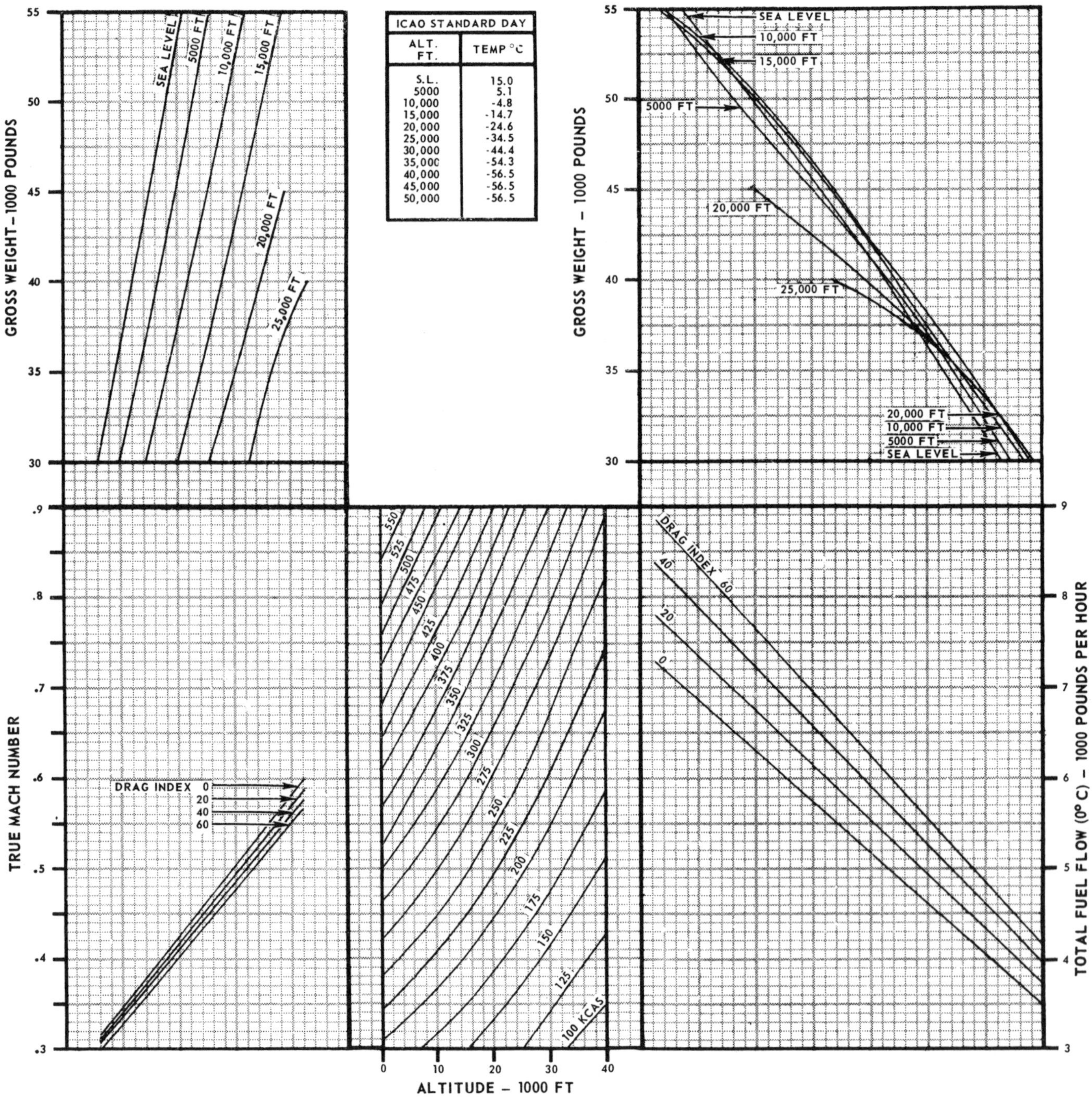

Figure 11-41

PART 6 AIR REFUELING

Note

Refer to NATOPS Air Refueling Manual.

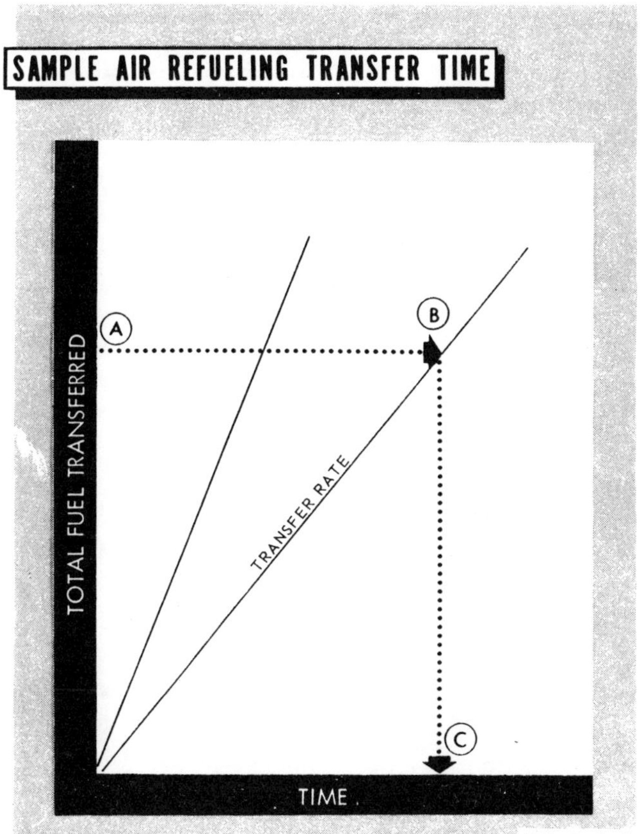

FDD-1-(166)

AIR REFUELING TRANSFER TIME CHART

This chart (figure 11-42) provides the capability of determining the amount of time required to take on a certain amount of fuel at a specified rate. This time segment should then be added to the planning profile.

USE

Enter the chart with a specified amount of fuel to be received and project horizontally to the right and intersect the applicable rate of transfer. From this point descend vertically to read the amount of time required for the transfer.

Sample Problem

D704 Buddy Tank

A.	Total fuel transferred	12, 000 Lbs.
B.	Transfer rate for D704 Buddy Tank	180 Gal/Min.
C.	Time of transfer	10 Min.

AIR REFUELING TRANSFER TIME

D-704 BUDDY TANK C-130 AND KA3B/EKA3B TANKER

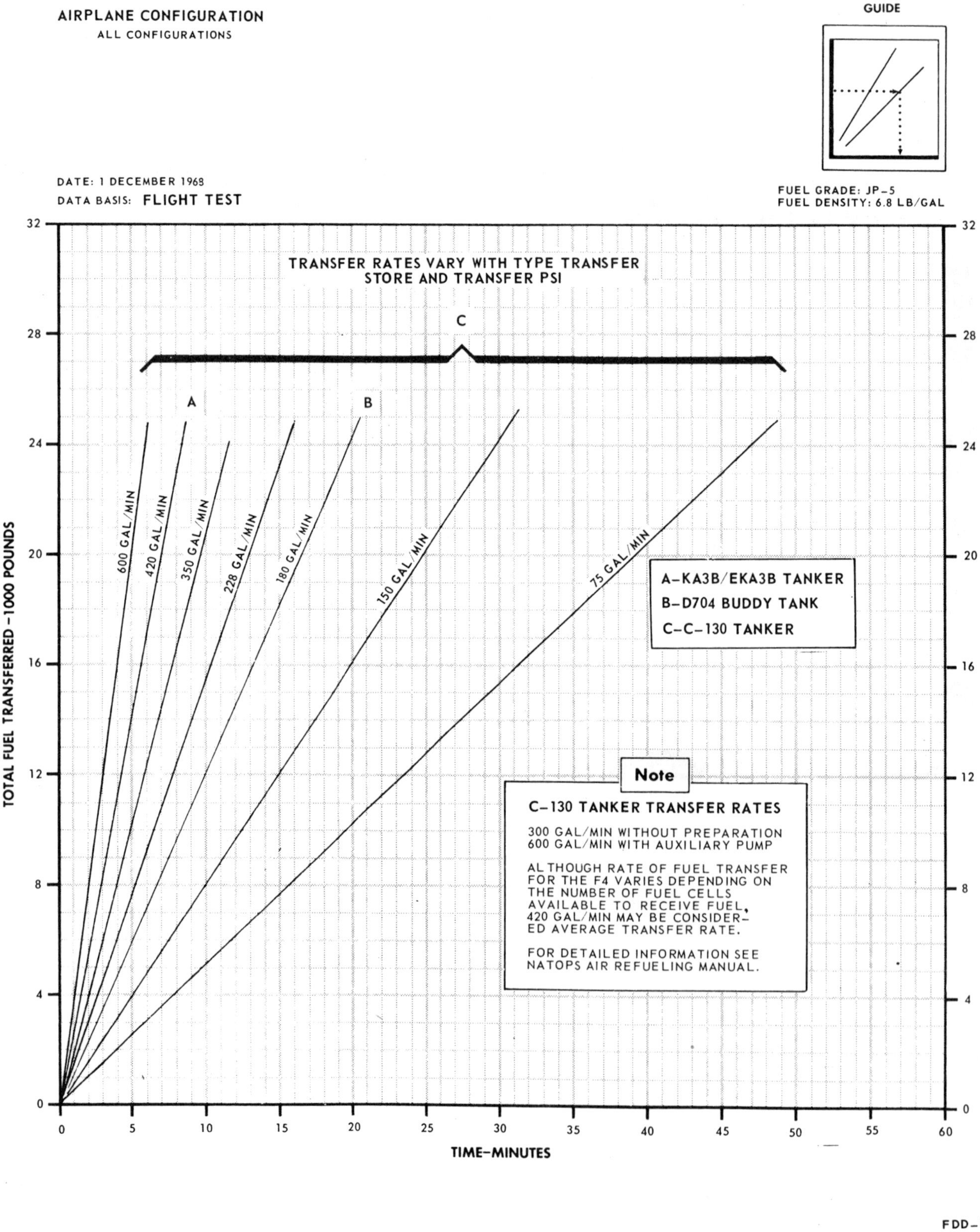

Figure 11-42

PART 7 DESCENT

DESCENT CHARTS

This chart (figure 11-43) presents the distance, time, and fuel required to make a 250-knot, idle thrust descent. Also included is a Mach number curve that corresponds to the 250 knots CAS maintained throughout the descent.

USE

Enter the upper graph with the flight altitude and project horizontally to intersect the applicable drag index within both plots. From the first intersection, project vertically downward to read distance traveled. From the second intersection, project vertically downward to read time to descend. Enter the lower graph with the flight altitude and project horizontally to intersect both plots. From the intersection of t' applicable drag index, project vertically downwai read total fuel used. From the intersection of the Mach number curve, project vertically downward to read Mach number corresponding to 250 knots CAS at the beginning of descent.

Sample Problem

A. Altitude	30, 000 Ft
B. Computed drag index	40
C. Distance traveled	45 Miles
D. Computed drag index	40
E. Time required	8.7 Mins
F. Altitude	30, 000 Ft
G. Computed drag index	40
H. Fuel used	202 Lbs
J. Single drag reflector	
K. Mach number at start of descent	0.675

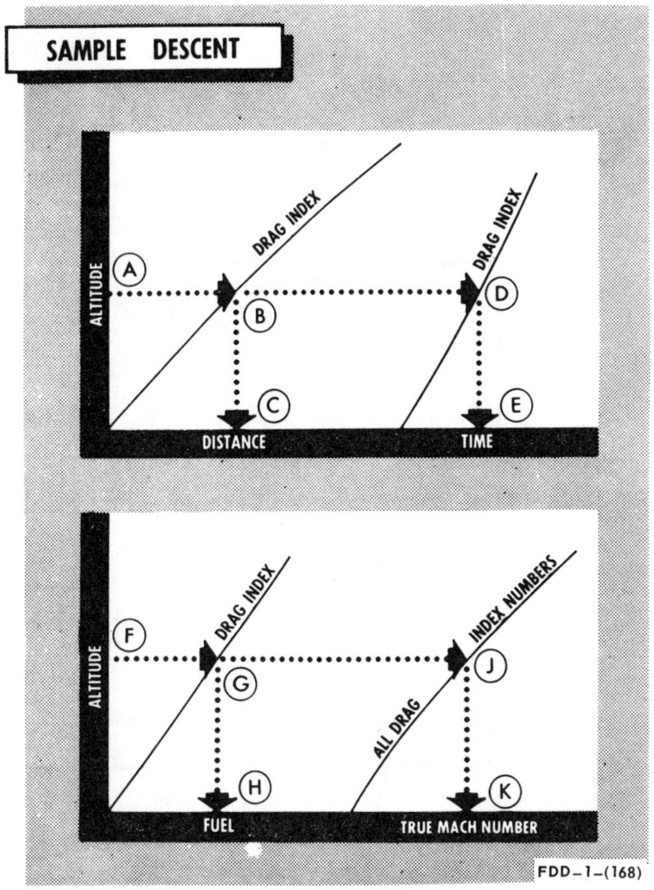

FDD-1-(168)

DESCENT

250 KCAS-IDLE THRUST
SPEED BRAKES RETRACTED

AIRPLANE CONFIGURATION
INDIVIDUAL DRAG INDEXES

REMARKS
ENGINES(S): (2) J79–GE–10
ALL GROSS WEIGHTS
ICAO STANDARD DAY

GUIDE

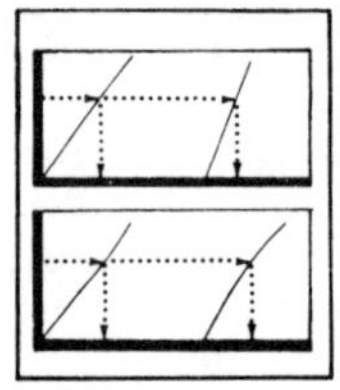

DATE: 1 JUNE 1971
DATA BASIS: **FLIGHT TEST**

FUEL GRADE: JP–5
FUEL DENSITY: 6.8 LB/GAL

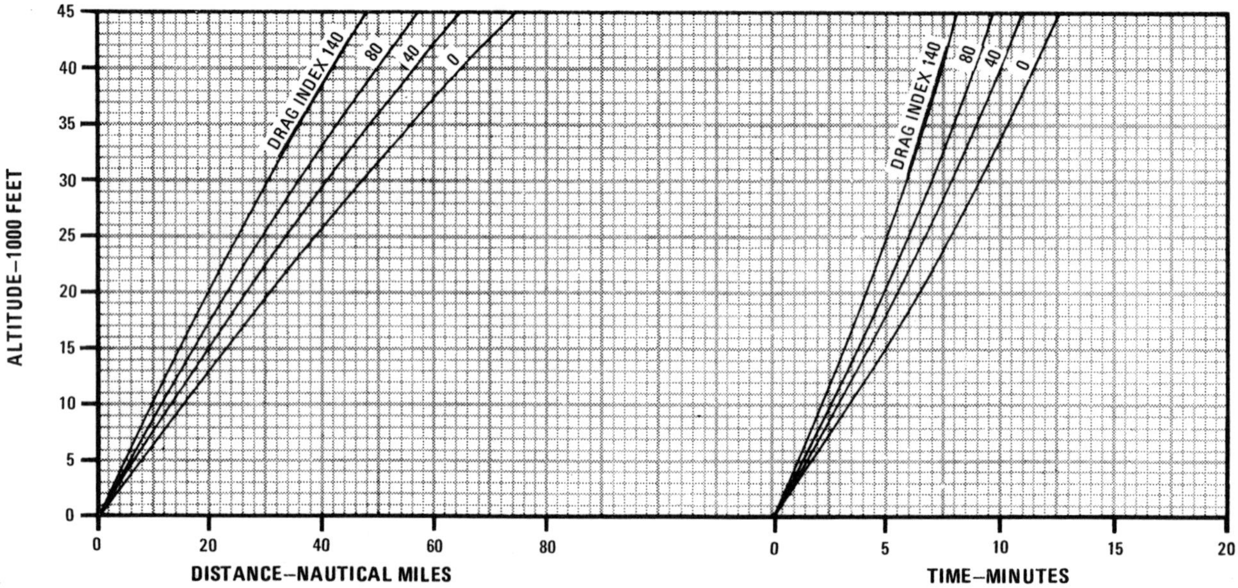

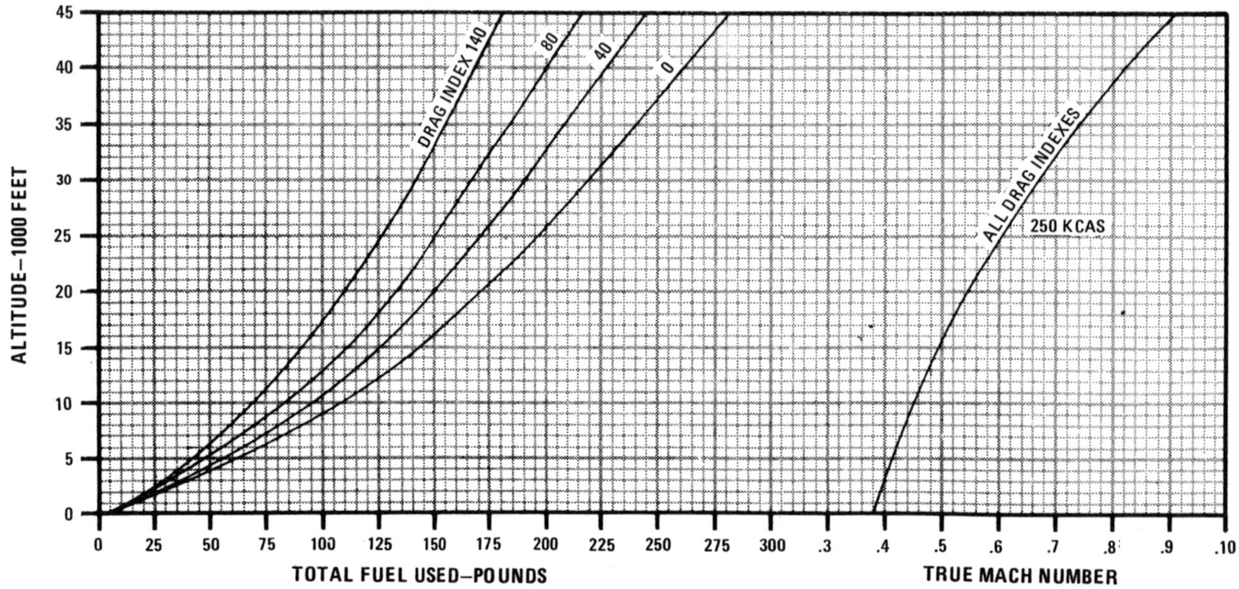

FDD–1–(169)A

Figure 11-43

PART 8 LANDING

Charts

MINIMUM LANDING ROLL DISTANCE CHART

NOTE

Landing Distance charts are based on making an on-speed AOA approach. For additional AOA/approach speeds, refer to Approach Speeds chart, section V.

This chart (figure 11-45) contains landing roll distance information. The variables of temperature, altitude, gross weight, wind, runway condition reading (RCR), and drag chute are taken into consideration.

USE

Enter the chart with the runway temperature and project vertically upward to the correct pressure altitude. From this point, proceed horizontally to the right to the landing gross weight. From this point, descend vertically to the wind base line. Parallel the nearest guide line down to the effective headwind or tailwind. From this point, descend vertically to the appropriate runway condition reading (RCR) and then horizontally to the left to read landing roll distance with drag chute. If the landing is to be made without the drag chute, continue further to the left to the appropriate RCR reflector and then proceed down to read the landing roll distance. If the landing is to be made over a 50-foot obstacle, allow 1900 feet for airborne distance required from the obstacle to the landing touchdown point. If field RCR factors are not available, use RCR 23 for dry, RCR 14 for wet and RCR for icy runway conditions.

Sample Problem

A.	Temperature	15°C
B.	Pressure altitude	2000 Ft

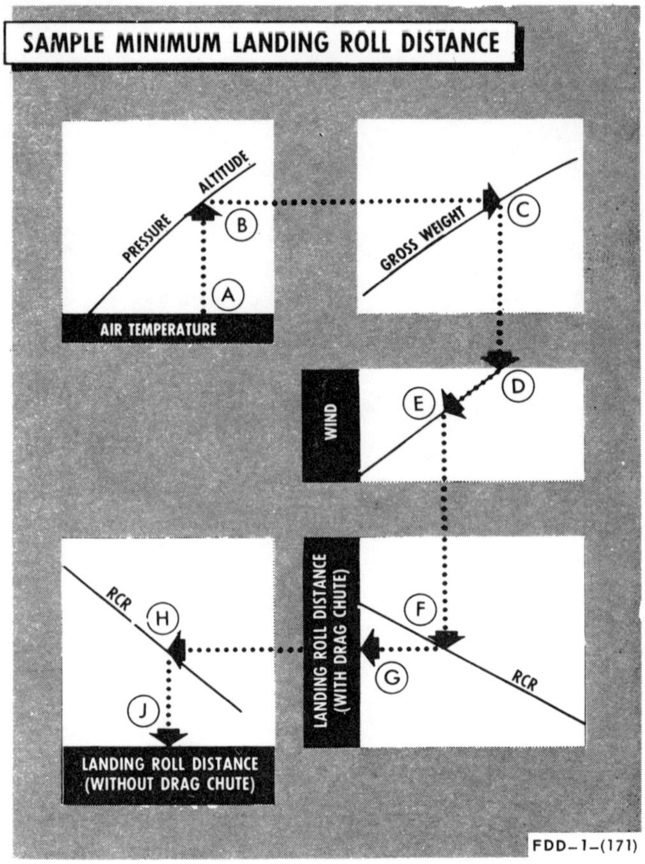

FDD-1-(171)

C.	Gross weight	30,000 Lb
D.	Wind base line	
E.	Effective headwind	20 Kt
F.	RCR	14
G.	Landing roll distance	4000 Ft

If operating without drag chute:

H.	RCR	14
J.	Landing roll distance	5800 Ft

 "Figure 11-44. Deleted."

☆ U. S. GOVERNMENT PRINTING OFFICE: 1976 — 665-672/2067

MINIMUM LANDING ROLL DISTANCE

IDLE THRUST

AIRPLANE CONFIGURATION
ALL DRAG INDEXES
FLAPS EXTENDED, GEAR DOWN
DRAG CHUTE DEPLOYED

REMARKS
ENGINE(S): (2) J79–GE–10

GUIDE

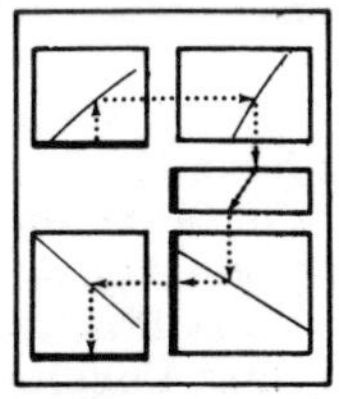

NOTE

FOR TOTAL DISTANCE FROM A 50 FT. HEIGHT, ADD 1900 FT. TO THE GROUND ROLL DISTANCE.

DATE: 1 MAY 1975
DATA BASIS: **FLIGHT TEST**

FUEL GRADE: JP–5
FUEL DENSITY: 6.8 LB/GAL

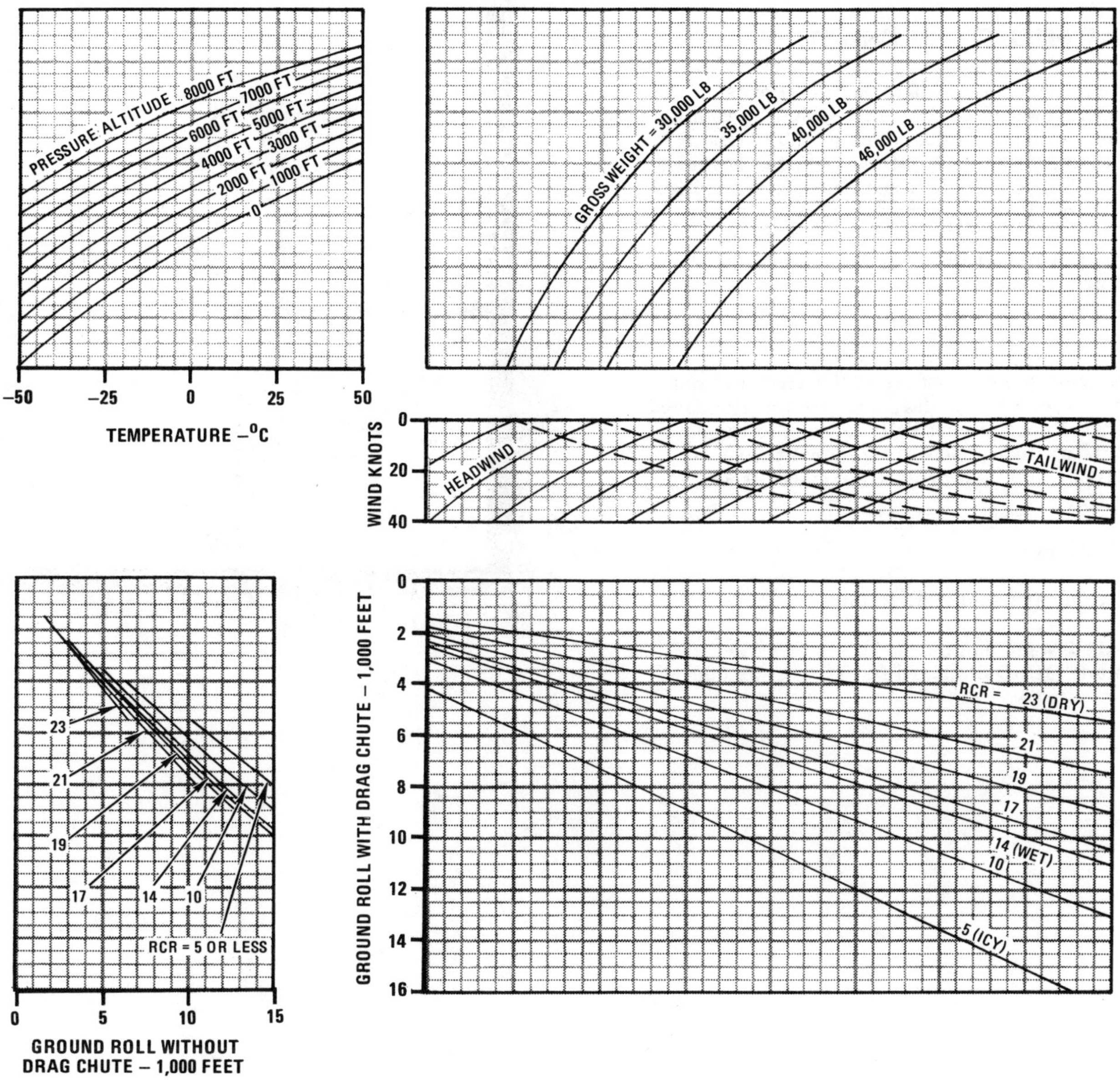

FDD–1–(322)

Figure 11-45

PART 9 COMBAT PERFORMANCE

CHARTS

COMBAT FUEL FLOW CHARTS

These charts (figures 11-46 thru 11-48) present the specific fuel flow and general thrust setting to maintain a constant Mach number for an ICAO standard day and standard day +10°C at all altitudes between sea level and 50, 000 feet. Each chart is plotted for a specific configuration. The fuel flow values are based on a stabilized level flight condition and do not represent the fuel flow required to accelerate to a given Mach number.

USE

Enter the chart corresponding to the aircraft configuration with the desired Mach number for stabilized level flight. Proceed vertically upward to the selected flight altitude. Note the general thrust setting required, and then project horizontally left to read fuel flow.

Sample Problem

Configuration: (4) AIM-7 Missiles

A.	Desired Mach number	1.5
B.	Altitude (ICAO Standard Day)	25,000 Ft
C.	Power setting required	Modulated After-burners
D.	Total fuel flow	890 PPM

SAMPLE COMBAT FUEL FLOW

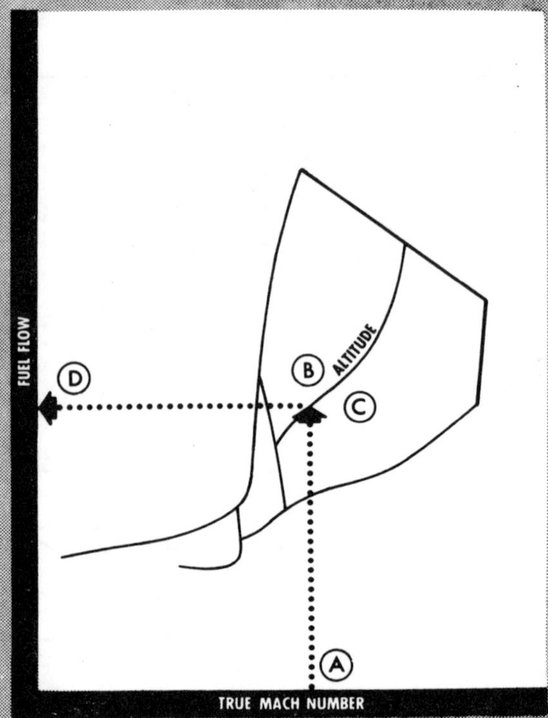

FDD-1-(174)

COMBAT SPECIFIC RANGE

These charts (figures 11-49 thru 11-51) present the specific range and the general power settings required to maintain a constant Mach number for an ICAO standard day and standard day +10°C at all altitudes from sea level to 50,000 feet. The specific range values are based on a stabilized level flight condition and do not represent the fuel flow required to accelerate to a given Mach number.

USE

Enter the chart corresponding to the aircraft configuration with the desired Mach number for stabilized level flight. Proceed vertically upward to the selected flight altitude. Note the general thrust setting required, and then project horizontally left to obtain the specific range.

Sample Problem

Configuration: (4) AIM-7 Missiles

A.	Desired Mach number	1.5
B.	Altitude	30,000 Ft
C.	Thrust required	Modulated After-burners
D.	Specific range	0.016 NMPP

SAMPLE COMBAT SPECIFIC RANGE

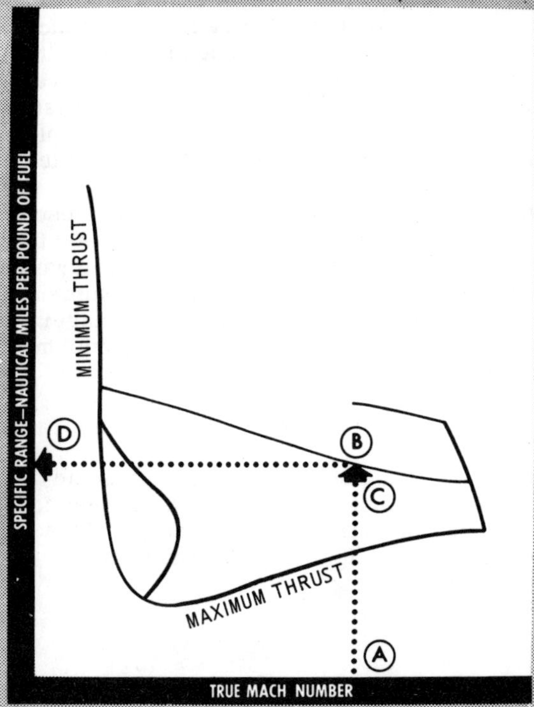

FDD-1-(175)

SUPERSONIC MAXIMUM THRUST CLIMB CHARTS

These charts (figures 11-52 thru 11-54) are plotted for supersonic maximum thrust climb from 35,000 feet to the supersonic combat ceiling. Distance traveled in the climb is plotted against gross weight, with guide lines provided to show the weight reduction as the climb progresses. The time to distance/altitude relationship is superimposed on the plot. Level flight acceleration data is provided which includes time, fuel used (gross weight change), and distance required to accelerate from the subsonic to the supersonic climb Mach number at 35,000 feet. If supersonic climb is contemplated, acceleration at 35,000 feet followed by the climb is recommended, since acceleration to supersonic Mach numbers at this altitude provides for the optimum performance capability.

Note

> If ramp cycling occurs during supersonic climb, the climb schedule Mach number can be increased until the cycling stops. This produces an insignificant degradation in climb performance.

USE

Enter the chart with the gross weight and proceed vertically to the initial Mach number and note the corresponding distance and time. Proceed parallel to the guide lines to the desired supersonic climb Mach number (end of acceleration). Project both vertically downward and horizontally left from this point to read gross weight and distance traveled; also note the time. From these values, subtract the distance, weight, and time corresponding to the initial Mach number to determine the distance, fuel, and time required to accelerate. From the climb Mach number gross weight intersection, (start of climb) proceed parallel along the guide lines to the desired altitude. Obtain the distance, gross weight, and time for this point. Subtract from this data the corresponding values at the start of climb to obtain the distance traveled, the weight change (fuel used), and the time required to complete the climb. If total distance, fuel, and time are desired, add the climb and acceleration values together.

Sample Problem

Configuration: (4) AIM-7 Missiles

A.	Initial gross weight	40,000 Lb
B.	Initial Mach number	1.2
C.	Time corresponding to initial Mach number	0.5 Min
D.	Distance corresponding to initial Mach number	6.7 Miles

SAMPLE SUPERSONIC MAXIMUM THRUST CLIMB

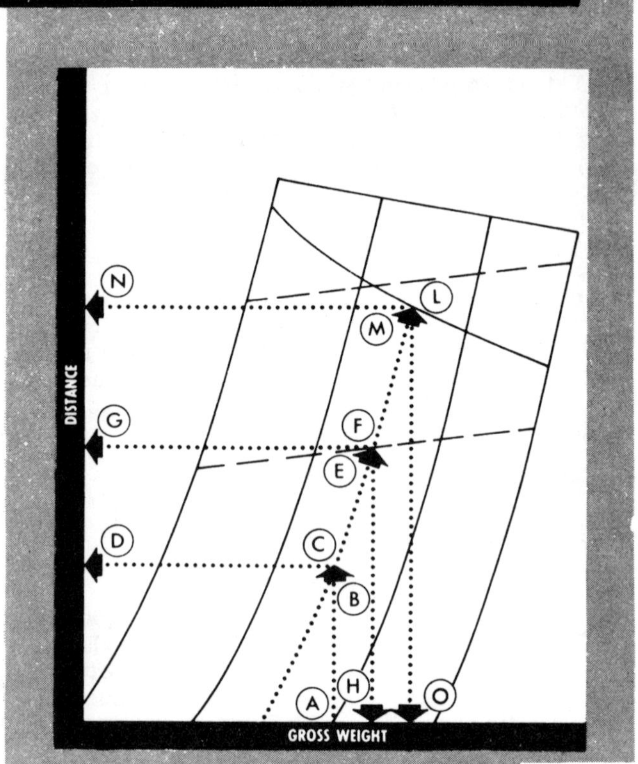

FDD-1-(176)

E.	Climb Mach number	1.745
F.	Time at end of acceleration	2 Min
G.	Distance at end of acceleration	26 Miles
H.	Gross weight at end of acceleration	38,000 Lb
I.	Time required for acceleration (F-C)	1.5 Min
J.	Fuel required for acceleration (A-H)	1200 Lb
K.	Distance required for acceleration (G-D)	19.3 Miles
L.	Altitude at end of climb	55,000 Ft
M.	Time at end of climb	4.4 Min
N.	Distance at end of climb	66 Miles
O.	Gross weight at end of climb	37,200 Lb
P.	Time required for climb (M-F)	2.4 Min
Q.	Distance required for climb (N-G)	40 Miles
R.	Fuel required for climb (H-O)	1600 Lb
S.	Total time required to accelerate and climb (I+P)	3.9 Min
T.	Total distance required to accelerate and climb (K+Q)	59.3 Miles
U.	Total fuel required to accelerate and climb (J+R)	2800 Lb

LOW ALTITUDE ACCELERATIONS

These charts (figures 11-55 and 11-60) present time and fuel required to accelerate from 0.5 to 0.9 Mach at altitudes of Sea Level, 2000, 4000, and 6000 feet. Separate charts are provided for several gross weights and for both maximum and military thrust. The time and fuel values are tabulated for ICAO Standard Day conditions; however, correction factors are given for non-standard temperature.

USE

After selecting the applicable chart for thrust, gross weight, and altitude, enter with the Mach number desired at end of acceleration and project horizontally to the applicable drag index column. Read time/fuel required to accelerate from 0.5 Mach.

ACCELERATION CHARTS

These charts (figures 11-61 thru 11-85) show the relationship of time, distance, and fuel required for level flight maximum or military thrust accelerations. The data is presented for various altitudes and configurations.

WARNING

Refer to section V for external stores operating limitations.

USE

Enter the applicable chart with the aircraft gross weight. Proceed vertically upward to the initial Mach number and note the time. Project horizontally and note the distance. From the initial Mach number, proceed parallel to the guide lines to the Mach number desired at the end of acceleration. At this point, note the time, then project horizontally and vertically and note the distance and gross weight. From this data, subtract the time, distance, and weight corresponding to the initial Mach number to determine the time, distance, and fuel required for acceleration.

Sample Problem

Configuration: Maximum Thrust, (4) AIM-7 Missiles and (2) Wing Tanks, 45,000 Ft

A.	Gross weight	45,000 Lb
B.	Initial Mach number	1.0

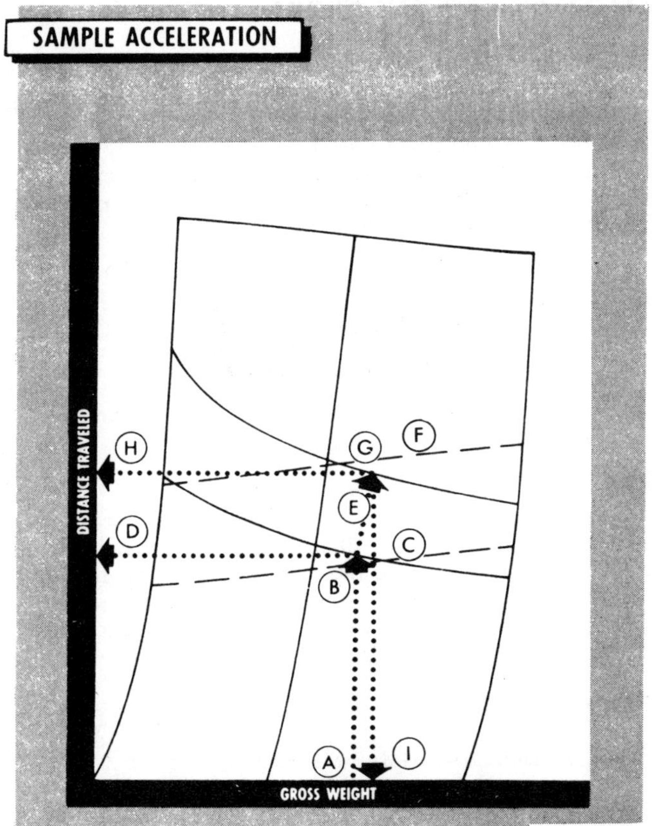

FDD-1-(177)

C.	Time	2 Min
D.	Distance	10 Miles
E.	Parallel guide lines	
F.	Desired Mach number	1.30
G.	Time corresponding to new Mach number	8 Min
H.	Distance corresponding new Mach number	75 Miles
I.	Gross weight corresponding to new Mach number	42,600 Lb
J.	Time required for acceleration (G-C)	0.26 Min
K.	Distance required for acceleration (H-D)	65 Miles
L.	Fuel required for acceleration (A-I)	2400 Lb

LEVEL FLIGHT ENVELOPE

This chart (figure 11-86) presents the aircraft level flight speed envelope for various configurations and average combat gross weights. Parameters of the envelopes extend from buffet onset to V_{max} throughout the altitude range. Maximum Mach number curves for additional aircraft configurations are plotted within the envelopes.

WARNING

Refer to section V for external stores operating limitations.

USE

Enter the chart with the desired combat altitude. Proceed horizontally to intersect the applicable configuration power curve. From this point, proceed vertically downward to read the maximum attainable Mach number in level flight.

Sample Problem

Configuration: (4) AIM-7 Missiles

A.	Combat altitude	36,000 Ft
B.	Airplane complete load	(4) AIM-7 and (1) ℄ Tank
C.	Maximum attainable Mach number (Curve 2)	2.0

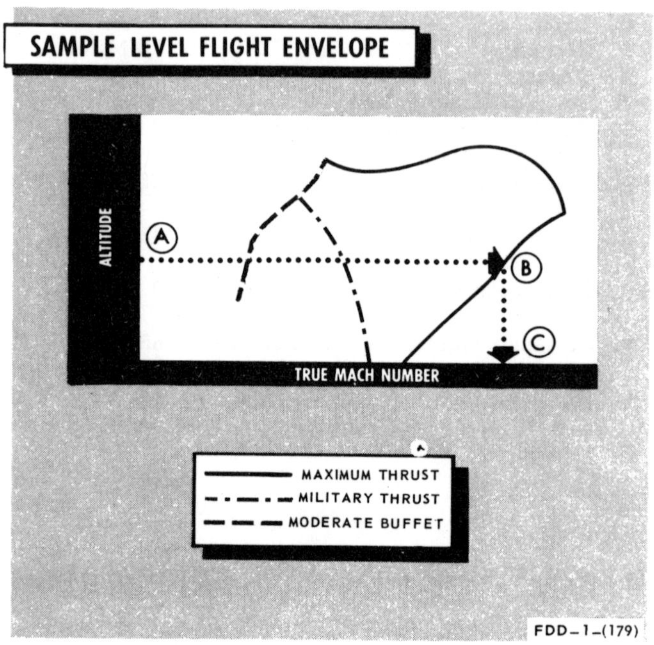

FDD-1-(179)

V-N

The symmetrical flight V-M Envelopes (figures 11-87 thru 11-90) are a graphical presentation of airspeed versus acceleration with lines of indicated angle of attack superimposed. The data are supplied for one gross weight at four altitudes. The charts may be used to determine the allowable maximum symmetrical maneuvering capability of the airplane as well as the indicated angle of attack for any desired G. The charts may be considered to be linear between altitudes for all practical purposes, provided the interpolation is carried out for a constant airspeed.

USE

To find the allowable maximum symmetrical performance capability, enter the chart with the calibrated airspeed and proceed vertically to the stall boundary (positive or negative G) or the maximum allowable acceleration (upper and lower) as applicable. From these intersections, project horizontally to the left to read the positive and negative G obtainable in the case of the stall boundaries, or the upper and lower maximum allowable G for the selected gross weight. To find the angle of attack for a given condition of G and airspeed, enter the appropriate chart with these parameters. Project horizontally to the right from the load factor and vertically upward from the airspeed. At the intersection of these two projections, read the indicated angle of attack.

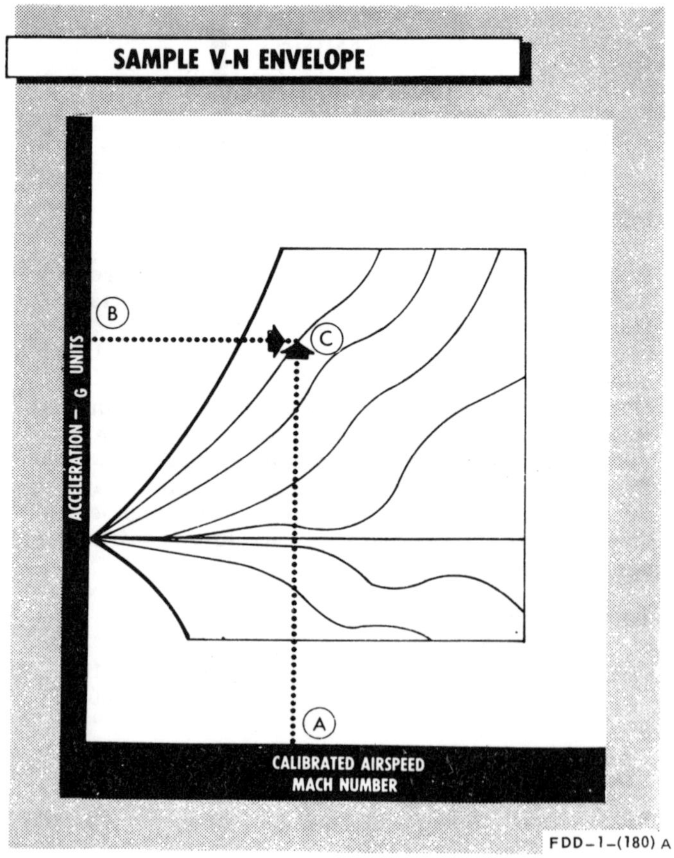

FDD-1-(180) A

Sample Problem

Altitude: 5000 Ft, gross weight 37,500 pounds.

A.	Speed (KCAS)	550
B.	Load factor	5G
C.	Angle-of-attack	8.6 Units

TEMPERATURE EFFECT ON MAXIMUM SPEED

This chart (figure 11-91) shows the effect of non-standard day temperatures on the maximum speed at maximum thrust. The speed variation is read out as the change in Mach number (Δ Mach) for a 10°C variation in temperature (hot or cold) from standard day.

USE

Determine the temperature variation from standard day for the desired altitude. M_{max} may be obtained from the Maximum Thrust Acceleration charts. Enter the chart at the desired Mach number on the standard day M_{max} line. Proceed vertically into either the Hot or Cold Day plot depending on the temperature variation. Continue vertically to the selected altitude, then proceed horizontally to the left to read Δ Mach. When the temperature variation differs from 10°C, simply divide the variation by 10 to reduce it to a decimal. Then multiply the Δ Mach by the decimal to obtain the Mach for a specific situation.

Sample Problem

Find Δ Mach for a standard day M_{max} of 1.8 at 30,000 feet. Forecast flight level temperature is -46.8°C.

A.	Temperature variation	-2.4°C
B.	Standard day M_{max}	1.8 Mach
C.	Altitude	30,000 Ft
D.	Mach/10°C variation	0.159
E.	Mach/2.4°C variation	0.04
F.	Mach number (B + E)	1.84 Mach

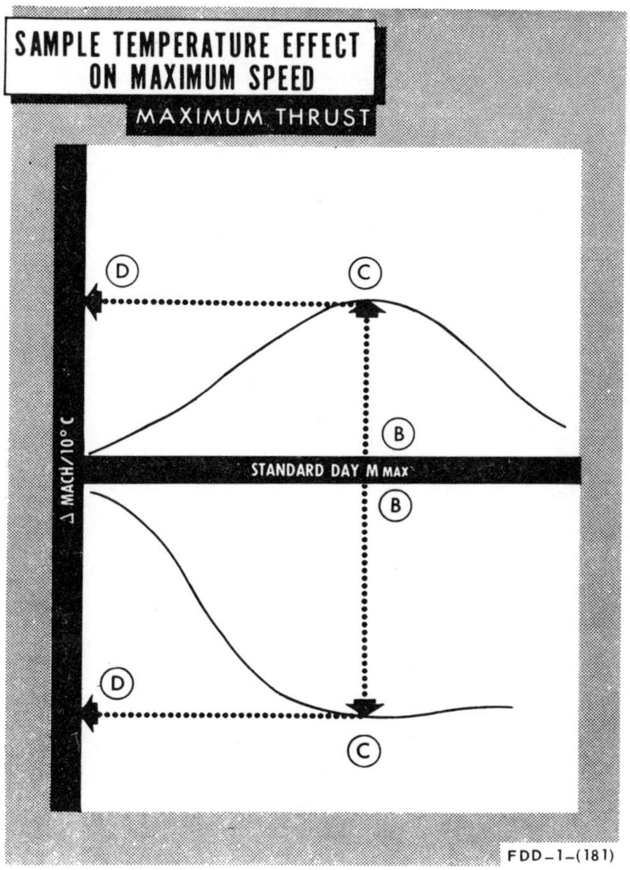

DIVE RECOVERY CHARTS

These charts, (figures 11-92 thru 11-95) present the airplanes dive recovery capability for various speeds (subsonic and supersonic), altitudes and dive angles at 16 units and 19 units AOA.

USE

Enter the applicable chart at the start of the pull-out, and project horizontally to intersect the Mach number at the start of the pull-out. From this point, descend vertically and intersect the dive angle at the start of pull-out, then proceed horizontally to the left to read altitude lost during pull-out.

Sample Problem

Configuration: (4) AIM-7 Missiles; 16 Units AOA; Supersonic

A.	Altitude at start of pull-out	40,000 Ft
B.	Mach number at start of pull-out	1.5 Mach
C.	Dive angle at start of pull-out	70°
D.	Altitude loss during constant 16 unit AOA pull-out	13,200 Ft.

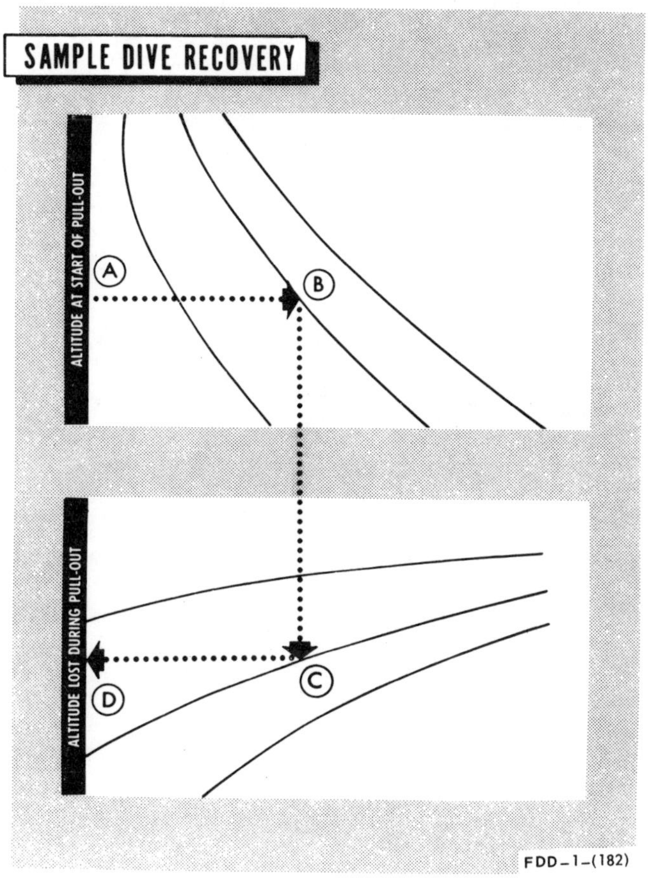

TURN CAPABILITIES

This chart (figure 11-96) presents the radius of turn and the rate of turn for a constant altitude, constant speed turn. Turn data is available for various speeds and bank angles. Load factor is also included for each bank angle.

USE

Enter the radius of turn plot with the true airspeed. Proceed horizontally to the right to the desired bank angle. Note the load factor, then proceed vertically downward and read the radius of turn. Enter the rate of turn plot with the true airspeed. Proceed horizontally to the right to the bank angle, note the load factor and then proceed vertically downward to read the rate of turn.

Sample Problem

Radius of Turn

A. True airspeed	400 Kts.
B. Bank angle	20°
C. Load factor	1.07 G
D. Radius of turn	40,000 Ft.

Rate of Turn

A. True airspeed	500 Kts.
B. Bank angle	30°
C. Load factor	1.15 G
D. Rate of turn	1.25 Deg/sec

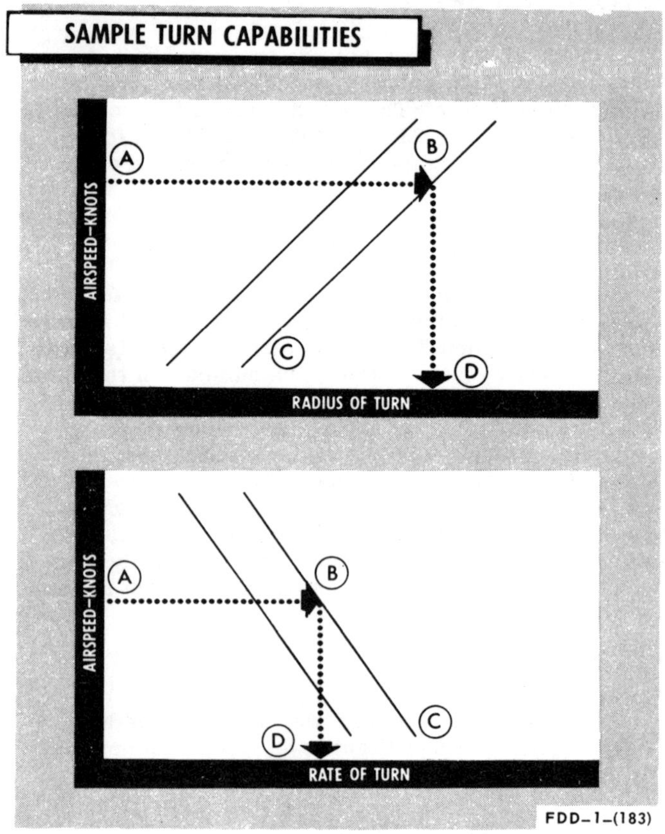

FDD–1–(183)

COMBAT FUEL FLOW

STABILIZED LEVEL FLIGHT

AIRPLANE CONFIGURATION
(4) AIM-7

REMARKS
ENGINE(S): (2) J79-GE-10

GUIDE

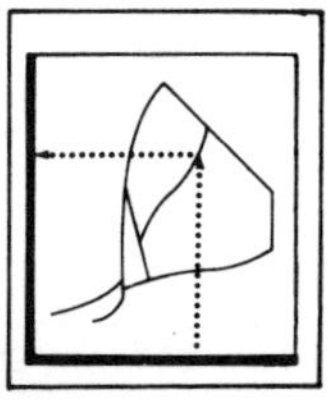

DATE: 1 JANUARY 1972
DATA BASIS: **ESTIMATED** (BASED ON FLIGHT TEST)

FUEL GRADE: JP-5
FUEL DENSITY: 6.8 LB/GAL

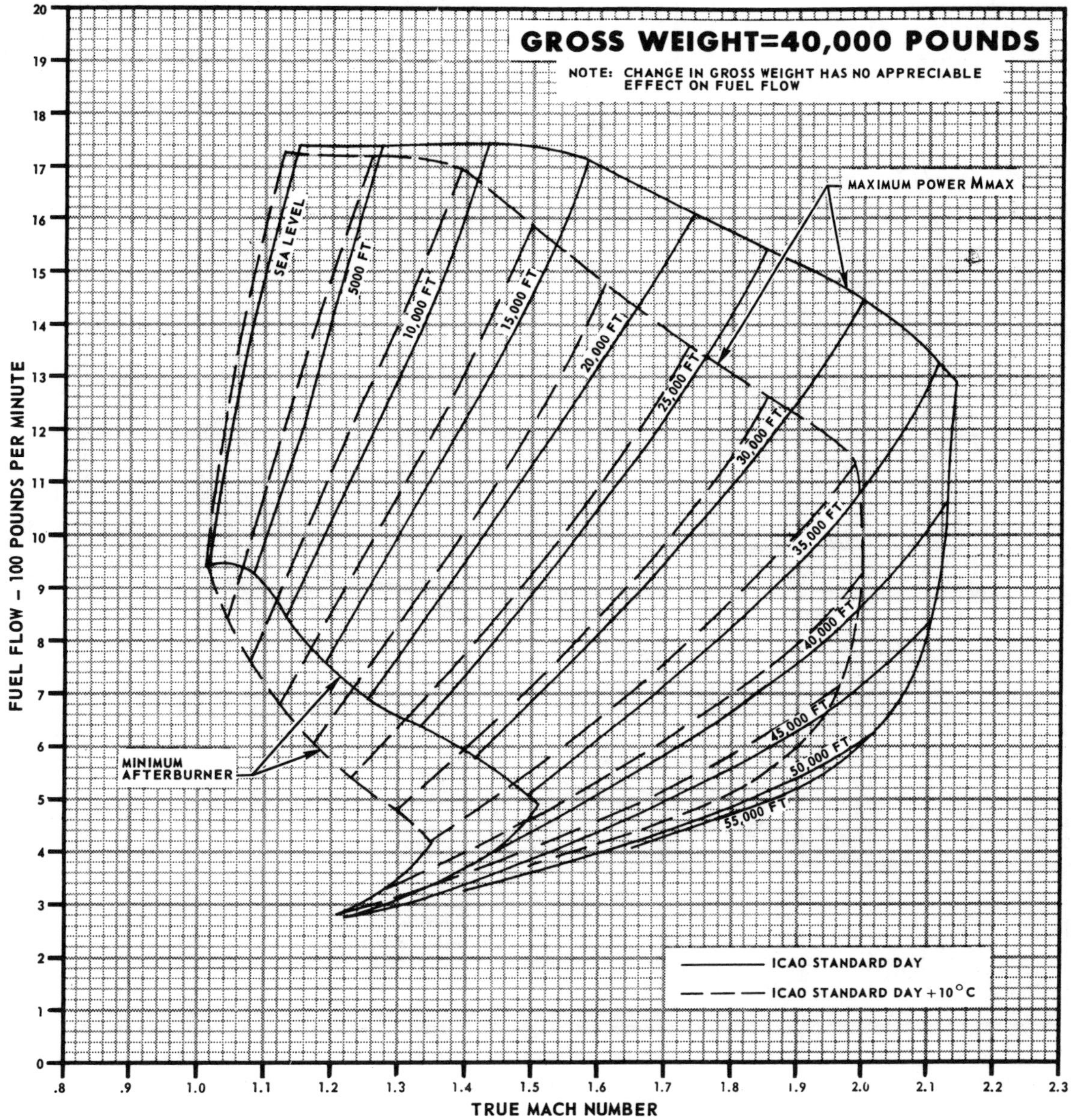

FDD-1-(184) A

Figure 11-46

COMBAT FUEL FLOW
STABILIZED LEVEL FLIGHT

AIRPLANE CONFIGURATION

(4) AIM-7 AND (1) ℄ TANK

REMARKS

ENGINES: (2) J79-GE-10
ICAO STANDARD DAY

GUIDE

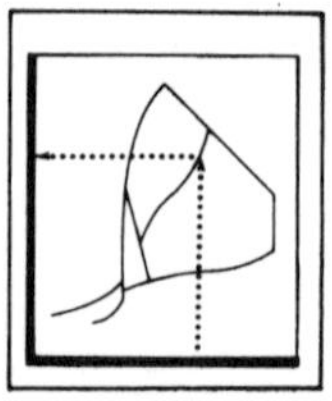

DATE: 1 JANUARY 1972
DATA BASIS: ESTIMATED (BASED ON FLIGHT TEST)

FUEL GRADE: JP-5
FUEL DENSITY: 6.8 LB/GAL

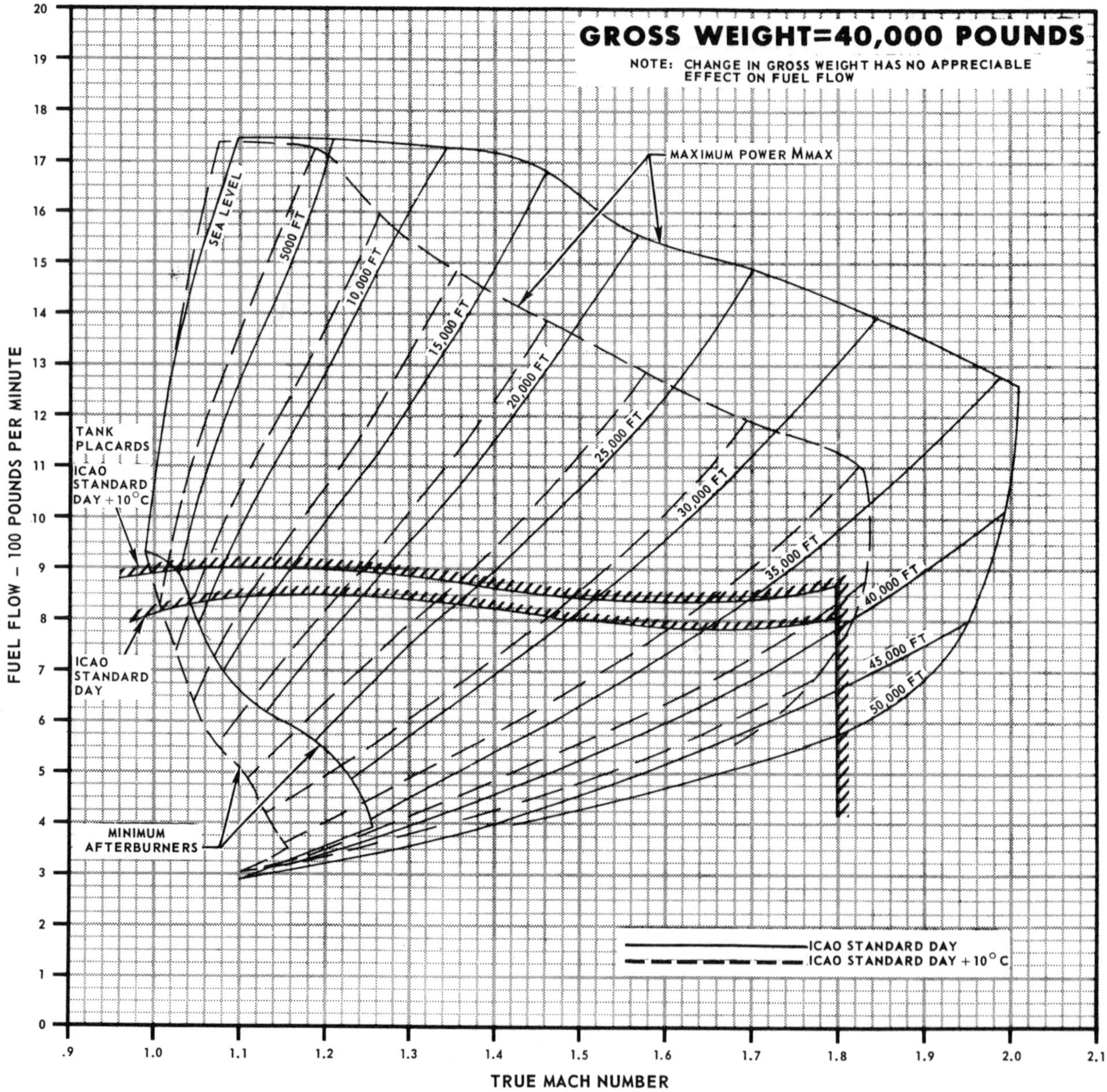

FDD-1-(185)A

Figure 11-47

COMBAT FUEL FLOW

STABILIZED LEVEL FLIGHT

AIRPLANE CONFIGURATION
(4) AIM–7 AND (2) WING TANKS

REMARKS
ENGINE(S): (2) J79-GE-10

GUIDE

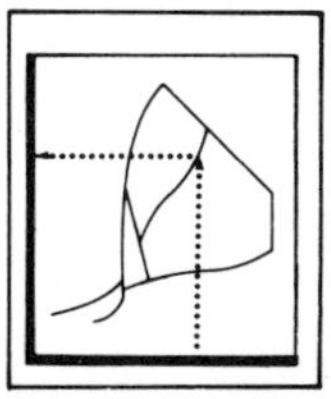

DATE: 1 JANUARY 1972
DATA BASIS: **ESTIMATED** (BASED ON FLIGHT TEST)

FUEL GRADE: JP-5
FUEL DENSITY: 6.8 LB/GAL

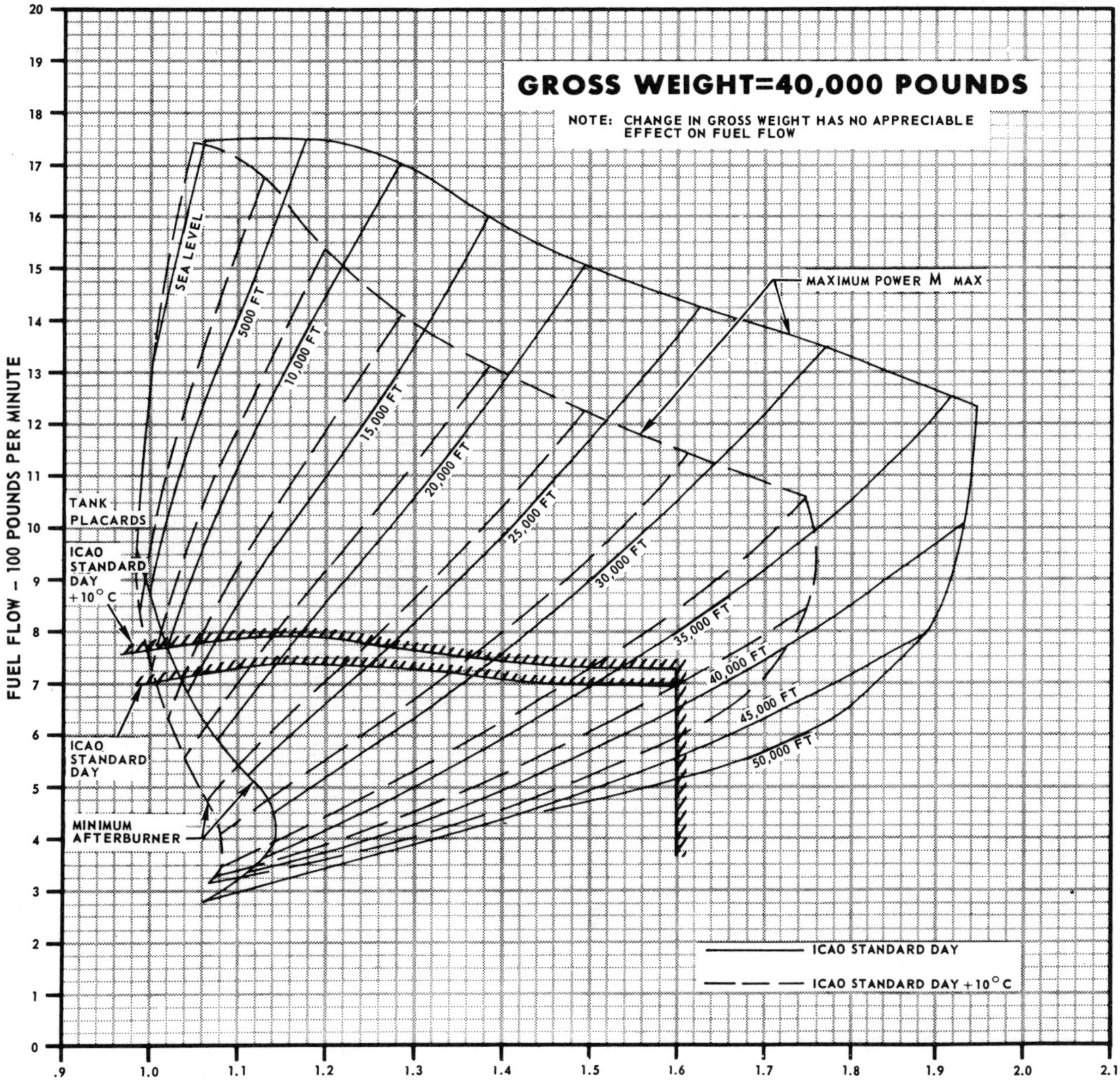

FDD-1-(186)A

Figure 11-48

COMBAT SPECIFIC RANGE
STABILIZED LEVEL FLIGHT

AIRPLANE CONFIGURATION
(4) AIM-7

REMARKS
ENGINE(S): (2) J79-GE-10
ICAO STANDARD DAY

GUIDE

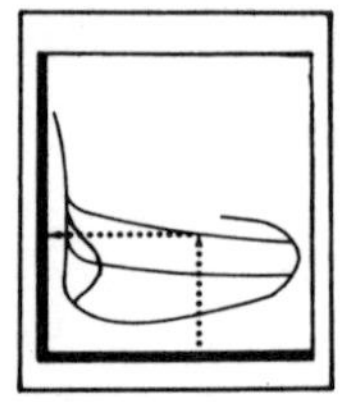

DATE: 1 JANUARY 1972
DATA BASIS: ESTIMATED (BASED ON FLIGHT TEST)

FUEL GRADE: JP-5
FUEL DENSITY: 6.8 LB/GAL

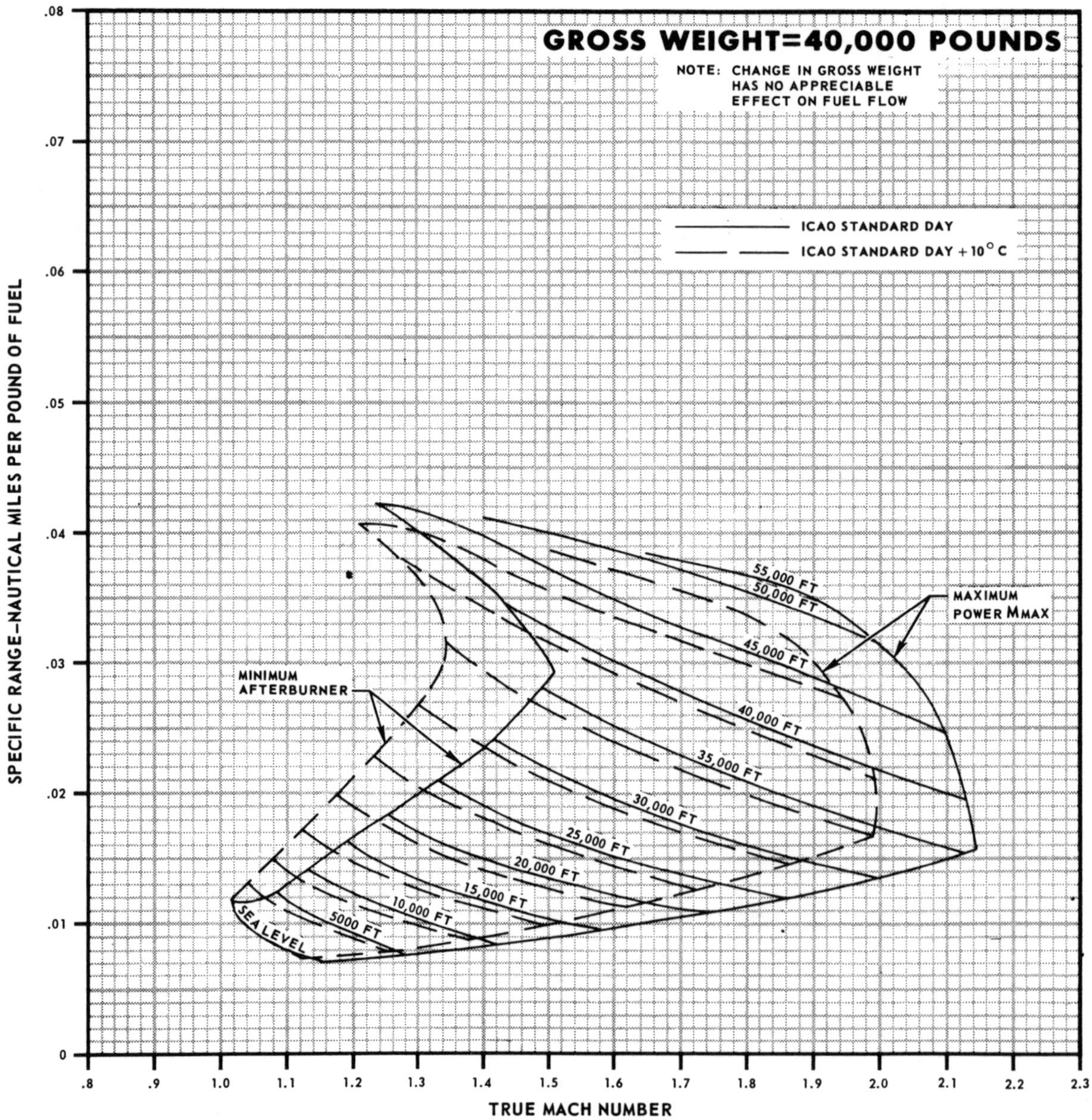

FDD-1-(190)A

Figure 11-49

COMBAT SPECIFIC RANGE

STABILIZED LEVEL FLIGHT

AIRPLANE CONFIGURATION
(4) AIM-7 AND (1) ℄ TANK

REMARKS
ENGINE(S): (2) J79-GE-10
ICAO STANDARD DAY

GUIDE

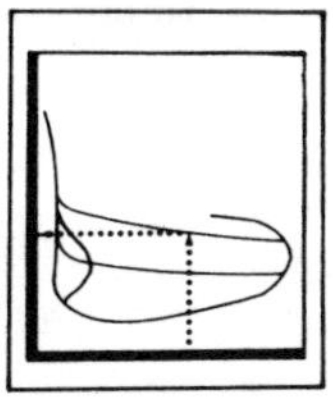

DATE: 1 JANUARY 1972
DATA BASIS: ESTIMATED (BASED ON FLIGHT TEST)

FUEL GRADE: JP-5
FUEL DENSITY: 6.8 LB/GAL

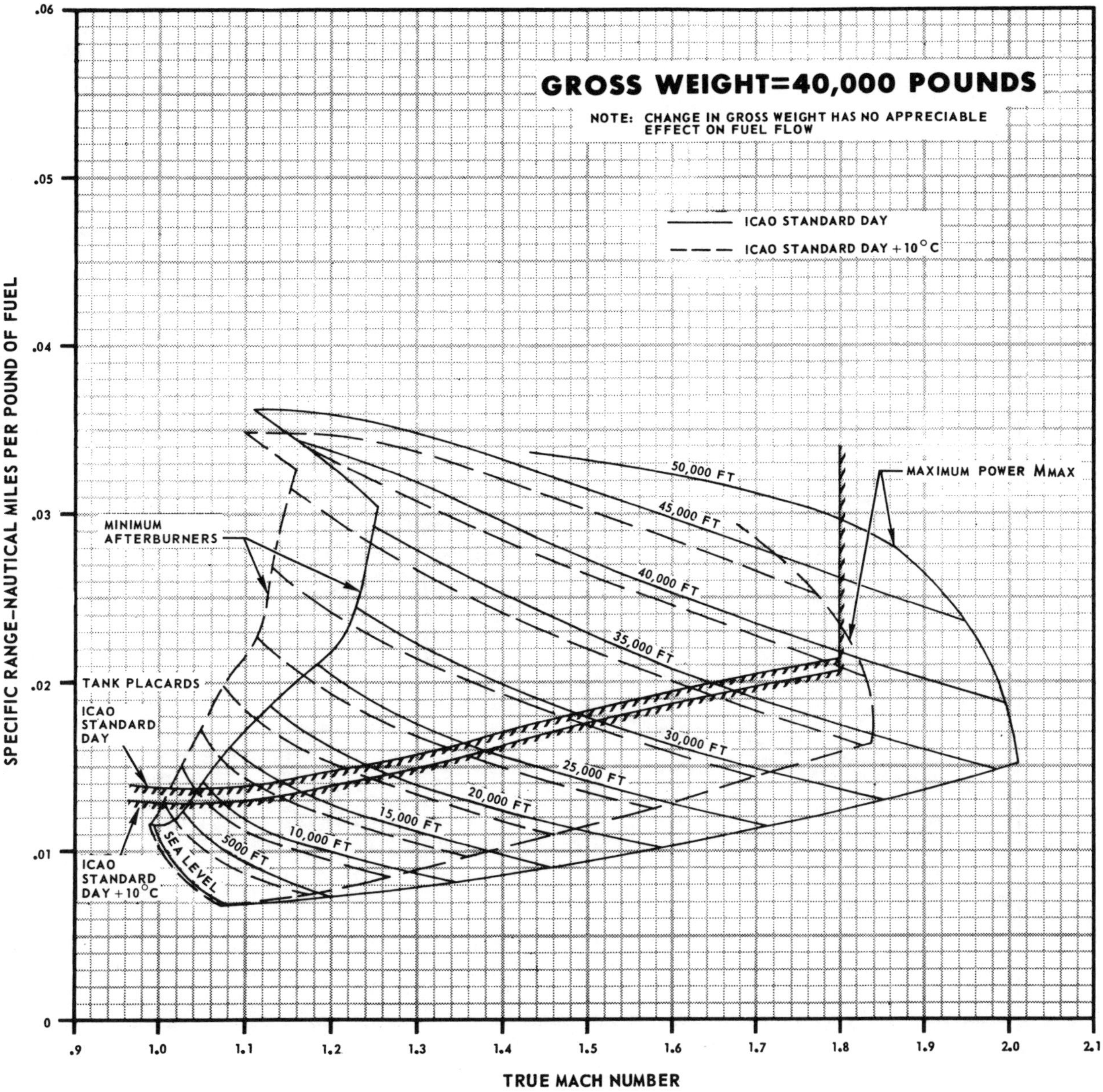

FDD-1-(191)A

Figure 11-50

COMBAT SPECIFIC RANGE
STABILIZED LEVEL FLIGHT

AIRPLANE CONFIGURATION
(4) AIM-7
AND (2) WING TANKS

REMARKS
ENGINE(S): (2) J79-GE-10
ICAO STANDARD DAY

GUIDE

DATE: 1 JANUARY 1972
DATA BASIS: **ESTIMATED** (BASED ON FLIGHT TEST)

FUEL GRADE: JP-5
FUEL DENSITY: 6.8 LB/GAL

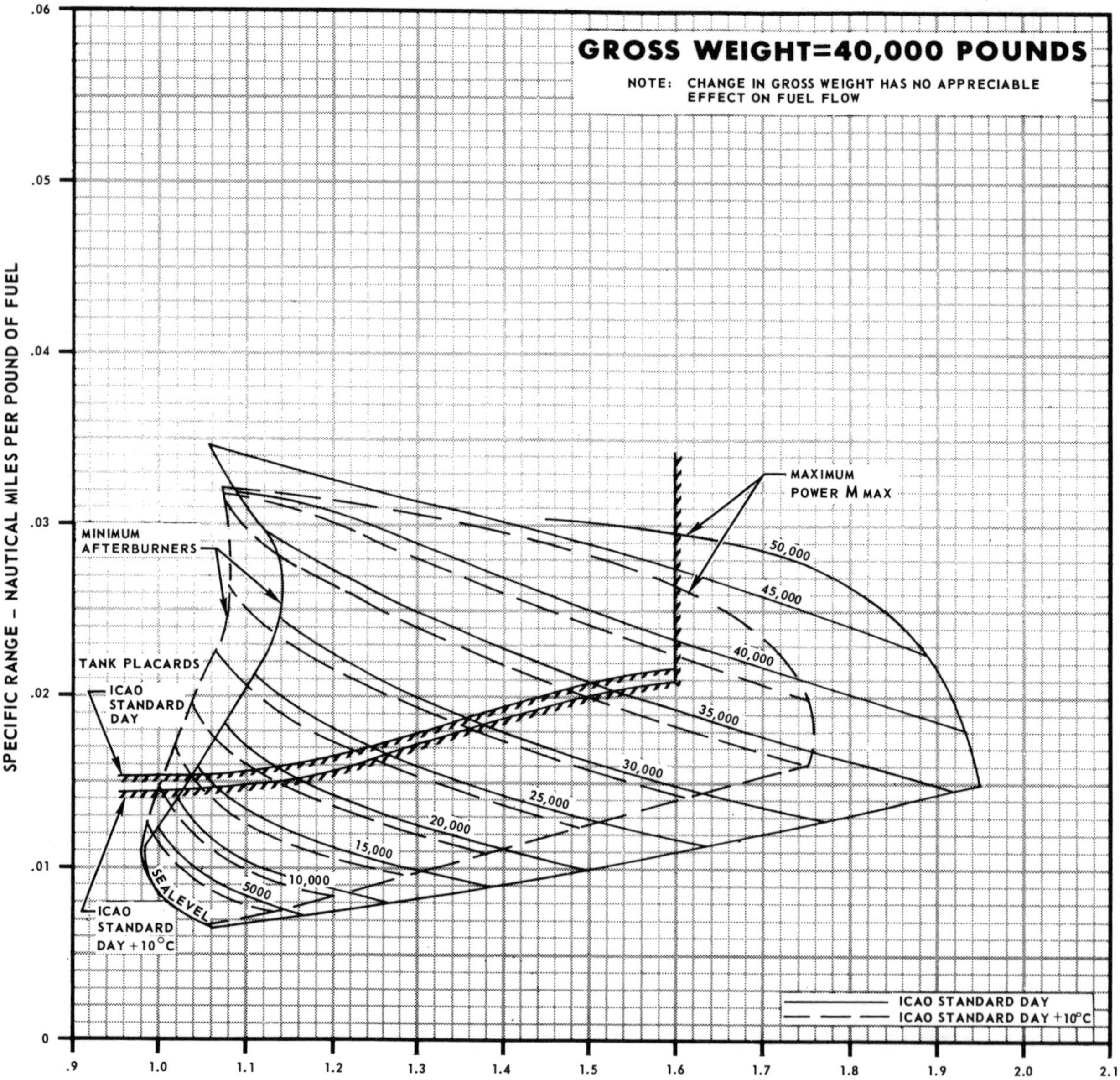

FDD-1-(192)A

Figure 11-51

SUPERSONIC MAXIMUM THRUST CLIMB

AIRPLANE CONFIGURATION

(4) AIM-7

REMARKS

ENGINE(S): (2) J79-GE-10
ICAO STANDARD DAY

GUIDE

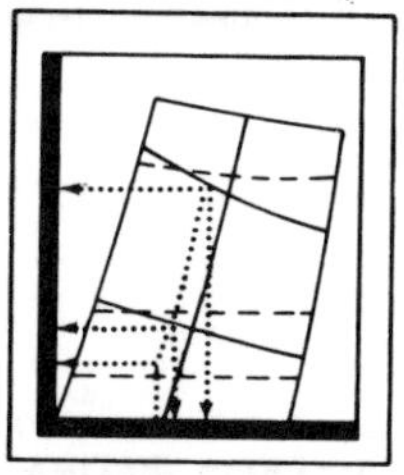

DATE: 1 OCTOBER 1971
DATA BASIS: **ESTIMATED** (BASED ON FLIGHT TEST)

FUEL GRADE: JP-5
FUEL DENSITY: 6.8 LB/GAL

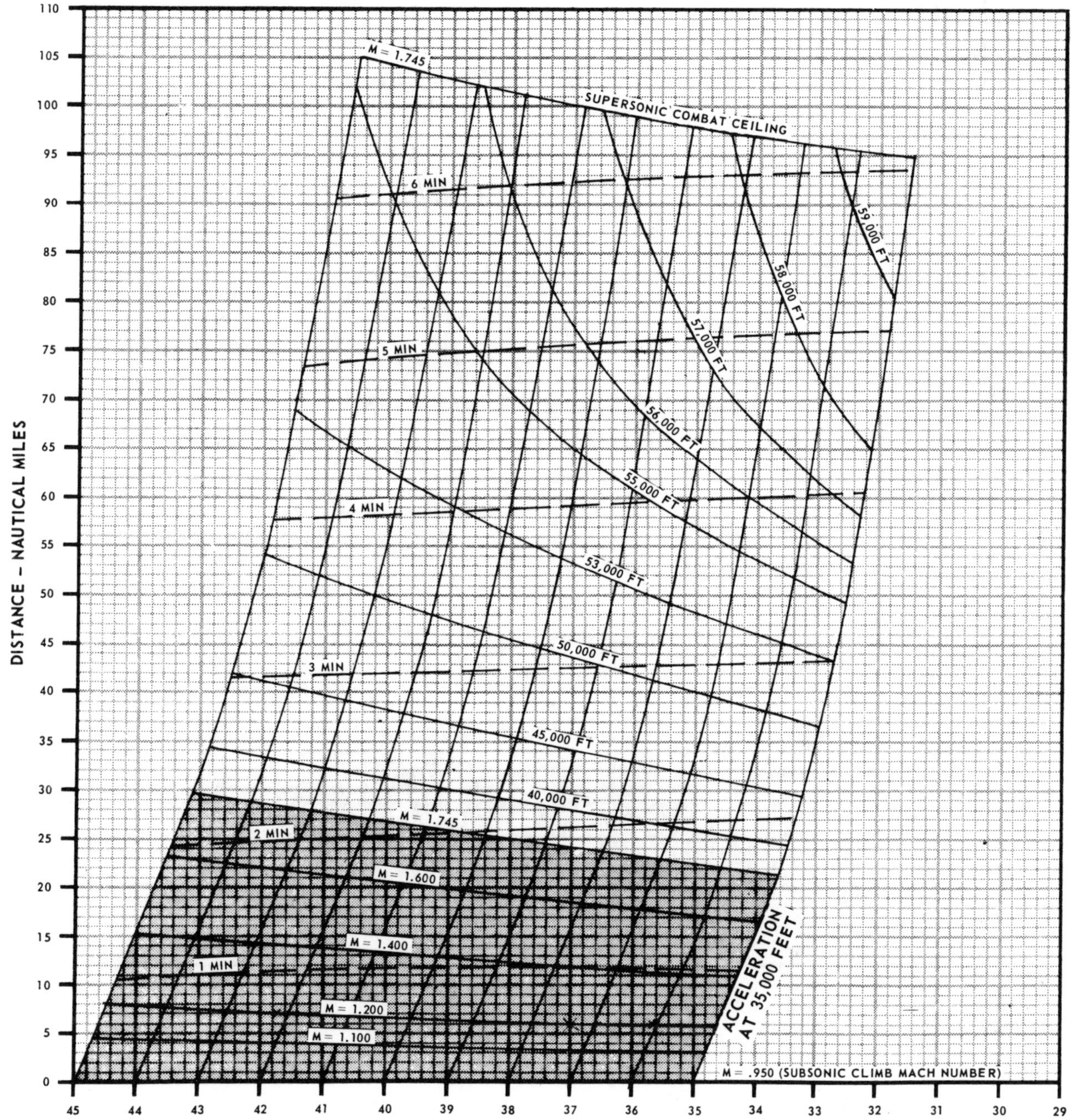

FDD-1-(196)A

Figure 11-52

SUPERSONIC MAXIMUM THRUST CLIMB

GUIDE

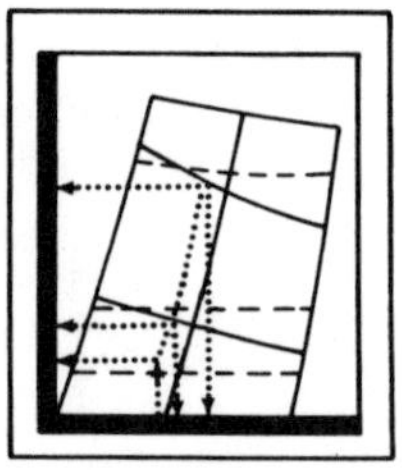

AIRPLANE CONFIGURATION
(4) AIM-7 AND (1) ℄ TANK

REMARKS

ENGINE(S): (2) J79-GE-10
ICAO STANDARD DAY

DATE: 1 JANUARY 1972
DATA BASIS: **ESTIMATED** (BASED ON FLIGHT TEST)

FUEL GRADE: JP-5
FUEL DENSITY: 6.8 LB/GAL

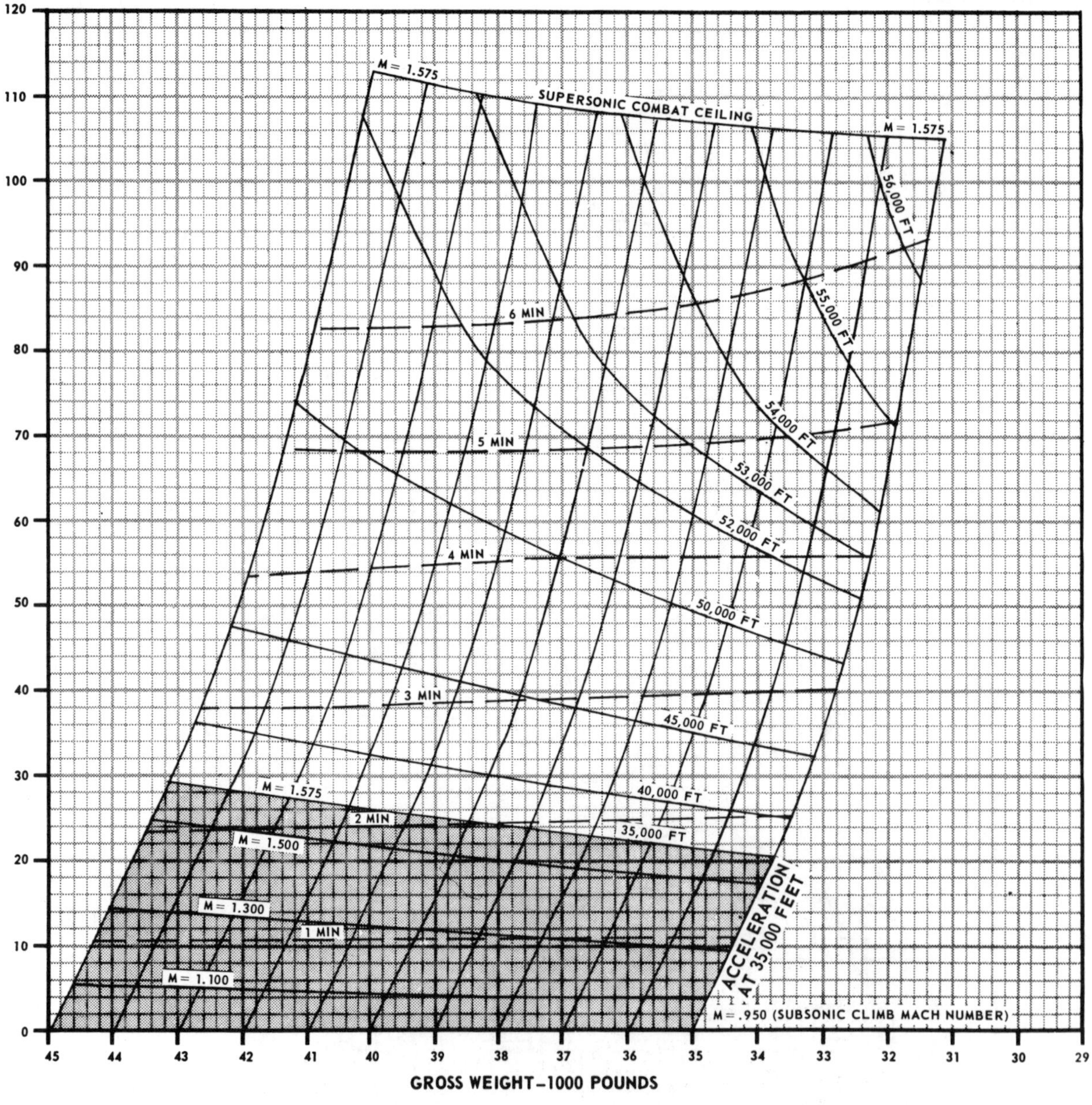

FDD-1-(197)A

Figure 11-53

SUPERSONIC MAXIMUM THRUST CLIMB

AIRPLANE CONFIGURATION
(4) AIM-7 AND (2) WING TANKS

REMARKS
ENGINE(S): (2) J79-GE-10
ICAO STANDARD DAY

GUIDE

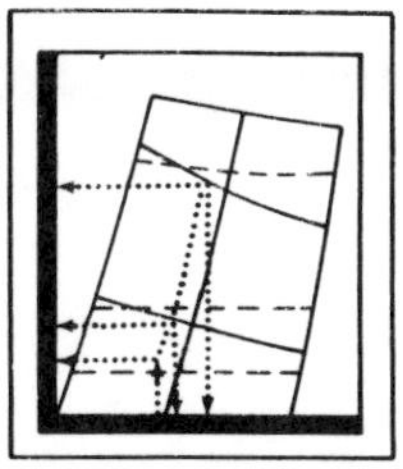

DATE: 1 JANUARY 1972
DATA BASIS: **ESTIMATED** (BASED ON FLIGHT TEST)

FUEL GRADE: JP-5
FUEL DENSITY: 6.8 LB/GAL

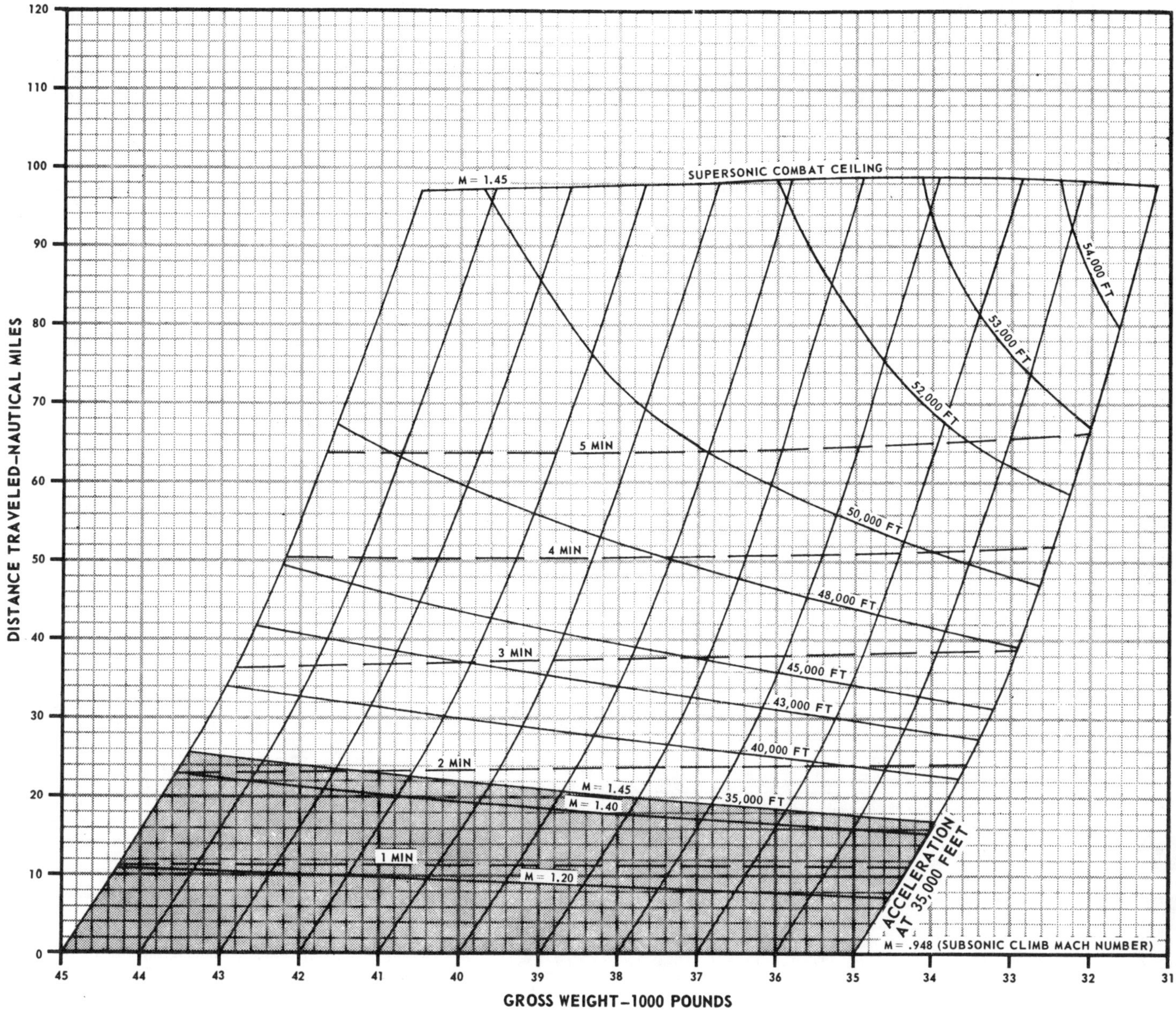

FDD-1-(198)A

Figure 11-54

LOW ALTITUDE ACCELERATION

MAXIMUM THRUST

AIRPLANE CONFIGURATION
INDIVIDUAL DRAG INDEXES

GROSS WEIGHT – 35,000 POUNDS

REMARKS
ENGINES: (2) J79-GE-10

DATE: 1 JULY 1971
DATA BASIS: FLIGHT TEST

FUEL GRADE: JP-5
FUEL DENSITY: 6.8 LB/GAL

	MACH	TIME TO ACCELERATE (MIN.)/FUEL TO ACCELERATE (LBS)							TEMP. EFFECTS FACTOR	
		DRAG INDEX 0	20	40	60	80	100	120	+10°C	−10°C
SEA LEVEL (15°C)	.5	.0 / 0	.0 / 0	.0 / 0	.0 / 0	.0 / 0	.0 / 0	.0 / 0	.0 / .0	.0 /.0
	.55	.04/ 48	.04/ 50	.04/ 51	.04/ 53	.04/ 55	.04/ 56	.04/ 58	1.10/1.07	.90/.93
	.6	.07/ 98	.07/101	.08/105	.08/109	.08/112	.08/117	.09/121	1.10/1.07	.90/.93
	.65	.11/150	.11/155	.11/161	.12/167	.12/174	.13/182	.14/190	1.10/1.07	.89/.92
	.7	.14/203	.15/211	.15/220	.16/229	.17/240	.18/252	.19/266	1.10/1.08	.89/.92
	.75	.18/258	.19/269	.19/282	.20/295	.21/312	.23/330	.24/352	1.11/1.08	.89/.91
	.8	.21/314	.22/329	.23/346	.25/366	.26/389	.28/417	.30/451	1.11/1.08	.88/.91
	.85	.25/373	.26/392	.28/414	.29/441	.31/475	.34/516	.38/571	1.12/1.09	.88/.90
	.9	.28/432	.30/457	.32/486	.34/525	.37/573	.41/638	.47/733	1.14/1.10	.87/.90
2000 FEET (11°C)	.5	.0 / 0	.0 / 0	.0 / 0	.0 / 0	.0 / 0	.0 / 0	.0 / 0	.0 / .0	.0 /.0
	.55	.04/ 48	.04/ 49	.04/ 50	.04/ 52	.04/ 53	.04/ 55	.04/ 57	1.10/1.07	.90/.93
	.6	.07/ 96	.08/100	.08/103	.08/106	.08/110	.09/114	.09/118	1.10/1.07	.90/.93
	.65	.11/147	.11/152	.12/157	.12/163	.13/170	.13/177	.14/185	1.10/1.07	.89/.92
	.7	.15/199	.15/206	.16/215	.17/224	.17/234	.18/245	.19/258	1.10/1.08	.89/.92
	.75	.18/252	.19/263	.20/275	.21/288	.22/303	.23/320	.25/340	1.11/1.08	.89/.91
	.8	.22/307	.23/322	.24/337	.25/355	.27/377	.29/402	.31/434	1.11/1.08	.88/.91
	.85	.26/364	.27/382	.28/403	.30/428	.32/459	.35/497	.38/547	1.12/1.09	.88/.90
	.90	.29/422	.31/445	.33/473	.35/508	.38/553	.42/611	.42/695	1.14/1.10	.87/.90
4000 FEET (7°C)	.5	.0 / 0	.0 / 0	.0 / 0	.0 / 0	.0 / 0	.0 / 0	.0 / 0	.0 / .0	.0 /.0
	.55	.04/ 47	.04/ 48	.04/ 49	.04/ 51	.04/ 52	.04/ 54	.05/ 56	1.10/1.07	.90/.93
	.6	.08/ 95	.08/ 98	.08/101	.08/104	.09/107	.09/111	.09/115	1.10/1.07	.90/.93
	.65	.11/144	.12/149	.12/154	.13/160	.13/166	.14/172	.14/180	1.10/1.07	.89/.92
	.7	.15/195	.16/202	.16/210	.17/218	.18/228	.19/238	.20/250	1.10/1.08	.89/.92
	.75	.19/247	.20/257	.21/268	.22/280	.23/294	.24/310	.25/328	1.11/1.08	.89/.91
	.8	.23/300	.24/314	.25/328	.26/345	.28/365	.29/389	.31/418	1.11/1.08	.88/.91
	.85	.27/355	.28/373	.29/392	.31/416	.33/444	.35/479	.39/524	1.12/1.09	.88/.90
	.9	.30/412	.32/434	.34/459	.36/493	.39/534	.43/587	.48/661	1.14/1.10	.87/.90
6000 FEET (3°C)	.5	.0 / 0	.0 / 0	.0 / 0	.0 / 0	.0 / 0	.0 / 0	.0 / 0	.0 / .0	.0 /.0
	.55	.04/ 46	.04/ 47	.04/ 49	.04/ 50	.04/ 52	.05/ 53	.05/ 55	1.10/1.07	.09/.93
	.6	.08/ 93	.08/96	.08/ 99	.09/102	.09/105	.09/109	.10/113	1.10/1.07	.90/.93
	.65	.12/141	.12/146	.13/151	.13/156	.14/162	.14/168	.15/175	1.10/1.07	.89/.92
	.7	.16/191	.16/198	.17/205	.18/213	.18/222	.19/232	.20/243	1.10/1.08	.89/.92
	.75	.20/242	.21/251	.21/261	.22/273	.23/286	.25/301	.26/318	1.11/1.08	.89/.91
	.8	.24/294	.25/306	.26/320	.27/336	.28/355	.30/377	.32/403	1.11/1.08	.88/.91
	.85	.28/347	.29/364	.30/382	.32/404	.34/430	.36/462	.40/503	1.12/1.09	.88/.90
	.9	.31/402	.33/423	.35/447	.37/478	.40/515	.43/564	.48/629	1.14/1.10	.87/.90

FDD-1-(201)A

Figure 11-55

LOW ALTITUDE ACCELERATION

MAXIMUM THRUST

AIRPLANE CONFIGURATION
INDIVIDUAL DRAG INDEXES

GROSS WEIGHT – 45,000 POUNDS

REMARKS
ENGINES: (2) J79–GE–10

DATE: 1 JULY 1971
DATA BASIS: FLIGHT TEST

FUEL GRADE: JP–5
FUEL DENSITY: 6.8 LB/GAL

	MACH	DRAG INDEX 0	20	40	60	80	100	120	TEMP. EFFECTS FACTOR +10°C	−10°C
		TIME TO ACCELERATE (MIN.)/FUEL TO ACCELERATE (LBS).								
SEA LEVEL (15°C)	.5	.0 / 0	.0 / 0	.0 / 0	.0 / 0	.0 / 0	.0 / 0	.0 / 0	1.0 / .0	.0 /.0
	.55	.05/ 63	.05/ 65	.05 /67	.05/ 69	.05/ 72	.05/ 74	.06/ 77	1.10/1.07	.90/.93
	.6	.09/129	.10/133	.10/137	.10/142	.11/147	.11/153	.12/159	1.10/1.07	.90/.93
	.65	.14/196	.14/203	.15/211	.16/219	.16/228	.17/238	.18/250	1.10/1.07	.89/.92
	.7	.19/265	.19/276	.20/287	.21/300	.22/314	.23/331	.24/349	1.10/1.08	.89/.92
	.75	.23/336	.24/351	.25/368	.27/386	.28/407	.30/432	.32/461	1.11/1.08	.89/.91
	.8	.28/410	.29/430	.31/451	.32/477	.34/508	.37/545	.40/591	1.11/1.08	.88/.91
	.85	.32/485	.34/511	.36/540	.38/576	.41/620	.45/675	.49/748	1.12/1.09	.88/.90
	.9	.37/562	.39/595	.41/633	.45/684	.49/748	.54/834	.62/960	1.14/1.10	.87/.90
2000 FEET (11°C)	.5	.0 / 0	.0 / 0	.0 / 0	.0 / 0	.0 / 0	.0 / 0	.0 / 0	.0 / .0	.0 /.0
	.55	.05/ 62	.05/ 64	.05/ 66	.05/ 68	.05/ 70	.06/ 73	.06/ 75	1.10/1.07	.90/.93
	.6	.10/126	.10/131	.10/135	.11/139	.11/144	.11/150	.12/156	1.10/1.07	.90/.93
	.65	.14/192	.15/199	.15/206	.16/214	.17/223	.17/232	.18/243	1.10/1.07	.89/.92
	.7	.19/260	.20/270	.21/281	.22/293	.23/306	.24/322	.25/339	1.10/1.08	.89/.92
	.75	.24/330	.25/344	.26/359	.27/376	.29/396	.30/419	.32/446	1.11/1.08	.89/.91
	.8	.29/401	.30/420	.31/440	.33/465	.35/493	.38/527	.40/569	1.11/1.08	.88/.91
	.85	.33/475	.35/499	.37/526	.39/559	.42/600	.45/651	.50/717	1.12/1.09	.88/.90
	.9	.38/550	.40/580	.43/616	.46/664	.50/722	.55/800	.62/911	1.14/1.10	.87/.90
4000 FEET (7°C)	.5	.0 / 0	.0 / 0	.0 / 0	.0 / 0	.0 / 0	.0 / 0	.0 / 0	.0 / .0	.0 /.0
	.55	.05/ 61	.05/ 63	.05/ 65	.05/ 67	.06/ 69	.06/ 71	.06/ 74	1.10/1.07	.90/.93
	.6	.10/124	.10/128	.11/132	.11/137	.11/142	.12/147	.12/152	1.10/1.07	.90/.93
	.65	.15/189	.16/195	.16/202	.17/210	.17/218	.18/227	.19/237	1.10/1.07	.89/.92
	.7	.20/255	.21/264	.21/275	.22/286	.23/299	.24/313	.26/329	1.10/1.08	.89/.92
	.75	.25/323	.26/336	.27/351	.28/367	.30/386	.31/407	.33/432	1.11/1.08	.89/.91
	.8	.30/393	.31/410	.33/430	.34/452	.36/479	.39/511	.41/549	1.11/1.08	.88/.91
	.85	.35/464	.36/487	.38/512	.40/544	.43/582	.47/628	.51/688	1.12/1.09	.88/.90
	.9	.39/537	.41/566	.44/600	.47/644	.51/698	.56/769	.63/867	1.14/1.10	.87/.90
6000 FEET (3°C)	.5	.0 / 0	.0 / 0	.0 / 0	.0 / 0	.0 / 0	.0 / 0	.0 / 0	.0 / .0	.0 /.0
	.55	.05/ 61	.05/ 62	.06/ 64	.06/ 66	.06/ 68	.06/ 70	.06/ 73	1.10/1.07	.90/.93
	.6	.10/123	.11/126	.11/130	.12/135	.12/139	.12/144	.13/149	1.10/1.07	.90/.93
	.65	.16/186	.16/192	.17/199	.17/206	.18/214	.19/222	.19/232	1.10/1.07	.89/.92
	.7	.21/251	.22/260	.22/270	.23/280	.24/292	.25/306	.27/321	1.10/1.08	.89/.92
	.75	.26/317	.27/329	.28/343	.29/359	.31/376	.32/396	.34/419	1.11/1.08	.89/.91
	.8	.31/385	.32/402	.34/420	.35/441	.37/466	.40/495	.42/531	1.11/1.08	.88/.91
	.85	.36/454	.38/476	.40/500	.42/529	.44/565	.48/607	.52/662	1.12/1.09	.88/.90
	.9	.41/526	.43/553	.45/585	.48/626	.52/676	.57/740	.63/828	1.14/1.10	.87/.90

FDD–1–(202)A

Figure 11-56

LOW ALTITUDE ACCELERATION
MAXIMUM THRUST

AIRPLANE CONFIGURATION
INDIVIDUAL DRAG INDEXES

GROSS WEIGHT – 55,000 POUNDS

REMARKS
ENGINES: (2) J79-GE-10

DATE: 1 JULY 1971
DATA BASIS: FLIGHT TEST

FUEL GRADE: JP-5
FUEL DENSITY: 6.8 LB/GAL

SEA LEVEL (15°C)

MACH	DRAG INDEX 0	20	40	60	80	100	120	TEMP. EFFECTS FACTOR +10°C	TEMP. EFFECTS FACTOR −10°C
.5	.0 / 0	.0 / 0	.0 / 0	.0 / 0	.0 / 0	.0 / 0	.0 / 0	0 /0	0 / 0
.55	.06/ 79	.06/ 82	.06/ 84	.06/ 87	.07/ 90	.07/ 93	.07/ 96	1.10/1.07	.90/.93
.6	.12/161	.12/166	.12/172	.13/178	.13/185	.14/ 192	.14/ 200	1.10/1.07	.90/.93
.65	.17/244	.18/253	.19/263	.19/274	.20/285	.21/ 298	.22/ 313	1.10/1.07	.89/.92
.7	.23/330	.24/343	.25/358	.26/374	.27/393	.29/ 413	.31/ 437	1.10/1.08	.89/.92
.75	.29/418	.30/437	.32/458	.33/481	.35/508	.37/ 540	.40/ 577	1.11/1.08	.89/.91
.8	.35/509	.36/534	.38/562	.40/594	.43/633	.46/ 680	.50/ 738	1.11/1.08	.88/.91
.85	.40/602	.42/634	.45/671	.48/716	.51/772	.56/ 842	.62/ 935	1.12/1.09	.88/.90
.9	.46/697	.48/738	.51/786	.55/850	.61/931	.67/1039	.77/1200	1.14/1.10	.87/.90

2000 FEET (11°C)

MACH	DRAG INDEX 0	20	40	60	80	100	120	TEMP. EFFECTS FACTOR +10°C	TEMP. EFFECTS FACTOR −10°C
.5	.0 / 0	.0 / 0	.0 / 0	.0 / 0	.0 / 0	.0 / 0	.0 / 0	0 /0	0 / 0
.55	.06/ 78	.06/ 80	.06/ 83	.07/ 86	.07/ 88	.07/ 91	.07/ 95	1.10/1.07	.90/.93
.6	.12/158	.12/163	.13/169	.13/175	.14/181	.14/ 188	.15/ 196	1.10/1.07	.90/.93
.65	.18/240	.19/249	.19/258	.20/268	.21/279	.22/ 292	.23/ 305	1.10/1.07	.89/.92
.7	.24/324	.25/337	.26/351	.27/366	.28/384	.30/ 403	.31/ 425	1.10/1.08	.89/.92
.75	.30/411	.31/428	.33/448	.34/470	.36/495	.38/ 574	.41/ 559	1.11/1.08	.89/.91
.8	.36/499	.37/523	.39/549	.41/580	.44/616	.47/ 659	.51/ 713	1.11/1.08	.88/.91
.85	.42/590	.44/620	.46/654	.49/697	.52/749	.57/ 813	.63/ 898	1.12/1.09	.88/.90
.9	.47/683	.50/721	.53/767	.57/826	.62/900	.68/ 998	.78/1140	1.14/1.10	.87/.90

4000 FEET (7°C)

MACH	DRAG INDEX 0	20	40	60	80	100	120	TEMP. EFFECTS FACTOR +10°C	TEMP. EFFECTS FACTOR −10°C
.5	.0 / 0	.0 / 0	.0 / 0	.0 / 0	.0 / 0	.0 / 0	.0 / 0	0 /0	0 / 0
.55	.06/ 77	.07/ 80	.07/ 82	.07/ 84	.07/ 87	.07/ 90	.08/ 93	1.10/1.07	.90/.93
.6	.13/156	.13/161	.13/166	.14/172	.14/178	.15/ 185	.15/ 192	1.10/1.07	.90/.93
.65	.19/236	.19/245	.20/254	.21/263	.22/274	.23/ 285	.24/ 298	1.10/1.07	.89/.92
.7	.25/319	.26/331	.27/344	.28/359	.29/375	.31/ 393	.32/ 414	1.10/1.08	.89/.92
.75	.31/403	.32/420	.34/439	.35/459	.37/483	.39/ 510	.42/ 543	1.11/1.08	.89/.91
.8	.37/490	.39/512	.41/537	.43/566	.45/600	.48/ 640	.52/ 689	1.11/1.08	.88/.91
.85	.43/578	.45/607	.48/639	.51/679	.54/727	.58/ 787	.64/ 864	1.12/1.09	.88/.90
.9	.49/669	.52/705	.55/748	.59/803	.63/872	.70/ 962	.79/1088	1.14/1.10	.87/.90

6000 FEET (3°C)

MACH	DRAG INDEX 0	20	40	60	80	100	120	TEMP. EFFECTS FACTOR +10°C	TEMP. EFFECTS FACTOR −10°C
.5	.0 / 0	.0 / 0	.0 / 0	.0 / 0	.0 / 0	.0 / 0	.0 / 0	0 /0	0 / 0
.55	.07/ 77	.07/ 79	.07/ 81	.07/ 84	.07/ 86	.08/ 89	.08/ 92	1.10/1.07	.90/.93
.6	.13/154	.14/159	.14/164	.15/170	.15/176	.16/ 182	.16/ 189	1.10/1.07	.90/.93
.65	.20/233	.20/242	.21/250	.22/259	.23/269	.24/ 281	.25/ 293	1.10/1.07	.89/.92
.7	.26/314	.27/326	.28/339	.29/352	.30/368	.32/ 385	.34/ 405	1.10/1.08	.89/.92
.75	.32/397	.34/413	.35/430	.37/450	.38/473	.41/ 498	.43/ 529	1.11/1.08	.89/.91
.8	.39/481	.40/502	.42/526	.44/553	.47/585	.50/ 623	.53/ 669	1.11/1.08	.88/.91
.85	.45/567	.47/595	.49/625	.52/663	.56/708	.60/ 763	.65/ 833	1.12/1.09	.88/.90
.9	.51/655	.54/690	.57/730	.61/782	.65/846	.71/ 928	.80/1041	1.14/1.10	.87/.90

FDD-1-(203)A

Figure 11-57

LOW ALTITUDE ACCELERATION
MILITARY THRUST

AIRPLANE CONFIGURATION
INDIVIDUAL DRAG INDEXES

GROSS WEIGHT – 35,000 POUNDS

REMARKS
ENGINES: (2) J79–GE–10

DATE: 1 JULY 1971
DATA BASIS: FLIGHT TEST

FUEL GRADE: UP–5
FUEL DENSITY: 6.8 LB/GAL

	MACH	TIME TO ACCELERATE (MIN.)/FUEL TO ACCELERATE (LBS). DRAG INDEX 0	20	40	60	80	100	120	TEMP. EFFECTS FACTOR +10°C	−10°C
SEA LEVEL (15°C)	.5	.0/0	.0/0	.0/0	.0/0	.0/0	.0/0	.0/0	0/0	0/0
	.55	.07/25	.07/26	.07/28	.08/30	.08/31	.09/34	.10/36	1.17/1.11	.87/.93
	.6	.13/51	.14/55	.15/58	.16/63	.18/68	.19/74	.21/81	1.18/1.11	.86/.91
	.65	.21/80	.22/86	.24/92	.26/100	.28/110	.32/123	.36/139	1.19/1.13	.84/.90
	.7	.28/111	.31/120	.33/131	.37/145	.41/163	.48/187	.57/224	1.23/1.16	.83/.88
	.75	.36/144	.40/158	.44/176	.50/199	.58/233	.72/290	1.05/426	1.24/1.18	.79/.83
	.8	.45/181	.50/202	.57/228	.66/269	.84/343	1.41/586		1.26/1.20	.80/.85
	.85	.54/223	.62/253	.72/296	.93/385				1.30/1.24	.83/.87
	.9	.66/275	.78/325	1.02/435					1.22/1.17	.81/.85

	MACH	0	20	40	60	80	100	120	+10°C	−10°C
2000 FEET (11°C)	.5	.0/0	.0/0	.0/0	.0/0	.0/0	.0/0	.0/0	0/0	0/0
	.55	.07/25	.07/26	.08/27	.08/29	.09/31	.09/33	.10/35	1.17/1.11	.87/.93
	.6	.14/51	.15/54	.16/57	.17/61	.18/66	.20/71	.21/78	1.18/1.11	.86/.91
	.65	.21/78	.23/84	.24/90	.26/98	.29/107	.32/118	.36/132	1.19/1.13	.84/.90
	.7	.29/108	.31/117	.34/127	.37/140	.42/156	.47/178	.56/209	1.23/1.16	.83/.88
	.75	.37/141	.41/154	.45/170	.50/191	.58/220	.70/265	.94/361	1.24/1.18	.79/.83
	.8	.46/176	.51/195	.57/220	.66/255	.81/314	1.18/463		1.26/1.20	.80/.85
	.85	.56/216	.62/244	.72/282	.90/354	1.48/596			1.30/1.24	.83/.87
	.9	.67/266	.78/310	.99/399					1.22/1.17	.81/.85

	MACH	0	20	40	60	80	100	120	+10°C	−10°C
4000 FEET (7°C)	.5	.0/0	.0/0	.0/0	.0/0	.0/0	.0/0	.0/0	0/0	0/0
	.55	.07/24	.07/26	.08/27	.08/29	.09/30	.09/32	.10/35	1.17/1.11	.87/.93
	.6	.14/50	.15/53	.16/56	.17/60	.18/64	.20/69	.22/76	1.18/1.11	.86/.91
	.65	.22/77	.23/82	.25/88	.27/95	.29/103	.32/114	.36/127	1.19/1.13	.84/.90
	.7	.30/106	.32/114	.35/124	.38/135	.42/150	.47/169	.55/196	1.23/1.16	.83/.88
	.75	.38/137	.42/150	.46/164	.51/183	.58/209	.68/248	.88/320	1.24/1.18	.79/.83
	.8	.47/172	.52/189	.58/211	.66/243	.80/293	1.08/401		1.26/1.20	.80/.85
	.85	.57/210	.64/236	.73/270	.88/331	1.28/486			1.30/1.24	.83/.87
	.9	.68/258	.79/297	.97/371					1.22/1.17	.81/.85

	MACH	0	20	40	60	80	100	120	+10°C	−10°C
6000 FEET (3°C)	.5	.0/0	.0/0	.0/0	.0/0	.0/0	.0/0	.0/0	0/0	0/0
	.55	.07/24	.08/25	.08/27	.09/28	.09/30	.10/32	.10/34	1.17/1.11	.87/.93
	.6	.15/49	.16/52	.17/55	.18/59	.19/63	.21/68	.22/74	1.18/1.11	.86/.91
	.65	.23/76	.24/81	.26/87	.28/93	.30/101	.33/110	.37/122	1.19/1.13	.84/.90
	.7	.31/104	.33/112	.36/121	.39/132	.43/146	.48/163	.55/187	1.23/1.16	.83/.88
	.75	.39/135	.43/146	.47/160	.52/177	.59/201	.68/234	.85/294	1.24/1.18	.79/.83
	.8	.49/168	.53/184	.59/205	.67/233	.79/277	1.02/359		1.26/1.20	.80/.85
	.85	.58/205	.65/228	.74/260	.88/312	1.19/425			1.30/1.24	.83/.87
	.9	.70/250	.80/286	.97/349					1.22/1.17	.81/.85

FDD–1–(204)A

Figure 11-58

LOW ALTITUDE ACCELERATION

MILITARY THRUST

AIRPLANE CONFIGURATION
INDIVIDUAL DRAG INDEXES

GROSS WEIGHT – 45,000 POUNDS

REMARKS
ENGINES: (2) J79-GE-10

DATE: 1 JULY 1971
DATA BASIS: FLIGHT TEST

FUEL GRADE: JP-5
FUEL DENSITY: 6.8 LB/GAL

	MACH	DRAG INDEX 0	20	40	60	80	100	120	TEMP. EFFECTS FACTOR +10°C	TEMP. EFFECTS FACTOR −10°C
		TIME TO ACCELERATE (MIN.)/FUEL TO ACCELERATE (LBS).								
SEA LEVEL (15°C)	.5	.0 / 0	.0 / 0	.0 / 0	.0 / 0	.0 / 0	.0 / 0	.0 / 0	.0 / .0	.0 / .0
	.55	.09/ 33	.09/ 35	.10/ 37	.10/ 40	.11/ 42	.12/ 46	.13/ 49	1.17/1.11	.87/.93
	.6	.18/ 68	.19/ 73	.20/ 78	.22/ 84	.24/ 91	.26/ 99	.29/109	1.18/1.11	.86/.91
	.65	.27/106	.29/114	.32/123	.35/134	.38/148	.43/165	.49/188	1.19/1.13	.84/.90
	.7	.37/147	.41/159	.45/175	.49/194	.56/219	.65/254	.79/311	1.23/1.16	.83/.88
	.75	.48/191	.53/210	.59/234	.67/266	.79/313	.98/393	1.53/619	1.24/1.18	.79/.83
	.8	.59/239	.66/267	.75/304	.89/360	1.11/464	2.17/903		1.26/1.20	.80/.85
	.85	.72/294	.82/335	.96/395	1.24/516				1.30/1.24	.83/.87
	.9	.87/364	1.03/431	1.38/587					1.22/1.17	.81/.85
2000 FEET (11°C)	.5	.0 / 0	.0 / 0	.0 / 0	.0 / 0	.0 / 0	.0 / 0	.0 / 0	.0 / .0	.0 / .0
	.55	.09/ 33	.10/ 35	.10/ 37	.11/ 39	.12/ 42	.12/ 45	.13/ 48	1.17/1.11	.87/.93
	.6	.18/ 67	.20/ 72	.21/ 77	.23/ 82	.24/ 89	.26/ 97	.29/106	1.18/1.11	.86/.91
	.65	.28/104	.30/112	.33/121	.35/131	.39/144	.43/159	.49/180	1.19/1.13	.84/.90
	.7	.39/144	.42/156	.45/170	.50/188	.56/210	.64/241	.78/289	1.23/1.16	.83/.88
	.75	.49/187	.54/205	.60/227	.67/256	.78/297	.96/366	1.34/513	1.24/1.18	.79/.83
	.8	.61/233	.68/259	.76/293	.89/343	1.10/428	1.70/669		1.26/1.20	.80/.85
	.85	.73/286	.83/324	.96/377	1.21/478	2.29/926			1.30/1.24	.83/.87
	.9	.89/352	1.04/412	1.33/537					1.22/1.17	.81/.85
4000 FEET (7°C)	.5	.0 / 0	.0 / 0	.0 / 0	.0 / 0	.0 / 0	.0 / 0	.0 / 0	.0 / .0	.0 / .0
	.55	.09/ 33	.10/ 34	.11/ 36	.11/ 39	.12/ 41	.13/ 44	.14/ 47	1.17/1.11	.87/.93
	.6	.19/ 67	.20/ 71	.22/ 76	.23/ 81	.25/ 87	.27/ 95	.30/103	1.18/1.11	.86/.91
	.65	.29/103	.31/110	.34/118	.36/128	.40/140	.44/155	.49/173	1.19/1.13	.84/.90
	.7	.40/141	.43/153	.47/166	.51/182	.57/203	.65/231	.76/273	1.23/1.16	.83/.88
	.75	.51/183	.55/200	.61/220	.68/247	.79/284	.94/342	1.24/452	1.24/1.18	.79/.83
	.8	.63/228	.69/252	.77/283	.89/328	1.09/403	1.53/567		1.26/1.20	.80/.85
	.85	.75/279	.85/314	.97/362	1.21/451	1.83/697			1.30/1.24	.83/.87
	.9	.91/342	1.05/396	1.31/500					1.22/1.17	.81/.85
6000 FEET (3°C)	.5	.0 / 0	.0 / 0	.0 / 0	.0 / 0	.0 / 0	.0 / 0	.0 / 0	.0 / .0	.0 / .0
	.55	.10/ 33	.11/ 34	.11/ 36	.12/ 38	.13/ 41	.13/ 44	.14/ 47	1.17/1.11	.87/.93
	.6	.20/ 66	.21/ 70	.23/ 75	.24/ 80	.26/ 86	.28/ 93	.31/102	1.18/1.11	.86/.91
	.65	.31/102	.33/109	.35/117	.38/126	.41/137	.45/151	.50/168	1.19/1.13	.84/.90
	.7	.41/140	.45/151	.48/163	.53/178	.58/198	.66/223	.76/259	1.23/1.16	.83/.88
	.75	.53/180	.57/196	.63/215	.70/240	.80/273	.94/325	1.20/414	1.24/1.18	.79/.83
	.8	.65/224	.71/247	.79/275	.91/316	1.09/381	1.44/506		1.26/1.20	.80/.85
	.85	.78/273	.87/305	.99/349	1.20/426	1.66/596			1.30/1.24	.83/.87
	.9	.93/333	1.07/382	1.30/470					1.22/1.17	.81/.85

FDD-1-(205) A

Figure 11-59

LOW ALTITUDE ACCELERATION

MILITARY THRUST

AIRPLANE CONFIGURATION
INDIVIDUAL DRAG INDEXES

GROSS WEIGHT – 55,000 POUNDS

REMARKS
ENGINES: (2) J79-GE-10

DATE: 1 JULY 1971
DATA BASIS: FLIGHT TEST

FUEL GRADE: JP-5
FUEL DENSITY: 6.8 LB/GAL

	MACH	DRAG INDEX / TIME TO ACCELERATE (MIN)/FUEL TO ACCELERATE (LBS)							TEMP EFFECTS FACTOR	
		0	20	40	60	80	100	120	+ 10°C	−10°C
SEA LEVEL (15°C)	.5	.0 / 0	.0 / 0	.0 / 0	.0 / 0	.0 / 0	.0 / 0	.0 / 0	.0 / .0	.0 / .0
	.55	.11/ 42	.12/ 45	.13/ 48	.13/ 51	.14/ 55	.16/ 59	.17/ 64	1.17/1.11	.87/.93
	.6	.23/ 87	.24/ 93	.26/100	.28/108	.31/117	.34/129	.37/143	1.18/1.11	.86/.91
	.65	.35/135	.38/145	.41/158	.45/172	.49/191	.55/215	.64/247	1.19/1.13	.84/.90
	.7	.47/186	.52/ 203	.57/223	.63/248	.72/283	.84/331	1.05/414	1.23/1.16	.83/.88
	.75	.61/241	.67/267	.75/298	.86/341	1.02/407	1.30/523	2.26/918	1.24/1.18	.79/.83
	.8	.75/302	.84/339	.96/388	1.14/464	1.49/608			1.26/1.20	.80/.85
	.85	.91/372	1.04/426	1.25/515	1.61/671				1.30/1.24	.83/.87
	.9	1.10/459	1.31/547	1.79/762					1.22/1.17	.81/.85
2000 FEET (11°C)	.5	.0 / 0	.0 / 0	.0 / 0	.0 / 0	.0 / 0	.0 / 0	.0 / 0	.0 / .0	.0 / .0
	.55	.12/ 42	.12/ 45	.13/ 47	.14/ 50	.15/ 54	.16/ 58	.17/ 63	1.17/1.11	.87/.93
	.6	.24/ 86	.25/ 92	.27/ 99	.29/106	.31/115	.34/126	.38/139	1.18/1.11	.86/.91
	.65	.37/133	.39/143	.42/155	.46/169	.50/186	.56/208	.64/237	1.19/1.13	.84/.90
	.7	.49/183	.53/199	.58/218	.65/242	.73/273	.84/316	1.02/385	1.23/1.16	.83/.88
	.75	.63/237	.69/261	.77/290	.87/329	1.02/387	1.27/485	1.87/722	1.24/1.18	.79/.83
	.8	.77/296	.86/331	.98/375	1.15/443	1.45/564	2.47/974		1.26/1.20	.80/.85
	.85	.93/363	1.06/413	1.23/483	1.53/629				1.30/1.24	.83/.87
	.9	1.12/446	1.32/525	1.72/694					1.22/1.17	.81/.85
4000 FEET (7°C)	.5	.0 / 0	.0 / 0	.0 / 0	.0 / 0	.0 / 0	.0 / 0	.0 / 0	.0 / .0	.0 / .0
	.55	.12/ 42	.13/ 45	.14/ 47	.15/ 50	.16/ 54	.17/ 58	.18/ 63	1.17/1.11	.87/.93
	.6	.25/ 86	.26/ 92	.28/ 98	.30/105	.33/114	.36/124	.39/137	1.18/1.11	.86/.91
	.65	.38/132	.40/142	.43/153	.47/166	.52/183	.58/203	.65/230	1.19/1.13	.84/.90
	.7	.51/181	.55/196	.60/214	.66/236	.74/265	.85/304	1.01/362	1.23/1.16	.83/.88
	.75	.65/234	.71/256	.79/283	.89/319	1.03/371	1.25/454	1.72/626	1.24/1.18	.79/.83
	.8	.80/291	.88/323	1.00/364	1.16/425	1.48/544	2.14/795		1.26/1.20	.80/.85
	.85	.96/355	1.08/401	1.25/466	1.57/588	2.62/1001			1.30/1.24	.83/.87
	.9	1.15/434	1.34/506	1.70/647					1.22/1.17	.81/.85
6000 FEET (3°C)	.5	.0 / 0	.0 / 0	.0 / 0	.0 / 0	.0 / 0	.0 / 0	.0 / 0	.0 / .0	.0 / .0
	.55	.13/ 42	.14/ 45	.15/ 47	.15/ 50	.17/ 54	.18/ 58	.19/ 63	1.17/1.11	.87/.93
	.6	.26/ 86	.28/ 91	.30/ 98	.32/105	.34/113	.37/123	.41/135	1.18/1.11	.86/.91
	.65	.39/132	.42/141	.45/152	.49/165	.54/181	.60/200	.67/225	1.19/1.13	.84/.90
	.7	.53/180	.58/194	.63/212	.69/233	.77/260	.87/296	1.03/348	1.23/1.16	.83/.88
	.75	.68/231	.74/253	.81/279	.91/312	1.05/360	1.26/433	1.64/569	1.24/1.18	.79/.83
	.8	.83/287	.92/317	1.03/356	1.18/411	1.44/504	1.98/698		1.26/1.20	.80/.85
	.85	.99/349	1.12/392	1.28/451	1.58/559	2.30/828			1.30/1.24	.83/.87
	.9	1.19/425	1.37/491	1.70/612					1.22/1.17	.81/.85

FDD-1-(206)A

Figure 11-60

MAXIMUM THRUST ACCELERATION

10,000 FEET

AIRPLANE CONFIGURATION
(4) AIM-7

REMARKS
ENGINE(S): (2) J79-GE-10
ICAO STANDARD DAY

GUIDE

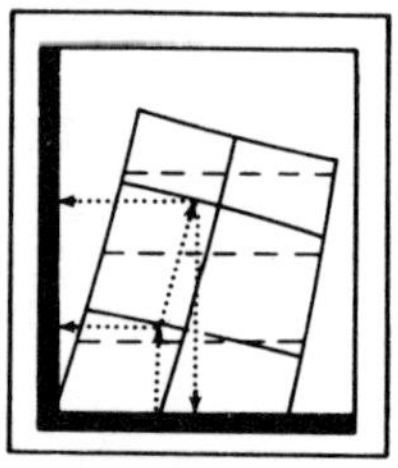

DATE: 1 JANUARY 1972
DATA BASIS: FLIGHT TEST

FUEL GRADE: JP-5
FUEL DENSITY: 6.8 LB/GAL

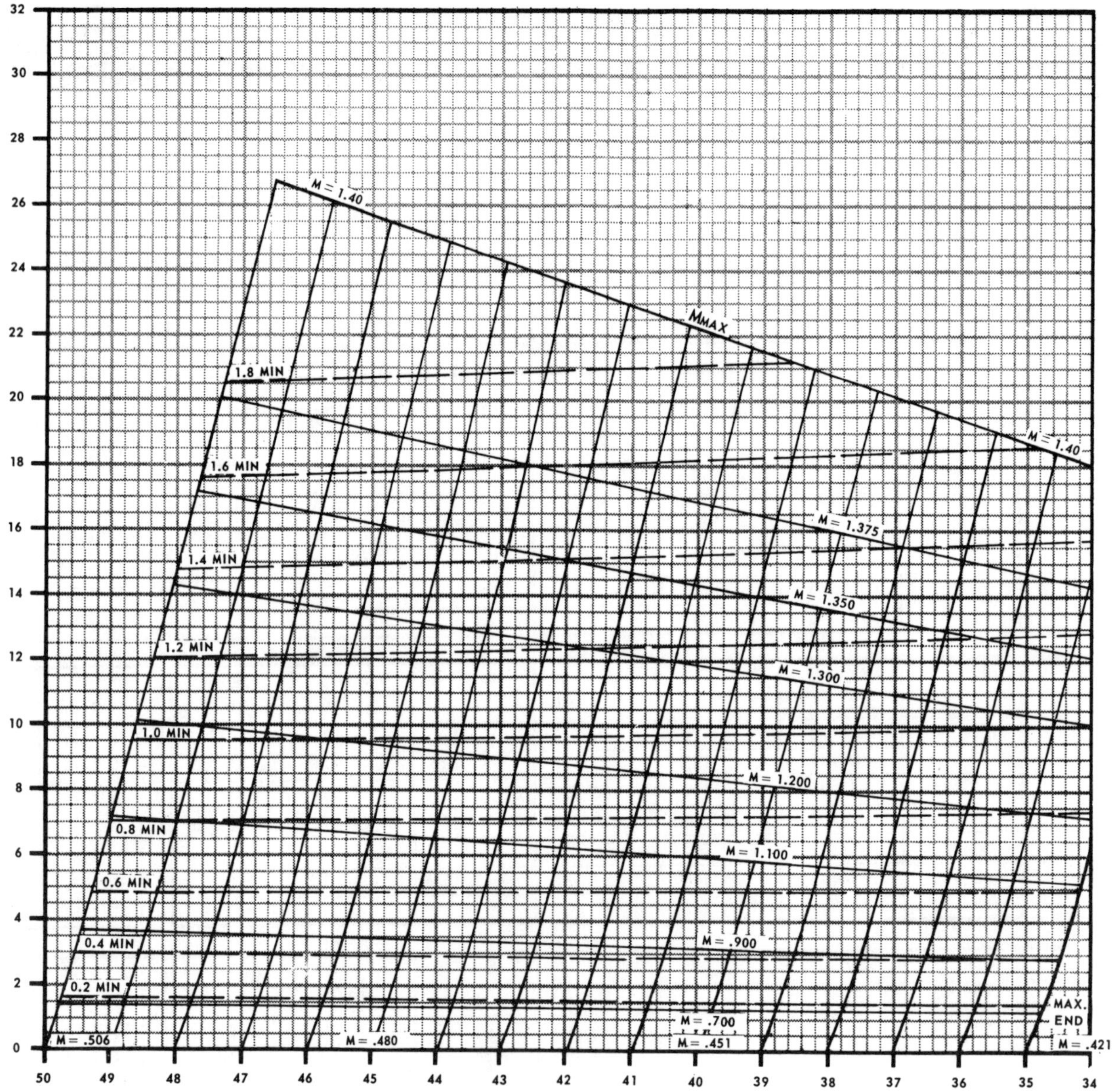

FDD-1-(207)A

Figure 11-61

MAXIMUM THRUST ACCELERATION
30,000 FEET

AIRPLANE CONFIGURATION
(4) AIM-7

REMARKS
ENGINE(S): (2) J79-GE-10
ICAO STANDARD DAY

GUIDE

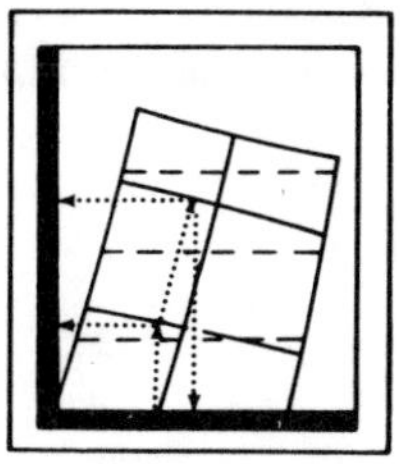

DATE: 1 JANUARY 1972
DATA BASIS: FLIGHT TEST

FUEL GRADE: JP-5
FUEL DENSITY: 6.8 LB/GAL

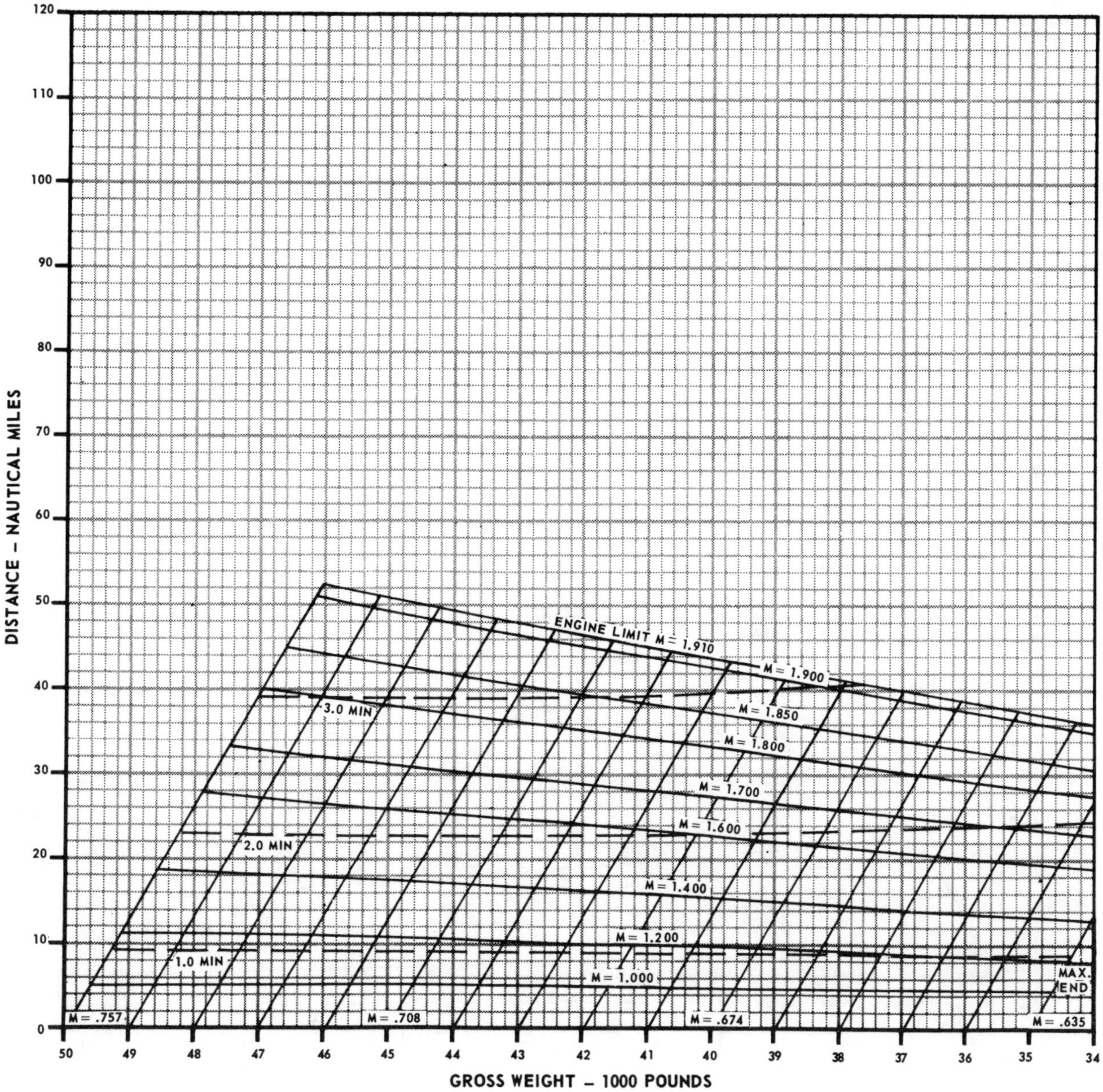

FDD-1-(209)A

Figure 11-62

MAXIMUM THRUST ACCELERATION

35,000 FEET

AIRPLANE CONFIGURATION

(4) AIM-7

REMARKS

ENGINE(S): (2) J79-GE-10
ICAO STANDARD DAY

GUIDE

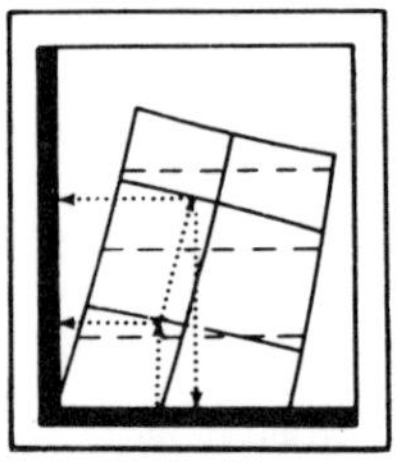

DATE: 1 JANUARY 1972
DATA BASIS: FLIGHT TEST

FUEL GRADE: JP-5
FUEL DENSITY: 6.8 LB/GAL

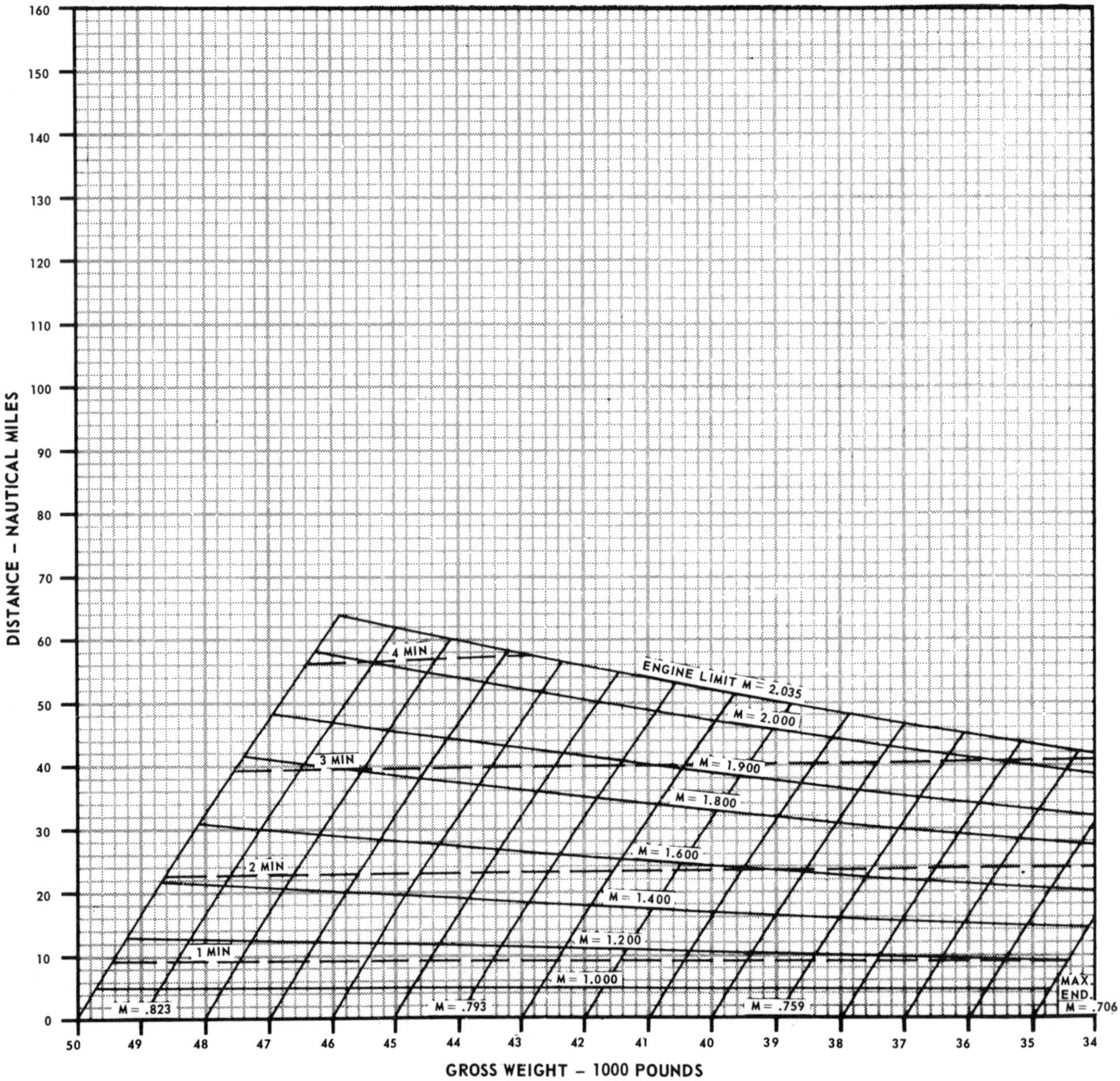

FDD-1-(210)A

Figure 11-63

MAXIMUM THRUST ACCELERATION

40,000 FEET

AIRPLANE CONFIGURATION
(4) AIM-7

REMARKS
ENGINE(S): (2) J79-GE-10
ICAO STANDARD DAY

GUIDE

DATE: 1 JANUARY 1972
DATA BASIS: **FLIGHT TEST**

FUEL GRADE: JP-5
FUEL DENSITY: 6.8 LB/GAL

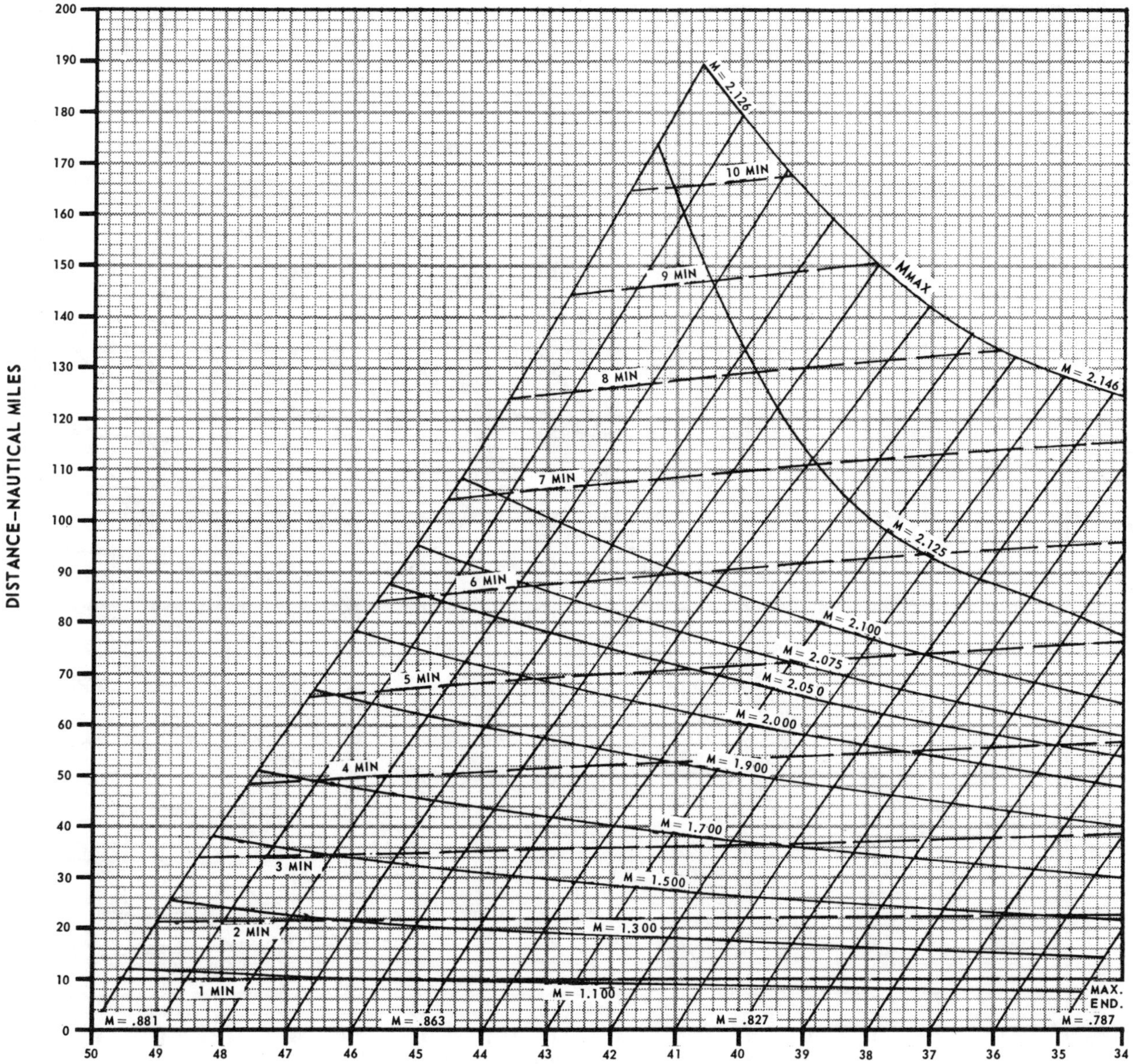

FDD-1-(211)A

Figure 11-64

MAXIMUM THRUST ACCELERATION
45,000 FEET

AIRPLANE CONFIGURATION
(4) AIM-7

REMARKS
ENGINE(S): (2) J79-GE-10
ICAO STANDARD DAY

GUIDE

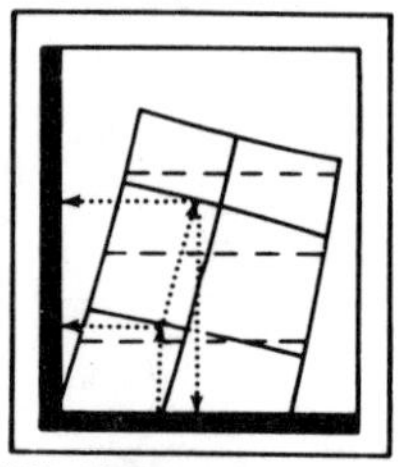

DATE: 1 JANUARY 1972
DATA BASIS: FLIGHT TEST

FUEL GRADE: JP-5
FUEL DENSITY: 6.8 LB/GAL

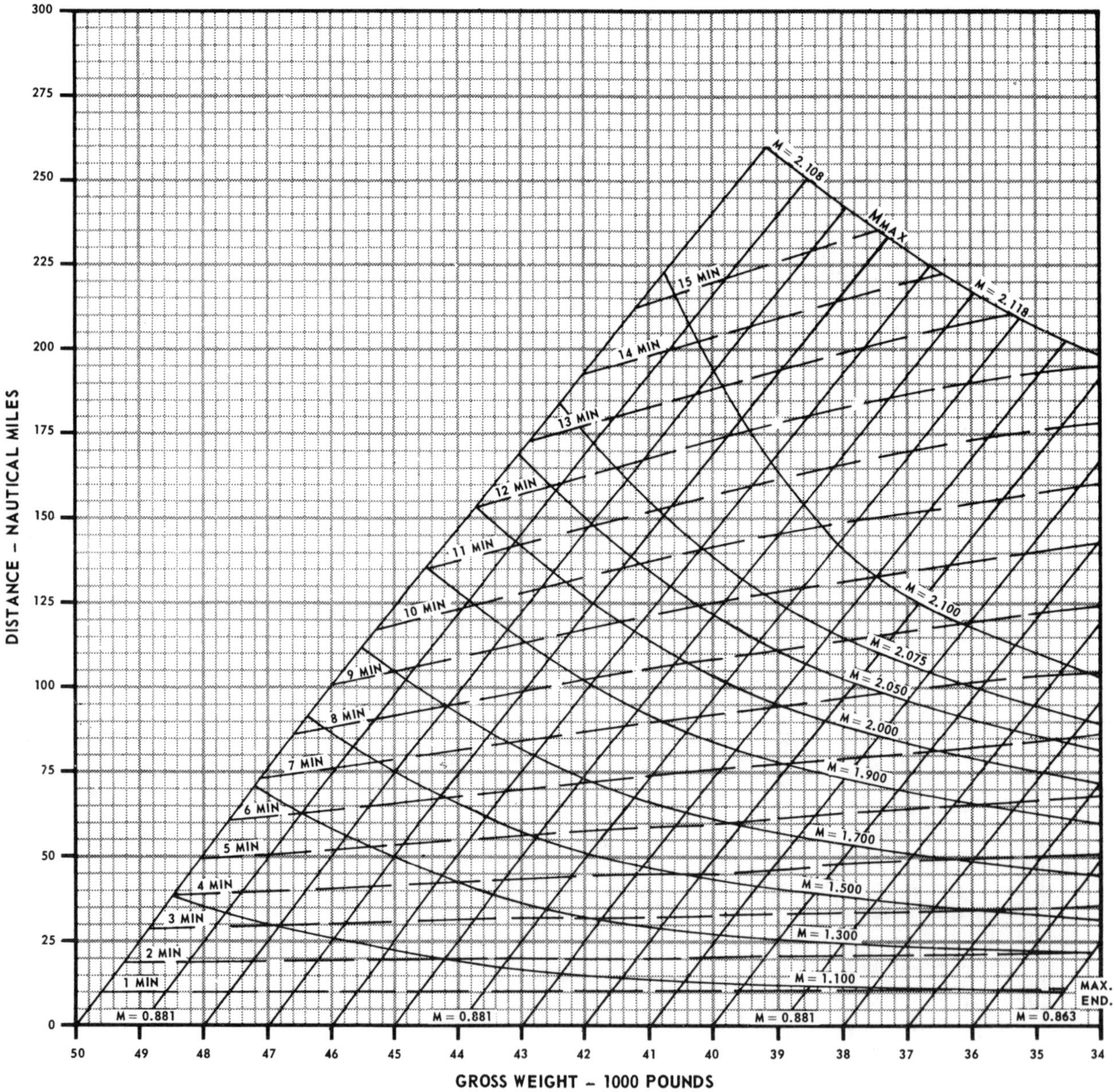

FDD-1-(212)A

Figure 11-65

MAXIMUM THRUST ACCELERATION

30,000 FEET

AIRPLANE CONFIGURATION
(4) AIM-7 AND (1) ℄ TANK

REMARKS

ENGINE(S): (2) J79-GE-10
ICAO STANDARD DAY

GUIDE

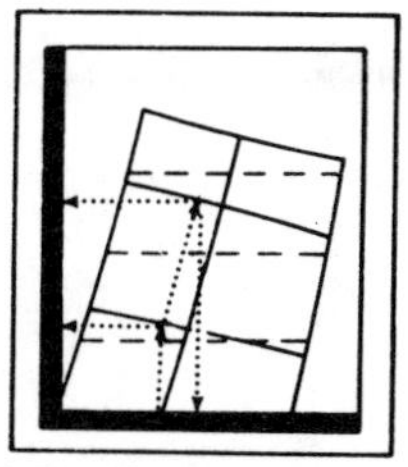

DATE: 1 JANUARY 1972
DATA BASIS: **FLIGHT TEST**

FUEL GRADE: JP-5
FUEL DENSITY: 6.8 LB/GAL

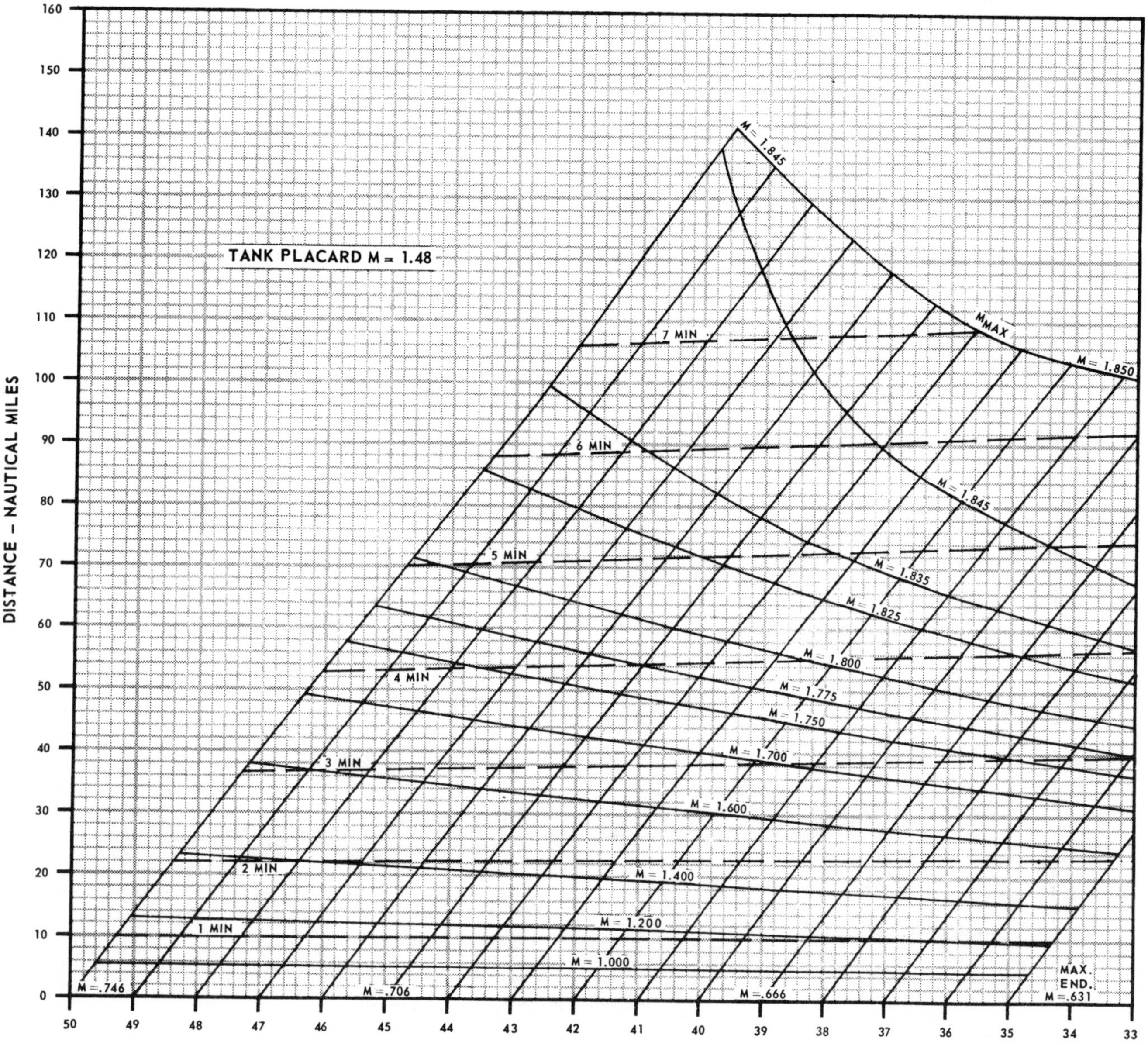

FDD-1-(215)A

Figure 11-66

MAXIMUM THRUST ACCELERATION

35,000 FEET

AIRPLANE CONFIGURATION
(4) AIM-7 AND (1) ℄ TANK

REMARKS

ENGINE(S): (2) J79-GE-10
ICAO STANDARD DAY

GUIDE

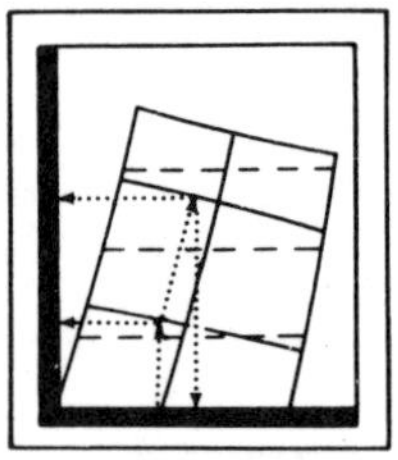

DATE: 1 JANUARY 1972
DATA BASIS: FLIGHT TEST

FUEL GRADE: JP-5
FUEL DENSITY: 6.8 LB/GAL

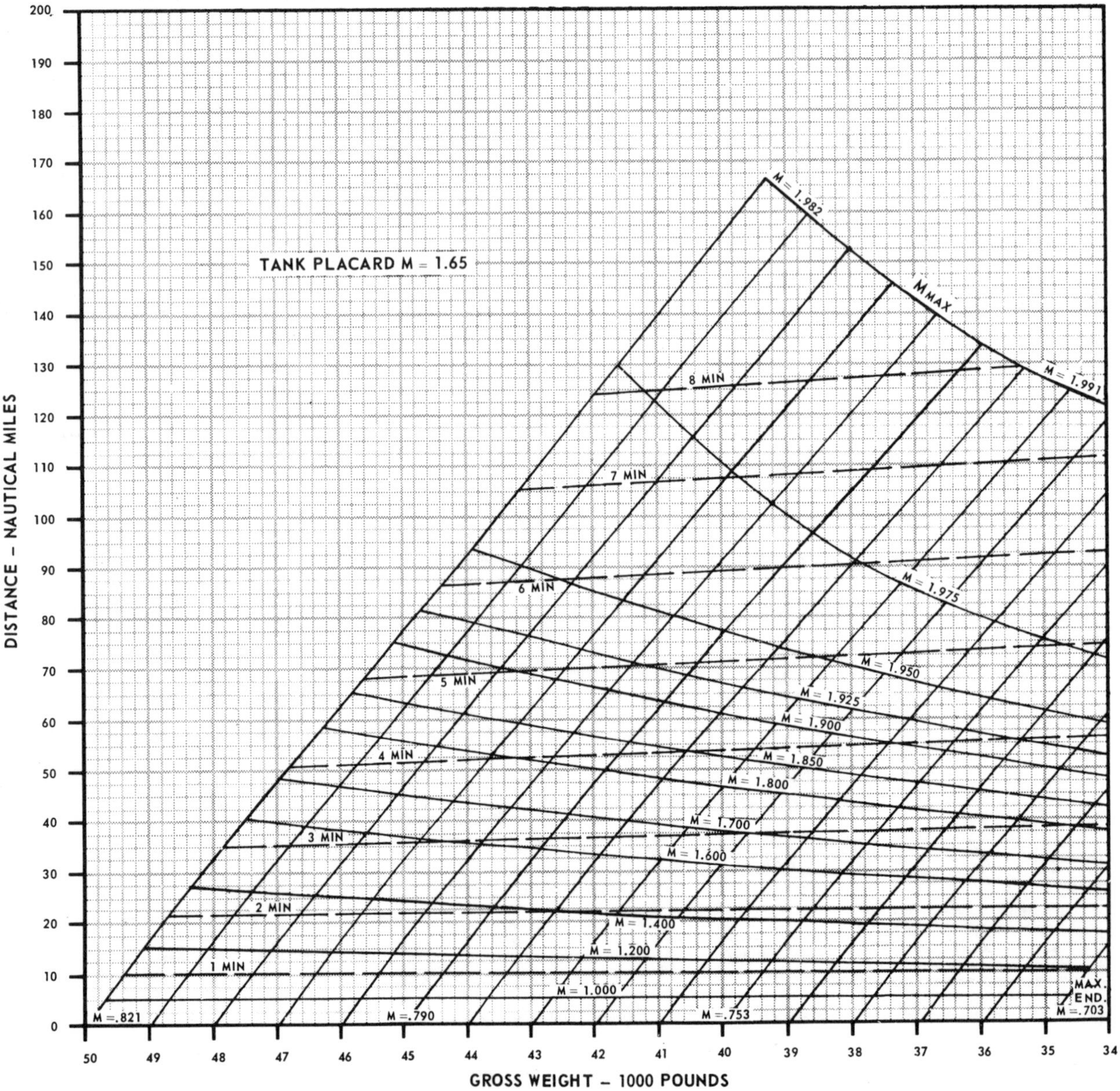

FDD-1-(216)A

Figure 11-67

MAXIMUM THRUST ACCELERATION
40,000 FEET

AIRPLANE CONFIGURATION
(4) AIM-7 AND (1) ℄ TANK

REMARKS
ENGINE(S): (2) J79-GE-10
ICAO STANDARD DAY

GUIDE

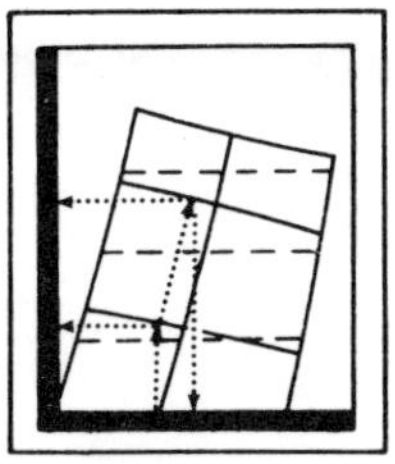

DATE: 1 JANUARY 1972
DATA BASIS: **FLIGHT TEST**

FUEL GRADE: JP-5
FUEL DENSITY: 6.8 LB/GAL

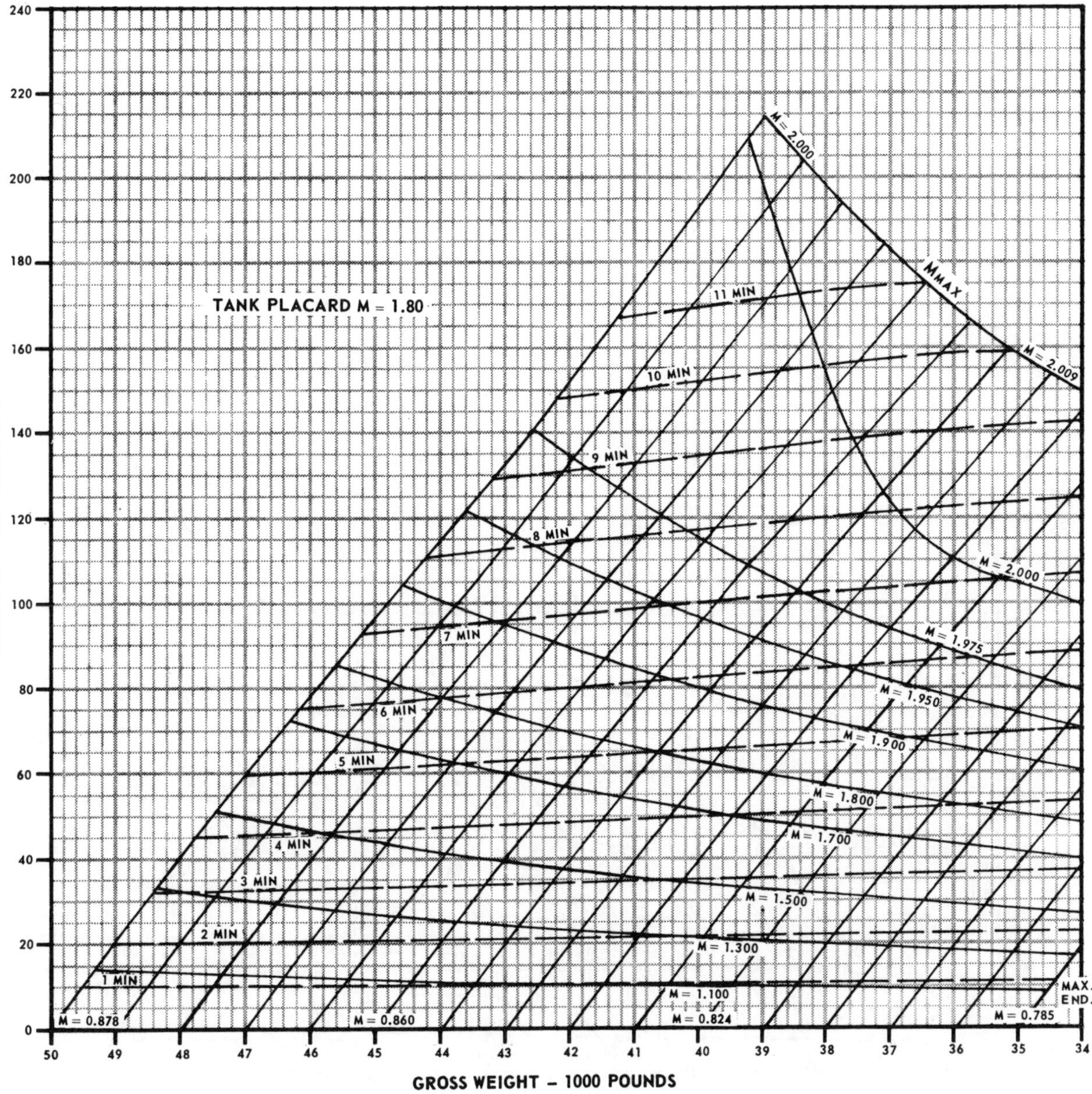

FDD-1-(217)A

Figure 11-68

MAXIMUM THRUST ACCELERATION
45,000 FEET

AIRPLANE CONFIGURATION
(4) AIM-7 AND (1) ℄ TANK

REMARKS
ENGINE(S): (2) J79-GE-10
ICAO STANDARD DAY

GUIDE

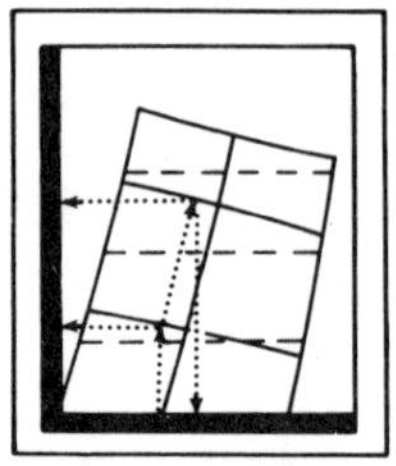

DATE: 1 JANUARY 1972
DATA BASIS: FLIGHT TEST

FUEL GRADE: JP-5
FUEL DENSITY: 6.8 LB/GAL.

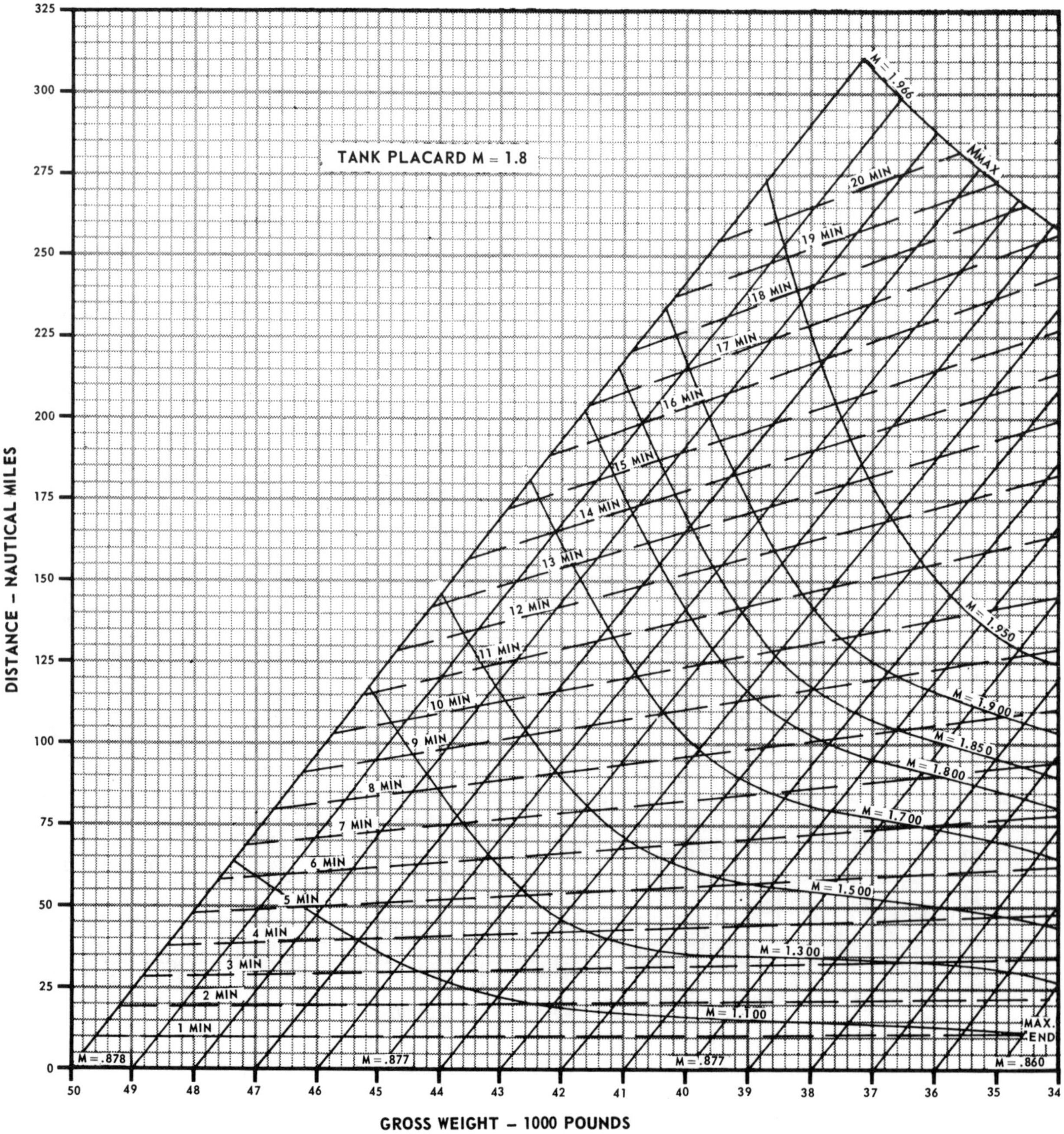

FDD-1-(218)A

Figure 11-69

MAXIMUM THRUST ACCELERATION

30,000 FEET

AIRPLANE CONFIGURATION
(4) AIM-7 AND (2) WING TANKS

REMARKS

ENGINE(S): (2) J79-GE-10
ICAO STANDARD DAY

GUIDE

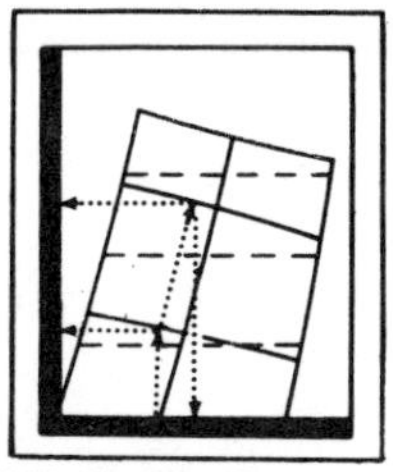

DATE: 1 JANUARY 1972
DATA BASIS: **FLIGHT TEST**

FUEL GRADE: JP-5
FUEL DENSITY: 6.8 LB/GAL

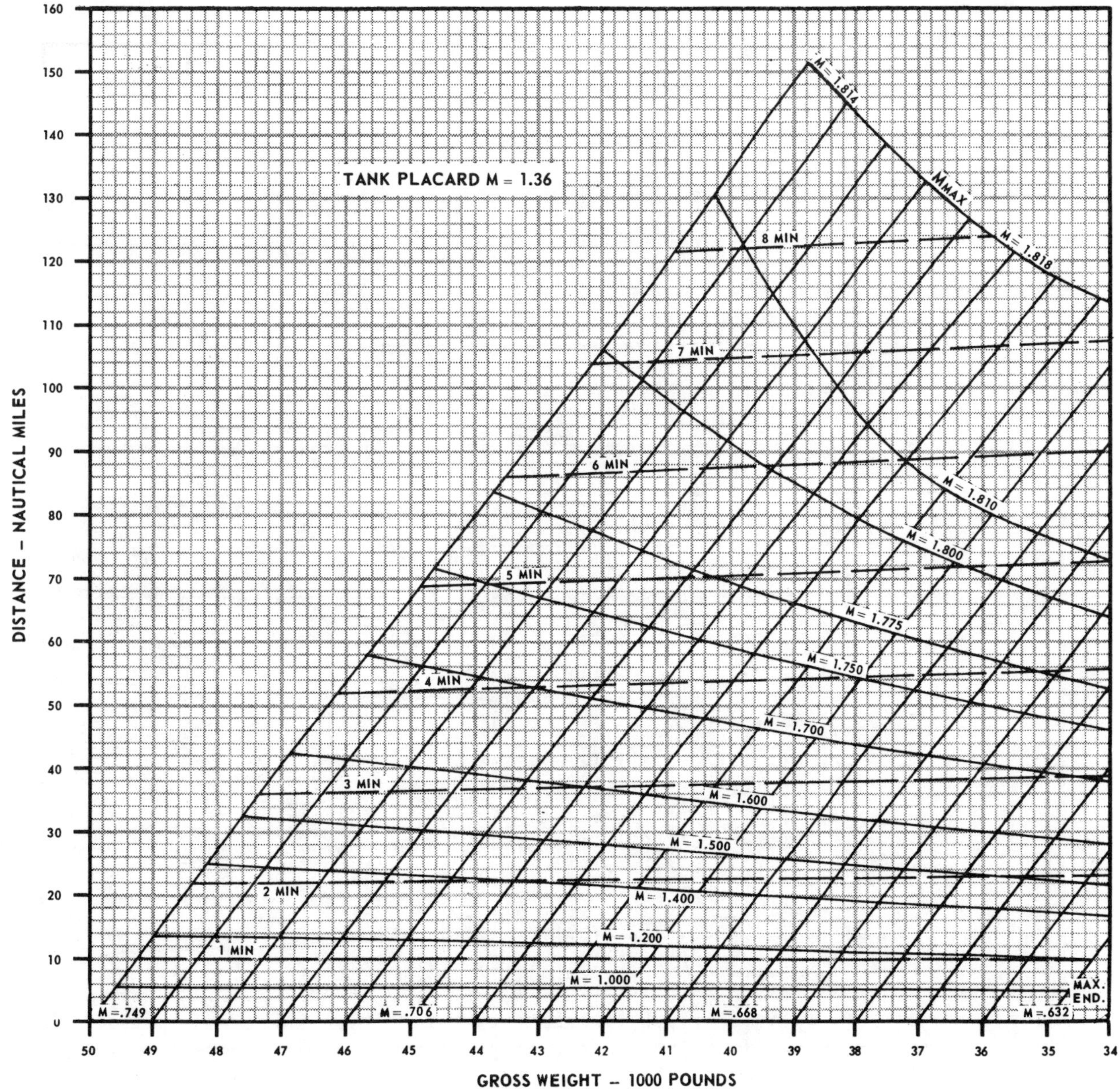

FDD-1-(221)A

Figure 11-70

MAXIMUM THRUST ACCELERATION

35,000 FEET

AIRPLANE CONFIGURATION
(4) AIM-7 AND (2) WING TANKS

REMARKS
ENGINE(S): (2) J79-GE-10
ICAO STANDARD DAY

GUIDE

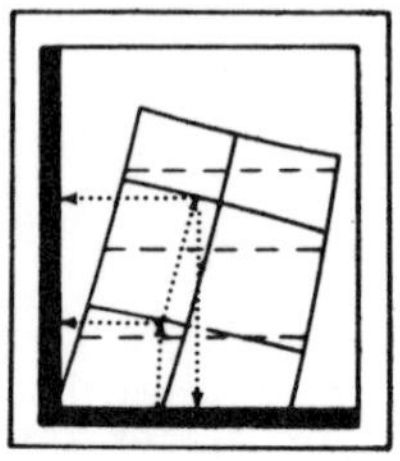

DATE: 1 JANUARY 1972
DATA BASIS: FLIGHT TEST

FUEL GRADE: JP-5
FUEL DENSITY: 6.8 LB/GAL

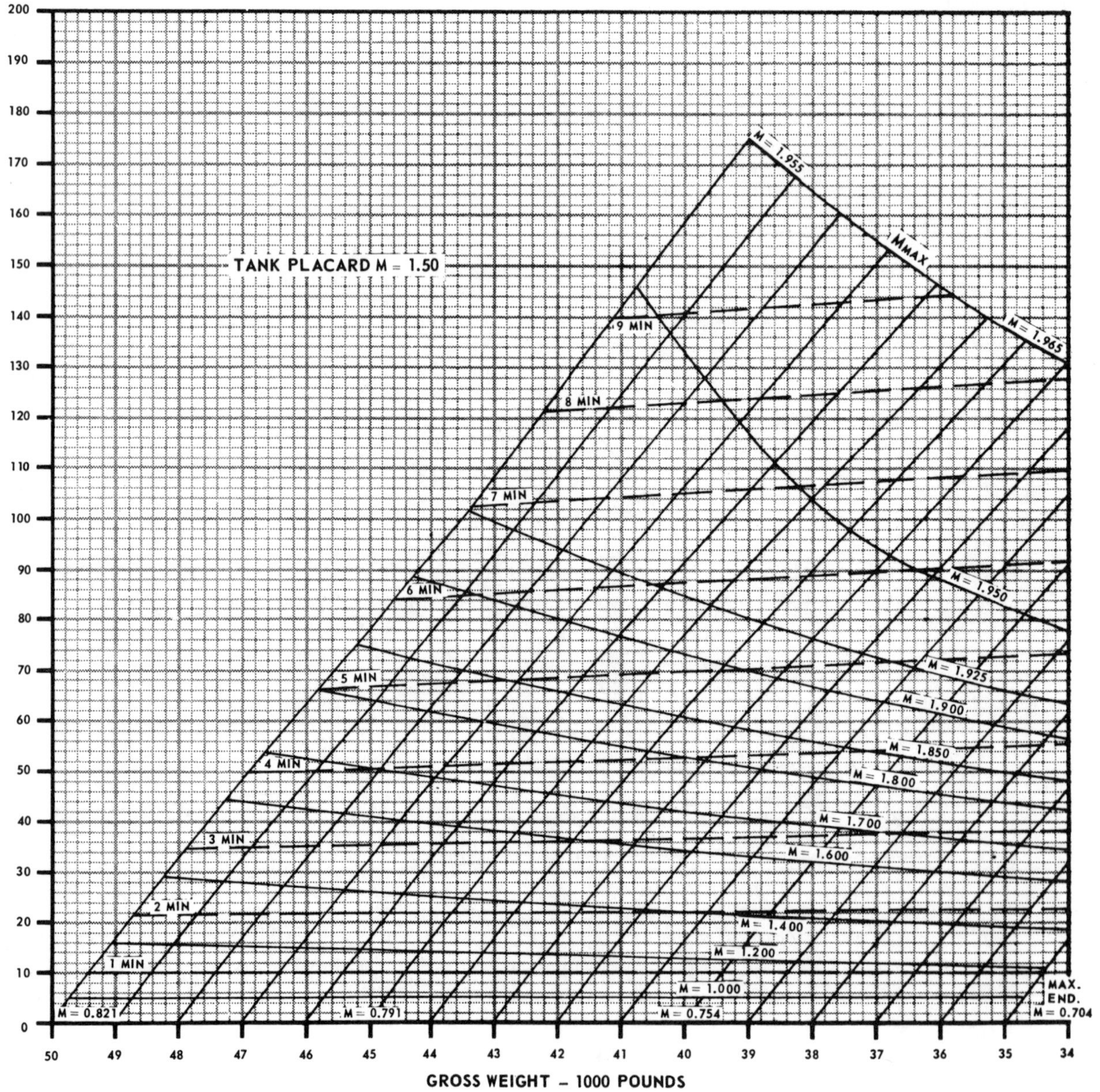

FDD-1-(222)A

Figure 11-71

MAXIMUM THRUST ACCELERATION

40,000FEET

AIRPLANE CONFIGURATION
(4) AIM-7 AND (2) WING TANKS

REMARKS

ENGINE(S): (2) J79-GE-10
ICAO STANDARD DAY

GUIDE

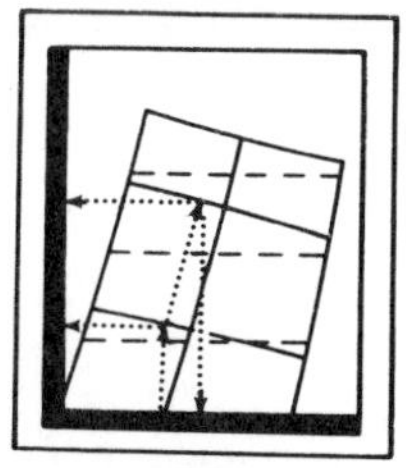

FUEL GRADE: JP-5
FUEL DENSITY: 6.8 LB/GAL

DATE: 1 JANUARY 1972
DATA BASIS: FLIGHT TEST

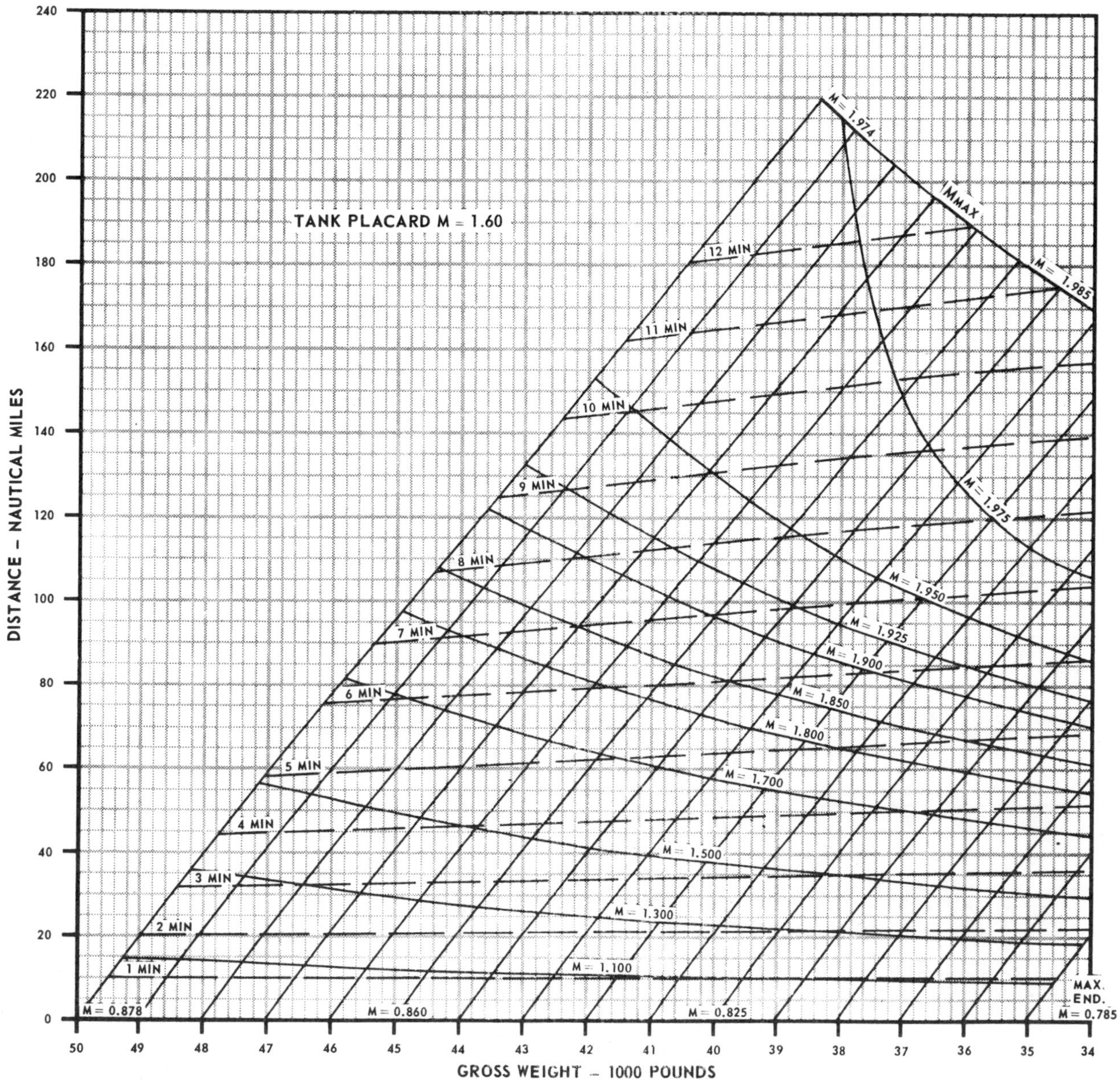

FDD-1-(223)A

Figure 11-72

MAXIMUM THRUST ACCELERATION

45,000 FEET

AIRPLANE CONFIGURATION
(4) AIM-7 AND (2) WING TANKS

REMARKS

ENGINE(S): (2) J79-GE-10
ICAO STANDARD DAY

GUIDE

DATE: 1 JANUARY 1972
DATA BASIS: FLIGHT TEST

FUEL GRADE: JP-5
FUEL DENSITY: 6.8 LB/GAL

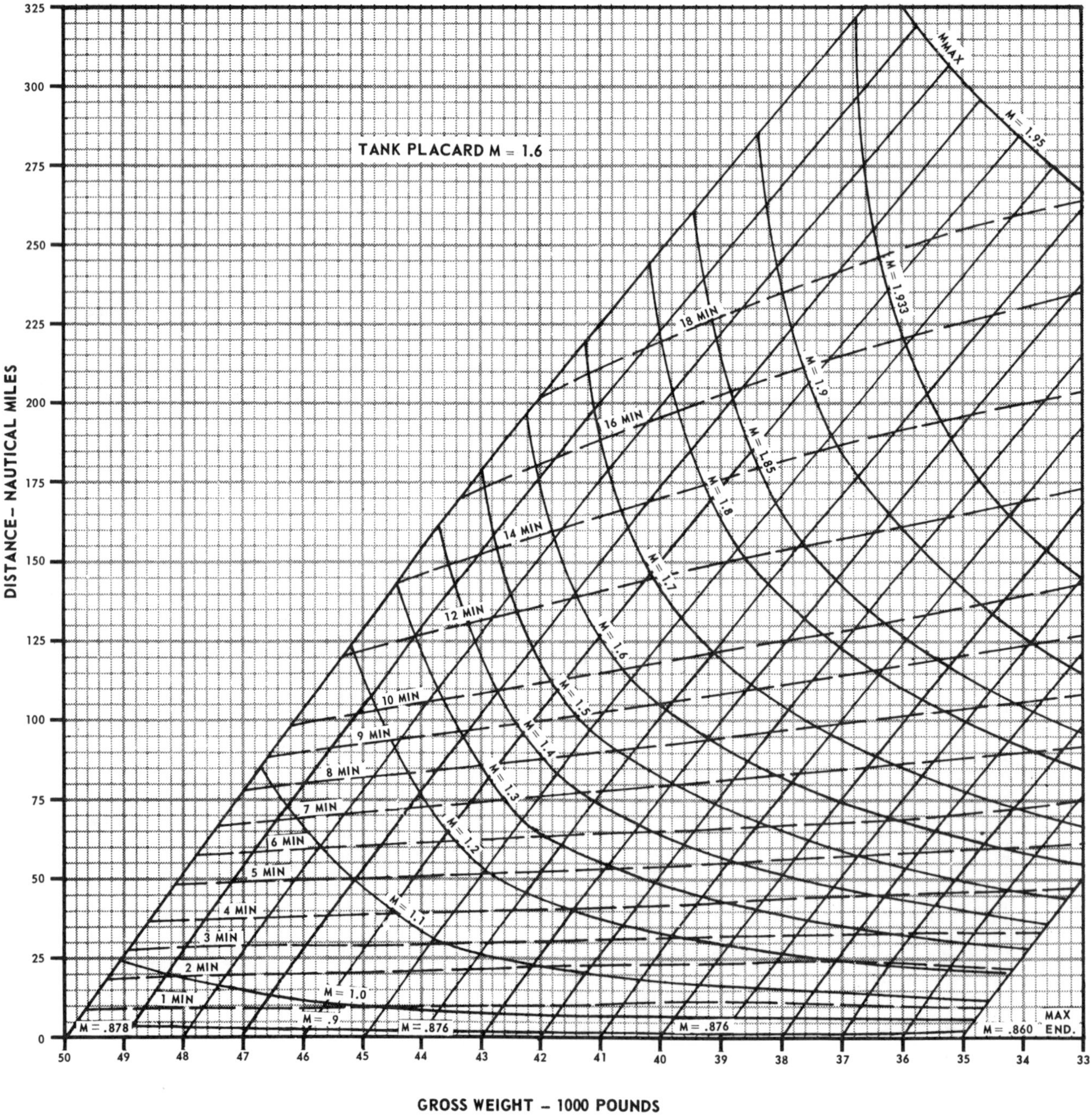

FDD-1-(224)A

Figure 11-73

MILITARY THRUST ACCELERATION

15,000 FEET

AIRPLANE CONFIGURATION
(4) AIM-7

REMARKS
ENGINE(S): (2) J79-GE-10
ICAO STANDARD DAY

DATE: 1 JANUARY 1972
DATA BASIS: FLIGHT TEST

FUEL GRADE: JP-5
FUEL DENSITY: 6.8 LB/GAL

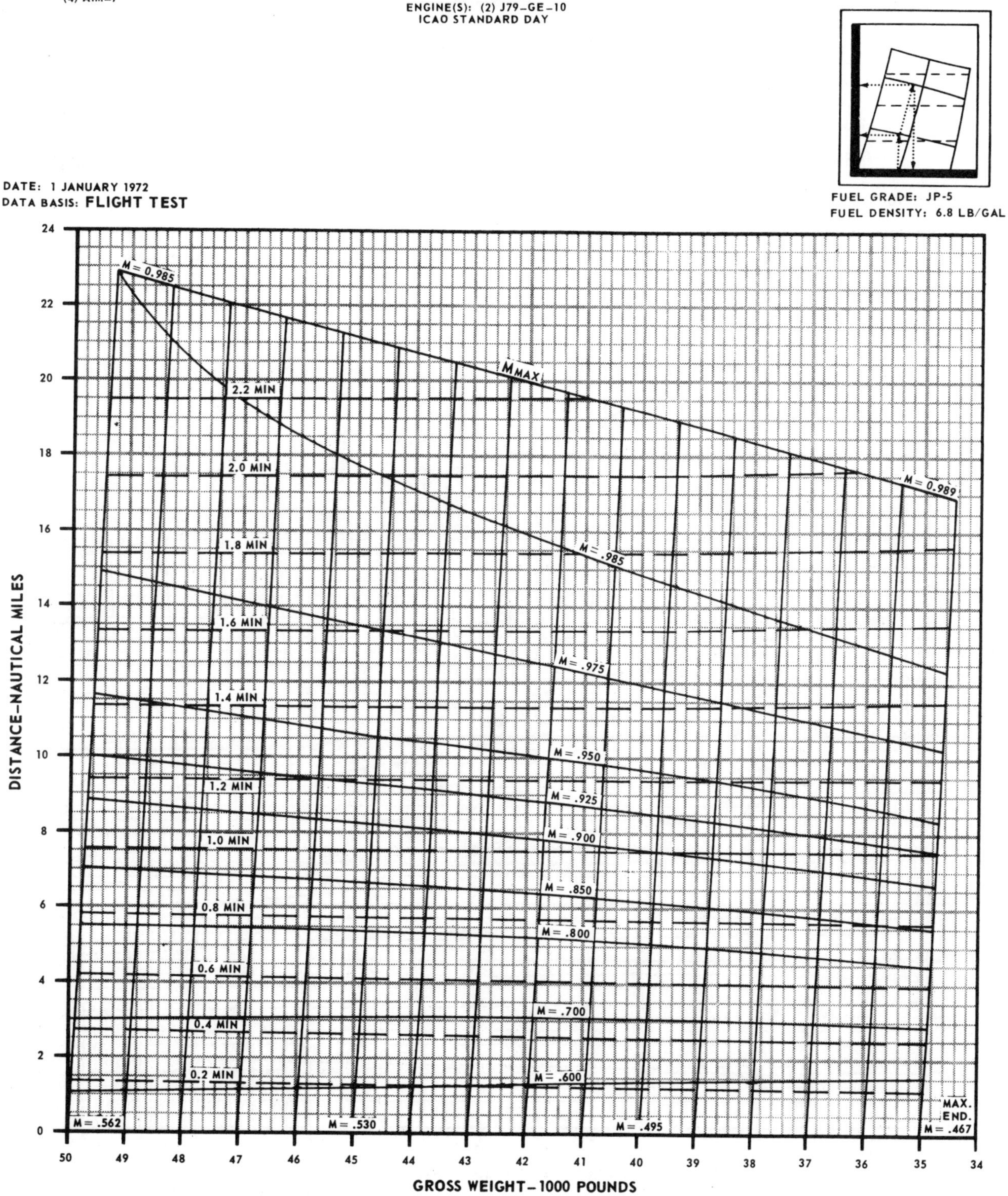

FDD-1-(240)A

Figure 11-74

MILITARY THRUST ACCELERATION

25,000 FEET

AIRPLANE CONFIGURATION
(4) AIM-7

REMARKS
ENGINE(S): (2) J79-GE-10
ICAO STANDARD DAY

GUIDE

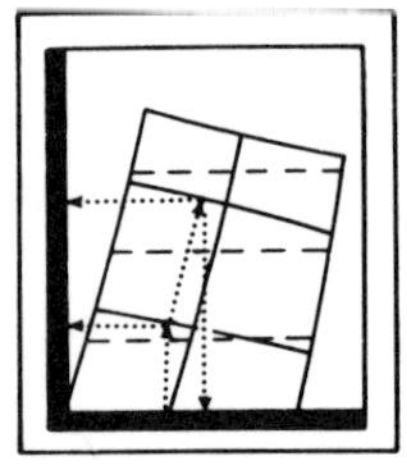

DATE: 1 JANUARY 1972
DATA BASIS: **FLIGHT TEST**

FUEL GRADE: JP-5
FUEL DENSITY: 6.8 LB/GAL

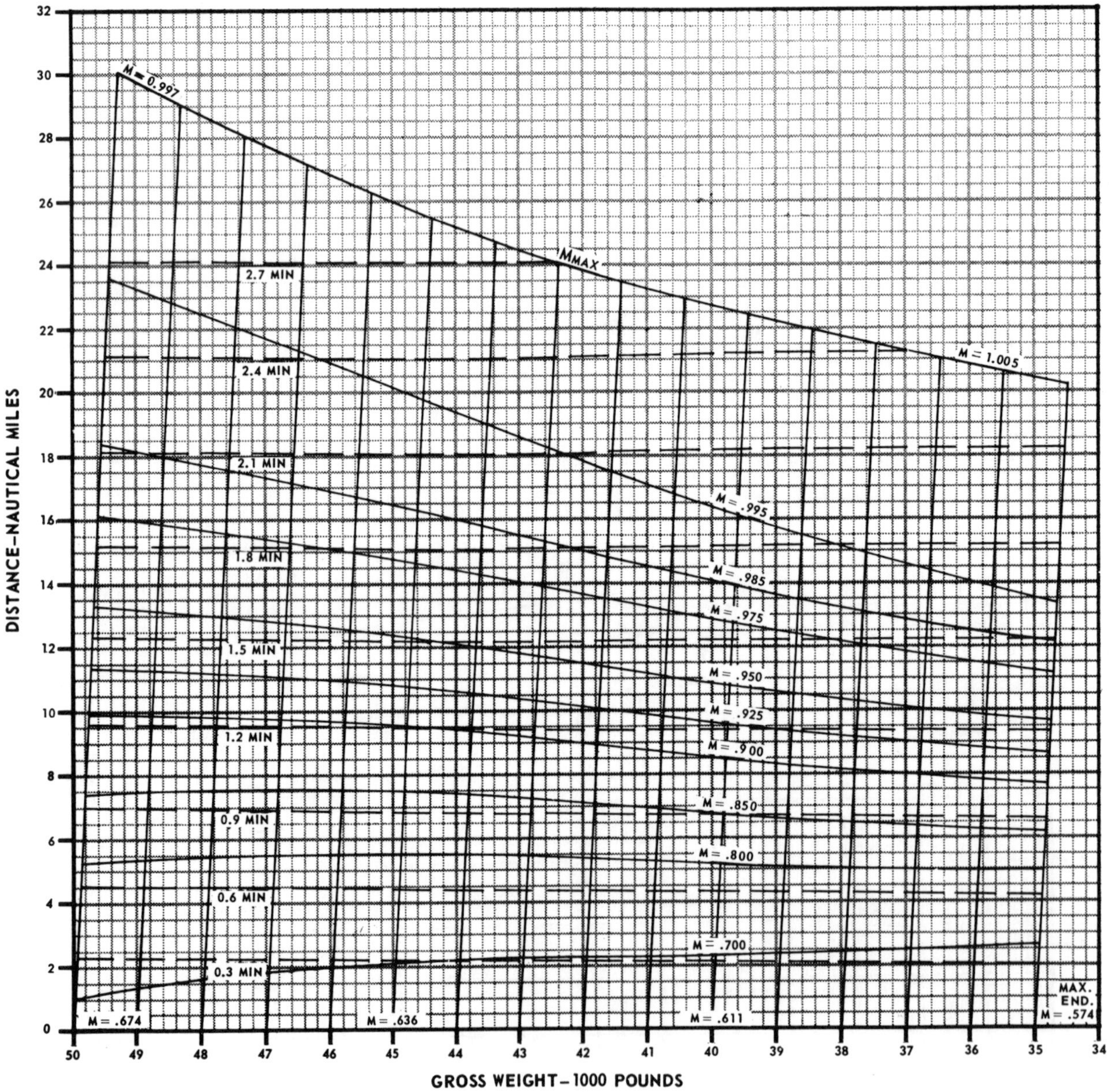

FDD-1-(241)A

Figure 11-75

MILITARY THRUST ACCELERATION

35,000 FEET

AIRPLANE CONFIGURATION
(4) AIM-7

REMARKS
ENGINE(S): (2) J79-GE-10
ICAO STANDARD DAY

GUIDE

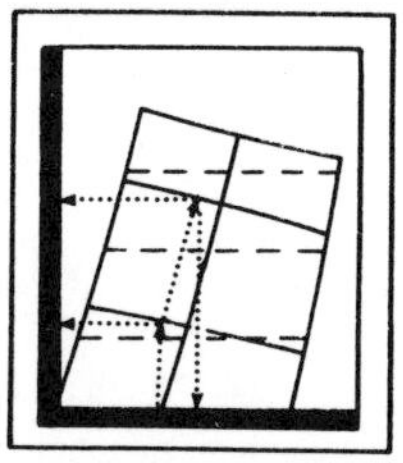

DATE: 1 JANUARY 1972
DATA BASIS: FLIGHT TEST

FUEL GRADE: JP-5
FUEL DENSITY: 6.8 LB/GAL

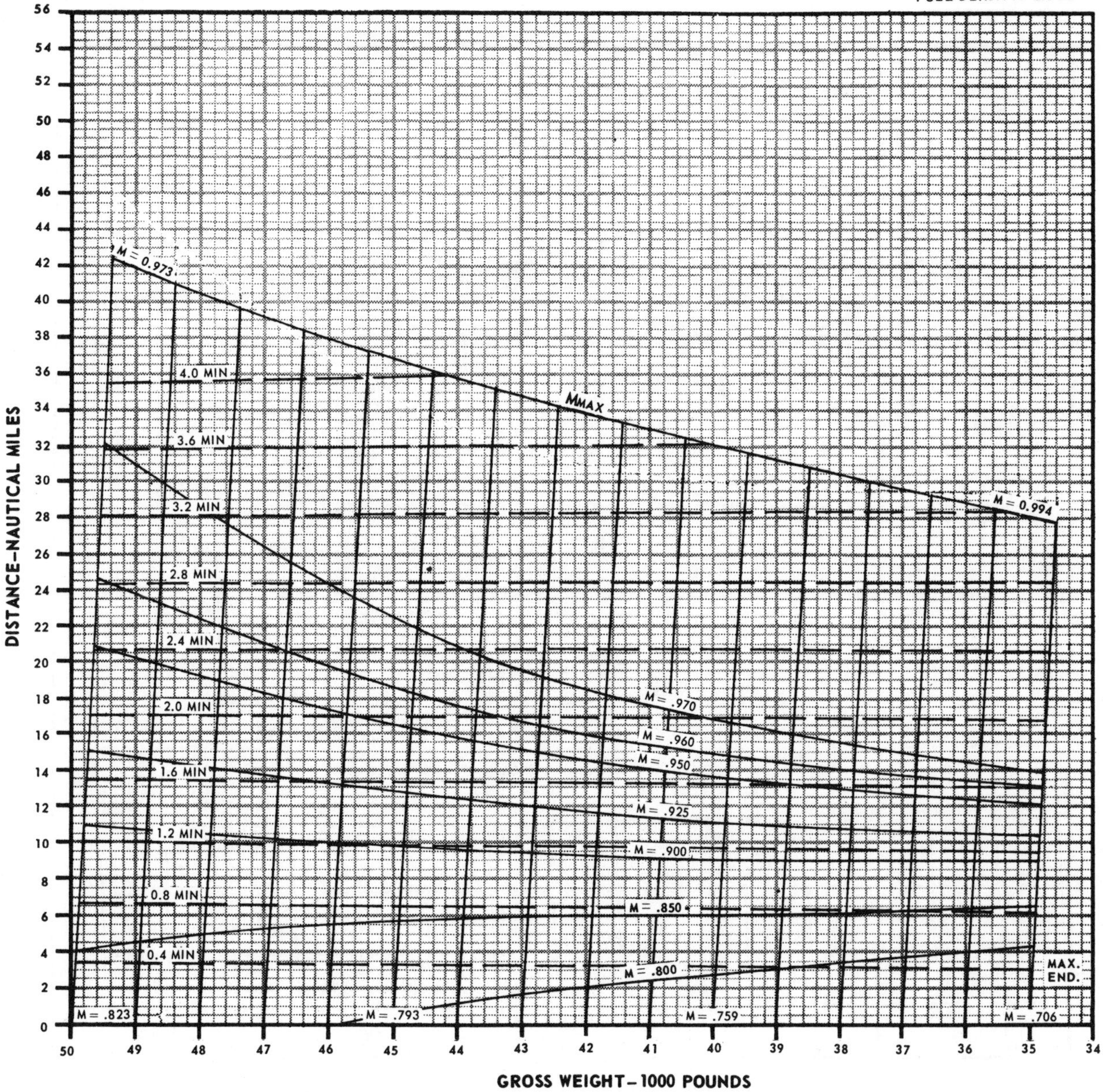

FDD-1-(242)A

Figure 11-76

MILITARY THRUST ACCELERATION

15,000 FEET

AIRPLANE CONFIGURATION
(4) AIM-7, (1) ℄ TANK

REMARKS
ENGINE(S): (2) J79-GE-10
ICAO STANDARD DAY

GUIDE

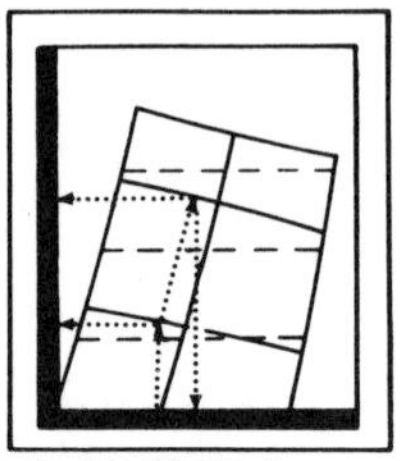

DATE: 1 JANUARY 1972
DATA BASIS: FLIGHT TEST

FUEL GRADE: JP-4
FUEL DENSITY: 6.5 LB/GAL

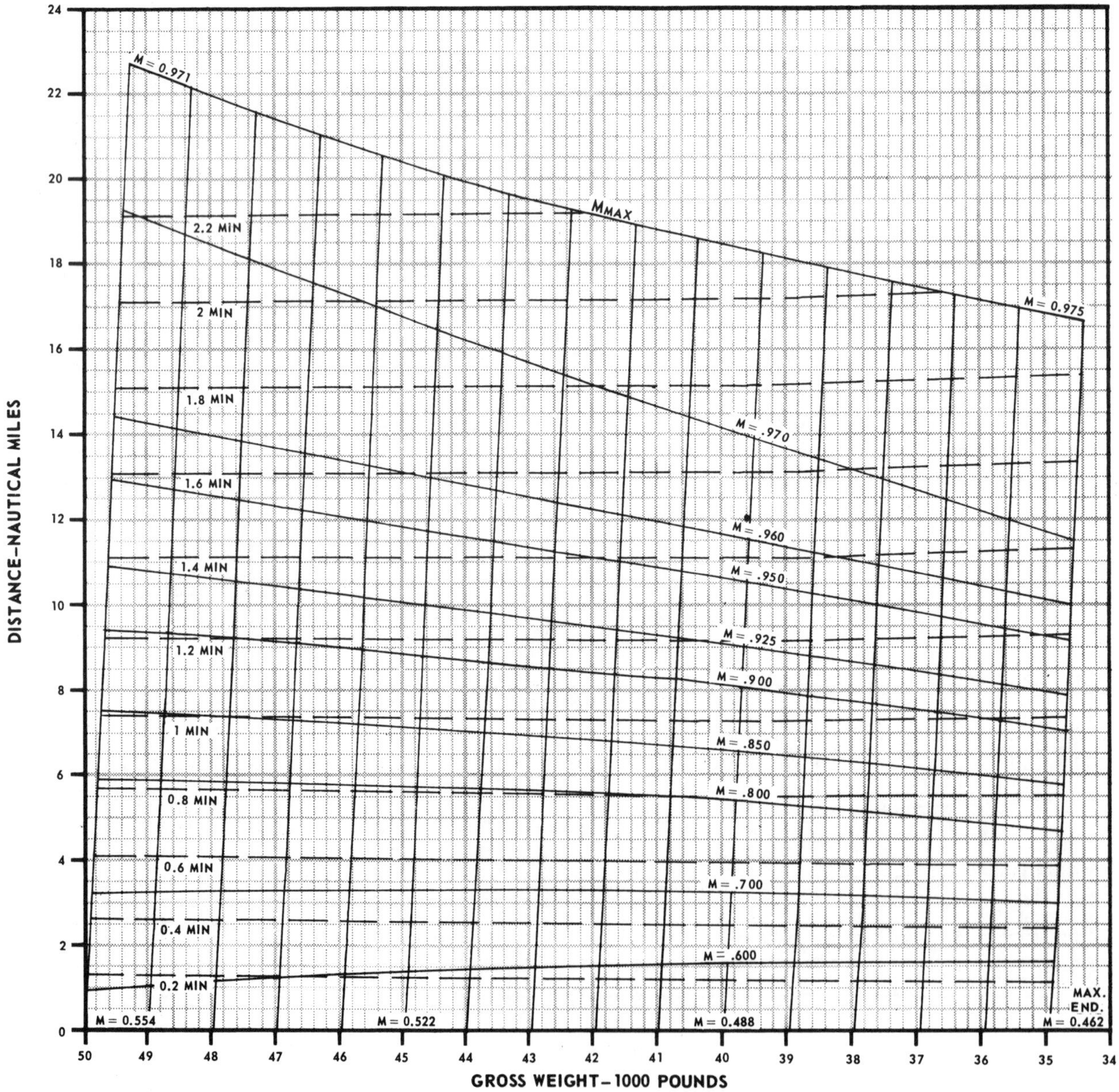

FDD-1-(246)A

Figure 11-77

MILITARY THRUST ACCELERATION

25,000 FEET

AIRPLANE CONFIGURATION
(4) AIM-7, (1) ℄ TANK

REMARKS

ENGINE(S): (2) J79-GE-10
ICAO STANDARD DAY

DATE: 1 JANUARY 1972
DATA BASIS: FLIGHT TEST

FUEL GRADE: JP-5
FUEL DENSITY: 6.8 LB/GAL

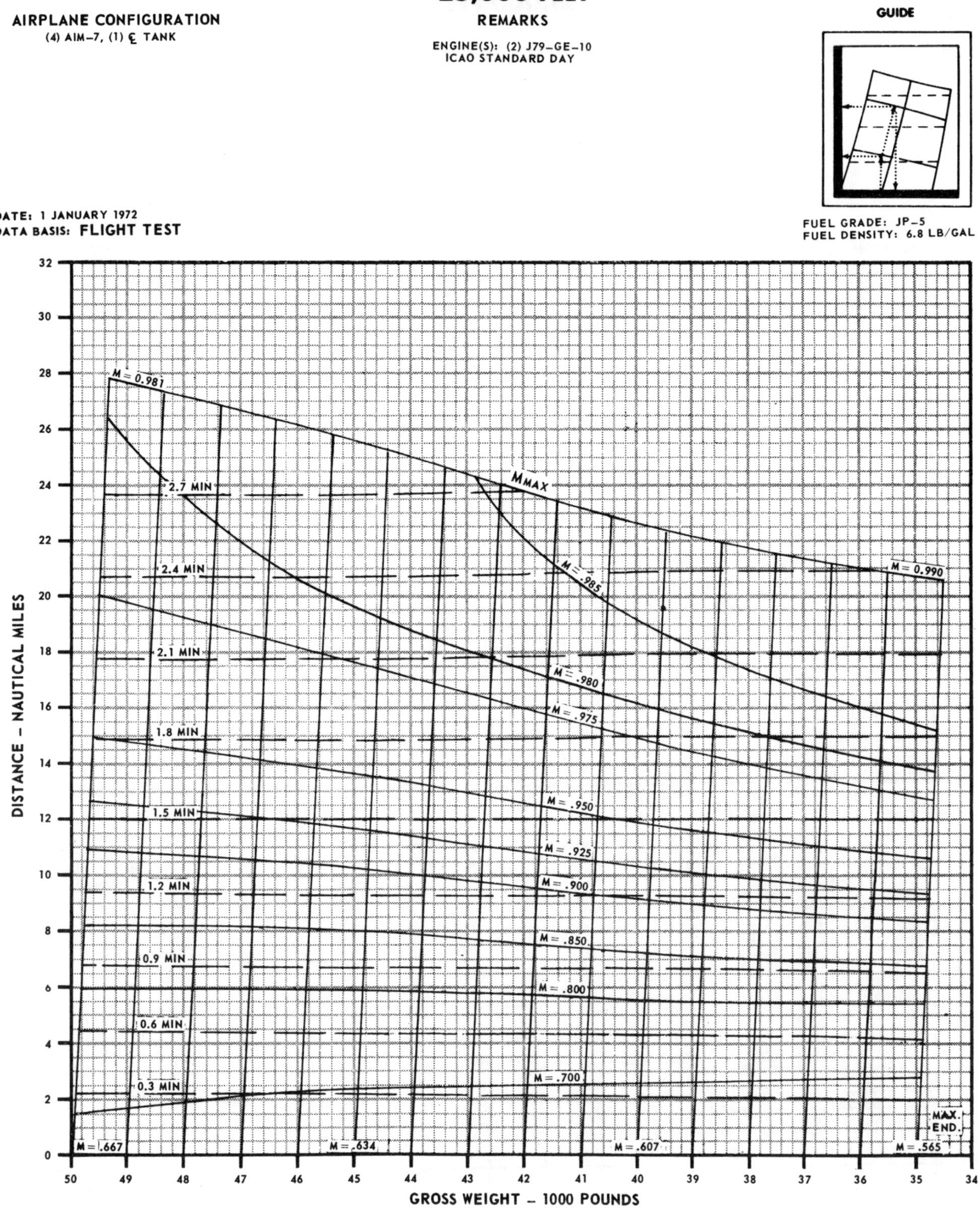

FDD-1-(247)A

Figure 11-78

MILITARY THRUST ACCELERATION

35,000 FEET

AIRPLANE CONFIGURATION
(4) AIM-7, (1) ℄ TANK

REMARKS
ENGINE(S) J79-GE-10
ICAO STANDARD DAY

GUIDE

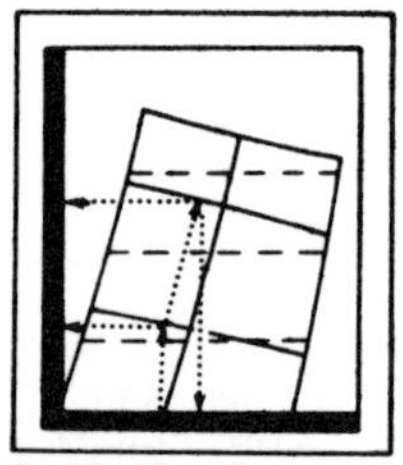

DATE: 1 JANUARY 1972
DATA BASIS: FLIGHT TEST

FUEL GRADE: JP-5
FUEL DENSITY: 6.8 LB/GAL

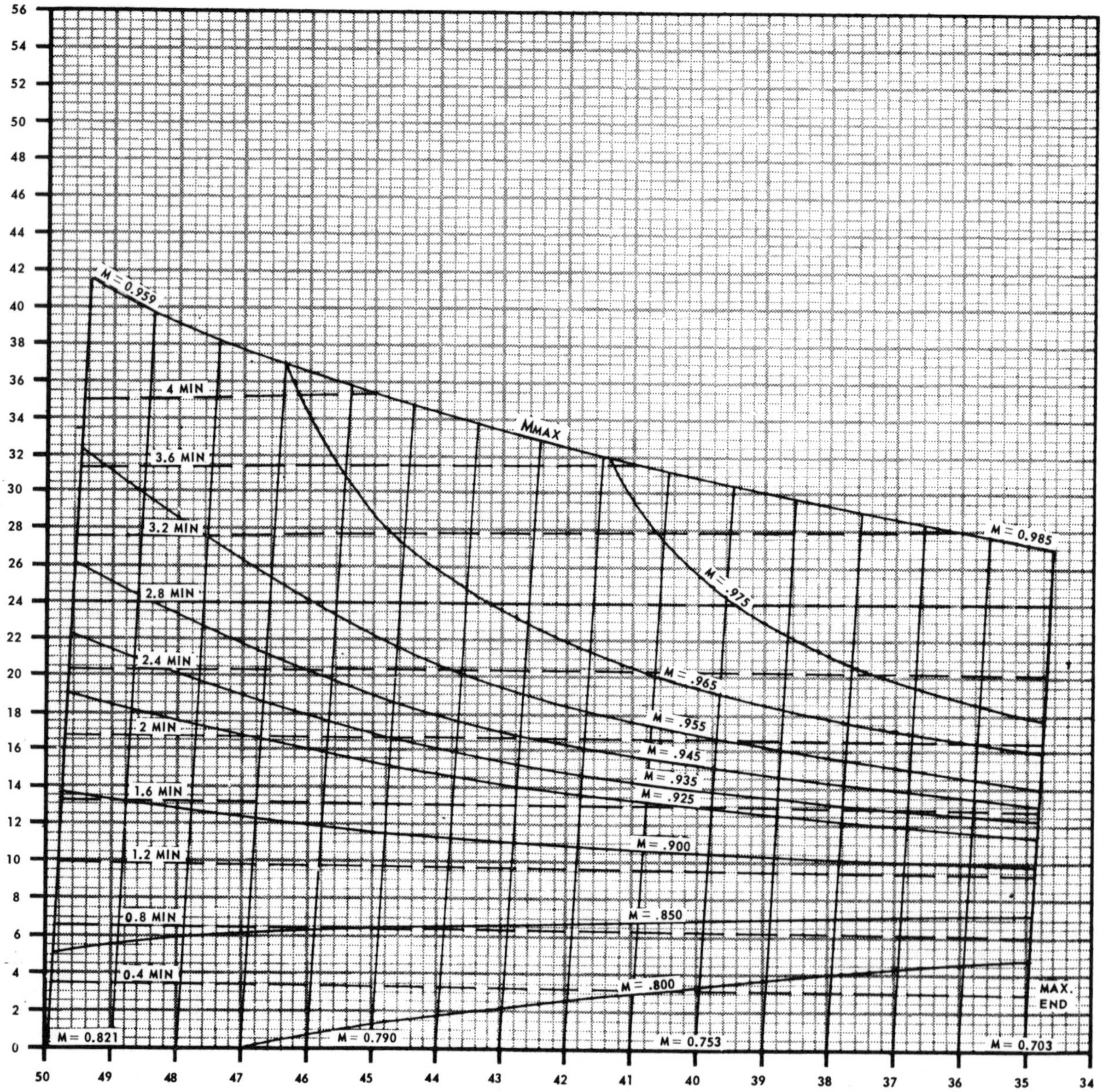

FDD-1-(248)A

Figure 11-79

MILITARY THRUST ACCELERATION

15,000 FEET

AIRPLANE CONFIGURATION
(4) AIM-7 AND (2) WING TANKS

REMARKS

ENGINE(S): (2) J79-GE-10
ICAO STANDARD DAY

GUIDE

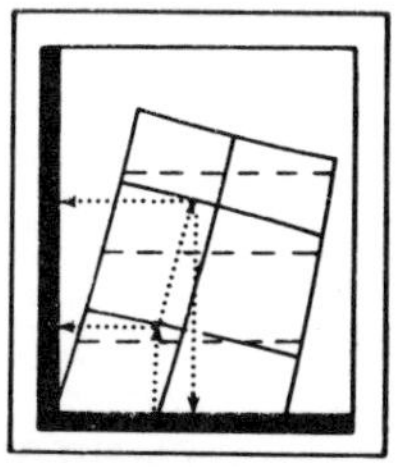

DATE: 1 JANUARY 1972
DATA BASIS: **FLIGHT TEST**

FUEL GRADE: JP-5
FUEL DENSITY: 6.8 LB/GAL

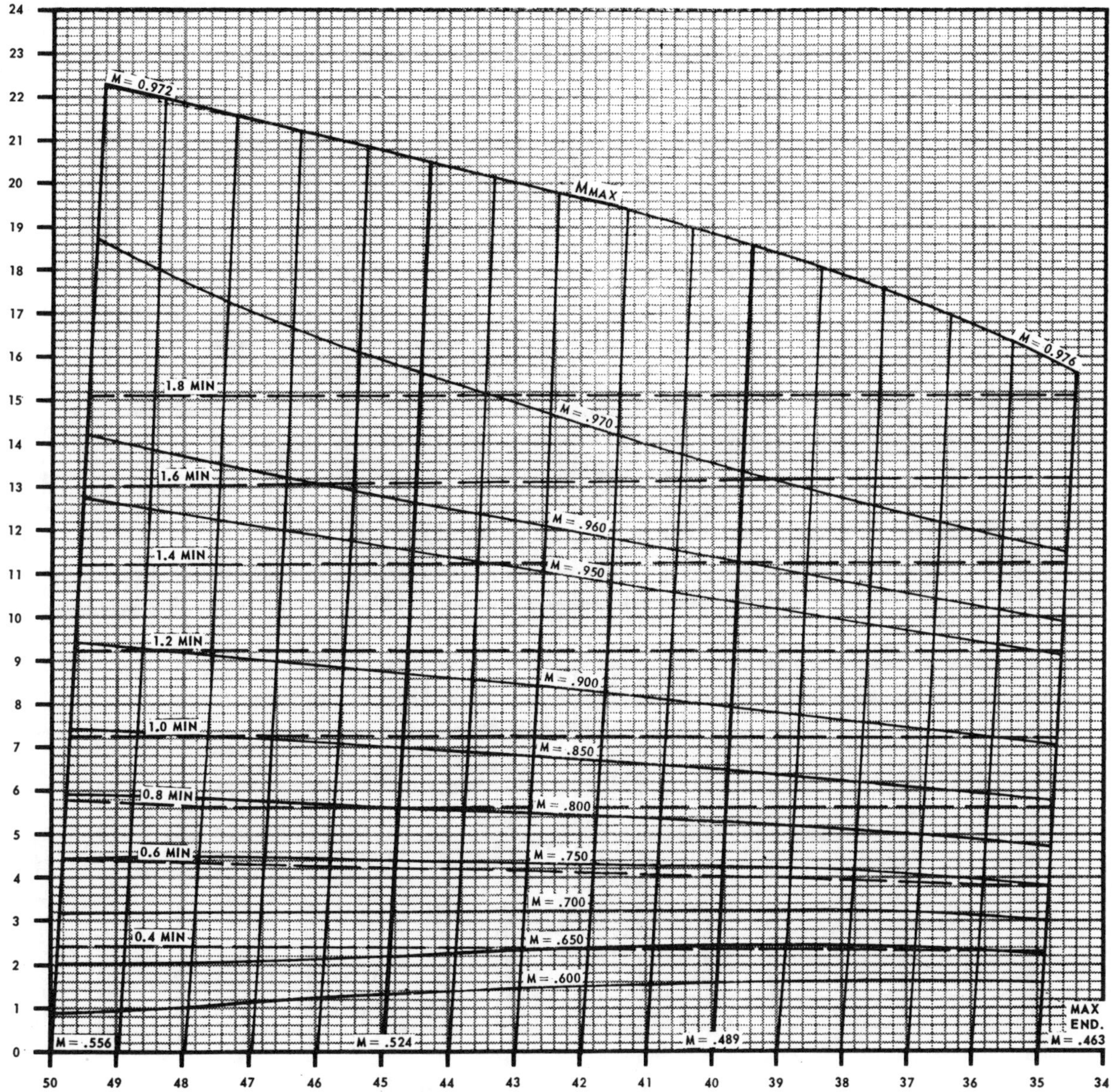

FDD-1-(249) A

Figure 11-80

MILITARY THRUST ACCELERATION

25,000 FEET

AIRPLANE CONFIGURATION
(4) AIM-7E AND (2) WING TANKS

REMARKS
ENGINE(S): (2) J79-GE-10
ICAO STANDARD DAY

GUIDE

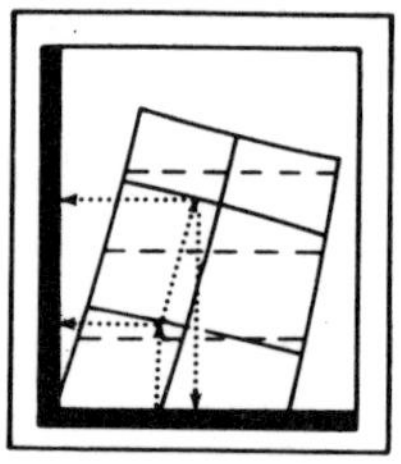

DATE: 1 JANUARY 1972
DATA BASIS: FLIGHT TEST

FUEL GRADE: JP-5
FUEL DENSITY: 6.8 LB/GAL

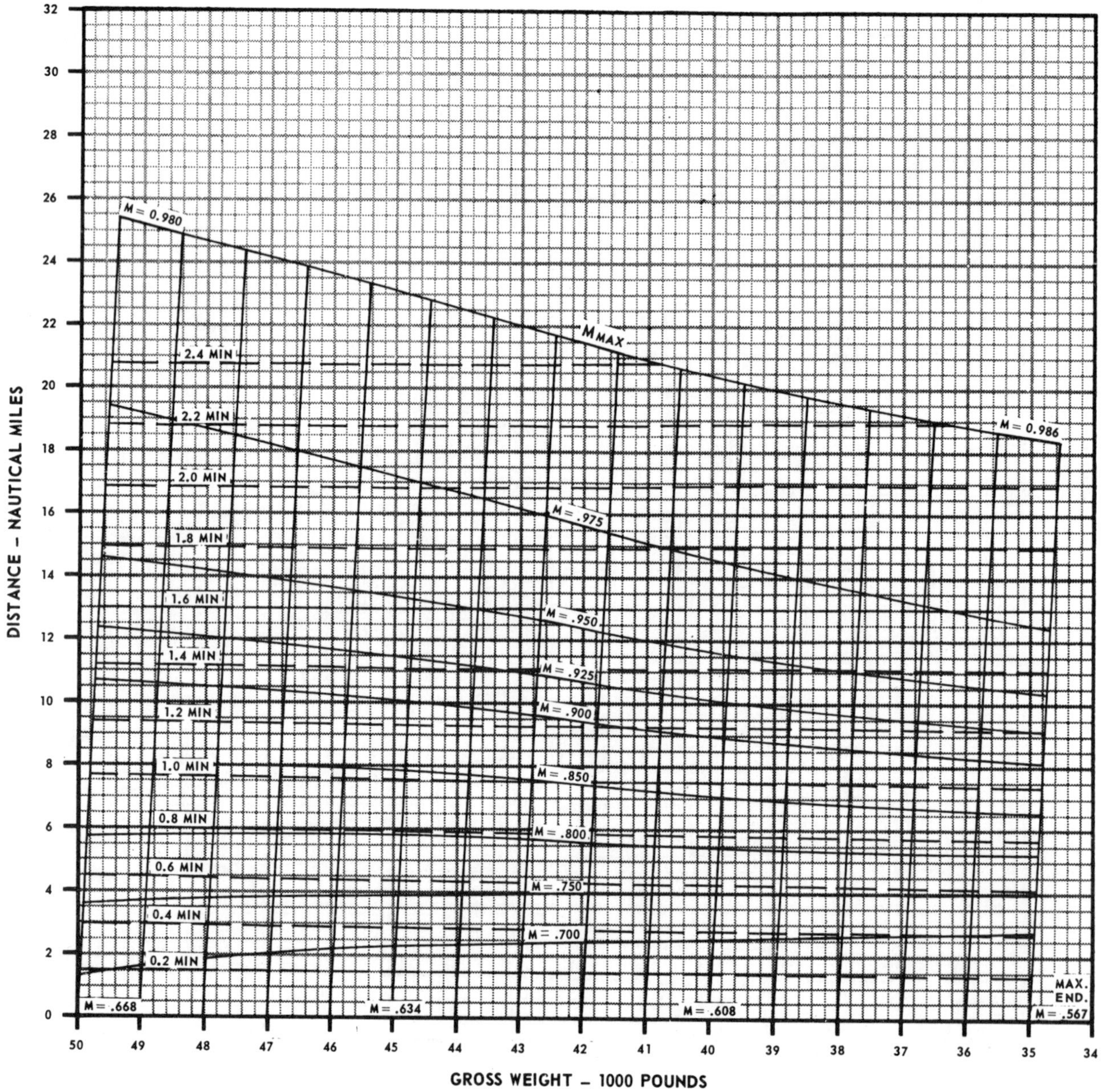

FDD-1-(250)A

Figure 11-81

MILITARY THRUST ACCELERATION

35,000 FEET

AIRPLANE CONFIGURATION
(4) AIM-7 AND (2) WING TANKS

REMARKS
ENGINES: (2) J79-GE-10
ICAO STANDARD DAY

GUIDE

DATE: 1 JANUARY 1972
DATA BASIS: FLIGHT TEST

FUEL GRADE: JP-5
FUEL DENSITY: 6.8 LB/GAL

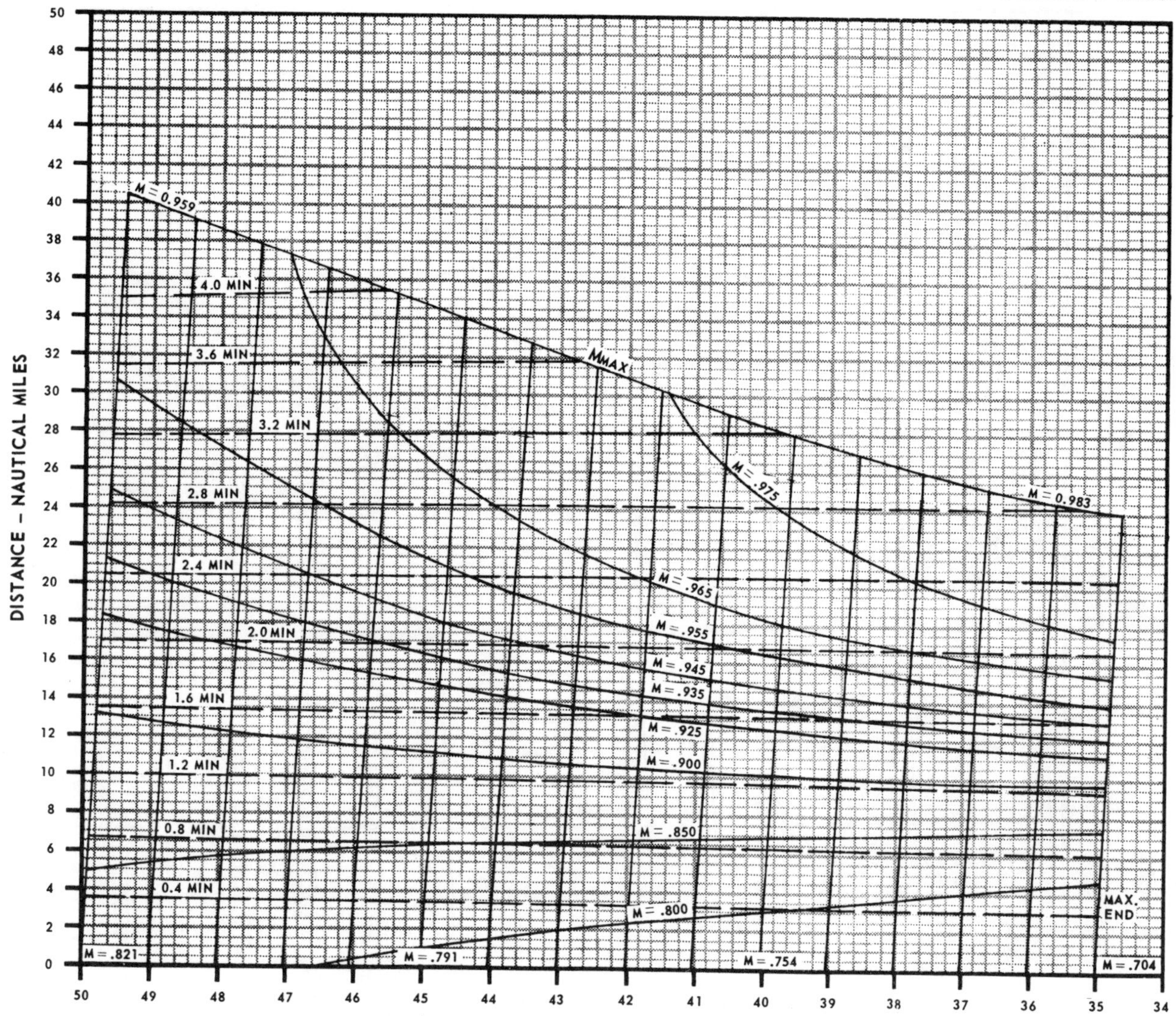

FDD-1-(251)A

Figure 11-82

MILITARY THRUST ACCELERATION

15,000 FEET

AIRPLANE CONFIGURATION
(4) AIM-7, (1) ℄ TANK
AND (2) WING TANKS

REMARKS
ENGINE(S): (2) J79-GE-10
ICAO STANDARD DAY

GUIDE

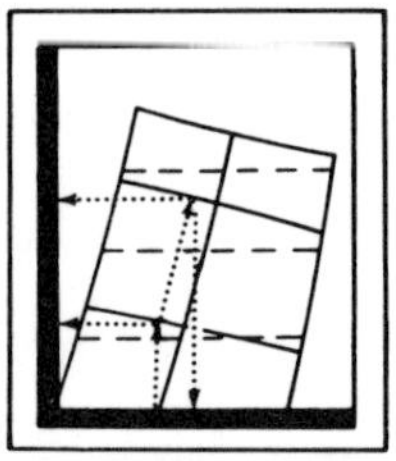

DATE: 1 JANUARY 1972
DATA BASIS: FLIGHT TEST

FUEL GRADE: JP-5
FUEL DENSITY: 6.8 LB/GAL

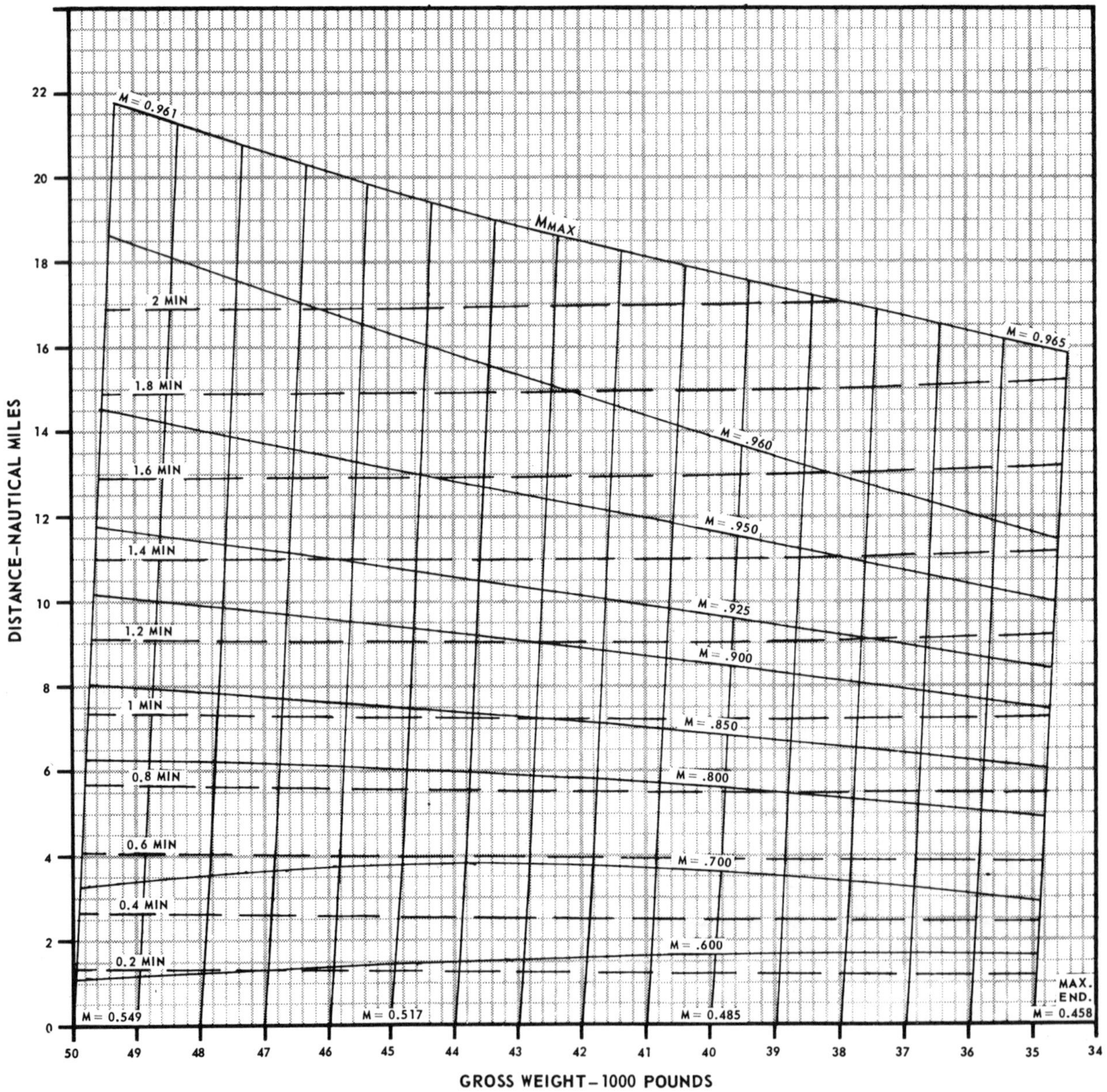

FDD-1-(300)

Figure 11-83

MILITARY THRUST ACCELERATION

25,000 FEET

AIRPLANE CONFIGURATION

(4) AIM-7, (1) ℄ TANK
AND (2) WING TANKS

REMARKS

ENGINE(S): (2) J79-GE-10
ICAO STANDARD DAY

GUIDE

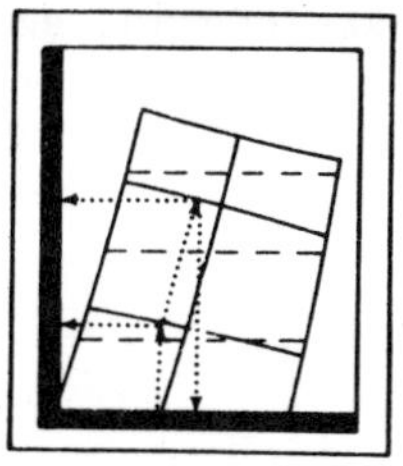

DATE: 1 JANUARY 1972
DATA BASIS: FLIGHT TEST

FUEL GRADE: JP-5
FUEL DENSITY: 6.8 LB/GAL

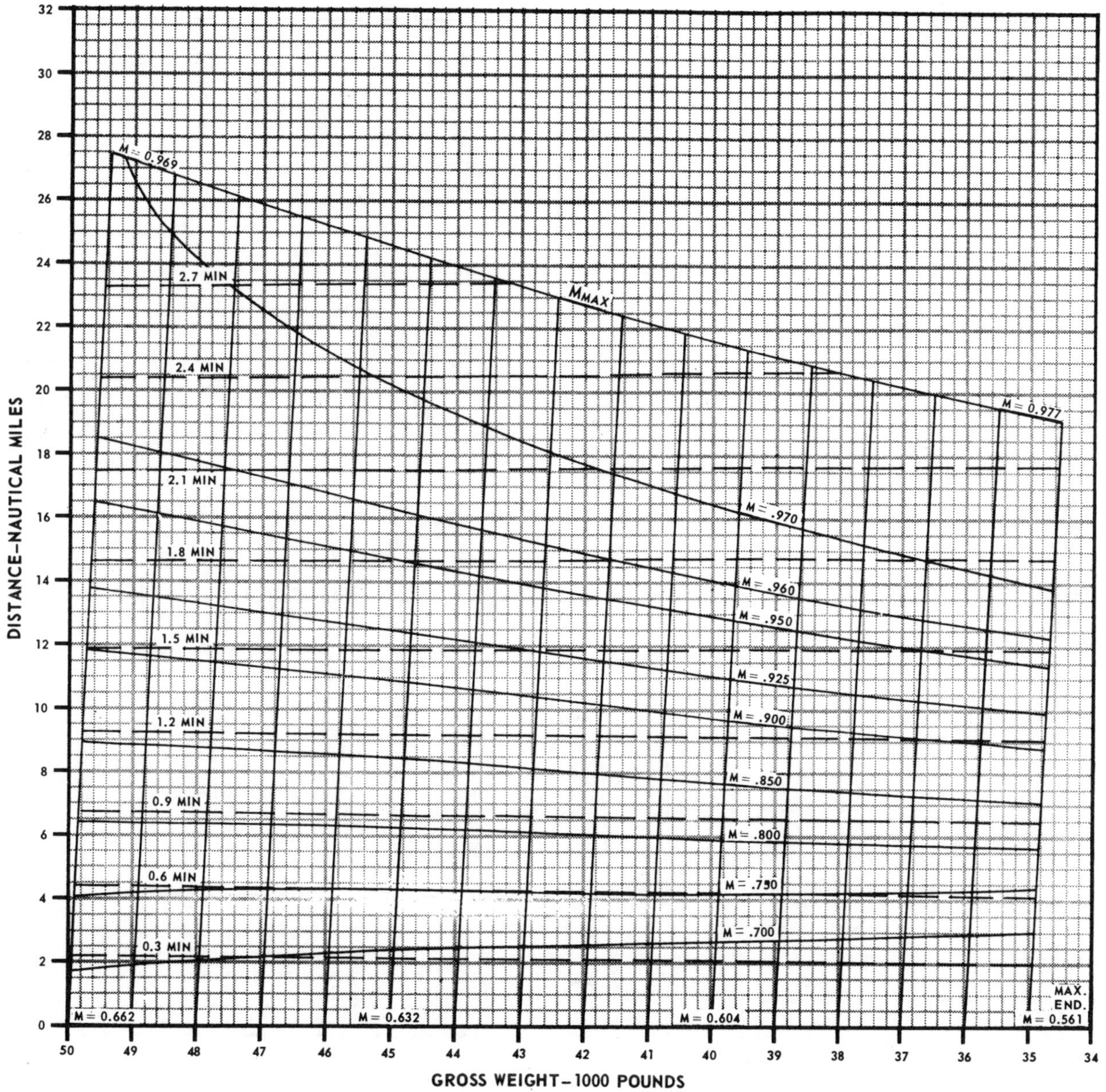

FDD-1-(301)

Figure 11-84

MILITARY THRUST ACCELERATION

35,000 FEET

AIRPLANE CONFIGURATION

(4) AIM-7, (1) ℄ TANK
AND (2) WING TANKS

REMARKS

ENGINE(S): (2) J79-GE-10
ICAO STANDARD DAY

GUIDE

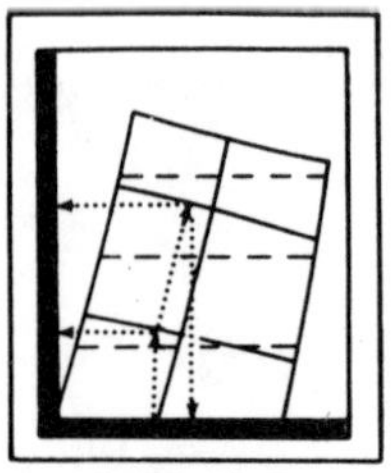

DATE: 1 JANUARY 1972
DATA BASIS: FLIGHT TEST

FUEL GRADE: JP-5
FUEL DENSITY: 6.8 LB/GAL

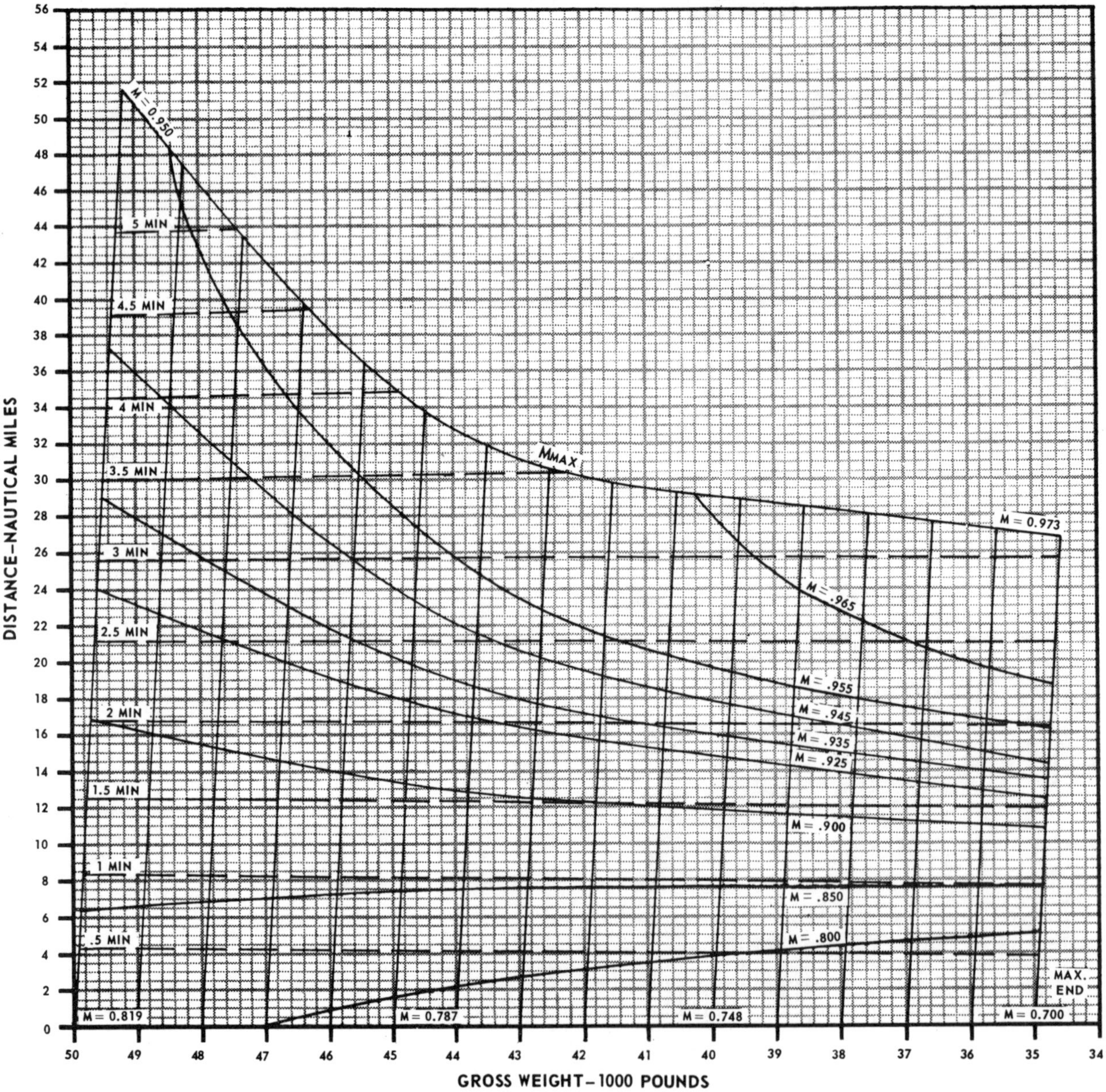

FDD-1-(302)

Figure 11-85

LEVEL FLIGHT ENVELOPE

ENGINE(S): (2) J79-GE-10
ICAO STANDARD DAY
COMBAT GROSS WEIGHTS

DATE: 1 SEPTEMBER 1971
DATA BASIS: FLIGHT TEST

FUEL GRADE: JP-5
FUEL DENSITY: 6.8 LB/GAL

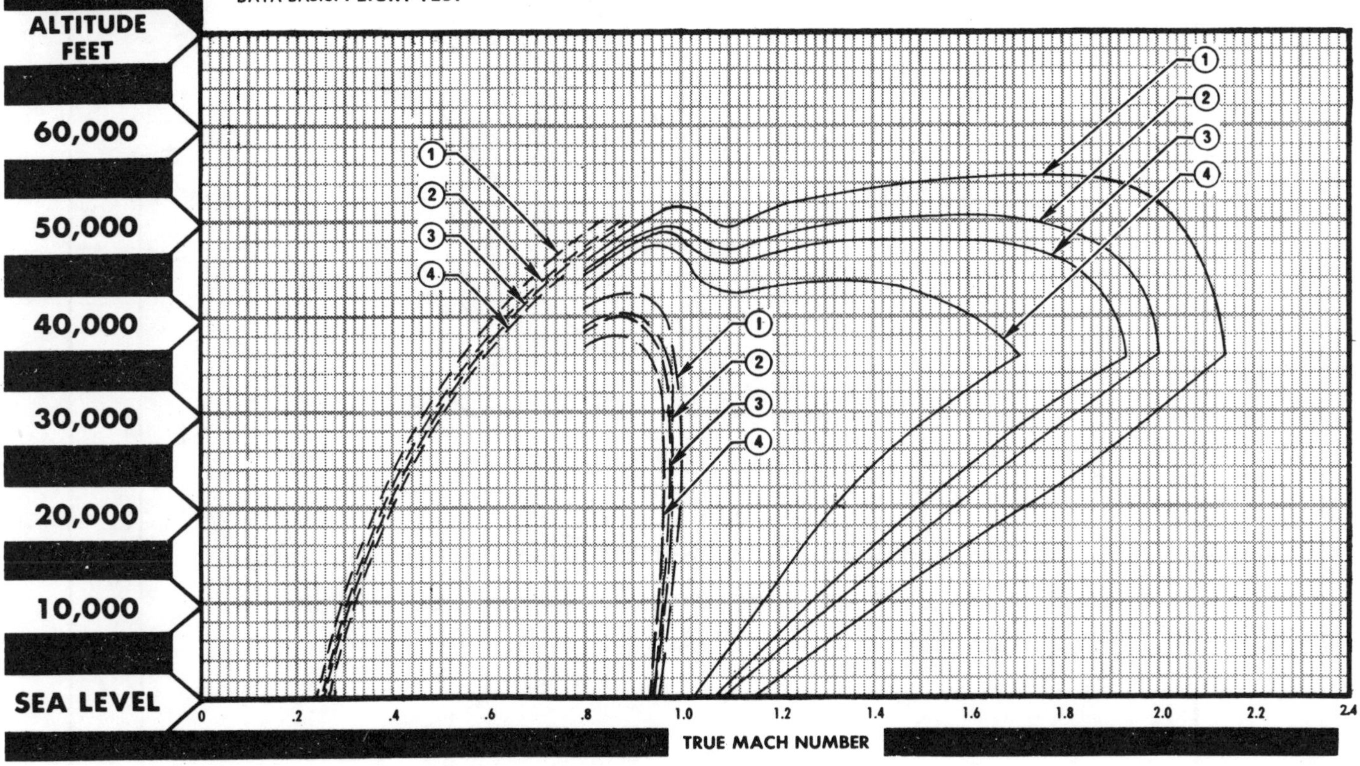

CURVE NO.	CONFIGURATION	GROSS WEIGHT
①	(4) AIM – 7	41,487 LBS
②	(4) AIM – 7, AND (1) ℄ TANK	44,182 LBS
③	(4) AIM – 7, AND (2) WING TANKS	45,120 LBS
④	(4) AIM – 7, (1) ℄ TANK, AND (2) WING TANKS	47,684 LBS

LEGEND

———————— MAXIMUM THRUST

— — — — — MILITARY THRUST

– – – – – MODERATE BUFFET

FDD-1-(303)

Figure 11-86

V-N ENVELOPE
SYMMETRICAL FLIGHT

AIRPLANE CONFIGURATION
CLEAN
OR
(4) AIM-7

GROSS WEIGHT – 37,500 POUNDS

REMARKS
ENGINES: (2) J79-GE-10
ICAO STANDARD DAY

GUIDE

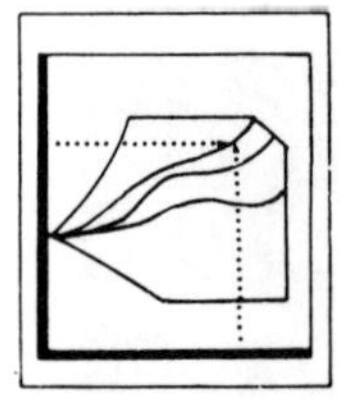

DATE: 15 AUGUST 1969
DATA BASIS: **FLIGHT TEST**

FUEL GRADE: JP-5
FUEL DENSITY: 6.8 LB/GAL

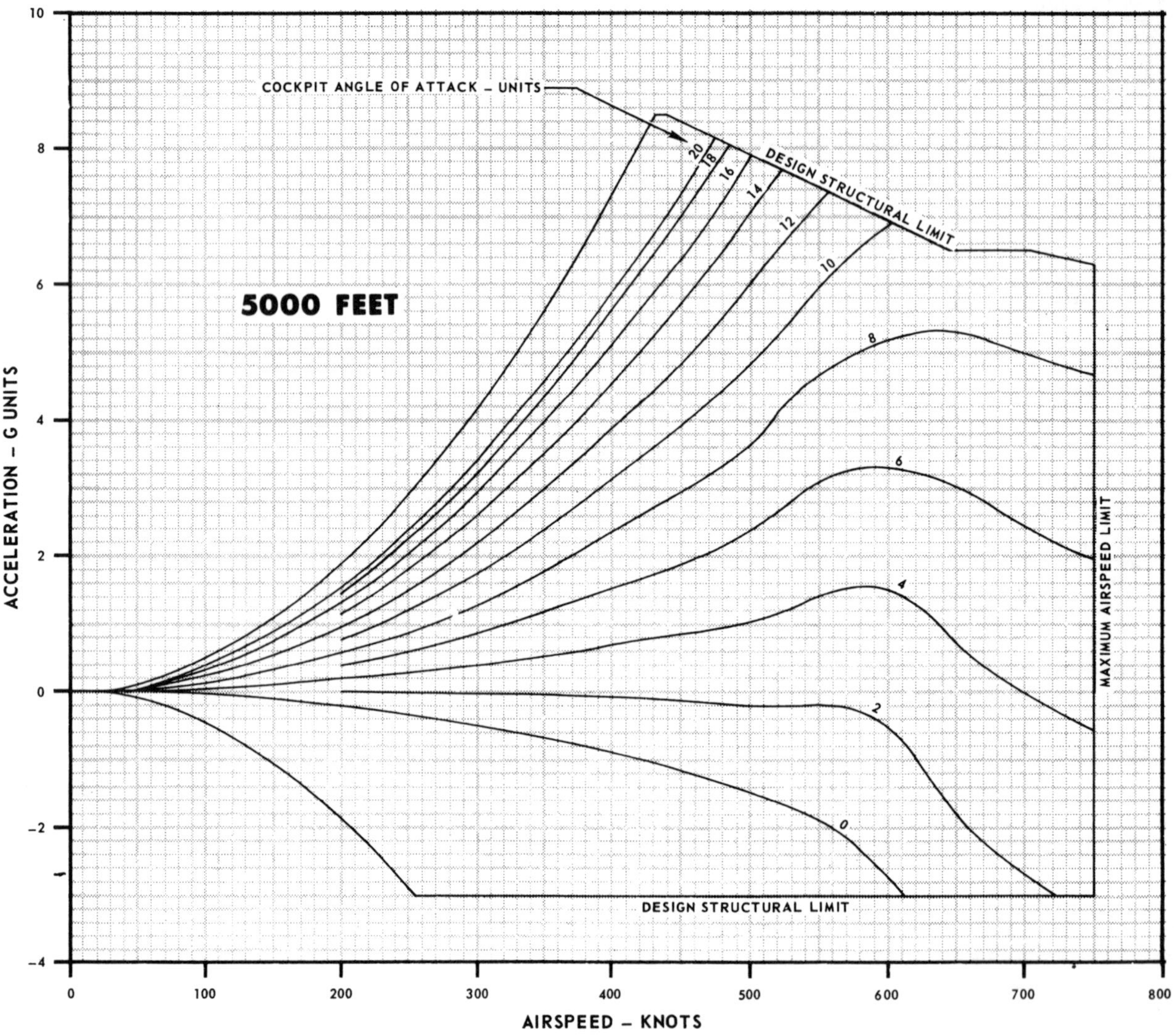

FDD-1-(305)

Figure 11-87

V-N ENVELOPE
SYMMETRICAL FLIGHT

AIRPLANE CONFIGURATION
CLEAN
OR
(4) AIM-7

GROSS WEIGHT – 37,500 POUNDS

REMARKS
ENGINES: (2) J79-GE-10
ICAO STANDARD DAY

GUIDE

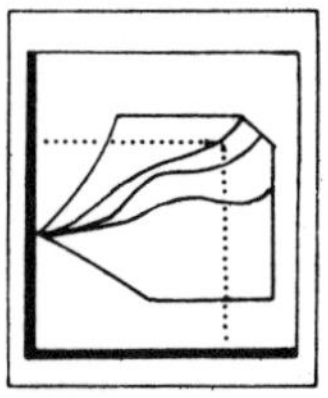

DATE: 15 AUGUST 1969
DATA BASIS: FLIGHT TEST

FUEL GRADE: JP-5
FUEL DESNITY: 6.8 LB/GAL

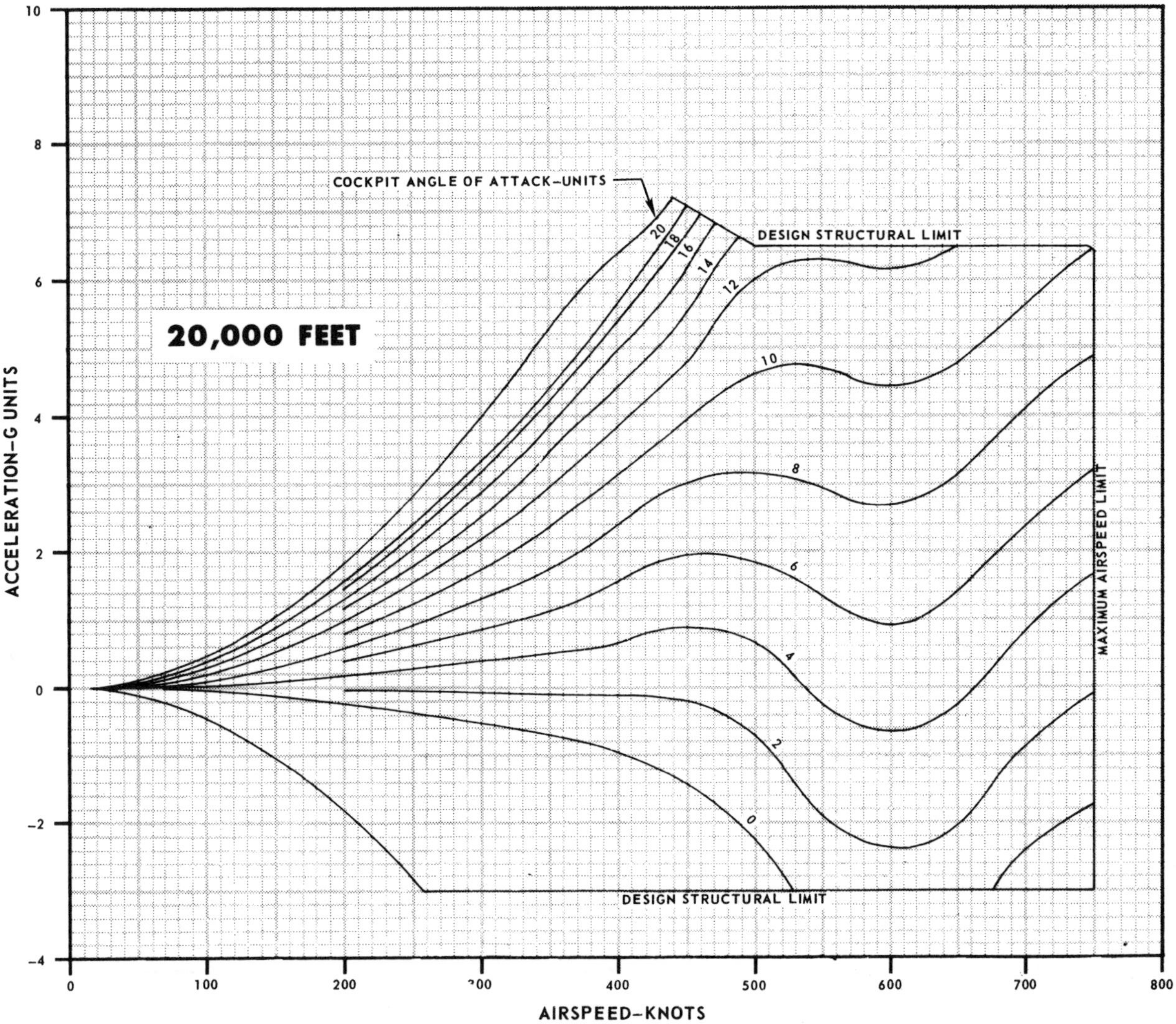

FDD-1-(304)

Figure 11-88

V-N ENVELOPE
SYMMETRICAL FLIGHT

AIRPLANE CONFIGURATION
CLEAN
OR
(4) AIM-7

GROSS WEIGHT-37,500 POUNDS

REMARKS
ENGINE(S): (2) J79-GE-10
ICAO STANDARD DAY

GUIDE

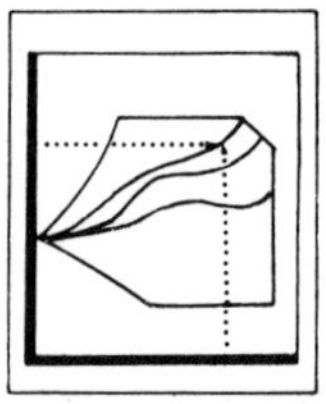

DATE: 15 AUGUST 1969
DATA BASIS: FLIGHT TEST

FUEL GRADE: JP-5
FUEL DENSITY: 6.8 LB/GAL

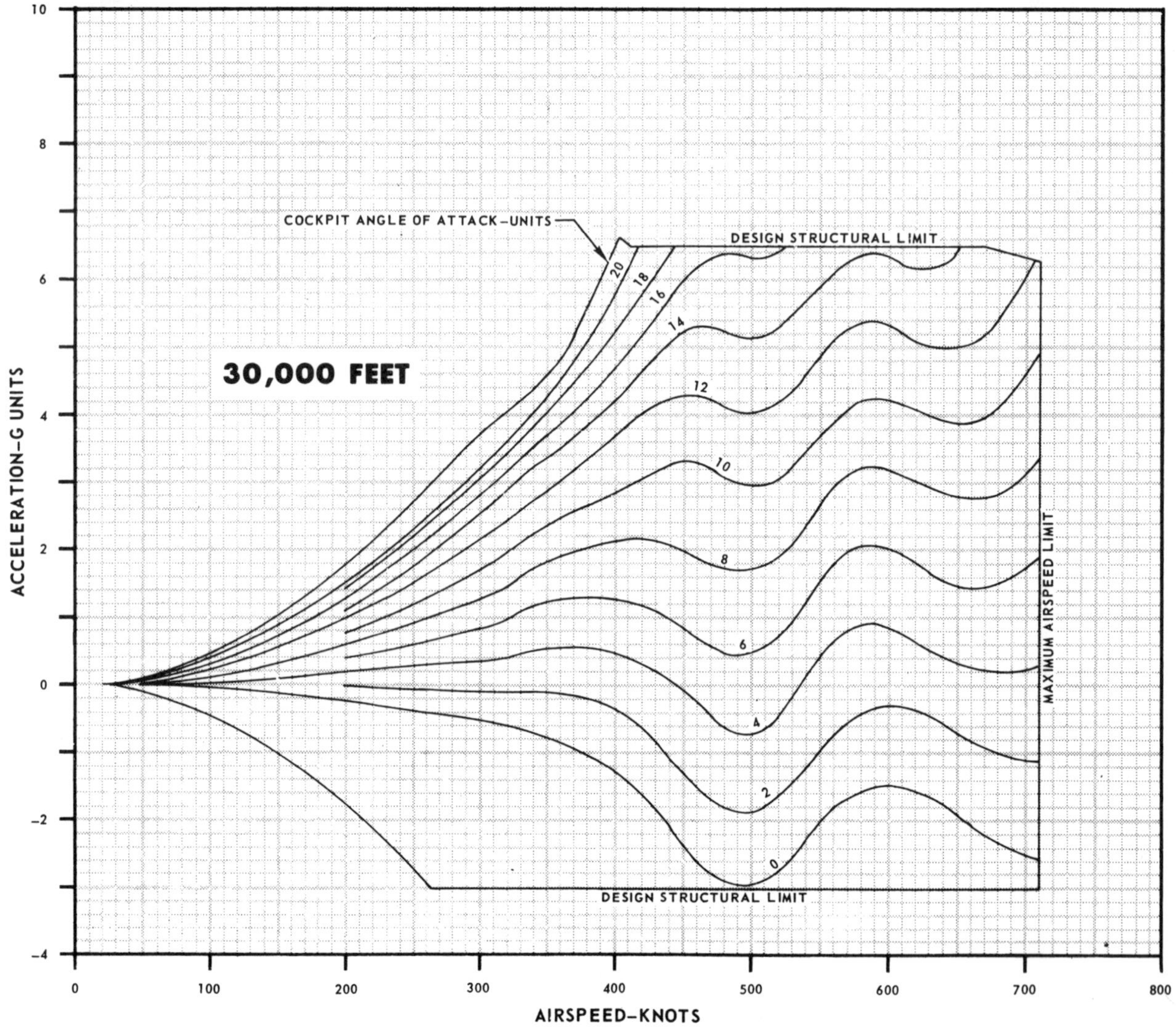

FDD-1-(306)

Figure 11-89

V-N ENVELOPE
SYMMETRICAL FLIGHT

AIRPLANE CONFIGURATION
CLEAN
OR
(4) AIM-7

GROSS WEIGHT – 37,500 POUNDS

REMARKS
ENGINE(S): (2) J79-GE-10
ICAO STANDARD DAY

GUIDE

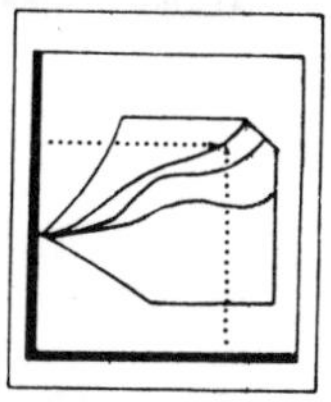

DATE: 15 AUGUST 1969
DATA BASIS: FLIGHT TEST

FUEL GRADE: JP-5
FUEL DENSITY: 6.8 LB/GAL

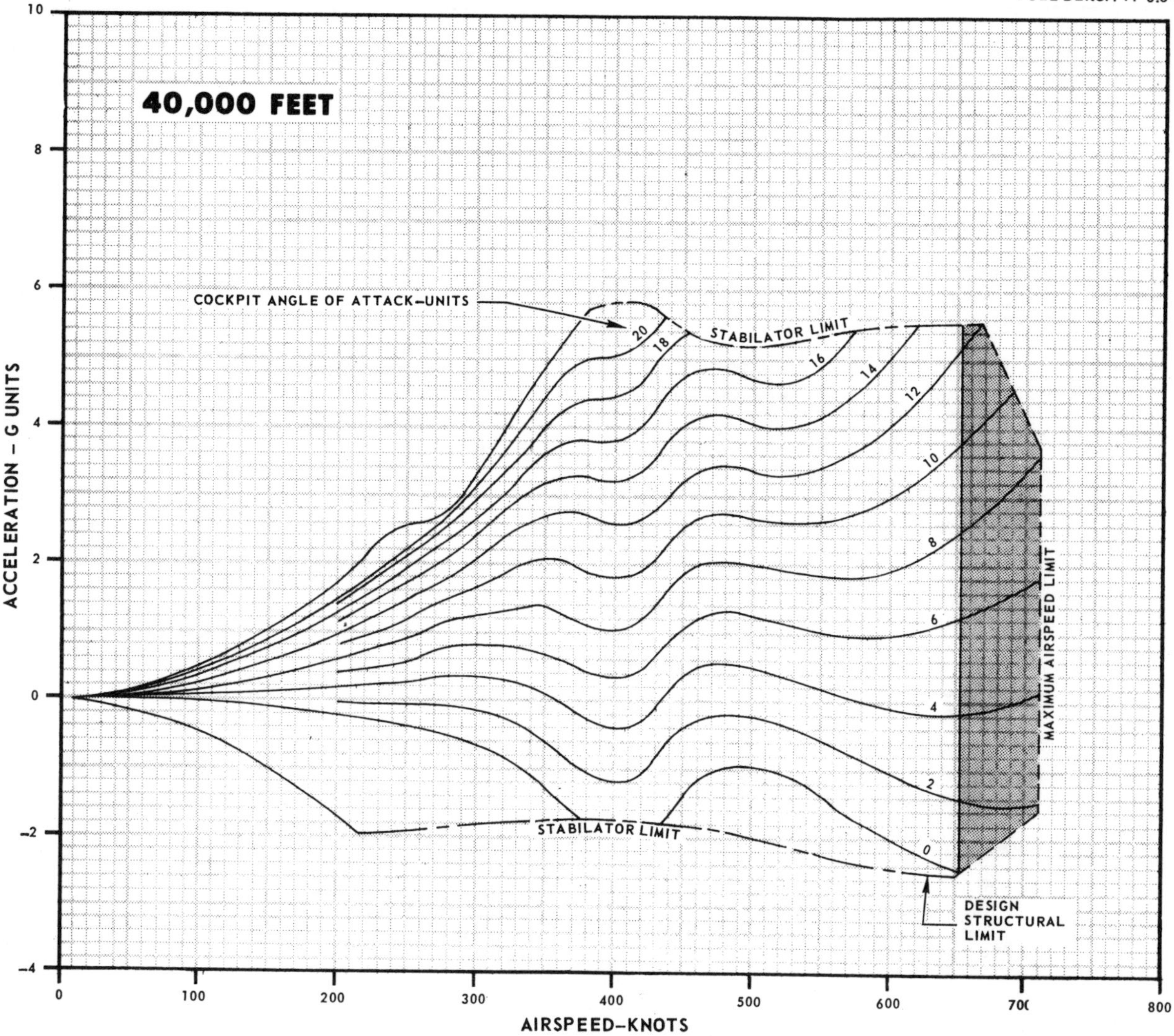

FDD-1-(307)

Figure 11-90

TEMPERATURE EFFECT ON MAXIMUM SPEED

MAXIMUM THRUST

AIRPLANE CONFIGURATION
ALL DRAG INDEXES

REMARKS
ENGINE(S): (2) J79-GE-10

GUIDE

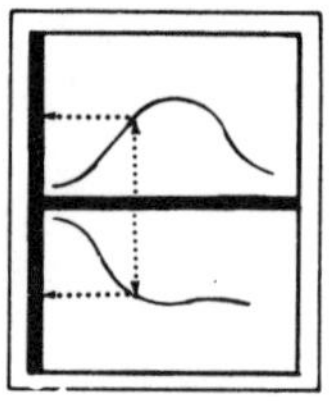

DATE: 1 AUGUST 1968
DATA BASIS: FLIGHT TEST

FUEL GRADE: JP-5
FUEL DENSITY: 6.8 LB/GAL

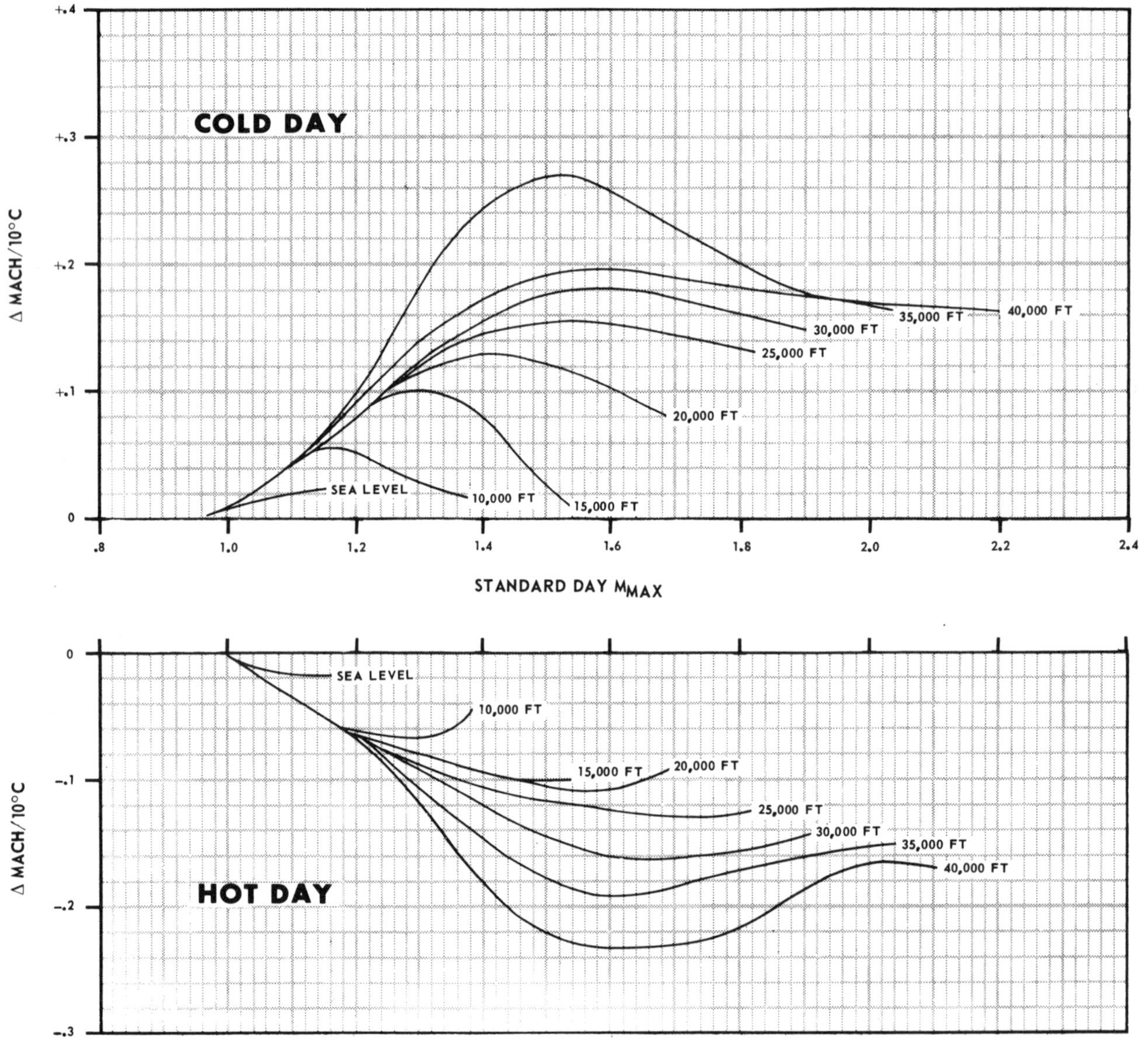

FDD-1-(267)A

Figure 11-91

DIVE RECOVERY -16 UNITS AOA

SUBSONIC-SPEED BRAKES RETRACTED

AIRPLANE CONFIGURATION
(4) AIM-7

GROSS WEIGHT 40,000 POUNDS

REMARKS

ENGINE(S): (2) J79-GE-10
ICAO STANDARD DAY

NOTES

1. ALTITUDE LOSS WITH MAXIMUM THRUST IS ESSENTIALLY THE SAME WITH MILITARY THRUST.
2. PULLOUT BASED ON 1.0G PER SECOND ACCELERATION BUILDUP TO 19 UNITS (AOA), STABILATOR LIMIT OR 6.0G WHICHEVER OCCURS FIRST.

GUIDE

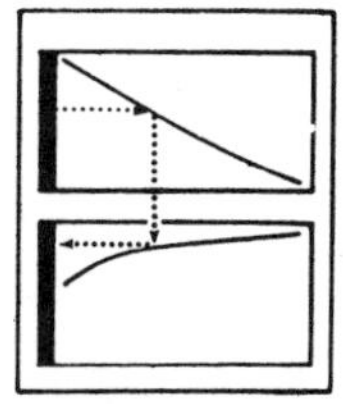

DATE: 1 AUGUST 1968
DATA BASIS: FLIGHT TEST

FUEL GRADE: JP-5
FUEL DENSITY: 6.8 LB/GAL

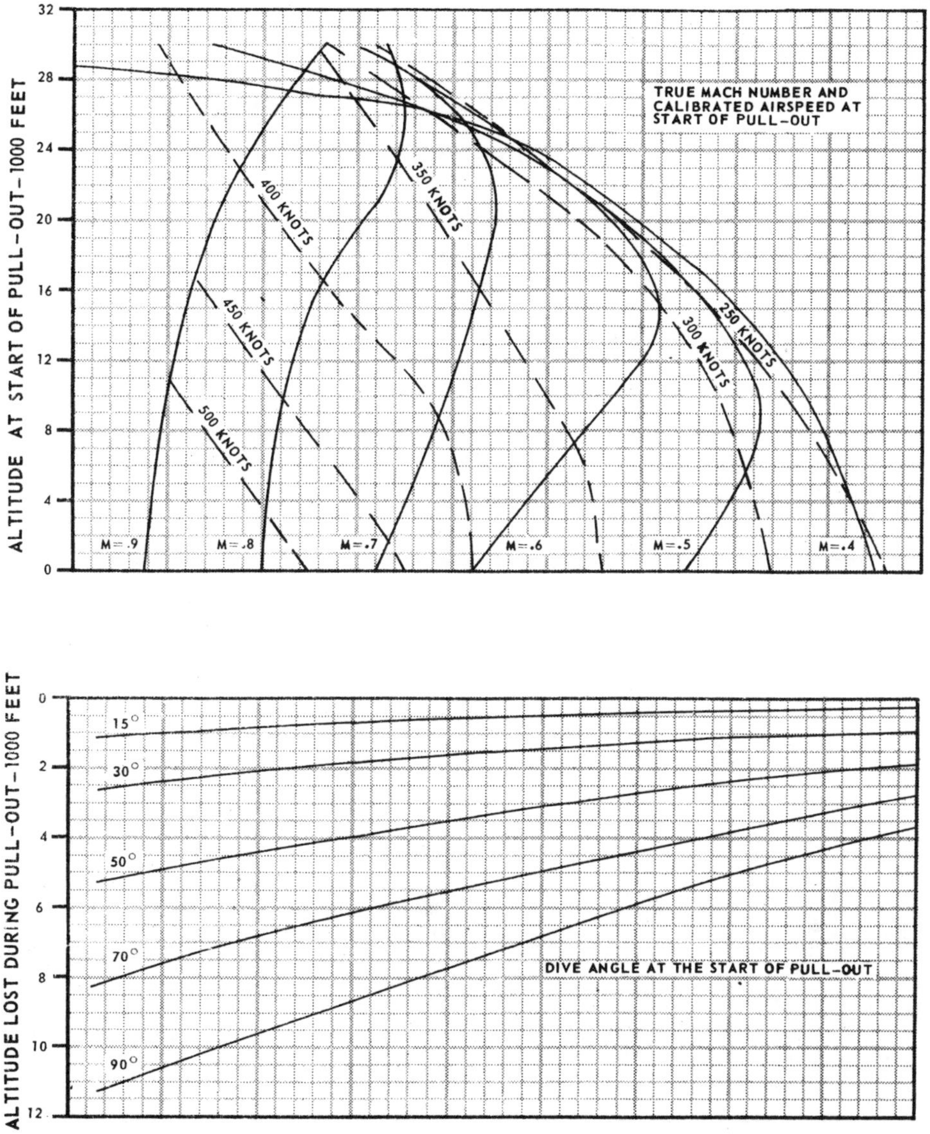

FDD-1-(309)

Figure 11-92

DIVE RECOVERY -19 UNITS AOA

SUBSONIC-SPEED BRAKES RETRACTED

AIRPLANE CONFIGURATION
(4) AIM-7

GROSS WEIGHT 40,000 POUNDS

GUIDE

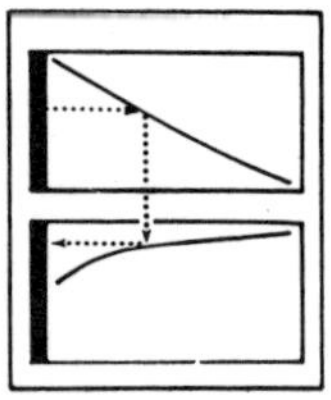

REMARKS

ENGINE(S): (2) J79-GE-10
ICAO STANDARD DAY

NOTES

1. ALTITUDE LOSS WITH MAXIMUM THRUST IS ESSENTIALLY THE SAME WITH MILITARY THRUST.
2. PULLOUT BASED ON 1.0G PER SECOND ACCELERATION BUILDUP TO 19 UNITS (AOA), STABILATOR LIMIT OR 6.0G WHICHEVER OCCURS FIRST.

DATE: 1 AUGUST 1968
DATA BASIS: FLIGHT TEST

FUEL GRADE: JP-5
FUEL DENSITY: 6.8 LB/GAL

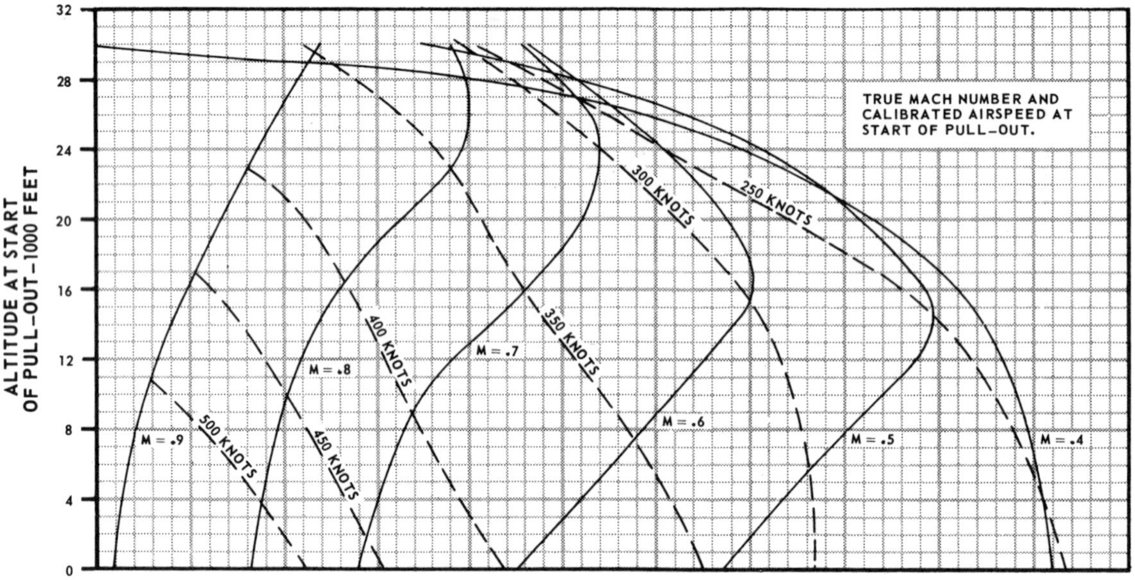

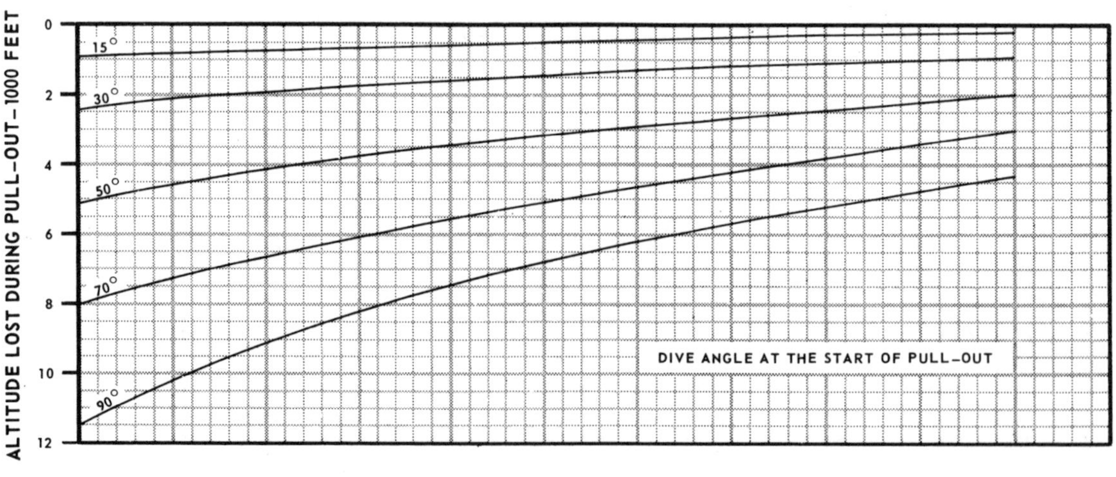

FDD-1-(310)

Figure 11-93

DIVE RECOVERY -16 UNITS AOA

SUPERSONIC-SPEED BRAKES RETRACTED

AIRPLANE CONFIGURATION
(4) AIM-7

GROSS WEIGHT 40,000 POUNDS

REMARKS
ENGINE(S): (2) J79-GE-10
ICAO STANDARD DAY

NOTES
1. ALTITUDE LOSS WITH MAXIMUM THRUST IS ESSENTIALLY THE SAME WITH MILITARY THRUST.
2. PULLOUT BASED ON 1.0G PER SECOND ACCELERATION BUILDUP TO 16 UNITS (AOA), STABILATOR LIMIT OR 6.0G WHICHEVER OCCURS FIRST.

GUIDE

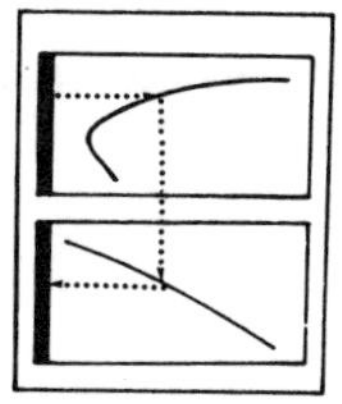

DATE: 1 AUGUST 1968
DATA BASIS: FLIGHT TEST

FUEL GRADE: JP-5
FUEL DENSITY: 6.8 LB/GAL

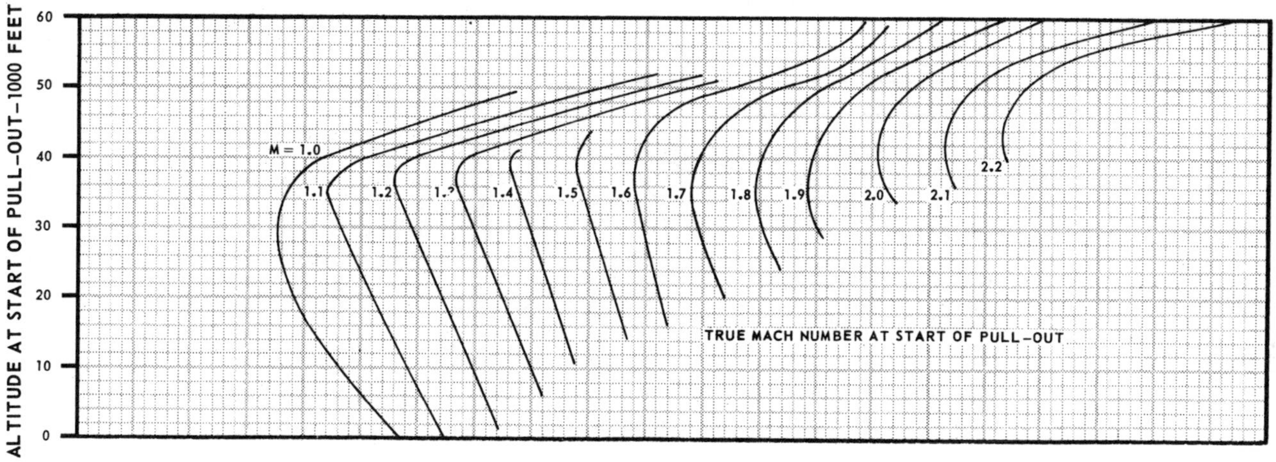

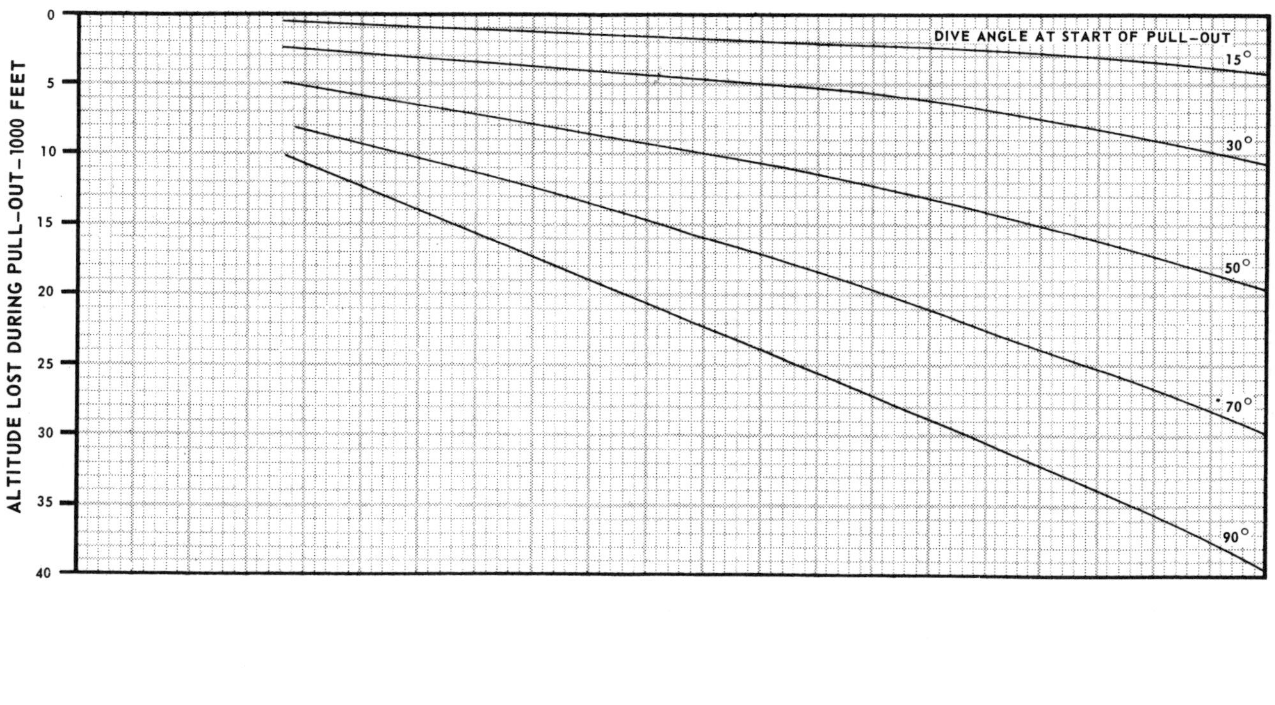

FDD-1-(311)

Figure 11-94

DIVE RECOVERY -19 UNITS AOA

SUPERSONIC-SPEED BRAKES RETRACTED

AIRPLANE CONFIGURATION
(4) AIM-7

GROSS WEIGHT 40,000 POUNDS

REMARKS

ENGINE(S): (2) J79-GE-10
ICAO STANDARD DAY

NOTES

1. ALTITUDE LOSS WITH MAXIMUM THRUST IS ESSENTIALLY THE SAME WITH MILITARY THRUST.
2. PULL-OUT BASED ON 1.0G PER SECOND ACCELERATION BUILDUP TO 19 UNITS (AOA), STABILATOR LIMIT OR 6.0G WHICHEVER OCCURS FIRST.

GUIDE

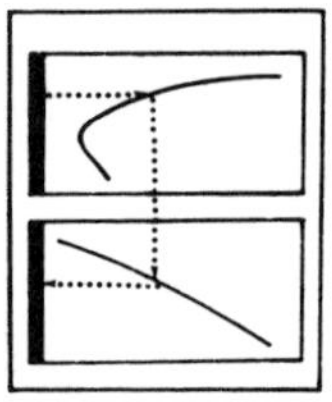

DATE: 1 AUGUST 1968
DATA BASIS: FLIGHT TEST

FUEL GRADE: JP-5
FUEL DENSITY: 6.8 LB/GAL

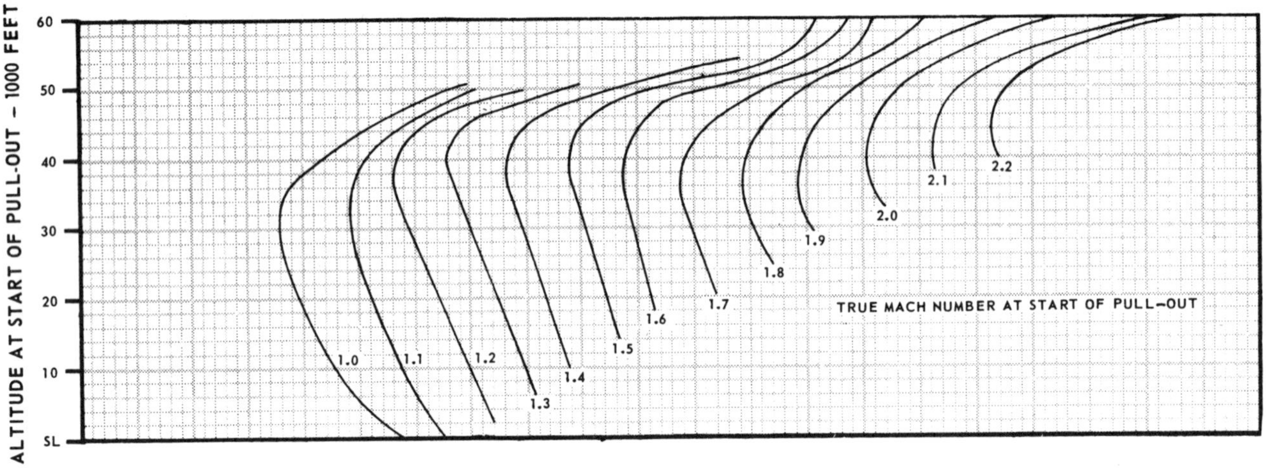

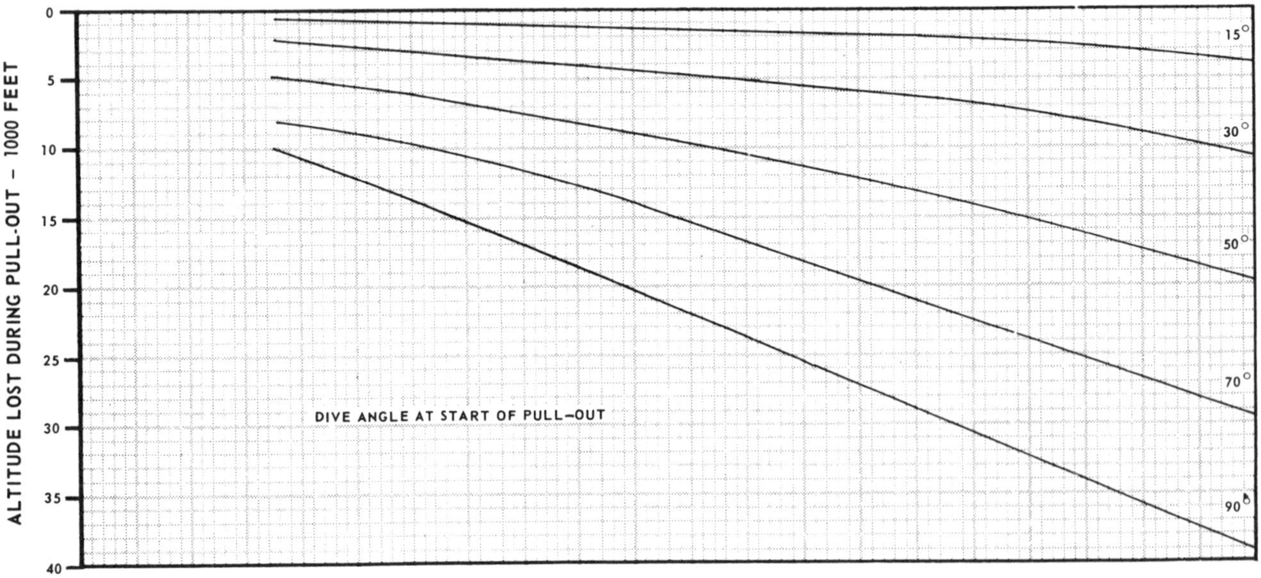

FDD-1-(308)

Figure 11-95

TURN CAPABILITIES
CONSTANT SPEED AND ALTITUDE

REMARKS

ENGINE(S): (2) J79-GE -10
ICAO STANDARD DAY

DATE: 1 MAY 1968
DATA BASIS: ESTIMATED (BASED ON FLIGHT TEST)

FUEL GRADE: JP-5
FUEL DENSITY: 6.8 LB/GAL

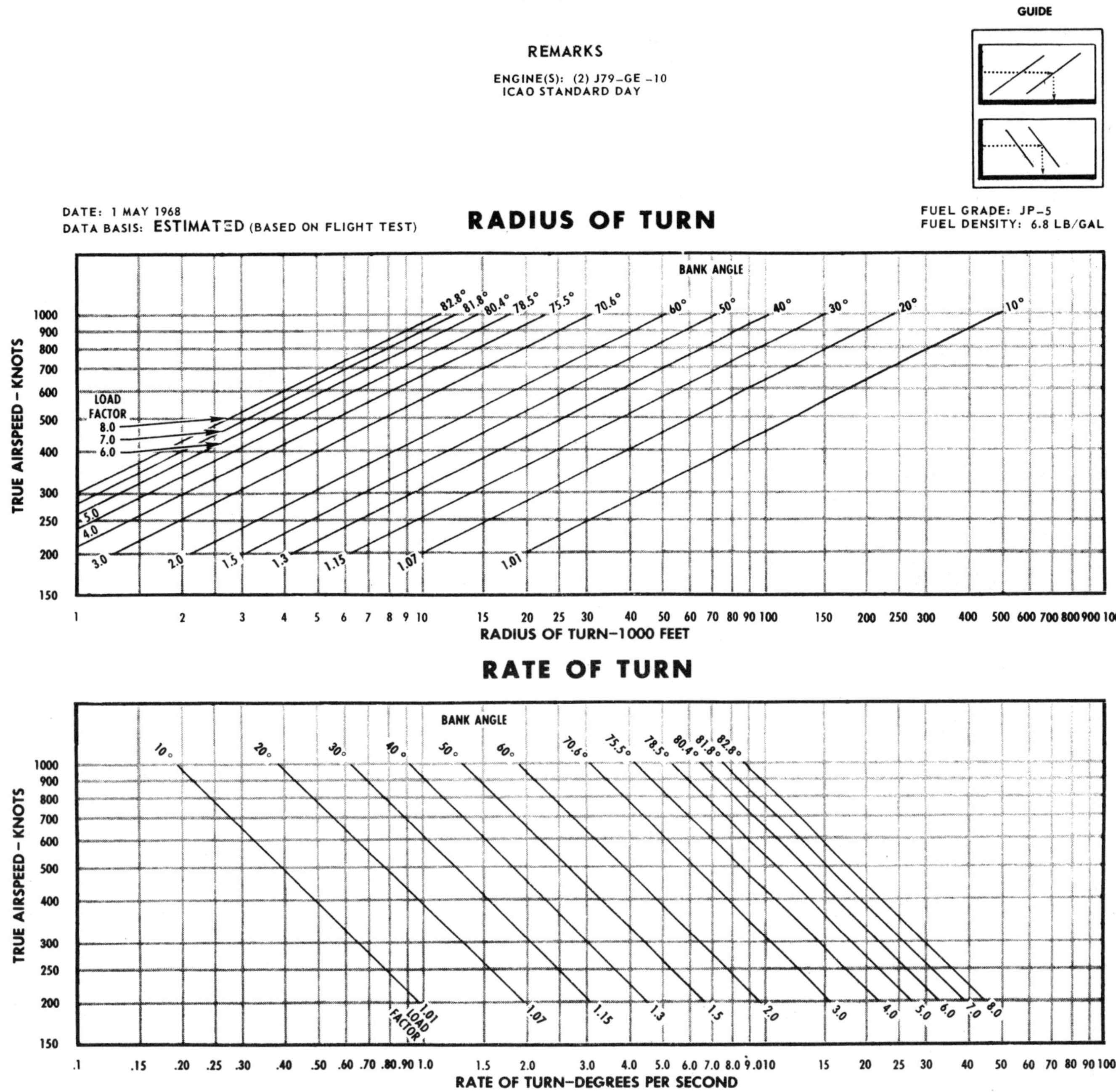

FDD-1-(272)A

Figure 11-96

PART 10 MISSION PLANNING

This information will be supplied when available.

APPENDIX A
FOLDOUT ILLUSTRATIONS

TABLE OF CONTENTS

GENERAL

The purpose of the Foldout Section is to make the subject illustrations always available for quick and ready reference while reading the associated text. The subject illustrations are referenced from several sections throughout the manual and are referred to in the text as (figure A- , appendix A).

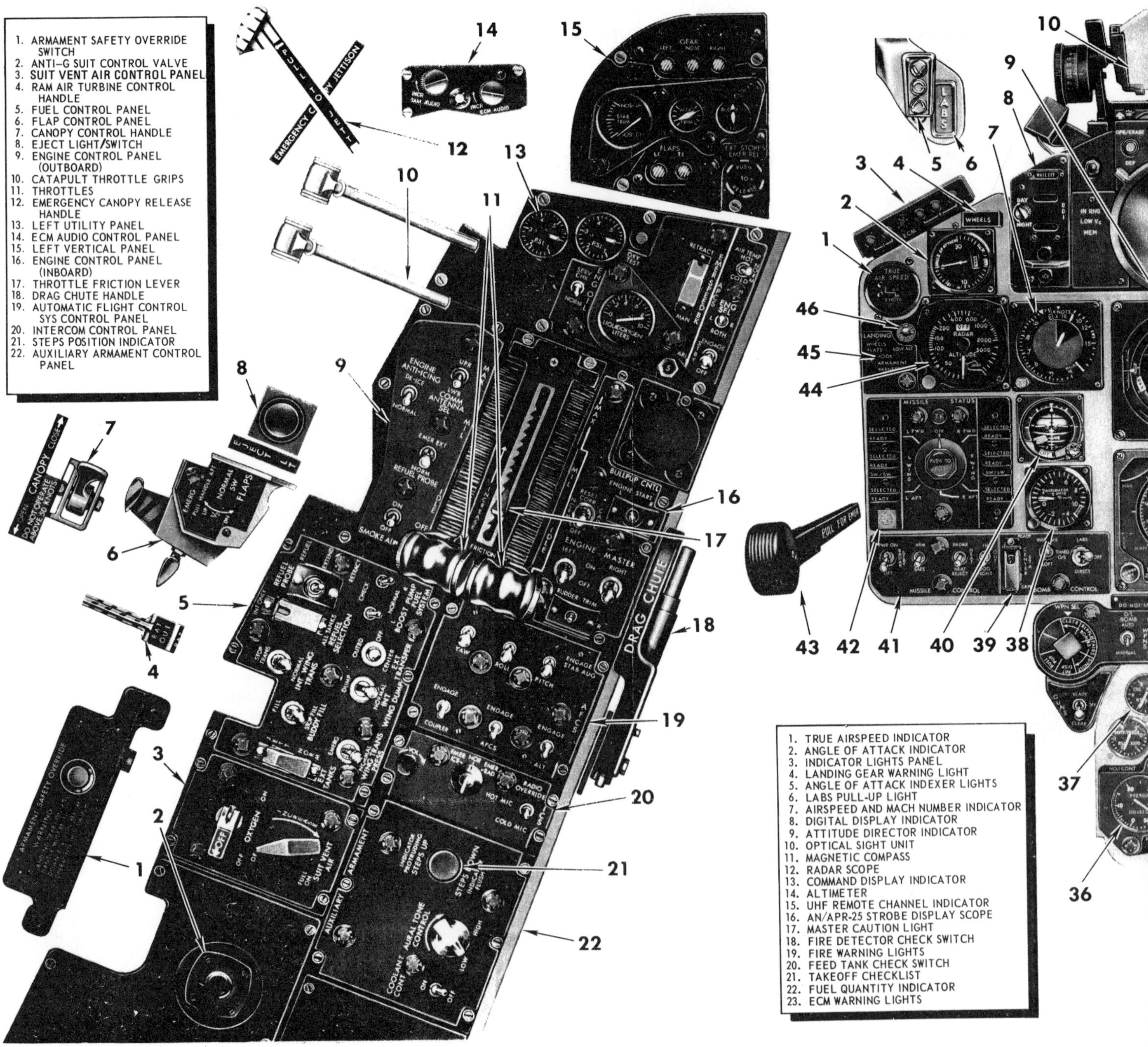

1. ARMAMENT SAFETY OVERRIDE SWITCH
2. ANTI-G SUIT CONTROL VALVE
3. SUIT VENT AIR CONTROL PANEL
4. RAM AIR TURBINE CONTROL HANDLE
5. FUEL CONTROL PANEL
6. FLAP CONTROL PANEL
7. CANOPY CONTROL HANDLE
8. EJECT LIGHT/SWITCH
9. ENGINE CONTROL PANEL (OUTBOARD)
10. CATAPULT THROTTLE GRIPS
11. THROTTLES
12. EMERGENCY CANOPY RELEASE HANDLE
13. LEFT UTILITY PANEL
14. ECM AUDIO CONTROL PANEL
15. LEFT VERTICAL PANEL
16. ENGINE CONTROL PANEL (INBOARD)
17. THROTTLE FRICTION LEVER
18. DRAG CHUTE HANDLE
19. AUTOMATIC FLIGHT CONTROL SYS CONTROL PANEL
20. INTERCOM CONTROL PANEL
21. STEPS POSITION INDICATOR
22. AUXILIARY ARMAMENT CONTROL PANEL

1. TRUE AIRSPEED INDICATOR
2. ANGLE OF ATTACK INDICATOR
3. INDICATOR LIGHTS PANEL
4. LANDING GEAR WARNING LIGHT
5. ANGLE OF ATTACK INDEXER LIGHTS
6. LABS PULL-UP LIGHT
7. AIRSPEED AND MACH NUMBER INDICATOR
8. DIGITAL DISPLAY INDICATOR
9. ATTITUDE DIRECTOR INDICATOR
10. OPTICAL SIGHT UNIT
11. MAGNETIC COMPASS
12. RADAR SCOPE
13. COMMAND DISPLAY INDICATOR
14. ALTIMETER
15. UHF REMOTE CHANNEL INDICATOR
16. AN/APR-25 STROBE DISPLAY SCOPE
17. MASTER CAUTION LIGHT
18. FIRE DETECTOR CHECK SWITCH
19. FIRE WARNING LIGHTS
20. FEED TANK CHECK SWITCH
21. TAKEOFF CHECKLIST
22. FUEL QUANTITY INDICATOR
23. ECM WARNING LIGHTS

TYPICAL FRONT COCKPIT
BLOCK 33

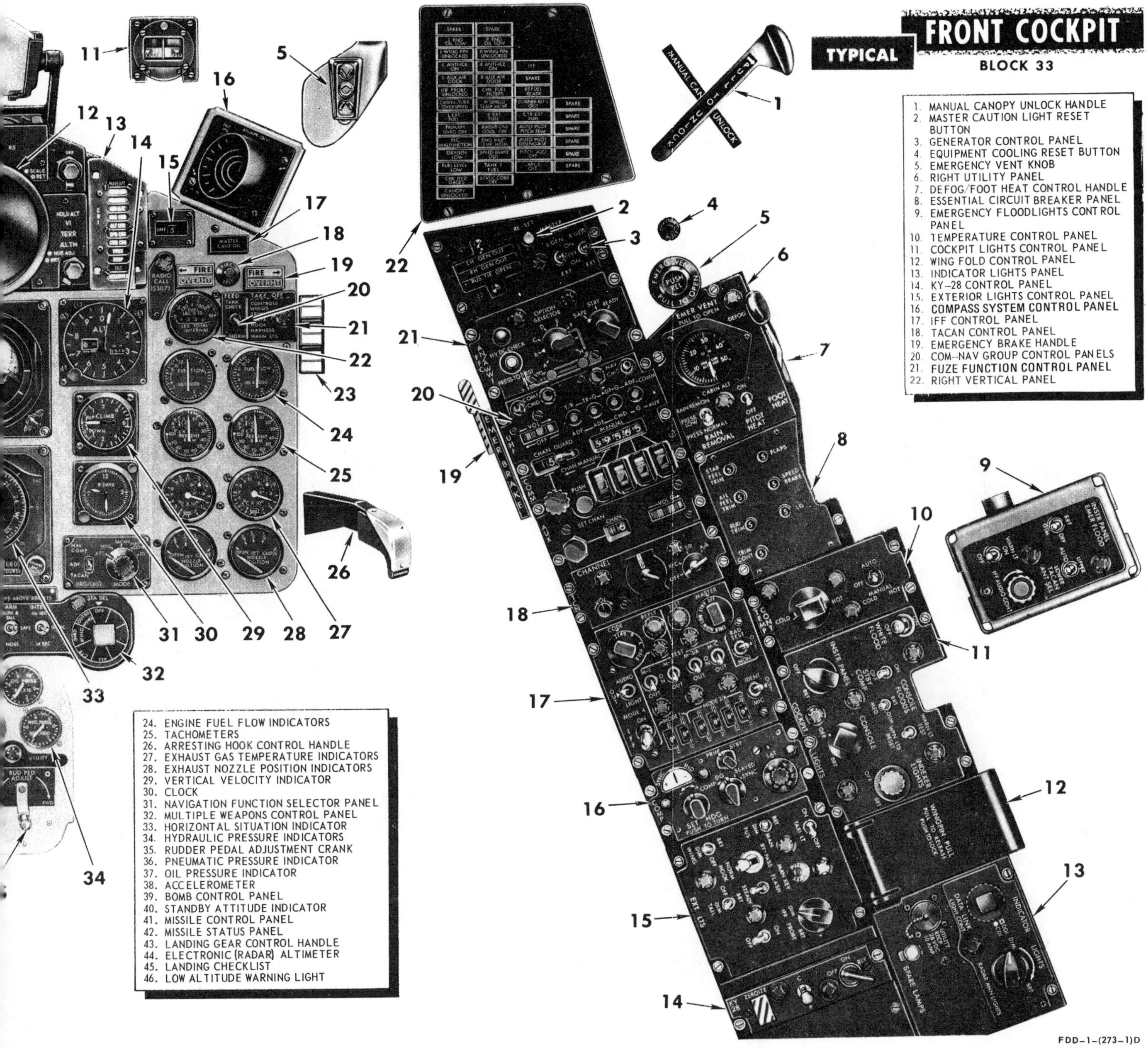

1. MANUAL CANOPY UNLOCK HANDLE
2. MASTER CAUTION LIGHT RESET BUTTON
3. GENERATOR CONTROL PANEL
4. EQUIPMENT COOLING RESET BUTTON
5. EMERGENCY VENT KNOB
6. RIGHT UTILITY PANEL
7. DEFOG/FOOT HEAT CONTROL HANDLE
8. ESSENTIAL CIRCUIT BREAKER PANEL
9. EMERGENCY FLOODLIGHTS CONTROL PANEL
10. TEMPERATURE CONTROL PANEL
11. COCKPIT LIGHTS CONTROL PANEL
12. WING FOLD CONTROL PANEL
13. INDICATOR LIGHTS PANEL
14. KY-28 CONTROL PANEL
15. EXTERIOR LIGHTS CONTROL PANEL
16. COMPASS SYSTEM CONTROL PANEL
17. IFF CONTROL PANEL
18. TACAN CONTROL PANEL
19. EMERGENCY BRAKE HANDLE
20. COM-NAV GROUP CONTROL PANELS
21. FUZE FUNCTION CONTROL PANEL
22. RIGHT VERTICAL PANEL

24. ENGINE FUEL FLOW INDICATORS
25. TACHOMETERS
26. ARRESTING HOOK CONTROL HANDLE
27. EXHAUST GAS TEMPERATURE INDICATORS
28. EXHAUST NOZZLE POSITION INDICATORS
29. VERTICAL VELOCITY INDICATOR
30. CLOCK
31. NAVIGATION FUNCTION SELECTOR PANEL
32. MULTIPLE WEAPONS CONTROL PANEL
33. HORIZONTAL SITUATION INDICATOR
34. HYDRAULIC PRESSURE INDICATORS
35. RUDDER PEDAL ADJUSTMENT CRANK
36. PNEUMATIC PRESSURE INDICATOR
37. OIL PRESSURE INDICATOR
38. ACCELEROMETER
39. BOMB CONTROL PANEL
40. STANDBY ATTITUDE INDICATOR
41. MISSILE CONTROL PANEL
42. MISSILE STATUS PANEL
43. LANDING GEAR CONTROL HANDLE
44. ELECTRONIC (RADAR) ALTIMETER
45. LANDING CHECKLIST
46. LOW ALTITUDE WARNING LIGHT

FDD-1-(273-1)D

Figure A-1 (Sheet 1 of 2)

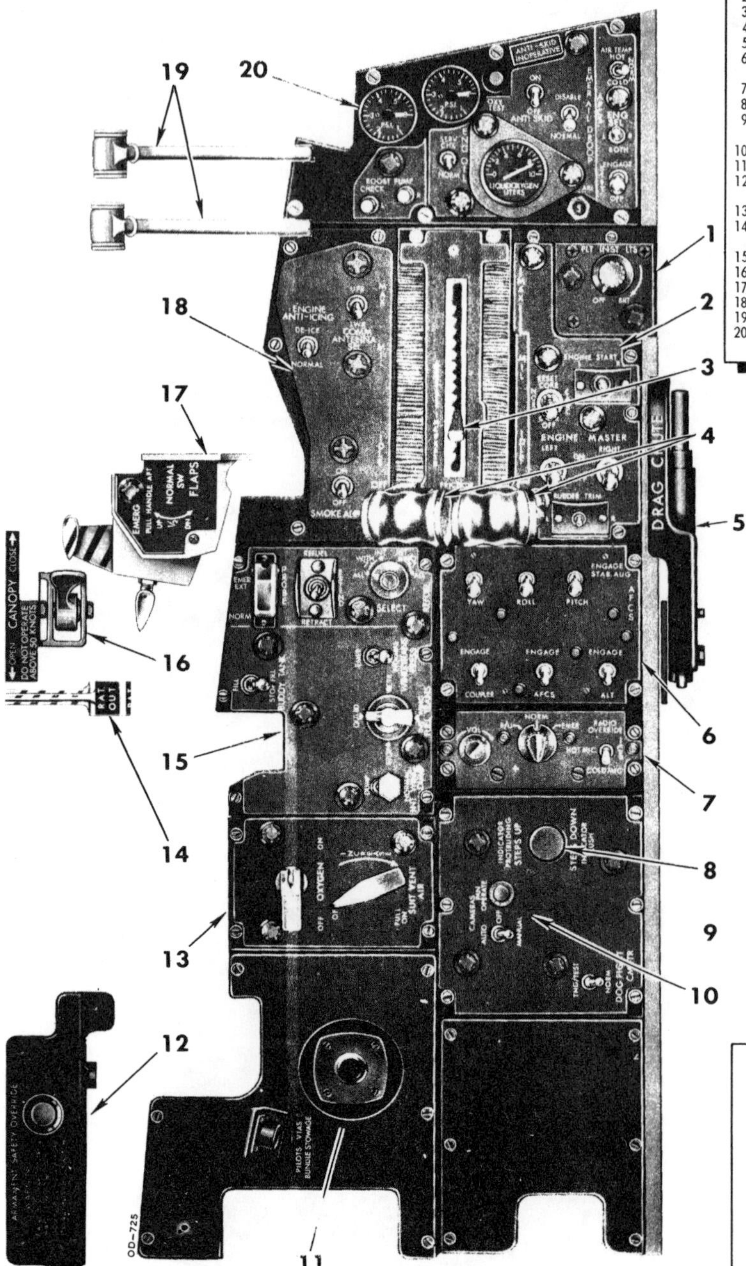

1. FLIGHT INSTRUMENT LIGHTS CONTROL
2. ENGINE CONTROL PANEL (INBOARD)
3. THROTTLE FRICTION LEVER
4. THROTTLES
5. DRAG CHUTE HANDLE
6. AUTOMATIC FLIGHT CONTROL SYS CONTROL PANEL
7. INTERCOM CONTROL PANEL
8. STEPS POSITION INDICATOR
9. AUXILIARY ARMAMENT CONTROL PANEL
10. LB 30A CAMERA POD CONTROL
11. ANTI–G SUIT CONTROL VALVE
12. ARMAMENT SAFETY OVERRIDE SWITCH
13. SUIT VENT AIR CONTROL PANEL
14. RAM AIR TURBINE CONTROL HANDLE
15. FUEL CONTROL PANEL
16. CANOPY CONTROL HANDLE
17. FLAP CONTROL PANEL
18. ENGINE CONTROL PANEL (OUTBOARD)
19. CATAPULT THROTTLE GRIPS
20. LEFT UTILITY PANEL

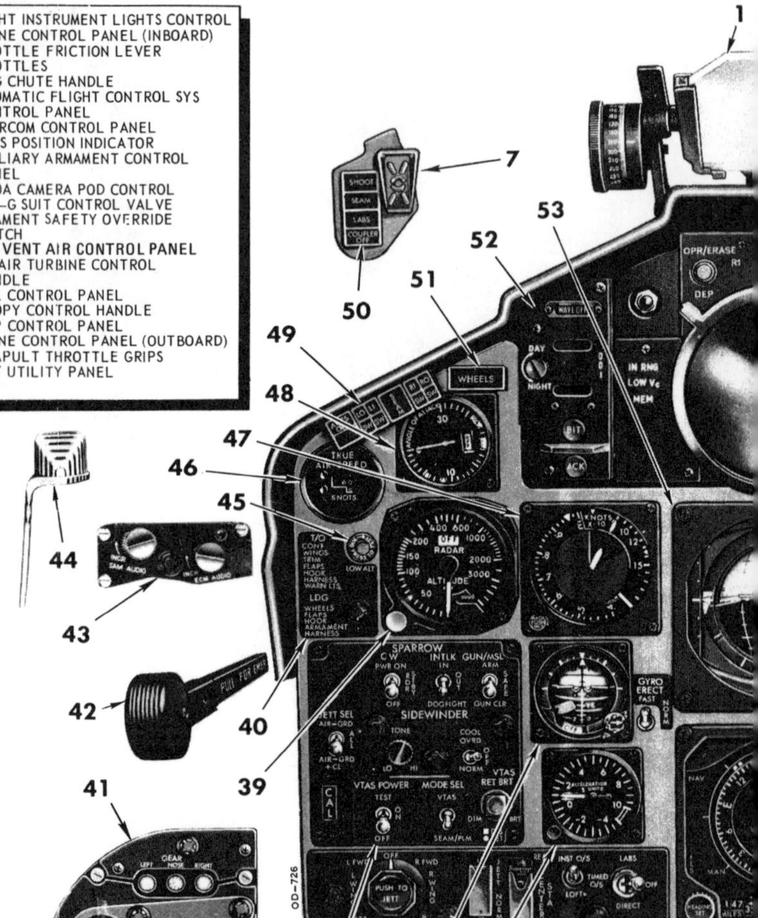

1. OPTICAL SIGHT UNIT
2. MAGNETIC COMPASS
3. RADAR SCOPE
4. COMMAND DISPLAY INDICATOR
5. MASTER CAUTION LIGHT
6. FEED TANK CHECK SWITCH
7. ANGLE OF ATTACK INDEXER LIGHTS
8. STROBE DISPLAY SCOPE
9. WEAPON STATUS PANEL
10. AIM–7 STATUS PANEL
11. UHF REMOTE CHANNEL
12. FIRE DETECTOR CHECK SWITCH
13. FUEL QUANTITY INDICATOR
14. ECM WARNING LIGHTS
15. FIRE WARNING LIGHTS
16. ALTIMETER
17. ENGINE FUEL FLOW INDICATORS
18. MANUAL CANOPY UNLOCK HANDLE
19. ARRESTING HOOK CONTROL HANDLE
20. RIGHT VERTICAL PANEL
21. TACHOMETERS
22. EXHAUST GAS TEMPERATURE INDICATORS
23. EMERGENCY BRAKE PRESSURE INDICATOR
24. EXHAUST NOZZLE POSITION INDICATORS
25. VERTICAL VELOCITY INDICATOR

TYPICAL FRONT COCKPIT

AIRCRAFT 158355at AND UP
ALL OTHERS AFTER AFC 506

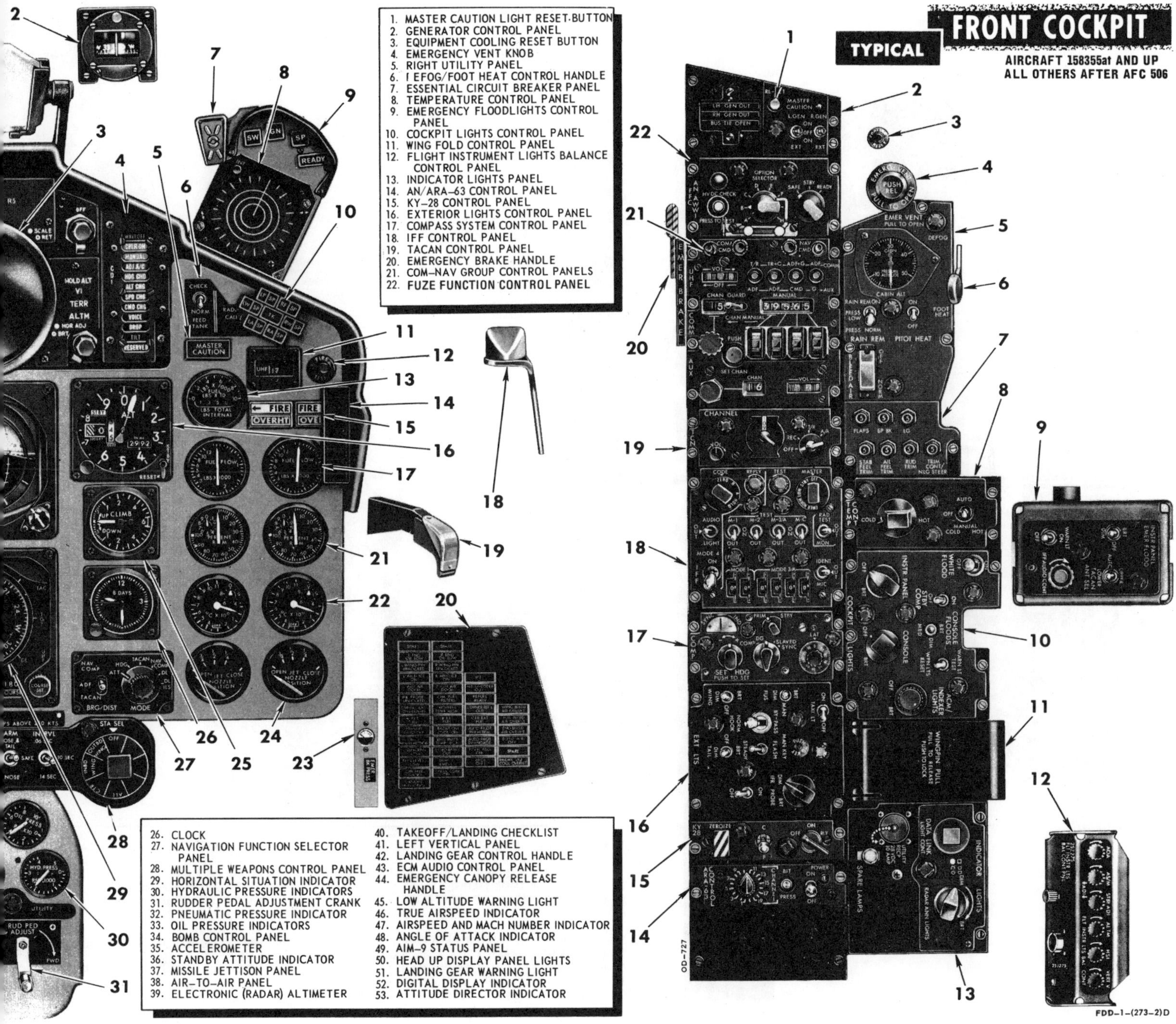

1. MASTER CAUTION LIGHT RESET BUTTON
2. GENERATOR CONTROL PANEL
3. EQUIPMENT COOLING RESET BUTTON
4. EMERGENCY VENT KNOB
5. RIGHT UTILITY PANEL
6. I EFOG/FOOT HEAT CONTROL HANDLE
7. ESSENTIAL CIRCUIT BREAKER PANEL
8. TEMPERATURE CONTROL PANEL
9. EMERGENCY FLOODLIGHTS CONTROL PANEL
10. COCKPIT LIGHTS CONTROL PANEL
11. WING FOLD CONTROL PANEL
12. FLIGHT INSTRUMENT LIGHTS BALANCE CONTROL PANEL
13. INDICATOR LIGHTS PANEL
14. AN/ARA-63 CONTROL PANEL
15. KY-28 CONTROL PANEL
16. EXTERIOR LIGHTS CONTROL PANEL
17. COMPASS SYSTEM CONTROL PANEL
18. IFF CONTROL PANEL
19. TACAN CONTROL PANEL
20. EMERGENCY BRAKE HANDLE
21. COM-NAV GROUP CONTROL PANELS
22. FUZE FUNCTION CONTROL PANEL

26. CLOCK
27. NAVIGATION FUNCTION SELECTOR PANEL
28. MULTIPLE WEAPONS CONTROL PANEL
29. HORIZONTAL SITUATION INDICATOR
30. HYDRAULIC PRESSURE INDICATORS
31. RUDDER PEDAL ADJUSTMENT CRANK
32. PNEUMATIC PRESSURE INDICATOR
33. OIL PRESSURE INDICATORS
34. BOMB CONTROL PANEL
35. ACCELEROMETER
36. STANDBY ATTITUDE INDICATOR
37. MISSILE JETTISON PANEL
38. AIR-TO-AIR PANEL
39. ELECTRONIC (RADAR) ALTIMETER
40. TAKEOFF/LANDING CHECKLIST
41. LEFT VERTICAL PANEL
42. LANDING GEAR CONTROL HANDLE
43. ECM AUDIO CONTROL PANEL
44. EMERGENCY CANOPY RELEASE HANDLE
45. LOW ALTITUDE WARNING LIGHT
46. TRUE AIRSPEED INDICATOR
47. AIRSPEED AND MACH NUMBER INDICATOR
48. ANGLE OF ATTACK INDICATOR
49. AIM-9 STATUS PANEL
50. HEAD UP DISPLAY PANEL LIGHTS
51. LANDING GEAR WARNING LIGHT
52. DIGITAL DISPLAY INDICATOR
53. ATTITUDE DIRECTOR INDICATOR

OD-727

FDD-1-(273-2)D

Figure A-1. (Sheet 2 of 2)

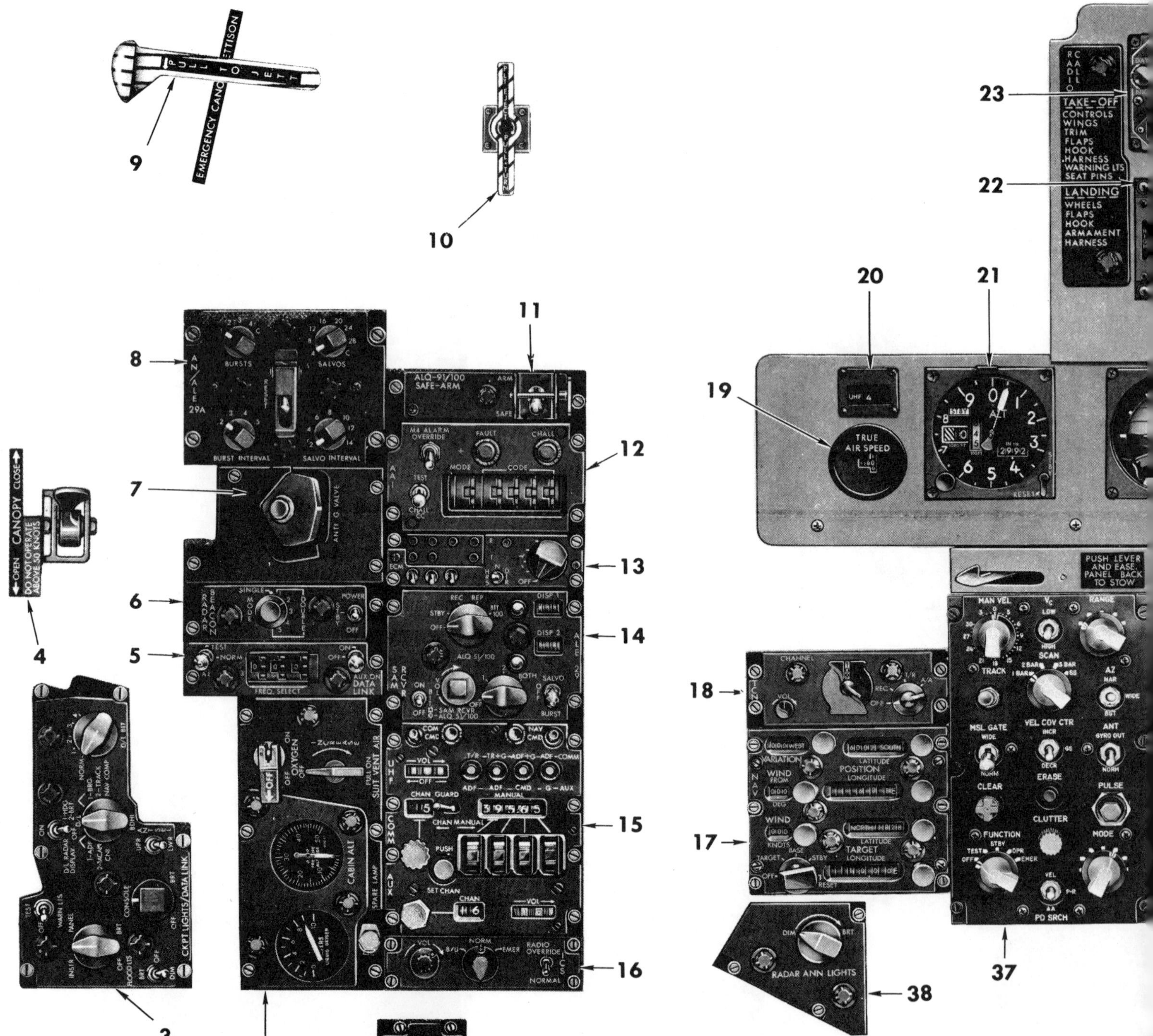

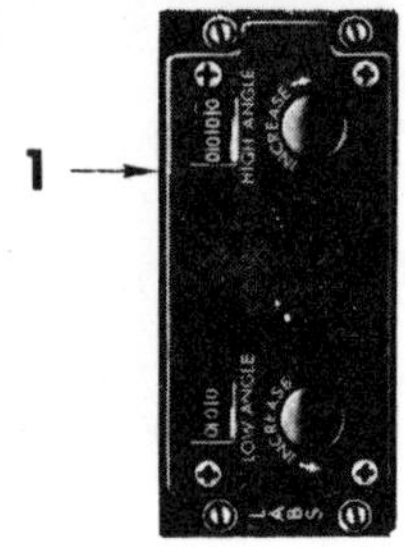

1. BOMB RELEASE ANGLE COMPUTER
2. UTILITY PANEL
3. COCKPIT LIGHTS DATA LINK CONTROL PANEL
4. CANOPY CONTROL HANDLE
5. COMMUNICATIONS SET CONTROL
6. AN/APN-154 RADAR BEACON CONTROL PANEL
7. ANTI-G VALVE
8. AN/ALE-29 DISPENSER PROGRAMMER
9. EMERGENCY CANOPY RELEASE HANDLE
10. COMMAND SELECTOR VALVE HANDLE
11. AN/ALQ-91/100 SAFE-ARM CONTROL PANEL
12. AN/APX-76 INTERROGATOR SET CONTROL PANEL
13. AN/ALQ-91 CONTROL-INDICATOR PANEL
14. INTEGRATED CONTROL PANEL
15. COMM-NAV GROUP CONTROL PANEL
16. INTERCOM CONTROL PANEL
17. NAVIGATION COMPUTER CONTROL PANEL
18. TACAN CONTROL PANEL
19. TRUE AIRSPEED INDICATOR
20. UHF REMOTE CHANNEL INDICATOR
21. ALTIMETER
22. COMMAND TARGET ALTITUDE INDICATOR
23. DIGITAL DISPLAY INDICATOR
24. AN/APR-25 STROBE DISPLAY SCOPE

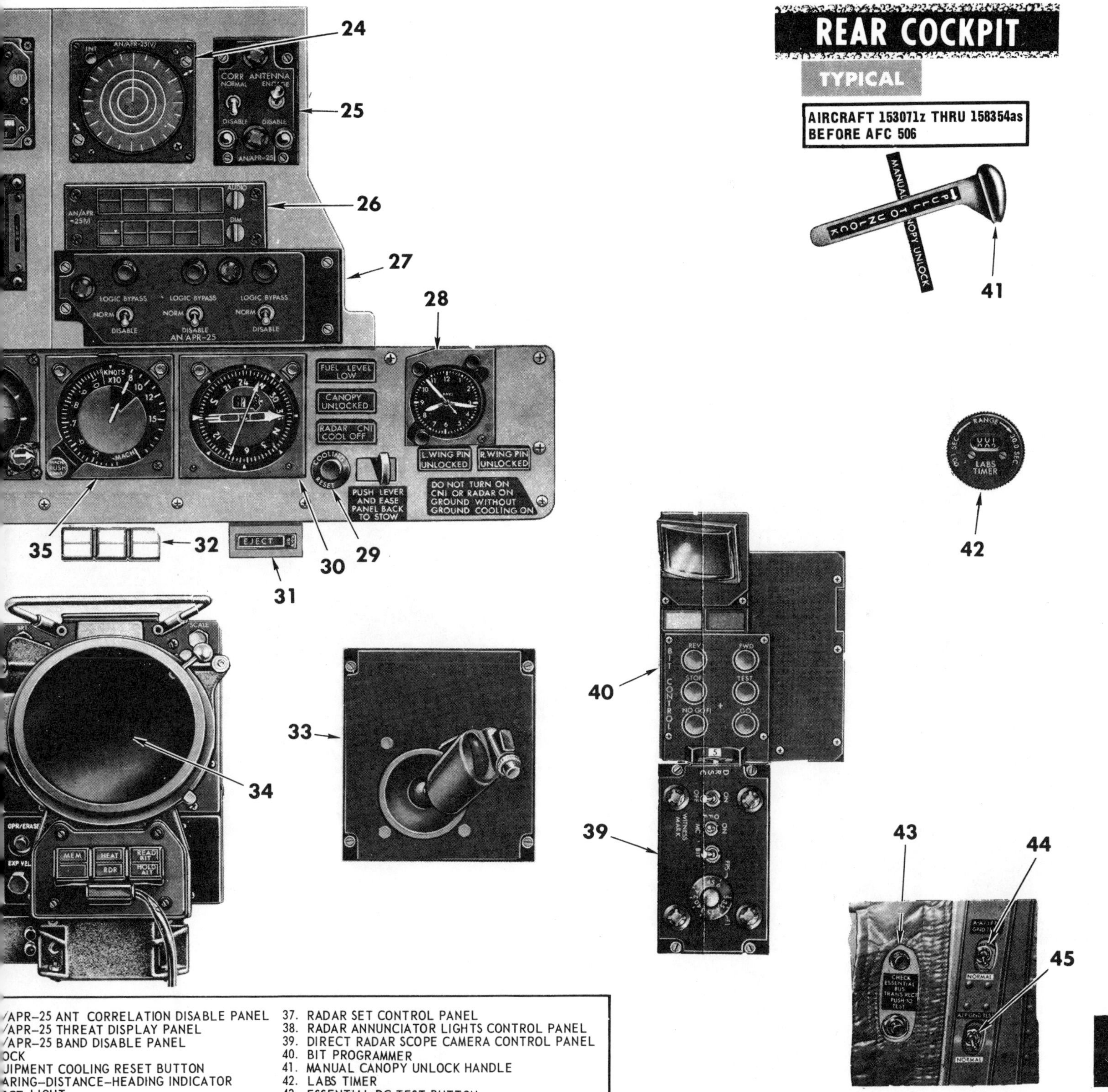

/APR-25 ANT CORRELATION DISABLE PANEL
/APR-25 THREAT DISPLAY PANEL
/APR-25 BAND DISABLE PANEL
OCK
UIPMENT COOLING RESET BUTTON
ARING-DISTANCE-HEADING INDICATOR
ECT LIGHT
M WARNING LIGHTS PANEL
TENNA HAND CONTROL PANEL
DAR SCOPE
SPEED AND MACH INDICATOR
MOTE ATTITUDE INDICATOR

37. RADAR SET CONTROL PANEL
38. RADAR ANNUNCIATOR LIGHTS CONTROL PANEL
39. DIRECT RADAR SCOPE CAMERA CONTROL PANEL
40. BIT PROGRAMMER
41. MANUAL CANOPY UNLOCK HANDLE
42. LABS TIMER
43. ESSENTIAL DC TEST BUTTON
44. IFF GROUND TEST SWITCH
45. AUTOPILOT GROUND POWER TEST SWITCH

FDD-1-(274-1)B

Figure A-2 (Sheet 1 of 2)

1. BOMB RELEASE ANGLE COMPUTER
2. UTILITY PANEL
3. COCKPIT LIGHTS/DATA LINK CONTROL PANEL
4. CANOPY CONTROL HANDLE
5. COMMUNICATIONS SET CONTROL
6. AN/APN-154 RADAR BEACON CONTROL PANEL
7. ANTI-G VALVE
8. AN/ALE-39 DISPENSER PROGRAMMER
9. EMERGENCY CANOPY RELEASE HANDLE
10. AN/ALQ-91/100 SAFE-ARM CONTROL PANEL
11. FLARE/CHAFF CONTROL PANEL
12. RADAR SET CONTROL PANEL
13. COMMAND SELECTOR VALVE HANDLE

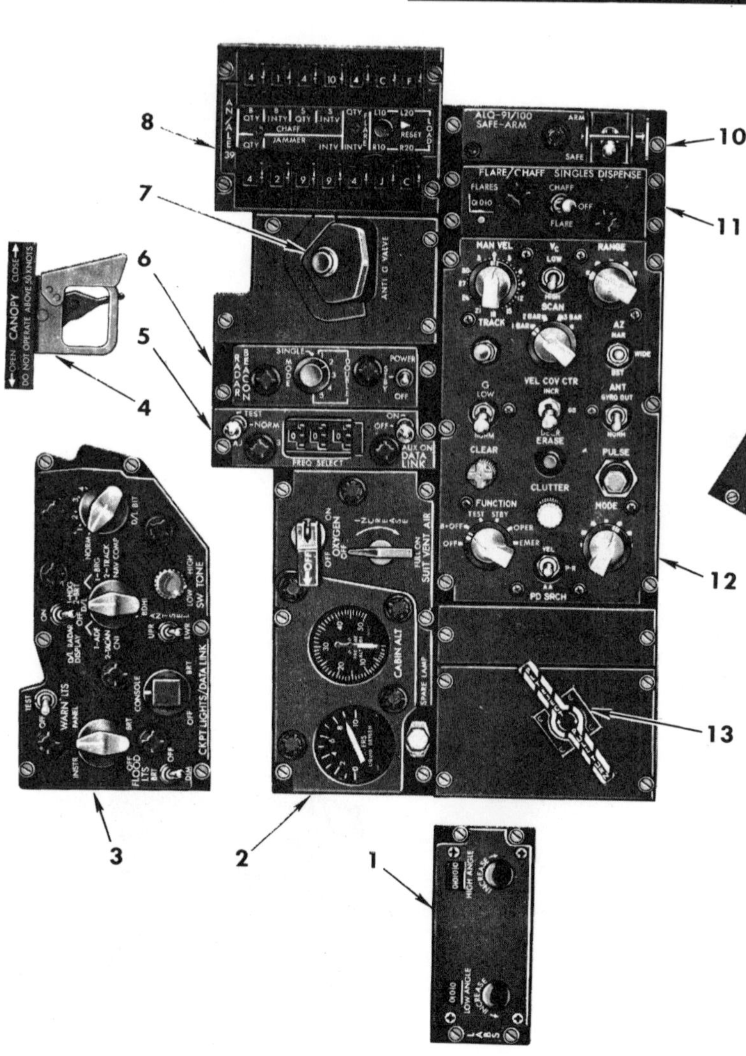

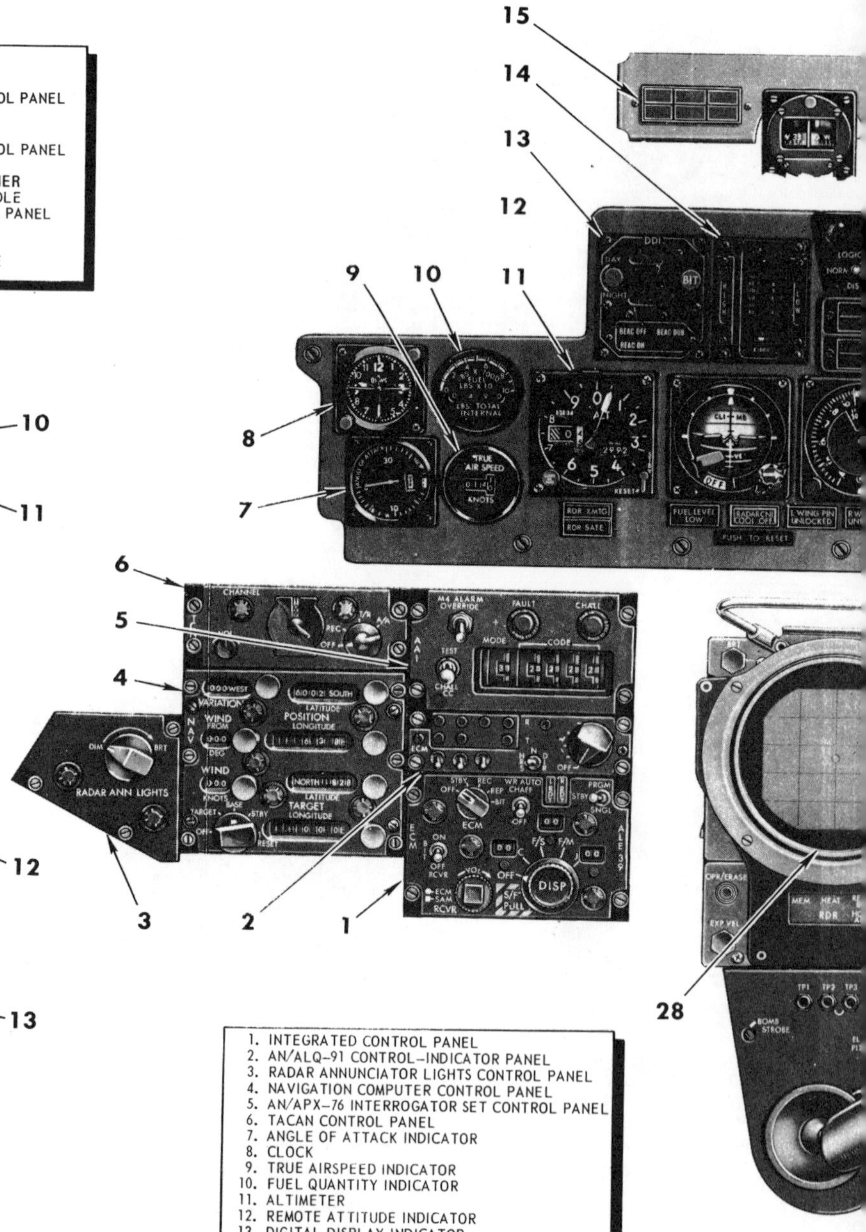

1. INTEGRATED CONTROL PANEL
2. AN/ALQ-91 CONTROL-INDICATOR PANEL
3. RADAR ANNUNCIATOR LIGHTS CONTROL PANEL
4. NAVIGATION COMPUTER CONTROL PANEL
5. AN/APX-76 INTERROGATOR SET CONTROL PANEL
6. TACAN CONTROL PANEL
7. ANGLE OF ATTACK INDICATOR
8. CLOCK
9. TRUE AIRSPEED INDICATOR
10. FUEL QUANTITY INDICATOR
11. ALTIMETER
12. REMOTE ATTITUDE INDICATOR
13. DIGITAL DISPLAY INDICATOR
14. COMMAND TARGET ALTITUDE INDICATOR
15. ECM WARNING LIGHTS PANEL

1. BIT PROGRAMMER
2. LABS TIMER
3. MANUAL CANOPY UNLOCK HANDLE
4. AUTOPILOT GROUND POWER TEST SWITCH
5. IFF GROUND TEST SWITCH
6. ESSENTIAL DC TEST BUTTON

REAR COCKPIT

TYPICAL

AIRCRAFT 158355at AND UP
ALL OTHERS AFTER AFC 506

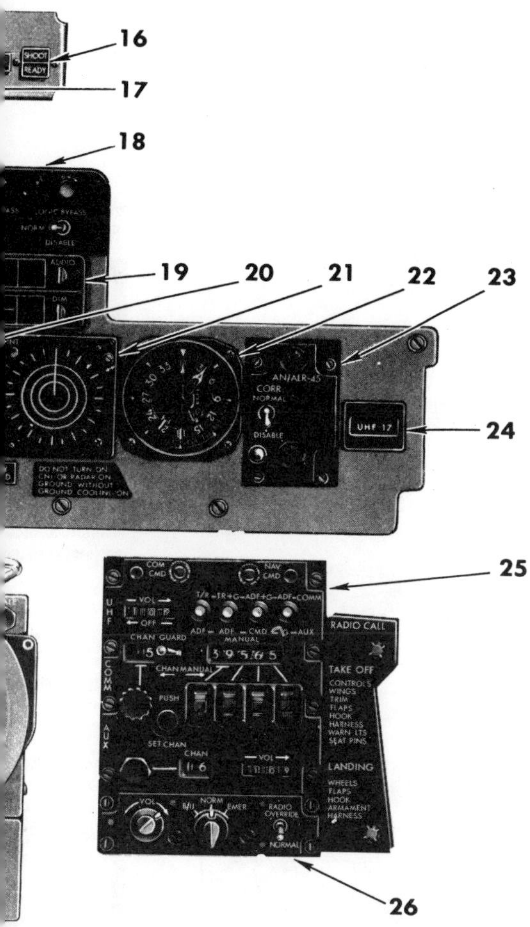

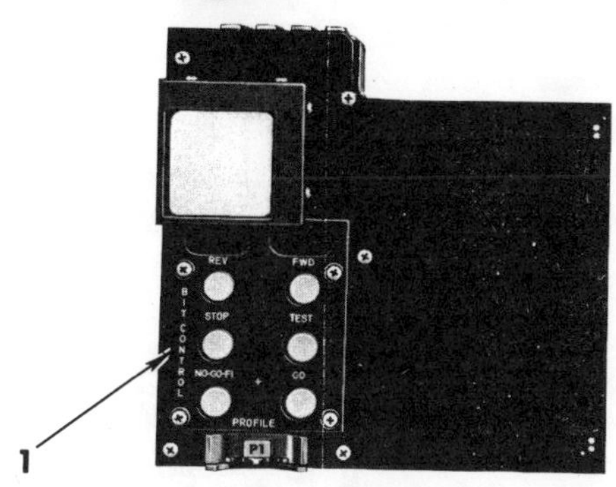

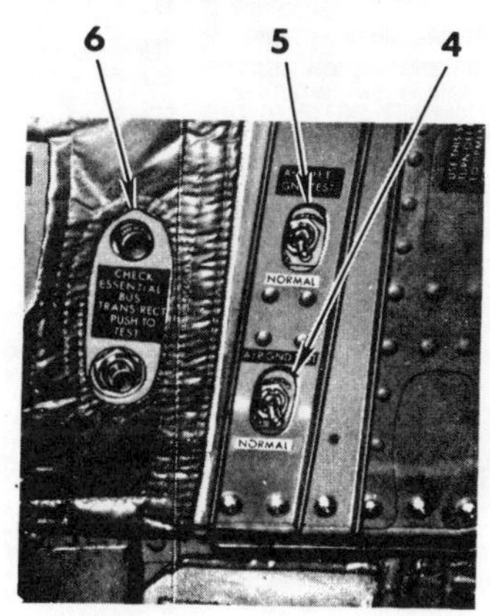

16. WEAPON STATUS PANEL
17. MAGNETIC COMPASS
18. BAND DISABLE PANEL
19. THREAT DISPLAY PANEL
20. AIRSPEED AND MACH INDICATOR
21. STROBE DISPLAY SCOPE
22. BEARING-DISTANCE-HEADING INDICATOR
23. ANTENNA CORRELATION DISABLE PANEL
24. UHF REMOTE CHANNEL INDICATOR
25. COMM-NAV GROUP CONTROL PANEL
26. INTERCOM CONTROL PANEL
27. ANTENNA HAND CONTROL
28. RADAR SCOPE

FDD-1-(274-2)A

Figure A-2 (Sheet 2 of 2)

ELECTRICAL SYSTEM

EXTERNAL POWER CONNECTED

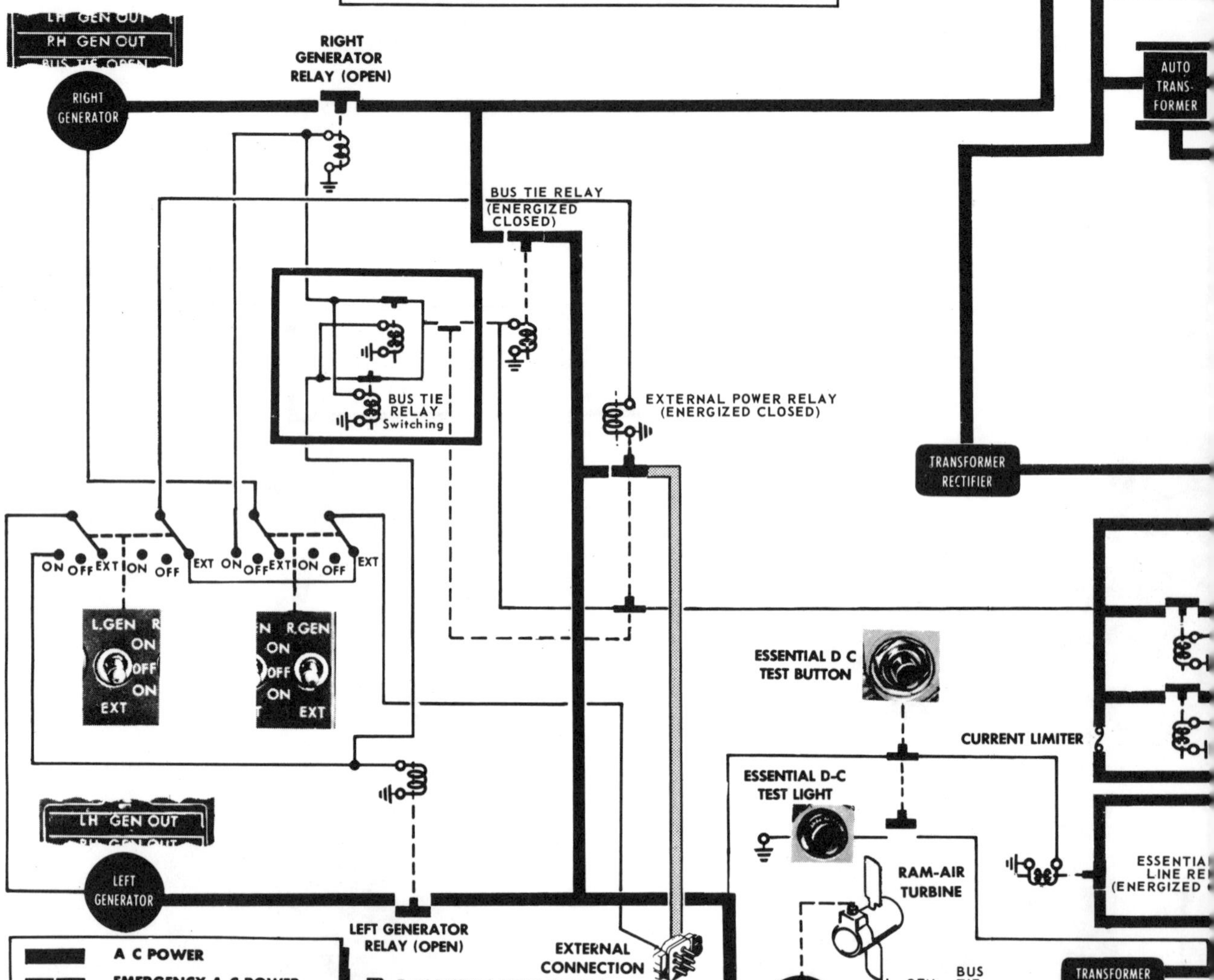

▬▬	A C POWER
▬ ▬	EMERGENCY A C POWER
▬▬	D-C POWER
▬ ▬	EMERGENCY D C POWER
▒▒	EXTERNAL A C POWER
——	RELAY CONTROL
- - -	MECHANICAL CONNECTION

* ARMAMENT BUS RELAY ENERGIZED BY THE ARMAMENT SAFETY SWITCH IN THE LANDING GEAR CONTROL HANDLE.

** MISSILE MASTER POWER RELAY ENERGIZED BY THE POWER SWITCH ON THE MISSILE CONTROL PANEL.

1. F-4J AIRCRAFT 153071z THRU 153799ab
2. F-4J AIRCRAFT 153780ac AND UP, AND ALL OTHERS AFTER AFC 388
3. F-4J AIRCRAFT 153076ac AND UP
4. F-4J AIRCRAFT 154786aq AND UP
5. F-4J AIRCRAFT 153071z THRU 153088aa, 153769ab, 153770ab, AND 153773ab
6. F-4J AIRCRAFT 153768ab, 153771ab, 153772ab, 153774ab AND UP
7. F-4J 155529ag AND UP, AND 153768ab THRU 155528aa AFTER AFC 388
8. F-4J 153071z THRU 155843ai BEFORE AFC 399
9. F-4J 155844aj AND UP, AND ALL OTHERS AFTER AFC 399
10. F-4J 155867ak AND UP, AND ALL OTHERS AFTER AFC 370
11. F-4J 153768ab THRU 155874ak BEFORE AFC 463
12. F-4J 155875al AND UP, AND 153768ab THRU 155874ak AFTER AFC 463
13. F-4J 157242an AND UP, AND ALL OTHERS AFTER AFC 486
14. F-4J 153071z THRU 155902am BEFORE AFC 486
15. F-4J 155529ag THRU 155569ag, AND 153768ab THRU 155528ag AFTER AFC 388
16. F-4J 155570ah AND UP BEFORE AFC 541
17. F-4J 155529ag THRU 155784ah, AND 153768ab THRU 155528ag AFTER AFC 388
18. F-4J 155785ai AND UP BEFORE AFC 574
19. F-4J 155529ag AND UP, AND 153768ab THRU 155528ag AFTER AFC 331
20. F-4J 155903ap, AND 157274ap AND UP
21. F-4J 157274ap AND UP, AND ALL OTHERS AFTER AFC 440
22. F-4J 155903ap, AND 157274ap AND UP, AND ALL OTHERS AFTER AFC 439
23. F-4J 158355at AND UP, AND 153768ab THRU 158354as AFTER AFC 506 AND BEFORE AFC 574
24. F-4J 153071z THRU 158354as.
25. F-4J 153768ab THRU 158354as BEFORE AFC 506 AND 153071z.
26. F-4J 157309ar AND UP; AND ALL OTHERS AFTER AFC 470.
27. F-4J 158355at AND UP.
28. F-4J 153071z, 153531ae, 153768ab THRU 157309ar
29. F-4J 158346as AND UP
30. F-4J AFTER AFC 518
31. F-4J AFTER AFC 524
32. F-4J 155529ag AND UP, AND 153768ab THRU 155528ag AFTER AFC 388 AND BEFORE AFC 524
33. F-4J 153768ab THRU 153838ad AND 153840ae THRU 154785af AFTER AFC 534
34. F-4J AFTER AFC 541
35. F-4J 155570ah AND UP
36. 153768ab THRU 158379au AFTER AFC 574
37. F-4J 15376… 158379au A… AFC 576 PT…

RIGHT MAIN 115/200V AC BUS

- AFTERBURNER IGNITION
- 27 AFT COOLING FAN
- 4 AILERON DROOP ACT
- AILERON FEEL TRIM
- AIRSPEED PITOT HEATER
- 1 ALTIMETER VIBRATOR
- 28 AN/AWW-1 BOMB FUZING
- ANTI-ICE
- 13 ANTI-SKID
- APN-141 ELECTRONIC ALT
- AUTO THROTTLE
- BELLMOUTH PITOT HEATER
- 22 BLEED AIR OVERHEAT DETECTION
- 29 BOMB FUZING
- BUDDY TANK HYD. PUMP & GUN POD
- 30 CAMERA AC (LB–30A STRIKE)
- CKPT HEAT & VENT
- ℄ GUN POD
- 24 EQUIP. COOLING
- FLIGHT REFUEL PROBE LT
- 27 FWD COOLING FAN
- LH RADAR MISSILE
- 27 LIQUID LOOP COOLING FAN
- NO. 4 FUEL TRANSFER PUMP (PWR)
- 20 NO. 7 TANK FUEL TRANSFER PUMP
- OXYGEN GAGE
- RIGHT FUEL BOOST PUMP (PWR)
- RH 14/28V AUTO XMFR
- RH RADAR MISSILE
- RH XMFR RECTIFIER
- SIDEWINDER

RIGHT MAIN 28V AC BUS

- ANTI-COLLISION LIGHT
- APPROACH LIGHT (PWR)
- FUS AND ANTI-COLLISION LT
- TAXI LIGHT
- UTILITY FLOOD LIGHT
- UTILITY HYD. PRESSURE INDICATOR
- WHITE FLOOD LT
- WING AND TAIL LIGHTS (BRT)

RIGHT MAIN 14V AC BUS

- AFT CKPT FLOODLIGHT (DIM)
- FWD CKPT RED CSL FLOOD LT (MED)
- FWD CKPT RED INST FLOOD LT (DIM)
- WING AND TAIL LIGHTS (DIM)

WARNING LIGHTS 28/14V AC BUS

- 35 AN/ALQ-91A ECM CONT LT
- 13 ANTI-SKID WARN LT
- ARR HOOK WARNING LT
- AUTOPILOT
- AUTO THROTTLE
- AUX AIR DOOR CAUTION LT
- 22 BLEED AIR OVERHEAT WARN LTS.
- BLC WARNING LIGHT
- CANOPY UNLOCK WARNING LT
- COR REFL WARNING LT
- DUCT TEMP WARNING LT
- 15 ECM LT SHOEHORN
- ELECT OV HT AND TURB OV SP
- 2 ENCODER IFF WARN LT
- EXT FUEL TRANSFER WARNING LT
- F LEVEL TK PRESS WARNING LT
- HYD PRESS WARNING LT
- IFR PROBE UNLOCK WARNING LT
- LANDING GEAR WARNING LT
- MAIN CONT WARNING LT
- MASTER CAUTION LT
- MISSILE LIGHTS
- OXYGEN LOW WARNING LT
- PRIM GYRO WARNING LT
- 14 SCOPE LTS
- SPEED BRAKE WARNING LT
- TAKE COMMAND LT
- WINDSHIELD TEMP WARNING LT
- WING LOCK WARNING LT

WARNING RELAY

RIGHT MAIN 28V DC BUS

- 12 AWG-10 & CONV WPNS
- AN/ARW-73
- 11 AN/AWW-1 BOMBING FUZING
- AN/AWW-1 BOMB FUZING
- ANGLE OF ATTACK PROBE HEATER
- APN-141 ELECTRONIC ALT
- APPROACH LT CONTROL
- ARMT BUS CONT
- ARR HOOK CONT
- AUTO THROTTLE
- 22 BLEED AIR OVERHEAT DETECTION
- BOMB AND RKT RELEASE PWR
- BUDDY TANK FUEL DUMP
- 30 CAMERA DC (LB–30A STRIKE)
- CKPT HEAT AND VENT
- CNI
- CTR STORE PWR
- 21 ENG BLEED AIR
- EQUIP COOLING
- EXTERIOR LT CONT
- FLAPS
- INTERNAL WING FUEL DUMP
- LANDING GEAR
- NO. 4 FUEL TRANSFER PUMP CONT
- 20 NO. 7 TANK FUEL TRANSFER CONT
- 25 NOSE WHEEL STEERING
- RADAR MISSILE FAIRING
- R FUEL BOOST PUMP CONT
- 6 R AND L ENGINE AUX AIR DOOR
- 5 R AND L ENGINE BELLMOUTH CONT
- RUDDER TRIM
- SIDEWINDER
- SPEED BRAKE
- 23 SP SW GUN VTAS SEL CONT
- UTILITY RECEPTACLE (DC)
- WARNING LIGHTS CONT

ARMAMENT 28V DC BUS

- 18 AN/ALE–29A FIRE SWS
- 36 AN/ALE–39 DOOR
- AN/AWW–1 BOMB FUZING
- BOMB AND RKT RELEASE CONT
- MISSILE FIRING
- 17 SHOE HORN
- 23 SP SW GUN FIRE POWER
- STORES RELEASE CONT

RADAR 28V DC BUS

- AWG-10
- MISSILE ARM

LEFT MAIN 28V DC BUS

- 7 A–A/IFF
- AILERON RUDDER INTERCONNECT
- 31 ALR–45
- 2 ALT ENCODER
- 13 ANTI-SKID
- 18 AN/ALE-29A DOOR SWS
- 23 AN/ALE–29A FIRE SYS
- 36 AN/ALE–39 sys
- 16 AN/ALQ–51/100 CONTROL
- 35 AN/ALQ–91A CONTROL
- 7 APN-154
- 13 ARI/ANTI-SKID
- 32 APR-25
- AUTOPILOT
- 7 AWG-10
- 7 DATA LINK
- 34 ECM CONTROL
- GVR–10 (VERTICAL FLIGHT REF)
- L BOOST PUMP NORMAL CONTROL
- NAV COMPUTER
- NO. 6 FUEL TRANSFER PUMP
- 37 OPTICAL SIGHT
- OUTBD JETT
- 25 RADAR SCOPE CAMERA
- RAIN REMOVAL
- 26 RECEIVING DECODING GRP AN/ARA–63
- 15 SHOE HORN
- 17 SHOE HORN
- STAB POSITION INDICATOR
- STATIC ACCEL
- 23 VTAS
- WING FOLD CONT

ESSENTIAL 28V DC BUS

- AJB–7
- BUDDY TANK HOSE JETTISON
- CADC
- CNI
- CTR STORE SAFETY
- 33 EMER AIL DROOP
- 10 EMER REFUEL PROBE
- 3 ENGINE OIL LEVEL
- EXT STORES EMER JETT
- EXTERNAL WING FUEL CONT
- FUEL LOW WARNING LT
- FUEL VALVE POWER
- FWD CKPT RED INST FLOOD LT (BRT)
- INTERCOM
- 19 JULIET–28
- LNDG GEAR AND FLAP POS. IND.
- L ENGINE FUEL SHUTOFF
- LH BOOST PUMP EMER CONT
- MAIN FUEL CONTROL
- MASTER CAUTION LIGHT RESET
- NOZZLE POSITION INDICATOR
- OUTBD JETT
- PNEU SYSTEM CONT
- RADAR MISSILE JETT
- REFUEL PROBE
- 17 RUDDER FEEL TRIM
- 18 RUDDER FEEL AND BELLMOUTH
- STAB FEEL TRIM
- TRIM CONT
- WING TRANSFER CONT

ESSENTIAL 28V AC BUS

- AFT CKPT FLOOD LT (BRT)
- AILERON POSITION INDICATOR
- AJB–7
- AWG–10 AND CADC
- CNI
- FUEL FLOW INDICATOR
- FUEL PRESSURE INDICATOR
- FWD CKPT RED CSL FLOOD LT (BRT)
- HYD PRESSURE INDICATORS (NO. 1 & 2)
- OIL PRESSURE INDICATORS
- PNEU PRESS INDICATORS
- RUDDER POSITION INDICATOR

ESSENTIAL 115/200V AC BUS

- AFT CKPT INST LT XMFR
- AJB–7
- CADC
- CNI
- ESS 28V AUTO XMFR
- FIRE DETECTOR
- FUEL QUANTITY
- 23 FUEL QTY ADAPTER
- FWD CKPT INST LT XMFR
- LH ESS XMFR RECTIFIER
- LH FUEL BOOSTER PUMP (PWR)
- L AND R ENGINE TURBINE OUTLET TEMP
- 8 L AND R MAIN IGNITION
- WINDSHIELD TEMP SENSOR

LEFT MAIN 28V AC BUS

- AILERON RUDDER INTERCONNECT
- 2 ALT ENCODER
- AUTOPILOT
- 37 OPTICAL SIGHT
- 13 SCOPE LTS

LEFT MAIN 115/200V AC BUS

- 7 A–A/IFF
- AFT CKPT EQUIP LT
- AILERON RUDDER INTERCONNECT
- 2 ALT ENCODER
- 16 AN/ALQ-51/100 RCVR XMTR
- 35 AN/ALQ-91A RCVR XMTR
- ANGLE-OF-ATTACK XMTR HEATER
- 7 APR-25
- AUTOPILOT
- 18 BLANKING UNIT
- CNI
- 7 DATA LINK
- 34 ECM
- FWD CSL LIGHT XMFR
- GVR-10 (VERTICAL FLIGHT REF)
- LH 28V AUTO XMFR
- L AND R RAMP CONTROL
- NAV COMPUTER
- NO. 6 FUEL TRANSFER PUMP
- NOSEWHEEL STEERING
- RADAR POWER
- 25 RADAR SCOPE CAMERA
- RADAR STANDBY
- 26 RECEIVING DECODING GRP AN/ARA–63
- SEAT ADJUST
- 15 SHOE HORN
- 17 SHOE HORN
- UTILITY RECEPTACLE (AC)
- 23 VTAS

IGNITION 115V AC BUS

- 9 L AND R MAIN IGNITION

Figure A-3

FDD–1–(275)D

Figure A-3

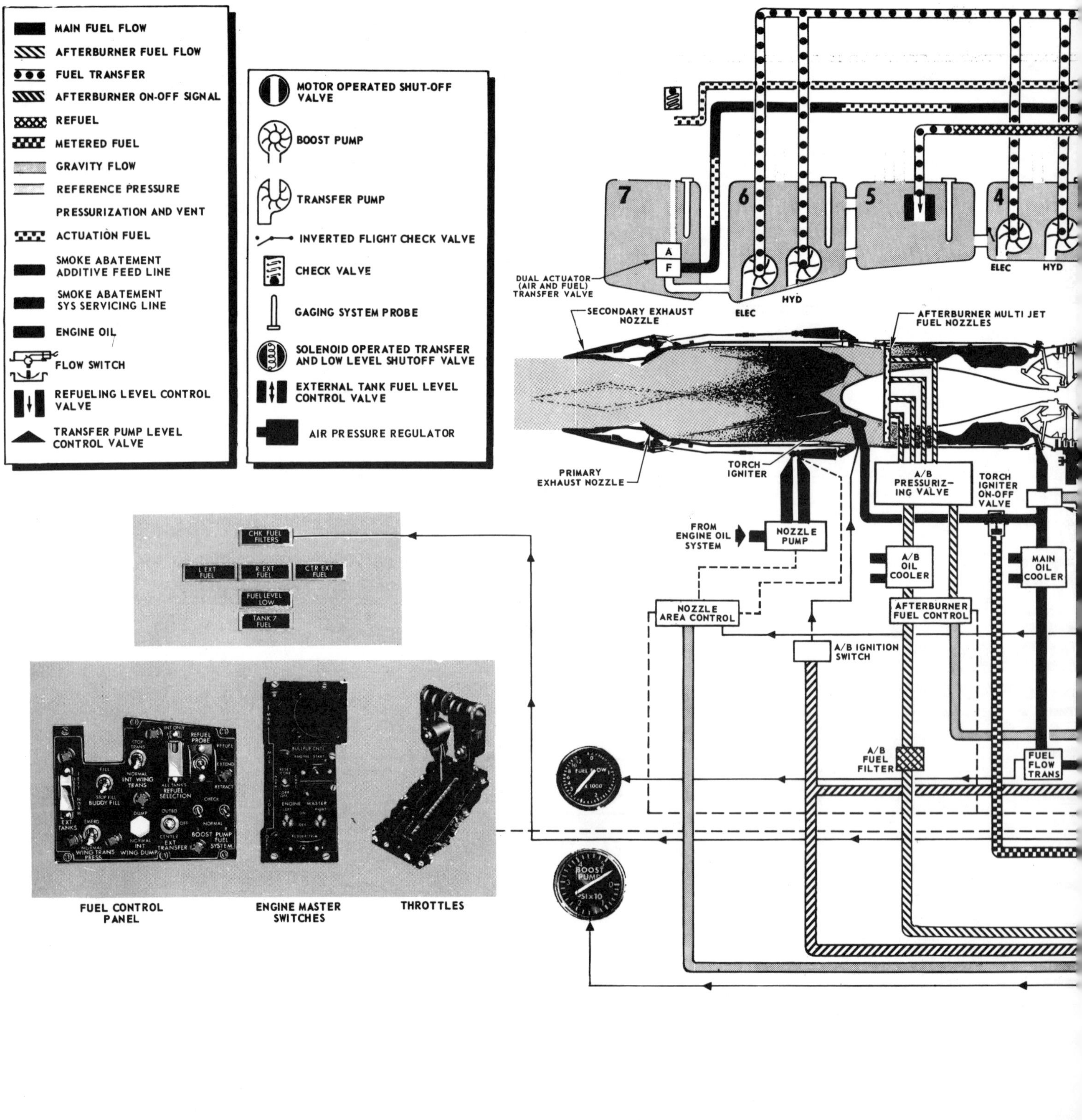
MAIN FUEL FLOW
AFTERBURNER FUEL FLOW
FUEL TRANSFER
AFTERBURNER ON-OFF SIGNAL
REFUEL
METERED FUEL
GRAVITY FLOW
REFERENCE PRESSURE
PRESSURIZATION AND VENT
ACTUATION FUEL
SMOKE ABATEMENT ADDITIVE FEED LINE
SMOKE ABATEMENT SYS SERVICING LINE
ENGINE OIL
FLOW SWITCH
REFUELING LEVEL CONTROL VALVE
TRANSFER PUMP LEVEL CONTROL VALVE
MOTOR OPERATED SHUT-OFF VALVE
BOOST PUMP
TRANSFER PUMP
INVERTED FLIGHT CHECK VALVE
CHECK VALVE
GAGING SYSTEM PROBE
SOLENOID OPERATED TRANSFER AND LOW LEVEL SHUTOFF VALVE
EXTERNAL TANK FUEL LEVEL CONTROL VALVE
AIR PRESSURE REGULATOR
7
6
5
4
A
F
DUAL ACTUATOR (AIR AND FUEL) TRANSFER VALVE
ELEC
HYD
ELEC
HYD
SECONDARY EXHAUST NOZZLE
AFTERBURNER MULTI JET FUEL NOZZLES
PRIMARY EXHAUST NOZZLE
TORCH IGNITER
A/B PRESSURIZ-ING VALVE
TORCH IGNITER ON-OFF VALVE
FROM ENGINE OIL SYSTEM
NOZZLE PUMP
A/B OIL COOLER
MAIN OIL COOLER
CHK FUEL FILTERS
L EXT FUEL
R EXT FUEL
CTR EXT FUEL
FUEL LEVEL LOW
TANK 7 FUEL
NOZZLE AREA CONTROL
AFTERBURNER FUEL CONTROL
A/B IGNITION SWITCH
A/B FUEL FILTER
FUEL FLOW TRANS
FUEL FLOW
X 1000
BOOST PUMP
PSI x 10
FUEL CONTROL PANEL
ENGINE MASTER SWITCHES
THROTTLES

AIRPLANE AND ENGINE FUEL SYSTEM

AIRCRAFT 153071z THRU 155902am AND 157242an THRU 157273ao

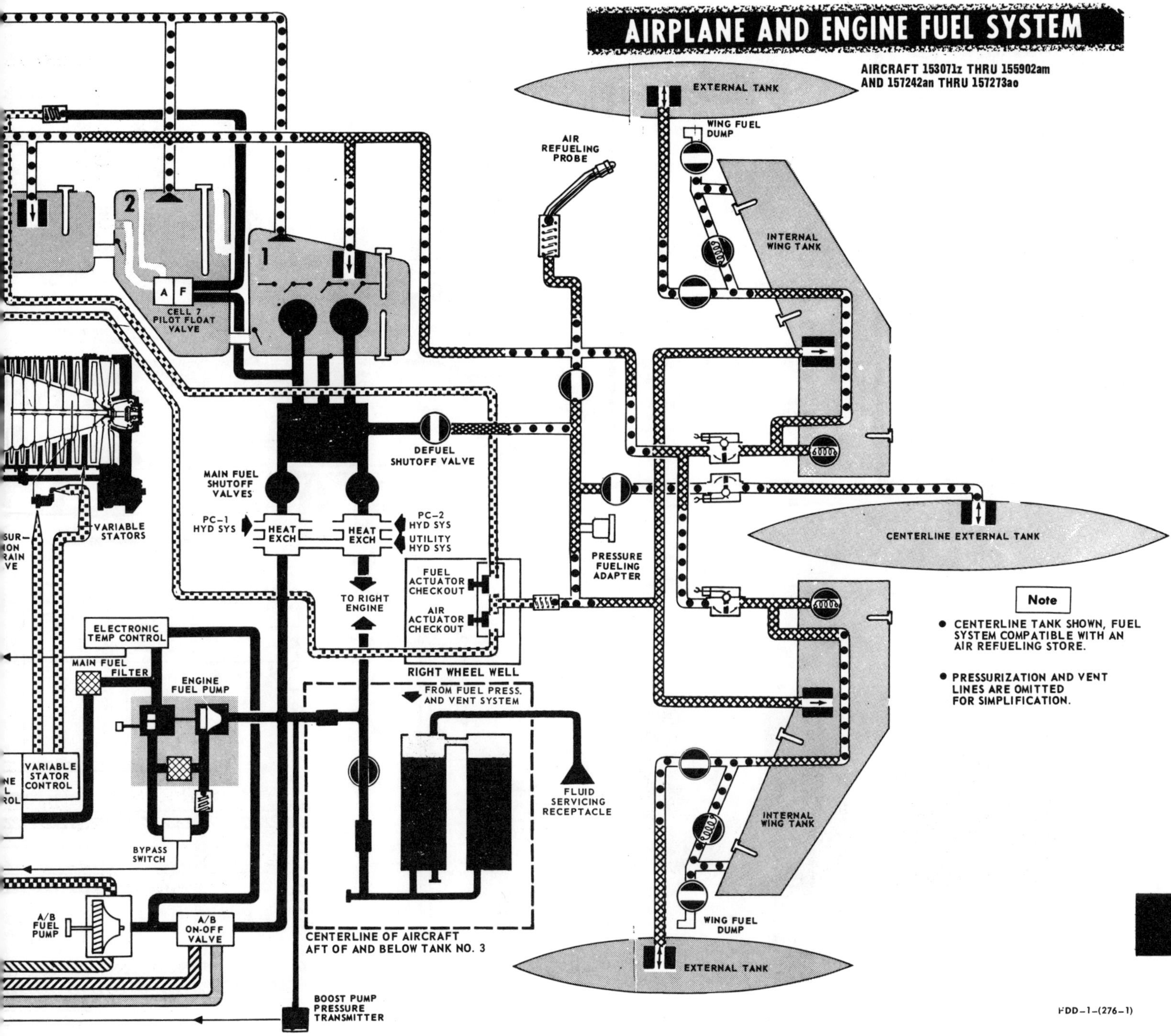

Note

- CENTERLINE TANK SHOWN, FUEL SYSTEM COMPATIBLE WITH AN AIR REFUELING STORE.
- PRESSURIZATION AND VENT LINES ARE OMITTED FOR SIMPLIFICATION.

FDD-1-(276-1)

Figure A-4 (Sheet 1 of 2)

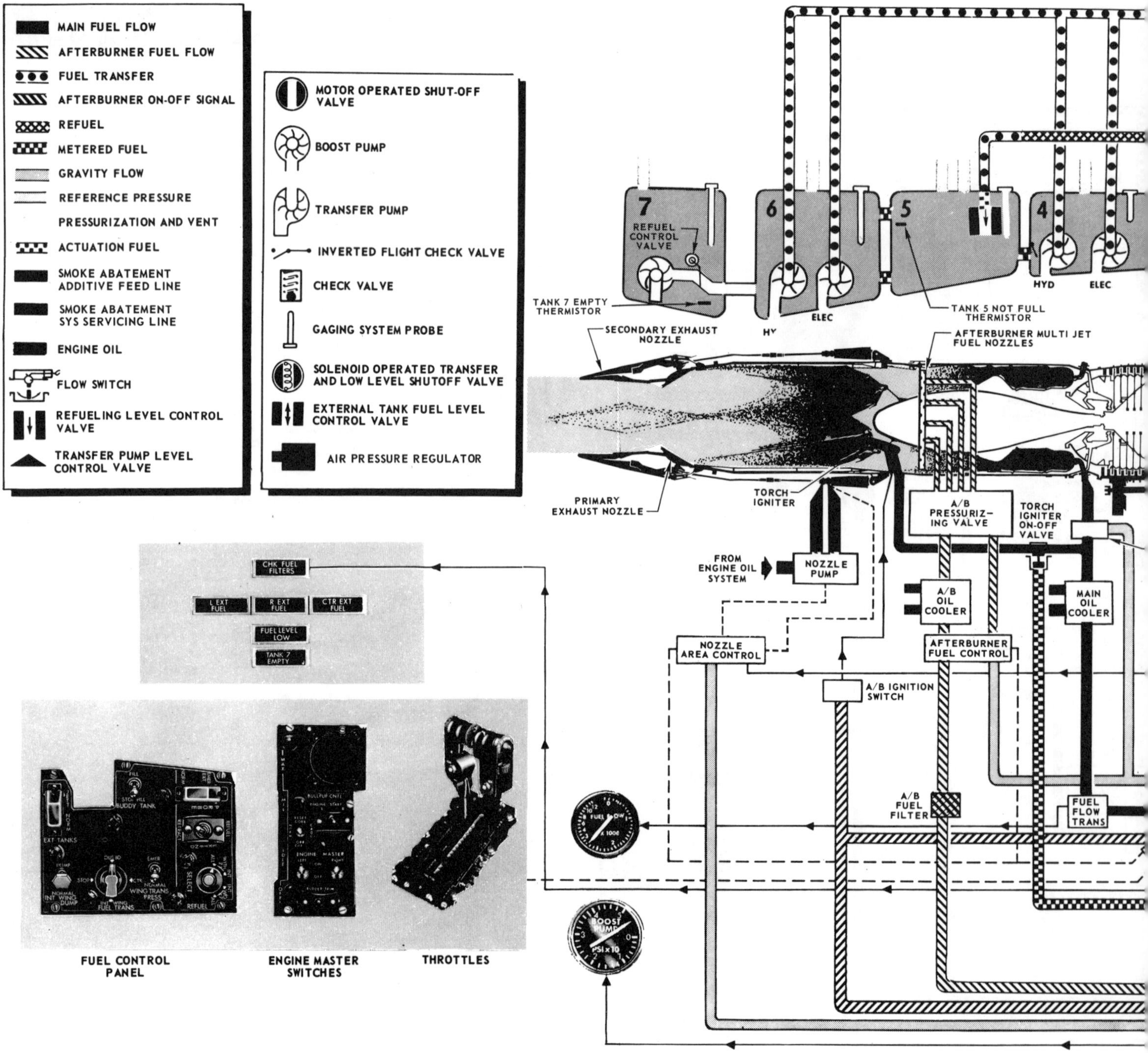
MAIN FUEL FLOW
AFTERBURNER FUEL FLOW
FUEL TRANSFER
AFTERBURNER ON-OFF SIGNAL
REFUEL
METERED FUEL
GRAVITY FLOW
REFERENCE PRESSURE
PRESSURIZATION AND VENT
ACTUATION FUEL
SMOKE ABATEMENT ADDITIVE FEED LINE
SMOKE ABATEMENT SYS SERVICING LINE
ENGINE OIL
FLOW SWITCH
REFUELING LEVEL CONTROL VALVE
TRANSFER PUMP LEVEL CONTROL VALVE
MOTOR OPERATED SHUT-OFF VALVE
BOOST PUMP
TRANSFER PUMP
INVERTED FLIGHT CHECK VALVE
CHECK VALVE
GAGING SYSTEM PROBE
SOLENOID OPERATED TRANSFER AND LOW LEVEL SHUTOFF VALVE
EXTERNAL TANK FUEL LEVEL CONTROL VALVE
AIR PRESSURE REGULATOR
7
REFUEL CONTROL VALVE
6
5
4
TANK 7 EMPTY THERMISTOR
ELEC
HYD
ELEC
TANK 5 NOT FULL THERMISTOR
SECONDARY EXHAUST NOZZLE
AFTERBURNER MULTI JET FUEL NOZZLES
PRIMARY EXHAUST NOZZLE
TORCH IGNITER
A/B PRESSURIZING VALVE
TORCH IGNITER ON-OFF VALVE
FROM ENGINE OIL SYSTEM
NOZZLE PUMP
A/B OIL COOLER
MAIN OIL COOLER
NOZZLE AREA CONTROL
AFTERBURNER FUEL CONTROL
A/B IGNITION SWITCH
A/B FUEL FILTER
FUEL FLOW TRANS
CHK FUEL FILTERS
L EXT FUEL
R EXT FUEL
CTR EXT FUEL
FUEL LEVEL LOW
TANK 7 EMPTY
FUEL FLOW x 1000
BOOST PUMP PSI x 10
FUEL CONTROL PANEL
ENGINE MASTER SWITCHES
THROTTLES

AIRPLANE AND ENGINE FUEL SYSTEM

AIRCRAFT 155903ap AND
AIRCRAFT 157274ap AND UP

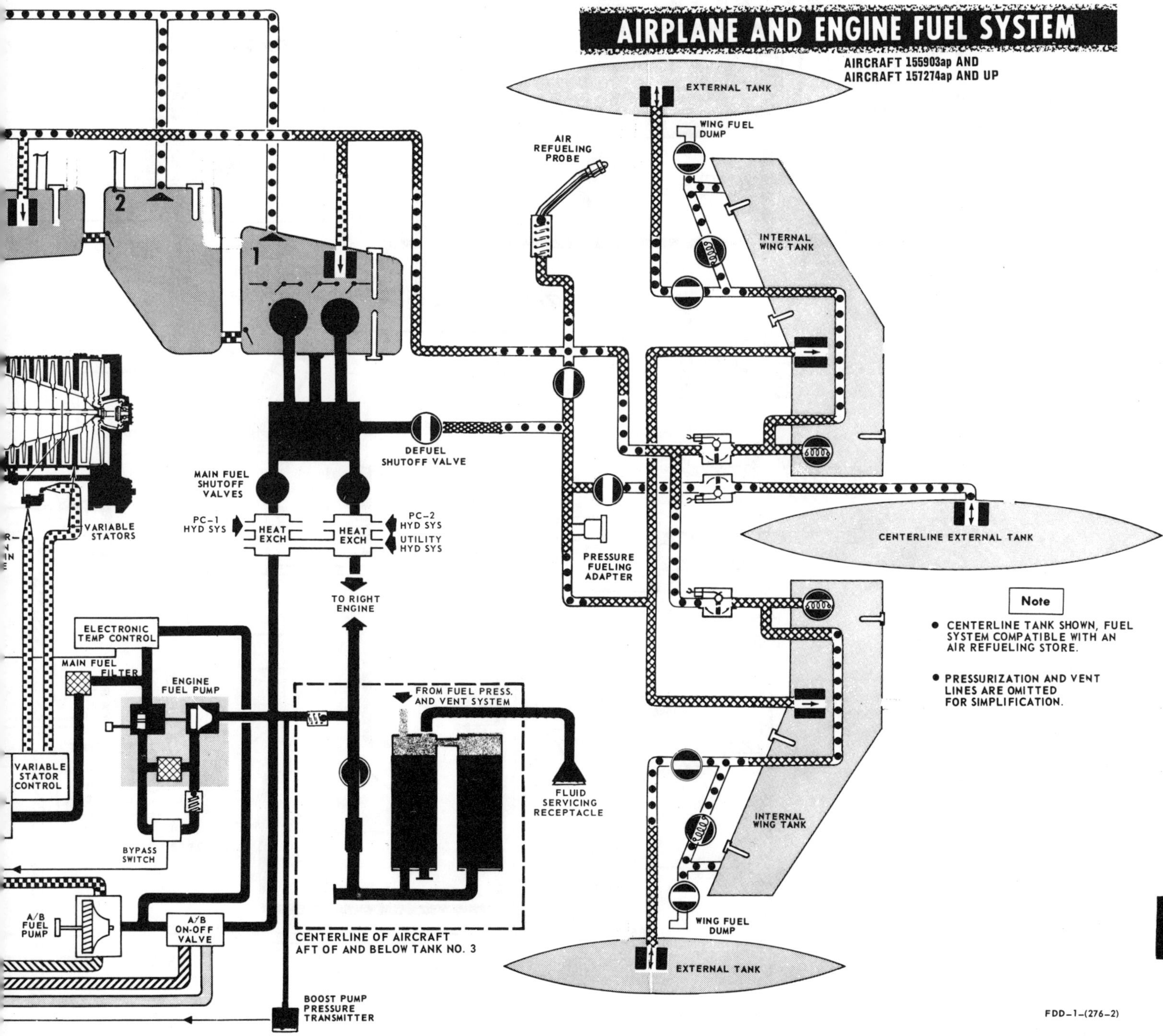

Note

- CENTERLINE TANK SHOWN, FUEL SYSTEM COMPATIBLE WITH AN AIR REFUELING STORE.
- PRESSURIZATION AND VENT LINES ARE OMITTED FOR SIMPLIFICATION.

FDD-1-(276-2)

Figure A-4 (Sheet 2 of 2)

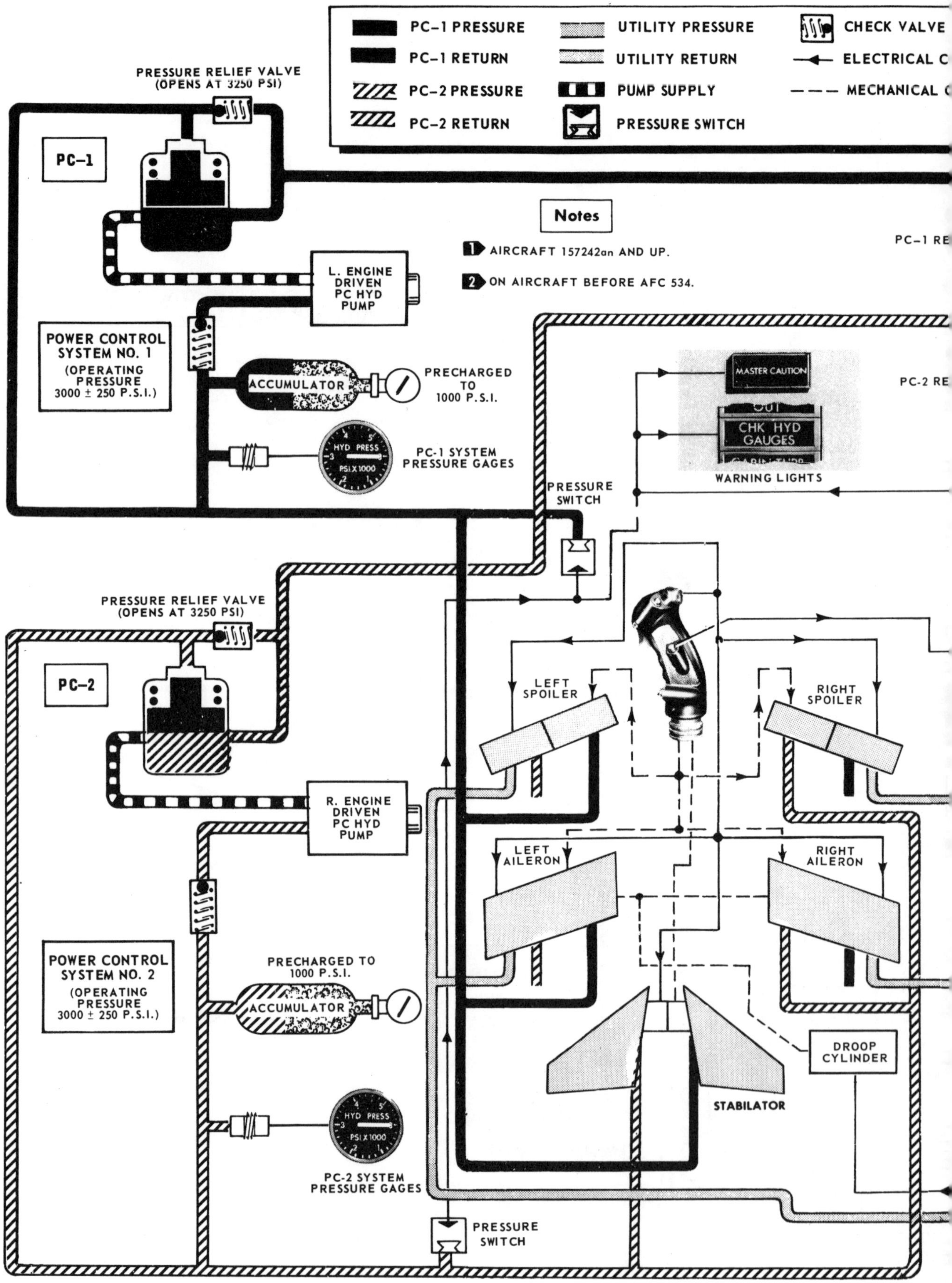

PC–1 PRESSURE
PC–1 RETURN
PC–2 PRESSURE
PC–2 RETURN
UTILITY PRESSURE
UTILITY RETURN
PUMP SUPPLY
PRESSURE SWITCH
CHECK VALVE
ELECTRICAL C
MECHANICAL C
PRESSURE RELIEF VALVE (OPENS AT 3250 PSI)
PC–1
Notes
1 AIRCRAFT 157242an AND UP.
2 ON AIRCRAFT BEFORE AFC 534.
PC–1 RE
L. ENGINE DRIVEN PC HYD PUMP
POWER CONTROL SYSTEM NO. 1
(OPERATING PRESSURE 3000 ± 250 P.S.I.)
ACCUMULATOR
PRECHARGED TO 1000 P.S.I.
PC-1 SYSTEM PRESSURE GAGES
HYD PRESS
PSI X 1000
MASTER CAUTION
CHK HYD GAUGES
WARNING LIGHTS
PC-2 RE
PRESSURE SWITCH
PRESSURE RELIEF VALVE (OPENS AT 3250 PSI)
PC–2
LEFT SPOILER
RIGHT SPOILER
R. ENGINE DRIVEN PC HYD PUMP
LEFT AILERON
RIGHT AILERON
POWER CONTROL SYSTEM NO. 2
(OPERATING PRESSURE 3000 ± 250 P.S.I.)
PRECHARGED TO 1000 P.S.I.
ACCUMULATOR
DROOP CYLINDER
STABILATOR
PC-2 SYSTEM PRESSURE GAGES
PRESSURE SWITCH

HYDRAULIC SYSTEMS

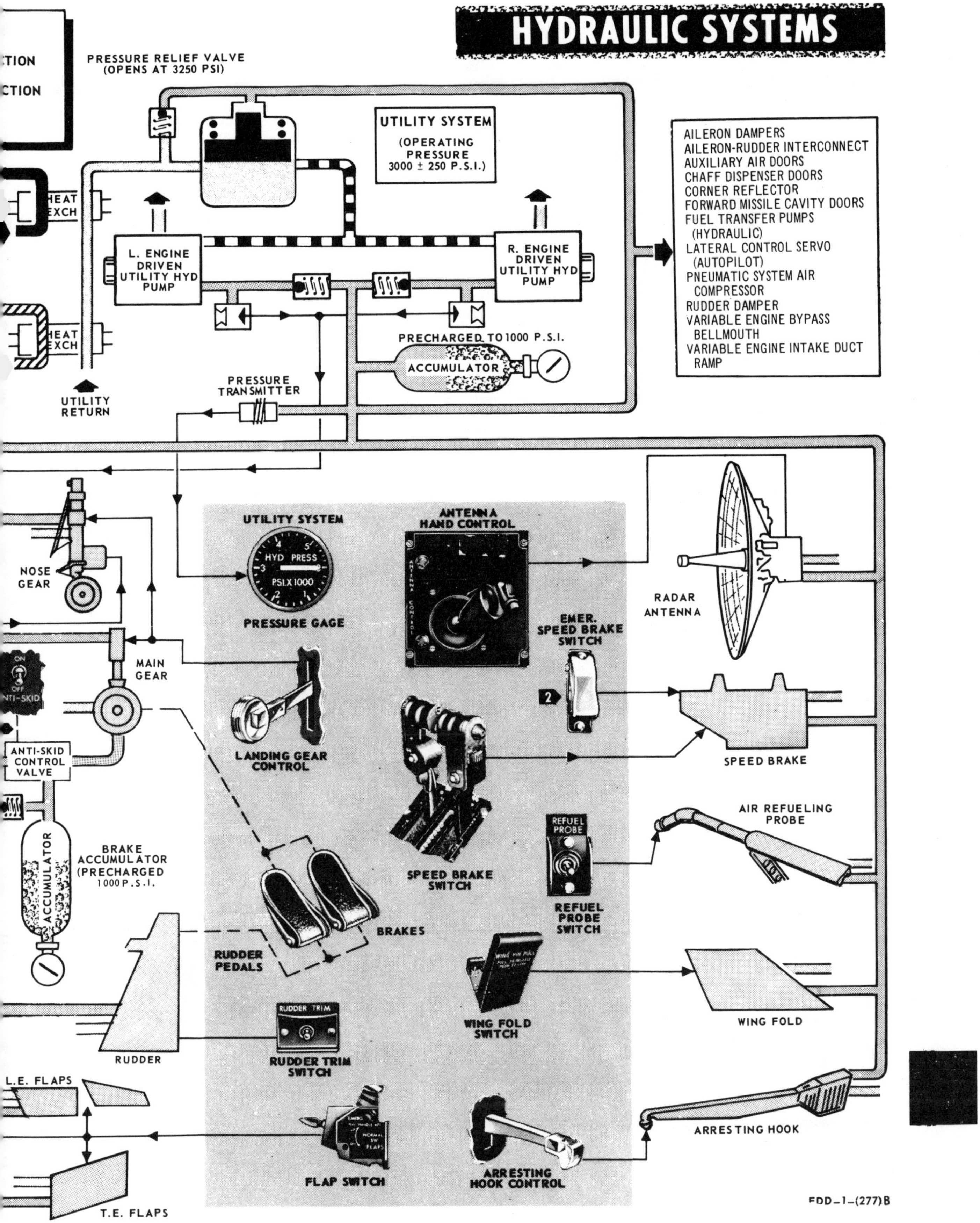

Figure A-5

Figure A-5

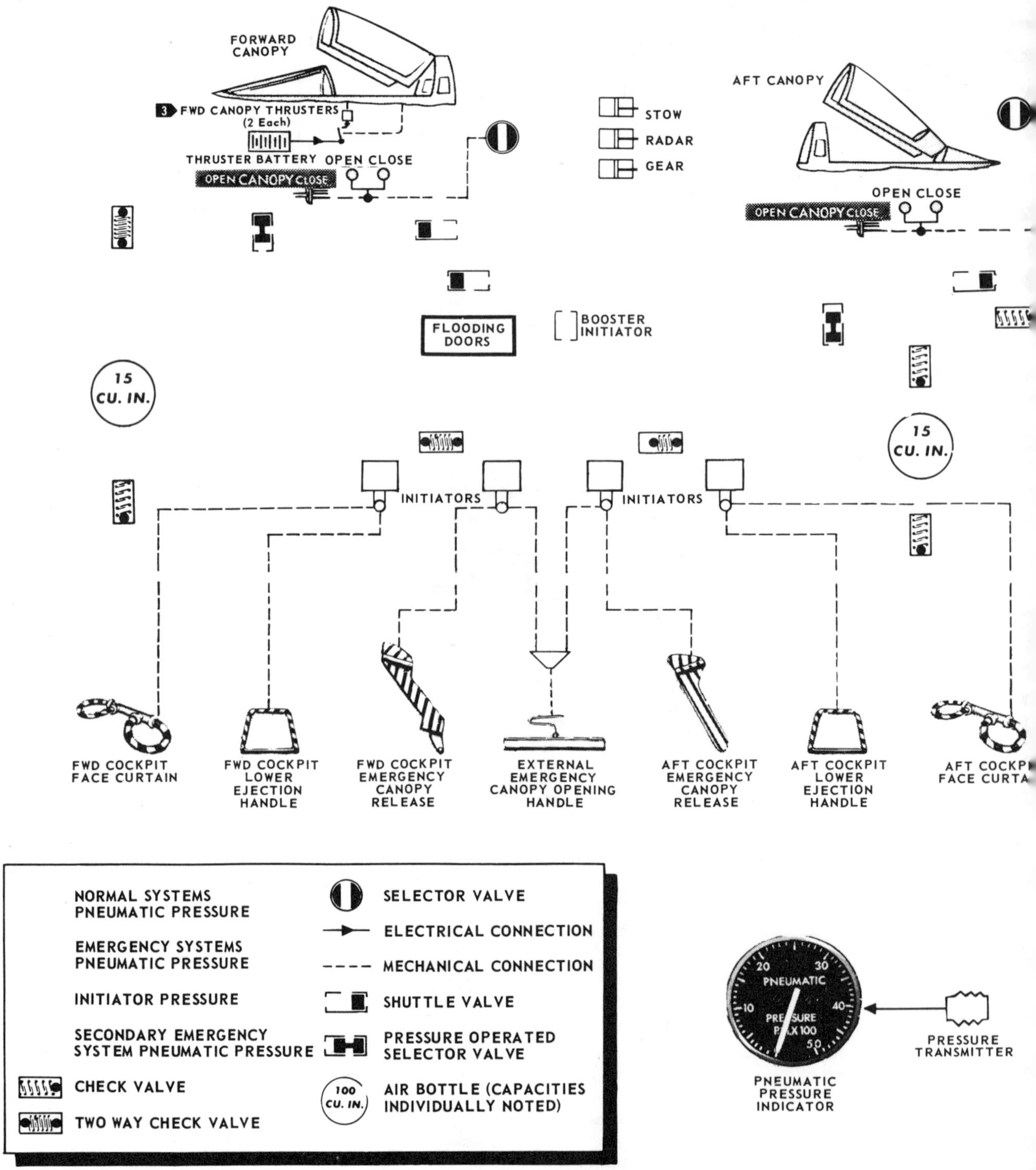

FORWARD CANOPY
3 FWD CANOPY THRUSTERS (2 Each)
THRUSTER BATTERY
OPEN CLOSE
OPEN CANOPY CLOSE
STOW
RADAR
GEAR
AFT CANOPY
OPEN CLOSE
OPEN CANOPY CLOSE
FLOODING DOORS
BOOSTER INITIATOR
15 CU. IN.
15 CU. IN.
INITIATORS
INITIATORS
FWD COCKPIT FACE CURTAIN
FWD COCKPIT LOWER EJECTION HANDLE
FWD COCKPIT EMERGENCY CANOPY RELEASE
EXTERNAL EMERGENCY CANOPY OPENING HANDLE
AFT COCKPIT EMERGENCY CANOPY RELEASE
AFT COCKPIT LOWER EJECTION HANDLE
AFT COCKPIT FACE CURTAIN
NORMAL SYSTEMS PNEUMATIC PRESSURE
EMERGENCY SYSTEMS PNEUMATIC PRESSURE
INITIATOR PRESSURE
SECONDARY EMERGENCY SYSTEM PNEUMATIC PRESSURE
CHECK VALVE
TWO WAY CHECK VALVE
SELECTOR VALVE
ELECTRICAL CONNECTION
MECHANICAL CONNECTION
SHUTTLE VALVE
PRESSURE OPERATED SELECTOR VALVE
100 CU. IN.
AIR BOTTLE (CAPACITIES INDIVIDUALLY NOTED)
10 20 30 40 50
PNEUMATIC PRESSURE PSI X 100
PRESSURE TRANSMITTER
PNEUMATIC PRESSURE INDICATOR

PNEUMATIC SYSTEM

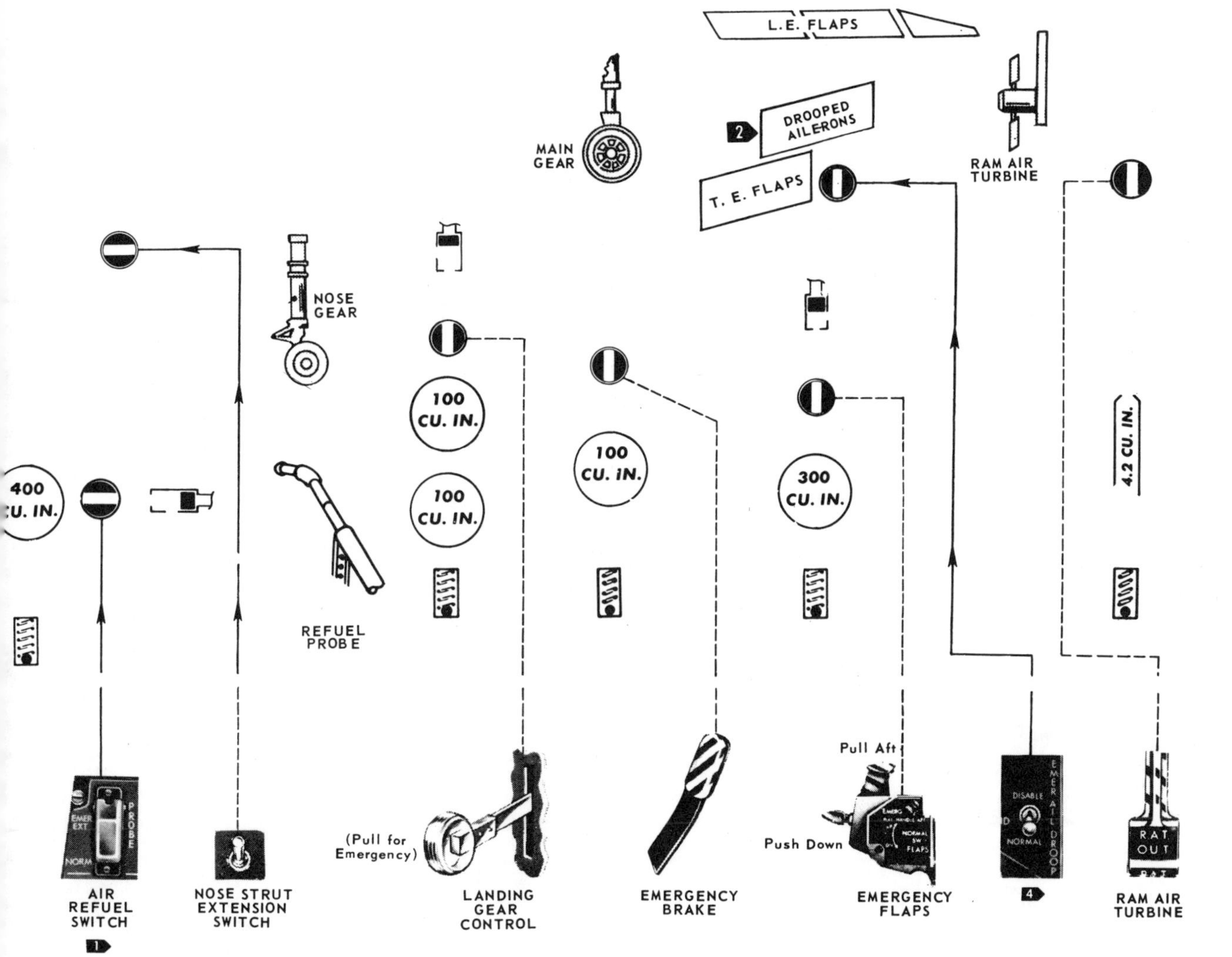

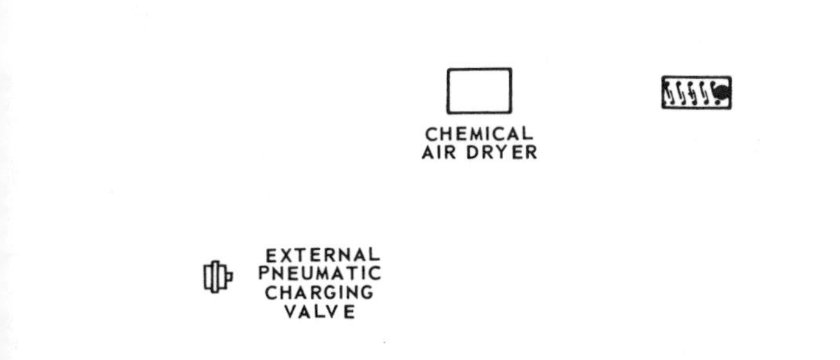

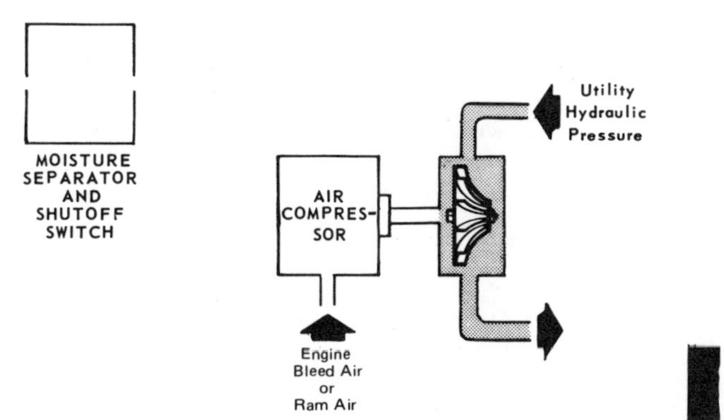

1 On aircraft 155867ak and up; and all others after AFC 370.

2 On aircraft before AFC 534, the ailerons will not droop when the flaps are lowered by the emergency system.

3 On aircraft 157258an and up; and others after AFC 497.

4 On aircraft through 154785af after AFC 534 (pneumatic emergency aileron droop).

FDD-1-(278)B

Figure A-6

Figure A-6

ALPHABETICAL LISTING

	C/B Panel	Zone
A/B IGN. L. ENG	1	4A
A/B IGN. R. ENG	1	3A
AFT CKPT. EQUIP. LT.	2	3L
AFT CKPT. FLOOD LT. BRT	2	3N
AFT CKPT. FLOOD LT. DIM	2	3M
AFT CKPT. INST. LT. XMFR	2	3P
AIL. FEEL TRIM	RUP	–
AILERON DROOP ACT [4]	1	12J
AILERON POS. IND.	1	5F
AILERON RUDDER INT.	1	7J
AILERON RUDDER INT.	1	8J
AIRSPEED PITOT HEATER	1	16J
AJB-7	1	14A
AJB-7	1	14B
AJB-7	1	14C
AJB-7	1	14D
AJB-7	1	14F
AJB-7	1	15B
ALT. ENCODER [2]	1	8F
ALT. ENCODER [2]	1	7G
ALT. ENCODER [2]	1	12H
ALT. VIBRATOR [1]	1	7G
AN/ARW-73	2	5E
AN/AWW-1 BOMB FUZING	2	7A
AN/AWW-1 BOMB FUZING	2	7B
AN/AWW-1 BOMB FUZING	2	7C
AN/AWW-1 BOMB FUZING	2	7D
AN/AWW-1 BOMB FUZING	2	4D
ANTI-COLLISION LT.	2	9F
ANTI-ICE	1	15E
A.O.A. PROBE HEATER	1	8G
A.O.A. XMTR. HEATER	1	13H
APN-141	1	13E
APN-141	1	14E
APPROACH LT. CONT.	2	8H
APPROACH LT. POWER	2	8G
ARI	LUP	–
ARMT. BUS CONT.	1	7E
ARRESTING HOOK CONT.	2	7H
ARRESTING HOOK WARN. LT.	2	7J
AUTOPILOT	1	9G
AUTOPILOT	1	9H
AUTOPILOT	1	9J
AUTOPILOT	1	6H
AUTOPILOT	1	7H
AUTOPILOT	1	8H
AUTO. THROTTLE	2	4K
AUTO. THROTTLE	2	5K
AUTO. THROTTLE	2	3J
AUX. AIR DOOR CAUTION LT.	2	5H
AUX AIR DOOR, L. ENG.	1	4D
AUX AIR DOOR, R. ENG.	1	3D
AWG-10	2	2M
AWG-10	2	2P
AWG-10 AND CADC	1	13F
BELLMOUTH PITOT HEATER	1	15J
BLC WARN. LT.	1	4J
BLEED AIR OFF VALVE [8]	1	15C
BLEED AIR OVERHEAT DETECTION [9]	1	16F
BLEED AIR OVERHEAT DETECTION [9]	1	16G
BLEED AIR OVERHEAT WARN LT [9]	1	16E
BOMB AND RKT. REL. CONT.	2	4E
BOMB AND RKT. REL. PWR.	2	5D
BUDDY TANK FUEL DUMP	2	6E
BUDDY TANK HOSE JETT.	2	6D
BUDDY TANK HYD. PUMP	2	6C
CADC	1	13A
CADC	1	13B
CADC	1	13C
CADC	1	13D
CANOPY UNLOCK WARN. LT.	2	8J
CKPT. HEAT AND VENT	2	9L
CKPT. HEAT AND VENT	2	9M
℄ GUN POD	1	16H
CNI	1	12A
CNI	1	12B
CNI	1	12C
CNI	1	12D
CNI	1	12E
CNI	1	12F
CNI	1	12G
CORNER REFL. WARN LT.	2	10J
CTR. STORE PWR.	2	8B
CTR. STORE SAFETY	2	9B
DUCT TEMP. WARN. LT.	2	6J
ELECT. OVHT. AND TURBINE OVSP.	2	8M
EMER. REFUEL PROBE [7]	1	15A
ENCODER IFF WARN. LT. [2]	1	4F
ENGINE OIL LEVEL [3]	1	13G
EQUIP. COOLING	2	8L
EQUIP. COOLING	2	8N
ESS. 28V AUTO. XMFR.	2	6N
EXT. FUEL TRANS. WARN LT.	1	3F
EXT. LT. CONT.	2	8C
EXT. STORES EMER. JETT.	1	3E
EXT. WING FUEL CONT.	1	3H
FIRE DETECTOR	1	5A
FLAPS	RUP	–
FLT. REFUEL PROBE LT.	2	4H
FUEL FLOW IND.	1	9F
FUEL LEVEL TK. PRESS. WARN. LT.	1	2F
FUEL LOW WARN. LT.	1	2G
FUEL PRESS. IND.	1	10F
FUEL QTY.	1	1H
FUEL VALVE POWER L.H.	1	4G
FUS. AND ANTI-COLL LT.	2	9G
FWD. CKPT. INST. LT. XMFR.	2	4P
FWD. CKPT. RED CSL FLOOD LT.	2	4N
FWD. CKPT. RED CSL FLOOD LT.	2	5M
FWD. CKPT. RED INST. FLOOD LT.	2	5N
FWD. CKPT. RED INST. FLOOD LT.	2	4M
FWD. CSL LT. XMFR.	2	5P
GVR-10	1	16A
GVR-10	1	16B
GVR-10	1	16C
GVR-10	1	16D
HYD. PRESS. IND. NO. 1	2	6F
HYD. PRESS. IND. NO. 2	2	6G
HYD. PRESS. WARN. LT.	2	6H
IFR PROBE UNLOCK WARN LT.	1	3J
IGN. L. MAIN	1	3B
IGN. R. MAIN	1	2B
INTERCOM	1	14G
INT. WING FUEL DUMP	1	2H
LG	RUP	–
LDG. GEAR AND FLAP POS. IND.	1	5H
LDG. GEAR WARN LT.	1	4B
L. ENG. BELLMOUTH CONT. [5]	1	4D
L. ENG. FUEL SHUTOFF	1	4E
L.H. BOOST PUMP EMER. CONT.	1	11A
L.H. BOOST PUMP NORM CONT.	1	11E
L.H. ESS. XMFR. RECT.	1	17D
L.H. ESS. XMFR. RECT.	1	17E
L.H. ESS. XMFR. RECT.	1	17F
L.H. FUEL BOOST PUMP	1	11B
L.H. FUEL BOOST PUMP	1	11C
L.H. FUEL BOOST PUMP	1	11D
L.H. RADAR MISSILE	1	6A
L.H. RADAR MISSILE	1	6B
L.H. RADAR MISSILE	1	6C
L.H. 28V AUTO. XMFR.	2	6M
MAIN CONT. WARN. LT.	2	4J
MAIN FUEL CONT.	1	3G
MASTER CAUTION LT.	1	2J
MASTER CAUTION LT. RESET	1	1J
MISSILE ARM	1	6D
MISSILE FIRING	1	6E
MISSILE LTS.	1	5B
NAV COMPUTER	1	8E
NAV COMPUTER	1	9E
NOSE WHEEL STEER.	2	6L
NOSE WHEEL STEER.	2	7L
NOZZLE POS. IND.	1	4C
NO. 4 FUEL TRANS. PUMP	1	9A
NO. 4 FUEL TRANS. PUMP	1	9B
NO. 4 FUEL TRANS. PUMP	1	9C
NO. 4 FUEL TRANS. PUMP	1	9D
NO. 6 FUEL TRANS. PUMP	1	7F
NO. 6 FUEL TRANS. PUMP	1	11G
NO. 6 FUEL TRANS. PUMP	1	11H
NO. 6 FUEL TRANS. PUMP	1	11J
OIL PRESS. IND.	1	11F
OUTBD. JETT.	1	2D
OUTBD. JETT.	1	2E
OXYGEN GAGE	1	6J
OXYGEN LOW WARN. LT.	1	5J
PNEU. PRESS. IND.	2	7F
PNEU. SYS. CONT.	2	7E
PRIM. GYRO WARN. LT.	2	10M
RADAR MISSILE FAIRING	1	5D
RADAR MISSILE JETT.	1	5C
RADAR POWER	2	1P
RADAR SCOPE CAMERA	2	5F
RADAR SCOPE CAMERA	2	4F
RADAR STANDBY	2	1N
RAIN REMOVAL	2	7P
RAMP CONT. L. ENG.	1	14J
RAMP CONT. R. ENG.	1	13J
REFUEL PROBE	1	5G
R. ENG. BELLMOUTH CONT. [5]	1	3D
R.H. FUEL BOOST PUMP CONT.	1	8D
R.H. FUEL BOOST PUMP	1	8A
R.H. FUEL BOOST PUMP	1	8B
R.H. FUEL BOOST PUMP	1	8C
R.H. RADAR MISSILE	2	6A
R.H. RADAR MISSILE	2	6B
R.H. RADAR MISSILE	2	5C
R.H. XMFR. RECT.	1	17G
R.H. XMFR. RECT.	1	17H
R.H. XMFR. RECT.	1	17J
RUDDER FEEL TRIM	1	15G
RUDDER POS. IND.	1	6F
R.H. 14/28V AUTO. XMFR.	2	6P
RUD. TRIM	RUP	–
SCOPE LTS.	1	15H
SEAT ADJUST	2	9N
SIDEWINDER	1	10A
SIDEWINDER	1	10B
SIDEWINDER	1	10C
SIDEWINDER	1	10D
SIDEWINDER	1	10E
SIDEWINDER	1	10G
SIDEWINDER	1	10H
SIDEWINDER	1	10J
SPEED BRAKE	RUP	–
SPEED BRAKE WARN. LT.	2	5J
STAB. FEEL TRIM	RUP	–
STAB. POS. IND.	1	6G
STATIC ACCEL.	2	8P
STORES REL. CONT.	1	5E
TAKE COMMAND LT.	1	14H
TAXI LT.	2	8F
TRIM CONT.	RUP	–
TURBINE OUTLET TEMP. L. ENG.	1	2C
TURBINE OUTLET TEMP. R. ENG.	1	3C
UTILITY FLOOD LT.	2	5L
UTILITY HYD. PRESS. IND.	2	7G
UTILITY RECEP.	2	9C
UTILITY RECEP. AC	2	9D
WARNING LT. CONT.	2	3K
WHITE FLOOD LT.	2	4L
WING AND TAIL LT. DIM	2	8E
WING AND TAIL LT. BRT.	2	8D
WING FOLD CONT.	2	9H
WING LOCK WARNING LT.	2	9J
WING TRANS. CONT.	1	4H
WSHLD. TEMP. SENSOR	2	7N
WSHLD. TEMP. WARN. LT.	2	7M

ZONE LISTING

Zone	C/B Panel	
A3	1	A/B IGN R ENG
A4	1	A/B IGN L ENG
A5	1	FIRE DETECTOR
A6	1	LH RADAR MISSILE
A8	1	RH FUEL BOOSTER PUMP
A9	1	NO 4 FUEL TRANS PUMP
A10	1	SIDEWINDER
A11	1	LH BOOST PUMP EMER CONT
A12	1	CNI
A13	1	CADC
A14	1	AJB-7
A15	1	EMER REFUEL PROBE [7]
A16	1	GVR-10
B2	1	IGN R MAIN
B3	1	IGN L MAIN
B4	1	LDG GEAR WARN LT
B5	1	MISSILE LTS
B6	1	LH RADAR MISSILE
B8	1	RH FUEL BOOST PUMP
B9	1	NO 4 FUEL TRANS PUMP
B10	1	SIDEWINDER
B11	1	LH FUEL BOOST PUMP
B12	1	CNI
B13	1	CADC
B14	1	AJB-7
B15	1	AJB-7
B16	1	GVR-10
C2	1	TURBINE OUTLET TEMP L ENG
C3	1	TURBINE OUTLET TEMP R ENG
C4	1	NOZZLE POSITION IND
C5	1	RADAR MISSILE JETT
C6	1	LH RADAR MISSILE
C8	1	RH FUEL BOOST PUMP
C9	1	NO 4 FUEL TRANS PUMP
C10	1	SIDEWINDER
C11	1	LH FUEL BOOST PUMP
C12	1	CNI
C13	1	CADC
C14	1	AJB-7
C15	1	BLEED AIR OFF VALVE [8]
C16	1	GVR-10
D2	1	OUTBD JETT
D3	1	R ENG AUX AIR DOOR [6]
D3	1	R ENG BELLMOUTH CONT [5]
D4	1	L ENG AUX AIR DOOR [6]
D4	1	L ENG BELLMOUTH CONT [5]
D5	1	RADAR MISSILE FAIRING
D6	1	MISSILE ARM
D8	1	RH FUEL BOOST PUMP CONT
D9	1	NO 4 FUEL TRANS PUMP
D10	1	SIDEWINDER
D11	1	LH FUEL BOOST PUMP
D12	1	CNI
D13	1	CADC
D14	1	AJB-7
D16	1	GVR-10
D17	1	LH ESS XMFR RECT
E2	1	OUTBD JETT
E3	1	EXT STORES EMER JETT
E4	1	L ENG FUEL SHUTOFF
E5	1	STORES RELEASE CONT
E6	1	MISSILE FIRING
E7	1	ARMT BUS CONT
E8	1	NAV CMPTR
E9	1	NAV CMPTR
E10	1	SIDEWINDER
E11	1	LH BOOST PUMP NORM CONT
E12	1	CNI
E13	1	APN 141
E14	1	APN 141
E15	1	ANTI-ICE
E16	1	BLEED AIR OVERHEAT WARN LT [9]
E17	1	LH ESS XMFR RECT
F2	1	F LEVEL TK PRESS LT
F3	1	EXT FUEL TRANS WA
F4	1	ENCODER IFF WARN
F5	1	AILERON POS IND
F6	1	RUDDER POSIT IND
F7	1	NO 6 FUEL TRANS P
F8	1	ALT ENCODER [2]
F9	1	FUEL FLOW IND
F10	1	FUEL PRESS IND
F11	1	OIL PRESS IND
F12	1	CNI
F13	1	AWG-10 & CADC
F14	1	AJB-7
F16	1	BLEED AIR OVERHE DETECTION [9]
F17	1	LH ESS XMFR RECT
G2	1	FUEL LOW WARN LT
G3	1	MAIN FUEL CONT
G4	1	FUEL VALVE PWR L
G5	1	REFUEL PROBE
G6	1	STAB POSIT IND
G7	1	ALT ENCODER
G7	1	ALT VIBRATOR
G8	1	A.O.A. PROBE HEAT
G9	1	AUTO PILOT
G10	1	SIDEWINDER
G11	1	NO 6 FUEL TRANS F
G12	1	CNI
G13	1	ENG OIL LEVEL
G14	1	INTERCOM
G15	1	RUDDER FEEL TRIM
G16	1	BLEED AIR OVERHE DETECTION [9]
G17	1	RH XMFR RECT
H1	1	FUEL QTY
H2	1	INT WING FUEL DUM
H3	1	EXT WING FUEL CO
H4	1	WING TRANS CONT
H5	1	LDG GEAR AND FLA IND
H6	1	AUTO PILOT
H7	1	AUTO PILOT
H8	1	AUTO PILOT
H9	1	AUTO PILOT
H10	1	SIDEWINDER
H11	1	NO 6 FUEL TRANS F
H12	1	ALT ENCODER
H13	1	A.O.A. XMTR HEATE
H14	1	TAKE COMMAND LT
H15	1	SCOPE LTS
H16	1	℄ GUN POD
H17	1	RH XMFR RECT
J1	1	MASTER CAUTION L RESET
J2	1	MASTER CAUTION L
J3	1	IFR PROBE UNLOCK LT
J4	1	BLC WARN LT
J5	1	OXYGEN LOW WARN
J6	1	OXYGEN GAGE
J7	1	AILERON RUDDER I CONNECT
J8	1	AILERON RUDDER I CONNECT
J9	1	AUTO PILOT
J10	1	SIDEWINDER
J11	1	NO 6 FUEL TRANS F
J12	1	AILERON DROOP AC
J13	1	RAMP CONT R ENG
J14	1	RAMP CONT L ENG
J15	1	BELLMOUTH PITOT
J16	1	AIRSPEED PITOT H
J17	1	RH XMFR RECT
A2	2	RH RADAR MISSILE
A7	2	AN/AWW-1 BOMB FU

CIRCUIT BREAKER LOCATION

Zone	C/B Panel	
B6	2	RH RADAR MISSILE
B7	2	AN/AWW-1 BOMB FUZING
B8	2	CTR STORE PWR
B9	2	CTR STORE SAFETY
C5	2	RH RADAR MISSILE
C6	2	BUDDY TANK HYD PUMP
C7	2	AN/AWW-1 BOMB FUZING
C8	2	EXT LT CONT
C9	2	UTILITY RECEP
D4	2	AN/AWW-1 BOMB FUZING
D5	2	BOMB & RKT REL PWR
D6	2	BUDDY TANK HOSE JETTI-SON
D7	2	AN/AWW-1 BOMB FUZING
D8	2	WING AND TAIL LT BRT
D9	2	UTILITY RECEP AC
E4	2	BOMB & RKT REL CONT
E5	2	AN/ARW-73
E6	2	BUDDY TANK FUEL DUMP
E7	2	PNEU SYS CONT
E8	2	WING & TAIL LT DIM
F4	2	RADAR SCOPE CAMERA
F5	2	RADAR SCOPE CAMERA
F6	2	HYD PRESS IND NO 1
F7	2	PNEU PRESS IND
F8	2	TAXI LT
F9	2	ANTI-COLLISION LT
G6	2	HYD PRESS IND NO 2
G7	2	UTILITY HYD PRESS IND
G8	2	APPROACH LT PWR
G9	2	FUS AND ANTI-COLLISION LT
H4	2	FLT REFUEL PROBE LT
H5	2	AUX AIR DOOR CAUTION LT
H6	2	HYD PRESS WARNING LT
H7	2	ARR HOOK CONT
H8	2	APPROACH LT CONT
H9	2	WING FOLD CONT
J3	2	AUTO THROT
J4	2	MAIN CONT WARNING LT
J5	2	SPEED BRAKE WARNING LT
J6	2	DUCT TEMP WARNING LT
J7	2	ARR HOOK WARNING LT
J8	2	CANOPY UNLOCK WARNING LT
J9	2	WING LOCK WARNING LT
J10	2	COR REFL WARNING LT
K3	2	WARNING LT CONT
K4	2	AUTO THROT
K5	2	AUTO THROT
L3	2	AFT CKPT EQUIP LT
L4	2	WHT FLOOD LT
L5	2	UTILITY FLOOD LT
L6	2	NOSE WHEEL STEER
L7	2	NOSE WHEEL STEER
L8	2	EQUIP COOLING
L9	2	CKPT HEAT AND VENT
M2	2	AWG-10
M3	2	AFT CKPT FLOOD LT DIM
M4	2	FWD CKPT RED INST/FLOOD LT DIM
M5	2	FWD CKPT RED CSL FLOOD LT DIM
M6	2	LH 28V AUTO XMFR
M7	2	WSHLD TEMP WARNING LT
M8	2	ELECT OVHT AND TURB OVSP
M9	2	CKPT HEAT AND VENT
M10	2	PRIM GYRO WARNING LT
N1	2	RADAR STANDBY
N3	2	AFT CKPT FLOOD LT BRT
N4	2	FWD CKPT RED CSL FLOOD LT BRT
N5	2	FWD CKPT RED INST FLOOD LT BRT
N6	2	ESS 28V AUTO XMFR
N7	2	WSHLD TEMP SENSOR

Zone	C/B Panel	
N8	2	EQUIP COOLING
N9	2	SEAT ADJUST
P1	2	RADAR PWR
P2	2	AWG-10
P3	2	AFT CKPT INST LT XMFR
P4	2	FWD CKPT INST LT XMFR
P5	2	FWD CSL LT XMFR
P6	2	RH 14/28V AUTO XMFR
P7	2	RAIN REM
P8	2	STATIC ACCEL
–	LUP	ARI
–	RUP	AIL FEEL TRIM
–	RUP	FLAPS
–	RUP	LG
–	RUP	RUD TRIM
–	RUP	SPEED BRAKE
–	RUP	TRIM CONT

1 153072z THRU 153779ab
2 153780ac AND UP
3 153076aa AND UP
4 154786ag AND UP
5 153072z THRU 153088aa, 153769ab 153770ab AND 153773ab
6 153768ab, 153771ab, 153772ab, 153774ab AND UP
7 AFTER AFC 370
8 153072z THRU 153088aa AFTER AFC 440
9 153072z THRU 153088aa AFTER AFC 439

LUP – LEFT UTILITY PANEL (FWD CKPT)
RUP – RIGHT UTILITY PANEL (FWD CKPT)

153072z THRU 153088aa, AND 153768ab THRU 155528ag BEFORE AFC 388.

NO. 2 CIRCUIT BREAKER PANEL

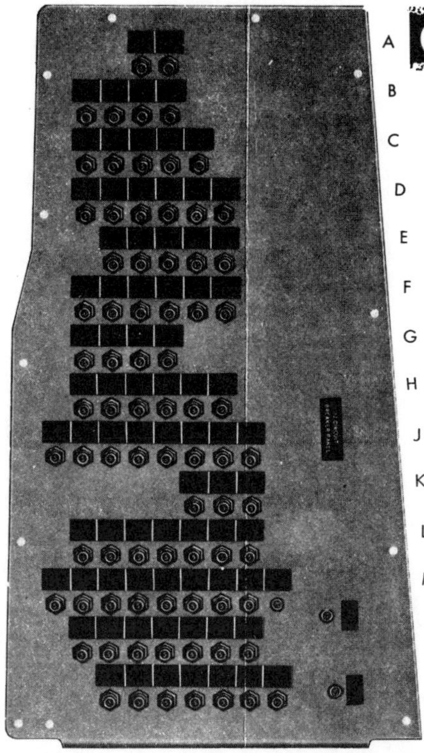

NO. 1 CIRCUIT BREAKER PANEL

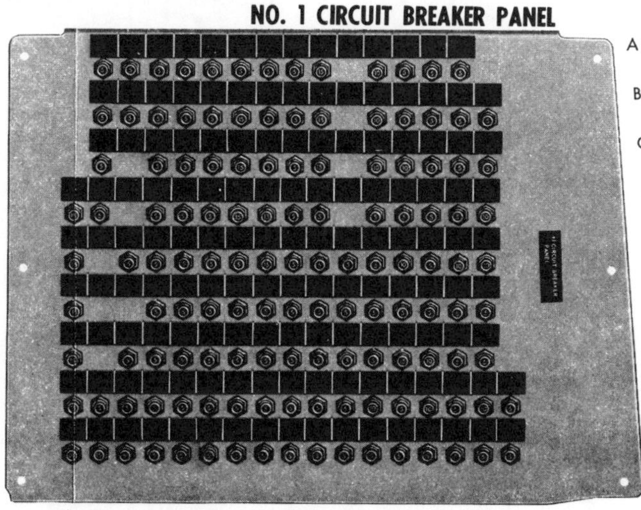

FDD-1-(279-1)

Figure A-7 (Sheet 1 of 3)

ALPHABETICAL LISTING

	C/B Panel	Zone
A-A/IFF [13]	1	13A
A/A IFF [27]	**1**	**13A**
A-A/IFF [13][26]	1	14A
AA/IFF [12]	1	15A
A/B IGN. L. ENG.	1	14E
A/B IGN. R. ENG.	1	13E
AFT CKPT. EQUIP. LT.	2	3N
AFT CKPT. FLOOD LT. BRT.	2	3Q
AFT CKPT. FLOOD LT. DIM.	2	3P
AFT CKPT. INST. LT. XMFR.	2	3R
AILERON DROOP ACT.	1	15H
AILERON FEEL TRIM	RUP	–
AILERON POS. IND.	1	5K
AILERON RUDDER INT.	1	11K
AILERON RUDDER INT.	1	12K
AIRSPEED PITOT HEATER	1	15D
AJB-7	1	16E
AJB-7	1	16G
AJB-7	1	17E
AJB-7	1	17F
AJB-7	1	17G
AJB-7	1	17H
ALR-45 [21]	1	11G
ALR-45 [21]	1	11H
ALT. ENCODER	1	10G
ALT. ENCODER	1	10H
ALT. ENCODER	1	10K
AN/ALE-29A DOOR SWS [8]	1	12A
AN/ALE-29A FIRE SWS [8]	1	11A
AN/ALE-29A FIRE SYS [26]	1	13A
AN/ALE –39 DOOR [27]	**1**	**11A**
AN/ALE –39 SYS [27]	**1**	**12A**
AN/ALQ–51/100 CONTROL [6]	2	5E
AN/ALQ–51/100 RCVR XMTR [6]	2	5B
AN/ALQ–51/100 RCVR XMTR [6]	2	5C
AN/ALQ–51/100 RCVR XMTR [6]	2	5D
AN/ALQ–91A CONTROL [24]	2	4E
AN/ALQ–91A ECM LT CONT [24]	2	4F
AN/ALQ–91A RCVR XMTR [24]	2	4B
AN/ALQ–91A RCVR XMTR [24]	2	4C
AN/ALQ–91A RCVR XMTR [24]	2	4D
AN/ARW-73	2	5F
AN/AWW-1 BOMB FUZING [3]	2	6E
AN/AWW-1 BOMB FUZING [18]	2	7B
AN/AWW-1 BOMB FUZING [18]	2	7C
AN/AWW-1 BOMB FUZING [18]	2	7D
AN/AWW-1 BOMB FUZING [18]	2	7E
ANTI-COLLISION LT.	2	10H
ANTI-ICE	1	14H
ANTI-SKID [2]	1	2B
ANTI-SKID [2]	1	2C
ANTI-SKID [2]	1	2D
ANTI-SKID [2]	1	2E
ANTI-SKID WARN LT [2]	1	2G
A.O.A. PROBE HEATER	1	14K
A.O.A. XMTR. HEATER	1	4J
APN-141	1	13G
APN-141	1	13H
APN-154	1	8J
APPROACH LT. CONT.	2	8K
APPROACH LT. POWER	2	7J
APR-25 [20]	1	11G
APR-25 [20]	1	11H
ARI	LUP	–
ARI/ANTI-SKID [2]	1	2F
ARMT. BUS CONT.	1	5A
ARRESTING HOOK CONT.	2	7K
ARRESTING HOOK WARN. LT.	2	7L
AUTO PILOT	1	9F
AUTO PILOT	1	9G
AUTO PILOT	1	9H
AUTO PILOT PITCH	1	9J
AUTO PILOT	1	9E
AUTO PILOT	1	9K
AUTO THROTTLE	2	3M

	C/B Panel	Zone
AUTO THROTTLE	2	3K
AUTO THROTTLE	2	3L
AUX. AIR DOOR CAUTION LT	2	5K
AUX AIR DOOR, L. ENG.	1	16K
AUX AIR DOOR, R. ENG.	2	15K
AWG-10	2	2Q
AWG-10	2	2R
AWG-10 & CONV WPNS [4]	2	6E
BLANKING UNIT [8]	1	10F
BELLMOUTH PITOT HEATER	1	14D
BLC WARNING LT.	1	10C
BLEED AIR OVERHEAT DETECTION [9]	1	1J
BLEED AIR OVERHEAT DETECTION [11]	1	1E
BLEED AIR OVERHEAT DETECTION [9]	1	1H
BLEED AIR OVERHEAT DETECTION [11]	1	1C
BLEED AIR OVERHEAT WARN LT [9]	1	1G
BLEED AIR OVERHEAT WARN LT [11]	1	1D
BOMB AND RKT. REL. CONT.	2	5G
BOMB AND RKT. REL. PWR.	2	6F
BOMB FUZING [19]	2	7B
BOMB FUZING [19]	2	7C
BOMB FUZING [19]	2	7D
BOMB FUZING [19]	2	7E
BUDDY TANK FUEL DUMP	2	9C
BUDDY TANK HOSE JETT.	2	8B
BUDDY TANK HYD PUMP	2	8D
CADC	1	4E
CADC	1	4H
CADC	1	4G
CADC	1	4F
CADC	1	4K
CAMERA AC [25]	**2**	**9D**
CAMERA DC [25]	**2**	**10G**
CANOPY UNLOCK WARN. LT.	2	8L
CKPT. HEAT AND VENT	2	9M
CKPT. HEAT AND VENT	2	9N
CL GUN POD	1	13K
CNI	1	3E
CNI	1	3F
CNI	1	3G
CNI	1	2K
CNI	1	3K
CNI	1	2J
CNI	1	2H
CORNER REFL. WARN LT.	2	10L
CTR. STORE PWR	2	8C
CTR. STORE SAFETY	2	9B
DATA LINK	1	8E
DATA LINK	1	8F
DATA LINK	1	8G
DATA LINK	1	8H
DUCT TEMP WARNING LT.	2	6L
ECM [23]	**2**	**5B**
ECM [23]	**2**	**5C**
ECM [23]	**2**	**5D**
ECM CONTROL [23]	**2**	**5E**
ECM LT. SHOEHORN [7]	2	4F
ELECT. OVHT. AND TURBINE OVSP.	2	8P
EMER AIL DROOP [22]	1	1F
EMER. REFUEL PROBE [1]	1	14G
ENCODER IFF WARN. LT.	1	10J
ENGINE BLEED AIR [10]	1	2A
ENGINE OIL LEVEL	1	14B
EQUIP. COOLING [14]	2	8M
EQUIP. COOLING	2	8N
ESS. 28V AUTO XMFR	2	6Q
EXT. FUEL TRANS. WARN LT.	1	14C
EXT. LT. CONT.	2	8F
EXT. STORES EMER. JETT.	1	10B
EXT. WING FUEL CONT.	1	12J

	C/B Panel	Zone
FIRE DETECTOR	1	15E
FLAPS	RUP	
FUEL FLOW IND.	1	7K
FUEL LEVEL TK. PRESS. WARN. LT.	1	11C
FUEL LOW WARN. LT.	1	11B
FUEL PRESS. IND.	1	8K
FUEL QTY.	1	11D
FUEL QTY ADAPTER [12]	1	**11E**
FUEL VALVE POWER L.H.	1	15J
FUS. AND ANTI-COLL. LT.	2	9H
FWD. CKPT. INST. LT. XMFR	2	4R
FWD. CKPT. RED CSL FLOOD LT. DIM.	2	5P
FWD. CKPT. RED CSL FLOOD LT. BRT.	2	4O
FWD. CKPT. RED INST. FLOOD LT. DIM.	2	4P
FWD. CKPT. RED INST. FLOOD LT. BRT.	2	5Q
FWD. CSL LT. XMFR	2	5R
GVR-10	1	12E
GVR-10	1	12F
GVR-10	1	12G
GVR-10	1	12H
HYD. PRESS. IND. NO. 1	2	6G
HYD. PRESS. IND. NO. 2	2	6J
HYD. PRESS. WARNING LT.	2	6K
INTERCOM	1	3H
IFR PROBE LT.	2	9G
IFR PROBE UNLOCK WARN LT	1	11J
IGN. L. MAIN	1	16F
IGN. R. MAIN	1	15F
INT. WING FUEL DUMP	1	17K
JULIET - 28 (KY-28)	1	5H
LG	RUP	
LDG. GEAR AND FLAP POS. IND.	1	15B
LDG. GEAR WARN LT.	1	13C
L. ENG. FUEL SHUTOFF	1	17J
L.H. BOOST PUMP EMER. CONT.	1	6H
L.H. BOOST PUMP NORM CONT.	1	7H
L.H. ESS. XMFR. RECT	1	5E
L.H. ESS. XMFR. RECT	1	5F
L.H. ESS. XMFR. RECT	1	5G
L.H. FUEL BOOST PUMP	1	6E
L.H. FUEL BOOST PUMP	1	6F
L.H. FUEL BOOST PUMP	1	6G
L.H. RADAR MISSILE	1	8B
L.H. RADAR MISSILE	1	8C
L.H. RADAR MISSILE	1	8D
L.H. 28V AUTO. XMFR	2	6P
MAIN CONT. WARN. LT.	2	4L
MAIN FUEL CONT.	1	16J
MASTER CAUTION LT.	1	12C
MASTER CAUTION LT. RESET	1	12B
MISSILE ARM	1	9D
MISSILE FIRING	1	9A
MISSILE LTS.	1	9C
NAV COMPUTER	1	10E
NAV COMPUTER	1	10D
NOSE WHEEL STEER.	2	7M
NOSE WHEEL STEER. [13]	2	7N
NOZZLE POS. IND.	1	15G
NO. 4 FUEL TRANS PUMP	1	7A
NO. 4 FUEL TRANS PUMP	1	7B
NO. 4 FUEL TRANS PUMP	1	7C
NO. 4 FUEL TRANS. PUMP	1	7D
NO. 6 FUEL TRANS. PUMP	1	7E
NO. 6 FUEL TRANS. PUMP	1	7F
NO. 6 FUEL TRANS. PUMP	1	7G
NO. 6 FUEL TRANS. PUMP	1	7J

NO. 7 TANK F TRANS F
NO. 7 TANK F TRANS F
NO. 7 TANK F TRANS F
NO. 7 TANK F TRANS F
OIL PRESS. IND.
OPTICAL SIGHT [28]
OPTICAL SIGHT [28]
OUTBD. JETT.
OUTBD. JETT.
OXYGEN GAGE
OXYGEN LOW WARN. LT
PNEU. PRESS. IND.
PNEU. SYS. CONT.
PRIM. GYRO WARN. LT.
RADAR MISSILE FAIRIN
RADAR MISSILE JETT.
RADAR POWER
RADAR SCOPE CAMERA
RADAR SCOPE CAMERA
RADAR STANDBY
RAIN REMOVAL
RAMP CONT. L. ENG.
RAMP CONT. R. ENG.
RECEIVING DECODING GROUP AN/ARA–63
RECEIVING DECODING GROUP AN/ARA–63
RECEIVING DECODING GROUP AN/ARA–63
RECEIVING DECODING GROUP AN/ARA- 63
REFUEL PROBE [16]
REFUEL PROBE [17]
R.H. FUEL BOOST PUMP CONT.
R.H. FUEL BOOST PUMP
R.H. FUEL BOOST PUMP
R.H. FUEL BOOST PUMP
R.H. RADAR MISSILE
R.H. RADAR MISSILE
R.H. RADAR MISSILE
R.H. XMFR. RECT.
R.H. XMFR. RECT.
R.H. XMFR. RECT.
RUDDER FEEL TRIM
RUDDER FEEL TRIM AN BELLMOUTH [8]
RUDDER POS IND.
R.H. 14/28V AUTO. XMF
RUDDER TRIM
SCOPE LTS.
SEAT ADJUST
SHOE HORN [7]
SHOE HORN [7]
SHOE HORN [7]
SHOE HORN [7]
SHOE HORN [7]
SHOE HORN [7]
SHOE HORN [7]
SHOE HORN [7]
SHOE HORN [7]
SHOE HORN [7]
SHOE HORN [7]
SIDEWINDER
SIDEWINDER
SIDEWINDER
SIDEWINDER
SIDEWINDER
SIDEWINDER
SIDEWINDER
SIDEWINDER
SPEED BRAKE
SPEED BRAKE WARN. L
SP SW GUN VTAS SEL CONT
SP SW GUN FIRE POWER
STAB. FEEL TRIM

C/B Panel	Zone
2	G4
2	H4
2	J4
2	K4
1	16H
1	**K1**
2	**M10**
1	16A
1	16B
1	16D
1	16C
2	7G
2	7F
2	9P
1	8A
1	9B
2	1R
2	10K
2	10J
2	1P
2	7P
1	13D
1	12D
2	4A
2	5A
2	6A
2	7A
1	14J
1	5J
1	6A
1	6B
1	6C
1	6D
2	6B
2	6C
2	6D
1	5B
1	5C
1	5D
1	13B
1	13B
1	6K
2	6R
RUP	–
1	15C
2	8E
1	11A
1	12A
1	10F
2	4B
2	4C
2	4D
2	4E
2	5B
2	5C
2	5D
2	5E
1	3A
1	3B
1	3C
1	3D
1	4A
1	4B
1	4C
1	4D
RUP	-
2	5L
1	17C
1	17D
RUP	

	C/B Panel	Zone
STAB. POS. IND.	1	6J
STATIC ACCEL.	2	5M
STORES REL. CONT.	1	10A
TAKE COMMAND LT.	1	3J
TAXI LT.	2	7H
TRIM CONT. 13	RUP	–
TRIM CONT NLG STEER 12	RUP	–
TURBINE OUTLET TEMP. L. ENG.	1	13F
TURBINE OUTLET TEMP R. ENG.	1	14F
UTILITY FLOOD LT.	2	5N
UTILITY HYD. PRESS. IND.	2	6H
UTILITY RECEP.	2	9F
UTILITY RECEP. AC	2	9E
VTAS 12	2	8J
VTAS 12	2	9J
VTAS 12	2	10J
VTAS 12	2	10K
WARNING LT. CONT.	2	4M
WHITE FLOOD LT.	2	4N
WING AND TAIL LT. DIM	2	8G
WING AND TAIL LT. BRT.	2	8H
WING FOLD CONT.	2	9K
WING LOCK WARNING LT.	2	9L
WING TRANS. CONT.	1	13J
WSHLD. TEMP. SENSOR	2	6M
WSHLD. TEMP. WARN. LT.	2	6N

1. 155867ak AND UP, AND ALL OTHERS AFTER AFC 370
2. 155903ap, 157242an AND UP
3. 153768ab THRU 155874ak BEFORE AFC 463
4. 155875al AND UP, AND 153768ab THRU 155874ak AFTER AFC 463
5. 155529ag THRU 155569ag AND 153768ab THRU 155528ag AFTER AFC 388 BEFORE AFC 541
6. 155570ah AND UP BEFORE AFC 541
7. 155529ag THRU 155784ah, AND 153768ab THRU 155528ag AFTER AFC 388
8. 155785ai AND UP
9. 155903ap AND 157274ap AND UP
10. 157274ap AND UP, AND ALL OTHERS AFTER AFC 440
11. 153071z, 153851ae, 155529ag THRU 157273ao AFTER AFC 439
12. 158355at AND UP, AND 153768ab THRU 158354as AFTER AFC 506
13. 153768ab THRU 158354as BEFORE AFC 506, AND 153071z
14. 153071z AND 153768ab THRU 158354as
15. 157309ar AND UP, AND ALL OTHERS AFTER AFC 470
16. 153071z, 153768ab THRU 158365at BEFORE AFC 545
17. 158366au THRU 158379au, AND ALL OTHERS AFTER AFC 545
18. 153071z, 153531ae, 153768ab THRU 157309ar
19. 158346as AND UP
20. BEFORE AFC 524
21. AFTER AFC 524
22. 153768ab THRU 153838ab AND 153840ae THRU 154785af AFTER AFC 534
23. AFTER AFC 541
24. 155570ah AND UP
25. 153768ab THRU 158379au AFTER AFC 518
26. 158355at AND UP, AND 153768ab THRU 158354as AFTER AFC 506, AND BEFORE AFC 574
27. 153768ab THRU 158379au AFTER AFC 574
28. 153768ab THRU 158379au AFTER AFC 576 PT 1.

LUP–LEFT UTILITY PANEL (FWD CKPT)
RUP–RIGHT UTILITY PANEL (FWD CKPT)

CIRCUIT BREAKER LOCATION

ALPHABETICAL LISTING

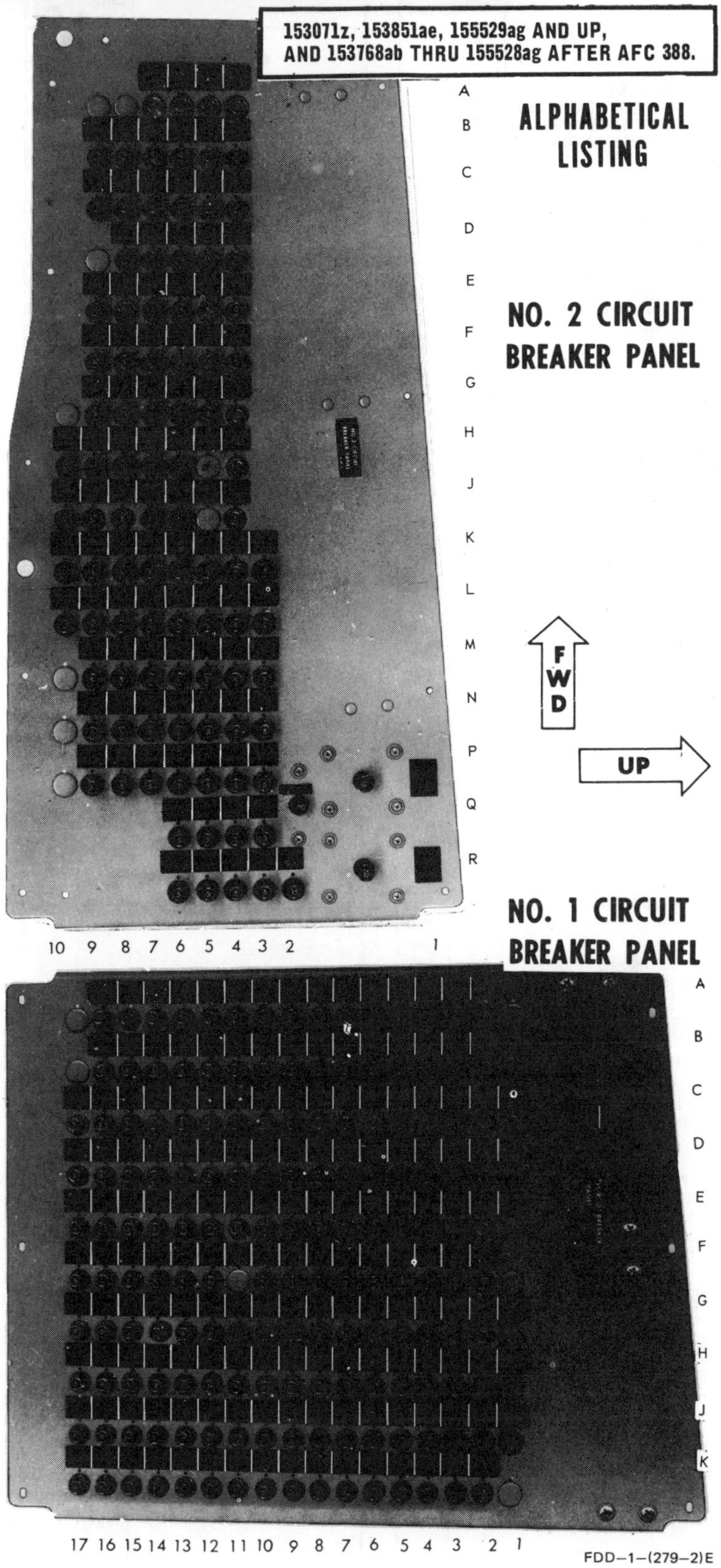

Figure A-7 (Sheet 2 of 3)

Figure A-7

ZONE LISTING

	ZONE	C/B Panel
ENGINE BLEED AIR (10)	**A2**	**1**
SIDEWINDER	**A3**	**1**
SIDEWINDER	**A4**	**1**
ARMT BUS CONT	**A5**	**1**
RH FUEL BOOST PUMP/CONT	**A6**	**1**
NO. 4 FUEL TRANS PUMP	**A7**	**1**
RADAR MISSILE FAIRING	**A8**	**1**
MISSILE FIRING	**A9**	**1**
STORES RELEASE CONT	**A10**	**1**
SHOEHORN (7)	**A11**	**1**
AN/ALE–29A FIRE SW (8)	**A11**	**1**
AN/ALE–39 DOOR (27)	**A11**	**1**
SHOEHORN (7)	**A12**	**1**
AN/ALE–29A DOOR SW (8)	**A12**	**1**
AN/ALE–39 SYS (27)	**A12**	**1**
A–A IFF (13)	**A13**	**1**
AN/ALE–29A FIRE SYS (26)	**A13**	**1**
A–A IFF (27)	**A13**	**1**
A–A IFF (13)(26)	**A14**	**1**
AA/IFF (12)	A15	1
OUTBD JETT	A16	1
ANTI-SKID (2)	B2	1
SIDEWINDER	B3	1
SIDEWINDER	B4	1
RH XMFR RECT	B5	1
RH FUEL BOOST PUMP	B6	1
NO. 4 FUEL TRANS PUMP	B7	1
LH RADAR MISSILE	B8	1
RADAR MISSILE JETT	B9	1
EXT STORES EMER JETT	B10	1
FUEL LOW WARN LT	B11	1
MASTER CAUTION LT RESET	B12	1
RUDDER FEEL TRIM (7)	B13	1
RUDDER FEEL AND BELL-MOUTH (8)	B13	1
ENG OIL LEVEL	B14	1
LG AND FLAP POS IND	B15	1
OUTBD JETT	B16	1
ENGINE OVERHEAT DETECTION (11)	C1	1
ANTI-SKID (2)	C2	1
SIDEWINDER	C3	1
SIDEWINDER	C4	1
RH XMFR RECT	C5	1
RH FUEL BOOST PUMP	C6	1
NO. 4 FUEL TRANS PUMP	C7	1
LH RADAR MISSILE	C8	1
MISSILE LTS	C9	1
BLC WARN LT	C10	1
F LEVEL TK PRESS WARN LT	C11	1
MASTER CAUTION LT	C12	1
LDG GEAR WARN LT	C13	1
EXT FUEL TRANS WARN LT	C14	1
SCOPE LTS	C15	1
OXYGEN LOW WARN LT	C16	1
SP SW GUN VTAS SEL CONT (12)	C17	1
BLEED AIR OVERHEAT WARN LT (11)	D1	1
ANTI-SKID (2)	D2	1
SIDEWINDER	D3	1
SIDEWINDER	D4	1
RH XMFR RECT	D5	1
RH FUEL BOOST PUMP	D6	1
NO. 4 FUEL TRANS PUMP	D7	1
LH RADAR MISSILE	D8	1
MISSILE ARM	D9	1
NAV CMTR	D10	1
FUEL QTY	D11	1
RAMP CONT R ENG	D12	1
RAMP CONT L ENG	D13	1
BELLMOUTH PITOT HEATER	D14	1
AIRSPEED PITOT HEATER	D15	1
OXYGEN GAGE	D16	1
SP SW GUN FIRE POWER (12)	D17	1
BLEED AIR OVFRHEAT DETECTION (11)	E1	1

	ZONF	C/B Panel
ANTİ-SKID (2)	E2	1
CNI	E3	1
CADC	E4	1
LH ESS XMFR RECT	E5	1
LH FUFL BOOST PUMP	E6	1
NO. 6 FUEL TRANS PUMP	E7	1
DATA LINK	E8	1
AUTO PILOT	E9	1
NAV CMTR	E10	1
FUEL QTY ADAPTER (12)	E11	1
GVR-10	E12	1
A/B IGN R ENG	E13	1
A/B IGN L ENG	E14	1
FIRE DETECTOR	E15	1
AJB-7	E16	1
AJB-7	E17	1
EMER AIL DROOP (22)	F1	1
ARI/ANTI-SKID (2)	F2	1
CNI	F3	1
CADC	F4	1
LH ESS XMFR RECT	F5	1
LH FUEL BOOST PUMP	F6	1
NO. 6 FUEL TRANS PUMP	F7	1
DATA LINK	F8	1
AUTO PILOT	F9	1
SHOEHORN (7)	F10	1
BLANKING UNIT (8)	F10	1
GVR-10	F12	1
TURBINE OUTLET TEMP L ENG	F13	1
TURBINE OUTLET TEMP R ENG	F14	1
IGN R MAIN	F15	1
IGN L MAIN	F16	1
AJB-7	F17	1
BLEED AIR OVERHEAT WARN LT (9)	G1	1
ANTI-SKID WARN LT (2)	G2	1
CNI	G3	1
CADC	G4	1
LH ESS XMFR RECT	G5	1
LH FUEL BOOST PUMP	G6	1
NO. 6 FUEL TRANS PUMP	G7	1
DATA LINK	G8	1
AUTO PILOT	G9	1
ALT ENCODER	G10	1
APR-25 (20)	G11	1
ALR–45 (21)	G11	1
GVR-10	G12	1
APN-141	G13	1
EMER REFUEL PROBE (1)	G14	1
NOZZLE POS IND	G15	1
AJB-7	G16	1
AJB-7	G17	1
BLEED AIR OVERHEAT DETECTION (9)	H1	1
CNI	H2	1
INTERCOM	H3	1
CADC	H4	1
JULIET-28	H5	1
LH BOOST PUMP EMER CONT	H6	1
LH BOOST PUMP NORM CONT	H7	1
DATA LINK	H8	1
AUTO PILOT	H9	1
ALT ENCODER	H10	1
APR-25 (20)	H11	1
ALR–45 (21)	H11	1
GVR-10	H12	1
APN-141	H13	1
ANTI-ICE	H14	1
AILERON DROOP ACTR	H15	1
OIL PRESS	H16	1
AJB-7	H17	1
BLEED AIR OVERHEAT DETECTION (9)	J1	1
CNI	J2	1

	ZONE	C/B Panel
TAKE COMMAND LT	J3	1
A.O.A. XMTR HEATER	J4	1
REFUEL PROBE (17)	J5	1
STAB POS IND	J6	1
NO. 6 FUEL TRANS PUMP	J7	1
APN-154	J8	1
AUTO PILOT PITCH	J9	1
ENCODER IFF WARN LT	J10	1
IFR PROBE UNLOCK WARN LT	J11	1
EXT WING FUEL CONT	J12	1
WING TRANS CONT	J13	1
REFUEL PROBE (16)	J14	1
FUEL VALVE PWR LH	J15	1
MAIN FUEL CONT	J16	1
L ENG FUEL SHUTOFF	J17	1
OPTICAL SIGHT (28)	**K1**	**1**
CNI	K2	1
CNI	K3	1
CADC	K4	1
AILERON POS IND	K5	1
RUDDER POS IND	K6	1
FUEL FLOW IND	K7	1
FUEL PRESS IND	K8	1
AUTO PILOT	K9	1
ALT ENCODER	K10	1
AILERON RUDDER INTER-CONNECT	K11	1
AILERON RUDDER INTER-CONNECT	K12	1
CL GUN POD	K13	1
A.O.A. PROBE HEATER	K14	1
R ENG AUX AIR DOOR	K15	1
L ENG AUX AIR DOOR	K16	1
INT WING FUEL DUMP	K17	1
RECEIVING DECODING (8) GROUP AN/ARA–63	A4	2
RECEIVING DECODING (15) GROUP AN/ARA–63	A5	2
RECEIVING DECODING (15) GROUP AN/ARA–63	A6	2
RECEIVING DECODING (15) GROUP AN/ARA–63	A7	2
SHOEHORN (5)	B4	2
AN/ALQ-91A RCVR XMTR (24)	B4	2
SHOEHORN (5)	B5	2
AN/ALQ-51/100 RCVR XMTR (6)	B5	2
ECM (23)	**B5**	**2**
RH RADAR MISSILE	B6	2
AN/AWW-1 BOMB FUZING (18)	B7	2
BOMB FUZING (19)	B7	2
BUDDY TANK HOSE JETT	B8	2
CTR STORE SAFETY	B9	2
SHOEHORN (5)	C4	2
AN/ALQ-91A RCVR XMTR (24)	C4	2
SHOEHORN (5)	C5	2
AN/ALQ-51/100 RCVR XMTR (6)	C5	2
ECM (23)	**C5**	**2**
RH RADAR MISSILE	C6	2
AN/AWW-1 BOMB FUZING (18)	C7	2
BOMB FUZING (19)	C7	2
CTR STORE PWR	C8	2
BUDDY TANK FUEL DUMP	C9	2
SHOEHORN (5)	D4	2
AN/ALQ-91A RCVR XMTR (24)	D4	2
SHOEHORN (5)	D5	2
AN/ALQ-51/100 RCVR XMTR (6)	D5	2
ECM (23)	**D5**	**2**
RH RADAR MISSILE	D6	2
AN/AWW-1 BOMB FUZING (18)	D7	2
BOMB FUZING (19)	D7	2
BUDDY TANK HYD PUMP	D8	2
CAMERA AC (25)	**D9**	**2**
SHOEHORN (5)	E4	2
AN/ALQ-91A CONTROL (24)	E4	2
SHOEHORN (5)	E5	2

AN/ALQ-51/100 CONT
ECM CONTROL (23)
AWG-10 & CONV WPNS
AN/AWW-1 BOMB FUZIN
AN/AWW-1 BOMB FUZIN
BOMB FUZING (19)
SEAT ADJUST
UTILITY RECEP AC
ECM LT SHOEHORN (5)
AN/ALQ-91A ECM LT CONT (24)
AN/ARW-73
BOMB & ROCKET REL
PNEU SYS CONT
EXT LTS CONTROL
UTILITY RECEP
NO. 7 TANK F TRANS PUMP (9)
BOMB & RKT REL CON
HYD PRESS IND NO. 1
PNEU PRESS IND
WING & TAIL LT DIM
IFR PROBE LT
CAMERA DC (25)
NO. 7 TANK F TRANS PUMP (9)
UTILITY HYD PRESS I
TAXI LT
WING AND TAIL LT BR
FUS & ANTI-COLLISIO
ANTI-COLLISION LT
NO. 7 TANK F TRANS PUMP (9)
HYD PRESS IND NO. 2
APPROACH LT PWR
VTAS (12)
VTAS (12)
VTAS (12)
RADAR SCOPE CAMER
AUTO THROT
NO. 7 TANK F TRANS PUMP (9)
AUX AIR DOOR CAUTI
HYD PRESS WARN LT
ARR HOOK CONT
APPROACH LT CONT
WING FOLD CONT
RADAR SCOPE CAMER
VTAS (12)
AUTO THROT
MAIN CONT WARNING
SPEED BRAKE WARNI
DUCT TEMP WARNING
ARR HOOK WARNING L
CANOPY UNLOCK WAR LT
WING LOCK WARNING
COR REFL WARNING L
AUTO THROT
WARNING LT CONT
STATIC ACCEL
WSHLD TEMP SENSOR
NOSE WHEEL STEER
EQUIP COOLING (14)
CKPT HEAT AND VEN
OPTICAL SIGHT (2
AFT CKPT EQUIP LT
WHT FLOOD LT
UTILITY FLOOD LT
WSHLD TEMP WARNIN
NOSE WHEEL STEER
EQUIP COOLING
CKPT HEAT AND VEN
RADAR STANDBY
AFT CKPT FLOOD LT
FWD CKPT RED INST L
FWD CKPT RED CSL F LT DIM

CIRCUIT BREAKER LOCATION

ZONE	C/B Panel
E5	2
E5	2
E6	2
E6	2
E7	2
E7	2
E8	2
E9	2
F4	2
F4	2
F5	2
F6	2
F7	2
F8	2
F9	2
G4	2
G5	2
G6	2
G7	2
G8	2
G9	2
G10	2
H4	2
H6	2
H7	2
H8	2
H9	2
H10	2
J4	2
J6	2
J7	2
J8	2
J9	2
J10	2
J10	2
K3	2
K4	2
K5	2
K6	2
K7	2
K8	2
K9	2
K10	2
K10	2
L3	2
L4	2
L5	2
L6	2
L7	2
L8	2
L9	2
L10	2
M3	2
M4	2
M5	2
M6	2
M7	2
M8	2
M9	2
M10	2
N3	2
N4	2
N5	2
N6	2
N7	2
N8	2
N9	2
P1	2
P3	2
P4	2
P5	2

	ZONE	C/B Panel
L28 V AUTO XMFR	P6	2
RAIN REMOVAL	P7	2
ELECT OVHT AND TURB OVSP	P8	2
PRIM GYRO WARNING LT	P9	2
AWG-10	Q2	2
AFT CKPT FLOOD LT BRT	Q3	2
FWD CKPT RED CSL FLOOD LT BRT	Q4	2
FWD CKPT RED INST FLOOD LT BRT	Q5	2
ESS 28V AUTO XMFR	Q6	2
RADAR POWER	R1	2
AWG-10	R2	2
AFT CKPT INST LT XMFR	R3	2
FWD CKPT INST LT XMFR	R4	2
FWD CSL LT XMFR	R5	2
RH 14/28V AUTO XMFR	R6	2
ARI	–	LUP
AILERON FEEL TRIM	–	RUP
FLAPS	–	RUP
LG	–	RUP
RUDDER TRIM	–	RUP
SPEED BRAKE	–	RUP
STAB FEEL TRIM	–	RUP
TRIM CONT 13	–	RUP
TRIM CONT NLG STEER 12		RUP

153071z, 153851ae, 155529ag AND UP, AND 153768ab THRU 155528ag AFTER AFC 388.

ZONE LISTING

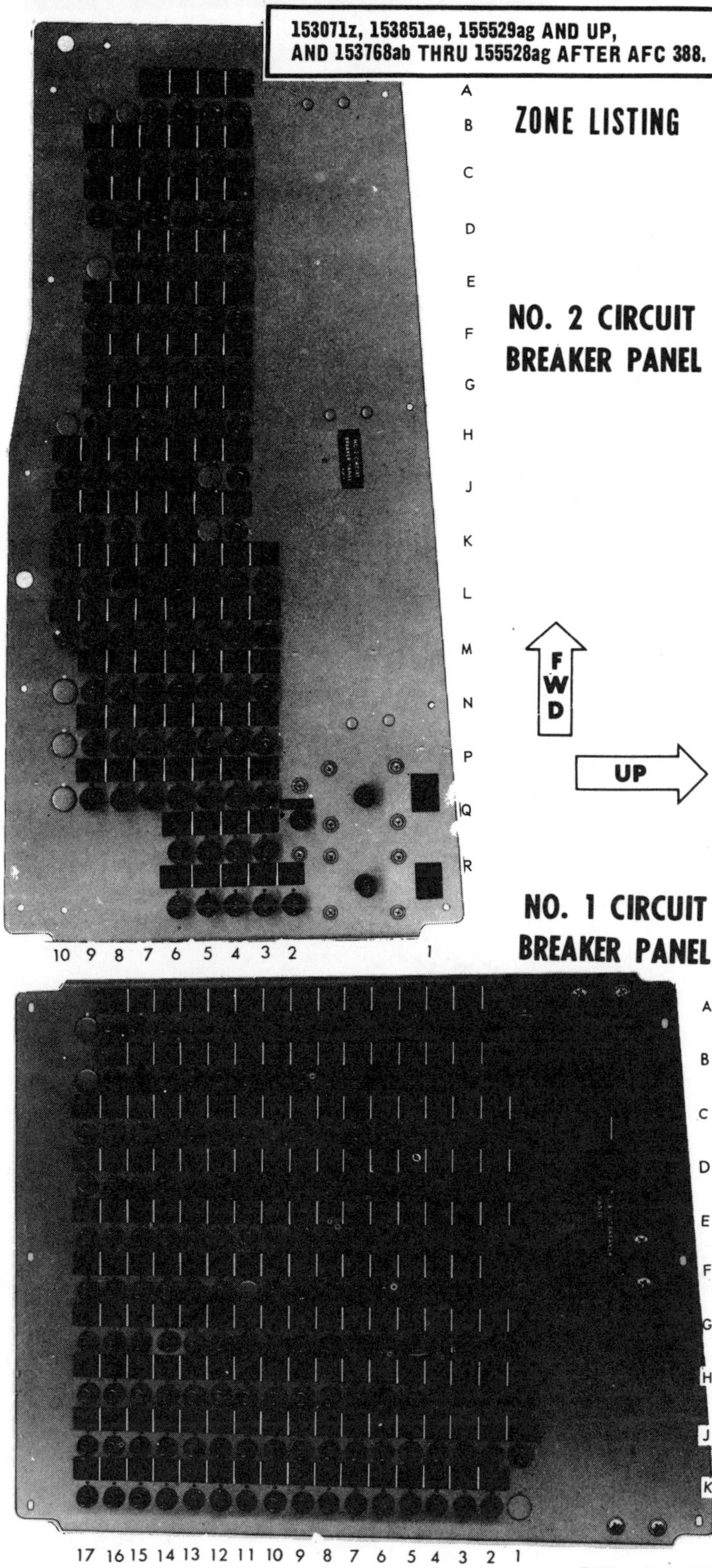

FDD-1-(279-3)E

1. 155867ak AND UP, AND ALL OTHERS AFTER AFC 370
2. 155903ap, 157242an AND UP
3. 153768ab THRU 155874ak BEFORE AFC 463
4. 155875al AND UP, AND 153768ab THRU 155874ak AFTER AFC 463
5. 155529ag THRU 155529ag THRU AND 153768ab THRU 155528ag AFTER AFC 388 BEFORE AFC 541
6. 155570ah AND UP BEFORE AFC 541
7. 155529ag THRU 155784ah,AND 153768ab THRU 155528ag AFTER AFC 388
8. 155785ai AND UP
9. 155903ap AND 157274ap AND UP
10. 157274ap AND UP, AND ALL OTHERS AFTER AFC 440
11. 153071z, 153851ae, 155529ag THRU 157273ao AFTER AFC 439
12. 158355at AND UP, AND 153768ab THRU 158354as AFTER AFC 506
13. 153768ab THRU 158354as BEFORE AFC 506, AND 153071z
14. 153071z AND 153768ab THRU 158354as
15. 157309ar AND UP, AND ALL OTHERS AFTER AFC 470
16. 153071z, 153768ab THRU 158365at BEFORE AFC 545
17. 158366au THRU 158379au AND ALL OTHERS AFTER AFC 545
18. 153071z, 153531ae, 153768ab THRU 157309ar
19. 158346as AND UP
20. BEFORE AFC 524
21. AFTER AFC 524
22. 153768ab THRU 153838ad AND 153840ae THRU 154785af AFTER AFC 534
23. AFTER AFC 541
24. 155570ah AND UP
25. 153768ab THRU 158379au AFTER AFC 518
26. 158355at AND UP, AND 153768ab THRU 158354as AFTER AFC 506, AND BEFORE AFC 574
27. 153768ab THRU 158379au AFTER AFC 574
28. 153768ab THRU 158379au AFTER AFC 576 PT 1.

LUP–LEFT UTILITY PANEL (FWD CKPT)
RUP–RIGHT UTILITY PANEL (FWD CKPT)

Figure A-7 (Sheet 3 of 3)

Figure A-7

ALPHABETICAL INDEX

Note

All numbers in this alphabetical index refer to page numbers, not paragraph and figure numbers.

☆ U.S. GOVERNMENT PRINTING OFFICE: 1975— 665-677/6002